DNA - The Blueprint For Life

" The Latest Information on Deoxyribonucleic Acid"

Edited by Paul F. Kisak

Contents

Chapter 1

DNA

For a non-technical introduction to the topic, see Introduction to genetics. For other uses, see DNA (disambiguation).

Deoxyribonucleic acid (◄ⁱ/diˌɒksiˌraɪbəˈnjuːˌkleɪ.ik ˈæsɪd/; **DNA**) is a molecule that carries most of the genetic instructions used in the development, functioning and reproduction of all known living organisms and many viruses. DNA is a nucleic acid; alongside proteins and carbohydrates, nucleic acids compose the three major macromolecules essential for all known forms of life. Most DNA molecules consist of two biopolymer strands coiled around each other to form a double helix. The two DNA strands are known as polynucleotides since they are composed of simpler units called nucleotides.[1] Each nucleotide is composed of a nitrogen-containing nucleobase—either cytosine (C), guanine (G), adenine (A), or thymine (T)—as well as a monosaccharide sugar called deoxyribose and a phosphate group. The nucleotides are joined to one another in a chain by covalent bonds between the sugar of one nucleotide and the phosphate of the next, resulting in an alternating sugar-phosphate backbone. According to base pairing rules (A with T, and C with G), hydrogen bonds bind the nitrogenous bases of the two separate polynucleotide strands to make double-stranded DNA. The total amount of related DNA base pairs on Earth is estimated at 5.0×10^{37}, and weighs 50 billion tonnes.[2] In comparison, the total mass of the biosphere has been estimated to be as much as 4 TtC (trillion tons of carbon).[3]

DNA stores biological information. The DNA backbone is resistant to cleavage, and both strands of the double-stranded structure store the same biological information. Biological information is replicated as the two strands are separated. A significant portion of DNA (more than 98% for humans) is non-coding, meaning that these sections do not serve as patterns for protein sequences.

The two strands of DNA run in opposite directions to each other and are therefore anti-parallel. Attached to each sugar is one of four types of nucleobases (informally, *bases*). It is the sequence of these four nucleobases along the backbone that encodes biological information. Under the genetic code, RNA strands are translated to specify the sequence of amino acids within proteins. These RNA strands are initially created using DNA strands as a template in a process called transcription.

Within cells, DNA is organized into long structures called chromosomes. During cell division these chromosomes are duplicated in the process of DNA replication, providing each cell its own complete set of chromosomes. Eukaryotic organisms (animals, plants, fungi, and protists) store most of their DNA inside the cell nucleus and some of their DNA in organelles, such as mitochondria or chloroplasts.[4] In contrast, prokaryotes (bacteria and archaea) store their DNA only in the cytoplasm. Within the chromosomes, chromatin proteins such as histones compact and organize DNA. These compact structures guide the interactions between DNA and other proteins, helping control which parts of the DNA are transcribed.

DNA was first isolated by Friedrich Miescher in 1869. Its molecular structure was identified by James Watson and Francis Crick in 1953, whose model-building efforts were guided by X-ray diffraction data acquired by Rosalind Franklin. DNA is used by researchers as a molecular tool to explore physical laws and theories, such as the ergodic theorem and the theory of elasticity. The unique material properties of DNA have made it an attractive molecule for material scientists and engineers interested in micro- and nano-fabrication. Among notable advances in this field are DNA origami and DNA-based hybrid materials.[5]

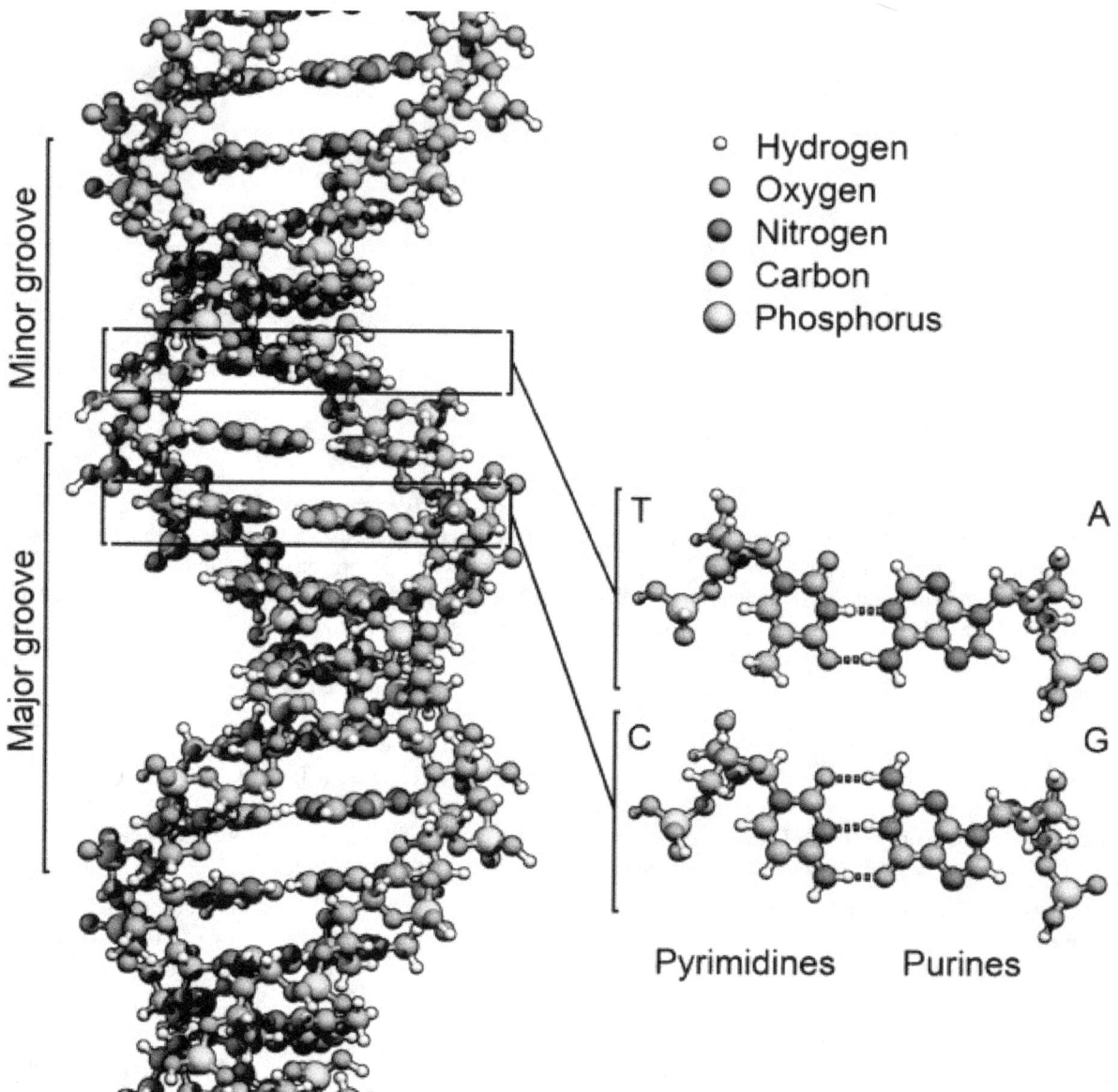

The structure of the DNA double helix. The atoms in the structure are colour-coded by element and the detailed structure of two base pairs are shown in the bottom right.

1.1   Properties

DNA is a long polymer made from repeating units called nucleotides.[6][7] DNA was first identified and isolated by Friedrich Miescher in 1869 at the University of Tübingen, a substance he called *nuclein*, and the double helix structure of DNA was first discovered in 1953 by Watson and Crick at the University of Cambridge, using experimental data collected by Rosalind Franklin and Maurice Wilkins. The structure of DNA is non-static,[8] all species comprises two helical chains each coiled round the same axis, and each with a pitch of 34 ångströms (3.4 nanometres) and a radius of 10 ångströms (1.0 nanometre).[9] According to another study, when measured in a particular solution, the DNA chain measured 22 to 26 ångströms wide (2.2 to 2.6 nanometres), and one nucleotide unit measured 3.3 Å (0.33 nm) long.[10] Although each individual repeating unit is very small, DNA polymers can be very large molecules containing millions of nucleotides. For instance, the DNA in the largest human chromosome, chromosome number 1, consists of approximately 220 million base pairs[11] and would be 85 mm long if straightened.

In living organisms DNA does not usually exist as a single molecule, but instead as a pair of molecules that are held tightly together.[12][13] These two long strands entwine like vines, in the shape of a double helix. The nucleotide repeats contain both the segment of the backbone of the molecule, which holds the chain together, and a nucleobase, which interacts with the other DNA strand in the helix. A nucleobase linked to a sugar is called a nucleoside and a base linked to a sugar and one or more phosphate groups is called a nucleotide. A polymer comprising multiple linked nucleotides (as in DNA) is called a polynucleotide.[14]

The backbone of the DNA strand is made from alternating phosphate and sugar residues.[15] The sugar in DNA is 2-deoxyribose, which is a pentose (five-carbon) sugar. The sugars are joined together by phosphate groups that form phosphodiester bonds between the third and fifth carbon atoms of adjacent sugar rings. These asymmetric bonds mean a strand of DNA has a direction. In a double helix the direction of the nucleotides in one strand is opposite to their direction in the other strand: the strands are *antiparallel*. The asymmetric ends of DNA strands are called the 5′ (*five prime*) and 3′ (*three prime*) ends, with the 5′ end having a terminal phosphate group and the 3′ end a terminal hydroxyl group. One major difference between DNA and RNA is the sugar, with the 2-deoxyribose in DNA being replaced by the alternative pentose sugar ribose in RNA.[13]

The DNA double helix is stabilized primarily by two forces: hydrogen bonds between nucleotides and base-stacking interactions among aromatic nucleobases.[17] In the aqueous environment of the cell, the conjugated π bonds of nucleotide bases align perpendicular to the axis of the DNA molecule, minimizing their interaction with the solvation shell and therefore, the Gibbs free energy. The four bases found in DNA are adenine (abbreviated A), cytosine (C), guanine (G) and thymine (T). These four bases are attached to the sugar/phosphate to form the complete nucleotide, as shown for adenosine monophosphate. Adenine pairs with thymine and guanine pairs with cytosine. It was represented by A-T base pairs and G-C base pairs.[18][19]

1.1.1 Nucleobase classification

The nucleobases are classified into two types: the purines, A and G, being fused five- and six-membered heterocyclic compounds, and the pyrimidines, the six-membered rings C and T.[13] A fifth pyrimidine nucleobase, uracil (U), usually takes the place of thymine in RNA and differs from thymine by lacking a methyl group on its ring. In addition to RNA and DNA a large number of artificial nucleic acid analogues have also been created to study the properties of nucleic acids, or for use in biotechnology.[20]

Uracil is not usually found in DNA, occurring only as a breakdown product of cytosine. However, in a number of bacteriophages – *Bacillus subtilis* bacteriophages PBS1 and PBS2 and *Yersinia* bacteriophage piR1-37 – thymine has been replaced by uracil.[21] Another phage - Staphylococcal phage S6 - has been identified with a genome where thymine has been replaced by uracil.[22]

Base J (beta-d-glucopyranosyloxymethyluracil), a modified form of uracil, is also found in a number of organisms: the flagellates *Diplonema* and *Euglena*, and all the kinetoplastid genera.[23] Biosynthesis of J occurs in two steps: in the first step a specific thymidine in DNA is converted into hydroxymethyldeoxyuridine; in the second HOMedU is glycosylated to form J.[24] Proteins that bind specifically to this base have been identified.[25][26][27] These proteins appear to be distant relatives of the Tet1 oncogene that is involved in the pathogenesis of acute myeloid leukemia.[28] J appears to act as a termination signal for RNA polymerase II.[29][30]

1.1.2 Grooves

Twin helical strands form the DNA backbone. Another double helix may be found tracing the spaces, or grooves, between the strands. These voids are adjacent to the base pairs and may provide a binding site. As the strands are not symmetrically located with respect to each other, the grooves are unequally sized. One groove, the major groove, is 22 Å wide and the other, the minor groove, is 12 Å wide.[31] The width of the major groove means that the edges of the bases are more accessible in the major groove than in the minor groove. As a result, proteins such as transcription factors that can bind to specific sequences in double-stranded DNA usually make contact with the sides of the bases exposed in the major groove.[32] This situation varies in unusual conformations of DNA within the cell (*see below*), but the major and minor grooves are always named to reflect the differences in size that would be seen if the DNA is twisted back into the ordinary B form.

1.1.3 Base pairing

Further information: Base pair

In a DNA double helix, each type of nucleobase on one strand bonds with just one type of nucleobase on the other strand. This is called complementary base pairing. Here, purines form hydrogen bonds to pyrimidines, with adenine bonding only to thymine in two hydrogen bonds, and cytosine bonding only to guanine in three hydrogen bonds. This arrangement of two nucleotides binding together across the double helix is called a base pair. As hydrogen bonds are not covalent, they can be broken and rejoined relatively easily. The two strands of DNA in a double helix can therefore be pulled apart like a zipper, either by a mechanical force or high temperature.[33] As a result of this complementarity, all the information in the double-stranded sequence of a DNA helix is duplicated on each strand, which is vital in DNA replication. Indeed, this reversible and specific interaction between complementary base pairs is critical for all the functions of DNA in living organisms.[7]

Top, a **GC** base pair with three hydrogen bonds. Bottom, an **AT** base pair with two hydrogen bonds. Non-covalent hydrogen bonds between the pairs are shown as dashed lines.

The two types of base pairs form different numbers of hydrogen bonds, AT forming two hydrogen bonds, and GC forming three hydrogen bonds (see figures, right). DNA with high GC-content is more stable than DNA with low GC-content.

As noted above, most DNA molecules are actually two polymer strands, bound together in a helical fashion by noncovalent bonds; this double stranded structure (**dsDNA**) is maintained largely by the intrastrand base stacking interactions, which are strongest for G,C stacks. The two strands can come apart – a process known as melting – to form two single-stranded DNA molecules (**ssDNA**) molecules. Melting occurs at high temperature, low salt and high pH (low pH also melts DNA, but since DNA is unstable due to acid depurination, low pH is rarely used).

The stability of the dsDNA form depends not only on the GC-content (% G,C basepairs) but also on sequence (since stacking is sequence specific) and also length (longer molecules are more stable). The stability can be measured in various ways; a common way is the "melting temperature", which is the temperature at which 50% of the ds molecules are converted to ss molecules; melting temperature is dependent on ionic strength and the concentration of DNA. As a result, it is both the percentage of GC base pairs and the overall length of a DNA double helix that determines the strength of the association between the two strands of DNA. Long DNA helices with a high GC-content have stronger-interacting strands, while short helices with high AT content have weaker-interacting strands.[34] In biology, parts of the DNA double helix that need to separate easily, such as the TATAAT Pribnow box in some promoters, tend to have a high AT content, making the strands easier to pull apart.[35]

In the laboratory, the strength of this interaction can be measured by finding the temperature necessary to break the hydrogen bonds, their melting temperature (also called *Tm* value). When all the base pairs in a DNA double helix melt, the strands separate and exist in solution as two entirely independent molecules. These single-stranded DNA molecules (*ssDNA*) have no single common shape, but some conformations are more stable than others.[36]

1.1.4 Sense and antisense

Further information: Sense (molecular biology)

A DNA sequence is called "sense" if its sequence is the same as that of a messenger RNA copy that is translated into protein.[37] The sequence on the opposite strand is called the "antisense" sequence. Both sense and antisense sequences can exist on different parts of the same strand of DNA (i.e. both strands can contain both sense and antisense sequences). In both prokaryotes and eukaryotes, antisense RNA sequences are produced, but the functions of these RNAs are not entirely clear.[38] One proposal is that antisense RNAs are involved in regulating gene expression through RNA-RNA base pairing.[39]

A few DNA sequences in prokaryotes and eukaryotes, and more in plasmids and viruses, blur the distinction between sense and antisense strands by having overlapping genes.[40] In these cases, some DNA sequences do double duty, encoding one protein when read along one strand, and a second protein when read in the opposite direction along the other strand.

In bacteria, this overlap may be involved in the regulation of gene transcription,[41] while in viruses, overlapping genes increase the amount of information that can be encoded within the small viral genome.[42]

1.1.5 Supercoiling

Further information: DNA supercoil

DNA can be twisted like a rope in a process called DNA supercoiling. With DNA in its "relaxed" state, a strand usually circles the axis of the double helix once every 10.4 base pairs, but if the DNA is twisted the strands become more tightly or more loosely wound.[43] If the DNA is twisted in the direction of the helix, this is positive supercoiling, and the bases are held more tightly together. If they are twisted in the opposite direction, this is negative supercoiling, and the bases come apart more easily. In nature, most DNA has slight negative supercoiling that is introduced by enzymes called topoisomerases.[44] These enzymes are also needed to relieve the twisting stresses introduced into DNA strands during processes such as transcription and DNA replication.[45]

1.1.6 Alternate DNA structures

Further information: Molecular Structure of Nucleic Acids: A Structure for Deoxyribose Nucleic Acid, Molecular models of DNA, and DNA structure

DNA exists in many possible conformations that include A-DNA, B-DNA, and Z-DNA forms, although, only B-DNA and Z-DNA have been directly observed in functional organisms.[15] The conformation that DNA adopts depends on the hydration level, DNA sequence, the amount and direction of supercoiling, chemical modifications of the bases, the type and concentration of metal ions, as well as the presence of polyamines in solution.[46]

The first published reports of A-DNA X-ray diffraction patterns—and also B-DNA—used analyses based on Patterson transforms that provided only a limited amount of structural information for oriented fibers of DNA.[47][48] An alternate analysis was then proposed by Wilkins *et al.*, in 1953, for the *in vivo* B-DNA X-ray diffraction/scattering patterns of highly hydrated DNA fibers in terms of squares of Bessel functions.[49] In the same journal, James Watson and Francis Crick presented their molecular modeling analysis of the DNA X-ray diffraction patterns to suggest that the structure was a double-helix.[9]

Although the "B-DNA form" is most common under the conditions found in cells,[50] it is not a well-defined conformation but a family of related DNA conformations[51] that occur at the high hydration levels present in living cells. Their corresponding X-ray diffraction and scattering patterns are characteristic of molecular paracrystals with a significant degree of disorder.[52][53]

Compared to B-DNA, the A-DNA form is a wider right-handed spiral, with a shallow, wide minor groove and a narrower, deeper major groove. The A form occurs under non-physiological conditions in partially dehydrated samples of DNA, while in the cell it may be produced in hybrid pairings of DNA and RNA strands, as well as in enzyme-DNA complexes.[54][55] Segments of DNA where the bases have been chemically modified by methylation may undergo a larger change in conformation and adopt the Z form. Here, the strands turn about the helical axis in a left-handed spiral, the opposite of the more common B form.[56] These unusual structures can be recognized by specific Z-DNA binding proteins and may be involved in the regulation of transcription.[57]

1.1.7 Alternative DNA chemistry

For a number of years exobiologists have proposed the existence of a shadow biosphere, a postulated microbial biosphere of Earth that uses radically different biochemical and molecular processes than currently known life. One of the proposals was the existence of lifeforms that use arsenic instead of phosphorus in DNA. A report in 2010 of the possibility in the bacterium GFAJ-1, was announced,[58][58][59] though the research was disputed,[59][60] and evidence suggests the bacterium actively prevents the incorporation of arsenic into the DNA backbone and other biomolecules.[61]

1.1.8   Quadruplex structures

Further information: G-quadruplex

At the ends of the linear chromosomes are specialized regions of DNA called telomeres. The main function of these regions is to allow the cell to replicate chromosome ends using the enzyme telomerase, as the enzymes that normally replicate DNA cannot copy the extreme 3′ ends of chromosomes.[62] These specialized chromosome caps also help protect the DNA ends, and stop the DNA repair systems in the cell from treating them as damage to be corrected.[63] In human cells, telomeres are usually lengths of single-stranded DNA containing several thousand repeats of a simple TTAGGG sequence.[64]

These guanine-rich sequences may stabilize chromosome ends by forming structures of stacked sets of four-base units, rather than the usual base pairs found in other DNA molecules. Here, four guanine bases form a flat plate and these flat four-base units then stack on top of each other, to form a stable G-quadruplex structure.[66] These structures are stabilized by hydrogen bonding between the edges of the bases and chelation of a metal ion in the centre of each four-base unit.[67] Other structures can also be formed, with the central set of four bases coming from either a single strand folded around the bases, or several different parallel strands, each contributing one base to the central structure.

In addition to these stacked structures, telomeres also form large loop structures called telomere loops, or T-loops. Here, the single-stranded DNA curls around in a long circle stabilized by telomere-binding proteins.[68] At the very end of the T-loop, the single-stranded telomere DNA is held onto a region of double-stranded DNA by the telomere strand disrupting the double-helical DNA and base pairing to one of the two strands. This triple-stranded structure is called a displacement loop or D-loop.[66]

Branched DNA can form networks containing multiple branches.

1.1.9   Branched DNA

Further information: Branched DNA and DNA nanotechnology

In DNA fraying occurs when non-complementary regions exist at the end of an otherwise complementary double-strand of DNA. However, branched DNA can occur if a third strand of DNA is introduced and contains adjoining regions able to hybridize with the frayed regions of the pre-existing double-strand. Although the simplest example of branched DNA involves only three strands of DNA, complexes involving additional strands and multiple branches are also possible.[69] Branched DNA can be used in nanotechnology to construct geometric shapes, see the section on uses in technology below.

1.2   Chemical modifications and altered DNA packaging

Structure of cytosine with and without the 5-methyl group. Deamination converts 5-methylcytosine into thymine.

1.2.1   Base modifications and DNA packaging

Further information: DNA methylation, Chromatin remodeling

The expression of genes is influenced by how the DNA is packaged in chromosomes, in a structure called chromatin. Base modifications can be involved in packaging, with regions that have low or no gene expression usually containing high levels of methylation of cytosine bases. DNA packaging and its influence on gene expression can also occur by covalent modifications of the histone protein core around which DNA is wrapped in the chromatin structure or else by remodeling carried out by chromatin remodeling complexes (see Chromatin remodeling). There is, further, crosstalk between DNA methylation and histone modification, so they can coordinately affect chromatin and gene expression.[70]

For one example, cytosine methylation, produces 5-methylcytosine, which is important for X-chromosome inactivation.[71] The average level of methylation varies between organisms – the worm *Caenorhabditis elegans* lacks cytosine methylation, while vertebrates have higher levels, with up to 1% of their DNA containing 5-methylcytosine.[72] Despite the importance of 5-methylcytosine, it can deaminate to leave a thymine base, so methylated cytosines are particularly prone to mutations.[73] Other base modifications include adenine methylation in bacteria, the presence of 5-hydroxymethylcytosine in the brain,[74] and the glycosylation of uracil to produce the "J-base" in kinetoplastids.[75][76]

1.2.2 Damage

Further information: DNA damage (naturally occurring), Mutation, DNA damage theory of aging
 DNA can be damaged by many sorts of mutagens, which change the DNA sequence. Mutagens include oxidizing agents, alkylating agents and also high-energy electromagnetic radiation such as ultraviolet light and X-rays. The type of DNA damage produced depends on the type of mutagen. For example, UV light can damage DNA by producing thymine dimers, which are cross-links between pyrimidine bases.[78] On the other hand, oxidants such as free radicals or hydrogen peroxide produce multiple forms of damage, including base modifications, particularly of guanosine, and double-strand breaks.[79] A typical human cell contains about 150,000 bases that have suffered oxidative damage.[80] Of these oxidative lesions, the most dangerous are double-strand breaks, as these are difficult to repair and can produce point mutations, insertions and deletions from the DNA sequence, as well as chromosomal translocations.[81] These mutations can cause cancer. Because of inherent limitations in the DNA repair mechanisms, if humans lived long enough, they would all eventually develop cancer.[82][83] DNA damages that are naturally occurring, due to normal cellular processes that produce reactive oxygen species, the hydrolytic activities of cellular water, etc., also occur frequently. Although most of these damages are repaired, in any cell some DNA damage may remain despite the action of repair processes. These remaining DNA damages accumulate with age in mammalian postmitotic tissues. This accumulation appears to be an important underlying cause of aging.[84][85][86]

Many mutagens fit into the space between two adjacent base pairs, this is called *intercalation*. Most intercalators are aromatic and planar molecules; examples include ethidium bromide, acridines, daunomycin, and doxorubicin. For an intercalator to fit between base pairs, the bases must separate, distorting the DNA strands by unwinding of the double helix. This inhibits both transcription and DNA replication, causing toxicity and mutations.[87] As a result, DNA intercalators may be carcinogens, and in the case of thalidomide, a teratogen.[88] Others such as benzo[*a*]pyrene diol epoxide and aflatoxin form DNA adducts that induce errors in replication.[89] Nevertheless, due to their ability to inhibit DNA transcription and replication, other similar toxins are also used in chemotherapy to inhibit rapidly growing cancer cells.[90]

1.3 Biological functions

DNA usually occurs as linear chromosomes in eukaryotes, and circular chromosomes in prokaryotes. The set of chromosomes in a cell makes up its genome; the human genome has approximately 3 billion base pairs of DNA arranged into 46 chromosomes.[91] The information carried by DNA is held in the sequence of pieces of DNA called genes. Transmission of genetic information in genes is achieved via complementary base pairing. For example, in transcription, when a cell uses the information in a gene, the DNA sequence is copied into a complementary RNA sequence through the attraction between the DNA and the correct RNA nucleotides. Usually, this RNA copy is then used to make a matching protein sequence in a process called translation, which depends on the same interaction between RNA nucleotides. In alternative fashion, a cell may simply copy its genetic information in a process called DNA replication. The details of these functions are covered in other articles; here the focus is on the interactions between DNA and other molecules that mediate the function of the genome.

1.3.1 Genes and genomes

Further information: Cell nucleus, Chromatin, Chromosome, Gene, Noncoding DNA

Genomic DNA is tightly and orderly packed in the process called DNA condensation to fit the small available volumes of the cell. In eukaryotes, DNA is located in the cell nucleus, as well as small amounts in mitochondria and chloroplasts. In prokaryotes, the DNA is held within an irregularly shaped body in the cytoplasm called the nucleoid.[92] The genetic information in a genome is held within genes, and the complete set of this information in an organism is called its genotype. A gene is a unit of heredity and is a region of DNA that influences a particular characteristic in an organism. Genes contain an open reading frame that can be transcribed, as well as regulatory sequences such as promoters and enhancers, which control the transcription of the open reading frame.

In many species, only a small fraction of the total sequence of the genome encodes protein. For example, only about 1.5% of the human genome consists of protein-coding exons, with over 50% of human DNA consisting of non-coding repetitive sequences.[93] The reasons for the presence of so much noncoding DNA in eukaryotic genomes and the extraordinary differences in genome size, or *C-value*, among species represent a long-standing puzzle known as the "C-value enigma".[94] However, some DNA sequences that do not code protein may still encode functional non-coding RNA molecules, which are involved in the regulation of gene expression.[95]

Some noncoding DNA sequences play structural roles in chromosomes. Telomeres and centromeres typically contain few genes, but are important for the function and stability of chromosomes.[63][97] An abundant form of noncoding DNA in humans are pseudogenes, which are copies of genes that have been disabled by mutation.[98] These sequences are usually just molecular fossils, although they can occasionally serve as raw genetic material for the creation of new genes through the process of gene duplication and divergence.[99]

1.3.2   Transcription and translation

Further information: Genetic code, Transcription (genetics), Protein biosynthesis

A gene is a sequence of DNA that contains genetic information and can influence the phenotype of an organism. Within a gene, the sequence of bases along a DNA strand defines a messenger RNA sequence, which then defines one or more protein sequences. The relationship between the nucleotide sequences of genes and the amino-acid sequences of proteins is determined by the rules of translation, known collectively as the genetic code. The genetic code consists of three-letter 'words' called *codons* formed from a sequence of three nucleotides (e.g. ACT, CAG, TTT).

In transcription, the codons of a gene are copied into messenger RNA by RNA polymerase. This RNA copy is then decoded by a ribosome that reads the RNA sequence by base-pairing the messenger RNA to transfer RNA, which carries amino acids. Since there are 4 bases in 3-letter combinations, there are 64 possible codons (4^3 combinations). These encode the twenty standard amino acids, giving most amino acids more than one possible codon. There are also three 'stop' or 'nonsense' codons signifying the end of the coding region; these are the TAA, TGA, and TAG codons.

1.3.3   Replication

Further information: DNA replication

Cell division is essential for an organism to grow, but, when a cell divides, it must replicate the DNA in its genome so that the two daughter cells have the same genetic information as their parent. The double-stranded structure of DNA provides a simple mechanism for DNA replication. Here, the two strands are separated and then each strand's complementary DNA sequence is recreated by an enzyme called DNA polymerase. This enzyme makes the complementary strand by finding the correct base through complementary base pairing, and bonding it onto the original strand. As DNA polymerases can only extend a DNA strand in a 5′ to 3′ direction, different mechanisms are used to copy the antiparallel strands of the double helix.[100] In this way, the base on the old strand dictates which base appears on the new strand, and the cell ends up with a perfect copy of its DNA.

1.3.4   Extracellular nucleic acids

Naked extracellular DNA (eDNA), most of it released by cell death, is nearly ubiquitous in the environment. Its concentration in soil may be as high as 2 μg/L, and its concentration in natural aquatic environments may be as high at 88 µg/L.[101] Various possible functions have been proposed for eDNA: it may be involved in horizontal gene transfer;[102] it may provide nutrients;[103] and it may act as a buffer to recruit or titrate ions or antibiotics.[104] Extracellular DNA acts as a functional extracellular matrix component in the biofilms of a number of bacterial species. It may act as a recognition factor to regulate the attachment and dispersal of specific cell types in the biofilm;[105] it may contribute to biofilm formation;[106] and it may contribute to the biofilm's physical strength and resistance to biological stress.[107]

1.4 Interactions with proteins

All the functions of DNA depend on interactions with proteins. These protein interactions can be non-specific, or the protein can bind specifically to a single DNA sequence. Enzymes can also bind to DNA and of these, the polymerases that copy the DNA base sequence in transcription and DNA replication are particularly important.

1.4.1 DNA-binding proteins

Further information: DNA-binding protein
Interaction of DNA (shown in orange) with histones (shown in blue). These proteins' basic amino acids bind to the acidic phosphate groups on DNA.

Structural proteins that bind DNA are well-understood examples of non-specific DNA-protein interactions. Within chromosomes, DNA is held in complexes with structural proteins. These proteins organize the DNA into a compact structure called chromatin. In eukaryotes this structure involves DNA binding to a complex of small basic proteins called histones, while in prokaryotes multiple types of proteins are involved.[108][109] The histones form a disk-shaped complex called a nucleosome, which contains two complete turns of double-stranded DNA wrapped around its surface. These non-specific interactions are formed through basic residues in the histones making ionic bonds to the acidic sugar-phosphate backbone of the DNA, and are therefore largely independent of the base sequence.[110] Chemical modifications of these basic amino acid residues include methylation, phosphorylation and acetylation.[111] These chemical changes alter the strength of the interaction between the DNA and the histones, making the DNA more or less accessible to transcription factors and changing the rate of transcription.[112] Other non-specific DNA-binding proteins in chromatin include the high-mobility group proteins, which bind to bent or distorted DNA.[113] These proteins are important in bending arrays of nucleosomes and arranging them into the larger structures that make up chromosomes.[114]

A distinct group of DNA-binding proteins are the DNA-binding proteins that specifically bind single-stranded DNA. In humans, replication protein A is the best-understood member of this family and is used in processes where the double helix is separated, including DNA replication, recombination and DNA repair.[115] These binding proteins seem to stabilize single-stranded DNA and protect it from forming stem-loops or being degraded by nucleases.

In contrast, other proteins have evolved to bind to particular DNA sequences. The most intensively studied of these are the various transcription factors, which are proteins that regulate transcription. Each transcription factor binds to one particular set of DNA sequences and activates or inhibits the transcription of genes that have these sequences close to their promoters. The transcription factors do this in two ways. Firstly, they can bind the RNA polymerase responsible for transcription, either directly or through other mediator proteins; this locates the polymerase at the promoter and allows it to begin transcription.[117] Alternatively, transcription factors can bind enzymes that modify the histones at the promoter. This changes the accessibility of the DNA template to the polymerase.[118]

As these DNA targets can occur throughout an organism's genome, changes in the activity of one type of transcription factor can affect thousands of genes.[119] Consequently, these proteins are often the targets of the signal transduction processes that control responses to environmental changes or cellular differentiation and development. The specificity of these transcription factors' interactions with DNA come from the proteins making multiple contacts to the edges of the DNA bases, allowing them to "read" the DNA sequence. Most of these base-interactions are made in the major groove, where the bases are most accessible.[32]

1.4.2 DNA-modifying enzymes

Nucleases and ligases

Nucleases are enzymes that cut DNA strands by catalyzing the hydrolysis of the phosphodiester bonds. Nucleases that hydrolyse nucleotides from the ends of DNA strands are called exonucleases, while endonucleases cut within strands. The most frequently used nucleases in molecular biology are the restriction endonucleases, which cut DNA at specific sequences. For instance, the EcoRV enzyme shown to the left recognizes the 6-base sequence 5′-GATATC-3′ and makes a cut at the vertical line. In nature, these enzymes protect bacteria against phage infection by digesting the phage DNA when it enters the bacterial cell, acting as part of the restriction modification system.[121] In technology, these sequence-specific nucleases are used in molecular cloning and DNA fingerprinting.

Enzymes called DNA ligases can rejoin cut or broken DNA strands.[122] Ligases are particularly important in lagging strand DNA replication, as they join together the short segments of DNA produced at the replication fork into a complete copy of the DNA template. They are also used in DNA repair and genetic recombination.[122]

Topoisomerases and helicases

Topoisomerases are enzymes with both nuclease and ligase activity. These proteins change the amount of supercoiling in DNA. Some of these enzymes work by cutting the DNA helix and allowing one section to rotate, thereby reducing its level of supercoiling; the enzyme then seals the DNA break.[44] Other types of these enzymes are capable of cutting one DNA helix and then passing a second strand of DNA through this break, before rejoining the helix.[123] Topoisomerases are required for many processes involving DNA, such as DNA replication and transcription.[45]

Helicases are proteins that are a type of molecular motor. They use the chemical energy in nucleoside triphosphates, predominantly ATP, to break hydrogen bonds between bases and unwind the DNA double helix into single strands.[124] These enzymes are essential for most processes where enzymes need to access the DNA bases.

Polymerases

Polymerases are enzymes that synthesize polynucleotide chains from nucleoside triphosphates. The sequence of their products are created based on existing polynucleotide chains—which are called *templates*. These enzymes function by repeatedly adding a nucleotide to the 3′ hydroxyl group at the end of the growing polynucleotide chain. As a consequence, all polymerases work in a 5′ to 3′ direction.[125] In the active site of these enzymes, the incoming nucleoside triphosphate base-pairs to the template: this allows polymerases to accurately synthesize the complementary strand of their template. Polymerases are classified according to the type of template that they use.

In DNA replication, DNA-dependent DNA polymerases make copies of DNA polynucleotide chains. In order to preserve biological information, it is essential that the sequence of bases in each copy are precisely complementary to the sequence of bases in the template strand. Many DNA polymerases have a proofreading activity. Here, the polymerase recognizes the occasional mistakes in the synthesis reaction by the lack of base pairing between the mismatched nucleotides. If a mismatch is detected, a 3′ to 5′ exonuclease activity is activated and the incorrect base removed.[126] In most organisms, DNA polymerases function in a large complex called the replisome that contains multiple accessory subunits, such as the DNA clamp or helicases.[127]

RNA-dependent DNA polymerases are a specialized class of polymerases that copy the sequence of an RNA strand into DNA. They include reverse transcriptase, which is a viral enzyme involved in the infection of cells by retroviruses, and telomerase, which is required for the replication of telomeres.[62][128] Telomerase is an unusual polymerase because it contains its own RNA template as part of its structure.[63]

Transcription is carried out by a DNA-dependent RNA polymerase that copies the sequence of a DNA strand into RNA. To begin transcribing a gene, the RNA polymerase binds to a sequence of DNA called a promoter and separates the DNA strands. It then copies the gene sequence into a messenger RNA transcript until it reaches a region of DNA called the terminator, where it halts and detaches from the DNA. As with human DNA-dependent DNA polymerases, RNA polymerase II, the enzyme that transcribes most of the genes in the human genome, operates as part of a large protein complex with multiple regulatory and accessory subunits.[129]

1.5   Genetic recombination

Structure of the Holliday junction intermediate in genetic recombination. The four separate DNA strands are coloured red, blue, green and yellow.[130]

Further information: Genetic recombination

 A DNA helix usually does not interact with other segments of DNA, and in human cells the different chromosomes even occupy separate areas in the nucleus called "chromosome territories".[131] This physical separation of different chromosomes is important for the ability of DNA to function as a stable repository for information, as one of the few times chromosomes interact is in chromosomal crossover which occurs during sexual reproduction, when genetic recombination occurs. Chromosomal crossover is when two DNA helices break, swap a section and then rejoin.

Recombination allows chromosomes to exchange genetic information and produces new combinations of genes, which increases the efficiency of natural selection and can be important in the rapid evolution of new proteins.[132] Genetic recombination can also be involved in DNA repair, particularly in the cell's response to double-strand breaks.[133]

The most common form of chromosomal crossover is homologous recombination, where the two chromosomes involved share very similar sequences. Non-homologous recombination can be damaging to cells, as it can produce chromosomal translocations and genetic abnormalities. The recombination reaction is catalyzed by enzymes known as recombinases, such as RAD51.[134] The first step in recombination is a double-stranded break caused by either an endonuclease or damage to the DNA.[135] A series of steps catalyzed in part by the recombinase then leads to joining of the two helices by at least one Holliday junction, in which a segment of a single strand in each helix is annealed to the complementary strand in the other helix. The Holliday junction is a tetrahedral junction structure that can be moved along the pair of chromosomes, swapping one strand for another. The recombination reaction is then halted by cleavage of the junction and re-ligation of the released DNA.[136]

1.6   Evolution

Further information: RNA world hypothesis

DNA contains the genetic information that allows all modern living things to function, grow and reproduce. However, it is unclear how long in the 4-billion-year history of life DNA has performed this function, as it has been proposed that the earliest forms of life may have used RNA as their genetic material.[137][138] RNA may have acted as the central part of early cell metabolism as it can both transmit genetic information and carry out catalysis as part of ribozymes.[139] This ancient RNA world where nucleic acid would have been used for both catalysis and genetics may have influenced the evolution of the current genetic code based on four nucleotide bases. This would occur, since the number of different bases in such an organism is a trade-off between a small number of bases increasing replication accuracy and a large number of bases increasing the catalytic efficiency of ribozymes.[140] However, there is no direct evidence of ancient genetic systems, as recovery of DNA from most fossils is impossible because DNA survives in the environment for less than one million years, and slowly degrades into short fragments in solution.[141] Claims for older DNA have been made, most notably a report of the isolation of a viable bacterium from a salt crystal 250 million years old,[142] but these claims are controversial.[143][144]

Building blocks of DNA (adenine, guanine and related organic molecules) may have been formed extraterrestrially in outer space.[145][146][147] Complex DNA and RNA organic compounds of life, including uracil, cytosine and thymine, have also been formed in the laboratory under conditions mimicking those found in outer space, using starting chemicals, such as pyrimidine, found in meteorites. Pyrimidine, like polycyclic aromatic hydrocarbons (PAHs), the most carbon-rich chemical found in the universe, may have been formed in red giants or in interstellar dust and gas clouds.[148]

1.7   Uses in technology

1.7.1   Genetic engineering

Further information: Molecular biology, nucleic acid methods and genetic engineering

Methods have been developed to purify DNA from organisms, such as phenol-chloroform extraction, and to manipulate it in the laboratory, such as restriction digests and the polymerase chain reaction. Modern biology and biochemistry make intensive use of these techniques in recombinant DNA technology. Recombinant DNA is a man-made DNA sequence that has been assembled from other DNA sequences. They can be transformed into organisms in the form of plasmids or in the appropriate format, by using a viral vector.[149] The genetically modified organisms produced can be used to produce products such as recombinant proteins, used in medical research,[150] or be grown in agriculture.[151][152]

1.7.2   DNA profiling

Further information: DNA profiling

Forensic scientists can use DNA in blood, semen, skin, saliva or hair found at a crime scene to identify a matching DNA of an individual, such as a perpetrator. This process is formally termed DNA profiling, but may also be called "genetic fingerprinting". In DNA profiling, the lengths of variable sections of repetitive DNA, such as short tandem repeats and minisatellites, are compared between people. This method is usually an extremely reliable technique for identifying a matching DNA.[153] However, identification can be complicated if the scene is contaminated with DNA from several people.[154] DNA profiling was developed in 1984 by British geneticist Sir Alec Jeffreys,[155] and first used in forensic science to convict Colin Pitchfork in the 1988 Enderby murders case.[156]

The development of forensic science, and the ability to now obtain genetic matching on minute samples of blood, skin, saliva or hair has led to a re-examination of a number of cases. Evidence can now be uncovered that was not scientifically possible at the time of the original examination. Combined with the removal of the double jeopardy law in some places, this can allow cases to be reopened where previous trials have failed to produce sufficient evidence to convince a jury. People charged with serious crimes may be required to provide a sample of DNA for matching purposes. The most obvious defence to DNA matches obtained forensically is to claim that cross-contamination of evidence has taken place. This has resulted in meticulous strict handling procedures with new cases of serious crime. DNA profiling is also used to identify victims of mass casualty incidents.[157] As well as positively identifying bodies or body parts in serious accidents, DNA profiling is being successfully used to identify individual victims in mass war graves – matching to family members.

DNA profiling is also used in DNA paternity testing in order to determine if someone is the biologicalparent or grandparent of a child with the probability of parentage is typically 99.99% when the alleged parent is biologically related to the child. Normal DNA sequencing methods happen after birth but there are new methods to test paternity while the mother is still pregnant.[158]

1.7.3   DNA enzymes or catalytic DNA

Further information: Deoxyribozyme

Deoxyribozymes, also called DNAzymes or catalytic DNA are first discovered in 1994.[159] They are mostly single stranded DNA sequences isolated from a large pool of random DNA sequences through a combinatorial approach called in vitro selection or SELEX. DNAzymes catalyze variety of chemical reactions including RNA/DNA cleavage, RNA/DNA ligation, amino acids phosphorylation/dephosphorylation, carbon-carbon bond formation, and etc. DNAzymes can enhance catalytic rate of chemical reactions up to 100,000,000,000-fold over the uncatalyzed reaction.[160] The most extensively studied class of DNAzymes are RNA-cleaving DNAzymes which have been used in detection of different metal ions and designing therapeutic agents. Several metal-specific DNAzymes have been reported including the GR-5 DNAzyme (lead-specific),[159] the CA1-3 DNAzymes (copper-specific),[161] the 39E DNAzyme (uranyl-specific) and the NaA43

DNAzyme (sodium-specific).[162] The NaA43 DNAzyme, which is reported to be more than 10,000-fold selective for sodium over other metal ions, was used to make a real-time sodium sensor in living cells.

1.7.4 Bioinformatics

Further information: Bioinformatics

Bioinformatics involves the development of techniques to store, data mine, search and manipulate biological data, including DNA nucleic acid sequence data. These have led to widely applied advances in computer science, especially string searching algorithms, machine learning and database theory.[163] String searching or matching algorithms, which find an occurrence of a sequence of letters inside a larger sequence of letters, were developed to search for specific sequences of nucleotides.[164] The DNA sequence may be aligned with other DNA sequences to identify homologous sequences and locate the specific mutations that make them distinct. These techniques, especially multiple sequence alignment, are used in studying phylogenetic relationships and protein function.[165] Data sets representing entire genomes' worth of DNA sequences, such as those produced by the Human Genome Project, are difficult to use without the annotations that identify the locations of genes and regulatory elements on each chromosome. Regions of DNA sequence that have the characteristic patterns associated with protein- or RNA-coding genes can be identified by gene finding algorithms, which allow researchers to predict the presence of particular gene products and their possible functions in an organism even before they have been isolated experimentally.[166] Entire genomes may also be compared, which can shed light on the evolutionary history of particular organism and permit the examination of complex evolutionary events.

1.7.5 DNA nanotechnology

Further information: DNA nanotechnology

DNA nanotechnology uses the unique molecular recognition properties of DNA and other nucleic acids to create self-assembling branched DNA complexes with useful properties.[167] DNA is thus used as a structural material rather than as a carrier of biological information. This has led to the creation of two-dimensional periodic lattices (both tile-based and using the "DNA origami" method) as well as three-dimensional structures in the shapes of polyhedra.[168] Nanomechanical devices and algorithmic self-assembly have also been demonstrated,[169] and these DNA structures have been used to template the arrangement of other molecules such as gold nanoparticles and streptavidin proteins.[170]

1.7.6 History and anthropology

Further information: Phylogenetics and Genetic genealogy

Because DNA collects mutations over time, which are then inherited, it contains historical information, and, by comparing DNA sequences, geneticists can infer the evolutionary history of organisms, their phylogeny.[171] This field of phylogenetics is a powerful tool in evolutionary biology. If DNA sequences within a species are compared, population geneticists can learn the history of particular populations. This can be used in studies ranging from ecological genetics to anthropology; For example, DNA evidence is being used to try to identify the Ten Lost Tribes of Israel.[172][173]

1.7.7 Information storage

Main article: DNA digital data storage

In a paper published in *Nature* in January 2013, scientists from the European Bioinformatics Institute and Agilent Technologies proposed a mechanism to use DNA's ability to code information as a means of digital data storage. The group was able to encode 739 kilobytes of data into DNA code, synthesize the actual DNA, then sequence the DNA and decode the information back to its original form, with a reported 100% accuracy. The encoded information consisted of text files and audio files. A prior experiment was published in August 2012. It was conducted by researchers at Harvard University, where the text of a 54,000-word book was encoded in DNA.[174][175]

1.8   History of DNA research

Further information: History of molecular biology

DNA was first isolated by the Swiss physician Friedrich Miescher who, in 1869, discovered a microscopic substance in the pus of discarded surgical bandages. As it resided in the nuclei of cells, he called it "nuclein".[176][177] In 1878, Albrecht Kossel isolated the non-protein component of "nuclein", nucleic acid, and later isolated its five primary nucleobases.[178][179] In 1919, Phoebus Levene identified the base, sugar and phosphate nucleotide unit.[180] Levene suggested that DNA consisted of a string of nucleotide units linked together through the phosphate groups. Levene thought the chain was short and the bases repeated in a fixed order. In 1937, William Astbury produced the first X-ray diffraction patterns that showed that DNA had a regular structure.[181]

In 1927, Nikolai Koltsov proposed that inherited traits would be inherited via a "giant hereditary molecule" made up of "two mirror strands that would replicate in a semi-conservative fashion using each strand as a template".[182][183] In 1928, Frederick Griffith in his experiment discovered that traits of the "smooth" form of *Pneumococcus* could be transferred to the "rough" form of the same bacteria by mixing killed "smooth" bacteria with the live "rough" form.[184][185] This system provided the first clear suggestion that DNA carries genetic information—the Avery–MacLeod–McCarty experiment— when Oswald Avery, along with coworkers Colin MacLeod and Maclyn McCarty, identified DNA as the transforming principle in 1943.[186] DNA's role in heredity was confirmed in 1952, when Alfred Hershey and Martha Chase in the Hershey–Chase experiment showed that DNA is the genetic material of the T2 phage.[187]

In 1953, James Watson and Francis Crick suggested what is now accepted as the first correct double-helix model of DNA structure in the journal *Nature*.[9] Their double-helix, molecular model of DNA was then based on a single X-ray diffraction image (labeled as "Photo 51")[188] taken by Rosalind Franklin and Raymond Gosling in May 1952, as well as the information that the DNA bases are paired—also obtained through private communications from Erwin Chargaff in the previous years.

Experimental evidence supporting the Watson and Crick model was published in a series of five articles in the same issue of *Nature*.[189] Of these, Franklin and Gosling's paper was the first publication of their own X-ray diffraction data and original analysis method that partially supported the Watson and Crick model;[48][190] this issue also contained an article on DNA structure by Maurice Wilkins and two of his colleagues, whose analysis and *in vivo* B-DNA X-ray patterns also supported the presence *in vivo* of the double-helical DNA configurations as proposed by Crick and Watson for their double-helix molecular model of DNA in the previous two pages of *Nature*.[49] In 1962, after Franklin's death, Watson, Crick, and Wilkins jointly received the Nobel Prize in Physiology or Medicine.[191] Nobel Prizes were awarded only to living recipients at the time. A debate continues about who should receive credit for the discovery.[192]

In an influential presentation in 1957, Crick laid out the central dogma of molecular biology, which foretold the relationship between DNA, RNA, and proteins, and articulated the "adaptor hypothesis".[193] Final confirmation of the replication mechanism that was implied by the double-helical structure followed in 1958 through the Meselson–Stahl experiment.[194] Further work by Crick and coworkers showed that the genetic code was based on non-overlapping triplets of bases, called codons, allowing Har Gobind Khorana, Robert W. Holley and Marshall Warren Nirenberg to decipher the genetic code.[195] These findings represent the birth of molecular biology.

1.9   See also

- Autosome

- Crystallography

- DNA-encoded chemical library

- DNA microarray

- DNA sequencing

- DNA, RNA and proteins: The three essential macromolecules of life

- Genetic disorder

- Haplotype

- Nucleic acid modeling

- Meiosis

- Nucleic acid double helix

- Nucleic acid notation

- Nucleic acid sequence

- Pangenesis

- Phosphoramidite

- Southern blot

- X-ray scattering techniques

- Xeno nucleic acid

- *Proteopedia DNA*

- RNA

- Deoxyribozyme

1.10 References

[1] Purcell, Adam. "DNA". *Basic Biology.*

[2] Nuwer, Rachel (18 July 2015). "Counting All the DNA on Earth". *The New York Times* (New York: The New York Times Company). ISSN 0362-4331. Retrieved 2015-07-18.

[3] "The Biosphere: Diversity of Life". *Aspen Global Change Institute.* Basalt, CO. Retrieved 2015-07-19.

[4] Russell, Peter (2001). *iGenetics.* New York: Benjamin Cummings. ISBN 0-8053-4553-1.

[5] Mashaghi A, Katan A (2013). "A physicist's view of DNA". *De Physicus* **24e**(3): 59–61. arXiv:1311.2545v1. Bibcode:2013

[6] Saenger, Wolfram (1984). *Principles of Nucleic Acid Structure.* New York: Springer-Verlag. ISBN 0-387-90762-9.

[7] Alberts, Bruce; Johnson, Alexander; Lewis, Julian; Raff, Martin; Roberts, Keith; Walters, Peter (2002). *Molecular Biology of the Cell; Fourth Edition.* New York and London: Garland Science. ISBN 0-8153-3218-1. OCLC 145080076 48122761 57023651 69932405.

[8] Irobalieva, Rossitza N.; Fogg, Jonathan M.; Catanese Jr, Daniel J.; Sutthibutpong, Thana; Chen, Muyuan; Barker, Anna K.; Ludtke, Steven J.; Harris, Sarah A.; Schmid, Michael F. (2015-10-12). "Structural diversity of supercoiled DNA". *Nature Communications* **6**. doi:10.1038/ncomms9440. PMC 4608029. PMID 26455586.

[9] Watson JD, Crick FH (1953). "A Structure for Deoxyribose Nucleic Acid"(PDF). *Nature* **171**(4356): 737–738. Bibcode:1953 doi:10.1038/171737a0. PMID 13054692.

[10] Mandelkern M, Elias JG, Eden D, Crothers DM (1981). "The dimensions of DNA in solution". *J Mol Biol* **152** (1): 153–61. doi:10.1016/0022-2836(81)90099-1. PMID 7338906.

[11] Gregory SG, Barlow KF, McLay KE, Kaul R, Swarbreck D, Dunham A; et al. (2006). "The DNA sequence and biological annotation of human chromosome 1". *Nature* **441** (7091): 315–21. Bibcode:2006Natur.441..315G. doi:10.1038/nature04727. PMID 16710414.

[12] Watson JD, Crick FH (1953). "A Structure for Deoxyribose Nucleic Acid"(PDF).*Nature***171**(4356): 737–738. Bibcode:195 doi:10.1038/171737a0. PMID 13054692. Retrieved 4 May 2009.

[13] Berg J., Tymoczko J. and Stryer L. (2002) *Biochemistry.* W. H. Freeman and Company ISBN 0-7167-4955-6

[14] Abbreviations and Symbols for Nucleic Acids, Polynucleotides and their Constituents IUPAC-IUB Commission on Biochemical Nomenclature (CBN). Retrieved 3 January 2006.

[15] Ghosh A, Bansal M (2003). "A glossary of DNA structures from A to Z". *Acta Crystallogr D***59**(4): 620–6. doi:10.1107/S090 PMID 12657780.

[16] Created from PDB 1D65

[17] Yakovchuk P, Protozanova E, Frank-Kamenetskii MD (2006). "Base-stacking and base-pairing contributions into thermal stability of the DNA double helix". *Nucleic Acids Res.* **34** (2): 564–74. doi:10.1093/nar/gkj454. PMC 1360284. PMID 16449200.

[18] Burton E. Tropp - *"Molecular Biology"*- Jones and Barlett Learning, ISBN 978-0-7637-8663-2

[19] https://www.mun.ca - *Watson-Crick Structure of DNA - 1953*

[20] Verma S, Eckstein F (1998). "Modified oligonucleotides: synthesis and strategy for users". *Annu. Rev. Biochem.* **67**: 99–134. doi:10.1146/annurev.biochem.67.1.99. PMID 9759484.

[21] Kiljunen S, Hakala K, Pinta E, Huttunen S, Pluta P, Gador A, Lönnberg H, Skurnik M (2005). "Yersiniophage phiR1-37 is a tailed bacteriophage having a 270 kb DNA genome with thymidine replaced by deoxyuridine". *Microbiology* **151** (12): 4093–4102. doi:10.1099/mic.0.28265-0. PMID 16339954.

[22] Uchiyama J, Takemura-Uchiyama I, Sakaguchi Y, Gamoh K, Kato SI, Daibata M, Ujihara T, Misawa N, Matsuzaki S (2014) Intragenus generalized transduction in *Staphylococcus* spp. by a novel giant phage. ISME J. 2014 Mar 6. doi:10.1038/ismej.2014.29

[23] Simpson L (1998). "A base called J". *Proc Natl Acad Sci USA***95**(5): 2037–2038. Bibcode:1998PNAS...95.2037S.doi:10.1073/ PMC 33841. PMID 9482833.

[24] Borst P, Sabatini R (2008). "Base J: discovery, biosynthesis, and possible functions". *Annual review of microbiology* **62**: 235–51. doi:10.1146/annurev.micro.62.081307.162750. PMID 18729733.

[25] Cross M, Kieft R, Sabatini R, Wilm M, de Kort M, van der Marel GA, van Boom JH, van Leeuwen F, Borst P (1999). "The modified base J is the target for a novel DNA-binding protein in kinetoplastid protozoans". *The EMBO Journal* **18** (22): 6573–6581. doi:10.1093/emboj/18.22.6573. PMC 1171720. PMID 10562569.

[26] DiPaolo C, Kieft R, Cross M, Sabatini R (2005). "Regulation of trypanosome DNA glycosylation by a SWI2/SNF2-like protein". *Mol Cell* **17** (3): 441–451. doi:10.1016/j.molcel.2004.12.022. PMID 15694344.

[27] Vainio S, Genest PA, ter Riet B, van Luenen H, Borst P (2009). "Evidence that J-binding protein 2 is a thymidine hydroxylase catalyzing the first step in the biosynthesis of DNA base J". *Molecular and biochemical parasitology* **164** (2): 157–61. doi:10.1016/j.molbiopara.2008.12.001. PMID 19114062.

[28] Iyer LM, Tahiliani M, Rao A, Aravind L (2009). "Prediction of novel families of enzymes involved in oxidative and other complex modifications of bases in nucleic acids". *Cell Cycle* **8** (11): 1698–1710. doi:10.4161/cc.8.11.8580. PMC 2995806. PMID 19411852.

[29] van Luenen HG, Farris C, Jan S, Genest PA, Tripathi P, Velds A, Kerkhoven RM, Nieuwland M, Haydock A, Ramasamy G, Vainio S, Heidebrecht T, Perrakis A, Pagie L, van Steensel B, Myler PJ, Borst P (2012). "Leishmania". *Cell* **150** (5): 909–921. doi:10.1016/j.cell.2012.07.030. PMC 3684241. PMID 22939620.

[30] Hazelbaker DZ, Buratowski S (2012). "Transcription: base J blocks the way". *Curr Biol***22**(22): R960–2. doi:10.1016/j.cub.20 PMC 3648658. PMID 23174300.

[31] Wing R, Drew H, Takano T, Broka C, Tanaka S, Itakura K, Dickerson RE (1980). "Crystal structure analysis of a complete turn of B-DNA". *Nature* **287** (5784): 755–8. Bibcode:1980Natur.287..755W. doi:10.1038/287755a0. PMID 7432492.

[32] Pabo CO, Sauer RT (1984). "Protein-DNA recognition". *Annu Rev Biochem* **53**: 293–321. doi:10.1146/annurev.bi.53.070184 PMID 6236744.

[33] Clausen-Schaumann H, Rief M, Tolksdorf C, Gaub HE (2000). "Mechanical stability of single DNA molecules". *Biophys J* **78** (4): 1997–2007. Bibcode:2000BpJ....78.1997C. doi:10.1016/S0006-3495(00)76747-6. PMC 1300792. PMID 10733978.

[34] Chalikian TV, Völker J, Plum GE, Breslauer KJ (1999). "A more unified picture for the thermodynamics of nucleic acid duplex melting: A characterization by calorimetric and volumetric techniques". *Proc Natl Acad Sci USA* **96** (14): 7853–8. Bibcode:1999PNAS...96.7853C. doi:10.1073/pnas.96.14.7853. PMC 22151. PMID 10393911.

[35] deHaseth PL, Helmann JD (1995). "Open complex formation by Escherichia coli RNA polymerase: the mechanism of polymerase-induced strand separation of double helical DNA". *Mol Microbiol* **16** (5): 817–24. doi:10.1111/j.1365-2958.1995. PMID 7476180.

[36] Isaksson J, Acharya S, Barman J, Cheruku P, Chattopadhyaya J (2004). "Single-stranded adenine-rich DNA and RNA retain structural characteristics of their respective double-stranded conformations and show directional differences in stacking pattern". *Biochemistry* **43** (51): 15996–6010. doi:10.1021/bi048221v. PMID 15609994.

[37] Designation of the two strands of DNA JCBN/NC-IUB Newsletter 1989. Retrieved 7 May 2008

[38] Hüttenhofer A, Schattner P, Polacek N (2005). "Non-coding RNAs: hope or hype?". *Trends Genet* **21** (5): 289–97. doi:10.1016/ PMID 15851066.

[39] Munroe SH (2004). "Diversity of antisense regulation in eukaryotes: multiple mechanisms, emerging patterns". *J Cell Biochem* **93** (4): 664–71. doi:10.1002/jcb.20252. PMID 15389973.

[40] Makalowska I, Lin CF, Makalowski W (2005). "Overlapping genes in vertebrate genomes". *Comput Biol Chem* **29** (1): 1–12. doi:10.1016/j.compbiolchem.2004.12.006. PMID 15680581.

[41] Johnson ZI, Chisholm SW (2004). "Properties of overlapping genes are conserved across microbial genomes". *Genome Res* **14** (11): 2268–72. doi:10.1101/gr.2433104. PMC 525685. PMID 15520290.

[42] Lamb RA, Horvath CM (1991). "Diversity of coding strategies in influenza viruses". *Trends Genet* **7** (8): 261–6. doi:10.1016/0168-9525(91)90326-L. PMID 1771674.

[43] Benham CJ, Mielke SP (2005). "DNA mechanics". *Annu Rev Biomed Eng* **7**: 21–53. doi:10.1146/annurev.bioeng.6.062403.1320 PMID 16004565.

[44] Champoux JJ (2001). "DNA topoisomerases: structure, function, and mechanism". *Annu Rev Biochem* **70**: 369–413. doi:10.1146 PMID 11395412.

[45] Wang JC (2002). "Cellular roles of DNA topoisomerases: a molecular perspective". *Nature Reviews Molecular Cell Biology* **3** (6): 430–40. doi:10.1038/nrm831. PMID 12042765.

[46] Basu HS, Feuerstein BG, Zarling DA, Shafer RH, Marton LJ (1988). "Recognition of Z-RNA and Z-DNA determinants by polyamines in solution: experimental and theoretical studies". *J Biomol Struct Dyn* **6** (2): 299–309. doi:10.1080/07391102.1988. PMID 2482766.

[47] Franklin RE, Gosling RG (6 March 1953). "The Structure of Sodium Thymonucleate Fibres I. The Influence of Water Content" (PDF). *Acta Crystallogr* **6** (8–9): 673–7. doi:10.1107/S0365110X53001939. Archived from the original (PDF) on 2007-06-12.
Franklin RE, Gosling RG (1953). "The structure of sodium thymonucleate fibres. II. The cylindrically symmetrical Patterson function". *Acta Crystallogr* **6** (8–9): 678–85. doi:10.1107/S0365110X53001940.

[48] Franklin RE, Gosling RG (1953). "Molecular Configuration in Sodium Thymonucleate. Franklin R. and Gosling R.G" (PDF). *Nature* **171** (4356): 740–1. Bibcode:1953Natur.171..740F. doi:10.1038/171740a0. PMID 13054694.

[49] Wilkins MH, Stokes AR, Wilson HR (1953). "Molecular Structure of Deoxypentose Nucleic Acids" (PDF). *Nature* **171** (4356): 738–740. Bibcode:1953Natur.171..738W. doi:10.1038/171738a0. PMID 13054693.

[50] Leslie AG, Arnott S, Chandrasekaran R, Ratliff RL (1980). "Polymorphism of DNA double helices". *J. Mol. Biol.* **143** (1): 49–72. doi:10.1016/0022-2836(80)90124-2. PMID 7441761.

[51] Baianu, I.C. (1980). "Structural Order and Partial Disorder in Biological systems". *Bull. Math. Biol.* **42** (4): 137–141. doi:10.1016/s0092-8240(80)80083-8. http://cogprints.org/3822/

[52] Hosemann R., Bagchi R.N., *Direct analysis of diffraction by matter*, North-Holland Publs., Amsterdam – New York, 1962.

[53] Baianu, I.C. (1978). "X-ray scattering by partially disordered membrane systems". *Acta Crystallogr A* **34** (5): 751–753. Bibcode:1978AcCrA..34..751B. doi:10.1107/S0567739478001540.

[54] Wahl MC, Sundaralingam M (1997). "Crystal structures of A-DNA duplexes". *Biopolymers***44**(1): 45–63. doi:10.1002/(SICI) 0282(1997)44:1<45::AID-BIP4>3.0.CO;2-#. PMID 9097733.

[55] Lu XJ, Shakked Z, Olson WK (2000). "A-form conformational motifs in ligand-bound DNA structures". *J. Mol. Biol.* **300** (4): 819–40. doi:10.1006/jmbi.2000.3690. PMID 10891271.

[56] Rothenburg S, Koch-Nolte F, Haag F (2001). "DNA methylation and Z-DNA formation as mediators of quantitative differences in the expression of alleles". *Immunol Rev* **184**: 286–98. doi:10.1034/j.1600-065x.2001.1840125.x. PMID 12086319.

[57] Oh DB, Kim YG, Rich A (2002). "Z-DNA-binding proteins can act as potent effectors of gene expression in vivo". *Proc. Natl. Acad. Sci. U.S.A.* **99** (26): 16666–71. Bibcode:2002PNAS...9916666O. doi:10.1073/pnas.262672699. PMC 139201. PMID 12486233.

[58] Palmer, Jason (2 December 2010). "Arsenic-loving bacteria may help in hunt for alien life". *BBC News*. Retrieved 2 December 2010.

[59] Bortman, Henry (2 December 2010). "Arsenic-Eating Bacteria Opens New Possibilities for Alien Life". *Space.com*. Retrieved 2 December 2010.

[60] Katsnelson, Alla (2 December 2010). "Arsenic-eating microbe may redefine chemistry of life". *Nature News*. doi:10.1038/news

[61] Cressey, Daniel (3 October 2012). "'Arsenic-life' Bacterium Prefers Phosphorus after all". *Nature News*. doi:10.1038/nature.20

[62] Greider CW, Blackburn EH (1985). "Identification of a specific telomere terminal transferase activity in Tetrahymena extracts". *Cell* **43** (2 Pt 1): 405–13. doi:10.1016/0092-8674(85)90170-9. PMID 3907856.

[63] Nugent CI, Lundblad V (1998). "The telomerase reverse transcriptase: components and regulation". *Genes Dev* **12** (8): 1073–85. doi:10.1101/gad.12.8.1073. PMID 9553037.

[64] Wright WE, Tesmer VM, Huffman KE, Levene SD, Shay JW (1997). "Normal human chromosomes have long G-rich telomeric overhangs at one end". *Genes Dev* **11** (21): 2801–9. doi:10.1101/gad.11.21.2801. PMC 316649. PMID 9353250.

[65] Created from NDB UD0017

[66] Burge S, Parkinson GN, Hazel P, Todd AK, Neidle S (2006). "Quadruplex DNA: sequence, topology and structure". *Nucleic Acids Res* **34** (19): 5402–15. doi:10.1093/nar/gkl655. PMC 1636468. PMID 17012276.

[67] Parkinson GN, Lee MP, Neidle S (2002). "Crystal structure of parallel quadruplexes from human telomeric DNA". *Nature* **417** (6891): 876–80. Bibcode:2002Natur.417..876P. doi:10.1038/nature755. PMID 12050675.

[68] Griffith JD, Comeau L, Rosenfield S, Stansel RM, Bianchi A, Moss H, de Lange T (1999). "Mammalian telomeres end in a large duplex loop". *Cell* **97** (4): 503–14. doi:10.1016/S0092-8674(00)80760-6. PMID 10338214.

[69] Seeman NC (2005). "DNA enables nanoscale control of the structure of matter". *Q. Rev. Biophys.***38**(4): 363–71. doi:10.1017/ PMC 3478329. PMID 16515737.

[70] Hu Q, Rosenfeld MG (2012). "Epigenetic regulation of human embryonic stem cells". *Frontiers in Genetics***3**: 238. doi:10.3389/ PMC 3488762. PMID 23133442.

[71] Klose RJ, Bird AP (2006). "Genomic DNA methylation: the mark and its mediators". *Trends Biochem Sci* **31** (2): 89–97. doi:10.1016/j.tibs.2005.12.008. PMID 16403636.

[72] Bird A (2002). "DNA methylation patterns and epigenetic memory". *Genes Dev* **16** (1): 6–21. doi:10.1101/gad.947102. PMID 11782440.

[73] Walsh CP, Xu GL (2006). "Cytosine methylation and DNA repair". *Curr Top Microbiol Immunol*. Current Topics in Microbiology and Immunology **301**: 283–315. doi:10.1007/3-540-31390-7_11. ISBN 3-540-29114-8. PMID 16570853.

[74] Kriaucionis S, Heintz N (2009). "The nuclear DNA base 5-hydroxymethylcytosine is present in Purkinje neurons and the brain". *Science* **324** (5929): 929–30. Bibcode:2009Sci...324..929K. doi:10.1126/science.1169786. PMC 3263819. PMID 19372393.

[75] Ratel D, Ravanat JL, Berger F, Wion D (2006). "N6-methyladenine: the other methylated base of DNA". *BioEssays* **28** (3): 309–15. doi:10.1002/bies.20342. PMC 2754416. PMID 16479578.

[76] Gommers-Ampt JH, Van Leeuwen F, de Beer AL, Vliegenthart JF, Dizdaroglu M, Kowalak JA, Crain PF, Borst P (1993). "beta-D-glucosyl-hydroxymethyluracil: a novel modified base present in the DNA of the parasitic protozoan T. brucei". *Cell* **75** (6): 1129–36. doi:10.1016/0092-8674(93)90322-H. PMID 8261512.

[77] Created from PDB 1JDG

[78] Douki T, Reynaud-Angelin A, Cadet J, Sage E (2003). "Bipyrimidine photoproducts rather than oxidative lesions are the main type of DNA damage involved in the genotoxic effect of solar UVA radiation". *Biochemistry* **42** (30): 9221–6. doi:10.1021/bi034. PMID12885257.

[79] Cadet J, Delatour T, Douki T, Gasparutto D, Pouget JP, Ravanat JL, Sauvaigo S (1999). "Hydroxyl radicals and DNA base damage". *Mutat Res* **424** (1–2): 9–21. doi:10.1016/S0027-5107(99)00004-4. PMID 10064846.

[80] Beckman KB, Ames BN (1997). "Oxidative decay of DNA". *J. Biol. Chem.* **272** (32): 19633–6. doi:10.1074/jbc.272.32.19633. PMID 9289489.

[81] Valerie K, Povirk LF (2003). "Regulation and mechanisms of mammalian double-strand break repair". *Oncogene* **22** (37): 5792–812. doi:10.1038/sj.onc.1206679. PMID 12947387.

[82] Johnson, George (28 December 2010). "Unearthing Prehistoric Tumors, and Debate". *The New York Times*. If we lived long enough, sooner or later we all would get cancer.

[83] Alberts, B, Johnson A, Lewis J; et al. (2002). "The Preventable Causes of Cancer". *Molecular biology of the cell* (4th ed.). New York: Garland Science. ISBN 0-8153-4072-9. A certain irreducible background incidence of cancer is to be expected regardless of circumstances: mutations can never be absolutely avoided, because they are an inescapable consequence of fundamental limitations on the accuracy of DNA replication, as discussed in Chapter 5. If a human could live long enough, it is inevitable that at least one of his or her cells would eventually accumulate a set of mutations sufficient for cancer to develop.

[84] Bernstein H, Payne CM, Bernstein C, Garewal H, Dvorak K (2008). Cancer and aging as consequences of un-repaired DNA damage. In: New Research on DNA Damages (Editors: Honoka Kimura and Aoi Suzuki) Nova Science Publishers, Inc., New York, Chapter 1, pp. 1–47. open access, but read only https://www.novapublishers.com/catalog/product_info.php?products_id=43247 ISBN 978-1604565812

[85] Hoeijmakers JH (October 2009). "DNA damage, aging, and cancer". *N. Engl. J. Med.* **361**(15): 1475–85. doi:10.1056/NEJMr PMID 19812404.

[86] Freitas AA, de Magalhães JP (2011). "A review and appraisal of the DNA damage theory of ageing". *Mutat. Res.* **728** (1–2): 12–22. doi:10.1016/j.mrrev.2011.05.001. PMID 21600302.

[87] Ferguson LR, Denny WA (1991). "The genetic toxicology of acridines". *Mutat Res* **258** (2): 123–60. doi:10.1016/0165-1110(91)90006-H. PMID 1881402.

[88] Stephens TD, Bunde CJ, Fillmore BJ (2000). "Mechanism of action in thalidomide teratogenesis". *Biochem Pharmacol* **59** (12): 1489–99. doi:10.1016/S0006-2952(99)00388-3. PMID 10799645.

[89] Jeffrey AM (1985). "DNA modification by chemical carcinogens". *Pharmacol Ther* **28** (2): 237–72. doi:10.1016/0163-7258(85)90013-0. PMID 3936066.

[90] Braña MF, Cacho M, Gradillas A, de Pascual-Teresa B, Ramos A (2001). "Intercalators as anticancer drugs". *Curr Pharm Des* **7** (17): 1745–80. doi:10.2174/1381612013397113. PMID 11562309.

[91] Venter JC, Adams MD, Myers EW, Li PW, Mural RJ, Sutton GG; et al. (2001). "The sequence of the human genome". *Science* **291** (5507): 1304–51. Bibcode:2001Sci...291.1304V. doi:10.1126/science.1058040. PMID 11181995.

[92] Thanbichler M, Wang SC, Shapiro L (2005). "The bacterial nucleoid: a highly organized and dynamic structure". *J Cell Biochem* **96** (3): 506–21. doi:10.1002/jcb.20519. PMID 15988757.

[93] Wolfsberg TG, McEntyre J, Schuler GD (2001). "Guide to the draft human genome". *Nature* **409**(6822): 824–6. Bibcode:2001 doi:10.1038/35057000. PMID 11236998.

[94] Gregory TR (2005). "The C-value enigma in plants and animals: a review of parallels and an appeal for partnership". *Annals of Botany* **95** (1): 133–46. doi:10.1093/aob/mci009. PMID 15596463.

[95]   Birney E, Stamatoyannopoulos JA, Dutta A, Guigó R, Gingeras TR, Margulies EH; et al. (2007). "Identification and analysis of functional elements in 1% of the human genome by the ENCODE pilot project". *Nature* **447** (7146): 799–816. Bibcode:2007Natur.447..799B. doi:10.1038/nature05874. PMC 2212820. PMID 17571346.

[96]   Created from PDB 1MSW

[97]   Pidoux AL, Allshire RC (2005). "The role of heterochromatin in centromere function". *Philosophical Transactions of the Royal Society B* **360** (1455): 569–79. doi:10.1098/rstb.2004.1611. PMC 1569473. PMID 15905142.

[98]   Harrison PM, Hegyi H, Balasubramanian S, Luscombe NM, Bertone P, Echols N, Johnson T, Gerstein M (2002). "Molecular Fossils in the Human Genome: Identification and Analysis of the Pseudogenes in Chromosomes 21 and 22". *Genome Res* **12** (2): 272–80. doi:10.1101/gr.207102. PMC 155275. PMID 11827946.

[99]   Harrison PM, Gerstein M (2002). "Studying genomes through the aeons: protein families, pseudogenes and proteome evolution". *J Mol Biol* **318** (5): 1155–74. doi:10.1016/S0022-2836(02)00109-2. PMID 12083509.

[100] Albà M (2001). "Replicative DNA polymerases". *Genome Biol* **2** (1): reviews3002.1–reviews3002.4. doi:10.1186/gb-2001-2-1-reviews3002. PMC 150442. PMID 11178285.

[101] Tani, Katsuji; Nasu, Masao (2010). "Roles of Extracellular DNA in Bacterial Ecosystems". In Kikuchi, Yo; Rykova, Elena Y. *Extracellular Nucleic Acids*. Springer. pp. 25–38. ISBN 978-3-642-12616-1.

[102] Vlassov, V. V.; Laktionov, P. P.; Rykova, E. Y. (2007). "Extracellular nucleic acids". *Bioessays* **29**: 654–667. doi:10.1002/bies.

[103] Finkel, S. E.; Kolter, R. (2001). "DNA as a nutrient: novel role for bacterial competence gene homologs". *J. Bacteriol.* **183**: 6288–6293. doi:10.1128/JB.183.21.6288-6293.2001.

[104] Mulcahy, H.; Charron-Mazenod, L.; Lewenza, S. (2008). "Extracellular DNA chelates cations and induces antibiotic resistance in *Pseudomonas aeruginosa* biofilms". *PLoSPathog* **4**: e1000213. doi:10.1371/journal.ppat.1000213.

[105] Berne, C.; Kysela, D. T.; Brun, Y. V. (2010). "A bacterial extracellular DNA inhibits settling of motile progeny cells within a biofilm". *Mol. Microbiol.* **77**: 815–829. doi:10.1111/j.1365-2958.2010.07267.x.

[106] Whitchurch, C. B.; Tolker-Nielsen, T.; Ragas, P. C.; Mattick, J. S. (2002). "Extracellular DNA required for bacterial biofilm formation" (PDF). *Science* **295**: 1487. doi:10.1126/science.295.5559.1487.

[107] Hu, W.; Li, L.; Sharma, S.; Wang, J.; McHardy, I.; Lux, R.; Yang, Z.; He, X.; Gimzewski, J. K.; Li, Y.; Shi, W. (2012). "DNA Builds and Strengthens the Extracellular Matrix in Myxococcus xanthus Biofilms by Interacting with Exopolysaccharides". *PLoS ONE* **7** (12): e51905. Bibcode:2012PLoSO...751905H. doi:10.1371/journal.pone.0051905.

[108] Sandman K, Pereira SL, Reeve JN (1998). "Diversity of prokaryotic chromosomal proteins and the origin of the nucleosome". *Cell Mol Life Sci* **54** (12): 1350–64. doi:10.1007/s000180050259. PMID 9893710.

[109] Dame RT (2005). "The role of nucleoid-associated proteins in the organization and compaction of bacterial chromatin". *Mol. Microbiol.* **56** (4): 858–70. doi:10.1111/j.1365-2958.2005.04598.x. PMID 15853876.

[110] Luger K, Mäder AW, Richmond RK, Sargent DF, Richmond TJ (1997). "Crystal structure of the nucleosome core particle at 2.8 A resolution". *Nature* **389** (6648): 251–60. Bibcode:1997Natur.389..251L. doi:10.1038/38444. PMID 9305837.

[111] Jenuwein T, Allis CD (2001). "Translating the histone code". *Science* **293** (5532): 1074–80. doi:10.1126/science.1063127. PMID 11498575.

[112] Ito T (2003). "Nucleosome assembly and remodelling". *Curr Top Microbiol Immunol*. Current Topics in Microbiology and Immunology **274**: 1–22. doi:10.1007/978-3-642-55747-7_1. ISBN 978-3-540-44208-0. PMID 12596902.

[113] Thomas JO (2001). "HMG1 and 2: architectural DNA-binding proteins". *Biochem Soc Trans* **29**(Pt 4): 395–401. doi:10.1042/ PMID 11497996.

[114] Grosschedl R, Giese K, Pagel J (1994). "HMG domain proteins: architectural elements in the assembly of nucleoprotein structures". *Trends Genet* **10** (3): 94–100. doi:10.1016/0168-9525(94)90232-1. PMID 8178371.

[115] Iftode C, Daniely Y, Borowiec JA (1999). "Replication protein A (RPA): the eukaryotic SSB". *Crit Rev Biochem Mol Biol* **34** (3): 141–80. doi:10.1080/10409239991209255. PMID 10473346.

[116] Created from PDB 1LMB

[117] Myers LC, Kornberg RD (2000). "Mediator of transcriptional regulation". *Annu Rev Biochem* **69**: 729–49. doi:10.1146/annurev PMID 10966474.

[118] Spiegelman BM, Heinrich R (2004). "Biological control through regulated transcriptional coactivators". *Cell* **119** (2): 157–67. doi:10.1016/j.cell.2004.09.037. PMID 15479634.

[119] Li Z, Van Calcar S, Qu C, Cavenee WK, Zhang MQ, Ren B (2003). "A global transcriptional regulatory role for c-Myc in Burkitt's lymphoma cells". *Proc Natl Acad Sci USA* **100** (14): 8164–9. Bibcode:2003PNAS..100.8164L. doi:10.1073/pnas.1332 PMC166200. PMID12808131.

[120] Created from PDB 1RVA

[121] Bickle TA, Krüger DH (1993). "Biology of DNA restriction". *Microbiol Rev* **57** (2): 434–50. PMC 372918. PMID 8336674.

[122] Doherty AJ, Suh SW (2000). "Structural and mechanistic conservation in DNA ligases". *Nucleic Acids Res* **28** (21): 4051–8. doi:10.1093/nar/28.21.4051. PMC 113121. PMID 11058099.

[123] Schoeffler AJ, Berger JM (2005). "Recent advances in understanding structure-function relationships in the type II topoisomerase mechanism". *Biochem Soc Trans* **33** (Pt 6): 1465–70. doi:10.1042/BST20051465. PMID 16246147.

[124] Tuteja N, Tuteja R (2004). "Unraveling DNA helicases. Motif, structure, mechanism and function". *Eur J Biochem* **271** (10): 1849–63. doi:10.1111/j.1432-1033.2004.04094.x. PMID 15128295.

[125] Joyce CM, Steitz TA (1995). "Polymerase structures and function: variations on a theme?". *J Bacteriol* **177** (22): 6321–9. PMC 177480. PMID 7592405.

[126] Hubscher U, Maga G, Spadari S(2002). "Eukaryotic DNA polymerases". *Annu Rev Biochem* **71**: 133–63. doi:10.1146/annurev PMID 12045093.

[127] Johnson A, O'Donnell M (2005). "Cellular DNA replicases: components and dynamics at the replication fork". *Annu Rev Biochem* **74**: 283–315. doi:10.1146/annurev.biochem.73.011303.073859. PMID 15952889.

[128] Tarrago-Litvak L, Andréola ML, Nevinsky GA, Sarih-Cottin L, Litvak S (1 May 1994). "The reverse transcriptase of HIV-1: from enzymology to therapeutic intervention". *FASEB J* **8** (8): 497–503. PMID 7514143.

[129] Martinez E (2002). "Multi-protein complexes in eukaryotic gene transcription". *Plant Mol Biol* **50**(6): 925–47. doi:10.1023/A: PMID 12516863.

[130] Created from PDB 1M6G

[131] Cremer T, Cremer C (2001). "Chromosome territories, nuclear architecture and gene regulation in mammalian cells". *Nature Reviews Genetics* **2** (4): 292–301. doi:10.1038/35066075. PMID 11283701.

[132] Pál C, Papp B, Lercher MJ (2006). "An integrated view of protein evolution". *Nature Reviews Genetics* **7** (5): 337–48. doi:10.1038/nrg1838. PMID 16619049.

[133] O'Driscoll M, Jeggo PA (2006). "The role of double-strand break repair – insights from human genetics". *Nature Reviews Genetics* **7** (1): 45–54. doi:10.1038/nrg1746. PMID 16369571.

[134] Vispé S, Defais M (1997). "Mammalian Rad51 protein: a RecA homologue with pleiotropic functions". *Biochimie* **79** (9–10): 587–92. doi:10.1016/S0300-9084(97)82007-X. PMID 9466696.

[135] Neale MJ, Keeney S (2006). "Clarifying the mechanics of DNA strand exchange in meiotic recombination". *Nature* **442** (7099): 153–8. Bibcode:2006Natur.442..153N. doi:10.1038/nature04885. PMID 16838012.

[136] Dickman MJ, Ingleston SM, Sedelnikova SE, Rafferty JB, Lloyd RG, Grasby JA, Hornby DP (2002). "The RuvABC resolvasome". *Eur J Biochem* **269** (22): 5492–501. doi:10.1046/j.1432-1033.2002.03250.x. PMID 12423347.

[137] Joyce GF (2002). "The antiquity of RNA-based evolution". *Nature* **418** (6894): 214–21. Bibcode:2002Natur.418..214J. doi:10.1038/418214a. PMID 12110897.

[138] Orgel LE (2004). "Prebiotic chemistry and the origin of the RNA world". *Crit Rev Biochem Mol Biol* **39** (2): 99–123. doi:10.1080/10409230490460765. PMID 15217990.

[139] Davenport RJ (2001). "Ribozymes. Making copies in the RNA world". *Science* **292**(5520): 1278. doi:10.1126/science.292.55 PMID 11360970.

[140] Szathmáry E (1992). "What is the optimum size for the genetic alphabet?". *Proc Natl Acad Sci USA* **89** (7): 2614–8. Bibcode:1992PNAS...89.2614S. doi:10.1073/pnas.89.7.2614. PMC 48712. PMID 1372984.

[141] Lindahl T (1993). "Instability and decay of the primary structure of DNA". *Nature***362**(6422): 709–15. Bibcode:1993Natur.3 doi:10.1038/362709a0. PMID 8469282.

[142] Vreeland RH, Rosenzweig WD, Powers DW (2000). "Isolation of a 250 million-year-old halotolerant bacterium from a primary salt crystal". *Nature* **407** (6806): 897–900. doi:10.1038/35038060. PMID 11057666.

[143] Hebsgaard MB, Phillips MJ, Willerslev E (2005). "Geologically ancient DNA: fact or artefact?". *Trends Microbiol* **13** (5): 212–20. doi:10.1016/j.tim.2005.03.010. PMID 15866038.

[144] Nickle DC, Learn GH, Rain MW, Mullins JI, Mittler JE (2002). "Curiously modern DNA for a "250 million-year-old" bacterium". *J Mol Evol* **54** (1): 134–7. doi:10.1007/s00239-001-0025-x. PMID 11734907.

[145] Callahan MP, Smith KE, Cleaves HJ, Ruzicka J, Stern JC, Glavin DP, House CH, Dworkin JP (August 2011). "Carbonaceous meteorites contain a wide range of extraterrestrial nucleobases". *Proc. Natl. Acad. Sci. U.S.A.* **108** (34): 13995–8. Bibcode:201 1PNAS..10813995C.doi:10.1073/pnas.1106493108. PMC3161613. PMID21836052.

[146] Steigerwald, John (8 August 2011). "NASA Researchers: DNA Building Blocks Can Be Made in Space". NASA. Retrieved 10 August 2011.

[147] ScienceDaily Staff (9 August 2011). "DNA Building Blocks Can Be Made in Space, NASA Evidence Suggests". ScienceDaily. Retrieved 9 August 2011.

[148] Marlaire, Ruth (3 March 2015). "NASA Ames Reproduces the Building Blocks of Life in Laboratory". *NASA*. Retrieved 5 March 2015.

[149] Goff SP, Berg P (1976). "Construction of hybrid viruses containing SV40 and lambda phage DNA segments and their propagation in cultured monkey cells". *Cell* **9** (4 PT 2): 695–705. doi:10.1016/0092-8674(76)90133-1. PMID 189942.

[150] Houdebine LM (2007). "Transgenic animal models in biomedical research". *Methods Mol Biol* **360**: 163–202. doi:10.1385/1-59745-165-7:163. ISBN 1-59745-165-7. PMID 17172731.

[151] Daniell H, Dhingra A (2002). "Multigene engineering: dawn of an exciting new era in biotechnology". *Current Opinion in Biotechnology* **13** (2): 136–41. doi:10.1016/S0958-1669(02)00297-5. PMC 3481857. PMID 11950565.

[152] Job D (2002). "Plant biotechnology in agriculture". *Biochimie* **84** (11): 1105–10. doi:10.1016/S0300-9084(02)00013-5. PMID 12595138.

[153] Collins A, Morton NE (1994). "Likelihood ratios for DNA identification". *Proc Natl Acad Sci USA* **91** (13): 6007–11. Bibcode:1994PNAS...91.6007C. doi:10.1073/pnas.91.13.6007. PMC 44126. PMID 8016106.

[154] Weir BS, Triggs CM, Starling L, Stowell LI, Walsh KA, Buckleton J (1997). "Interpreting DNA mixtures". *J Forensic Sci* **42** (2): 213–22. PMID 9068179.

[155] Jeffreys AJ, Wilson V, Thein SL (1985). "Individual-specific 'fingerprints' of human DNA". *Nature* **316** (6023): 76–9. Bibcode:1985Natur.316...76J. doi:10.1038/316076a0. PMID 2989708.

[156] Colin Pitchfork — first murder conviction on DNA evidence also clears the prime suspect Forensic Science Service Accessed 23 December 2006

[157] "DNA Identification in Mass Fatality Incidents". National Institute of Justice. September 2006.

[158] "Paternity Blood Tests That Work Early in a Pregnancy" New York Times June 20, 2012

[159] Breaker, Ronald R.; Joyce, Gerald F. (1994-01-12). "A DNA enzyme that cleaves RNA". *Chemistry & Biology* **1** (4): 223–229. doi:10.1016/1074-5521(94)90014-0. ISSN 1074-5521. PMID 9383394.

[160] Chandra, Madhavaiah; Sachdeva, Amit; Silverman, Scott K. "DNA-catalyzed sequence-specific hydrolysis of DNA". *Nature Chemical Biology* **5** (10): 718–720. doi:10.1038/nchembio.201. PMC 2746877. PMID 19684594.

[161] Carmi, Nir; Shultz, Lisa A.; Breaker, Ronald R. (1996-01-12). "In vitro selection of self-cleaving DNAs". *Chemistry & Biology* **3** (12): 1039–1046. doi:10.1016/S1074-5521(96)90170-2. ISSN 1074-5521. PMID 9000012.

[162] Torabi, Seyed-Fakhreddin; Wu, Peiwen; McGhee, Claire E.; Chen, Lu; Hwang, Kevin; Zheng, Nan; Cheng, Jianjun; Lu, Yi (2015-05-12). "In vitro selection of a sodium-specific DNAzyme and its application in intracellular sensing". *Proceedings of the National Academy of Sciences* **112** (19): 5903–5908. doi:10.1073/pnas.1420361112. ISSN 0027-8424. PMC 4434688. PMID 25918425.

[163] Baldi, Pierre; Brunak, Soren (2001). *Bioinformatics: The Machine Learning Approach*. MIT Press. ISBN 978-0-262-02506-5. OCLC 45951728.

[164] Gusfield, Dan. *Algorithms on Strings, Trees, and Sequences: Computer Science and Computational Biology*. Cambridge University Press, 15 January 1997. ISBN 978-0-521-58519-4.

[165] Sjölander K (2004). "Phylogenomic inference of protein molecular function: advances and challenges". *Bioinformatics* **20** (2): 170–9. doi:10.1093/bioinformatics/bth021. PMID 14734307.

[166] Mount DM (2004). *Bioinformatics: Sequence and Genome Analysis* (2 ed.). Cold Spring Harbor, NY: Cold Spring Harbor Laboratory Press. ISBN 0-87969-712-1. OCLC 55106399.

[167] Rothemund PW (2006). "Folding DNA to create nanoscale shapes and patterns". *Nature***440**(7082): 297–302. Bibcode:2006N doi:10.1038/nature04586. PMID 16541064.

[168] Andersen ES, Dong M, Nielsen MM, Jahn K, Subramani R, Mamdouh W, Golas MM, Sander B, Stark H, Oliveira CL, Pedersen JS, Birkedal V, Besenbacher F, Gothelf KV, Kjems J (2009). "Self-assembly of a nanoscale DNA box with a controllable lid". *Nature* **459** (7243): 73–6. Bibcode:2009Natur.459...73A. doi:10.1038/nature07971. PMID 19424153.

[169] Ishitsuka Y, Ha T (2009). "DNA nanotechnology: a nanomachine goes live". *Nat Nanotechnol***4**(5): 281–2. Bibcode:2009NatN doi:10.1038/nnano.2009.101. PMID 19421208.

[170] Aldaye FA, Palmer AL, Sleiman HF (2008). "Assembling materials with DNA as the guide". *Science* **321** (5897): 1795–9. Bibcode:2008Sci...321.1795A. doi:10.1126/science.1154533. PMID 18818351.

[171] Wray GA (2002). "Dating branches on the Tree of Life using DNA". *Genome Biol* **3** (1): reviews0001.1–reviews0001.7. doi:10.1046/j.1525-142X.1999.99010.x. PMC 150454. PMID 11806830.

[172] *Lost Tribes of Israel*, NOVA, PBS airdate: 22 February 2000. Transcript available from PBS.org. Retrieved 4 March 2006.

[173] Kleiman, Yaakov. "The Cohanim/DNA Connection: The fascinating story of how DNA studies confirm an ancient biblical tradition". *aish.com* (13 January 2000). Retrieved 4 March 2006.

[174] Goldman N, Bertone P, Chen S, Dessimoz C, LeProust EM, Sipos B, Birney E (23 January 2013). "Towards practical, high-capacity, low-maintenance information storage in synthesized DNA". *Nature* **494** (7435): 77–80. Bibcode:2013Natur.494...77G. doi:10.1038/nature11875. PMC 3672958. PMID 23354052.

[175] Naik, Gautam (24 January 2013). "Storing Digital Data in DNA". *Wall Street Journal*. Retrieved 24 January 2013.

[176] Miescher, Friedrich (1871) "Ueber die chemische Zusammensetzung der Eiterzellen" (On the chemical composition of pus cells), *Medicinisch-chemische Untersuchungen*, 4 : 441–460. From p. 456: *"Ich habe mich daher später mit meinen Versuchen an die ganzen Kerne gehalten, die Trennung der Körper, die ich einstweilen ohne weiteres Präjudiz als lösliches und unlösliches Nuclein bezeichnen will, einem günstigeren Material überlassend."* (Therefore, in my experiments I subsequently limited myself to the whole nucleus, leaving to a more favorable material the separation of the substances, that for the present, without further prejudice, I will designate as soluble and insoluble nuclear material ("Nuclein").)

[177] Dahm R (2008). "Discovering DNA: Friedrich Miescher and the early years of nucleic acid research". *Hum. Genet.* **122** (6): 565–81. doi:10.1007/s00439-007-0433-0. PMID 17901982.

[178] See:

- Albrect Kossel (1879) "Ueber Nucleïn der Hefe" (On nuclein in yeast) *Zeitschrift für physiologische Chemie*, **3** : 284-291.
- Albrect Kossel (1880) "Ueber Nucleïn der Hefe II" (On nuclein in yeast, Part 2) *Zeitschrift für physiologische Chemie*, **4** : 290-295.
- Albrect Kossel (1881) "Ueber die Verbreitung des Hypoxanthins im Thier- und Pflanzenreich" (On the distribution of hypoxanthins in the animal and plant kingdoms) *Zeitschrift für physiologische Chemie*, **5** : 267-271.
- Albrect Kossel, *Untersuchungen über die Nucleine und ihre Spaltungsprodukte* [Investigations into nuclein and its cleavage products] (Strassburg, Germany: K.J. Trübne, 1881), 19 pages.

- Albrect Kossel (1882) "Ueber Xanthin und Hypoxanthin" (On xanthin and hypoxanthin), *Zeitschrift für physiologische Chemie*, **6** : 422-431.

- Albrect Kossel (1883) "Zur Chemie des Zellkerns" (On the chemistry of the cell nucleus), *Zeitschrift für physiologische Chemie*, **7** : 7-22.

- Albrect Kossel (1886) "Weitere Beiträge zur Chemie des Zellkerns" (Further contributions to the chemistry of the cell nucleus), *Zeitschrift für Physiologische Chemie*, **10** : 248-264. Available on-line at: Max Planck Institute for the History of Science, Berlin, Germany. On p. 264, Kossel remarked presciently: *"Der Erforschung der quantitativen Verhältnisse der vier stickstoffreichen Basen, der Abhängigkeit ihrer Menge von den physiologischen Zuständen der Zelle, verspricht wichtige Aufschlüsse über die elementaren physiologisch-chemischen Vorgänge."* (The study of the quantitative relations of the four nitrogenous bases — [and] of the dependence of their quantity on the physiological states of the cell — promises important insights into the fundamental physiological-chemical processes.)

[179] Jones ME (September 1953). "Albrecht Kossel, A Biographical Sketch". *Yale Journal of Biology and Medicine* (National Center for Biotechnology Information) **26** (1): 80–97. PMC 2599350. PMID 13103145.

[180] Levene P, (1 December 1919). "The structure of yeast nucleic acid". *J Biol Chem* **40** (2): 415–24.

[181] See:

- W. T. Astbury and Florence O. Bell (1938) "Some recent developments in the X-ray study of proteins and related structures," *Cold Spring Harbor Symposia on Quantitative Biology*, **6** : 109-121. Available on-line at: University of Leeds.

- Astbury, W. T., (1947) "X-ray studies of nucleic acids," *Symposia of the Society for Experimental Biology*, **1** : 66-76. Available on-line at: Oregon State University.

[182] Koltsov proposed that a cell's genetic information was encoded in a long chain of amino acids. See:

- Н. К. Кольцов, "Физико-химические основы морфологии" (The physical-chemical basis of morphology) -- speech given at the 3rd All-Union Meeting of Zoologist, Anatomists, and Histologists at Leningrad, U.S.S.R., December 12, 1927.

- Reprinted in: *Успехи экспериментальной биологии* (Advances in Experimental Biology), series B, 7 (1) : ?-? (1928).

- Reprinted in German as: Nikolaj K. Koltzoff (1928) "Physikalisch-chemische Grundlagen der Morphologie" (The physical-chemical basis of morphology), *Biologisches Zentralblatt*, **48** (6) : 345-369.

- In 1934, Koltsov contended that the proteins that contain a cell's genetic information replicate. See: N. K. Koltzoff (October 5, 1934) "The structure of the chromosomes in the salivary glands of Drosophila," *Science*, **80** (2075) : 312-313. From page 313: "I think that the size of the chromosomes in the salivary glands [of Drosophila] is determined through the multiplication of *genonemes*. By this term I designate the axial thread of the chromosome, in which the geneticists locate the linear combination of genes; … In the normal chromosome there is usually only one genoneme; before cell-division this genoneme has become divided into two strands."

[183] Soyfer VN (2001). "The consequences of political dictatorship for Russian science". *Nature Reviews Genetics* **2** (9): 723–729. doi:10.1038/35088598. PMID 11533721.

[184] Griffith F (January 1928). "The significance of pneumococcal types". *The Journal of Hygiene (London)* **27** (2): 113–59. doi:10.1017/S0022172400031879. PMC 2167760. PMID 20474956.

[185] Lorenz MG, Wackernagel W (1994). "Bacterial gene transfer by natural genetic transformation in the environment". *Microbiol. Rev.* **58** (3): 563–602. PMC 372978. PMID 7968924.

[186] Avery OT, Macleod CM, McCarty M (1944). "Studies on the Chemical Nature of the Substance Inducing Transformation of Pneumococcal Types: Induction of Transformation by a Desoxyribonucleic Acid Fraction Isolated from Pneumococcus Type Iii". *J Exp Med* **79** (2): 137–158. doi:10.1084/jem.79.2.137. PMC 2135445. PMID 19871359.

[187] Hershey AD, Chase M (1952). "Independent Functions of Viral Protein and Nucleic Acid in Growth of Bacteriophage". *J Gen Physiol* **36** (1): 39–56. doi:10.1085/jgp.36.1.39. PMC 2147348. PMID 12981234.

[188] The B-DNA X-ray pattern on the right of this linked image was obtained by Rosalind Franklin and Raymond Gosling in May 1952 at high hydration levels of DNA and it has been labeled as "Photo 51"

[189] Nature Archives Double Helix of DNA: 50 Years

[190] "Original X-ray diffraction image". Osulibrary.oregonstate.edu. Retrieved 6 February 2011.

[191] The Nobel Prize in Physiology or Medicine 1962 Nobelprize .org Accessed 22 December 06

[192] Maddox B (23 January 2003). "The double helix and the 'wronged heroine'"(PDF).*Nature***421**(6921): 407–408. Bibcode:200 doi:10.1038/nature01399. PMID 12540909.

[193] Crick, F.H.C. On degenerate templates and the adaptor hypothesis (PDF). genome.wellcome.ac.uk (Lecture, 1955). Retrieved 22 December 2006.

[194] Meselson M, Stahl FW (1958). "The replication of DNA in Escherichia coli". *Proc Natl Acad Sci USA* **44** (7): 671–82. Bibcode:1958PNAS...44..671M. doi:10.1073/pnas.44.7.671. PMC 528642. PMID 16590258.

[195] The Nobel Prize in Physiology or Medicine 1968 Nobelprize.org Accessed 22 December 06

1.11 Further reading

- Berry, Andrew; Watson, James. (2003). *DNA: the secret of life*. New York: Alfred A. Knopf. ISBN 0-375-41546-7.

- Calladine, Chris R.; Drew, Horace R.; Luisi, Ben F. and Travers, Andrew A. (2003). *Understanding DNA: the molecule & how it works*. Amsterdam: Elsevier Academic Press. ISBN 0-12-155089-3.

- Dennis, Carina; Julie Clayton (2003). *50 years of DNA*. Basingstoke: Palgrave Macmillan. ISBN 1-4039-1479-6.

- Judson, Horace F. 1979. *The Eighth Day of Creation: Makers of the Revolution in Biology*. Touchstone Books, ISBN 0-671-22540-5. 2nd edition: Cold Spring Harbor Laboratory Press, 1996 paperback: ISBN 0-87969-478-5.

- Olby, Robert C. (1994). *The path to the double helix: the discovery of DNA*. New York: Dover Publications. ISBN 0-486-68117-3., first published in October 1974 by MacMillan, with foreword by Francis Crick;the definitive DNA textbook,revised in 1994 with a 9-page postscript

- Micklas, David. 2003. *DNA Science: A First Course*. Cold Spring Harbor Press: ISBN 978-0-87969-636-8

- Ridley, Matt (2006). *Francis Crick: discoverer of the genetic code*. Ashland, OH: Eminent Lives, Atlas Books. ISBN 0-06-082333-X.

- Olby, Robert C. (2009). *Francis Crick: A Biography*. Plainview, N.Y: Cold Spring Harbor Laboratory Press. ISBN 0-87969-798-9.

- Rosenfeld, Israel. 2010. *DNA: A Graphic Guide to the Molecule that Shook the World*. Columbia University Press: ISBN 978-0-231-14271-7

- Schultz, Mark and Zander Cannon. 2009. *The Stuff of Life: A Graphic Guide to Genetics and DNA*. Hill and Wang: ISBN 0-8090-8947-5

- Stent, Gunther Siegmund; Watson, James. (1980). *The double helix: a personal account of the discovery of the structure of DNA*. New York: Norton. ISBN 0-393-95075-1.

- Watson, James. 2004. *DNA: The Secret of Life*. Random House: ISBN 978-0-09-945184-6

- Wilkins, Maurice (2003). *The third man of the double helix the autobiography of Maurice Wilkins*. Cambridge, Eng: University Press. ISBN 0-19-860665-6.

1.12 External links

- DNA at DMOZ

- DNA binding site prediction on protein

- DNA the Double Helix Game From the official Nobel Prize web site

- DNA under electron microscope

- Dolan DNA Learning Center

- Double Helix: 50 years of DNA, *Nature*

- *Proteopedia DNA*

- *Proteopedia Forms_of_DNA*

- ENCODE threads explorer ENCODE Home page. Nature (journal)

- Double Helix 1953–2003 National Centre for Biotechnology Education

- Genetic Education Modules for Teachers—*DNA from the Beginning* Study Guide

- PDB Molecule of the Month *pdb23_1*

- Rosalind Franklin's contributions to the study of DNA

- U.S. National DNA Day—watch videos and participate in real-time chat with top scientists

- Clue to chemistry of heredity found The New York Times June 1953. First American newspaper coverage of the discovery of the DNA structure

- Olby R (2003). "Quiet debut for the double helix". *Nature* **421** (6921): 402–5. Bibcode:2003Natur.421..402O. doi:10.1038/nature01397. PMID 12540907.

- DNA from the Beginning Another DNA Learning Center site on DNA, genes, and heredity from Mendel to the human genome project.

- The Register of Francis Crick Personal Papers 1938 – 2007 at Mandeville Special Collections Library, University of California, San Diego

- Seven-page, handwritten letter that Crick sent to his 12-year-old son Michael in 1953 describing the structure of DNA. See Crick's medal goes under the hammer, Nature, 5 April 2013.

- 3D map of DNA reveals hidden loops that allow genes to work together (11 December 2014), *Science (Daily News)*

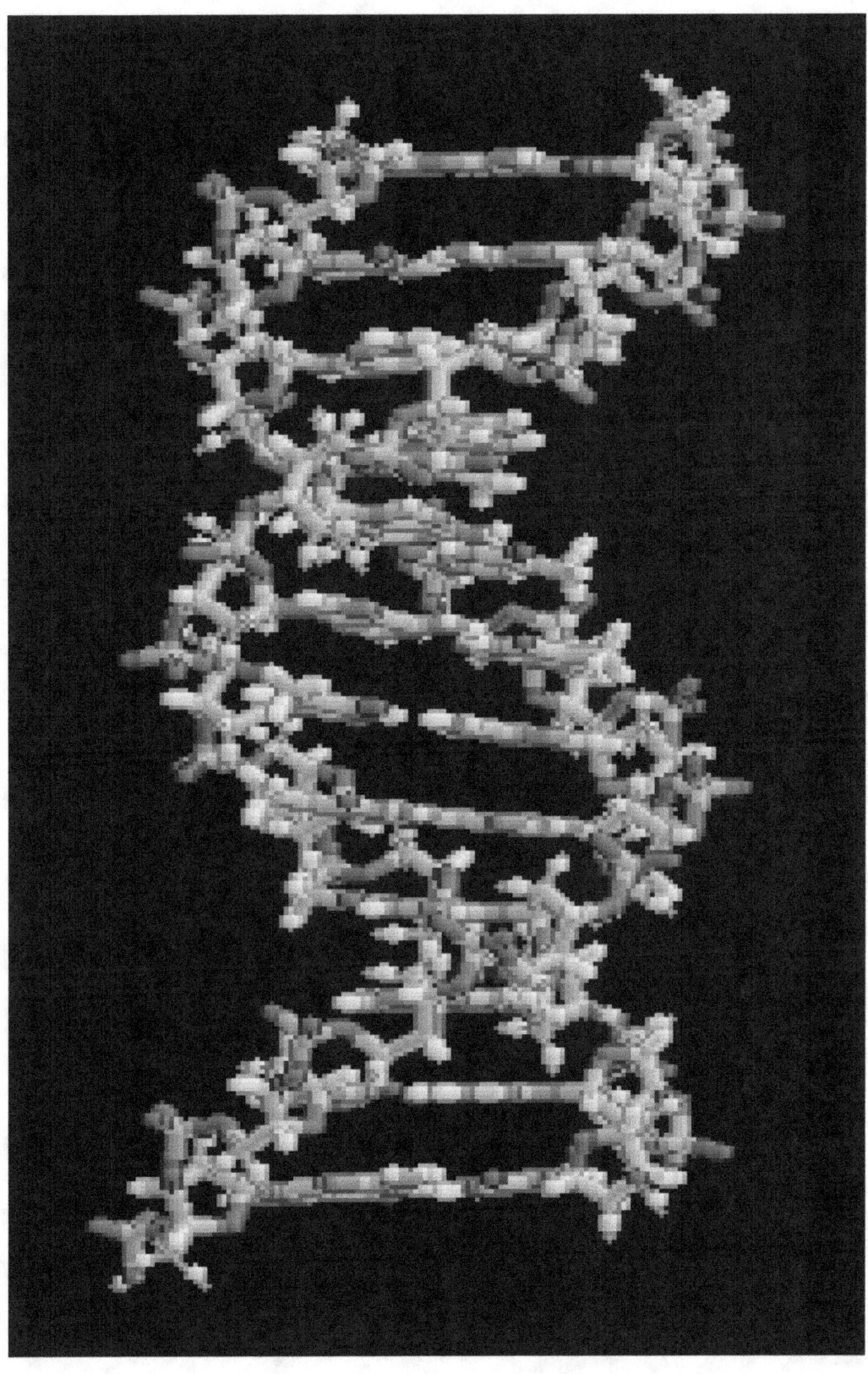

The structure of part of a DNA double helix

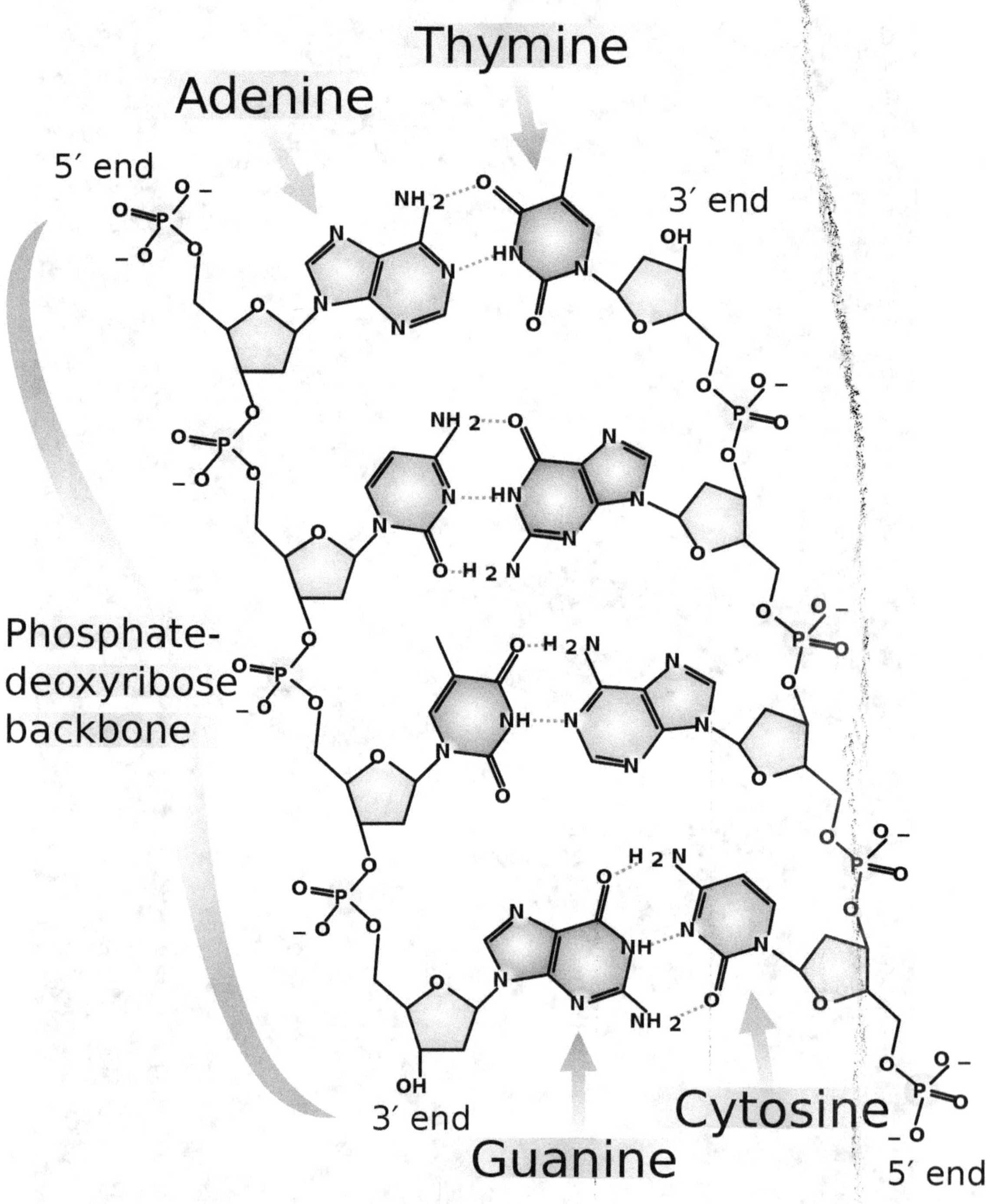

Adenine

Thymine

5′ end

3′ end

Phosphate-deoxyribose backbone

3′ end

Guanine

Cytosine

5′ end

Chemical structure of DNA; hydrogen bonds shown as dotted lines

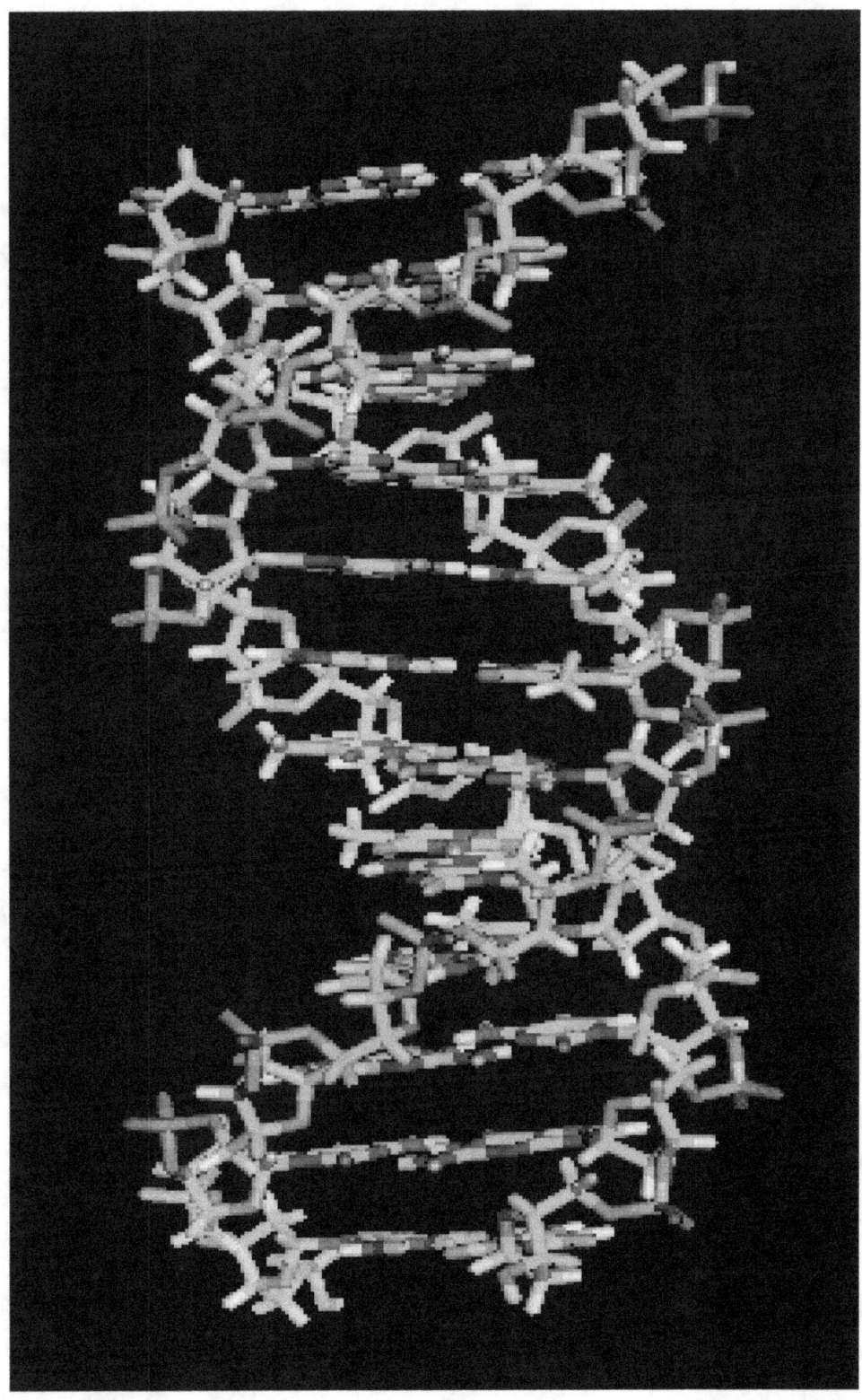

A section of DNA. The bases lie horizontally between the two spiraling strands.[16] (animated version).

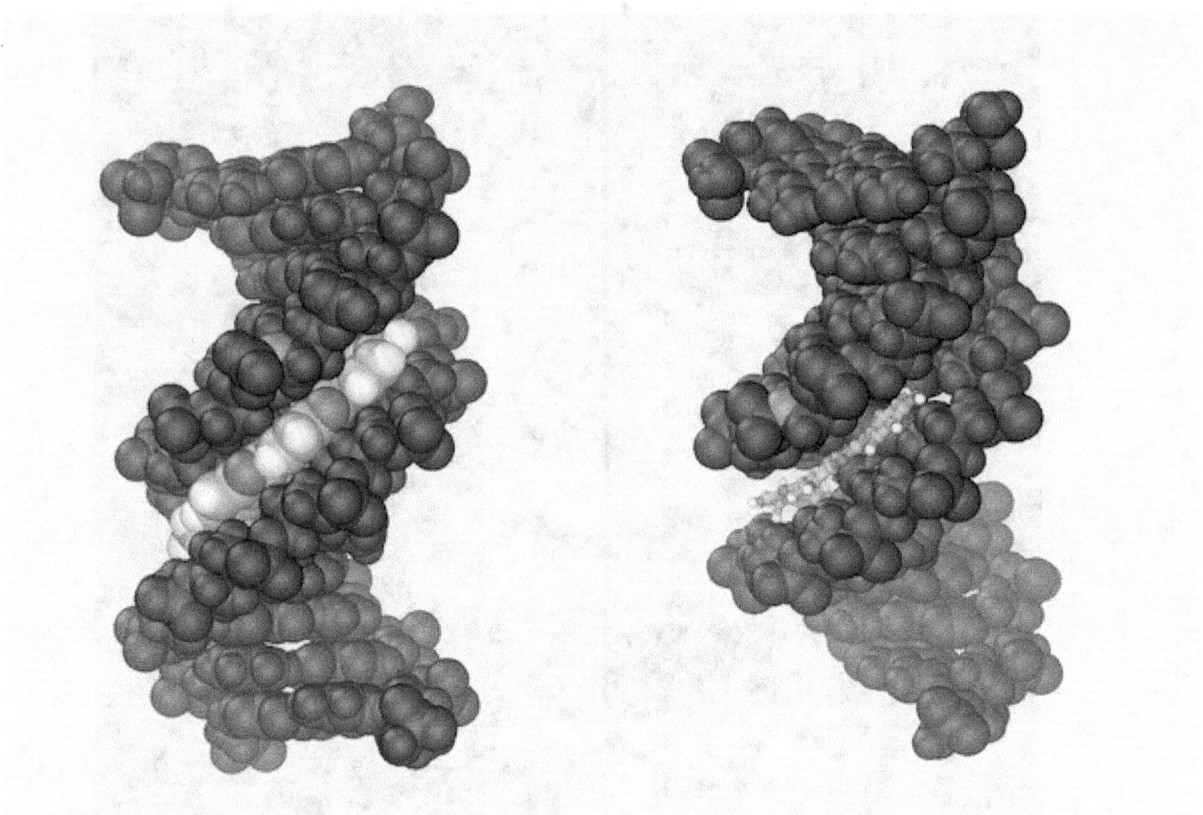

Major and minor grooves of DNA. Minor groove is a binding site for the dye Hoechst 33258.

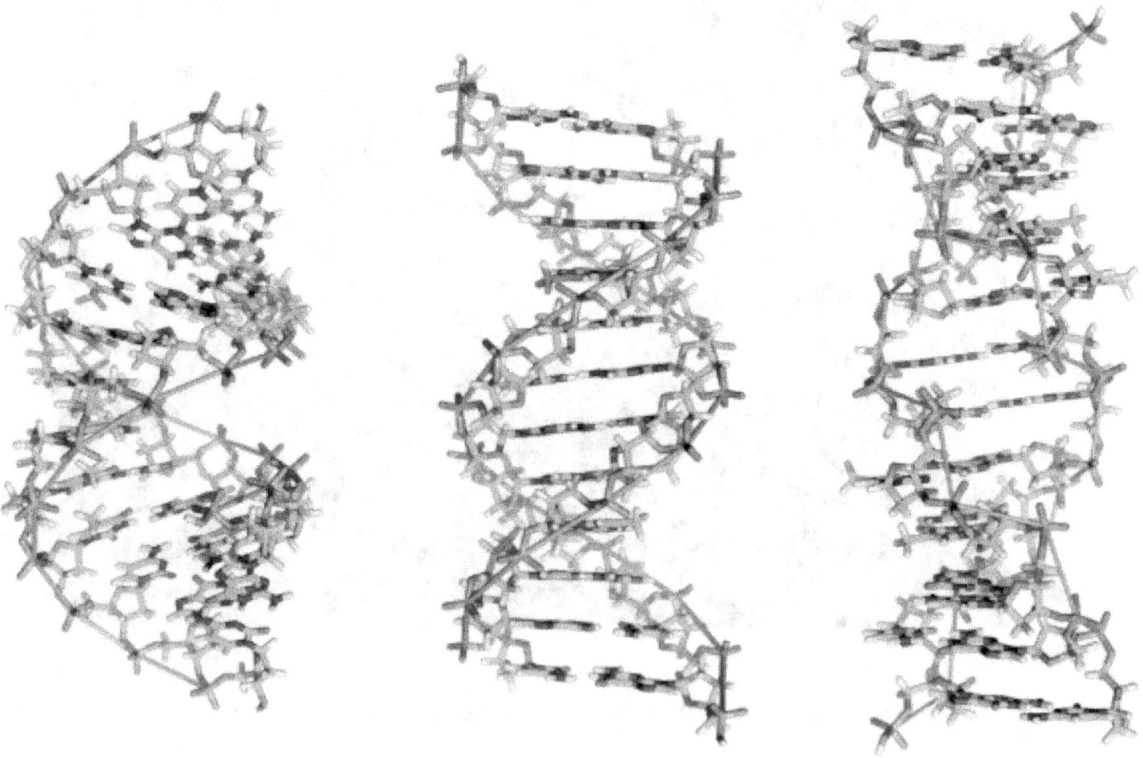

From left to right, the structures of A, B and Z DNA

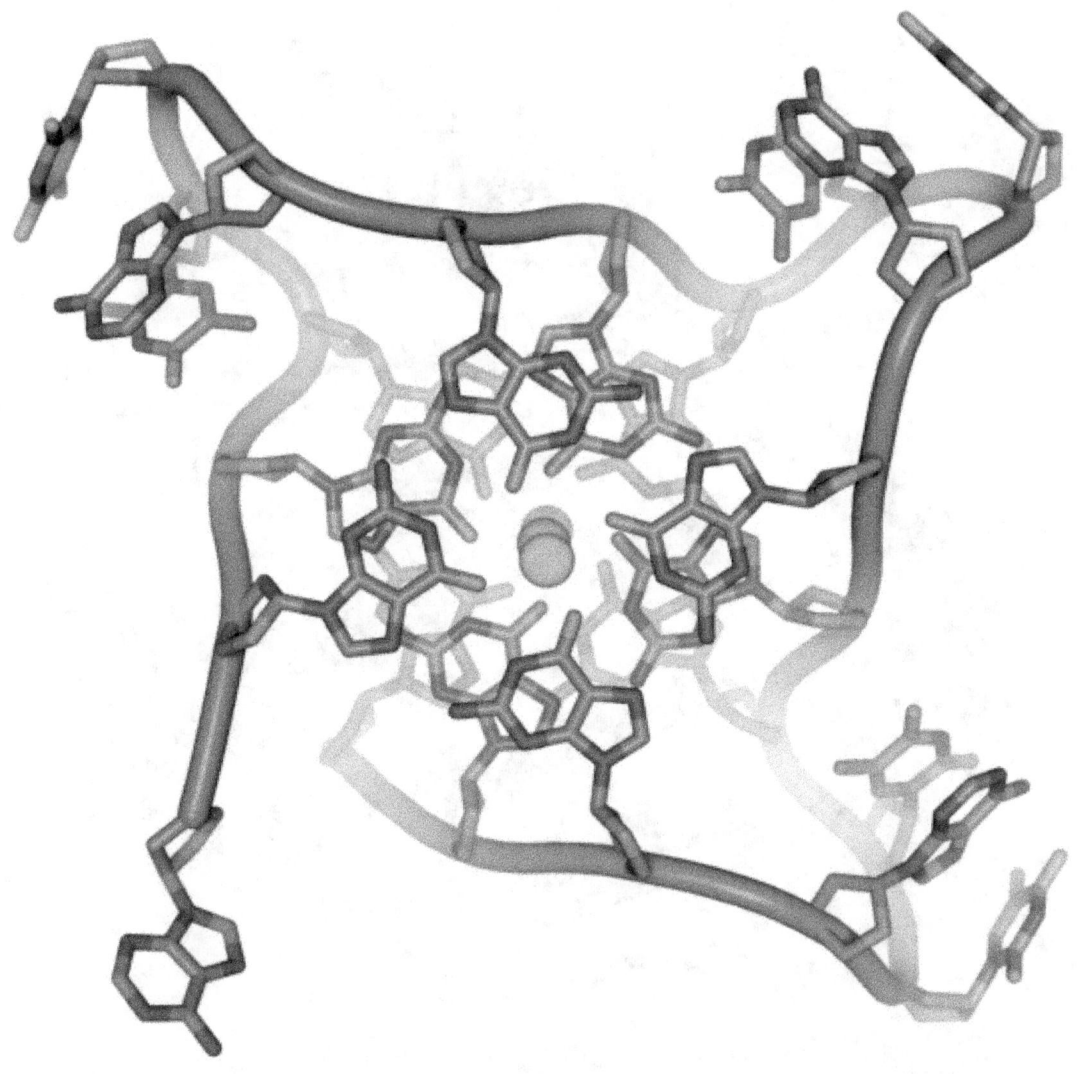

DNA quadruplex formed by telomere repeats. The looped conformation of the DNA backbone is very different from the typical DNA helix.[65]

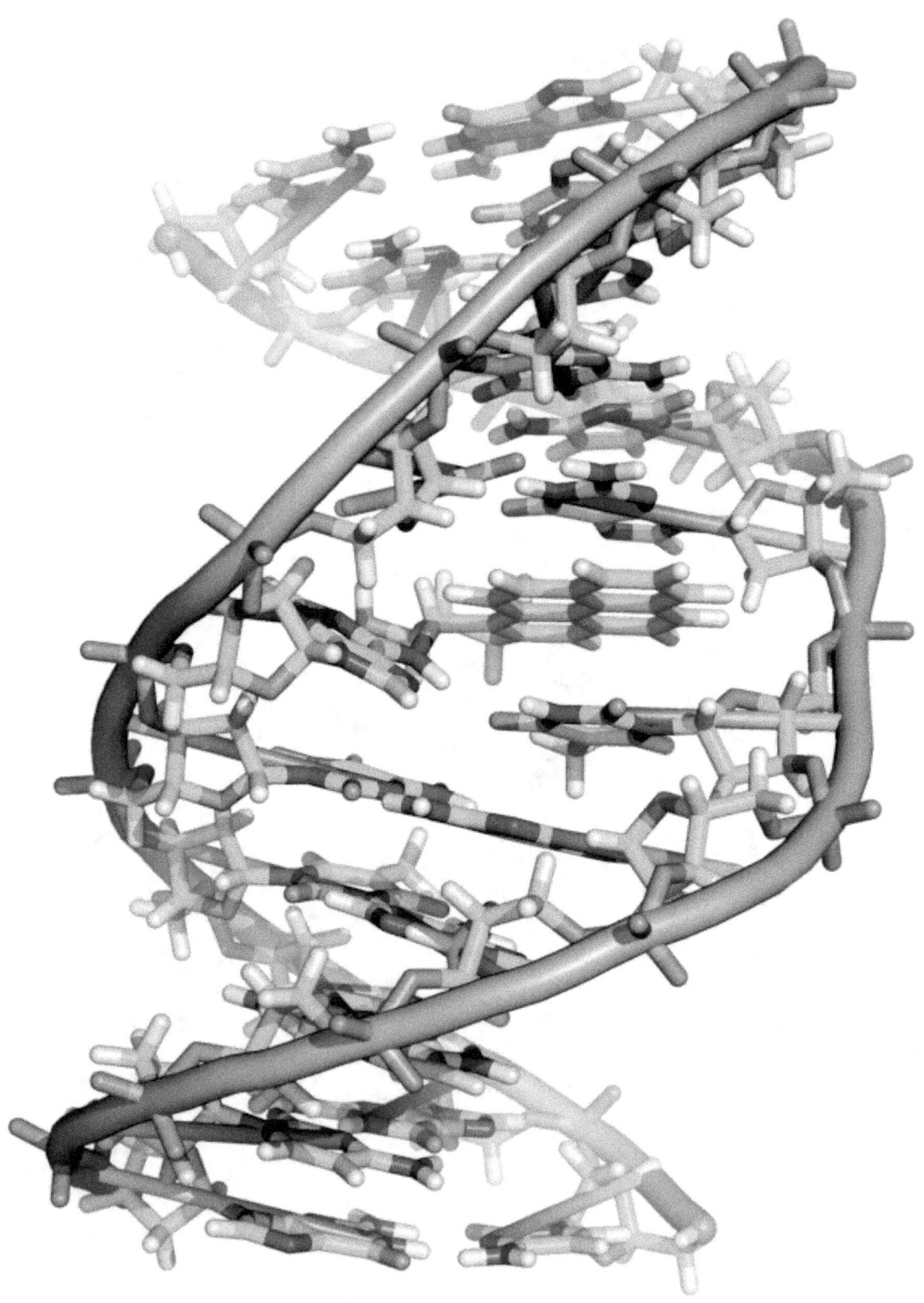

A covalent adduct between a metabolically activated form of benzo[a]pyrene, the major mutagen in tobacco smoke, and DNA[77]

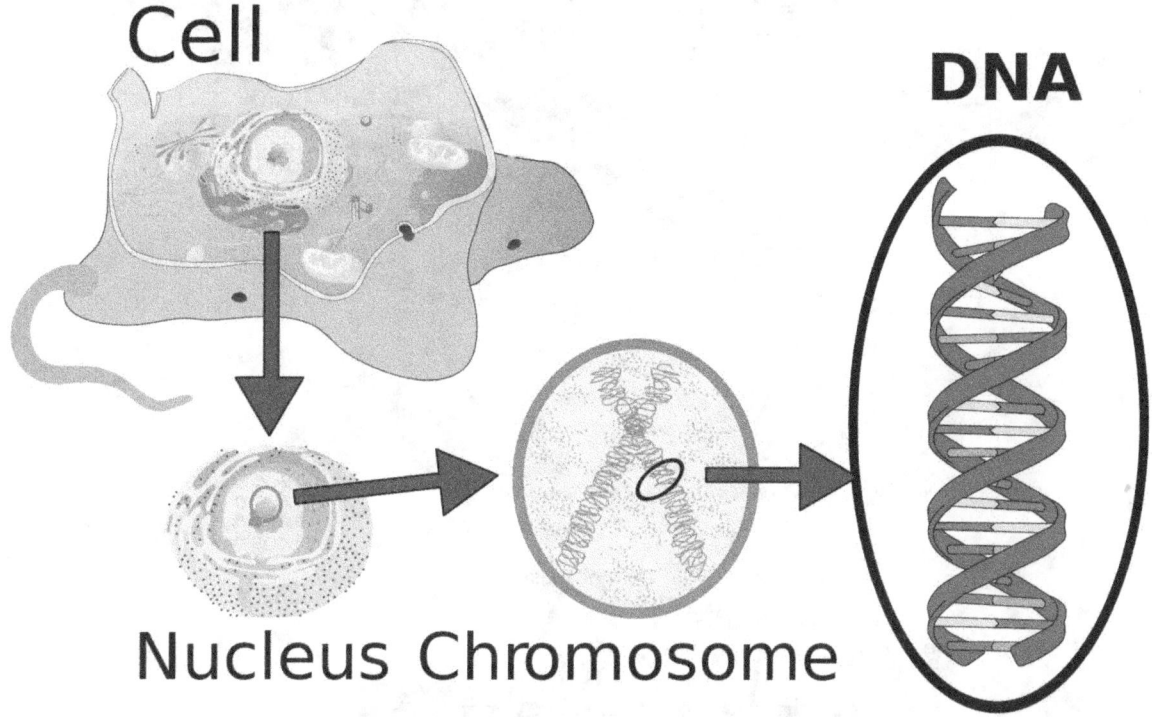

Location of eukaryote nuclear DNA within the chromosomes.

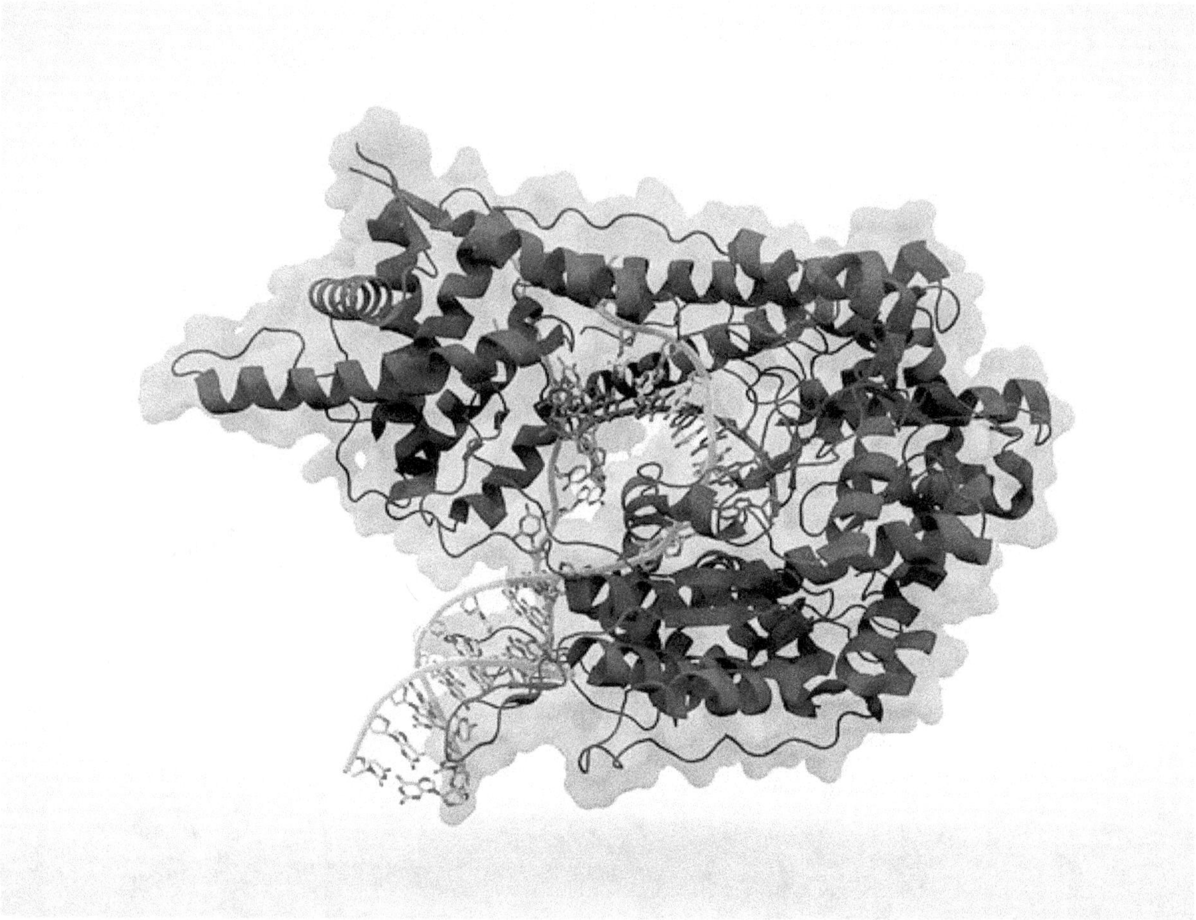

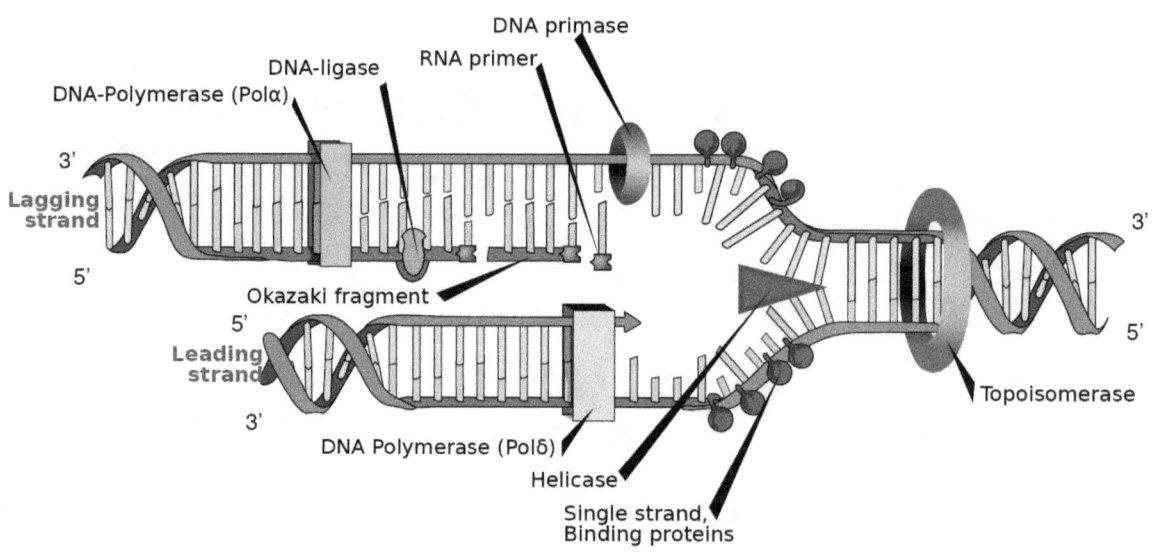

T7 RNA polymerase (blue) producing a mRNA (green) from a DNA template (orange).[96]

DNA replication. The double helix is unwound by a helicase and topoisomerase. Next, one DNA polymerase produces the leading strand copy. Another DNA polymerase binds to the lagging strand. This enzyme makes discontinuous segments (called Okazaki fragments) before DNA ligase joins them together.

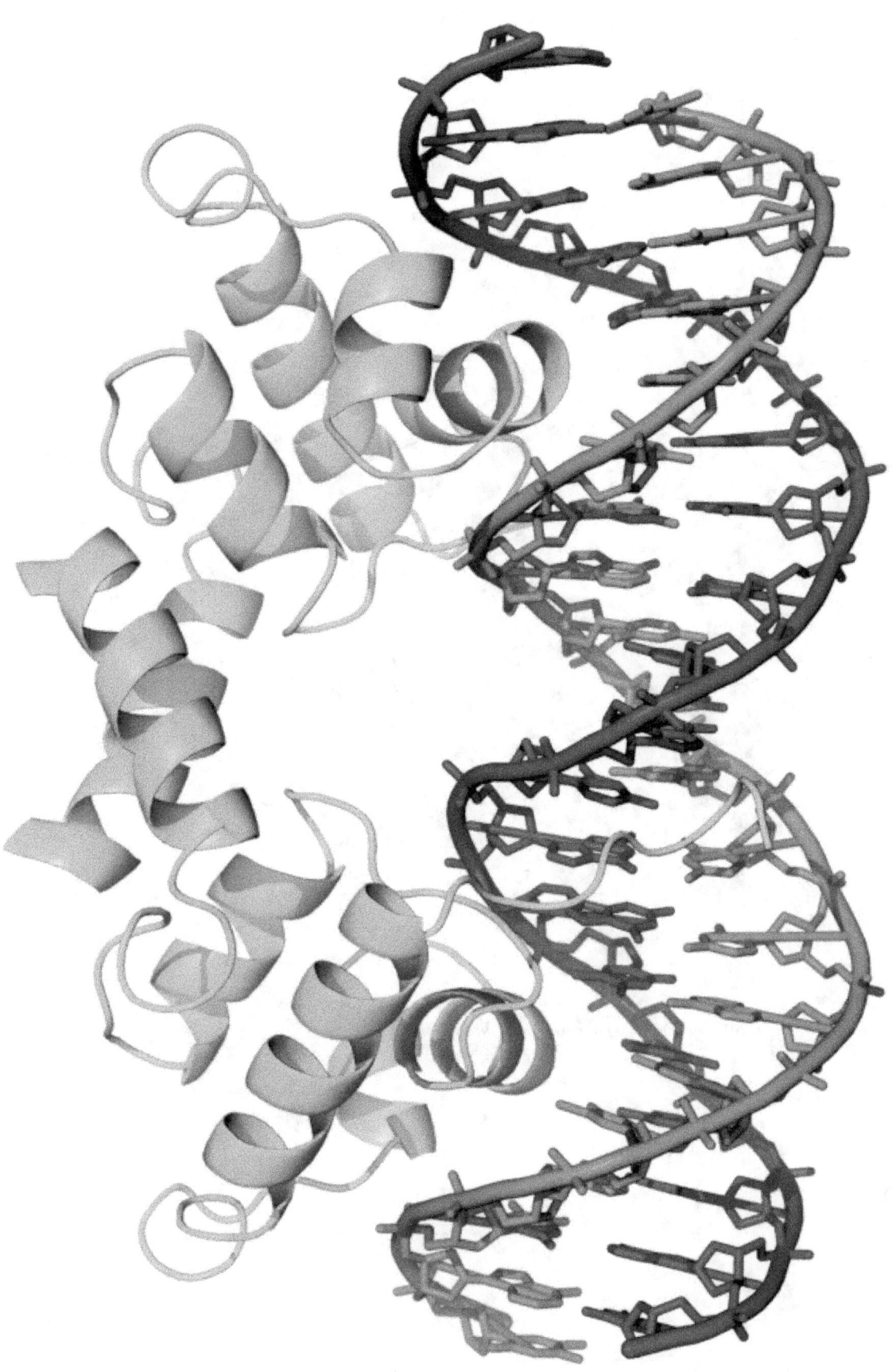

The lambda repressor helix-turn-helix transcription factor bound to its DNA target[116]

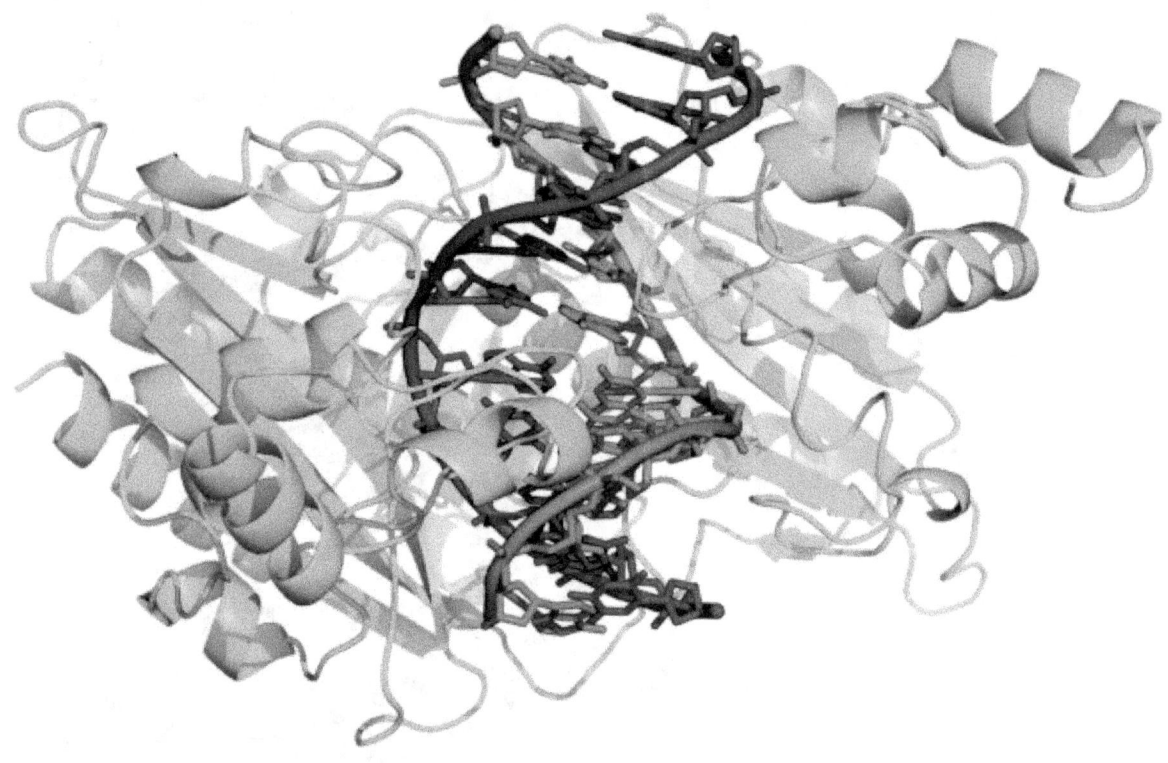

The restriction enzyme EcoRV (green) in a complex with its substrate DNA[120]

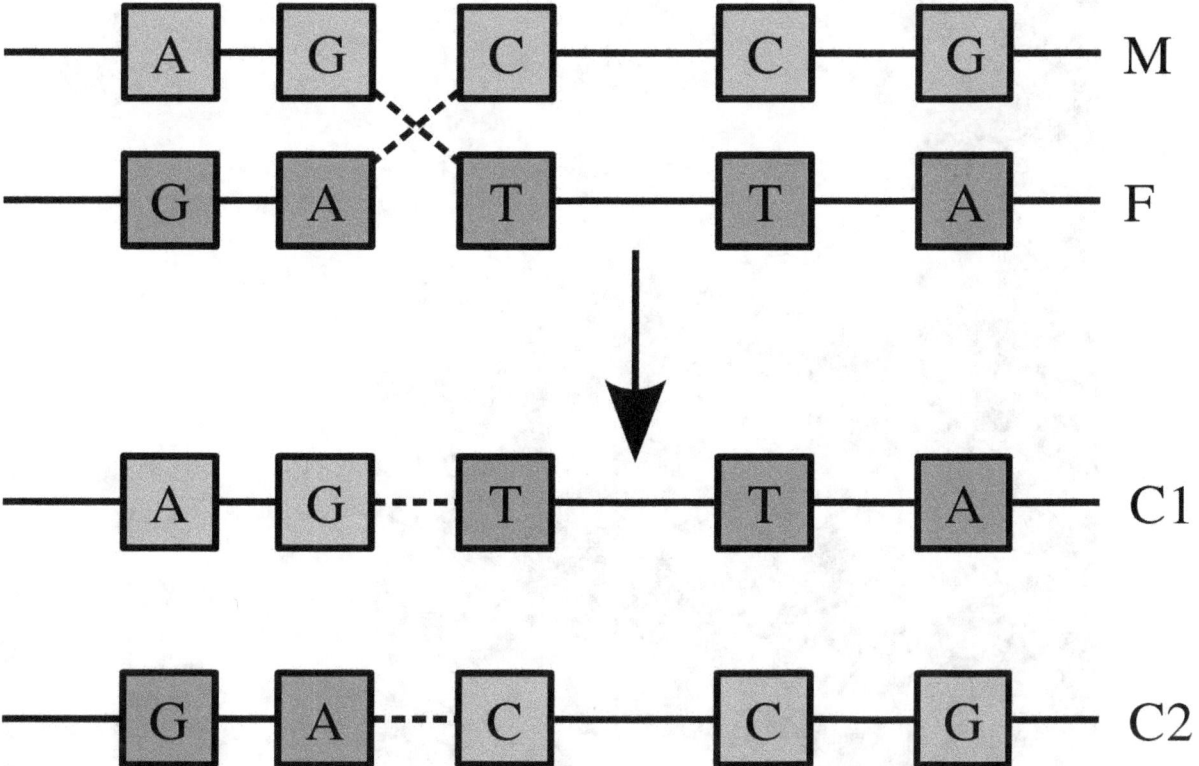

Recombination involves the breakage and rejoining of two chromosomes (M and F) to produce two re-arranged chromosomes (C1 and C2).

Sculpture of DNA made out of shopping carts

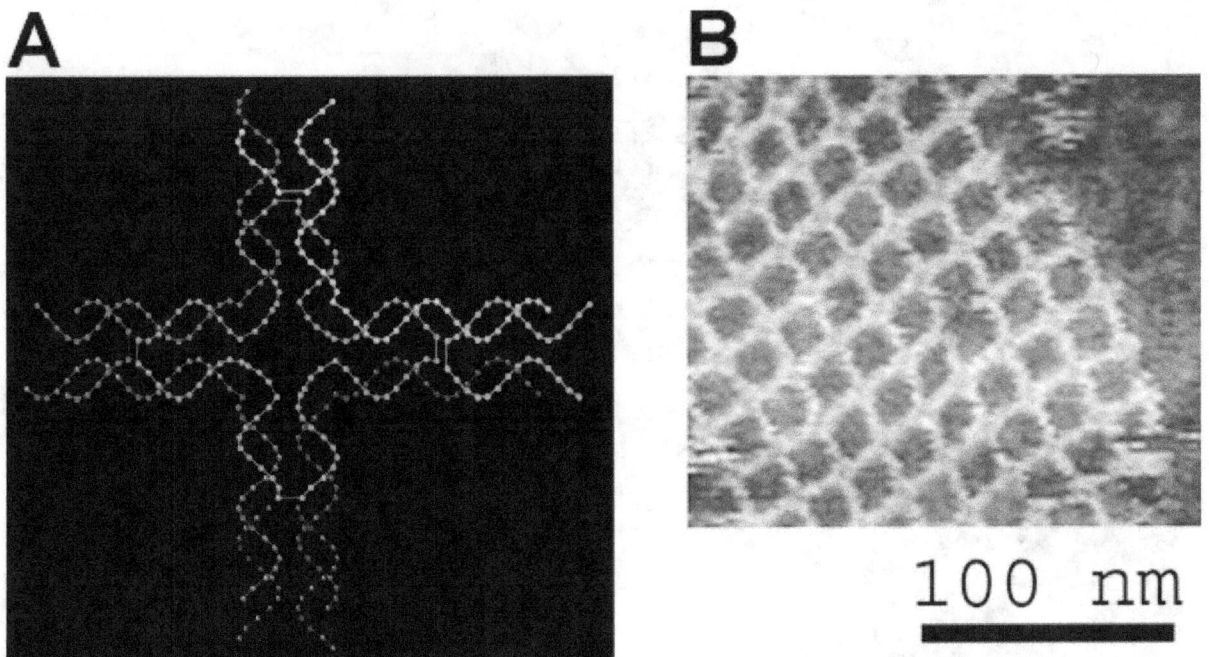

The DNA structure at left (schematic shown) will self-assemble into the structure visualized by atomic force microscopy at right. DNA nanotechnology is the field that seeks to design nanoscale structures using the molecular recognition properties of DNA molecules. Image from Strong, 2004.

James Watson and Francis Crick (right), co-originators of the double-helix model, with Maclyn McCarty (left).

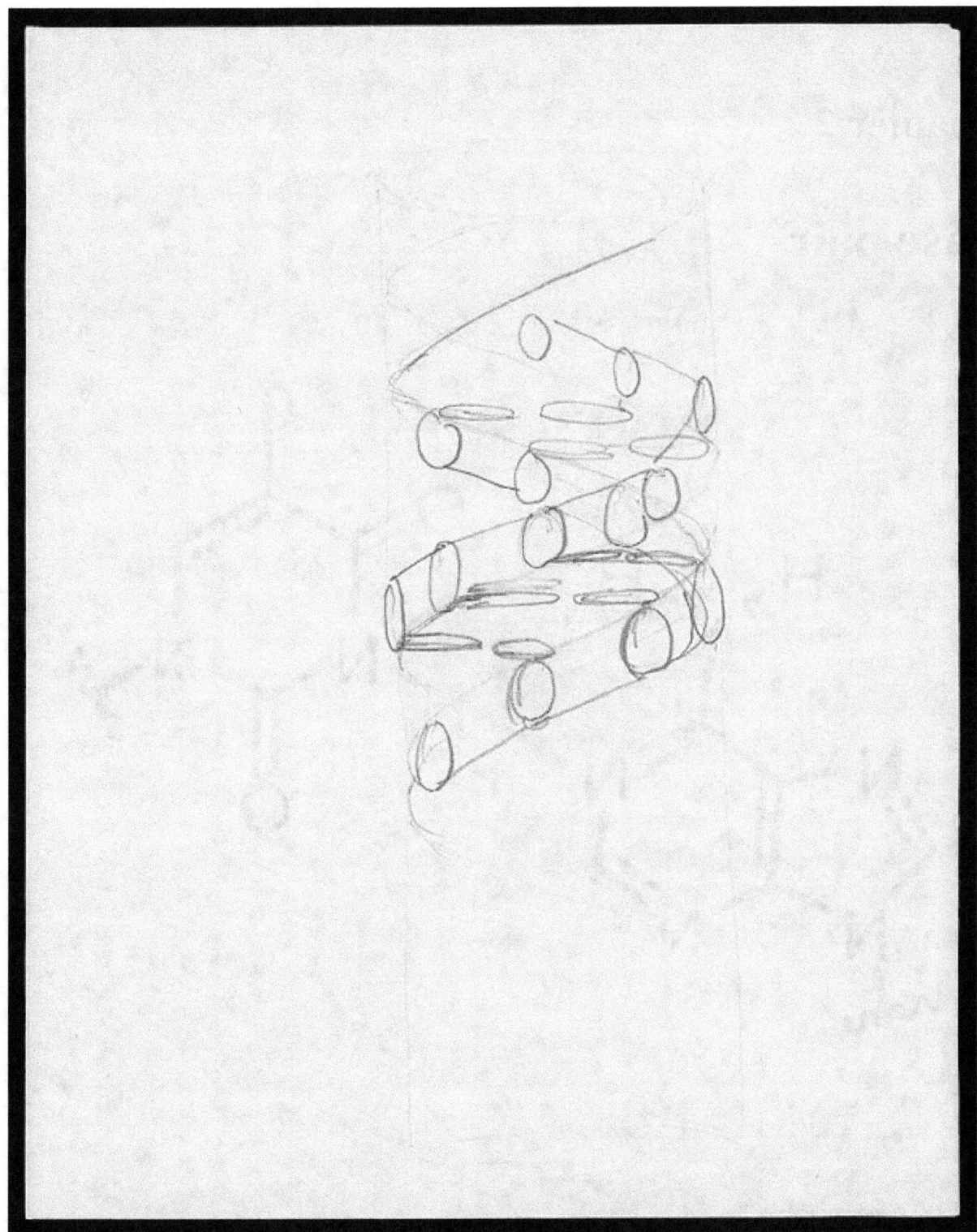

Pencil sketch of the DNA double helix by Francis Crick in 1953

Chapter 2

Base pair

Depiction of the adenine-thymine Watson-Crick base pair.

Base pairs (unit: **bp**), which form between specific nucleobases (also termed nitrogenous bases), are the building blocks of the DNA double helix and contribute to the folded structure of both DNA and RNA. Dictated by specific hydrogen bonding patterns, Watson-Crick base pairs (guanine-cytosine and adenine-thymine) allow the DNA helix to maintain a regular helical structure that is subtly dependent on its nucleotide sequence.[1] The complementary nature of this based-paired structure provides a backup copy of all genetic information encoded within double-stranded DNA. The regular structure and data redundancy provided by the DNA double helix make DNA well suited to the storage of genetic information, while base-pairing between DNA and incoming nucleotides provides the mechanism through which DNA polymerase replicates DNA, and RNA polymerase transcribes DNA into RNA. Many DNA-binding proteins can recognize specific base pairing patterns that identify particular regulatory regions of genes.

Intramolecular base pairs can occur within single-stranded nucleic acids. This is particularly important in RNA molecules (e.g., transfer RNA), where Watson-Crick base pairs (G-C and A-U) permit the formation of short double-stranded helices, and a wide variety of non-Watson-Crick interactions (e.g., G-U or A-A) allow RNAs to fold into a vast range of specific three-dimensional structures. In addition, base-pairing between transfer RNA (tRNA) and messenger RNA (mRNA) forms the basis for the molecular recognition events that result in the nucleotide sequence of mRNA becoming translated into the amino acid sequence of proteins.

The size of an individual gene or an organism's entire genome is often measured in base pairs because DNA is usually double-stranded. Hence, the number of total base pairs is equal to the number of nucleotides in one of the strands (with the exception of non-coding single-stranded regions of telomeres). The haploid human genome (23 chromosomes) is estimated to be about 3.2 billion bases long and to contain 20,000–25,000 distinct protein-coding genes.[2][3][4] A kilobase (kb) is a unit of measurement in molecular biology equal to 1000 base pairs of DNA or RNA.[5] The total amount of related DNA base pairs on Earth is estimated at 5.0×10^{37}, and weighs 50 billion tonnes.[6] In comparison, the total mass of the biosphere has been estimated to be as much as 4 TtC (trillion tons of carbon).[7]

2.1   Hydrogen bonding and stability

Top, a **GC** base pair with three hydrogen bonds. Bottom, an **AT** base pair with two hydrogen bonds. Non-covalent hydrogen bonds between the pairs are shown as dashed lines.

Hydrogen bonding is the chemical interaction that underlies the base-pairing rules described above. Appropriate geometrical correspondence of hydrogen bond donors and acceptors allows only the "right" pairs to form stably. DNA with high GC-content is more stable than DNA with low GC-content, but, contrary to popular belief, the hydrogen bonds do not stabilize the DNA significantly, and stabilization is mainly due to stacking interactions.[8]

The larger nucleobases, adenine and guanine, are members of a class of double-ringed chemical structures called purines; the smaller nucleobases, cytosine and thymine (and uracil), are members of a class of single-ringed chemical structures called pyrimidines. Purines are complementary only with pyrimidines: pyrimidine-pyrimidine pairings are energetically unfavorable because the molecules are too far apart for hydrogen bonding to be established; purine-purine pairings are energetically unfavorable because the molecules are too close, leading to overlap repulsion. Purine-pyrimidine base pairing of AT or GC or UA (in RNA) results in proper duplex structure. The only other purine-pyrimidine pairings would be AC and GT and UG (in RNA); these pairings are mismatches because the patterns of hydrogen donors and acceptors do not correspond. The GU pairing, with two hydrogen bonds, does occur fairly often in RNA (see wobble base pair).

Paired DNA and RNA molecules are comparatively stable at room temperature but the two nucleotide strands will separate above a melting point that is determined by the length of the molecules, the extent of mispairing (if any), and the GC content. Higher GC content results in higher melting temperatures; it is, therefore, unsurprising that the genomes of extremophile organisms such as *Thermus thermophilus* are particularly GC-rich. On the converse, regions of a genome that need to separate frequently — for example, the promoter regions for often-transcribed genes — are comparatively GC-poor (for example, see TATA box). GC content and melting temperature must also be taken into account when designing primers for PCR reactions.

2.1.1   Examples

The following DNA sequences illustrate pair double-stranded patterns. By convention, the top strand is written from the 5' end to the 3' end; thus, the bottom strand is written 3' to 5'.

A base-paired DNA sequence:

```
ATCGATTGAGCTCTAGCG
TAGCTAACTCGAGATCGC
```

The corresponding RNA sequence, in which uracil is substituted for thymine where uracil takes its place in the RNA strand:

```
AUCGAUUGAGCUCUAGCG
UAGCUAACUCGAGAUCGC
```

2.2   Base analogs and intercalators

Main article: Nucleic acid analogues

Chemical analogs of nucleotides can take the place of proper nucleotides and establish non-canonical base-pairing, leading to errors (mostly point mutations) in DNA replication and DNA transcription. This is due to their isosteric chemistry. One common mutagenic base analog is 5-bromouracil, which resembles thymine but can base-pair to guanine in its enol form.

Other chemicals, known as DNA intercalators, fit into the gap between adjacent bases on a single strand and induce frameshift mutations by "masquerading" as a base, causing the DNA replication machinery to skip or insert additional nucleotides at the intercalated site. Most intercalators are large polyaromatic compounds and are known or suspected carcinogens. Examples include ethidium bromide and acridine.

2.3   Unnatural base pair (UBP)

See also: Artificial gene synthesis, Expanded genetic code, Nucleic acid analogue and Synthetic genomics

An unnatural base pair (UBP) is a designed subunit (or nucleobase) of DNA which is created in a laboratory and does not occur in nature. DNA sequences have been described which use newly created nucleobases to form a third base pair, in addition to the two base pairs found in nature, A-T (adenine - thymine) and G-C (guanine - cytosine). A few research groups have been searching for a third base pair for DNA, including teams led by Steven A. Benner, Philippe Marliere, Floyd Romesberg and Ichiro Hirao.[9] Some new base pairs have been reported.[10][11][12]

In 1989 Steven Benner, then at the Swiss Federal Institute of Technology in Zurich, and his team led with modified forms of cytosine and guanine into DNA molecules *in vitro*.[13] The nucleotides, which encoded RNA and proteins, were successfully replicated *in vitro*. Since then, Benner's team has been trying to engineer cells that can make foreign bases from scratch, obviating the need for a feedstock.[14]

In 2002, Ichiro Hirao's group in Japan developed an unnatural base pair between 2-amino-8-(2-thienyl)purine (s) and pyridine-2-one (y) that functions in transcription and translation, for the site-specific incorporation of non-standard amino acids into proteins.[15] In 2006, they created 7-(2-thienyl)imidazo[4,5-b]pyridine (Ds) and pyrrole-2-carbaldehyde (Pa) as a third base pair for replication and transcription.[16] Afterward, Ds and 4-[3-(6-aminohexanamido)−1-propynyl]−2-nitropyrrole (Px) was discovered as a high fidelity pair in PCR amplification.[17][18] In 2013, they applied the Ds-Px pair to DNA aptamer generation by *in vitro* selection (SELEX) and demonstrated the genetic alphabet expansion significantly augment DNA aptamer affinities to target proteins.[19]

In 2012, a group of American scientists led by Floyd Romesberg, a chemical biologist at the Scripps Research Institute in San Diego, California, published that his team designed an unnatural base pair (UBP).[20] The two new artificial nucleotides or *Unnatural Base Pair* (UBP) were named **d5SICS** and **dNaM**. More technically, these artificial nucleotides bearing hydrophobic nucleobases, feature two fused aromatic rings that form a (d5SICS–dNaM) complex or base pair in DNA.[14][21] His team designed a variety of *in vitro* or "test tube" templates containing the unnatural base pair and they confirmed that it was efficiently replicated with high fidelity in virtually all sequence contexts using the modern standard *in vitro* techniques, namely PCR amplification of DNA and PCR-based applications.[20] Their results show that for PCR and PCR-based applications, the d5SICS–dNaM unnatural base pair is functionally equivalent to a natural base pair, and when combined with the other two natural base pairs used by all organisms, A–T and G–C, they provide a fully functional and expanded six-letter "genetic alphabet".[21]

In 2014 the same team from the Scripps Research Institute reported that they synthesized a stretch of circular DNA known as a plasmid containing natural T-A and C-G base pairs along with the best-performing UBP Romesberg's laboratory had designed, and inserted it into cells of the common bacterium *E. coli* that successfully replicated the unnatural base pairs through multiple generations.[9] The transfection did not hamper the growth of the *E. coli* cells, and showed no sign of losing its unnatural base pairs to its natural DNA repair mechanisms. This is the first known example of a living organism passing along an expanded genetic code to subsequent generations.[21][22] Romesberg said he and his colleagues created 300 variants to refine the design of nucleotides that would be stable enough and would be replicated as easily as the natural ones when the cells divide. This was in part achieved by the addition of a supportive algal gene that expresses a nucleotide triphosphate transporter which efficiently imports the triphosphates of both d5SICSTP and dNaMTP into *E. coli* bacteria.[21] Then, the natural bacterial replication pathways use them to accurately replicate a plasmid containing d5SICS–dNaM. Other researchers were surprised that the bacteria replicated these human-made DNA subunits.[23]

The successful incorporation of a third base pair is a significant breakthrough toward the goal of greatly expanding the number of amino acids which can be encoded by DNA, from the existing 20 amino acids to a theoretically possible 172, thereby expanding the potential for living organisms to produce novel proteins.[9] The artificial strings of DNA do not encode for anything yet, but scientists speculate they could be designed to manufacture new proteins which could have industrial or pharmaceutical uses.[24] Experts said the synthetic DNA incorporating the unnatural base pair raises the possibility of life forms based on a different DNA code.[23][24]

2.4 Length measurements

The following abbreviations are commonly used to describe the length of a D/RNA molecule:

- bp = base pair(s)— one bp corresponds to approximately 3.4 Å (340 pm) of length along the strand, and to roughly 618 or 643 daltons for DNA and RNA respectively.

- kb (= kbp) = kilo base pairs = 1,000 bp

- Mb (= Mbp) = mega base pairs = 1,000,000 bp

- Gb = giga base pairs = 1,000,000,000 bp.

For case of single-stranded DNA/RNA units of nucleotides are used, abbreviated nt (or knt, Mnt, Gnt), as they are not paired. For distinction between units of computer storage and bases kbp, Mbp, Gbp, etc. may be used for base pairs.

The centimorgan is also often used to imply distance along a chromosome, but the number of base pairs it corresponds to varies widely. In the Human genome, the centimorgan is about 1 million base pairs.[25][26]

2.5 See also

- List of Y-DNA single-nucleotide polymorphisms

2.6 References

[1] "Sequence-Dependent Variability of B-DNA". Springer.

[2] Moran, Laurence A. (2011-03-24). "The total size of the human genome is very likely to be ~3,200 Mb". Sandwalk.blogspot.com. Retrieved 2012-07-16.

[3] "The finished length of the human genome is 2.86 Gb". Strategicgenomics.com. 2006-06-12. Retrieved 2012-07-16.

[4] International Human Genome Sequencing Consortium (2004). "Finishing the euchromatic sequence of the human genome". *Nature* **431** (7011): 931–45. doi:10.1038/nature03001. PMID 15496913.

[5] Cockburn, Andrew F.; Jane Newkirk, Mary; Firtel, Richard A. (1976). "Organization of the ribosomal RNA genes of dictyostelium discoideum: Mapping of the nontrascribed spacer regions". *Cell* **9** (4): 605–613. doi:10.1016/0092-8674(76)90043-X.

[6] Nuwer, Rachel (18 July 2015). "Counting All the DNA on Earth". *The New York Times* (New York: The New York Times Company). ISSN 0362-4331. Retrieved 2015-07-18.

[7] "The Biosphere: Diversity of Life". *Aspen Global Change Institute*. Basalt, CO. Retrieved 2015-07-19.

[8] Peter Yakovchuk, Ekaterina Protozanova and Maxim D. Frank-Kamenetskii. Base-stacking and base-pairing contributions into thermal stability of the DNA double helix. Nucleic Acids Research 2006 34(2):564-574.

[9] Fikes, Bradley J. (May 8, 2014). "Life engineered with expanded genetic code". *San Diego Union Tribune*. Retrieved 8 May 2014.

[10] Yang, Zunyi; et al. (August 15, 2011). "Amplification, Mutation, and Sequencing of a Six-Letter Synthetic Genetic System". *J. Am. Chem. Soc* **133** (38): 15105–15112. doi:10.1021/ja204910n.

[11] Yamashige, Rie; et al. (March 2012). "Highly specific unnatural base pair systems as a third base pair for PCR amplification". *Nucl Acids Res* **40** (6): 2793–2806. doi:10.1093/nar/gkr1068.

[12] Malyashev, D. A.; et al. (July 24, 2012). "Efficient and sequence-independent replication of DNA containing a third base pair establishes a functional six-letter genetic alphabet". *Proc. Nat. Acad. Sci. USA* **109**(30): 12005–12010. doi:10.1073/pnas.120517

[13] Switzer, Christopher; Moroney, Simon E.; Benner, Steven A. (1989). "Enzymatic incorporation of a new base pair into DNA and RNA". *J. Am. Chem. Soc.* **111** (21): 8322–8323. doi:10.1021/ja00203a067.

[14] Callaway, Ewan (May 7, 2014). "Scientists Create First Living Organism With 'Artificial' DNA". *Nature News* (Huffington Post). Retrieved 8 May 2014.

[15] Hirao, I.; et al. (2002). "An unnatural base pair for incorporating amino acid analogs into proteins". *Nat. Biotechnol* **20**: 177–182. doi:10.1038/nbt0202-177.

[16] Hirao, I.; et al. (2006). "An unnatural hydrophobic base pair system: site-specific incorporation of nucleotide analogs into DNA and RNA". *Nat. Methods* **6**: 729–735.

[17] Kimoto, M. et al. (2009) An unnatural base pair system for efficient PCR amplification and functionalization of DNA molecules. Nucleic acids Res. 37, e14

[18] Yamashige, R.; et al. "Highly specific unnatural base pair systems as a third base pair for PCR amplification". *Nucleic Acids Res* **40**: 2793–2806. doi:10.1093/nar/gkr1068.

[19] Kimoto, M.; et al. (2013). "Generation of high-affinity DNA aptamers using an expanded genetic alphabet". *Nat. Biotechnol* **31**: 453–457. doi:10.1038/nbt.2556.

[20] Malyshev, Denis A.; Dhami, Kirandeep; Quach, Henry T.; Lavergne, Thomas; Ordoukhanian, Phillip (24 July 2012). "Efficient and sequence-independent replication of DNA containing a third base pair establishes a functional six-letter genetic alphabet". *Proceedings of the National Academy of Sciences of the United States of America (PNAS)* **109** (30): 12005–12010. doi:10.1073/pnas.1205176109. Retrieved 2014-05-11.

[21] Malyshev, Denis A.; Dhami, Kirandeep; Lavergne, Thomas; Chen, Tingjian; Dai, Nan; Foster, Jeremy M.; Corrêa, Ivan R.; Romesberg, Floyd E. (May 7, 2014). "A semi-synthetic organism with an expanded genetic alphabet". *Nature (journal)*. doi:10.1038/nature13314. Retrieved May 7, 2014.

[22] Sample, Ian (May 7, 2014). "First life forms to pass on artificial DNA engineered by US scientists". *The Guardian*. Retrieved 8 May 2014.

[23] "Scientists create first living organism containing artificial DNA". *The Wall Street Journal* (Fox News). May 8, 2014. Retrieved 8 May 2014.

[24] Pollack, Andrew (May 7, 2014). "Scientists Add Letters to DNA's Alphabet, Raising Hope and Fear". *New York Times*. Retrieved 8 May 2014.

[25] "NIH ORDR - Glossary - C". Rarediseases.info.nih.gov. Retrieved 2012-07-16.

[26] Matthew P Scott; Paul Matsudaira; Harvey Lodish; James Darnell; Lawrence Zipursky; Chris A Kaiser; Arnold Berk; Monty Krieger (2004). *Molecular Cell Biology* (Fifth ed.). San Francisco: W. H. Freeman. p. 396. ISBN 0-7167-4366-3. ...in humans 1 centimorgan on average represents a distance of about 7.5×10^5 base pairs.

2.7 Further reading

- Watson JD; Baker TA; Bell SP; Gann A; Levine M; Losick R (2004). *Molecular Biology of the Gene* (5th ed.). Pearson Benjamin Cummings: CSHL Press. (See esp. ch. 6 and 9)

- Astrid Sigel; Helmut Sigel; Roland K. O. Sigel, eds. (2012). *Interplay between Metal Ions and Nucleic Acids*. Metal Ions in Life Sciences **10**. Springer. doi:10.1007/978-94-007-2172-2. ISBN 978-9-4007-2171-5.

- Clever, Guido H.; Shionoya, Mitsuhiko (2012). "Chapter 10. Alternative DNA Base-Pairing through Metal Coordination". *Interplay between Metal Ions and Nucleic Acids*. pp. 269–294. doi:10.1007/978-94-007-2172-2_10.

- Megger, Dominik A.; Megger, Nicole; Mueller, Jens (2012). "Chapter 11. Metal-Mediated Base Pairs in Nucleic Acids with Purine and Pyrimidine-Derived Neucleosides". *Interplay between Metal Ions and Nucleic Acids*. pp. 295–317. doi:10.1007/978-94-007-2172-2_11.

2.8 External links

- DAN—webserver version of the EMBOSS tool for calculating melting temperatures

Chapter 3

Sense (molecular biology)

In molecular biology and genetics, **sense** is a concept used to compare the polarity of nucleic acid molecules, such as DNA or RNA, to other nucleic acid molecules. Depending on the context within molecular biology, sense may have slightly different meanings.

3.1   DNA sense

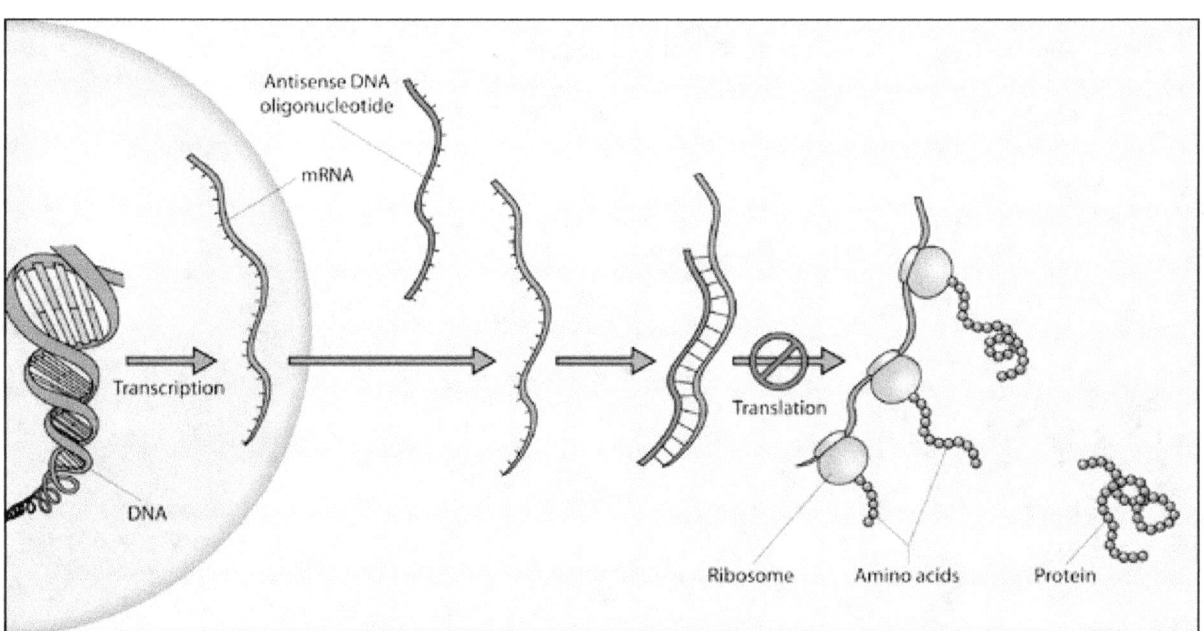

Schematic showing how antisense DNA strands can interfere with protein translation.

Molecular biologists call a single strand of DNA **sense** (or **positive (+)**) if an RNA version of the same sequence is translated or translatable into protein. Its complementary strand is called **antisense** (or **negative (-)** sense). Sometimes the phrase **coding strand** is encountered; however, protein coding and non-coding RNAs can be transcribed similarly from both strands, in some cases being transcribed in both directions from a common promoter region, or being transcribed from within introns, on both strands (see "ambisense" below).[1][2][3]

48

3.1.1   Antisense DNA

The two complementary strands of double-stranded DNA (dsDNA) are usually differentiated as the "sense" strand and the "antisense" strand. The DNA sense strand looks like the messenger RNA (mRNA) and can be used to read the expected protein code; for example, ATG in the sense DNA may correspond to an AUG codon in the mRNA, encoding the amino acid methionine. However, the DNA sense strand itself is not used to make protein by the cell. It is the DNA antisense strand which serves as the source for the protein code, because, with bases complementary to the DNA sense strand, it is used as a template for the mRNA. Since transcription results in an RNA product complementary to the DNA template strand, the mRNA is complementary to the DNA antisense strand. The mRNA is what is used for translation (protein synthesis).

Hence, a base triplet 3'-TAC-5' in the DNA antisense strand can be used as a template which will result in an 5'-AUG-3' base triplet in mRNA (AUG is the codon for methionine, the start codon). The DNA sense strand will have the triplet ATG which looks just like AUG but will not be used to make Methionine because it will not be used to make mRNA. The DNA sense strand is called a "sense" strand not because it will be used to make protein (it won't be), but because it has a sequence that looks like the protein codon sequence.

In biology and research, short antisense molecules can interact with complementary strands of nucleic acids, modifying expression of genes. See the section on "antisense oligonucleotides" below.

3.1.2   Example with double-stranded DNA

DNA strand 1: antisense strand (transcribed to)→ RNA strand (sense)

DNA strand 2: sense strand

Some regions within a double strand of DNA code for genes, which are usually instructions specifying the order of amino acids in a protein along with regulatory sequences, splicing sites, noncoding introns, and other complicating details. For a cell to use this information, one strand of the DNA serves as a template for the synthesis of a complementary strand of RNA. The template DNA strand is called the transcribed strand with antisense sequence and the mRNA transcript is said to be sense sequence (the complement of antisense). Because the DNA is double-stranded, the strand complementary to the antisense sequence is called non-transcribed strand and has the same sense sequence as the mRNA transcript (though T bases in DNA are substituted with U bases in RNA).

A note on the confusion between "sense" and "antisense" strands: The strand names actually depend on which direction you are writing the sequence that contains the information for proteins (the "sense" information), not on which strand is on the top or bottom (that is arbitrary). The only real biological information that is important for labeling strands is the location of the 5' phosphate group and the 3' hydroxyl group because these ends determine the direction of transcription and translation. A sequence 5' CGCTAT 3' is equivalent to a sequence written 3' TATCGC 5' as long as the 5' and 3' ends are noted. If the ends are not labeled, convention is to assume that the sequence is written in the 5' to 3' direction. Watson strand refers to 5' to 3' top strand (5' → 3'), whereas Crick strand refers to 5' to 3' bottom strand (3' ← 5').[4] Both Watson and Crick strands can be either sense or antisense strands depending on the gene whose sequences are displayed in the genome sequence database. For example, YEL021W, an alias of URA3 gene used in NCBI database, defines that this gene is located on the 21st open reading frame (ORF) from the centromere of the left arm (L) of Yeast (Y) chromosome number V (E), and that the expression coding strand is Watson strand (W). YKL074C defines the 74th ORF to the left of the centromere of chromosome XI and denotes coding strand from the Crick strand (C). Another confusing term referring to "Plus" and "Minus" strand is also widely used. Whether the strand is sense (positive) or antisense (negative), the default query sequence in NCBI BLAST alignment is "Plus" strand.

3.2   Ambisense

A single-stranded genome that contains both positive-sense and negative-sense is said to be **ambisense**. Bunyaviruses have 3 single-stranded RNA (ssRNA) fragments containing both positive-sense and negative-sense sections; arenaviruses are also ssRNA viruses with an ambisense genome, as they have 2 fragments that are mainly negative-sense except for part of the 5' ends of the large and small segments of their genome.

3.3   Antisense RNA

Main article: Antisense RNA

Antisense RNA is an RNA transcript that is complementary to endogenous mRNA. In other words, it is a non-coding strand complementary to the coding sequence of RNA; this is similar to negative-sense viral RNA. Introducing a transgene coding for antisense RNA is a technique used to block expression of a gene of interest. Radioactively-labelled antisense RNA can be used to show the level of transcription of genes in various cell types. Some alternative antisense structural types are being experimentally applied as antisense therapy, with at least one antisense therapy approved for use in humans.

When mRNA forms a duplex with a complementary antisense RNA sequence, translation is blocked. This process is related to RNA interference.

Antisense nucleic acid molecules have been used experimentally to bind to mRNA and prevent expression of specific genes. Antisense therapies are also in development; in the USA, the Food and Drug Administration (FDA) has approved phosphorothioate antisense oligos fomivirsen (Vitravene) and mipomersen (Kynamro)[5] for human therapeutic use.

Cells can produce antisense RNA molecules naturally, which interact with complementary mRNA molecules and inhibit their expression.

3.4   RNA sense in viruses

In virology, the genome of an RNA virus can be said to be either **positive-sense**, also known as a "plus-strand", or **negative-sense**, also known as a "minus-strand". In most cases, the terms *sense* and *strand* are used interchangeably, making such terms as *positive-strand* equivalent to *positive-sense*, and *plus-strand* equivalent to *plus-sense*. Whether a virus genome is positive-sense or negative-sense can be used as a basis for classifying viruses.

3.4.1   Positive-sense

Positive-sense (5' to 3') viral RNA signifies that a particular viral RNA sequence may be directly translated into the desired viral proteins. Therefore, in positive-sense RNA viruses, the viral RNA genome can be considered viral mRNA, and can be immediately translated by the host cell. Unlike negative-sense RNA, positive-sense RNA is of the same sense as mRNA. Some viruses (e.g., Coronaviridae) have positive-sense genomes that can act as mRNA and be used directly to synthesize proteins without the help of a complementary RNA intermediate. Because of this, these viruses do not need to have an RNA polymerase packaged into the virion.

3.4.2   Negative-sense

Negative-sense (3' to 5') viral RNA is complementary to the viral mRNA and thus from it a positive-sense RNA must be produced by an RNA-dependent RNA polymerase prior to translation. Negative-sense RNA (like DNA) has a nucleotide sequence complementary to the mRNA that it encodes. Like DNA, this RNA cannot be translated into protein directly. Instead, it must first be transcribed into a positive-sense RNA that acts as an mRNA. Some viruses (Influenza, for example) have negative-sense genomes and so must carry an RNA polymerase inside the virion.

3.5   Antisense oligonucleotides

Gene silencing can be achieved by introducing into cells a short "antisense oligonucleotide" that is complementary to an RNA target. This experiment was first done by Zamecnik and Stephenson in 1978[6] and continues to be a useful approach, both for laboratory experiments and potentially for clinical applications (antisense therapy).[7]

If the antisense oligonucleotide contains a stretch of DNA or a DNA mimic (phosphorothioate DNA, 2'F-ANA, or others) it can recruit RNase H to degrade the target RNA. This makes the mechanism of gene silencing catalytic. Double-stranded RNA can also act as a catalytic, enzyme-dependent antisense agent through the RNAi/siRNA pathway, involving target mRNA recognition through sense-antisense strand pairing followed by target mRNA degradation by the RNA-induced silencing complex (RISC). The R1 plasmid hok/sok system provides yet another example of an enzyme-dependent antisense regulation process through enzymatic degradation of the resulting RNA duplex.

Other antisense mechanisms are not enzyme-dependent, but involve steric blocking of their target RNA (e.g. to prevent translation or induce alternative splicing). Steric blocking antisense mechanisms often use oligonucleotides that are heavily modified. Since there is no need for RNase H recognition, this can include chemistries such as 2'-O-alkyl, peptide nucleic acid (PNA), locked nucleic acid (LNA), and Morpholino oligomers.

3.6 See also

- DNA
- DNA codon table
- Viral replication
- RNA virus
- Antisense therapy
- Transcription
- Translation
- Directionality (molecular biology)

3.7 References

[1] Anne-Lise Haenni (2003). "Expression strategies of ambisense viruses". *Virus Research* **93** (2): 141–150. doi:10.1016/S0168-1702(03)00094-7. PMID 12782362.

[2] Kakutani T, Hayano Y, Hayashi T, Minobe Y. (1991). "Ambisense segment 3 of rice stripe virus: the first instance of a virus containing two ambisense segments". *J Gen Virol.* **72**: 465–8. doi:10.1099/0022-1317-72-2-465. PMID 1993885.

[3] Zhu Y, Hayakawa T, Toriyama S, Takahashi M. (1991). "Complete nucleotide sequence of RNA 3 of rice stripe virus: an ambisense coding strategy". *J Gen Virol.* **72**: 763–7. doi:10.1099/0022-1317-72-4-763. PMID 2016591.

[4] Cartwright, Reed; Dan Graur (Feb 8, 2011). "The multiple personalities of Watson and Crick strands". *Biology Direct* **6**: 7. doi:10.1186/1745-6150-6-7. PMC 3055211. PMID 21303550.

[5] Staff (29 January 2013) FDA approves new orphan drug Kynamro to treat inherited cholesterol disorder U.S. Food and Drug Administration, Retrieved 31 January 2013

[6] Zamecnik, P.C.; Stephenson, M.L. (1978). "Inhibition of Rous sarcoma Virus Replication and Cell Transformation by a Specific Oligodeoxynucleotide". *Proc. Natl. Acad. Sci. USA* **75** (1): 280–284. doi:10.1073/pnas.75.1.280.

[7] Watts, J.K.; Corey, D.R. (2012). "Silencing Disease Genes in the Laboratory and in the Clinic". *J. Pathol* **226** (2): 365–379. doi:10.1002/path.2993.

Chapter 4

Messenger RNA

Messenger RNA (mRNA) is a large family of RNA molecules that convey genetic information from DNA to the ribosome, where they specify the amino acid sequence of the protein products of gene expression. Following transcription of primary transcript mRNA (known as pre-mRNA) by RNA polymerase, processed, mature mRNA is translated into a polymer of amino acids: a protein, as summarized in the central dogma of molecular biology.

As in DNA, mRNA genetic information is in the sequence of nucleotides, which are arranged into codons consisting of three bases each. Each codon encodes for a specific amino acid, except the stop codons, which terminate protein synthesis. This process of translation of codons into amino acids requires two other types of RNA: Transfer RNA (tRNA), that mediates recognition of the codon and provides the corresponding amino acid, and ribosomal RNA (rRNA), that is the central component of the ribosome's protein-manufacturing machinery.

The existence of mRNA was first suggested by Jacques Monod and François Jacob, and subsequently discovered by Jacob, Sydney Brenner and Matthew Meselson at the California Institute of Technology in 1961.

4.1   Synthesis, processing and function

The brief existence of an mRNA molecule begins with transcription, and ultimately ends in degradation. During its life, an mRNA molecule may also be processed, edited, and transported prior to translation. Eukaryotic mRNA molecules often require extensive processing and transport, while prokaryotic molecules do not.

4.1.1   Transcription

Main article: Transcription (genetics)

Transcription is when RNA is made from DNA. During transcription, RNA polymerase makes a copy of a gene from the DNA to mRNA as needed. This process is similar in eukaryotes and prokaryotes. One notable difference, however, is that eukaryotic RNA polymerase associates with mRNA-processing enzymes during transcription so that processing can proceed quickly after the start of transcription. The short-lived, unprocessed or partially processed product is termed *precursor mRNA*, or *pre-mRNA*; once completely processed, it is termed *mature mRNA*.

4.1.2   Eukaryotic pre-mRNA processing

Main article: Post-transcriptional modification

Processing of mRNA differs greatly among eukaryotes, bacteria, and archea. Non-eukaryotic mRNA is, in essence,

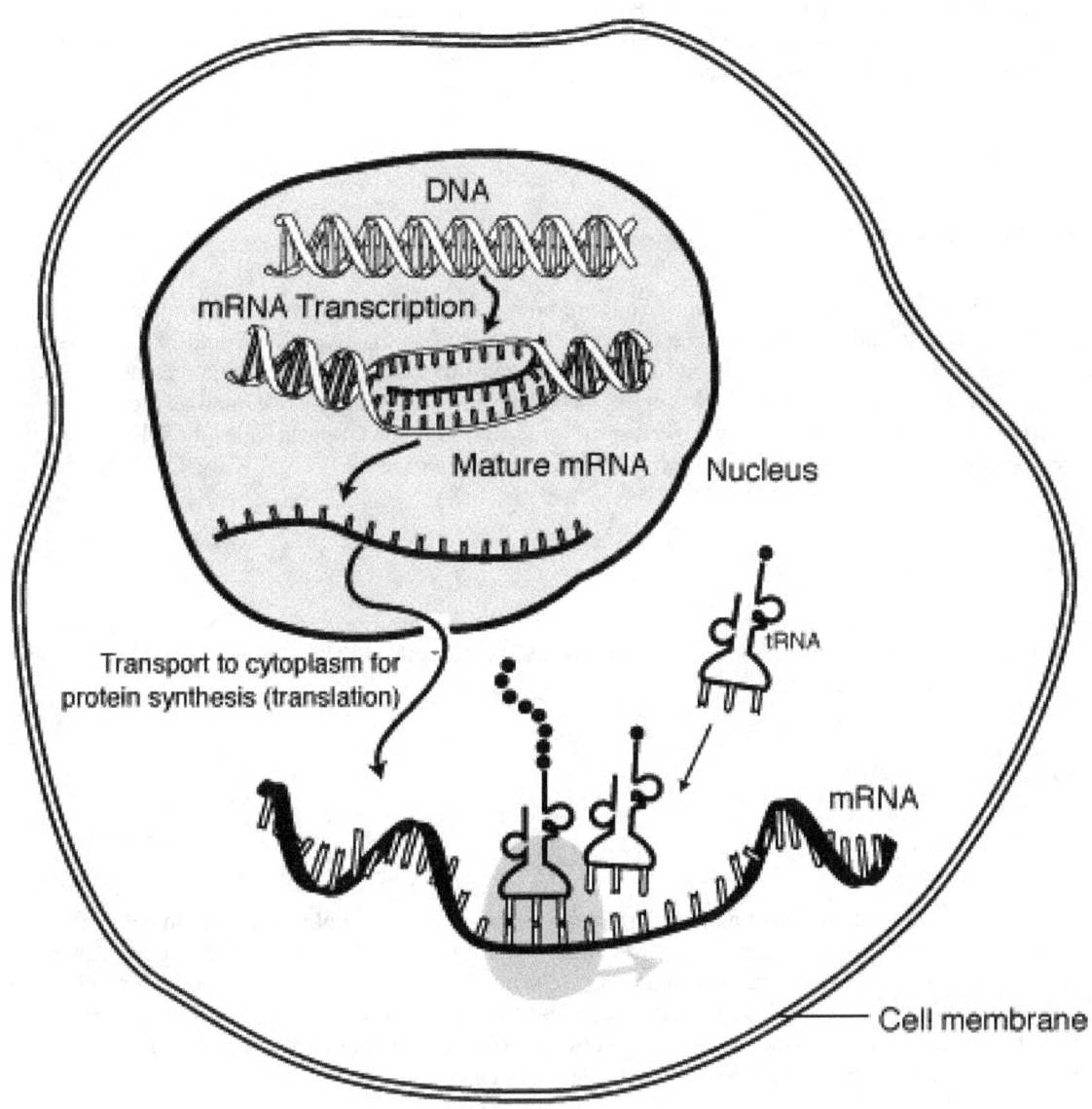

The "life cycle" of an **mRNA** in a eukaryotic cell. RNA is transcribed in the nucleus; processing, it is transported to the cytoplasm and translated by the ribosome. Finally, the mRNA is degraded.

mature upon transcription and requires no processing, except in rare cases. Eukaryotic pre-mRNA, however, requires extensive processing.

5' cap addition

Main article: 5' cap

A *5' cap* (also termed an RNA cap, an RNA 7-methylguanosine cap, or an RNA m^7G cap) is a modified guanine nucleotide that has been added to the "front" or 5' end of a eukaryotic messenger RNA shortly after the start of transcription. The 5' cap consists of a terminal 7-methylguanosine residue that is linked through a 5'−5'-triphosphate bond to the first transcribed nucleotide. Its presence is critical for recognition by the ribosome and protection from RNases.

Cap addition is coupled to transcription, and occurs co-transcriptionally, such that each influences the other. Shortly after the start of transcription, the 5' end of the mRNA being synthesized is bound by a cap-synthesizing complex associated with RNA polymerase. This enzymatic complex catalyzes the chemical reactions that are required for mRNA capping. Synthesis proceeds as a multi-step biochemical reaction.

Splicing

Main article: Splicing (genetics)

Splicing is the process by which pre-mRNA is modified to remove certain stretches of non-coding sequences called introns; the stretches that remain include protein-coding sequences and are called exons. To prevent confusion, use this mnemonic: **ex**ons contain base pairs that will be **ex**pressed in the protein, and **intr**ons are the **intr**agenic regions. Sometimes pre-mRNA messages may be spliced in several different ways, allowing a single gene to encode multiple proteins. This process is called alternative splicing. Splicing is usually performed by an RNA-protein complex called the spliceosome, but some RNA molecules are also capable of catalyzing their own splicing (*see ribozymes*).

Editing

In some instances, an mRNA will be edited, changing the nucleotide composition of that mRNA. An example in humans is the apolipoprotein B mRNA, which is edited in some tissues, but not others. The editing creates an early stop codon, which, upon translation, produces a shorter protein.

Polyadenylation

Main article: Polyadenylation

Polyadenylation is the covalent linkage of a polyadenylyl moiety to a messenger RNA molecule. In eukaryotic organisms most messenger RNA (mRNA) molecules are polyadenylated at the 3' end, but recent studies have shown that short stretches of uridine (oligouridylation) are also common.[1] The poly(A) tail and the protein bound to it aid in protecting mRNA from degradation by exonucleases. Polyadenylation is also important for transcription termination, export of the mRNA from the nucleus, and translation. mRNA can also be polyadenylated in prokaryotic organisms, where poly(A) tails act to facilitate, rather than impede, exonucleolytic degradation.

Polyadenylation occurs during and/or immediately after transcription of DNA into RNA. After transcription has been terminated, the mRNA chain is cleaved through the action of an endonuclease complex associated with RNA polymerase. After the mRNA has been cleaved, around 250 adenosine residues are added to the free 3' end at the cleavage site. This reaction is catalyzed by polyadenylate polymerase. Just as in alternative splicing, there can be more than one polyadenylation variant of an mRNA.

Polyadenylation site mutations also occur. The primary RNA transcript of a gene is cleaved at the poly-A addition site, and 100-200 A's are added to the 3' end of the RNA. If this site is altered, an abnormally long and unstable mRNA construct will be formed.

4.1.3 Transport

Another difference between eukaryotes and prokaryotes is mRNA transport. Because eukaryotic transcription and translation is compartmentally separated, eukaryotic mRNAs must be exported from the nucleus to the cytoplasm—a process that may be regulated by different signaling pathways.[2] Mature mRNAs are recognized by their processed modifications and then exported through the nuclear pore by binding to the cap-binding proteins CBP20 and CBP80,[3] as well as the transcription/export complex (TREX).[4][5] Multiple mRNA export pathways have been identified in eukaryotes.[6]

In neurons, mRNA must be transported from the soma to the dendrites where local translation occurs in response to external stimuli, such as β-actin mRNA.[7] Upon export from the nucleus, the mRNA associates with ZBP1 and the 40S subunit. The complex is bound by a motor protein and is transported to the target location (neurite extension) along the cytoskeleton. Eventually ZBP1 is phosphorylated by Src in order for translation to be initiated.[8] Many messages are marked with so-called "zip codes," which target their transport to a specific location.[9]

4.1.4 Translation

Main article: Translation (genetics)

Because prokaryotic mRNA does not need to be processed or transported, translation by the ribosome can begin immediately after the end of transcription. Therefore, it can be said that prokaryotic translation is *coupled* to transcription and occurs *co-transcriptionally*.

Eukaryotic mRNA that has been processed and transported to the cytoplasm (i.e., mature mRNA) can then be translated by the ribosome. Translation may occur at ribosomes free-floating in the cytoplasm, or directed to the endoplasmic reticulum by the signal recognition particle. Therefore, unlike in prokaryotes, eukaryotic translation *is not* directly coupled to transcription.[10]

4.2 Structure

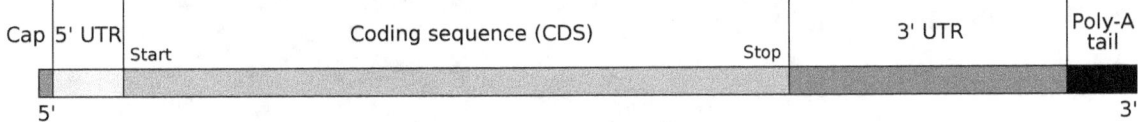

The structure of a mature eukaryotic mRNA. A fully processed mRNA includes a 5' cap, 5' UTR, coding region, 3' UTR, and poly(A) tail.

4.2.1 5' cap

Main article: 5' cap

The *5' cap* is a modified guanine nucleotide added to the "front" (5' end) of the pre-mRNA using a 5'−5'-triphosphate linkage. This modification is critical for recognition and proper attachment of mRNA to the ribosome, as well as protection from 5' exonucleases. It may also be important for other essential processes, such as splicing and transport.

4.2.2 Coding regions

Main article: Coding region

Coding regions are composed of codons, which are decoded and translated (in eukaryotes usually into one and in prokaryotes usually into several) into proteins by the ribosome. Coding regions begin with the start codon and end with a stop codon. In general, the start codon is an AUG triplet and the stop codon is UAA, UAG, or UGA. The coding regions tend to be stabilised by internal base pairs, this impedes degradation.[11][12] In addition to being protein-coding, portions of coding regions may serve as regulatory sequences in the pre-mRNA as exonic splicing enhancers or exonic splicing silencers.

4.2.3 Untranslated regions

Main articles: 5' UTR and 3' UTR

Untranslated regions (UTRs) are sections of the mRNA before the start codon and after the stop codon that are not translated, termed the five prime untranslated region (5' UTR) and three prime untranslated region (3' UTR), respectively. These regions are transcribed with the coding region and thus are exonic as they are present in the mature mRNA. Several roles in gene expression have been attributed to the untranslated regions, including mRNA stability, mRNA localization, and translational efficiency. The ability of a UTR to perform these functions depends on the sequence of the UTR and can differ between mRNAs. Genetic variants in 3' UTR have also been implicated in disease susceptibility because of the change in RNA structure and protein translation.[13]

The stability of mRNAs may be controlled by the 5' UTR and/or 3' UTR due to varying affinity for RNA degrading enzymes called ribonucleases and for ancillary proteins that can promote or inhibit RNA degradation. (See also, C-rich stability element.)

Translational efficiency, including sometimes the complete inhibition of translation, can be controlled by UTRs. Proteins that bind to either the 3' or 5' UTR may affect translation by influencing the ribosome's ability to bind to the mRNA. MicroRNAs bound to the 3' UTR also may affect translational efficiency or mRNA stability.

Cytoplasmic localization of mRNA is thought to be a function of the 3' UTR. Proteins that are needed in a particular region of the cell can also be translated there; in such a case, the 3' UTR may contain sequences that allow the transcript to be localized to this region for translation.

Some of the elements contained in untranslated regions form a characteristic secondary structure when transcribed into RNA. These structural mRNA elements are involved in regulating the mRNA. Some, such as the SECIS element, are targets for proteins to bind. One class of mRNA element, the riboswitches, directly bind small molecules, changing their fold to modify levels of transcription or translation. In these cases, the mRNA regulates itself.

4.2.4 Poly(A) tail

Main article: Polyadenylation

The 3' poly(A) tail is a long sequence of adenine nucleotides (often several hundred) added to the 3' end of the pre-mRNA. This tail promotes export from the nucleus and translation, and protects the mRNA from degradation.

4.2.5 Monocistronic versus polycistronic mRNA

See also: Cistron

An mRNA molecule is said to be monocistronic when it contains the genetic information to translate only a single protein chain (polypeptide). This is the case for most of the eukaryotic mRNAs.[14][15] On the other hand, polycistronic mRNA carries several open reading frames (ORFs), each of which is translated into a polypeptide. These polypeptides usually have a related function (they often are the subunits composing a final complex protein) and their coding sequence is grouped and regulated together in a regulatory region, containing a promoter and an operator. Most of the mRNA found in bacteria and archaea is polycistronic,[14] as is the human mitochondrial genome[16]. Dicistronic or bicistronic mRNA encodes only two proteins.

4.2.6 mRNA circularization

In eukaryotes mRNA molecules form circular structures due to an interaction between the eIF4E and poly(A)-binding protein, which both bind to eIF4G, forming an mRNA-protein-mRNA bridge.[17] Circularization is thought to promote cycling of ribosomes on the mRNA leading to time-efficient translation, and may also function to ensure only intact mRNA are translated (partially degraded mRNA characteristically have no m7G cap, or no poly-A tail).[18]

Other mechanisms for circularization exist, particularly in virus mRNA. Poliovirus mRNA uses a cloverleaf section towards its 5' end to bind PCBP2, which binds poly(A)-binding protein, forming the familiar mRNA-protein-mRNA circle. Barley yellow dwarf virus has binding between mRNA segments on its 5' end and 3' end (called kissing stem loops), circularizing the mRNA without any proteins involved.

RNA virus genomes (the + strands of which are translated as mRNA) are also commonly circularized. During genome replication the circularization acts to enhance genome replication speeds, cycling viral RNA-dependent RNA polymerase much the same as the ribosome is hypothesized to cycle.

4.3 Degradation

Different mRNAs within the same cell have distinct lifetimes (stabilities). In bacterial cells, individual mRNAs can survive from seconds to more than an hour; in mammalian cells, mRNA lifetimes range from several minutes to days.[19] The greater the stability of an mRNA the more protein may be produced from that mRNA. The limited lifetime of mRNA enables a cell to alter protein synthesis rapidly in response to its changing needs. There are many mechanisms that lead to the destruction of an mRNA, some of which are described below.

4.3.1 Prokaryotic mRNA degradation

In general, in prokaryotes the lifetime of mRNA is much shorter than in eukaryotes. Prokaryotes degrade messages by using a combination of ribonucleases, including endonucleases, 3' exonucleases, and 5' exonucleases. In some instances, small RNA molecules (sRNA) tens to hundreds of nucleotides long can stimulate the degradation of specific mRNAs by base-pairing with complementary sequences and facilitating ribonuclease cleavage by RNase III. It was recently shown that bacteria also have a sort of 5' cap consisting of a triphosphate on the 5' end.[20] Removal of two of the phosphates leaves a 5' monophosphate, causing the message to be destroyed by the exonuclease RNase J, which degrades 5' to 3'.

4.3.2 Eukaryotic mRNA turnover

Inside eukaryotic cells, there is a balance between the processes of translation and mRNA decay. Messages that are being actively translated are bound by ribosomes, the eukaryotic initiation factors eIF-4E and eIF-4G, and poly(A)-binding protein. eIF-4E and eIF-4G block the decapping enzyme (DCP2), and poly(A)-binding protein blocks the exosome complex, protecting the ends of the message. The balance between translation and decay is reflected in the size and abundance of cytoplasmic structures known as P-bodies[21] The poly(A) tail of the mRNA is shortened by specialized exonucleases that are targeted to specific messenger RNAs by a combination of cis-regulatory sequences on the RNA and trans-acting RNA-binding proteins. Poly(A) tail removal is thought to disrupt the circular structure of the message and destabilize the cap binding complex. The message is then subject to degradation by either the exosome complex or the decapping complex. In this way, translationally inactive messages can be destroyed quickly, while active messages remain intact. The mechanism by which translation stops and the message is handed-off to decay complexes is not understood in detail.

4.3.3 AU-rich element decay

The presence of AU-rich elements in some mammalian mRNAs tends to destabilize those transcripts through the action of cellular proteins that bind these sequences and stimulate poly(A) tail removal. Loss of the poly(A) tail is thought to promote mRNA degradation by facilitating attack by both the exosome complex[22] and the decapping complex.[23] Rapid mRNA degradation via AU-rich elements is a critical mechanism for preventing the overproduction of potent cytokines such as tumor necrosis factor (TNF) and granulocyte-macrophage colony stimulating factor (GM-CSF).[24] AU-rich elements also regulate the biosynthesis of proto-oncogenic transcription factors like c-Jun and c-Fos.[25]

4.3.4   Nonsense mediated decay

Main article: Nonsense mediated decay

Eukaryotic messages are subject to surveillance by nonsense mediated decay (NMD), which checks for the presence of premature stop codons (nonsense codons) in the message. These can arise via incomplete splicing, V(D)J recombination in the adaptive immune system, mutations in DNA, transcription errors, leaky scanning by the ribosome causing a frame shift, and other causes. Detection of a premature stop codon triggers mRNA degradation by 5' decapping, 3' poly(A) tail removal, or endonucleolytic cleavage.[26]

4.3.5   Small interfering RNA (siRNA)

Main article: siRNA

In metazoans, small interfering RNAs (siRNAs) processed by Dicer are incorporated into a complex known as the RNA-induced silencing complex or RISC. This complex contains an endonuclease that cleaves perfectly complementary messages to which the siRNA binds. The resulting mRNA fragments are then destroyed by exonucleases. siRNA is commonly used in laboratories to block the function of genes in cell culture. It is thought to be part of the innate immune system as a defense against double-stranded RNA viruses.[27]

4.3.6   MicroRNA (miRNA)

Main article: microRNA

MicroRNAs (miRNAs) are small RNAs that typically are partially complementary to sequences in metazoan messenger RNAs.[28] Binding of a miRNA to a message can repress translation of that message and accelerate poly(A) tail removal, thereby hastening mRNA degradation. The mechanism of action of miRNAs is the subject of active research.[29]

4.3.7   Other decay mechanisms

There are other ways by which messages can be degraded, including non-stop decay and silencing by Piwi-interacting RNA (piRNA), among others.

4.4   mRNA-based Therapeutics

mRNA is currently being investigated for its potential use in the treatment and prevention of diseases. mRNA-based vaccines are being developed as cancer immunotherapy and prophylactic vaccines for infectious diseases.[30] mRNA is also being studied as a source of therapeutic gene products and protein replacement therapies in vivo.[31]

4.5   References

[1] Choi et al. RNA. 2012. 18: 394-401

[2] Quaresma, Alexandre J.; Sievert, Jeffrey A.; Nickerson, J. A. (2013), "Regulation of mRNA export by the PI3 kinase/AKT signal transduction pathway", *Mol Biol Cell* **8** (8): 1208–21, doi:10.1091/mbc.E12-06-0450, PMC 3623641, PMID 23427269

[3] Kierzkowski, Daniel; Kmieciak, Maciej; Piontek, Paulina; Wojtaszek, Przemyslaw; Szweykowska-Kulinska, Zofia; Jarmolowski, Artur (September 2009). "The Arabidopsis CBP20 targets the cap-binding complex to the nucleus, and is stabilized by CBP80". *The Plant Journal* **59** (5): 814–825. doi:10.1111/j.1365-313X.2009.03915.x. Retrieved 12 December 2014.

[4] Sträßer, Katja; Masuda, Seiji; Mason, Paul; Pfannstiel, Jens; Oppizzi, Marisa; Rodriguez-Navarro, Susana; Rondón, Ana G.; Aguilera, Andres; Struhl, Kevin; Reed, Robin; Hurt, Ed (28 April 2002). "TREX is a conserved complex coupling transcription with messenger RNA export". *Nature* **417** (6886): 304–308. doi:10.1038/nature746. PMID 11979277. Retrieved 12 December 2014.

[5] Katahira, Jun; Yoneda, Yoshihiro (27 October 2014). "Roles of the TREX complex in nuclear export of mRNA". *RNA Biology* **6** (2): 149–152. doi:10.4161/rna.6.2.8046. Retrieved 12 December 2014.

[6] Cenik, Can; Chua, Hon Nian; Zhang, Hui; Tarnawsky, Stefan P.; Akef, Abdalla; Derti, Adnan; Tasan, Murat; Moore, Melissa J.; Palazzo, Alexander F.; Roth, Frederick P. (2011). "Genome Analysis Reveals Interplay between 5′UTR Introns and Nuclear mRNA Export for Secretory and Mitochondrial Genes". *PLoS Genetics* **7** (4): e1001366. doi:10.1371/journal.pgen.1001366. ISSN 1553-7404. PMC 3077370. PMID 21533221.

[7] Job, C.; Eberwine, J. (1912), "Localization and translation of mRNA in dendrites and axons", *Nat Rev Neurosci* **2001** (12): 889–98, doi:10.1038/35104069, PMID 11733796

[8] Spatial regulation of bold beta-actin translation by Src-dependent phosphorylation of ZBP1 Nature04115.

[9] Ainger, Kevin; Avossa, Daniela; Diana, Amy S.; Barry, Christopher; Barbarese, Elisa; Carson, John H. (1997), "Transport and Localization Elements in Myelin Basic Protein mRNA", *The Journal of Cell Biology* **138** (5): 1077–1087, doi:10.1083/jcb.138.5 PMC2136761, PMID9281585

[10] somalia

[11] Shabalina SA, Ogurtsov AY, Spiridonov NA (2006), "A periodic pattern of mRNA secondary structure created by the genetic code", *Nucleic Acids Res.* **34** (8): 2428–37, doi:10.1093/nar/gkl287, PMC 1458515, PMID 16682450

[12] Katz L, Burge CB (September 2003), "Widespread Selection for Local RNA Secondary Structure in Coding Regions of Bacterial Genes", *Genome Res.* **13** (9): 2042–51, doi:10.1101/gr.1257503, PMC 403678, PMID 12952875

[13] Lu, YF; Mauger, DM; Goldstein, DB; Urban, TJ; Weeks, KM; Bradrick, SS (4 November 2015). "IFNL3 mRNA structure is remodeled by a functional non-coding polymorphism associated with hepatitis C virus clearance.". *Scientific reports* **5**: 16037. PMID 26531896.

[14] Kozak, M. (March 1983), "Comparison of initiation of protein synthesis in procaryotes, eucaryotes, and organelles", *Microbiological Reviews* **47** (1): 1–45, PMC 281560, PMID 6343825

[15] Niehrs C, Pollet N (December 1999), "Synexpression groups in eukaryotes", *Nature* **402** (6761): 483–7, doi:10.1038/990025, PMID 10591207

[16] Mercer, Tim R.; Neph, Shane; Dinger, Marcel E.; Crawford, Joanna; Smith, Martin A.; Shearwood, Anne-Marie J.; Haugen, Eric; Bracken, Cameron P.; Rackham, Oliver; Stamatoyannopoulos, John A.; Filipovska, Aleksandra; Mattick, John S. (2011). "The Human Mitochondrial Transcriptome". *Cell* **146** (4): 645–658. doi:10.1016/j.cell.2011.06.051. ISSN 0092-8674.

[17] Wells, S.E.; Hillner, P.E.; Vale, R.D.; Sachs, A.B. (1998), "Circularization of mRNA by Eukaryotic Translation Initiation Factors" (w), *Molecular Cell* **2** (1): 135–140, doi:10.1016/S1097-2765(00)80122-7, PMID 9702200

[18] López-Lastra M, Rivas A, Barría MI (2005), "Protein synthesis in eukaryotes: The growing biological relevance of cap-independent translation initiation",*Biological Research***38**: 121–146,doi:10.4067/S0716-97602005000200003,PMID 16238092

[19] Yu, Jia; Russell, J. Eric. "Structural and Functional Analysis of an mRNP Complex That Mediates the High Stability of Human β-Globin mRNA" (PDF). *National Center for Biotechnology Information*. Retrieved 4 June 2014.

[20] Deana, Atilio; Celesnik, Helena; Belasco, Joel G. (2008), "The bacterial enzyme RppH triggers messenger RNA degradation by 5′ pyrophosphate removal", *Nature* **451** (7176): 355–8, doi:10.1038/nature06475, PMID 18202662

[21] Parker, R.; Sheth, U. (2007), "P Bodies and the Control of mRNA Translation and Degradation" (w), *Molecular Cell* **25** (5): 635–646, doi:10.1016/j.molcel.2007.02.011, PMID 17349952

[22] Chen, C.Y.; Gherzi, R.; Ong, S.E.; Chan, E.L.; Raijmakers, R.; Pruijn, G.J.M.; Stoecklin, G.; Moroni, C.; Mann, M.; Karin, Michael (2001), "AU Binding Proteins Recruit the Exosome to Degrade ARE-Containing mRNAs", *Cell* **107** (4): 451–464, doi:10.1016/S0092-8674(01)00578-5, PMID 11719186

[23] Fenger-Grøn M, Fillman C, Norrild B, Lykke-Andersen J (December 2005), "Multiple processing body factors and the ARE binding protein TTP activate mRNA decapping" (PDF), *Mol. Cell* **20** (6): 905–15, doi:10.1016/j.molcel.2005.10.031, PMID 16364915

[24] Shaw G, Kamen R (August 1986), "A conserved AU sequence from the 3' untranslated region of GM-CSF mRNA mediates selective mRNA degradation", *Cell* **46** (5): 659–67, doi:10.1016/0092-8674(86)90341-7, PMID 3488815

[25] Chen, C.Y.A.; Shyu, A.B. (1995), "AU-rich elements: characterization and importance in mRNA degradation", *Trends in Biochemical Sciences* **20** (11): 465–470, doi:10.1016/S0968-0004(00)89102-1, PMID 8578590

[26] Isken, O.; Maquat, L.E. (2007), "Quality control of eukaryotic mRNA: safeguarding cells from abnormal mRNA function", *Genes & Development* **21** (15): 1833–56, doi:10.1101/gad.1566807, PMID 17671086

[27] Obbard, D.J.; Gordon, K.H.J.; Buck, A.H.; Jiggins, F.M. (2009), "The evolution of RNAi as a defence against viruses and transposable elements", *Philosophical Transactions of the Royal Society B: Biological Sciences* **364** (1513): 99–115, doi:10.1098/rstb.2008.0168,PMC2592633, PMID18926973

[28] Brennecke J, Stark A, Russell RB, Cohen SM (March 2005), "Principles of MicroRNA–Target Recognition", *PLoS Biol.* **3** (3): e85, doi:10.1371/journal.pbio.0030085, PMC 1043860, PMID 15723116

[29] Eulalio, A.; Huntzinger, E.; Nishihara, T.; Rehwinkel, J.; Fauser, M.; Izaurralde, E. (2009), "Deadenylation is a widespread effect of miRNA regulation", *RNA* **15** (1): 21–32, doi:10.1261/rna.1399509, PMC 2612776, PMID 19029310

[30] "sp2 Inter-Active". September–October 2012.

[31] "BioWorld Today".

4.6 External links

- Life of mRNA Flash animation
- RNAi Atlas: a database of RNAi libraries and their target analysis results
- miRSearch: Tool for finding microRNAs that target mRNA
- Bio-Synthesis mRNA Synthesis: RNA transcript and mRNA synthesis
- How mRNA is coded?: YouTube video

Chapter 5

DNA supercoil

DNA supercoiling refers to the over- or under-winding of a DNA strand, and is an expression of the strain on that strand. Supercoiling is important in a number of biological processes, such as compacting DNA. Additionally, certain enzymes such as topoisomerases are able to change DNA topology to facilitate functions such as DNA replication or transcription. Mathematical expressions are used to describe supercoiling by comparing different coiled states to relaxed B-form DNA.

As a general rule, the DNA of most organisms is negatively supercoiled.[1]

5.1 Role of supercoiling

In a "relaxed" double-helical segment of **B-DNA**, the two strands twist around the helical axis once every 10.4–10.5 base pairs of sequence. Adding or subtracting twists, as some enzymes can do, imposes strain. If a DNA segment under twist strain were closed into a circle by joining its two ends and then allowed to move freely, the circular DNA would contort into a new shape, such as a simple figure-eight. Such a contortion is a **supercoil**.

The simple figure eight is the simplest supercoil, and is the shape a circular DNA assumes to accommodate one too many or one too few helical twists. The two lobes of the figure eight will appear rotated either clockwise or counterclockwise with respect to one another, depending on whether the helix is over- or underwound. For each additional helical twist being accommodated, the lobes will show one more rotation about their axis.

The noun form "supercoil" is rarely used in the context of DNA topology. Instead, global contortions of a circular DNA, such as the rotation of the figure-eight lobes above, are referred to as *writhe*. The above example illustrates that twist and writhe are interconvertible. "**Supercoiling**" is an abstract mathematical property representing the sum of twist and writhe. The twist is the number of helical turns in the DNA and the writhe is the number of times the double helix crosses over on itself (these are the supercoils).

Extra helical twists are positive and lead to positive supercoiling, while subtractive twisting causes negative supercoiling. Many topoisomerase enzymes sense supercoiling and either generate or dissipate it as they change DNA topology. DNA of most organisms is negatively supercoiled.

In part because chromosomes may be very large, segments in the middle may act as if their ends are anchored. As a result, they may be unable to distribute excess twist to the rest of the chromosome or to absorb twist to recover from underwinding—the segments may become *supercoiled*, in other words. In response to supercoiling, they will assume an amount of writhe, just as if their ends were joined.

Supercoiled DNA forms two structures; a plectoneme or a toroid, or a combination of both. A negatively supercoiled DNA molecule will produce either a one-start left-handed helix, the toroid, or a two-start right-handed helix with terminal loops, the plectoneme. Plectonemes are typically more common in nature, and this is the shape most bacterial plasmids will take. For larger molecules it is common for hybrid structures to form – a loop on a toroid can extend into a plectoneme. If all the loops on a toroid extend then it becomes a branch point in the plectonemic structure.

5.2   Occurrence of DNA supercoiling

DNA supercoiling is important for DNA packaging within all cells. Because the length of DNA can be thousands of times that of a cell, packaging this genetic material into the cell or nucleus (in eukaryotes) is a difficult feat. Supercoiling of DNA reduces the space and allows for much more DNA to be packaged. In prokaryotes, plectonemic supercoils are predominant, because of the circular chromosome and relatively small amount of genetic material. In eukaryotes, DNA supercoiling exists on many levels of both plectonemic and solenoidal supercoils, with the solenoidal supercoiling proving most effective in compacting the DNA. Solenoidal supercoiling is achieved with histones to form a 10 nm fiber. This fiber is further coiled into a 30 nm fiber, and further coiled upon itself numerous times more.

DNA packaging is greatly increased during nuclear division events such as mitosis or meiosis, where DNA must be compacted and segregated to daughter cells. Condensins and cohesins are *Structural Maintenance of Chromosome* proteins that aid in the condensation of sister chromatids and the linkage of the centromere in sister chromatids. These SMC proteins induce positive supercoils.

Supercoiling is also required for DNA/RNA synthesis. Because DNA must be unwound for DNA/RNA polymerase action, supercoils will result. The region ahead of the polymerase complex will be unwound; this stress is compensated with positive supercoils ahead of the complex. Behind the complex, DNA is rewound and there will be **compensatory** negative supercoils. Topoisomerases such as DNA gyrase (Type II Topoisomerase) play a role in relieving some of the stress during DNA/RNA synthesis.[2]

5.3   Mathematical description

In nature, circular DNA is always isolated as a higher-order helix-upon-a-helix, known as a *superhelix*. In discussions of this subject, the Watson-Crick twist is referred to as a "secondary" winding, and the superhelices as a "tertiary" winding. The sketch at right indicates a "relaxed", or "open circular" Watson-Crick double-helix, and, next to it, a right-handed superhelix. The "relaxed" structure on the left is not found unless the chromosome is nicked; the superhelix is the form usually found in nature.

For purposes of mathematical computations, a right-handed superhelix is defined as having a "negative" number of super-helical turns, and a left-handed superhelix is defined as having a "positive" number of superhelical turns. In the drawing (shown at the right), both the secondary (*i.e.,* "Watson-Crick") winding and the superhelical winding are right-handed, hence the supertwists are negative(-3 in this example).

The superhelicity is presumed to be a result of underwinding, meaning that there is a deficiency in the number of secondary Watson-Crick twists. Such a chromosome will be strained, just as a macroscopic metal spring is strained when it is either overwound or unwound. In DNA which is thusly strained, supertwists will appear.

DNA supercoiling can be described numerically by changes in the linking number Lk. The linking number is the most descriptive property of supercoiled DNA. Lk_o, the number of turns in the relaxed (B type) DNA plasmid/molecule, is determined by dividing the total base pairs of the molecule by the relaxed bp/turn which, depending on reference is 10.4–10.5.

$$Lk_o = bp/10.4$$

Lk is merely the number of crosses a single strand makes across the other . L_k, known as the "linking number", is the number of Watson-Crick twists found in a circular chromosome in a (usually imaginary) planar projection. This number is physically "locked in" at the moment of covalent closure of the chromosome, and cannot be altered without strand breakage.

The topology of the DNA is described by the equation below in which the linking number is equivalent to the sum of TW, which is the number of twists or turns of the double helix, and Wr which is the number of coils or 'writhes'. If there is a closed DNA molecule, the sum of Tw and Wr, or the linking number, does not change. However, there may be complementary changes in TW and Wr without changing their sum.

$$Lk = Tw + Wr$$

Tw, called "twist", refers to the number of Watson-Crick twists in the chromosome when it is not constrained to lie in a plane. We have already seen that native DNA is usually found to be superhelical. If one goes around the superhelically twisted chromosome, counting secondary Watson-Crick twists, that number will be different from the number counted when the chromosome is constrained to lie flat. In general, the number of secondary twists in the native, supertwisted chromosome is expected to be the "normal" Watson-Crick winding number, meaning a single 10-base-pair helical twist for every 34 Å of DNA length.

Wr, called "writhe", is the number of superhelical twists. Since biological circular DNA is usually underwound, $\mathbf{L_k}$ will generally be *less* than **Tw**, which means that **Wr** will typically be *negative.*

Now we can see that if DNA is underwound, it will be under strain, exactly as a metal spring is strained when forcefully unwound, and that the appearance of supertwists will allow the chromosome to relieve its strain by taking on negative supertwists, which correct the secondary underwinding in accordance with the topology equation above.

The topology equation teaches further that there is a one-to-one relationship between changes in **Tw** and **Wr**. For example, if a secondary "Watson-Crick" twist is removed, then a right-handed supertwist must have been removed simultaneously (or, if the chromosome is relaxed, with no supertwists, then a left-handed supertwist must be added).

The change in the linking number, ΔLk, is the actual number of turns in the plasmid/molecule, Lk, minus the number of turns in the relaxed plasmid/molecule Lk$_o$.

$$\Delta Lk = Lk - Lk_o$$

If the DNA is negatively supercoiled ΔLk < 0. The negative supercoiling implies that the DNA is underwound.

A standard expression independent of the molecule size is the "specific linking difference" or "superhelical density" denoted σ. σ represents the number of turns added or removed relative to the total number of turns in the relaxed molecule/plasmid, indicating the level of supercoiling.

$$\sigma = \Delta Lk / Lk_o$$

The Gibbs free energy associated with the coiling is given by the equation below[3]

$$\Delta G/N = 10RT\sigma^2$$

The difference in Gibbs free energy between the supercoiled circular DNA and uncoiled circular DNA with N > 2000 bp is approximated by:

$$\Delta G/N = 700 Kcal/bp * (\Delta Lk/N)$$

or, 16 cal/bp.

Since the linking number L of supercoiled DNA is the number of times the two strands are intertwined (and both strands remain covalently intact), L cannot change. The reference state (or parameter) L_0 of a circular DNA duplex is its relaxed state. In this state, its writhe $W = 0$. Since $L = T + W$, in a relaxed state $T = L$. Thus, if we have a 400 bp relaxed circular DNA duplex, $L \sim 40$ (assuming ~10 bp per turn in B-DNA). Then $T \sim 40$.

- Positively supercoiling:

    T = 0, W = 0, then L = 0
    T = +3, W = 0, then L = +3
    T = +2, W = +1, then L = +3

- Negatively supercoiling:

    $T = 0$, $W = 0$, then $L = 0$
    $T = -3$, $W = 0$, then $L = -3$
    $T = -2$, $W = -1$, then $L = -3$

Negative supercoils favor local unwinding of the DNA, allowing processes such as transcription, DNA replication, and recombination. Negative supercoiling is also thought to favour the transition between B-DNA and Z-DNA, and moderate the interactions of DNA binding proteins involved in gene regulation.[4]

5.4   Effects on sedimentation coefficient

The topological properties of circular DNA are complex, and only a brief introduction can be presented here. In standard texts, these properties are invariably explained in terms of a helical model for DNA, because the majority of scientists continue to believe that no other structure is possible.

When the sedimentation coefficient, s, of circular DNA is ascertained over a large range of pH, the following curves are seen.

Three curves are shown here, representing three species of DNA. From top-to-bottom they are: "Form IV" (green), "Form I" (blue) and "Form II" (red).

"Form I" (blue curve) is the traditional nomenclature used for the native form of duplex circular DNA, as recovered from viruses and intracellular plasmids. Form I is covalently closed, and any plectonemic winding which may be present is therefore locked in.

If one or more nicks are introduced to Form I, free rotation of one strand with respect to the other becomes possible, and Form II (red curve) is seen.

Form IV (green curve) is the product of alkali denaturation of Form I. Its structure is unknown, except that it is persistently duplex, and extremely dense.

Between pH 7 and pH 11.5, the sedimentation coefficient s, for Form I, is constant. Then it dips, and at a pH just below 12, reaches a minimum. With further increases in pH, s then returns to its former value. It doesn't stop there, however, but continues to increase relentlessly. By pH 13, the value of s has risen to nearly 50, two to three times its value at pH 7, indicating an extremely compact structure.

If the pH is then lowered, the s value is not restored. Instead, one sees the upper, green curve. The DNA, now in the state known as Form IV, remains extremely dense, even if the pH is restored to the original physiologic range. As stated previously, the structure of Form IV is almost entirely unknown, and there is no currently accepted explanation for its extraordinary density. About all that is known about the tertiary structure is that it is duplex, but has no hydrogen bonding between bases.

These behaviors of Forms I and IV are considered to be due to the peculiar properties of duplex DNA which has been covalently closed into a double-stranded circle. If the covalent integrity is disrupted by even a single nick in one of the strands, all such topological behavior ceases, and one sees the lower Form II curve (Δ). For Form II, alterations in pH have very little effect on s. Its physical properties are, in general, identical to those of linear DNA. At pH 13, the strands of Form II simply separate, just as the strands of linear DNA do. The separated single strands have slightly different s values, but display no significant changes in s with further increases in pH.

A complete explanation for these data is beyond the scope of this article. In brief, the alterations in s come about because of changes in the superhelicity of circular DNA. These changes in superhelicity are schematically illustrated by four little drawings which have been strategically superimposed upon the figure above.

Without going into great detail, let it simply be said that the alterations of s seen in the pH titration curve above are widely believed to be due to changes in the superhelical winding of DNA under conditions of increasing pH. Up to pH 11.5, the purported "underwinding" produces a right-handed ("negative") supertwist. But as the pH increases, and the secondary helical structure begins to denature and unwind, the chromosome (if we may speak anthropomorphically) no longer "wants" to have the full Watson-Crick winding, but rather "wants", increasingly, to be "underwound". Since there is less and less strain to be relieved by superhelical winding, the superhelices therefore progressively disappear as the pH increases. At a pH just below 12, all incentive for superhelicity has expired, and the chromosome will appear as a relaxed, open circle.

At higher pHs still, the chromosome, which is now denaturing in earnest, wishes to unwind entirely, which it cannot do so (because L_k is covalently locked in). Under these conditions, what was once treated as "underwinding" has actually now become "overwinding". Once again there is strain, and once again it is (in part at least) relieved by superhelicity, but this time in the opposite direction (*i.e.,* left-handed or "positive"). Each left-handed tertiary supertwist removes a single, now *undesirable* right-handed Watson-Crick secondary twist.

The titration ends at pH 13, where Form IV appears.

5.5 See also

- Jerome Vinograd

- Mechanical properties of DNA

- Ribbon theory

5.6 References

[1] Champoux J (2001). "DNA topoisomerases: structure, function, and mechanism". *Annu Rev Biochem* **70**: 369–413. doi:10.1146/ PMID 11395412.

[2] Albert A-C, Spirito F, Figueroa-Bossi N, Bossi L, Rahmouni AR (1996). "Hyper-negative template DNA supercoiling during transcription of the tetracycline-resistance gene in topA mutants is largely constrained in vivo". *Nucl Acids Res* **24** (15): 3093–3099. doi:10.1093/nar/24.15.3093. PMC 146055. PMID 8760899.

[3] Vologodskii AV, Lukashin AV, Anshelevich VV, et al. (1979). "Fluctuations in superhelical DNA". *Nucleic Acids Res* **6** (3): 967–682. doi:10.1093/nar/6.3.967. PMC 327745. PMID 155809.

[4] H. S. Chawla (2002). *Introduction to Plant Biotechnology*. Science Publishers. ISBN 1-57808-228-5.

5.6.1 General references

- Bloomfield, Victor A.; Crothers, Donald M.; Tinoco, Jr., Ignacio (2000). *Nucleic acids: structures, properties, and functions*. Sausalito, California: University Science Books. pp. 446–453. ISBN 0935702490.

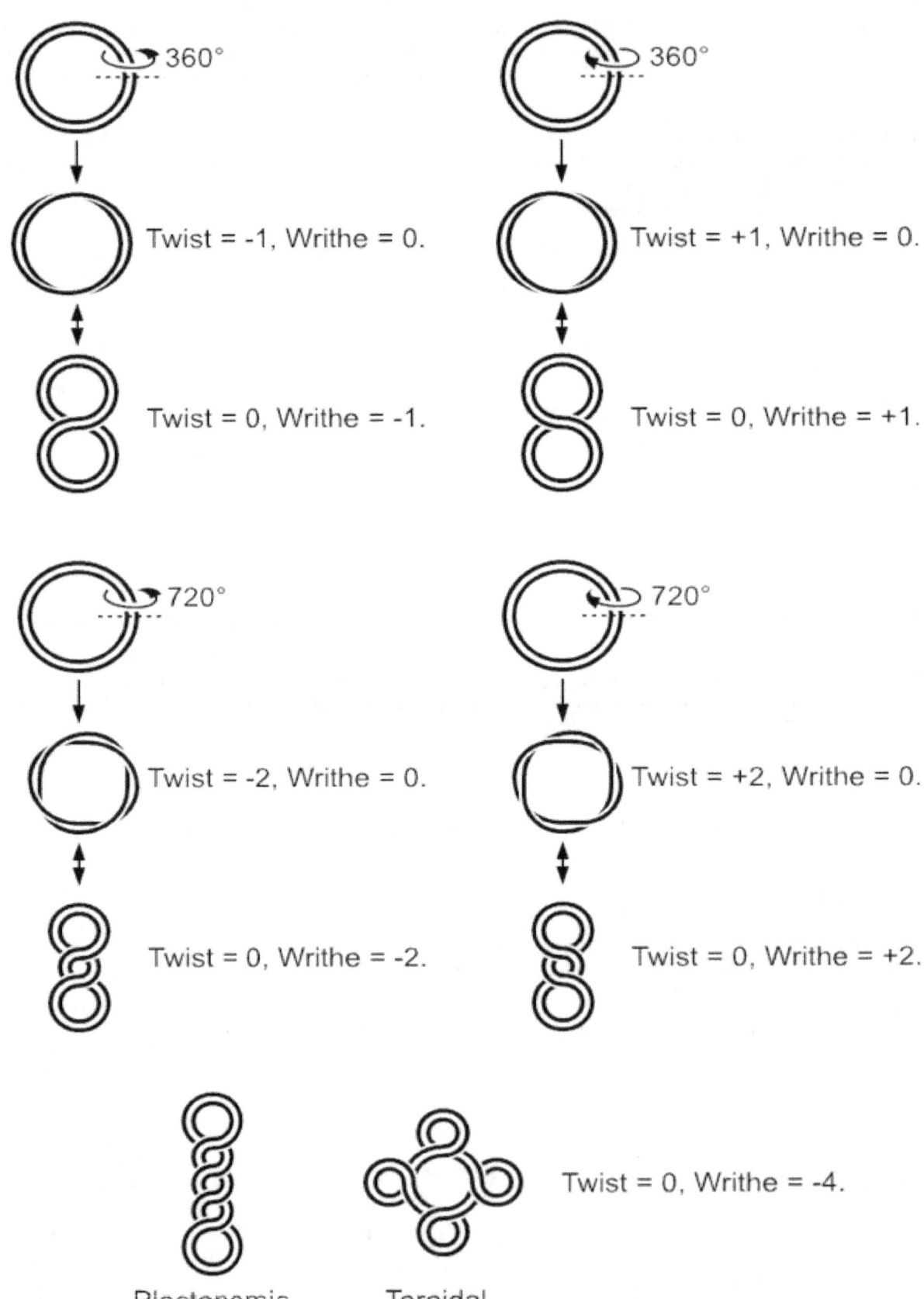

Supercoiled structure of circular DNA molecules with low writhe. The helical nature of the DNA duplex is omitted for clarity.

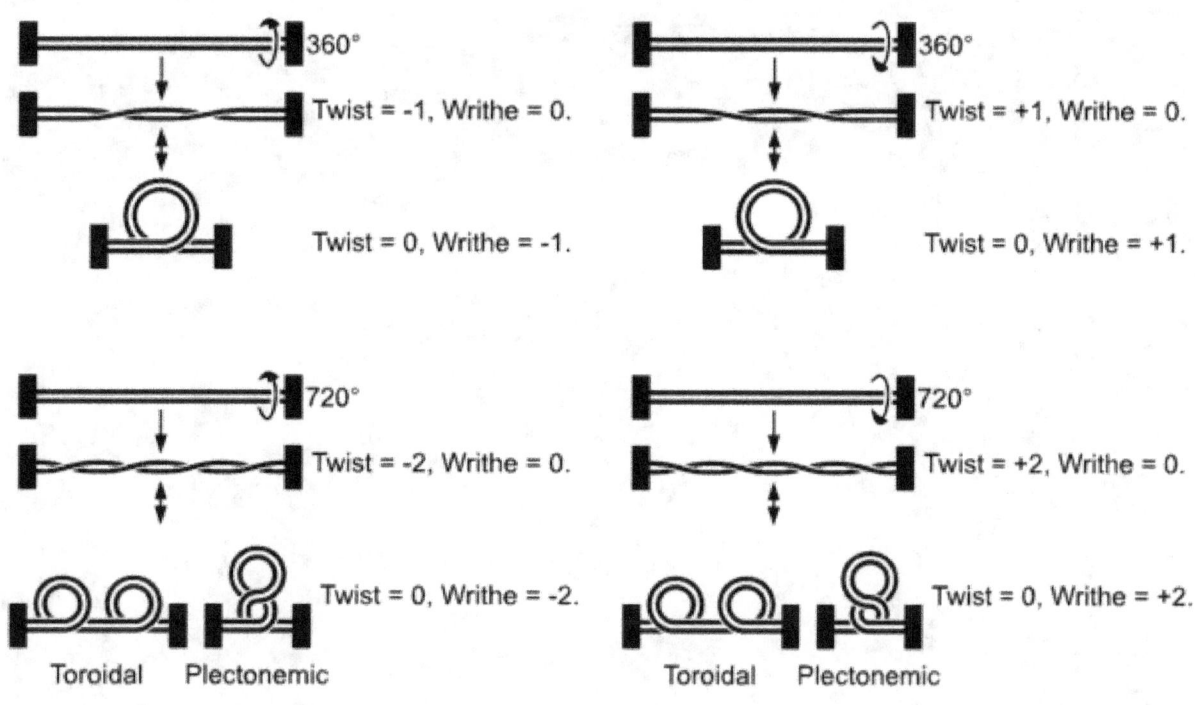

Supercoiled structure of linear DNA molecules with constrained ends. The helical nature of the DNA duplex is omitted for clarity.

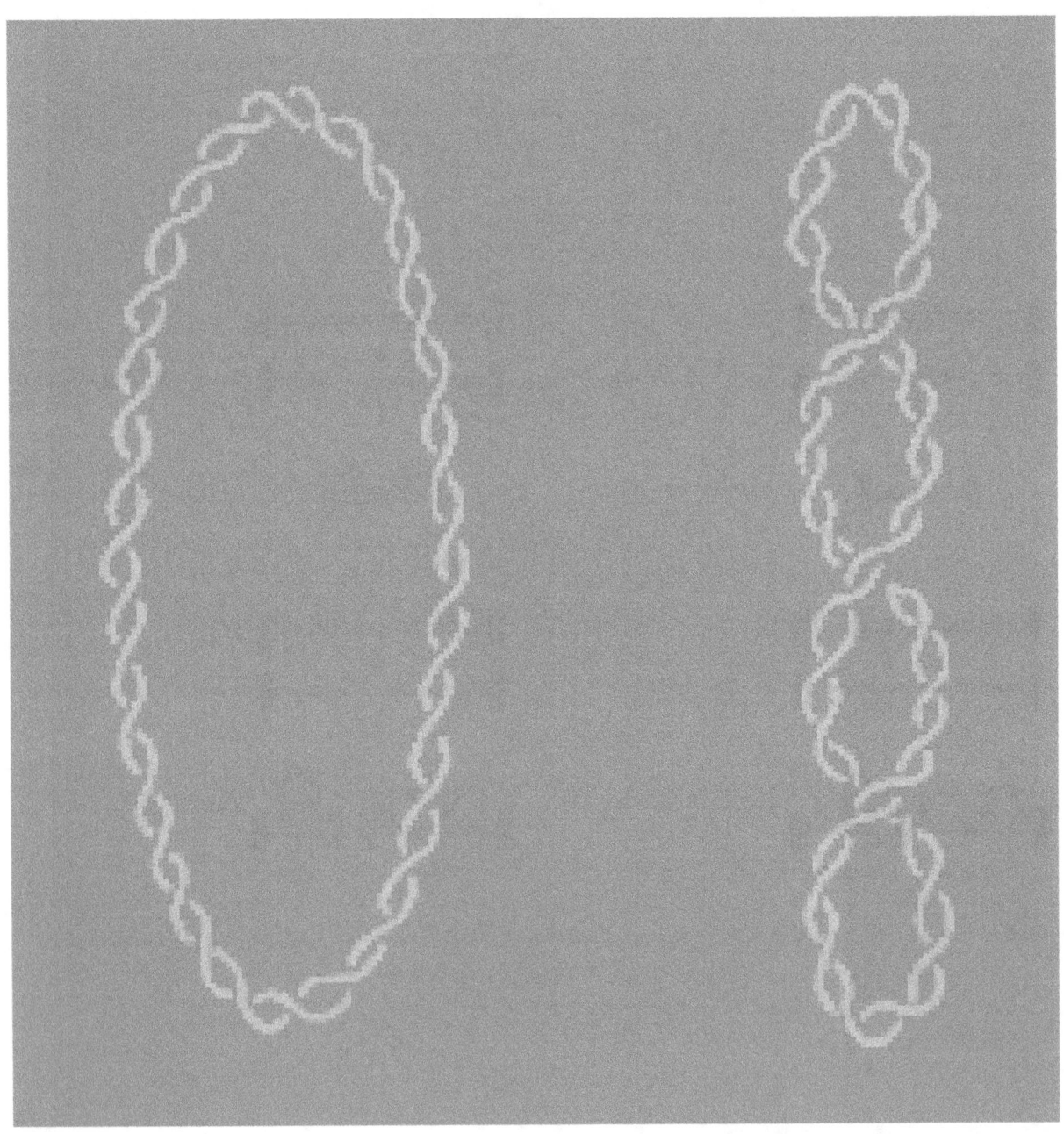

Drawing showing the difference between a circular DNA chromosome with a secondary helical twist only, and one containing an additional tertiary superhelical twist superimposed on the secondary helical winding.

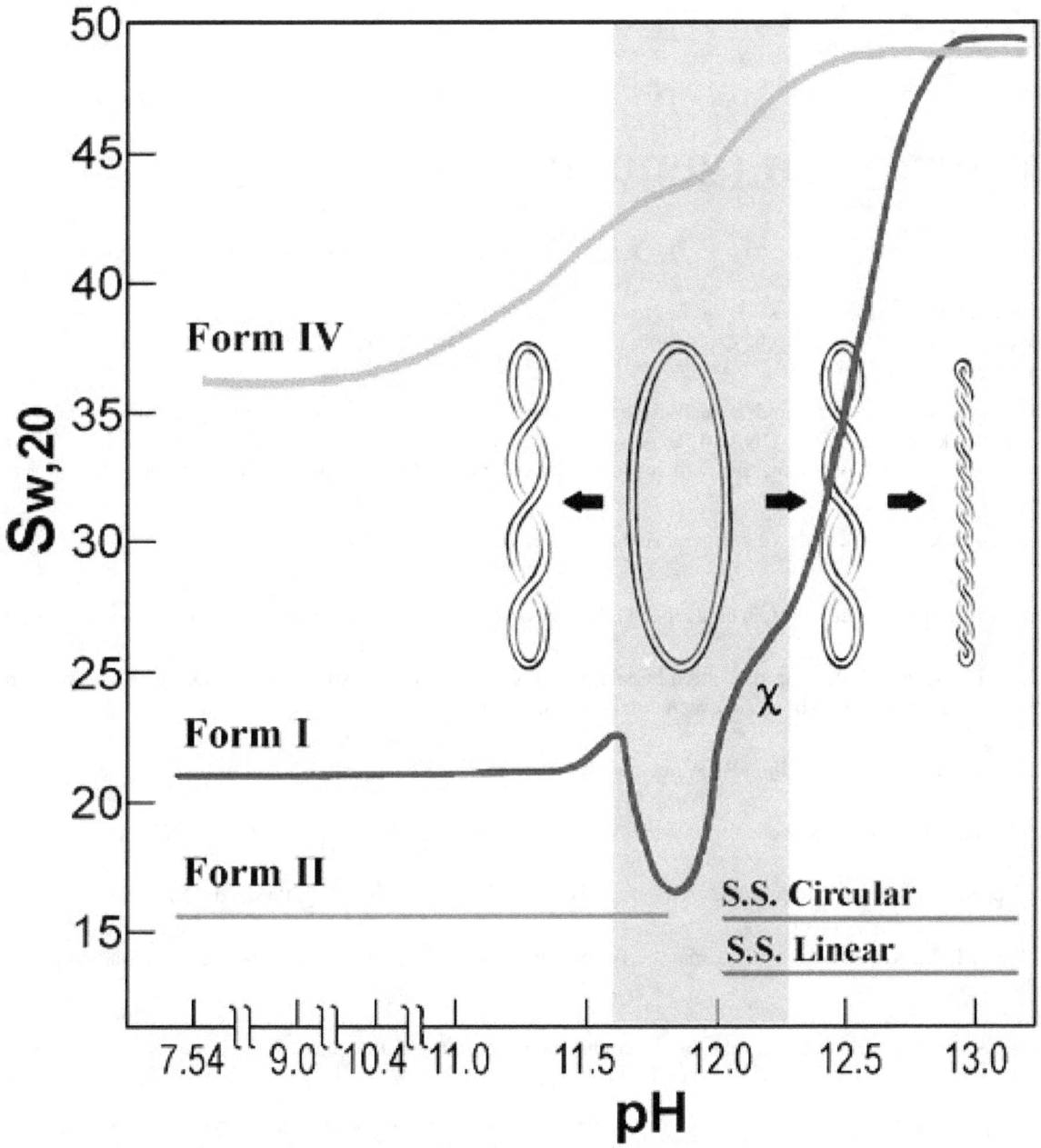

Figure showing the various conformational changes which are observed in circular DNA at different pHs. At a pH of about 12, there is a dip in the sedimentation coefficient, followed by a relentless increase up to a pH of about 13, at which pH the structure converts into the mysterious "Form IV".

Chapter 6

Transcription (genetics)

This article is about a genetic process. For other uses, see Transcription.

Transcription is the first step of gene expression, in which a particular segment of DNA is copied into RNA (mRNA) by the enzyme RNA polymerase.

Both RNA and DNA are nucleic acids, which use base pairs of nucleotides as a complementary language. The two can be converted back and forth from DNA to RNA by the action of the correct enzymes. During transcription, a DNA sequence is read by an RNA polymerase, which produces a complementary, antiparallel RNA strand called a primary transcript.

Transcription proceeds in the following general steps:

1. One or more sigma factor protein binds to the RNA polymerase holoenzyme, allowing it to bind to promoter DNA.

2. RNA polymerase creates a transcription bubble, which separates the two strands of the DNA helix. This is done by breaking the hydrogen bonds between complementary DNA nucleotides.

3. RNA polymerase adds matching RNA nucleotides to the complementary nucleotides of one DNA strand.

4. RNA sugar-phosphate backbone forms with assistance from RNA polymerase to form an RNA strand.

5. Hydrogen bonds of the untwisted RNA-DNA helix break, freeing the newly synthesized RNA strand.

6. If the cell has a nucleus, the RNA may be further processed. This may include polyadenylation, capping, and splicing.

7. The RNA may remain in the nucleus or exit to the cytoplasm through the nuclear pore complex.

The stretch of DNA transcribed into an RNA molecule is called a *transcription unit* and encodes at least one gene. If the gene transcribed encodes a protein, messenger RNA (mRNA) will be transcribed; the mRNA will in turn serve as a template for the protein's synthesis through translation. Alternatively, the transcribed gene may encode for either non-coding RNA (such as microRNA), ribosomal RNA (rRNA), transfer RNA (tRNA), or other enzymatic RNA molecules called ribozymes.[1] Overall, RNA helps synthesize, regulate, and process proteins; it therefore plays a fundamental role in performing functions within a cell.

In virology, the term may also be used when referring to mRNA synthesis from an RNA molecule (i.e., RNA replication). For instance, the genome of a negative-sense single-stranded RNA (ssRNA -) virus may be template for a positive-sense single-stranded RNA (ssRNA +). This is because the positive-sense strand contains the information needed to translate the viral proteins for viral replication afterwards. This process is catalysed by a viral RNA replicase.[2]

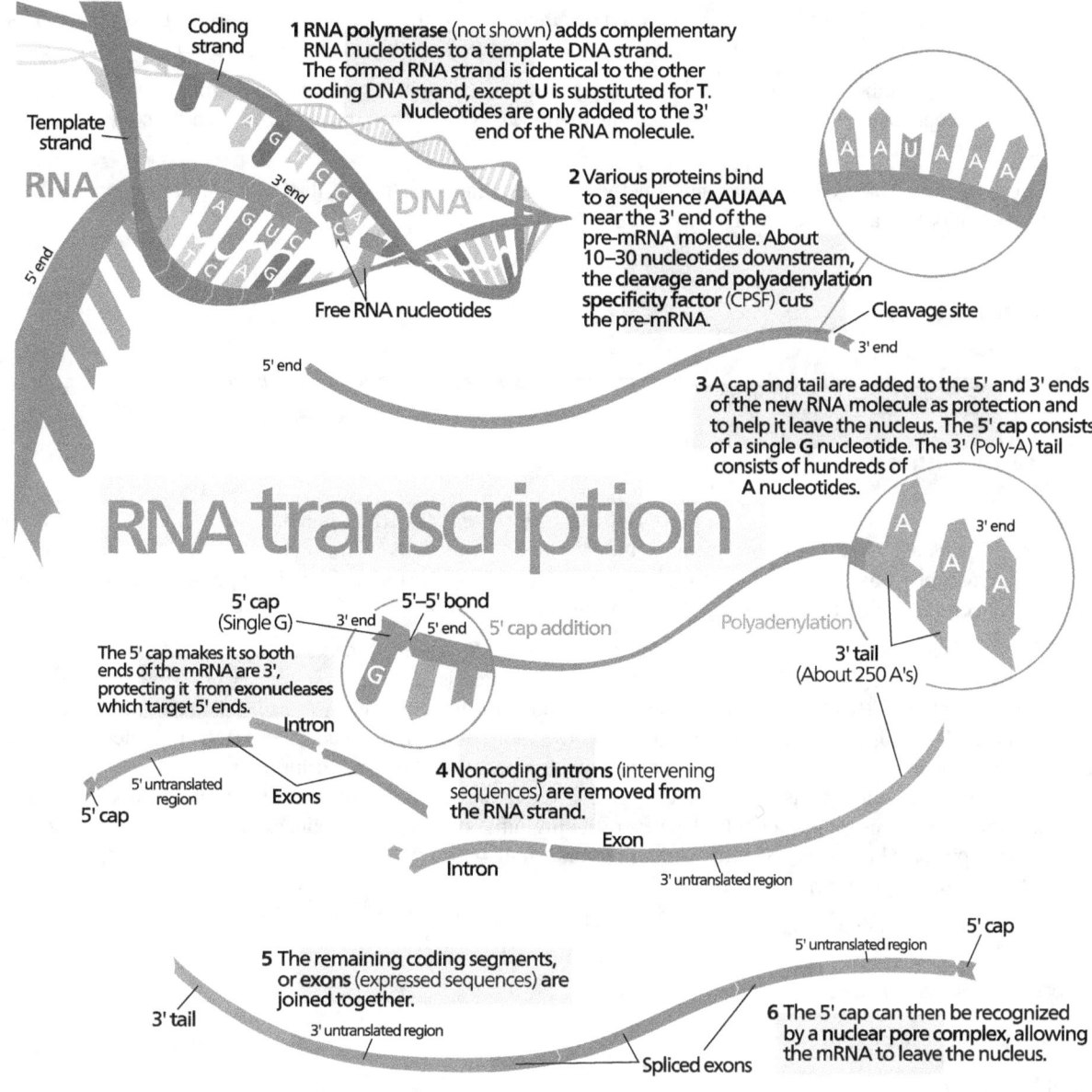

Simplified diagram of mRNA synthesis and processing. Enzymes not shown.

6.1 Background

A DNA transcription unit encoding for a protein may contain both a *coding sequence*, which will be translated into the protein, and *regulatory sequences*, which direct and regulate the synthesis of that protein. The regulatory sequence before ("upstream" from) the coding sequence is called the five prime untranslated region (5'UTR); the sequence after ("downstream" from) the coding sequence is called the three prime untranslated region (3'UTR).[1]

As opposed to DNA replication, transcription results in an RNA complement that includes the nucleotide uracil (U) in all instances where thymine (T) would have occurred in a DNA complement.

Only one of the two DNA strands serve as a template for transcription. The antisense strand of DNA is read by RNA polymerase from the 3' end to the 5' end during transcription (3' → 5'). The complementary RNA is created in the opposite direction, in the 5' → 3' direction, matching the sequence of the sense strand with the exception of switching uracil for thymine. This directionality is because RNA polymerase can only add nucleotides to the 3' end of the growing mRNA chain. This use of only the 3' → 5' DNA strand eliminates the need for the Okazaki fragments that are seen in DNA replication.[1] This removes the need for an RNA primer to initiate RNA synthesis, as is the case in DNA replication.

The *non*-template sense strand of DNA is called the coding strand, because its sequence is the same as the newly created RNA transcript (except for the substitution of uracil for thymine). This is the strand that is used by convention when presenting a DNA sequence.

Transcription has some proofreading mechanisms, but they are fewer and less effective than the controls for copying DNA; therefore, transcription has a lower copying fidelity than DNA replication.[3]

6.2   Major steps

Transcription is divided into *pre-initiation, initiation, promoter clearance, elongation* and *termination*.[1]

6.2.1   Pre-initiation

In **eukaryotes**, a core promoter sequence in the DNA must be present for RNA polymerase to initiate transcription. Promoters are regions of DNA that promote transcription - in eukaryotes, they are found at -30, -75, and -90 base pairs upstream from the transcription start site (abbreviated to TSS). Transcription factors are proteins that bind to these promoter sequences and facilitate the binding of RNA Polymerase.[4]

The most characterized type of core promoter in eukaryotes is a short DNA sequence known as a TATA box, found 25-30 base pairs upstream from the TSS.[4] The TATA box, as a core promoter, is the binding site for a transcription factor known as TATA-binding protein (TBP) (which is itself a subunit of another transcription factor, called Transcription Factor II D (TFIID)). After TFIID binds to the TATA box via the TBP, five more transcription factors and RNA polymerase combine around the TATA box in a series of stages to form a preinitiation complex. One transcription factor, Transcription factor II H, has two components with helicase activity and so is involved in the separating of opposing strands of double-stranded DNA to form the initial transcription bubble. However, the preinitiation complex produces only a low transcription rate on its own. Other proteins known as activators and repressors, along with any associated coactivators or corepressors, are responsible for modulating transcription rate.[4]

Thus, preinitiation complex contains:.

1. Core Promoter Sequence

2. Transcription Factors

3. RNA Polymerase

4. Activators and Repressors.

The transcription preinitiation in **archaea** is, in essence, homologous to that of eukaryotes, but is much less complex.[5] The archaeal preinitiation complex assembles at a TATA-box binding site; however, in archaea, this complex is composed of only RNA polymerase II, TBP, and TFB (the archaeal homologue of eukaryotic transcription factor II B (TFIIB)).[6][7]

6.2.2   Initiation

In **bacteria**, transcription begins with the binding of RNA polymerase to the promoter in DNA. RNA polymerase is a core enzyme consisting of five subunits: 2 α subunits, 1 β subunit, 1 β' subunit, and 1 ω subunit. At the start of initiation, the core enzyme is associated with a sigma factor that aids in finding the appropriate -35 and -10 base pairs upstream of promoter sequences.[8] When the sigma factor and RNA polymerase combine, they form a holoenzyme.

Transcription initiation is more complex in **eukaryotes**. Eukaryotic RNA polymerase does not directly recognize the core promoter sequences. Instead, a collection of proteins called transcription factors mediate the binding of RNA polymerase and the initiation of transcription. Only after certain transcription factors are attached to the promoter does the RNA

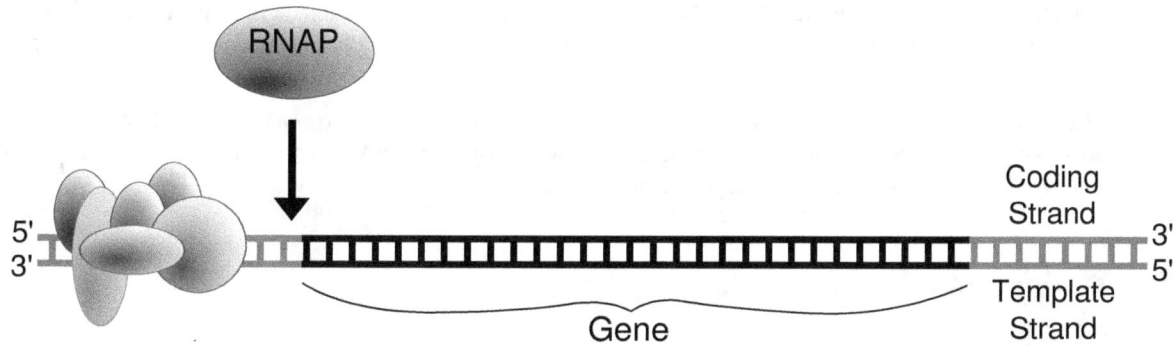

Simple diagram of transcription initiation. RNAP = RNA polymerase

polymerase bind to it. The completed assembly of transcription factors and RNA polymerase bind to the promoter, forming a transcription initiation complex. Transcription in the archaea domain is similar to transcription in eukaryotes.[9]

6.2.3   Promoter clearance

After the first bond is synthesized, the RNA polymerase must clear the promoter. During this time there is a tendency to release the RNA transcript and produce truncated transcripts. This is called *abortive initiation* and is common for both eukaryotes and prokaryotes.[10]

In **prokaryotes**, abortive initiation continues to occur until an RNA product of a threshold length of approximately 10 nucleotides is synthesized, at which point promoter escape occurs and a transcription elongation complex is formed. The σ factor is released according to a stochastic model.[11] Mechanistically, promoter escape occurs through a scrunching mechanism, where the energy built up by DNA scrunching provides the energy needed to break interactions between RNA polymerase holoenzyme and the promoter.[12]

In **eukaryotes**, after several rounds of 10nt abortive initiation, promoter clearance coincides with the TFIIH's phosphorylation of serine 5 on the carboxy terminal domain of RNAP II, leading to the recruitment of capping enzyme (CE).[13][14] The exact mechanism of how CE induces promoter clearance in eukaryotes is not yet known.

6.2.4   Elongation

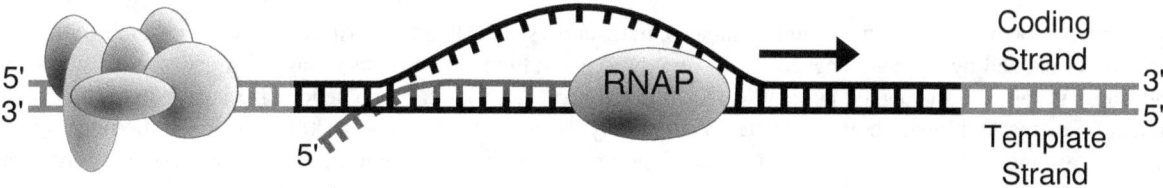

Simple diagram of transcription elongation

One strand of the DNA, the *template strand* (or noncoding strand), is used as a template for RNA synthesis. As transcription proceeds, RNA polymerase traverses the template strand and uses base pairing complementarity with the DNA template to create an RNA copy. Although RNA polymerase traverses the template strand from 3' → 5', the coding (non-template) strand and newly formed RNA can also be used as reference points, so transcription can be described as occurring 5' → 3'. This produces an RNA molecule from 5' → 3', an exact copy of the coding strand (except that thymines are replaced with uracils, and the nucleotides are composed of a ribose (5-carbon) sugar where DNA has deoxyribose (one fewer oxygen atom) in its sugar-phosphate backbone).

mRNA transcription can involve multiple RNA polymerases on a single DNA template and multiple rounds of transcription (amplification of particular mRNA), so many mRNA molecules can be rapidly produced from a single copy of a gene.

Elongation also involves a proofreading mechanism that can replace incorrectly incorporated bases. In eukaryotes, this may correspond with short pauses during transcription that allow appropriate RNA editing factors to bind. These pauses may be intrinsic to the RNA polymerase or due to chromatin structure.

6.2.5   Termination

Main article: Terminator (genetics)

Bacteria use two different strategies for transcription termination - *Rho-independent termination* and *Rho-dependent termination*. In Rho-independent transcription termination, also called intrinsic termination, RNA transcription stops when the newly synthesized RNA molecule forms a G-C-rich hairpin loop followed by a run of Us. When the hairpin forms, the mechanical stress breaks the weak rU-dA bonds, now filling the DNA-RNA hybrid. This pulls the poly-U transcript out of the active site of the RNA polymerase, in effect, terminating transcription. In the "Rho-dependent" type of termination, a protein factor called "Rho" destabilizes the interaction between the template and the mRNA, thus releasing the newly synthesized mRNA from the elongation complex.[15]

Transcription termination in **eukaryotes** is less understood but involves cleavage of the new transcript followed by template-independent addition of adenines at its new 3' end, in a process called polyadenylation.[16]

6.3   Inhibitors

Transcription inhibitors can be used as antibiotics against, for example, pathogenic bacteria (antibacterials) and fungi (antifungals). An example of such an antibacterial is rifampicin, which inhibits prokaryotic DNA transcription into mRNA by inhibiting DNA-dependent RNA polymerase by binding its beta-subunit. 8-Hydroxyquinoline is an antifungal transcription inhibitor.[17] The effects of histone methylation may also work to inhibit the action of transcription.

6.4   Transcription factories

Main article: Transcription factories

Active transcription units are clustered in the nucleus, in discrete sites called transcription factories or euchromatin. Such sites can be visualized by allowing engaged polymerases to extend their transcripts in tagged precursors (Br-UTP or Br-U) and immuno-labeling the tagged nascent RNA. Transcription factories can also be localized using fluorescence in situ hybridization or marked by antibodies directed against polymerases. There are ~10,000 factories in the nucleoplasm of a HeLa cell, among which are ~8,000 polymerase II factories and ~2,000 polymerase III factories. Each polymerase II factory contains ~8 polymerases. As most active transcription units are associated with only one polymerase, each factory usually contains ~8 different transcription units. These units might be associated through promoters and/or enhancers, with loops forming a 'cloud' around the factor.[18]

6.5   History

A molecule that allows the genetic material to be realized as a protein was first hypothesized by François Jacob and Jacques Monod. Severo Ochoa won a Nobel Prize in Physiology or Medicine in 1959 for developing a process for synthesizing RNA *in vitro* with polynucleotide phosphorylase, which was useful for cracking the genetic code. RNA synthesis by RNA polymerase was established *in vitro* by several laboratories by 1965; however, the RNA synthesized by these enzymes had properties that suggested the existence of an additional factor needed to terminate transcription correctly.

In 1972, Walter Fiers became the first person to actually prove the existence of the terminating enzyme.

Roger D. Kornbergwon the 2006Nobel Prize in Chemistry"for his studies of the molecular basis of eukaryotic transcription".

6.6 Measuring and detecting transcription

Transcription can be measured and detected in a variety of ways:

- Nuclear Run-on assay: measures the relative abundance of newly formed transcripts

- RNase protection assay and ChIP-Chip of RNAP: detect active transcription sites

- RT-PCR: measures the absolute abundance of total or nuclear RNA levels, which may however differ from transcription rates

- DNA microarrays: measures the relative abundance of the global total or nuclear RNA levels; however, these may differ from transcription rates

- In situ hybridization: detects the presence of a transcript

- MS2 tagging: by incorporating RNA stem loops, such as MS2, into a gene, these become incorporated into newly synthesized RNA. The stem loops can then be detected using a fusion of GFP and the MS2 coat protein, which has a high affinity, sequence-specific interaction with the MS2 stem loops. The recruitment of GFP to the site of transcription is visualised as a single fluorescent spot. This new approach has revealed that transcription occurs in discontinuous bursts, or pulses (see Transcriptional bursting). With the notable exception of in situ techniques, most other methods provide cell population averages, and are not capable of detecting this fundamental property of genes.[20]

- Northern blot: the traditional method, and until the advent of RNA-Seq, the most quantitative

- RNA-Seq: applies next-generation sequencing techniques to sequence whole transcriptomes, which allows the measurement of relative abundance of RNA, as well as the detection of additional variations such as fusion genes, post-transcriptional edits and novel splice sites

6.7 Reverse transcription

Some viruses (such as HIV, the cause of AIDS), have the ability to transcribe RNA into DNA. HIV has an RNA genome that is *reverse transcribed* into DNA. The resulting DNA can be merged with the DNA genome of the host cell. The main enzyme responsible for synthesis of DNA from an RNA template is called reverse transcriptase.

In the case of HIV, reverse transcriptase is responsible for synthesizing a complementary DNA strand (cDNA) to the viral RNA genome. The enzyme ribonuclease H then digests the RNA strand, and reverse transcriptase synthesises a complementary strand of DNA to form a double helix DNA structure ("cDNA"). The cDNA is integrated into the host cell's genome by the enzyme integrase, which causes the host cell to generate viral proteins that reassemble into new viral particles. In HIV, subsequent to this, the host cell undergoes programmed cell death, or apoptosis of T cells.[21] However, in other retroviruses, the host cell remains intact as the virus buds out of the cell.

Some eukaryotic cells contain an enzyme with reverse transcription activity called telomerase. Telomerase is a reverse transcriptase that lengthens the ends of linear chromosomes. Telomerase carries an RNA template from which it synthesizes a repeating sequence of DNA, or "junk" DNA. This repeated sequence of DNA is called a telomere and can be thought of as a "cap" for a chromosome. It is important because every time a linear chromosome is duplicated, it is shortened. With this "junk" DNA or "cap" at the ends of chromosomes, the shortening eliminates some of the non-essential, repeated sequence rather than the protein-encoding DNA sequence, that is farther away from the chromosome end.

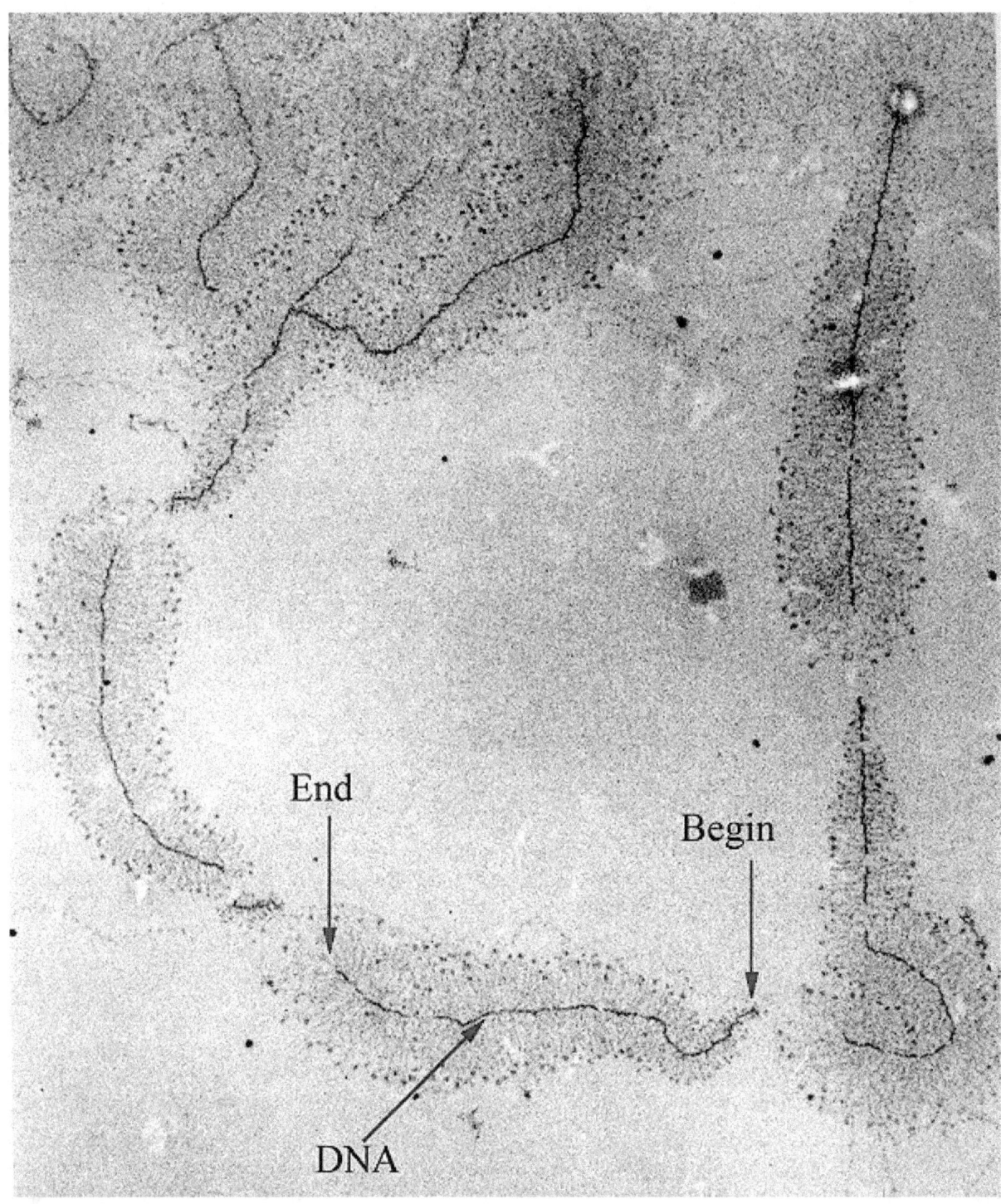

Electron micrograph of transcription of ribosomal RNA. The forming ribosomal RNA strands are visible as branches from the main DNA strand.

Telomerase is often activated in cancer cells to enable cancer cells to duplicate their genomes indefinitely without losing important protein-coding DNA sequence. Activation of telomerase could be part of the process that allows cancer cells to become *immortal*. The immortalizing factor of cancer via telomere lengthening due to telomerase has been proven to occur in 90% of all carcinogenic tumors *in vivo* with the remaining 10% using an alternative telomere maintenance route called ALT or Alternative Lengthening of Telomeres.[22]

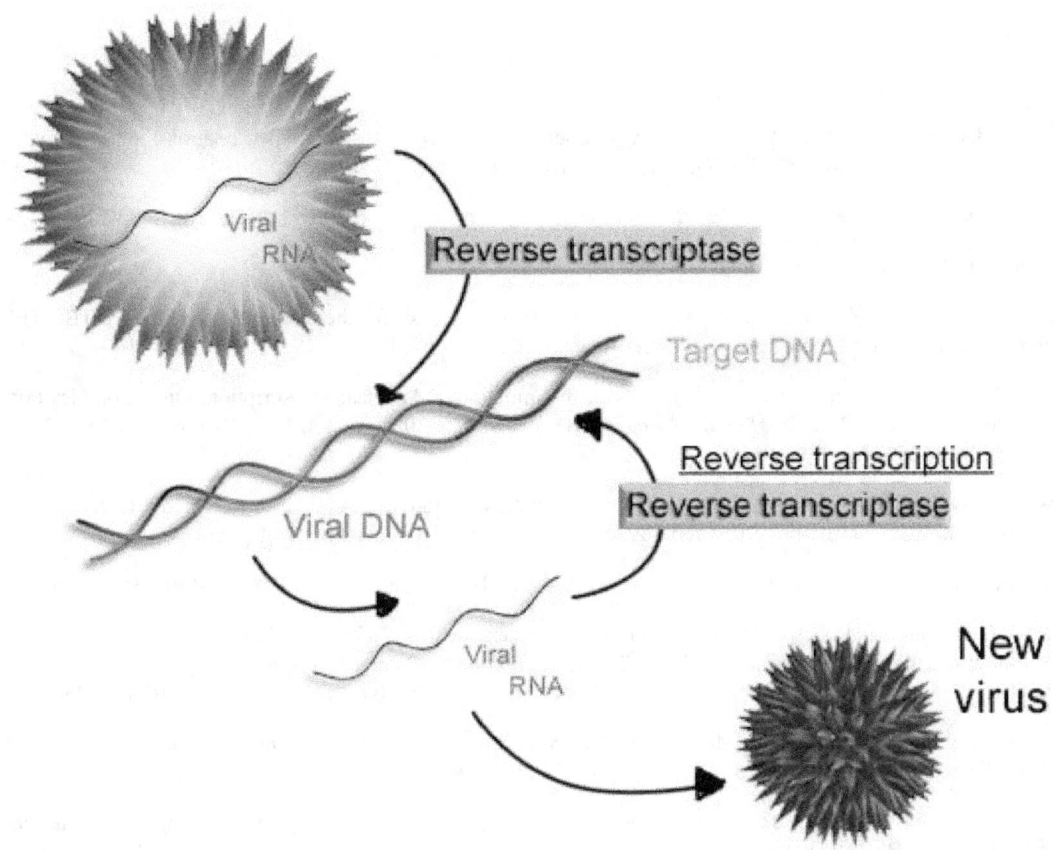

Scheme of reverse transcription

6.8   See also

- Crick's central dogma - DNA is transcribed to RNA, which is translated to polypeptides (polypeptides cannot "reverse translate" into RNA or DNA)

- Eukaryotic transcription

- Gene regulation

- Bacterial transcription

- RNA Polymerase

- Reverse transcription - process viruses use to make DNA from RNA

- Splicing - process of removing introns from precursor messenger RNA (pre-mRNA) to make messenger RNA (mRNA)

- Translation - process of decoding RNA to form polypeptides

- Transcription factor

6.9 References

[1] Eldra P. Solomon, Linda R. Berg, Diana W. Martin. *Biology, 8th Edition, International Student Edition*. Thomson Brooks/Cole. ISBN 978-0495317142

[2] "Tentative identification of RNA-dependent RNA polymerases of dsRNA viruses and their relationship to positive strand RNA viral polymerases". *FEBS Letters* **252**: 42–46. doi:10.1016/0014-5793(89)80886-5.

[3] Berg J, Tymoczko JL, Stryer L (2006). *Biochemistry* (6th ed.). San Francisco: W. H. Freeman. ISBN 0-7167-8724-5.

[4] Basic Medical Biochemistry, 4th edition, Marks. Chapter 14

[5] Littlefield, O., Korkhin, Y., and Sigler, P.B. (1999). "The structural basis for the oriented assembly of a TBP/TFB/promoter complex". *PNAS* **96** (24): 13668–13673. doi:10.1073/pnas.96.24.13668. PMC 24122. PMID 10570130.

[6] Hausner, W., Michael Thomm, M. (2001). "Events during Initiation of Archaeal Transcription: Open Complex Formation and DNA-Protein Interactions". *Journal of Bacteriology* **183** (10): 3025–3031. doi:10.1128/JB.183.10.3025-3031.2001. PMC 95201. PMID 11325929.

[7] Qureshi, SA; Bell, SD; Jackson, SP (1997). "Factor requirements for transcription in the archaeon Sulfolobus shibatae". *EMBO Journal* **16** (10): 2927–2936. doi:10.1093/emboj/16.10.2927. PMC 1169900. PMID 9184236.

[8] Raven, Peter H. (2011). *Biology* (9th ed.). New York: McGraw-Hill. pp. 278–301. ISBN 978-0-07-353222-6.

[9] Mohamed Ouhammouch, Robert E. Dewhurst, Winfried Hausner, Michael Thomm, and E. Peter Geiduschek (2003). "Activation of archaeal transcription by recruitment of the TATA-binding protein". *Proceedings of the National Academy of Sciences of the United States of America* **100** (9): 5097–5102. doi:10.1073/pnas.0837150100. PMC 154304. PMID 12692306.

[10] Goldman, S.; Ebright, R.; Nickels, B. (May 2009). "Direct detection of abortive RNA transcripts in vivo". *Science* **324** (5929): 927–928. doi:10.1126/science.1169237. PMC 2718712. PMID 19443781.

[11] Raffaelle, M.; Kanin, E. I.; Vogt, J.; Burgess, R. R.; Ansari, A. Z. (2005). "Holoenzyme Switching and Stochastic Release of Sigma Factors from RNA Polymerase in Vivo". *Molecular Cell* **20** (3): 357–366. doi:10.1016/j.molcel.2005.10.011. PMID 16285918.

[12] Revyakin, A.; Liu, C.; Ebright, R.; Strick, T. (2006). "Abortive initiation and productive initiation by RNA polymerase involve DNA scrunching". *Science* **314** (5802): 1139–1143. doi:10.1126/science.1131398. PMC 2754787. PMID 17110577.

[13] Mandal, S. S.; Chu, C.; Wada, T.; Handa, H.; Shatkin, A. J.; Reinberg, D. (2004). "Functional interactions of RNA-capping enzyme with factors that positively and negatively regulate promoter escape by RNA polymerase II". *Proceedings of the National Academy of Sciences* **101** (20): 7572–7577. doi:10.1073/pnas.0401493101. PMC 419647. PMID 15136722.

[14] Goodrich, J. A.; Tjian, R. (1994). "Transcription factors IIE and IIH and ATP hydrolysis direct promoter clearance by RNA polymerase II". *Cell* **77** (1): 145–156. doi:10.1016/0092-8674(94)90242-9. PMID 8156590.

[15] Richardson, J (2002). "Rho-dependent termination and ATPases in transcript termination". *Biochimica et Biophysica Acta* **1577** (2): 251–260. doi:10.1016/S0167-4781(02)00456-6.

[16] Lykke-Andersen, S; Jensen, TH (2007). "Overlapping pathways dictate termination of RNA polymerase II transcription". *Biochimie* **89** (10): 1177–82. doi:10.1016/j.biochi.2007.05.007.

[17] 8-Hydroxyquinoline info from SIGMA-ALDRICH. Retrieved Feb 2012

[18] Papantonis, A (2012-10-26). "TNFα signals through specialized factories where responsive coding and miRNA genes are transcribed". *Nature EMBO J.*

[19] "Chemistry 2006". *Nobel Foundation*. Retrieved March 29, 2007.

[20] Raj, A.; van Oudenaarden, A. (2008). "Nature, nurture, or chance: stochastic gene expression and its consequences". *Cell* **135**: 216–26. doi:10.1016/j.cell.2008.09.050.

[21] Kolesnikova I. N. (2000). "Some patterns of apoptosis mechanism during HIV-infection". *Dissertation* (in Russian). Retrieved February 20, 2011.

[22] ALT and Telomerase from Nature. Retrieved May 2010

6.10 External links

- Interactive Java simulation of transcription initiation. From Center for Models of Life at the Niels Bohr Institute.

- Interactive Java simulation of transcription interference--a game of promoter dominance in bacterial virus. From Center for Models of Life at the Niels Bohr Institute.

- Biology animations about this topic under Chapter 15 and Chapter 18

- Virtual Cell Animation Collection, Introducing Transcription

- Easy to use DNA transcription site

Chapter 7

DNA replication

DNA replication is the process of producing two identical replicas from one original DNA molecule. This biological process occurs in all living organisms and is the basis for biological inheritance. DNA is made up of two strands and each strand of the original DNA molecule serves as a template for the production of the complementary strand, a process referred to as semiconservative replication. Cellular proofreading and error-checking mechanisms ensure near perfect fidelity for DNA replication.[1][2]

In a cell, DNA replication begins at specific locations, or origins of replication, in the genome.[3] Unwinding of DNA at the origin and synthesis of new strands results in replication forks growing bidirectional from the origin. A number of proteins are associated with the replication fork which helps in terms of the initiation and continuation of DNA synthesis. Most prominently, DNA polymerase synthesizes the new DNA by adding complementary nucleotides to the template strand.

DNA replication can also be performed *in vitro* (artificially, outside a cell). DNA polymerases isolated from cells and artificial DNA primers can be used to initiate DNA synthesis at known sequences in a template DNA molecule. The polymerase chain reaction (PCR), a common laboratory technique, cyclically applies such artificial synthesis to amplify a specific target DNA fragment from a pool of DNA.

7.1 Background on DNA structure

DNA usually exists as a double-stranded structure, with both strands coiled together to form the characteristic double-helix. Each single strand of DNA is a chain of four types of nucleotides. Nucleotides in DNA contain a deoxyribose sugar, a phosphate, and a nucleobase. The four types of nucleotide correspond to the four nucleobases adenine, cytosine, guanine, and thymine, commonly abbreviated as A,C, G and T. Adenine and guanine are purine bases, while cytosine and thymine are pyrimidines. These nucleotides form phosphodiester bonds, creating the phosphate-deoxyribose backbone of the DNA double helix with the nucleobases pointing inward. Nucleotides (bases) are matched between strands through hydrogen bonds to form base pairs. Adenine pairs with thymine (two hydrogen bonds), and guanine pairs with cytosine (stronger: three hydrogen bonds).

DNA strands have a directionality, and the different ends of a single strand are called the "3' (three-prime) end" and the "5' (five-prime) end". By convention, if the base sequence of a single strand of DNA is given, the left end of the sequence is the 5' end, while the right end of the sequence is the 3' end. The strands of the double helix are anti-parallel with one being 5' to 3', and the opposite strand 3' to 5'. These terms refer to the carbon atom in deoxyribose to which the next phosphate in the chain attaches. Directionality has consequences in DNA synthesis, because DNA polymerase can synthesize DNA in only one direction by adding nucleotides to the 3' end of a DNA strand.

The pairing of complementary bases in DNA through hydrogen bonding means that the information contained within each strand is redundant. The nucleotides on a single strand can be used to reconstruct nucleotides on a newly synthesized partner strand.[4]

7.2 DNA polymerase

Main article: DNA polymerase

DNA polymerases are a family of enzymes that carry out all forms of DNA replication.[6] DNA polymerases in general cannot initiate synthesis of new strands, but can only extend an existing DNA or RNA strand paired with a template strand. To begin synthesis, a short fragment of RNA , called a primer, must be created and paired with the template DNA strand.

DNA polymerase adds a new strand of DNA by extending the 3' end of an existing nucleotide chain, adding new nucleotides matched to the template strand one at a time via the creation of phosphodiester bonds. The energy for this process of DNA polymerization comes from hydrolysis of the high-energy phosphate (phosphoanhydride) bonds between the three phosphates attached to each unincorporated base. (Free bases with their attached phosphate groups are called nucleotides; in particular, bases with three attached phosphate groups are called nucleoside triphosphates.) When a nucleotide is being added to a growing DNA strand, the formation of a phosphodiester bond between the proximal phosphate of the nucleotide to the growing chain is accompanied by hydrolysis of a high-energy phosphate bond with release of the two distal phosphates as a pyrophosphate. Enzymatic hydrolysis of the resulting pyrophosphate into inorganic phosphate consumes a second high-energy phosphate bond and renders the reaction effectively irreversible.[Note 1]

In general, DNA polymerases are highly accurate, with an intrinsic error rate of less than one mistake for every 10^7 nucleotides added.[7] In addition, some DNA polymerases also have proofreading ability; they can remove nucleotides from the end of a growing strand in order to correct mismatched bases. Finally, post-replication mismatch repair mechanisms monitor the DNA for errors, being capable of distinguishing mismatches in the newly synthesized DNA strand from the original strand sequence. Together, these three discrimination steps enable replication fidelity of less than one mistake for every 10^9 nucleotides added.[7]

The rate of DNA replication in a living cell was first measured as the rate of phage T4 DNA elongation in phage-infected E. coli.[8] During the period of exponential DNA increase at 37 °C, the rate was 749 nucleotides per second. The mutation rate per base pair per replication during phage T4 DNA synthesis is 1.7 per 10^8.[9] Thus DNA replication is both impressively fast and accurate.

7.3 Replication process

Main articles: Prokaryotic DNA replication and Eukaryotic DNA replication

DNA Replication, like all biological polymerization processes, proceeds in three enzymatically catalyzed and coordinated steps: initiation, elongation and termination.

7.3.1 Initiation

For a cell to divide, it must first replicate its DNA.[10] This process is initiated at particular points in the DNA, known as "origins", which are targeted by initiator proteins.[3] In *E. coli* this protein is DnaA; in yeast, this is the origin recognition complex.[11] Sequences used by initiator proteins tend to be "AT-rich" (rich in adenine and thymine bases), because A-T base pairs have two hydrogen bonds (rather than the three formed in a C-G pair) which are easier to unzip.[12] Once the origin has been located, these initiators recruit other proteins and form the pre-replication complex, which unzips the double-stranded DNA.

7.3.2 Elongation

DNA polymerase has 5'−3' activity. All known DNA replication systems require a free 3' hydroxyl group before synthesis can be initiated (Important note: DNA is read in 3' to 5' direction whereas a new strand is synthesized in the 5' to 3' direction—this is often confused). Four distinct mechanisms for initiation of synthesis are recognized:

1. All cellular life forms and many DNA viruses, phages and plasmids use a primase to synthesize a short RNA primer with a free 3' OH group which is subsequently elongated by a DNA polymerase.

2. The retroelements (including retroviruses) employ a transfer RNA that primes DNA replication by providing a free 3' OH that is used for elongation by the reverse transcriptase.

3. In the adenoviruses and the φ29 family of bacteriophages, the 3' OH group is provided by the side chain of an amino acid of the genome attached protein (the terminal protein) to which nucleotides are added by the DNA polymerase to form a new strand.

4. In the single stranded DNA viruses — a group that includes the circoviruses, the geminiviruses, the parvoviruses and others — and also the many phages and plasmids that use the rolling circle replication (RCR) mechanism, the RCR endonuclease creates a nick in the genome strand (single stranded viruses) or one of the DNA strands (plasmids). The 5' end of the nicked strand is transferred to a tyrosine residue on the nuclease and the free 3' OH group is then used by the DNA polymerase to synthesize the new strand.

The first is the best known of these mechanisms and is used by the cellular organisms. In this mechanism, once the two strands are separated, primase adds RNA primers to the template strands. The leading strand receives one RNA primer while the lagging strand receives several. The leading strand is continuously extended from the primer by a DNA polymerase with high processivity, while the lagging strand is extended discontinuously from each primer forming Okazaki fragments. RNase removes the primer RNA fragments, and a low processivity DNA polymerase distinct from the replicative polymerase enters to fill the gaps. When this is complete, a single nick on the leading strand and several nicks on the lagging strand can be found. Ligase works to fill these nicks in, thus completing the newly replicated DNA molecule.

The primase used in this process differs significantly between bacteria and archaea/eukaryotes. Bacteria use a primase belonging to the DnaG protein superfamily which contains a catalytic domain of the TOPRIM fold type.[13] The TOPRIM fold contains an α/β core with four conserved strands in a Rossmann-like topology. This structure is also found in the catalytic domains of topoisomerase Ia, topoisomerase II, the OLD-family nucleases and DNA repair proteins related to the RecR protein.

The primase used by archaea and eukaryotes in contrast contains a highly derived version of the RNA recognition motif (RRM). This primase is structurally similar to many viral RNA dependent RNA polymerases, reverse transcriptases, cyclic nucleotide generating cyclases and DNA polymerases of the A/B/Y families that are involved in DNA replication and repair. In eukaryotic replication, the primase forms a complex with Pol α.[14]

Multiple DNA polymerases take on different roles in the DNA replication process. In *E. coli*, DNA Pol III is the polymerase enzyme primarily responsible for DNA replication. It assembles into a replication complex at the replication fork that exhibits extremely high processivity, remaining intact for the entire replication cycle. In contrast, DNA Pol I is the enzyme responsible for replacing RNA primers with DNA. DNA Pol I has a 5' to 3' exonuclease activity in addition to its polymerase activity, and uses its exonuclease activity to degrade the RNA primers ahead of it as it extends the DNA strand behind it, in a process called nick translation. Pol I is much less processive than Pol III because its primary function in DNA replication is to create many short DNA regions rather than a few very long regions.

In eukaryotes, the low-processivity enzyme, Pol α, helps to initiate replication. The high-processivity extension enzymes are Pol δ and Pol ε.

As DNA synthesis continues, the original DNA strands continue to unwind on each side of the bubble, forming a replication fork with two prongs. In bacteria, which have a single origin of replication on their circular chromosome, this process creates a "theta structure" (resembling the Greek letter theta: θ). In contrast, eukaryotes have longer linear chromosomes and initiate replication at multiple origins within these.[15]

7.3.3   Replication fork

The replication fork is a structure that forms within the nucleus during DNA replication. It is created by helicases, which break the hydrogen bonds holding the two DNA strands together. The resulting structure has two branching "prongs", each one made up of a single strand of DNA. These two strands serve as the template for the leading and lagging strands, which will be created as DNA polymerase matches complementary nucleotides to the templates; the templates may be properly referred to as the leading strand template and the lagging strand template.

DNA is always synthesized in the 5' to 3' direction. Since the leading and lagging strand templates are oriented in opposite directions at the replication fork, a major issue is how to achieve synthesis of nascent (new) lagging strand DNA, whose direction of synthesis is opposite to the direction of the growing replication fork.

Leading strand

The leading strand is the strand of nascent DNA which is being synthesized in the same direction as the growing replication fork. A polymerase "reads" the leading strand *template* and adds complementary nucleotides to the nascent leading strand on a continuous basis.

The polymerase involved in leading strand synthesis is DNA polymerase III (DNA Pol III) in prokaryotes and Pol δ in eukaryotes.[16] Pol ε can substitute for Pol δ in special circumstances.[17]

Lagging strand

The lagging strand is the strand of nascent DNA whose direction of synthesis is opposite to the direction of the growing replication fork. Because of its orientation, replication of the lagging strand is more complicated as compared to that of the leading strand.

The lagging strand is synthesized in short, separated segments. On the lagging strand *template*, a primase "reads" the template DNA and initiates synthesis of a short complementary RNA primer. A DNA polymerase extends the primed segments, forming Okazaki fragments. The RNA primers are then removed and replaced with DNA, and the fragments of DNA are joined together by DNA ligase.

DNA polymerase III (in prokaryotes) or Pol δ (in eukaryotes) is responsible for extension of the primers added during replication of the lagging strand. Primer removal is performed by DNA polymerase I (in prokaryotes) and Pol δ (in eukaryotes).[18] Eukaryotic primase is intrinsic to Pol α.[19] In eukaryotes, pol ε helps with repair during DNA replication.

Dynamics at the replication fork

As helicase unwinds DNA at the replication fork, the DNA ahead is forced to rotate. This process results in a build-up of twists in the DNA ahead.[20] This build-up forms a torsional resistance that would eventually halt the progress of the replication fork. Topoisomerases are enzymes that temporarily break the strands of DNA, relieving the tension caused by unwinding the two strands of the DNA helix; topoisomerases (including DNA gyrase) achieve this by adding negative supercoils to the DNA helix.[21]

Bare single-stranded DNA tends to fold back on itself forming secondary structures; these structures can interfere with the movement of DNA polymerase. To prevent this, single-strand binding proteins bind to the DNA until a second strand is synthesized, preventing secondary structure formation.[22]

Clamp proteins form a sliding clamp around DNA, helping the DNA polymerase maintain contact with its template, thereby assisting with processivity. The inner face of the clamp enables DNA to be threaded through it. Once the polymerase reaches the end of the template or detects double-stranded DNA, the sliding clamp undergoes a conformational change that releases the DNA polymerase. Clamp-loading proteins are used to initially load the clamp, recognizing the junction between template and RNA primers.[2]:274-5

7.3.4 DNA replication proteins

At the replication fork, many replication enzymes assemble on the DNA into a complex molecular machine called the replisome. The following is a list of major DNA replication enzymes that participate in the replisome:[23]

7.3.5 Replication machinery

Replication machineries consist of factors involved in DNA replication and appearing on template ssDNAs. Replication machineries include primosomes also. The factors are replication enzymes; DNA polymerase, DNA helicases, DNA clamps and DNA topoisomerases, and replication proteins; e.g. single-stranded DNA binding proteins (SSB). In the replication machineries these components coordinate. In most of the bacteria, all of the factors involved in DNA replication are located on replication forks and the complexes stay on the forks during DNA replication. This replication machineries are called **replisomes** or **DNA replicase systems**, these terms mean originally a generic term for proteins located on replication forks. In eukaryotic and some bacterial cells the replisomes are not formed.

Since replication machineries do not move relatively to template DNAs such as factories, they are called **replication factory**.[25] In an alternative figure, DNA factories are similar to projectors and DNAs are like as cinematic films passing constantly into the projectors. In the replication factory model, after both DNA helicases for leading stands and lagging strands are loaded on the template DNAs, the helicases run along the DNAs into each other. The helicases remain associated for the remainder of replication process. Peter Meister et al. observed directly replication sites in budding yeast by monitoring green fluorescent protein(GFP)-tagged DNA polymerases α. They detected DNA replication of pairs of the tagged loci spaced apart symmetrically from a replication origin and found that the distance between the pairs decreased markedly by time.[26] This finding suggests that the mechanism of DNA replication goes with DNA factories. Suggesting, couples of replication factories are loaded on replication origins and the factories associated with each other. Also, template DNAs move into the factories, which bring extrusion of the template ssDNAs and nascent DNAs. Peter's finding is the first direct evidence of replication factory model. By later researches, it is revealed that DNA helicases form dimers in many eukaryotic cells and bacterial replication machineries stay in single intranuclear location during DNA synthesis.[25]

The replication factories perform disentanglement of sister chromatids. The disentanglement is essential to distribute the chromatids into daughter cells after DNA replication. Because sister chromatids after DNA replication hold each other by **Cohesin** rings, there is the only chance for the disentanglement in DNA replication. Fixing of replication machineries as replication factories can improve a success rate of DNA replication. If replication forks move freely in chromosomes, catenation of nuclei is aggravated and impedes mitotic segregation.[26]

7.3.6 Termination

Eukaryotes initiate DNA replication at multiple points in the chromosome, so replication forks meet and terminate at many points in the chromosome; these are not known to be regulated in any particular way. Because eukaryotes have linear chromosomes, DNA replication is unable to reach the very end of the chromosomes, but ends at the telomere region of repetitive DNA close to the end. This shortens the telomere of the daughter DNA strand. Shortening of the telomeres is a normal process in somatic cells. As a result, cells can only divide a certain number of times before the DNA loss prevents further division. (This is known as the Hayflick limit.) Within the germ cell line, which passes DNA to the next generation, telomerase extends the repetitive sequences of the telomere region to prevent degradation. Telomerase can become mistakenly active in somatic cells, sometimes leading to cancer formation. Increased telomerase activity is one of the Hallmarks of cancer.

Termination requires that the progress of the DNA replication fork must stop or be blocked. Termination at a specific locus, when it occurs, involves the interaction between two components: (1) a termination site sequence in the DNA, and (2) a protein which binds to this sequence to physically stop DNA replication. In various bacterial species, this is named the DNA replication terminus site-binding protein, or Ter protein.

Because bacteria have circular chromosomes, termination of replication occurs when the two replication forks meet each other on the opposite end of the parental chromosome. *E. coli* regulates this process through the use of termination sequences that, when bound by the Tus protein, enable only one direction of replication fork to pass through. As a result, the replication forks are constrained to always meet within the termination region of the chromosome.[27]

7.4 Regulation

7.4.1 Eukaryotes

Within eukaryotes, DNA replication is controlled within the context of the cell cycle. As the cell grows and divides, it progresses through stages in the cell cycle; DNA replication takes place during the S phase (synthesis phase). The progress of the eukaryotic cell through the cycle is controlled by cell cycle checkpoints. Progression through checkpoints is controlled through complex interactions between various proteins, including cyclins and cyclin-dependent kinases.[28] Unlike bacteria, eukaryotic DNA replicates in the confines of the nucleus.[29]

The G1/S checkpoint (or restriction checkpoint) regulates whether eukaryotic cells enter the process of DNA replication and subsequent division. Cells that do not proceed through this checkpoint remain in the G0 stage and do not replicate their DNA.

Replication of chloroplast and mitochondrial genomes occurs independently of the cell cycle, through the process of D-loop replication.

Replication focus

In vertebrate cells, replication sites concentrate into positions called **replication foci**.[26] Replication sites can be detected by immunostaining daughter strands and replication enzymes and monitoring GFP-tagged replication factors. By these methods it is found that replication foci of varying size and positions appear in S phase of cell division and their number per nucleus is far smaller than the number of genomic replication forks.

P. Heun et al.(2001) tracked GFP-tagged replication foci in budding yeast cells and revealed that replication origins move constantly in G1 and S phase and the dynamics decreased significantly in S phase.[26] Traditionally, replication sites were fixed on spatial structure of chromosomes by nuclear matrix or lamins. The Heun's results denied the traditional concepts, budding yeasts don't have lamins, and support that replication origins self-assemble and form replication foci.

By firing of replication origins, controlled spatially and temporally, the formation of replication foci is regulated. D. A. Jackson et al.(1998) revealed that neighboring origins fire simultaneously in mammalian cells.[26] Spatial juxtaposition of replication sites brings **clustering** of replication forks. The clustering do **rescue of stalled replication forks** and favors normal progress of replication forks. Progress of replication forks is inhibited by many factors; collision with proteins or with complexes binding strongly on DNA, deficiency of dNTPs, nicks on template DNAs and so on. If replication forks stall and remaining sequences from the stalled forks are not replicated, daughter strands have nick obtained un-replicated sites. The un-replicated sites on one parent's strand hold the other strand together but not daughter strands. Therefore, resulting sister chromatids cannot separate together and cannot divide into 2 daughter cells. When neighboring origins fire and a fork from one origin is stalled, fork from other origin access on an opposite direction of the stalled fork and duplicate the un-replicated sites. As other mechanism of the rescue there is application of **dormant replication origins** that excess origins don't fire in normal DNA replication.

7.4.2 Bacteria

Most bacteria do not go through a well-defined cell cycle but instead continuously copy their DNA; during rapid growth, this can result in the concurrent occurrence of multiple rounds of replication.[30] In E. coli, the best-characterized bacteria, DNA replication is regulated through several mechanisms, including: the hemimethylation and sequestering of the origin sequence, the ratio of adenosine triphosphate (ATP) to adenosine diphosphate (ADP), and the levels of protein DnaA. All these control the binding of initiator proteins to the origin sequences.

Because E. coli methylates GATC DNA sequences, DNA synthesis results in hemimethylated sequences. This hemimethylated DNA is recognized by the protein SeqA, which binds and sequesters the origin sequence; in addition, DnaA (required for initiation of replication) binds less well to hemimethylated DNA. As a result, newly replicated origins are prevented from immediately initiating another round of DNA replication.[31]

ATP builds up when the cell is in a rich medium, triggering DNA replication once the cell has reached a specific size. ATP competes with ADP to bind to DnaA, and the DnaA-ATP complex is able to initiate replication. A certain number of DnaA proteins are also required for DNA replication — each time the origin is copied, the number of binding sites for DnaA doubles, requiring the synthesis of more DnaA to enable another initiation of replication.

7.5   Polymerase chain reaction

Main article: Polymerase chain reaction

Researchers commonly replicate DNA *in vitro* using the polymerase chain reaction (PCR). PCR uses a pair of primers to span a target region in template DNA, and then polymerizes partner strands in each direction from these primers using a thermostable DNA polymerase. Repeating this process through multiple cycles amplifies the targeted DNA region. At the start of each cycle, the mixture of template and primers is heated, separating the newly synthesized molecule and template. Then, as the mixture cools, both of these become templates for annealing of new primers, and the polymerase extends from these. As a result, the number of copies of the target region doubles each round, increasing exponentially.[32]

7.6   Notes

[1] The energetics of this process may also help explain the directionality of synthesis—if DNA were synthesized in the 3' to 5' direction, the energy for the process would come from the 5' end of the growing strand rather than from free nucleotides. The problem is that if the high energy triphosphates were on the growing strand and not on the free nucleotides, proof-reading by removing a mismatched terminal nucleotide would be problematic: Once a nucleotide is added, the triphosphate is lost and a single phosphate remains on the backbone between the new nucleotide and the rest of the strand. If the added nucleotide were mismatched, removal would result in a DNA strand terminated by a monophosphate at the end of the "growing strand" rather than a high energy triphosphate. So strand would be stuck and wouldn't be able to grow anymore. In actuality, the high energy triphosphates hydrolyzed at each step originate from the free nucleotides, not the polymerized strand, so this issue does not exist.

7.7   References

[1] Imperfect DNA replication results in mutations. Berg JM, Tymoczko JL, Stryer L, Clarke ND (2002). "Chapter 27: DNA Replication, Recombination, and Repair". *Biochemistry*. W.H. Freeman and Company. ISBN 0-7167-3051-0. External link in |chapter= (help)

[2] Alberts B, Johnson A, Lewis J, Raff M, Roberts K, Walter P (2002). "Chapter 5: DNA Replication, Repair, and Recombination". *Molecular Biology of the Cell*. Garland Science. ISBN 0-8153-3218-1. External link in |chapter= (help)

[3] Berg JM, Tymoczko JL, Stryer L, Clarke ND (2002). "Chapter 27, Section 4: DNA Replication of Both Strands Proceeds Rapidly from Specific Start Sites". *Biochemistry*. W.H. Freeman and Company. ISBN 0-7167-3051-0. External link in |chapter= (help)

[4] Alberts, B., et al., *Molecular Biology of the Cell*, Garland Science, 4th ed., 2002, pp. 238–240 ISBN 0-8153-3218-1

[5] Allison, Lizabeth A. *Fundamental Molecular Biology*. Blackwell Publishing. 2007. p.112 ISBN 978-1-4051-0379-4

[6] Berg JM, Tymoczko JL, Stryer L, Clarke ND (2002). *Biochemistry*. W.H. Freeman and Company. ISBN 0-7167-3051-0. Chapter 27, Section 2: DNA Polymerases Require a Template and a Primer

[7] McCulloch SD, Kunkel TA (January 2008). "The fidelity of DNA synthesis by eukaryotic replicative and translesion synthesis polymerases". *Cell Research* **18** (1): 148–61. doi:10.1038/cr.2008.4. PMC 3639319. PMID 18166979.

[8] McCarthy D, Minner C, Bernstein H, Bernstein C (1976). "DNA elongation rates and growing point distributions of wild-type phage T4 and a DNA-delay amber mutant". *J Mol Biol* **106** (4): 963–81. doi:10.1016/0022-2836(76)90346-6. PMID 789903.

[9] Drake JW (1970) *The Molecular Basis of Mutation*. Holden-Day, San Francisco ISBN 0816224501 ISBN 978-0816224500

[10] Alberts B, Johnson A, Lewis J, Raff M, Roberts K, Walter P (2002). *Molecular Biology of the Cell*. Garland Science. ISBN 0-8153-3218-1. Chapter 5: DNA Replication Mechanisms

[11] Weigel C, Schmidt A, Rückert B, Lurz R, Messer W (November 1997). "DnaA protein binding to individual DnaA boxes in the Escherichia coli replication origin, oriC". *The EMBO Journal* **16** (21): 6574–83. doi:10.1093/emboj/16.21.6574. PMC 1170261. PMID 9351837.

[12] Lodish H, Berk A, Zipursky LS, Matsudaira P, Baltimore D, Darnell J (2000). *Molecular Cell Biology*. W. H. Freeman and Company. ISBN 0-7167-3136-3.12.1. General Features of Chromosomal Replication: Three Common Features of Replication Origins

[13] Aravind, L.; Leipe, D. D.; Koonin, E. V. (1998). "Toprim--a conserved catalytic domain in type IA and II topoisomerases, DnaG-type primases, OLD family nucleases and RecR proteins". *Nucleic acids research* **26** (18): 4205–4213. doi:10.1093/nar/26.18.4205.PMC147817. PMID9722641.

[14] Frick, David; Richardson, Charles (July 2001). "DNA Primases". *Annual Review of Biochemistry* **70**: 39–80. doi:10.1146/annurev.biochem.70.1.39. PMID 11395402.

[15] Huberman JA, Riggs AD (1968). "On the mechanism of DNA replication in mammalian chromosomes". *J Mol Biol* **32** (2): 327–341. doi:10.1016/0022-2836(68)90013-2. PMID 5689363.

[16] Johnson, RE; Klassen, R; Prakash, L; Prakash, S (July 2015). "A Major Role of DNA Polymerase δ in Replication of Both the Leading and Lagging DNA Strands.". *Molecular Cell* **59** (2): 163-175. PMID 26145172.

[17] Hansen, Barbara (2011). *Biochemistry and Medical Genetics: Lecture Notes*. Kaplan Medical. p. 21.

[18] Distinguishing the pathways of primer removal during Eukaryotic Okazaki fragment maturation Contributor Author Rossi, Marie Louise. Date Accessioned: 2009-02-23T17:05:09Z. Date Available: 2009-02-23T17:05:09Z. Date Issued: 2009-02-23T17:05:09Z. Identifier Uri: http://hdl.handle.net/1802/6537. Description: Dr. Robert A. Bambara, Faculty Advisor. Thesis (PhD) – School of Medicine and Dentistry, University of Rochester. UR only until January 2010. UR only until January 2010.

[19] Elizabeth R. Barry; Stephen D. Bell (December 2006). "DNA Replication in the Archaea". *Microbiology and Molecular Biology Reviews* **70** (4): 876–887. doi:10.1128/MMBR.00029-06. PMC 1698513. PMID 17158702.

[20] Alberts B, Johnson A, Lewis J, Raff M, Roberts K, Walter P (2002). *Molecular Biology of the Cell*. Garland Science. ISBN 0-8153-3218-1. DNA Replication Mechanisms: DNA Topoisomerases Prevent DNA Tangling During Replication

[21] Reece, R. J.; Maxwell, A.; Wang, J. C. (1991). "DNA Gyrase: Structure and Function". *Critical Reviews in Biochemistry and Molecular Biology* **26** (3–4): 335–375. doi:10.3109/10409239109114072. PMID 1657531.

[22] Alberts B, Johnson A, Lewis J, Raff M, Roberts K, Walter P (2002). *Molecular Biology of the Cell*. Garland Science. ISBN 0-8153-3218-1. DNA Replication Mechanisms: Special Proteins Help to Open Up the DNA Double Helix in Front of the Replication Fork

[23] Griffiths A.J.F., Wessler S.R., Lewontin R.C., Carroll S.B. (2008). *Introduction to Genetic Analysis*. W. H. Freeman and Company. ISBN 0-7167-6887-9.[Chapter 7: DNA: Structure and Replication. pg 283–290]

[24] "Will the Hayflick limit keep us from living forever?". *Howstuffworks*. Retrieved January 20, 2015.

[25] James D. Watson et al. (2008), "Molecular Biology of the gene", Pearson Education: 237

[26] Peter Meister, Angela Taddei1, Susan M. Gasser(June 2006), "In and out of the Replication Factory", *Cell* **125** (7): 1233–1235

[27] TA Brown (2002). *Genomes*. BIOS Scientific Publishers. ISBN 1-85996-228-9.13.2.3. Termination of replication

[28] Alberts B, Johnson A, Lewis J, Raff M, Roberts K, Walter P (2002). *Molecular Biology of the Cell*. Garland Science. ISBN 0-8153-3218-1. Intracellular Control of Cell-Cycle Events: S-Phase Cyclin-Cdk Complexes (S-Cdks) Initiate DNA Replication Once Per Cycle

[29] Brown, TA (2002). "13". *Genomes* (2nd ed.). Oxford: Wiley-Liss.

[30] Tobiason DM, Seifert HS (2006). "The Obligate Human Pathogen, Neisseria gonorrhoeae, Is Polyploid". *PLoS Biology* **4** (6): e185. doi:10.1371/journal.pbio.0040185. PMC 1470461. PMID 16719561.

[31] Slater S, Wold S, Lu M, Boye E, Skarstad K, Kleckner N (September 1995). "E. coli SeqA protein binds oriC in two different methyl-modulated reactions appropriate to its roles in DNA replication initiation and origin sequestration". *Cell* **82** (6): 927–36. doi:10.1016/0092-8674(95)90272-4. PMID 7553853.

[32] Saiki, RK; Gelfand DH; Stoffel S; Scharf SJ; Higuchi R; Horn GT; Mullis KB; Erlich HA (1988). "Primer-directed enzymatic amplification of DNA with a thermostable DNA polymerase". *Science* **239** (4839): 487–91. doi:10.1126/science.2448875. PMID 2448875.

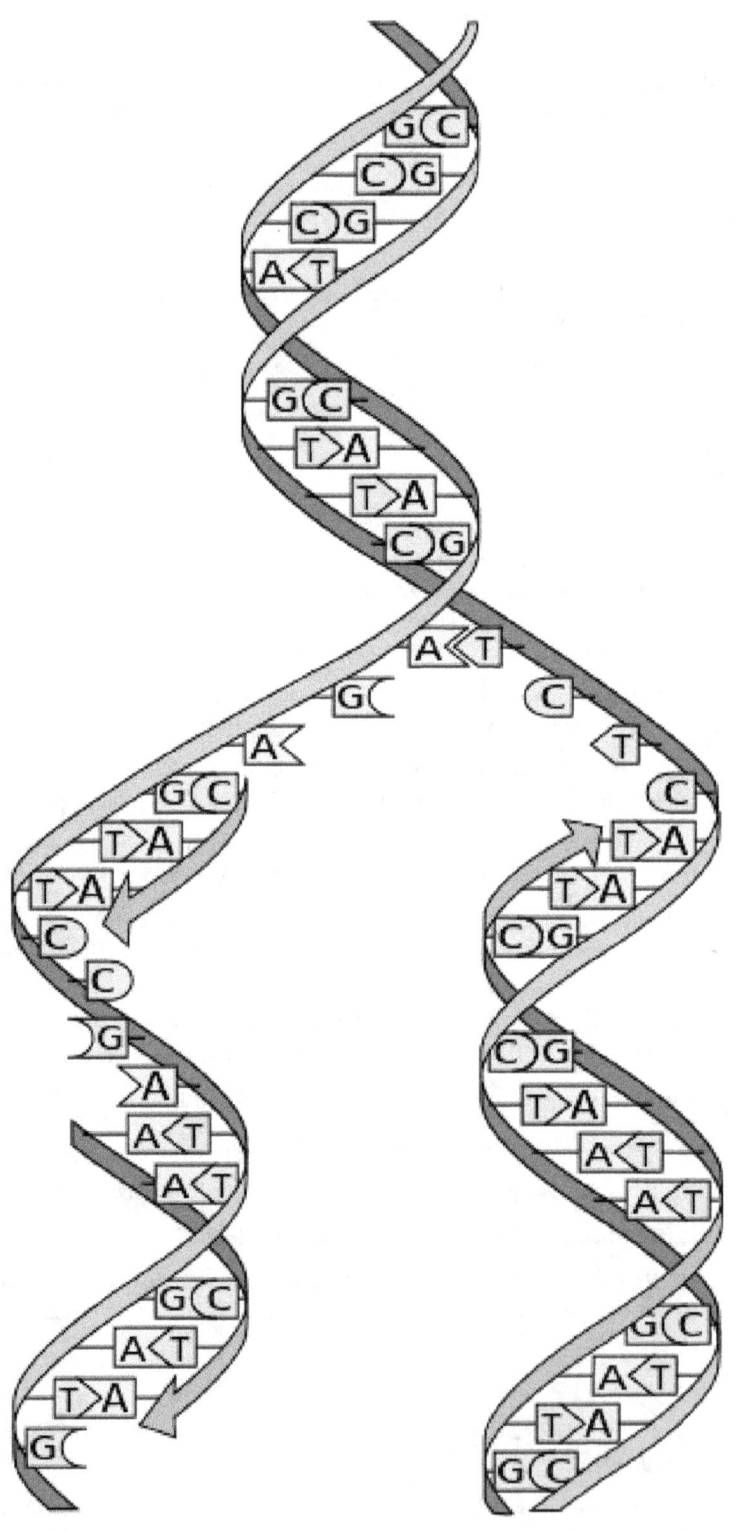

DNA replication. The double helix is unwound and each strand acts as a template for the next strand. Bases are matched to synthesize the new partner strands.

credit Madeleine Price Ball

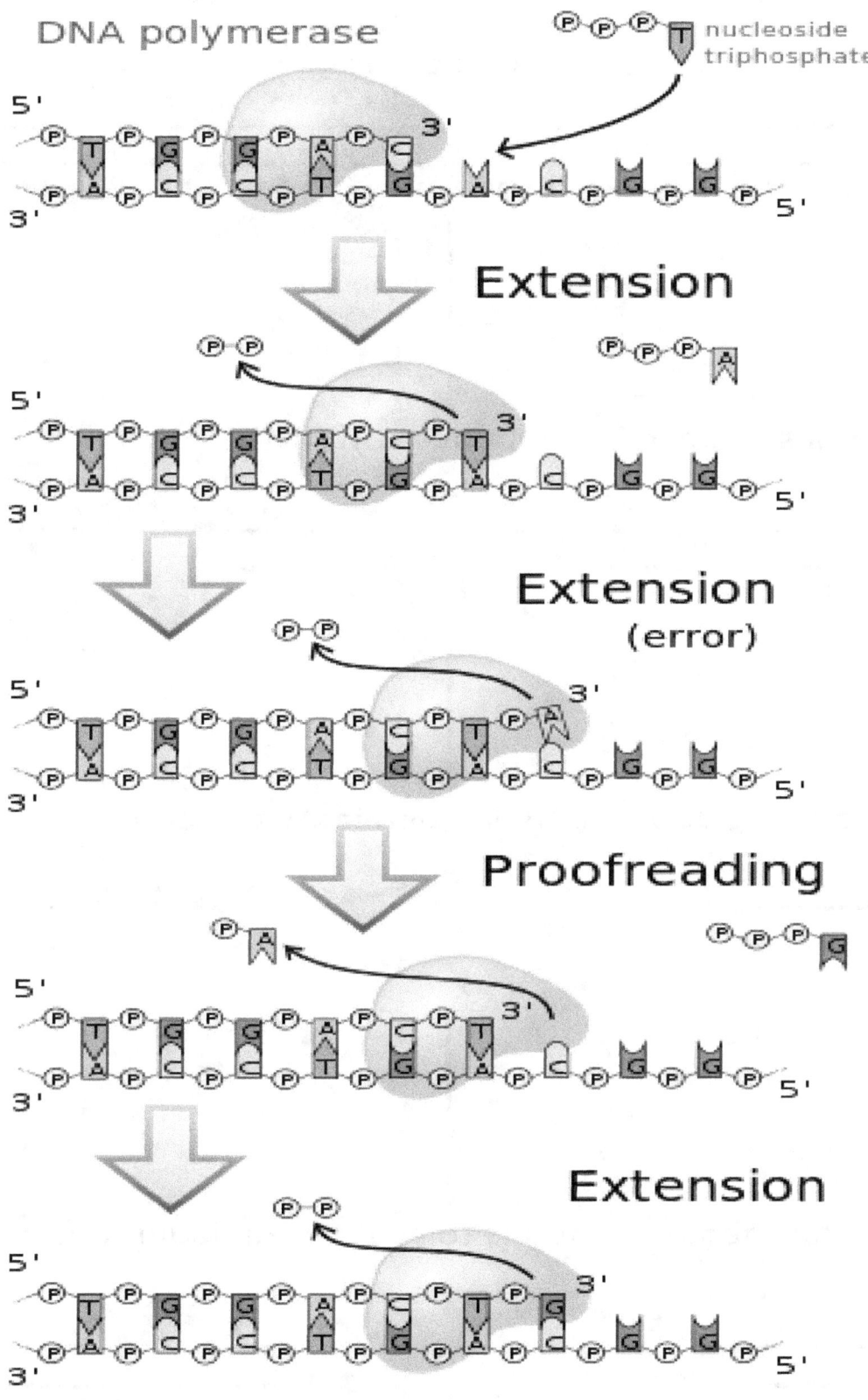

DNA polymerases adds nucleotides to the 3' end of a strand of DNA.[5] If a mismatch is accidentally incorporated, the polymerase is inhibited from further extension. Proofreading removes the mismatched nucleotide and extension continues.

Replicator

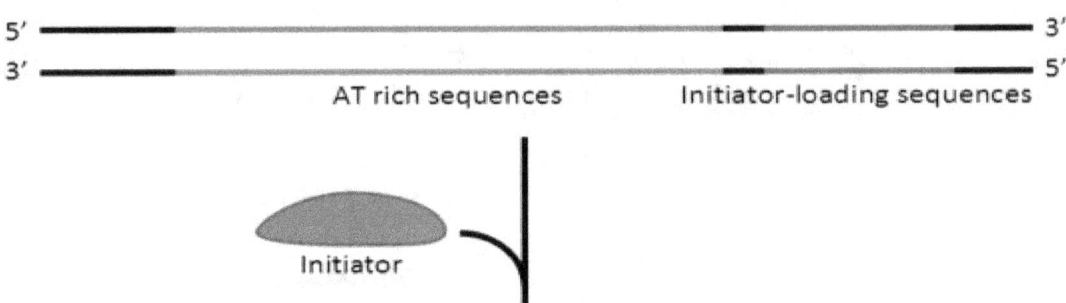

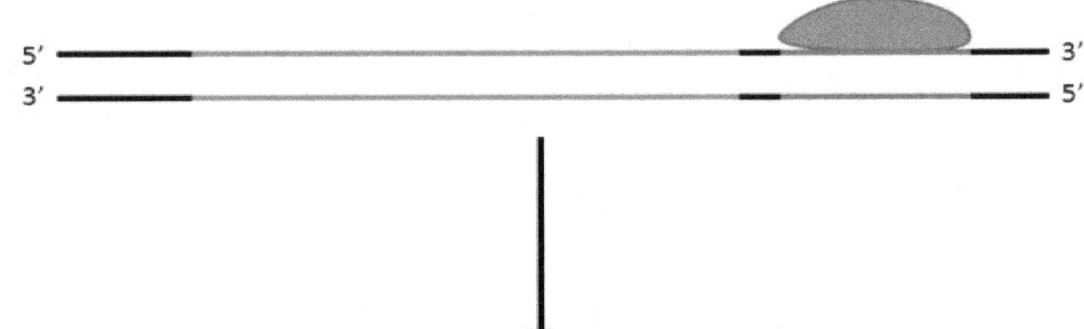

① Binding to specific sequences on DNAs

② Unzipping duplex DNA sequences with rich AT

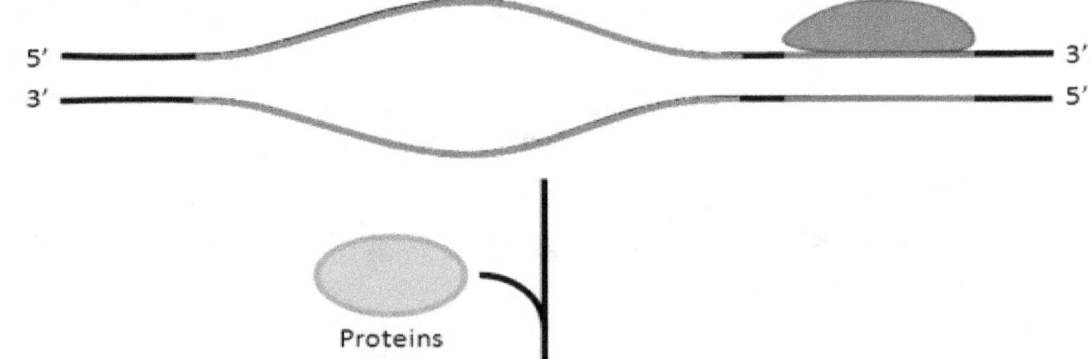

③ Recruitment of proteins to synthesize daughter strands

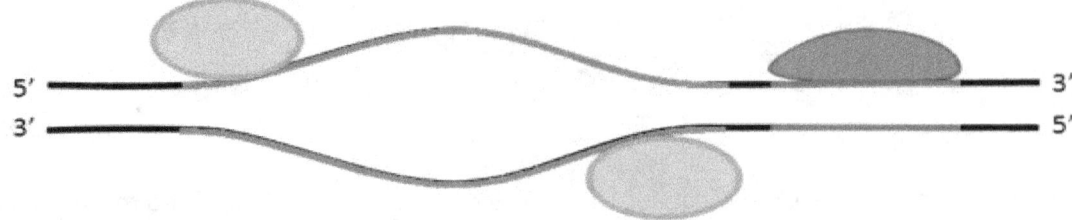

Role of initiators for initiation of DNA replication.

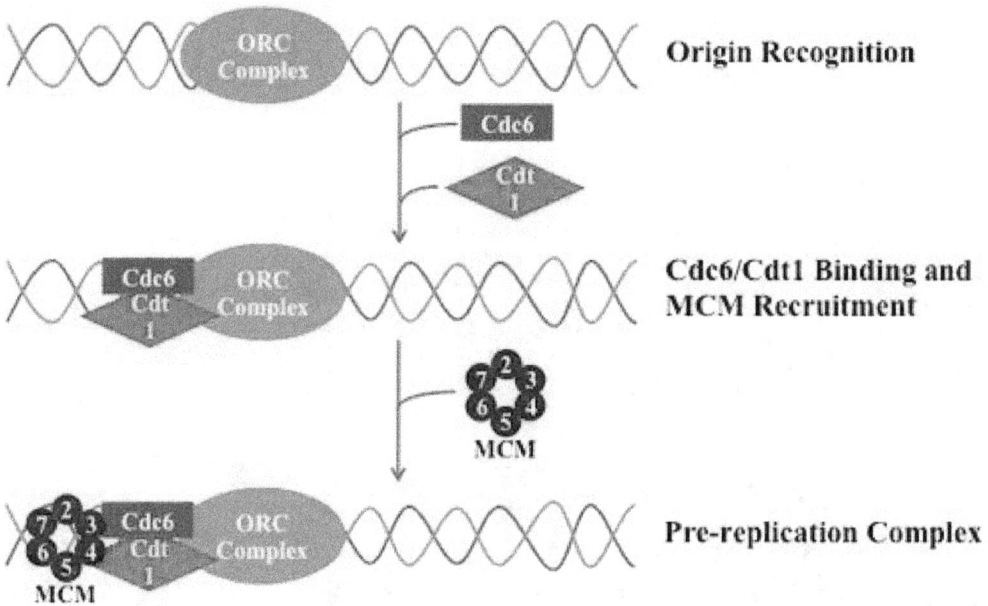

Formation of pre-replication complex.

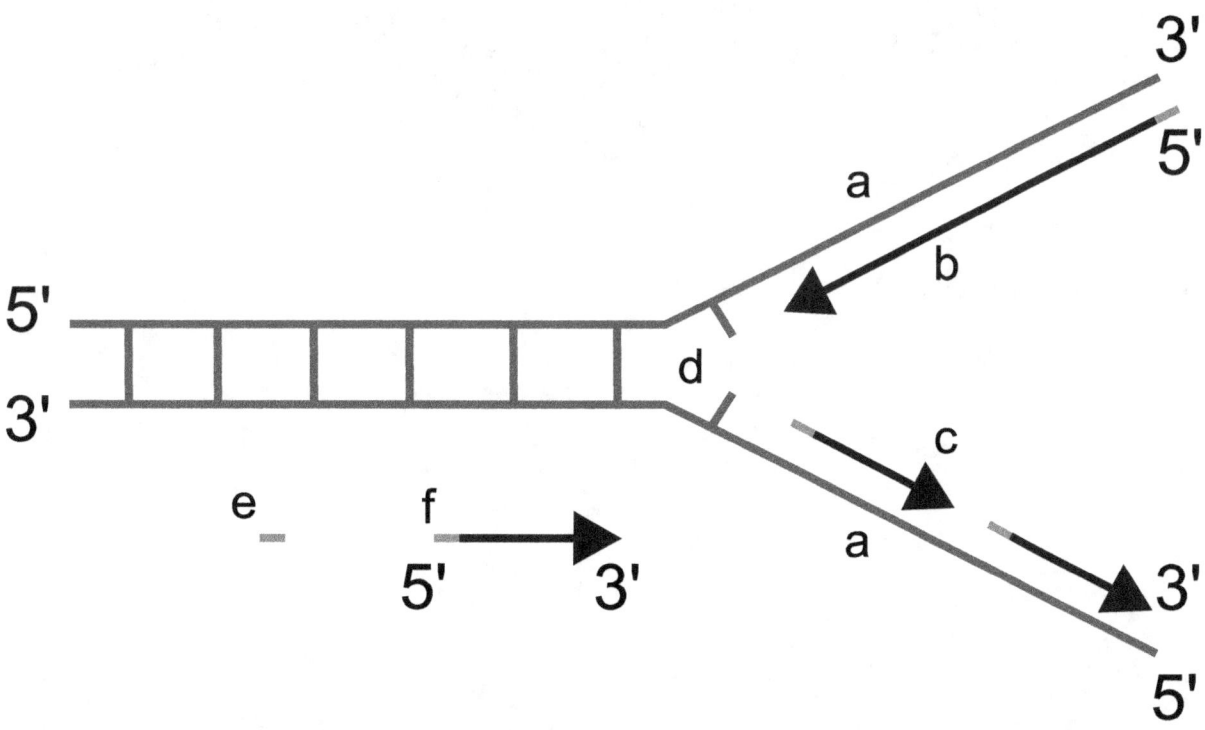

Scheme of the replication fork.
a: template, b: leading strand, c: lagging strand, d: replication fork, e: primer, f: Okazaki fragments

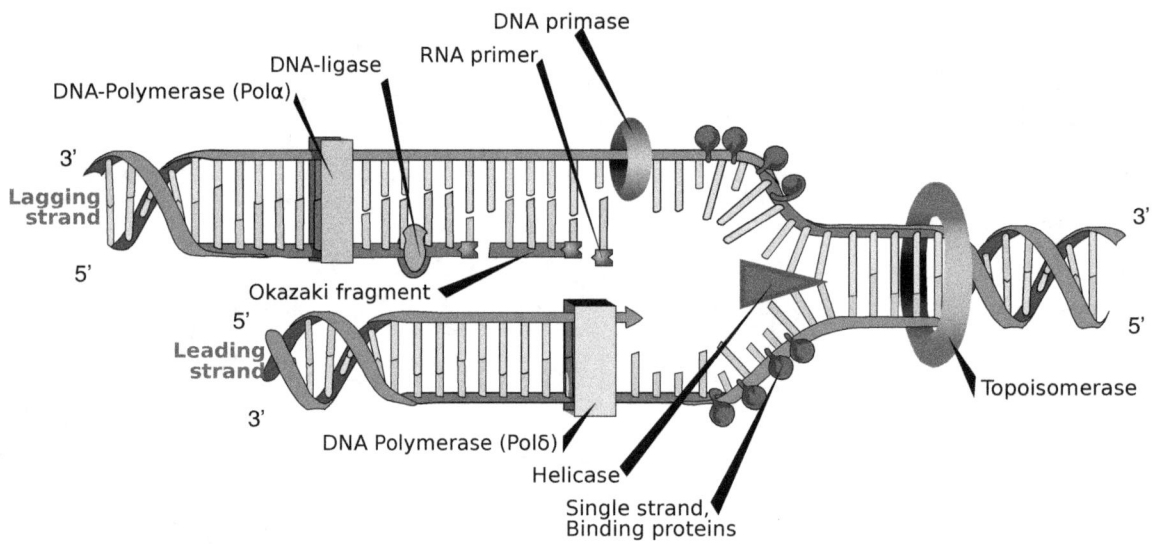

Many enzymes are involved in the DNA replication fork.

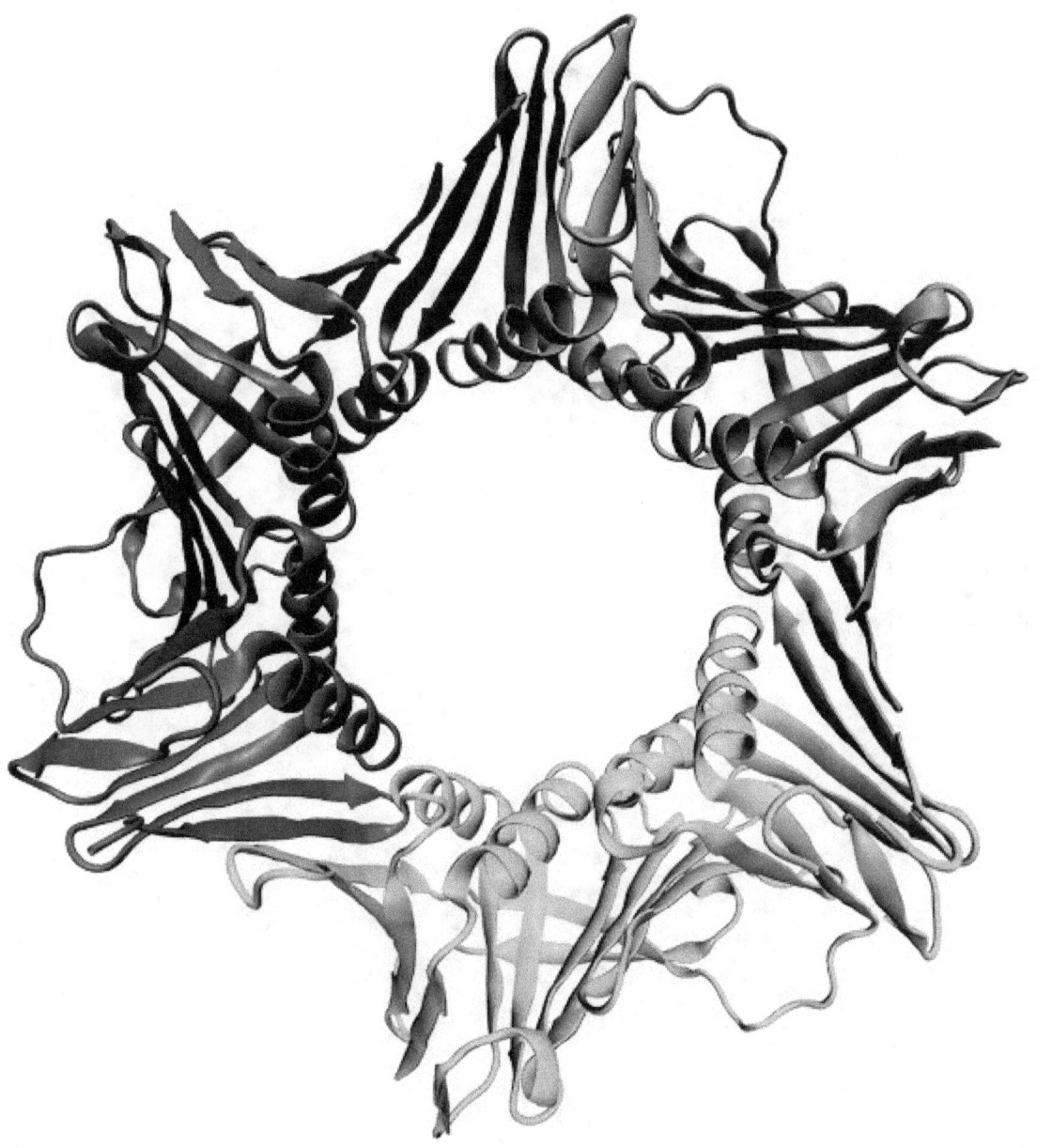

The assembled human DNA clamp, a trimer of the protein PCNA.

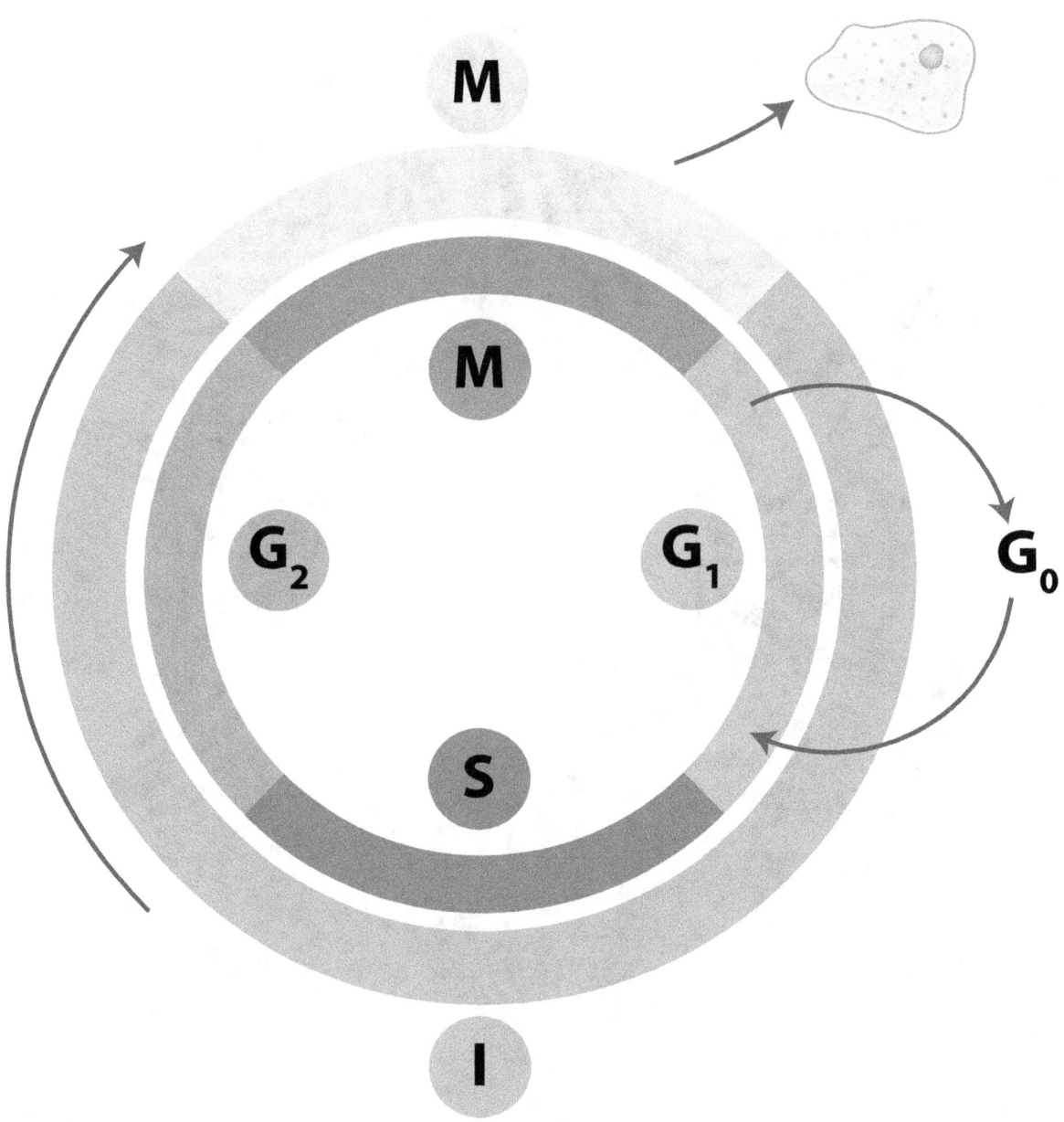

The cell cycle of eukaryotic cells.

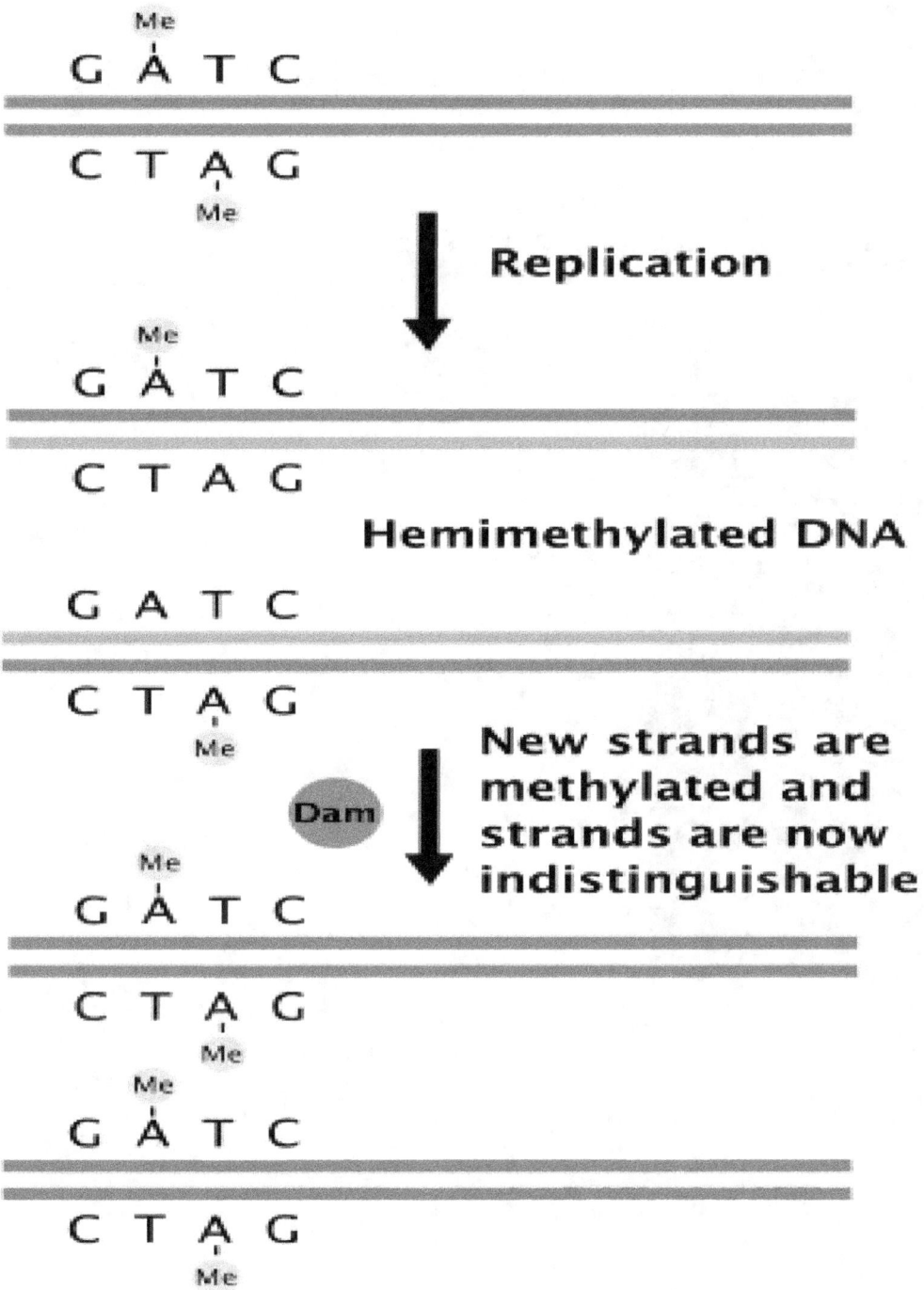

After DNA replication in E. coli, temporary hemimethylation occurs until the new strands are methylated by Dam methylase at the adenine of GATC sites.

Chapter 8

Molecular models of DNA

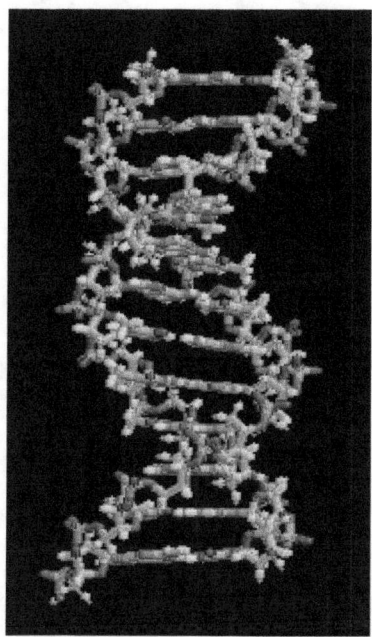

While this purified DNA precipitated in a water jug (left) appears to be a formless mass, nucleic acids actually possess intricate structure at the nanoscale (right).

Molecular models of DNA structures are representations of the molecular geometry and topology of Deoxyribonucleic acid (DNA) molecules using one of several means, with the aim of simplifying and presenting the essential, physical and chemical, properties of DNA molecular structures either *in vivo* or *in vitro*. These representations include closely packed spheres (CPK models) made of plastic, metal wires for 'skeletal models', graphic computations and animations by computers, artistic rendering. Computer molecular models also allow animations and molecular dynamics simulations that are very important for understanding how DNA functions *in vivo*.

The more advanced, computer-based molecular models of DNA involve molecular dynamics simulations as well as quantum mechanical computations of vibro-rotations, delocalized molecular orbitals (MOs), electric dipole moments, hydrogen-bonding, and so on. **DNA molecular dynamics modeling** involves simulations of DNA molecular geometry and topology changes with time as a result of both intra- and inter- molecular interactions of DNA. Whereas molecular models of Deoxyribonucleic acid (DNA) molecules such as closely packed spheres (CPK models) made of plastic or metal wires for 'skeletal models' are useful representations of static DNA structures, their usefulness is very limited for representing complex DNA dynamics. Computer molecular modeling allows both animations and molecular dynamics simulations that are very important for understanding how DNA functions *in vivo*.

8.1 History

Further information: History of molecular biology

From the very early stages of structural studies of DNA by X-ray diffraction and biochemical means, molecular models such as the Watson-Crick double-helix model were successfully employed to solve the 'puzzle' of DNA structure, and also find how the latter relates to its key functions in living cells. The first high quality X-ray diffraction patterns of A-DNA were reported by Rosalind Franklin and Raymond Gosling in 1953.[1] The first calculations of the Fourier transform of an atomic helix were reported one year earlier by Cochran, Crick and Vand,[2] and were followed in 1953 by the computation of the Fourier transform of a coiled-coil by Crick.[3]

Structural information is generated from X-ray diffraction studies of oriented DNA fibers with the help of molecular models of DNA that are combined with crystallographic and mathematical analysis of the X-ray patterns.

The first reports of a double-helix molecular model of B-DNA structure were made by James Watson and Francis Crick

The A-DNA double-helix molecular model of Crick and Watson (consistent with their X-ray data) for which they received a Nobel prize together with M.H.F. Wilkins.

in 1953.[4][5] Last-but-not-least, Maurice F. Wilkins, A. Stokes and H.R. Wilson, reported the first X-ray patterns of *in vivo* B-DNA in partially oriented salmon sperm heads.[6]

The development of the first correct double-helix molecular model of DNA by Crick and Watson may not have been possible without the biochemical evidence for the nucleotide base-pairing ([A---T]; [C---G]), or Chargaff's rules.[7][8][9][10][11][12] Although such initial studies of DNA structures with the help of molecular models were essentially static, their consequences for explaining the *in vivo* functions of DNA were significant in the areas of protein biosynthesis and the quasi-universality of the genetic code. Epigenetic transformation studies of DNA *in vivo* were however much slower to develop in spite of their importance for embryology, morphogenesis and cancer research. Such chemical dynamics and biochemical reactions of DNA are much more complex than the molecular dynamics of DNA physical interactions with water, ions and proteins/enzymes in living cells.

8.2 Importance

Further information: Nucleic acid structure

An old standing dynamic problem is how DNA "self-replication" takes place in living cells that should involve transient uncoiling of supercoiled DNA fibers. Although DNA consists of relatively rigid, very large elongated biopolymer molecules called "fibers" or chains (that are made of repeating nucleotide units of four basic types, attached to deoxyribose and phosphate groups), its molecular structure *in vivo* undergoes dynamic configuration changes that involve dynamically attached water molecules and ions. Supercoiling, packing with histones in chromosome structures, and other such supramolecular aspects also involve *in vivo* DNA topology which is even more complex than DNA molecular geometry, thus turning molecular modeling of DNA into an especially challenging problem for both molecular biologists and biotechnologists. Like other large molecules and biopolymers, DNA often exists in multiple stable geometries (that is, it exhibits conformational isomerism) and configurational, quantum states which are close to each other in energy on the potential energy surface of the DNA molecule.

Such varying molecular geometries can also be computed, at least in principle, by employing *ab initio* quantum chemistry methods that can attain high accuracy for small molecules, although claims that acceptable accuracy can be also achieved for polynuclelotides, as well as DNA conformations, were recently made on the basis of VCD spectral data. Such quantum geometries define an important class of *ab initio* molecular models of DNA whose exploration has barely started especially in connection with results obtained by VCD in solutions. More detailed comparisons with such *ab initio* quantum computations are in principle obtainable through 2D-FT NMR spectroscopy and relaxation studies of polynucleotide solutions or specifically labeled DNA, as for example with deuterium labels.

In an interesting twist of roles, the DNA molecule itself was proposed to be utilized for quantum computing. Both DNA nanostructures as well as DNA 'computing' biochips have been built.

8.3   Fundamental concepts

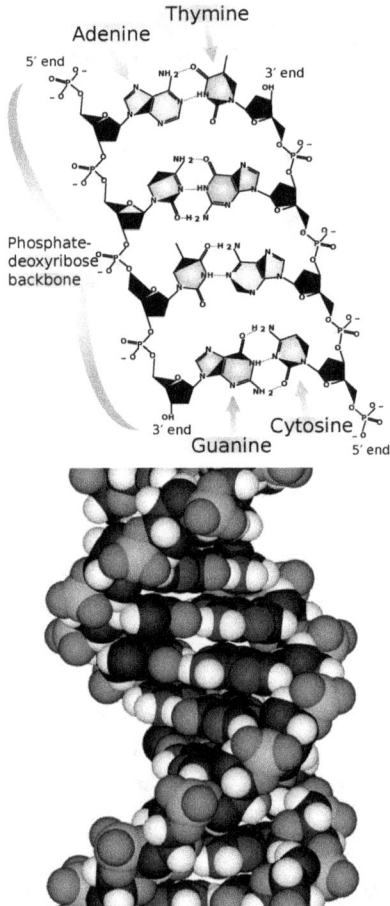

At left, the chemical structure of DNA showing the base-pairing. This depiction of a DNA duplex lacks information about the molecule's three-dimensional structure, at right.

The chemical structure of DNA is insufficient to understand the complexity of the 3D structures of DNA. On the other hand, animated molecular models allow one to visually explore the three-dimensional (3D) structure of DNA. The DNA model shown (far right) is a space-filling, or CPK, model of the DNA double-helix. Animated molecular models, such as the wire, or skeletal, type shown at the top of this article, allow one to visually explore the three-dimensional (3D) structure of DNA. Another type of DNA model is the space-filling, or CPK, model.

The hydrogen bonding dynamics and proton exchange is very different by many orders of magnitude between the two systems of fully hydrated DNA and water molecules in ice. Thus, the DNA dynamics is complex, involving nanosecond and several tens of picosecond time scales, whereas that of liquid ice is on the picosecond time scale, and that of proton exchange in ice is on the millisecond time scale. The proton exchange rates in DNA and attached proteins may vary from picosecond to nanosecond, minutes or years, depending on the exact locations of the exchanged protons in the large biopolymers.

A simple harmonic oscillator 'vibration' is only an oversimplified dynamic representation of the longitudinal vibrations of the DNA intertwined helices which were found to be anharmonic rather than harmonic as often assumed in quantum dynamic simulations of DNA.

8.3.1   DNA structure

The structure of DNA shows a variety of forms, both double-stranded and single-stranded. The mechanical properties of DNA, which are directly related to its structure, are a significant problem for cells. Every process which binds or reads DNA is able to use or modify the mechanical properties of DNA for purposes of recognition, packaging and modification. The extreme length (a chromosome may contain a 10 cm long DNA strand), relative rigidity and helical structure of DNA has led to the evolution of histones and of enzymes such as topoisomerases and helicases to manage a cell's DNA. The properties of DNA are closely related to its molecular structure and sequence, particularly the weakness of the hydrogen bonds and electronic interactions that hold strands of DNA together compared to the strength of the bonds within each strand.

Experimental techniques which can directly measure the mechanical properties of DNA are relatively new, and high-resolution visualization in solution is often difficult. Nevertheless, scientists have uncovered large amount of data on the mechanical properties of this polymer, and the implications of DNA's mechanical properties on cellular processes is a topic of active current research.

The DNA found in many cells can be macroscopic in length - a few centimetres long for each human chromosome. Consequently, cells must compact or "package" DNA to carry it within them. In eukaryotes this is carried by spool-like proteins known as histones, around which DNA winds. It is the further compaction of this DNA-protein complex which produces the well known mitotic eukaryotic chromosomes.

Alternate non-helical models were briefly considered in the late 1970s as a potential solution to problems in the replication of DNA in plasmids and chromatin. However, the models were set aside in favor of the double-helical model due to subsequent experimental advances such as X-ray crystallography of DNA duplexes and later the nucleosome core particle, as well as the discovery of topoisomerases, and these non-double-helical models are not currently accepted by the mainstream scientific community.[13][14]

8.4   DNA structure determination using molecular modeling and DNA X-ray patterns

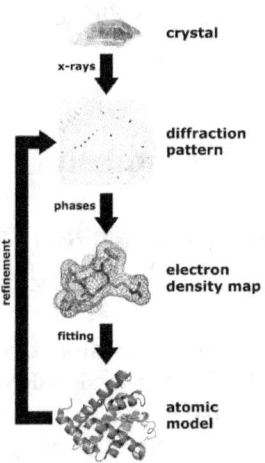

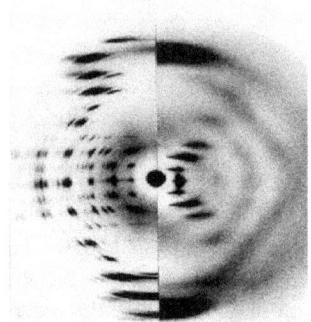

A-DNA   B-DNA

Left, the major steps involved in DNA structure determination by X-ray crystallography showing the important role played by molecular models of DNA structure in this iterative process. Right, an image of actual A- and B- DNA X-ray patterns obtained from oriented and hydrated DNA fibers (courtesy of Dr. Herbert R. Wilson, FRS- see refs. list). Main article: Nucleic acid structure determination

After DNA has been separated and purified by standard biochemical techniques one has a sample in a jar much like in the figure at the top of this article. Below are the main steps involved in generating structural information from X-ray diffraction studies of oriented DNA fibers that are drawn from the hydrated DNA sample with the help of molecular models of DNA that are combined with crystallographic and mathematical analysis of the X-ray patterns.

8.5   Paracrystalline lattice models of B-DNA structures

A paracrystalline lattice, or paracrystal, is a molecular or atomic lattice with significant amounts (e.g., larger than a few percent) of partial disordering of molecular arrangements. Limiting cases of the paracrystal model are nanostructures, such as glasses, liquids, etc., that may possess only local ordering and no global order. A simple example of a paracrystalline lattice is shown in the following figure for a silica glass:

Liquid crystals also have paracrystalline rather than crystalline structures.

Highly hydrated B-DNA occurs naturally in living cells in such a paracrystalline state, which is a dynamic one in spite of the relatively rigid DNA double-helix stabilized by parallel hydrogen bonds between the nucleotide base-pairs in the two complementary, helical DNA chains (see figures). For simplicity most DNA molecular models omit both water and ions dynamically bound to B-DNA, and are thus less useful for understanding the dynamic behaviors of B-DNA *in vivo*. The physical and mathematical analysis of X-ray[15][16] and spectroscopic data for paracrystalline B-DNA is therefore much more complicated than that of crystalline, A-DNA X-ray diffraction patterns. The paracrystal model is also important for DNA technological applications such as DNA nanotechnology. Novel techniques that combine X-ray diffraction of DNA with X-ray microscopy in hydrated living cells are now also being developed.[17]

8.6   Genomic and biotechnology applications of DNA molecular modeling

There are various uses of DNA molecular modeling in Genomics and Biotechnology research applications, from DNA repair to PCR and DNA nanostructures. Two-dimensional DNA junction arrays have been visualized by Atomic force microscopy.[18]

DNA molecular modeling has various uses in genomics and biotechnology, with research applications ranging from DNA repair to PCR and DNA nanostructures. These include computer molecular models of molecules as varied as RNA polymerase, an E. coli, bacterial DNA primase template suggesting very complex dynamics at the interfaces between the enzymes and the DNA template, and molecular models of the mutagenic, chemical interaction of potent carcinogen molecules with DNA. These are all represented in the gallery below.

Silica glass is another example of a material which is organized into a paracrystalline lattice.

Technological application include a DNA biochip and DNA nanostructures designed for DNA computing and other dynamic applications of DNA nanotechnology.[19][20][21][22][23][24] The image at right is of self-assembled DNA nanostructures. The DNA "tile" structure in this image consists of four branched junctions oriented at 90° angles. Each tile consists of nine DNA oligonucleotides as shown; such tiles serve as the primary "building block" for the assembly of the DNA nanogrids shown in the AFM micrograph.

Quadruplex DNA may be involved in certain cancers.[25][26] Images of quadruplex DNA are in the gallery below.

8.7 Gallery of DNA models

- Spinning DNA generic model.

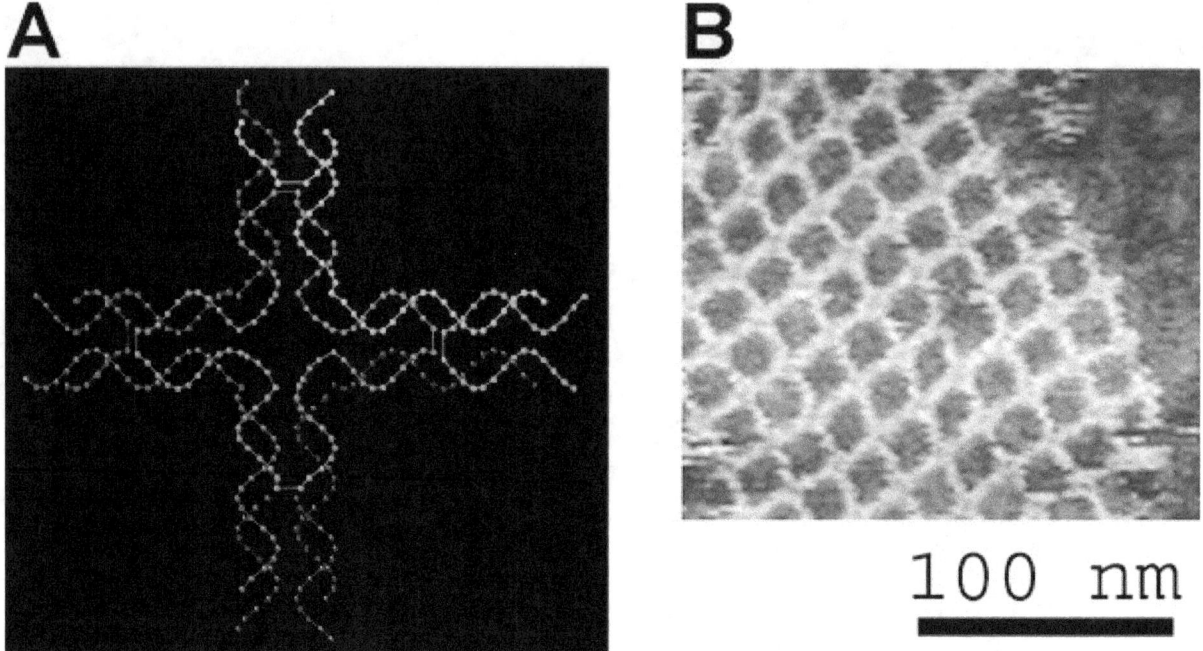

Molecular models are useful in the design of structures for DNA nanotechnology. Here, individual DNA tiles (model at left) self-assemble into a highly ordered DNA 2D-nanogrid (AFM image at right).

- An oversimplified sketch of the double-helix structure of A-DNA.

- A model of DNA replication based on the double-helix concept.

- Animated, space-filling moelcular model of the A-DNA double helix

- A large-scale Crick-Watson DNA Model shown in the Museum of Príncipe Felipe.

- Side view of molecular models of A-, B-, Z- DNA.

- Oversimplified model of the A-DNA double-helix.

- Molecular modeling of RNA Polymerase.

- Molecular modeling of a bacterial DNA Primase Template.

- Molecular modeling of DNA interactions with the carcinogen molecule MGMT.

- 3D Molecular model of DNA damaged by carcinogenic 2-aminofluorene(AF).

- Fig.6. Molecular modeling of DNA repair

- Animated skeletal model of A-DNA.

- Simplified models of chromatin.

- Simplified model of chromosome structure.

- A hypothetical quadruplex of guanine-rich DNA structures that may be involved in cancers.

- 3D Molecular Structure of the intramolecular human telomeric G-quadruplex in potassium solution.

- DNA spacefilling molecular model

- A model of a designed DNA tetrahedron.

8.8   See also

- G-quadruplex

- Crystallography

- Crystal lattices

- DiProDB Dinucleotide Property Database. The database is designed to collect and analyse thermodynamic, structural and other dinucleotide properties.

- X-ray microscopy

- X-ray scattering

- Neutron scattering

- Nucleic acid sequence

- Vibrational circular dichroism (VCD)

- Raman spectroscopy/microscopy and CARS

- Sir Lawrence Bragg, FRS

- List of nucleic acid simulation software

- AMBER

- CHARMM

- Abalone (molecular mechanics)

- Sirius visualization software

- QMC@Home

- FT-NMR

- NMR microscopy

- Microwave spectroscopy

- FT-IR

- FT-NIR

- Spectral, Hyperspectral, and Chemical imaging

- Fluorescence correlation spectroscopy

- Fluorescence cross-correlation spectroscopy and FRET

- Confocal microscopy

8.9   References

[1] Franklin, R.E., Gosling, R.G. (6 March 1953). "The Structure of Sodium Thymonucleate Fibres I. The Influence of Water Content". *Acta Cryst.* **6** (8): 673. doi:10.1107/S0365110X53001939.
Franklin, R.E., Gosling, R.G. (6 March 1953). "The Structure of Sodium Thymonucleate Fibres II. The Cylindrically Symmetrical Patterson Function". *Acta Cryst.* **6** (8): 678. doi:10.1107/S0365110X53001940.

[2] Cochran, W., Crick, F.H.C., Vand V. (1952). "The Structure of Synthetic Polypeptides. 1. The Transform of Atoms on a Helix". *Acta Cryst.* **5** (5): 581–6. doi:10.1107/S0365110X52001635.

[3] Crick, F.H.C. (1953a). "The Fourier Transform of a Coiled-Coil". *Acta Cryst.* **6** (8–9): 685–9. doi:10.1107/S0365110X53001952.

[4] Watson, James D., Crick, Francis H.C. (25 April 1953). "A structure for Deoxyribose Nucleic Acid" (PDF). *Nature* **171** (4356): 737–8. Bibcode:1953Natur.171..737W. doi:10.1038/171737a0. PMID 13054692., .

[5] Watson, J.D; Crick F.H.C. (1953b). "The Structure of DNA". *Cold Spring Harbor Symposia on Quantitative Biology* **18**: 123–31. doi:10.1101/SQB.1953.018.01.020. PMID 13168976.

[6] Wilkins M.H.F., A.R. Stokes A.R. & Wilson, H.R. (1953). "Molecular Structure of Deoxypentose Nucleic Acids" (PDF). *Nature* **171** (4356): 738–40. Bibcode:1953Natur.171..738W. doi:10.1038/171738a0. PMID 13054693.

[7] Elson D, Chargaff E (1952). "On the deoxyribonucleic acid content of sea urchin gametes". *Experientia* **8** (4): 143–5. doi:10.1007/BF02170221. PMID 14945441.

[8] Chargaff E, Lipshitz R, Green C (1952). "Composition of the deoxypentose nucleic acids of four genera of sea-urchin". *J Biol Chem* **195** (1): 155–160. PMID 14938364.

[9] Chargaff E, Lipshitz R, Green C, Hodes ME (1951). "The composition of the deoxyribonucleic acid of salmon sperm". *J Biol Chem* **192** (1): 223–230. PMID 14917668.

[10] Chargaff E (1951). "Some recent studies on the composition and structure of nucleic acids". *J Cell Physiol Suppl* **38** (Suppl).

[11] Magasanik B, Vischer E, Doniger R, Elson D, Chargaff E (1950). "The separation and estimation of ribonucleotides in minute quantities". *J Biol Chem* **186** (1): 37–50. PMID 14778802.

[12] Chargaff E (1950). "Chemical specificity of nucleic acids and mechanism of their enzymatic degradation". *Experientia* **6** (6): 201–9. doi:10.1007/BF02173653. PMID 15421335.

[13] Stokes, T. D. (1982). "The double helix and the warped zipper—an exemplary tale". *Social Studies of Science* **12** (2): 207–240. doi:10.1177/030631282012002002.

[14] Gautham, N. (25 May 2004). "Response to "Variety in DNA secondary structure"" (PDF). *Current Science* **86** (10): 1352–1353. Retrieved 25 May 2012. However, the discovery of topoisomerases took "the sting" out of the topological objection to the plectonaemic double helix. The more recent solution of the single crystal X-ray structure of the nucleosome core particle showed nearly 150 base pairs of the DNA (i.e. about 15 complete turns), with a structure that is in all essential respects the same as the Watson–Crick model. This dealt a death blow to the idea that other forms of DNA, particularly double helical DNA, exist as anything other than local or transient structures.

[15] Hosemann R., Bagchi R.N., *Direct analysis of diffraction by matter*, North-Holland Publs., Amsterdam – New York, 1962.

[16] Baianu, I.C. (1978). "X-ray scattering by partially disordered membrane systems". *Acta Cryst.* **A34** (5): 751–3. Bibcode:1978A doi:10.1107/S0567739478001540.

[17] Yamamoto Y, Shinohara K (October 2002). "Application of X-ray microscopy in analysis of living hydrated cells". *Anat. Rec.* **269** (5): 217–23. doi:10.1002/ar.10166. PMID 12379938.

[18] Mao, Chengde; Sun, Weiqiong; Seeman, Nadrian C. (16 June 1999). "Designed Two-Dimensional DNA Holliday Junction Arrays Visualized by Atomic Force Microscopy". *Journal of the American Chemical Society* **121** (23): 5437–43. doi:10.1021/ja990

[19] Robinson, Bruche H.; Seeman, Nadrian C. (August 1987). "The Design of a Biochip: A Self-Assembling Molecular-Scale Memory Device". *Protein Engineering* **1** (4): 295–300. doi:10.1093/protein/1.4.295. ISSN 0269-2139. PMID 3508280. Link

[20] Rothemund, Paul W. K.; Ekani-Nkodo, Axel; Papadakis, Nick; Kumar, Ashish; Fygenson, Deborah Kuchnir & Winfree, Erik (22 December 2004). "Design and Characterization of Programmable DNA Nanotubes". *Journal of the American Chemical Society* **126** (50): 16344–52. doi:10.1021/ja0443191. PMID 15600335.

[21] Keren, K.; Kinneret Keren, Rotem S. Berman, Evgeny Buchstab, Uri Sivan, Erez Braun (November 2003). "DNA-Templated Carbon Nanotube Field-Effect Transistor". *Science* **302** (6549): 1380–2. Bibcode:2003Sci...302.1380K. doi:10.1126/science.1091022.PMID14631035.

[22] Zheng, Jiwen; Constantinou, Pamela E.; Micheel, Christine; Alivisatos, A. Paul; Kiehl, Richard A. & Seeman Nadrian C. (2006). "2D Nanoparticle Arrays Show the Organizational Power of Robust DNA Motifs". *Nano Letters* **6** (7): 1502–4. Bibcode:2006NanoL...6.1502Z. doi:10.1021/nl060994c. PMC 3465979. PMID 16834438.

[23] Cohen, Justin D.; Sadowski, John P.; Dervan, Peter B. (2007). "Addressing Single Molecules on DNA Nanostructures". *Angewandte Chemie* **46** (42): 7956–9. doi:10.1002/anie.200702767. PMID 17763481.

[24] Constantinou, Pamela E.; Wang, Tong; Kopatsch, Jens; Israel, Lisa B.; Zhang, Xiaoping; Ding, Baoquan; Sherman, William B.; Wang, Xing; Zheng, Jianping; Sha, Ruojie & Seeman, Nadrian C. (2006). "Double cohesion in structural DNA nanotechnology". *Organic and Biomolecular Chemistry* **4** (18): 3414–9. doi:10.1039/b605212f. PMC 3491902. PMID 17036134.

[25] http://www.phy.cam.ac.uk/research/bss/molbiophysics.php

[26] http://planetphysics.org/encyclopedia/TheoreticalBiophysics.html

8.10 Further reading

- *Applications of Novel Techniques to Health Foods, Medical and Agricultural Biotechnology.*(June 2004) I. C. Baianu, P. R. Lozano, V. I. Prisecaru and H. C. Lin., q-bio/0406047.

- F. Bessel, *Untersuchung des Theils der planetarischen Störungen*, Berlin Abhandlungen (1824), article 14.

- Sir Lawrence Bragg, FRS. *The Crystalline State, A General survey.* London: G. Bells and Sons, Ltd., vols. 1 and 2., 1966., 2024 pages.

- Cantor, C. R. and Schimmel, P.R. *Biophysical Chemistry, Parts I and II.*, San Franscisco: W.H. Freeman and Co. 1980. 1,800 pages.

- Voet, D. and J.G. Voet. *Biochemistry*, 2nd Edn., New York, Toronto, Singapore: John Wiley & Sons, Inc., 1995, ISBN 0-471-58651-X., 1361 pages.

- Watson, G. N. *A Treatise on the Theory of Bessel Functions.*, (1995) Cambridge University Press. ISBN 0-521-48391-3.

- Watson, James D. *Molecular Biology of the Gene.* New York and Amsterdam: W.A. Benjamin, Inc. 1965., 494 pages.

- Wentworth, W.E. *Physical Chemistry. A short course.*, Malden (Mass.): Blackwell Science, Inc. 2000.

- Herbert R. Wilson, FRS. *Diffraction of X-rays by proteins, Nucleic Acids and Viruses.*, London: Edward Arnold (Publishers) Ltd. 1966.

- Kurt Wuthrich. *NMR of Proteins and Nucleic Acids.*, New York, Brisbane,Chicester, Toronto, Singapore: J. Wiley & Sons. 1986., 292 pages.

- Hallin PF, David Ussery D (2004). "CBS Genome Atlas Database: A dynamic storage for bioinformatic results and DNA sequence data". *Bioinformatics* **20** (18): 3682–6. doi:10.1093/bioinformatics/bth423. PMID 15256401.

- Zhang CT, Zhang R, Ou HY (2003). "The Z curve database: a graphic representation of genome sequences". *Bioinformatics* **19** (5): 593–599. doi:10.1093/bioinformatics/btg041. PMID 12651717.

8.11   External links

- DNA the Double Helix Game From the official Nobel Prize web site

- MDDNA: Structural Bioinformatics of DNA

- Double Helix 1953–2003 National Centre for Biotechnology Education

- DNAlive: a web interface to compute DNA physical properties. Also allows cross-linking of the results with the UCSC Genome browser and DNA dynamics.

- Further details of mathematical and molecular analysis of DNA structure based on X-ray data

- Bessel functions corresponding to Fourier transforms of atomic or molecular helices.

- overview of STM/AFM/SNOM principles with educative videos

8.11.1   Databases for DNA molecular models and sequences

X-ray diffraction

- NDB ID: UD0017 Database

- X-ray Atlas -database

- PDB files of coordinates for nucleic acid structures from X-ray diffraction by NA (incl. DNA) crystals

- Structure factors downloadable files in CIF format

Neutron scattering

- ISIS neutron source: ISIS pulsed neutron source:A world centre for science with neutrons & muons at Harwell, near Oxford, UK.

X-ray microscopy

Electron microscopy

- DNA under electron microscope

NMR databases

- NMR Atlas--database

- mmcif downloadable coordinate files of nucleic acids in solution from 2D-FT NMR data

- NMR constraints files for NAs in PDB format

Genomic and structural databases

- CBS Genome Atlas Database — contains examples of base skews.

- The Z curve database of genomes — a 3-dimensional visualization and analysis tool of genomes.

- DNA and other nucleic acids' molecular models: Coordinate files of nucleic acids molecular structure models in PDB and CIF formats

Atomic force microscopy

- How SPM Works

- SPM Image Gallery - AFM STM SEM MFM NSOM and more.

Chapter 9

Nucleic acid structure

Nucleic acid structure refers to the structure of nucleic acids such as DNA and RNA. Chemically speaking, DNA and RNA are very similar. Nucleic acid structure is often divided into four different levels: primary, secondary, tertiary and quaternary.

9.1 Primary structure

Main article: Nucleic acid sequence
 Primary structure consists of a linear sequence of nucleotides that are linked together by phosphodiester bonds. It is this linear sequence of nucleotides that make up the primary structure of DNA or RNA. Nucleotides consist of 3 components:

1. Nitrogenous base

    (a) Adenine

    (b) Guanine

    (c) Cytosine

    (d) Thymine (present in DNA only)

    (e) Uracil (present in RNA only)

2. 5-carbon sugar which is called deoxyribose (found in DNA) and ribose (found in RNA).

3. One or more phosphate groups.[1]

The nitrogen bases adenine and guanine are purine in structure and form a glycosidic bond between their 9' nitrogen and the 1' -OH group of the deoxyribose. Cytosine, thymine and uracil are pyrimidines, hence the glycosidic bonds forms between their 1' nitrogen and the 1' -OH of the deoxyribose. For both the purine and pyrimidine bases, the phosphate group forms a bond with the deoxyribose sugar through an ester bond between one of its negatively charged oxygen groups and the 5' -OH of the sugar.[2] The polarity in DNA and RNA is derived from the oxygen and nitrogen atoms in the backbone. Nucleic acids are formed when nucleotides come together through phosphodiester linkages between the 5' and 3' carbon atoms.[3] A Nucleic acid sequence is the order of nucleotides within a DNA (GACT) or RNA (GACU) molecule that is determined by a series of letters. Sequences are presented from the 5' to 3' end and determine the covalent structure of the entire molecule. Sequences can be complementary to another sequence in that the base on each position is complementary as well as in the reverse order. An example of a complementary sequence to AGCT is TCGA. DNA is double-stranded containing both a sense strand and an antisense strand. Therefore, the complementary sequence will be to the sense strand.[4]

Chemical structure of DNA

9.2   Secondary structure

Main article: Nucleic acid secondary structure

Secondary structure is the set of interactions between bases, i.e., which parts of strands are bound to each other. In DNA

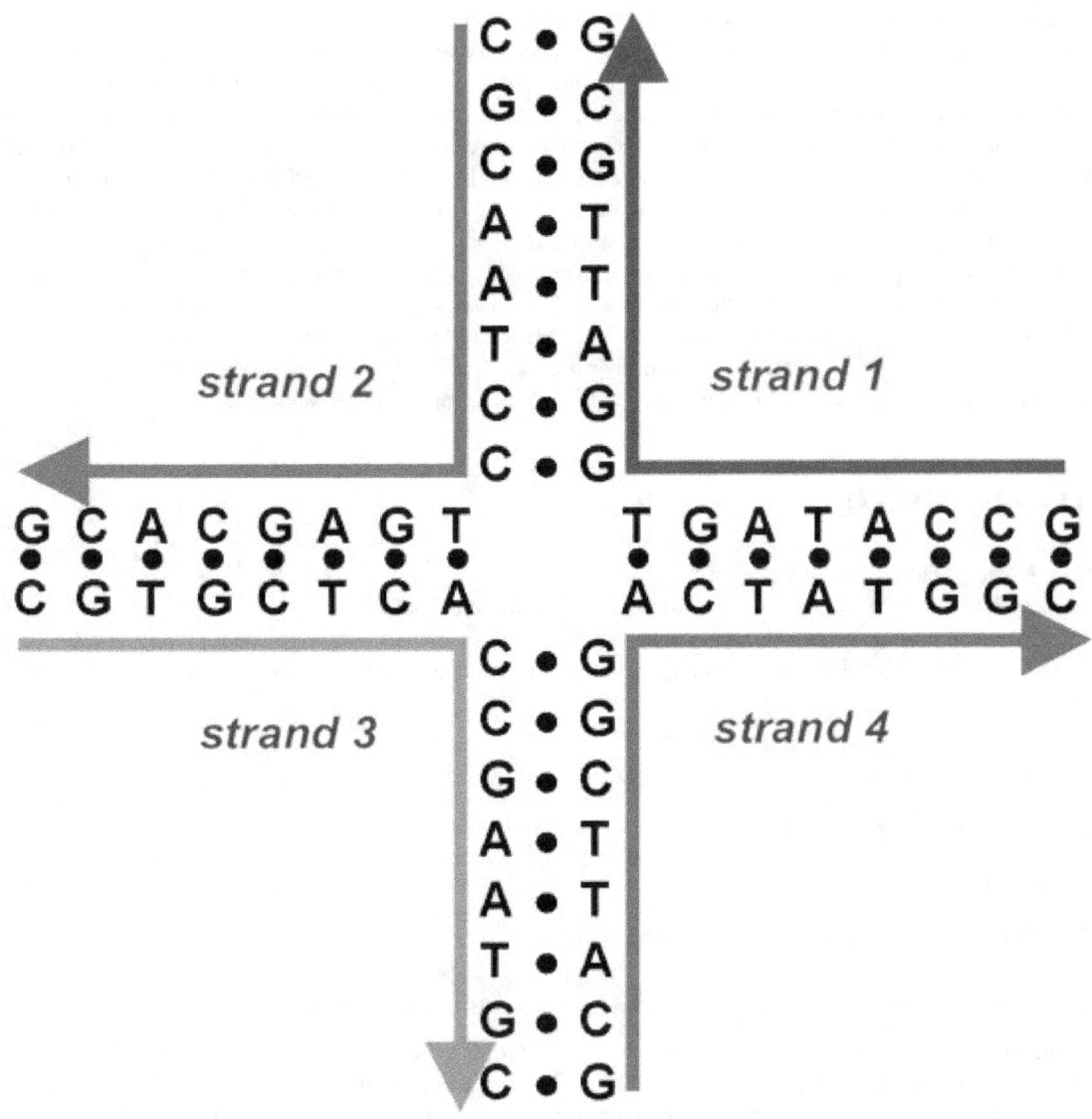

Nucleic acid design can be used to create nucleic acid complexes with complicated secondary structures such as this four-arm junction. These four strands associate into this structure because it maximizes the number of correct base pairs, with A's matched to T's and C's matched to G's. Image from Mao, 2004.[5]

double helix, the two strands of DNA are held together by hydrogen bonds. The nucleotides on one strand base pairs with the nucleotide on the other strand. The secondary structure is responsible for the shape that the nucleic acid assumes. The bases in the DNA are classified as Purines and Pyrimidines. The purines are Adenine and Guanine. Purines consist of a double ring structure, a six membered and a five membered ring containing nitrogen. The pyrimidines are Cytosine and Thymine. It has a single ringed structure, a six membered ring containing nitrogen. A purine base always pairs with a pyrimidine base (Guanosine (G) pairs with Cytosine(C) and Adenine(A) pairs with Thymine (T) or Uracil (U)). DNA's secondary structure is predominantly determined by base-pairing of the two polynucleotide strands wrapped around each other to form a double helix. There is also a major groove and a minor groove on the double helix.

The secondary structure of RNA consists of a single polynucleotide. Base pairing in RNA occurs when RNA folds between complementarity regions. Both single- and double-stranded regions are often found in RNA molecules. The antiparallel strands form a helical shape.[3] The four basic elements in the secondary structure of RNA are helices, loops, bulges, and junctions. Stem-loop or hairpin loop is the most common element of RNA secondary structure.[6] Stem-loop is formed when the RNA chains fold back on themselves to form a double helical tract called the stem, the unpaired nucleotides forms single stranded region called the loop.[7] Secondary structure of RNA can be predicted by experimental data on the secondary structure elements, helices, loops and bulges. Bulges and internal loops are formed by separation of the double helical tract on either one strand (bulge) or on both strands (internal loops) by unpaired nucleotides. A Tetraloop is a four-base pairs hairpin RNA structure. There are three common families of tetraloop in ribosomal RNA: UNCG, GNRA, and CUUG (N is one of the four nucleotides and R is a purine).UNCG is the most stable tetraloop.[8] Pseudoknot is a RNA secondary structure first identified in turnip yellow mosaic virus.[9] Pseudoknots are formed when nucleotides from the hairpin loop pairs with a single stranded region outside of the hairpin to form a helical segment. H-type fold pseudoknots are best characterized. In H-type fold, nucleotides in the hairpin loop pairs with the bases outside the hairpin stem forming second stem and loop. This causes formation of pseudoknots with two stems and two loops.[10] Pseudoknots are functional elements in RNA structure having diverse function and found in most classes of RNA. DotKnot-PW method is used for comparative pseudoknots prediction .The main points in the DotKnot-PW method is scoring the similarities found in stems, secondary elements and H-type pseudoknots.[11]

9.3   Tertiary structure

Main article: Nucleic acid tertiary structure
 Tertiary structure refers to the locations of the atoms in three-dimensional space, taking into consideration geometrical and steric constraints. It is a higher order than the secondary structure, in which large-scale folding in a linear polymer occurs and the entire chain is folded into a specific 3-dimensional shape. There are 4 areas in which the structural forms of DNA can differ.

1. Handedness - right or left

2. Length of the helix turn

3. Number of base pairs per turn

4. Difference in size between the major and minor grooves[3]

The tertiary arrangement of DNA's double helix in space includes B-DNA, A-DNA and Z-DNA.

B-DNA is the most common form of DNA in vivo and is a more narrow, elongated helix than A-DNA. Its wide major groove makes it more accessible to proteins. On the other hand, it has a narrow minor groove. B-DNA's favored conformations occur at high water concentrations; the hydration of the minor groove appears to favor B-DNA. B-DNA base pairs are nearly perpendicular to the helix axis. The sugar pucker which determines the shape of the a-helix, whether the helix will exist in the A-form or in the B-form, occurs at the C2'-endo.[12]

A-DNA, is a form of the DNA duplex observed under dehydrating conditions. It is shorter and wider than B-DNA. RNA adopts this double helical form, and RNA-DNA duplexes are mostly A-form, but B-form RNA-DNA duplexes have been observed.[13] In localized single strand dinucleotide contexts, RNA can also adopt the B-form without pairing to DNA.[14] A-DNA has a deep, narrow major groove which does not make it easily accessible to proteins. On the other hand, its wide, shallow minor groove makes it accessible to proteins but with lower information content than the major groove. Its favored conformation is at low water concentrations. A-DNAs base pairs are tilted relative to the helix axis, and are displaced from the axis. The sugar pucker occurs at the C3'-endo and in RNA 2'-OH inhibits C2'-endo conformation.[12] Long considered little more than a laboratory artifice, A-DNA is now known to have several biological functions.

Z-DNA is a relatively rare left-handed double-helix. Given the proper sequence and superhelical tension, it can be formed in vivo but its function is unclear. It has a more narrow, more elongated helix than A or B. Z-DNA's major groove is not really a groove, and it has a narrow minor groove. The most favored conformation occurs when there are high salt concentrations. There are some base substitutions but they require an alternating purine-pyrimidine sequence. The N2-amino of G H-bonds to 5' PO, which explains the slow exchange of protons and the need for the G purine. Z-DNA base pairs are nearly perpendicular to the helix axis. Z-DNA does not contain single base-pairs but rather a GpC repeat with P-P distances varying for GpC and CpG. On the GpC stack there is good base overlap, whereas on the CpG stack there is less overlap. Z-DNA's zigzag backbone is due to the C sugar conformation compensating for G glycosidic bond conformation. The conformation of G is syn, C2'-endo and for C it is anti, C3'-endo.[12]

A linear DNA molecule having free ends can rotate, to adjust to changes of various dynamic processes in the cell, by changing how many times the two chains of its double helix twist around each other. Some DNA molecules are circular and are topologically constrained. A covalently closed, circular DNA (also known as cccDNA) is topologically constrained as the number of times the chains coiled around one other cannot change. This cccDNA can be supercoiled, which is the tertiary structure of DNA. Supercoiling is characterized by the linking number, twist and writhe. The linking number (Lk) for circular DNA is defined as the number of times one strand would have to pass through the other strand to completely separate the two strands. The linking number for circular DNA can only be changed by breaking of a covalent bond in one of the two strands. Always an integer, the linking number of a cccDNA is the sum of two components: twists (Tw) and writhes (Wr).[15]

$$Lk = Tw + Wr$$

Twists are the number of times the two strands of DNA are twisted around each other. Writhes are number of times the DNA helix crosses over itself. DNA in cells is negatively supercoiled and has the tendency to unwind. Hence the separation of strands is easier in negatively supercoiled DNA than in relaxed DNA. The two components of supercoiled DNA are solenoid and plectonemic. The plectonemic supercoil is found in prokaryotes, while the solenoidal supercoiling is mostly seen in eukaryotes.

9.4 Quaternary structure

Main article: Nucleic acid quaternary structure
The quaternary structure of nucleic acids is similar to that of protein quaternary structure. Although some of the concepts are not exactly the same, the quaternary structure refers to a higher-level of organization of nucleic acids. Moreover, it refers to interactions of the nucleic acids with other molecules. The most commonly seen form of higher-level organization of nucleic acids is seen in the form of chromatin which leads to its interactions with the small proteins histones. Also, the quaternary structure refers to the interactions between separate RNA units in the ribosome or spliceosome.[16]

9.5 See also

- Nucleic acid double helix
- Nucleic acid structure determination (experimental)
- Nucleic acid structure prediction (computational)
- DNA nanotechnology
- Nucleic acid design
- Nucleic acid thermodynamics
- Crosslinking of DNA
- Non-helical models of DNA structure
- DNA supercoil

9.6   References

[1] Krieger M, Scott MP, Matsudaira PT, Lodish HF, Darnell JE, Lawrence Z, Kaiser C, Berk A (2004). "Section 4.1: Structure of Nucleic Acids". *Molecular cell biology*. New York: W.H. Freeman and CO. ISBN 0-7167-4366-3.

[2] "Structure of Nucleic Acids". *SparkNotes*.

[3] Anthony-Cahill SJ; Mathews CK, van Holde KE, Appling DR (2012). *Biochemistry (4th Edition)*. Englewood Cliffs, N.J: Prentice Hall. ISBN 0-13-800464-1.

[4] Alberts B, Johnson A, Lewis J, Raff M, Roberts K & Wlater P (2002). *Molecular Biology of the Cell (4th ed.)*. New York NY: Garland Science. ISBN 0-8153-3218-1.

[5] Mao, Chengde (December 2004). "The Emergence of Complexity: Lessons from DNA". *PLOS Biology* **2** (12): 2036–2038. doi:10.1371/journal.pbio.0020431. ISSN 1544-9173. PMC 535573. PMID 15597116.

[6] Tinoco I, Jr; Bustamante, C (Oct 22, 1999). "How RNA folds.". *Journal of Molecular Biology* **293**(2): 271–81. doi:10.1006/jm PMID 10550208.

[7] "RNA structure (Molecular Biology)".

[8] Hollyfield, JG; Besharse, JC; Rayborn, ME (December 1976). "The effect of light on the quantity of phagosomes in the pigment epithelium.". *Experimental eye research* **23** (6): 623–35. doi:10.1016/0014-4835(76)90221-9. PMID 1087245.

[9] Rietveld, K; Van Poelgeest, R; Pleij, CW; Van Boom, JH; Bosch, L (Mar 25, 1982). "The tRNA-like structure at the 3' terminus of turnip yellow mosaic virus RNA. Differences and similarities with canonical tRNA". *Nucleic Acids Research* **10** (6): 1929–46. doi:10.1093/nar/10.6.1929. PMC 320581. PMID 7079175.

[10] Staple, DW; Butcher, SE (June 2005). "Pseudoknots: RNA structures with diverse functions". *PLoS Biology* **3** (6): e213. doi:10.1371/journal.pbio.0030213. PMC 1149493. PMID 15941360.

[11] Sperschneider, J; Datta, A; Wise, MJ (Dec 1, 2012). "Predicting pseudoknotted structures across two RNA sequences". *Bioinformatics (Oxford, England)* **28** (23): 3058–65. doi:10.1093/bioinformatics/bts575. PMC 3516145. PMID 23044552.

[12] Dickerson RE, Drew HR, Conner BN, Wing RM, Fratini AV, Kopka ML (April 1982). "The anatomy of A-, B-, and Z-DNA". *Science* **216** (4545): 475–85. doi:10.1126/science.7071593. PMID 7071593.

[13] Chen X; Ramakrishnan B; Sundaralingam M (1995). "Crystal structures of B-form DNA-RNA chimers complexed with distamycin". *Nature Structural Biology* **2** (9): 733–735.

[14] Sedova A; Banavali NK (2016). "RNA approaches the B-form in stacked single strand dinucleotide contexts". *Biopolymers*. in press. doi:10.1002/bip.22750. PMID 26443416.

[15] Mirkin SM (2001). "DNA Topology: Fundamentals". *Encyclopedia of Life Sciences*. doi:10.1038/npg.els.0001038. ISBN 0470016175.

[16] "Strucual Biochemistry/Nucleic Acid/DNA/DNA Structure". Retrieved 11 December 2012.

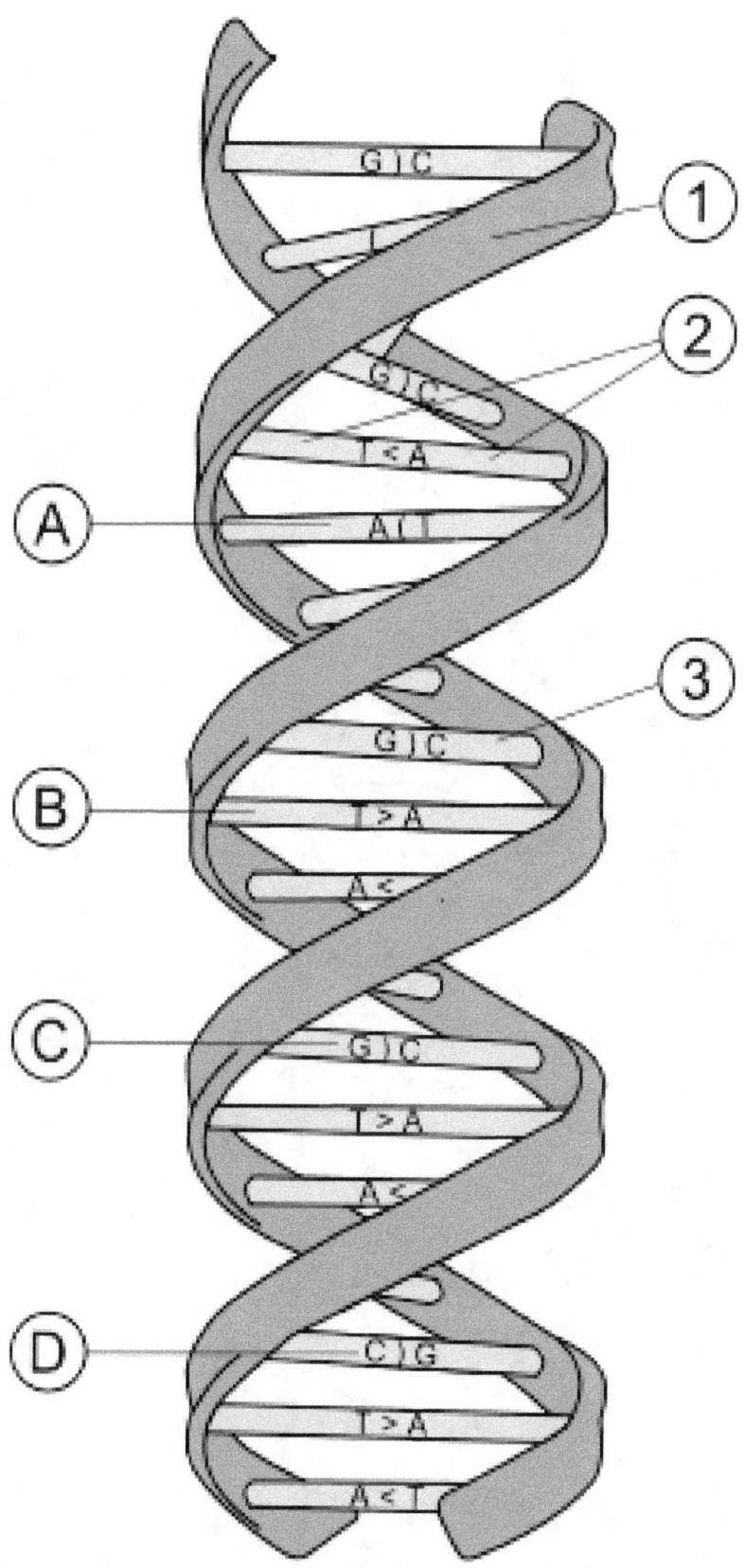

The structure of DNA, based on Image:DNA-structure-and-bases.png: A. Adenine B. Thymine C. Guanine D. Cytosine 1. Sugar, Phosphate, Backbone 2. Base pair 3. Nitrogeous base

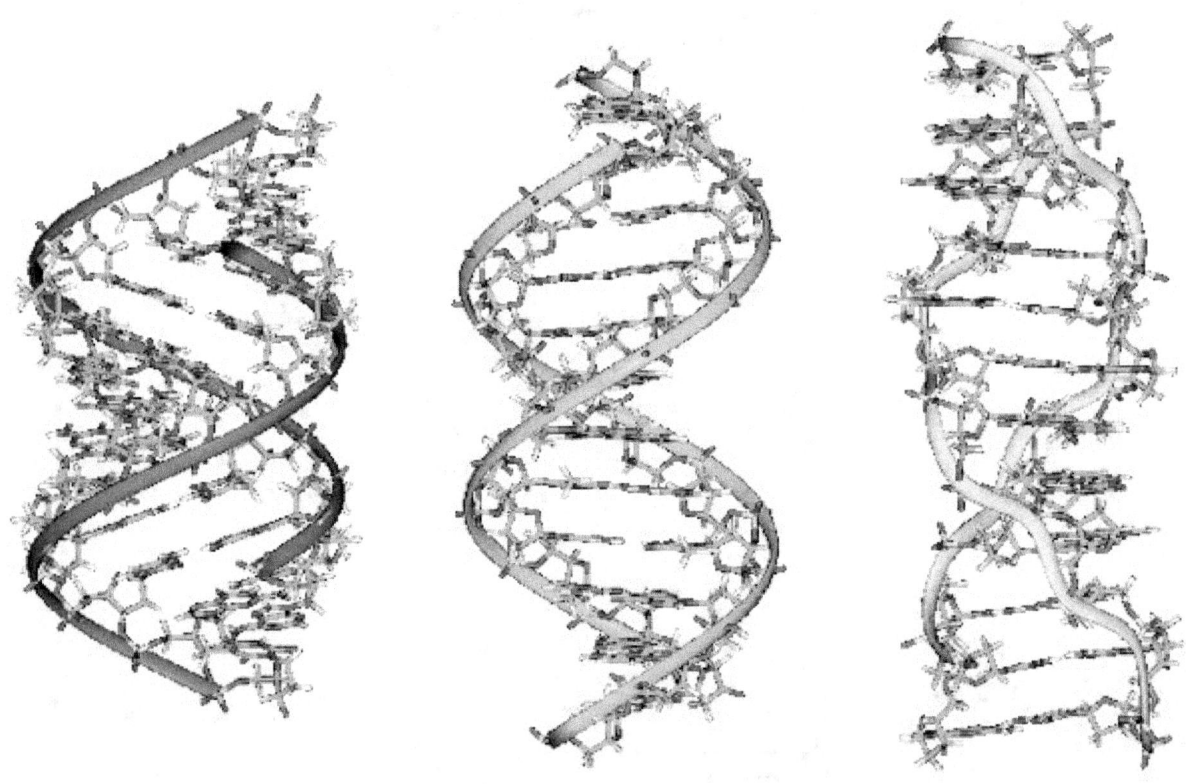

A-B-Z-DNA Side View

naked duplex DNA

"beads-on-a-string" created by formation of nucleosomes

30nm solenoid

extended form of chromosome

condensed section of chromatin

mitotic chromosome

DNA to Chromatin

Chapter 10

Molecular Structure of Nucleic Acids: A Structure for Deoxyribose Nucleic Acid

"**Molecular Structure of Nucleic Acids: A Structure for Deoxyribose Nucleic Acid**" was an article published by Francis Crick and James D. Watson in the scientific journal *Nature* in its 171st volume on pages 737–738 (dated 25 April 1953).[1] It was the first publication which described the discovery of the double helix structure of DNA, using X-ray diffraction and the mathematics of a helix transform.[2]

This article is often termed a "pearl" of science because it is brief and contains the answer to a fundamental mystery about living organisms. This mystery was the question of how it is possible that genetic instructions are held inside organisms and how they are passed from generation to generation. The article presents a simple and elegant solution, which surprised many biologists at the time who believed that DNA transmission was going to be more difficult to deduce and understand. The discovery had a major impact on biology, particularly in the field of genetics, enabling later researchers to understand the genetic code.

10.1   Origins of molecular biology

The application of physics and chemistry to biological problems led to the development of molecular biology. Molecular biology is particularly concerned with the flow and consequences of biological information at the level of genes and proteins. The discovery of the DNA double helix made clear that genes are functionally defined parts of DNA molecules and that there must be a way for cells to make use of their DNA genes in order to make proteins.

Linus Pauling was a chemist who was very influential in developing an understanding of the structure of biological molecules. In 1951, Pauling published the structure of the alpha helix, a fundamentally important structural component of proteins. In early 1953, Pauling published an incorrect triple helix model of DNA.[3] Both Crick, and particularly Watson, thought that they were racing against Pauling to discover the structure of DNA.

Max Delbrück was a physicist who recognized some of the biological implications of quantum physics. Delbruck's thinking about the physical basis of life stimulated Erwin Schrödinger to write, *What Is Life?* Schrödinger's book was an important influence on Crick, Watson, and Maurice Wilkins who won the Nobel Prize for Medicine in recognition of their discovery of the DNA double helix. Delbruck's efforts to promote the "Phage Group" (exploring genetics by way of the viruses that infect bacteria) was important in the early development of molecular biology in general and the development of Watson's scientific interests in particular.[4]

10.2 DNA structure and function

It is not always the case that the structure of a molecule is easy to relate to its function. What makes the structure of DNA so obviously related to its function was described modestly at the end of the article: "It has not escaped our notice that the specific pairing we have postulated immediately suggests a possible copying mechanism for the genetic material".

The "specific pairing" is a key feature of the Watson and Crick model of DNA, the pairing of nucleotide subunits.[5] In DNA, the amount of guanine is equal to cytosine and the amount of adenine is equal to thymine. The A:T and C:G pairs are structurally similar. In particular, the length of each base pair is the same and they fit equally between the two sugar-phosphate backbones (Figure 2). The base pairs are held together by hydrogen bonds, a type of chemical attraction that is easy to break and easy to reform. After realizing the structural similarity of the A:T and C:G pairs, Watson and Crick soon produced their double helix model of DNA with the hydrogen bonds at the core of the helix providing a way to unzip the two complementary strands for easy replication: the last key requirement for a likely model of the genetic molecule.

Indeed, the base-pairing did suggest a way to copy a DNA molecule. Just pull apart the two sugar-phosphate backbones, each with its hydrogen bonded A, T, G, and C components. Each strand could then be used as a template for assembly of a new base-pair complementary strand.

10.2.1 Future considerations

When Watson and Crick produced their double helix model of DNA, it was known that most of the specialized features of the many different life forms on Earth are made possible by proteins. Structurally, proteins are long chains of amino acid subunits. In some way, the genetic molecule, DNA, had to contain instructions for how to make the thousands of proteins found in cells. From the DNA double helix model, it was clear that there must be some correspondence between the linear sequences of nucleotides in DNA molecules to the linear sequences of amino acids in proteins. The details of how sequences of DNA instruct cells to make specific proteins was worked out by molecular biologists during the period from 1953 to 1965. Francis Crick played an integral role in both the theory and analysis of the experiments that led to an improved understanding of the genetic code.[6]

10.2.2 Consequences

Other advances in molecular biology stemming from the discovery of the DNA double helix eventually led to ways to sequence genes. James Watson directed the Human Genome Project at the National Institutes of Health.[7] The ability to sequence and manipulate DNA is now central to the biotechnology industry and modern medicine. The austere beauty of the structure and the practical implications of the DNA double helix combined to make *Molecular structure of Nucleic Acids; A Structure for Deoxyribose Nucleic Acid* one of the most prominent biology articles of the twentieth century.

10.3 Collaborators' controversy

Main article: King's College London DNA Controversy

Although Watson and Crick were first to put together all the scattered fragments of information that were required to produce a successful molecular model of DNA, their findings had been based on data collected by researchers in several other laboratories. The discovery of the DNA double helix used a considerable amount of material from the unpublished work of Maurice Wilkins, Rosalind Franklin, A.R. Stokes, and H.R. Wilson at King's College London. Key data from Wilkins, Stokes, and Wilson, and, separately, by Franklin and Gosling, were published in two separate additional articles in the same issue of *Nature* with the article by Watson and Crick.[8][9] The article by Watson and Crick acknowledged that they had been "stimulated" by experimental results from the King's College researchers, and a similar acknowledgment was published by Wilkins, Stokes, and Wilson in the following three-page article.

In 1968, Watson published a highly controversial autobiographical account of the discovery of the double-helical, molecular structure of DNA called *The Double Helix*, which was not publicly accepted either by Crick or Wilkins.[10] Furthermore, Erwin Chargaff also printed a rather "unsympathetic review" of Watson's booklet in the March 29, 1968 issue of *Science*. In the booklet, Watson stated among other things that he and Crick had access to some of Franklin's data from a source that she was not aware of, and also that he had seen—without her permission—the B-DNA X-ray diffraction pattern obtained by Franklin and Gosling in May 1952 at King's in London. In particular, in late 1952, Franklin had submitted a progress report to the Medical Research Council, which was reviewed by Max Perutz, then at the Cavendish Laboratory of the University of Cambridge. Watson and Crick also worked in the MRC-supported Cavendish Laboratory in Cambridge whereas Wilkins and Franklin were in the MRC supported laboratory at King's in London. Such MRC reports were not usually widely circulated, but Crick read a copy of Franklin's research summary in early 1953.[10][11]

Perutz's justification for passing Franklin's report about the crystallographic unit of the B-DNA and A-DNA structures to both Crick and Watson was that the report contained information which Watson had previously heard in November 1951 when Franklin talked about her unpublished results with Raymond Gosling during a meeting arranged by M.H.F. Wilkins at King's College, following a request from Crick and Watson;[12] Perutz said he had not acted unethically because the report had been part of an effort to promote wider contact between different MRC research groups and was not confidential.[13] This justification would exclude Crick, who was not present at the November 1951 meeting, yet Perutz also gave him access to Franklin's MRC report data. Crick and Watson then sought permission from Cavendish Laboratory head Lawrence Bragg, to publish their double-helix molecular model of DNA based on data from Franklin and Wilkins.

By November 1951 Watson had acquired—by his own admission—little training in X-ray crystallography, and therefore had not fully understood (again, according to his own admission, in *The Double Helix*) what Franklin was saying about the structural symmetry of the DNA molecule. Crick, however, knowing the Fourier transforms of Bessel functions that represent the X-ray diffraction patterns of helical structures of atoms, correctly interpreted further one of Franklin's experimental findings as indicating that DNA was most likely to be a double helix with the two polynucleotide chains running in opposite directions. Crick was thus in a unique position to make this interpretation because he had previously worked on the X-ray diffraction data for other large molecules that had helical symmetry similar to that of DNA. Franklin, on the other hand, rejected the first molecular model building approach proposed by Crick and Watson: the first DNA model (which in 1952 Watson presented to her and to Wilkins in London) had an obviously incorrect structure with hydrated charged groups on the inside of the model, rather than on the outside. Watson explicitly admitted this in his "Double Helix" booklet.[11]

10.4 See also

- Crystallography

- DNA

- List of nucleic acid simulation software: nucleic acid modeling

- *Miles from Tomorrowland*, a TV series with twin admirals named Watson and Crick

- Paracrystal model and theory

- X-ray scattering

10.5 Works

- Judson, Horace Freeland (1979). *The Eighth Day of Creation. Makers of the Revolution in Biology*. Simon and Schuster. ISBN 0-671-22540-5.

- Maddox, Brenda (2002). *Rosalind Franklin: The Dark Lady of DNA*. ISBN 0-060-98508-9.

- Olby, Robert (1974). *The Path to The Double Helix: Discovery of DNA*. MacMillan. ISBN 0-486-68117-3. (with foreword by Francis Crick; revised in 1994, with a 9 page postscript.)

- Watson, James D. (1980). *The Double Helix: A Personal Account of the Discovery of the Structure of DNA.* Atheneum. ISBN 0-689-70602-2. (first published in 1968)

- Wilkins, Maurice (2003). *The Third Man of the Double Helix: The Autobiography of Maurice Wilkins.* ISBN 0-198-60665-6.

- Life Story (TV film) a BBC dramatization about the scientific race to discover the DNA double-helix.

10.6 References

[1] Watson JD, Crick FH (April 1953). "Molecular structure of nucleic acids; a structure for deoxyribose nucleic acid" (PDF). *Nature* **171** (4356): 737–738. Bibcode:1953Natur.171..737W. doi:10.1038/171737a0. PMID 13054692.

[2] Cochran W, Crick FHC and Vand V. (1952) "The Structure of Synthetic Polypeptides. I. The Transform of Atoms on a Helix", Acta Cryst., 5, 581–586.

[3] Pauling L, Corey RB (1953). "A Proposed Structure for the Nucleic Acids". *PNAS* **39** (2): 84–97. Bibcode:1953PNAS...39...84P. doi:10.1073/pnas.39.2.84. PMC 1063734. PMID 16578429.

[4] Judson, Horace Freeland (1979). *Eighth Day of Creation: Makers of the Revolution in Biology.* New York: Simon & Schuster. ISBN 978-8-796-94785-3.

[5] Discover the rules of DNA base pairing with an online simulator.

[6] Perutz MF, Randall JT, Thomson L, Wilkins MH, Watson JD (June 1969). "DNA helix". *Science* **164** (3887): 1537–9. Bibcode:1969Sci...164.1537W. doi:10.1126/science.164.3887.1537. PMID 5796048.

[7] "History - Historic Figures: Watson and Crick (1928-)". BBC. Retrieved 2014-06-15.

[8] Franklin R, Gosling RG (1953-04-25). "Molecular configuration in sodium thymonucleate" (PDF). *Nature* **171** (4356): 740–741. Bibcode:1953Natur.171..740F. doi:10.1038/171740a0. PMID 13054694.

[9] Wilkins MHF, Stokes AR, Wilson HR (25 April 1953). "Molecular structure of deoxypentose nucleic acids" (PDF). *Nature* **171** (4356): 738–740. Bibcode:1953Natur.171..738W. doi:10.1038/171738a0. PMID 13054693.

[10] Beckwith, Jon (2003). "Double Take on the Double Helix". In Victor K. McElheny. *Watson and DNA: Making a Scientific Revolution.* Cambridge, MA: Perseus Publishing. p. 363. ISBN 978-0-738-20341-6. OCLC 51440191.

[11] Watson, James D. (1980). *The Double Helix: A Personal Account of the Discovery of the Structure of DNA.* Atheneum. ISBN 0-689-70602-2. (first published in 1968)

[12] Sayre, Anne (1975). *Rosalind Franklin and DNA.* New York: Norton.

[13] Perutz MF, Randall JT, Thomson L, Wilkins MH, Watson JD (1969-06-27). "DNA helix". *Science* **164** (3887): 1537–1539. Bibcode:1969Sci...164.1537W. doi:10.1126/science.164.3887.1537. PMID 5796048.

10.7 External links

- Annotated copy of the article from San Francisco's Exploratorium

- Access Excellence Classic Collection article on DNA structure.

- Linus Pauling and the Race for DNA: A Documentary History

10.7.1 Online versions

- Online version (Original text) at nature.com

- National Library of Medicine's PDF copy in the Francis Crick Documents Collection.

- Commemorative HTML version Am J Psychiatry 160:623-624, April 2003.

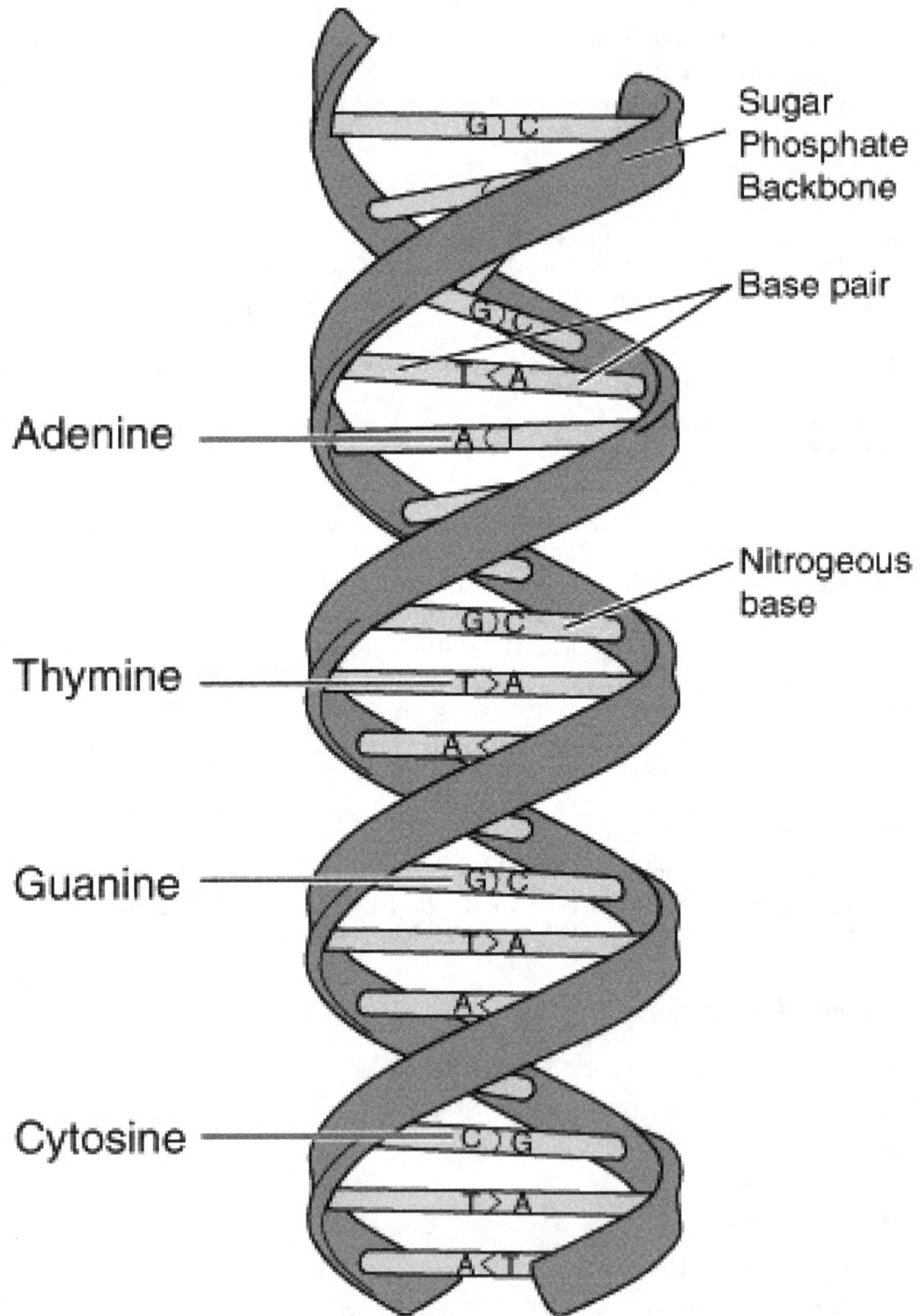

Figure 2. Diagramatic representation of the key structural features of the DNA double helix. This figure does not depict B-DNA.

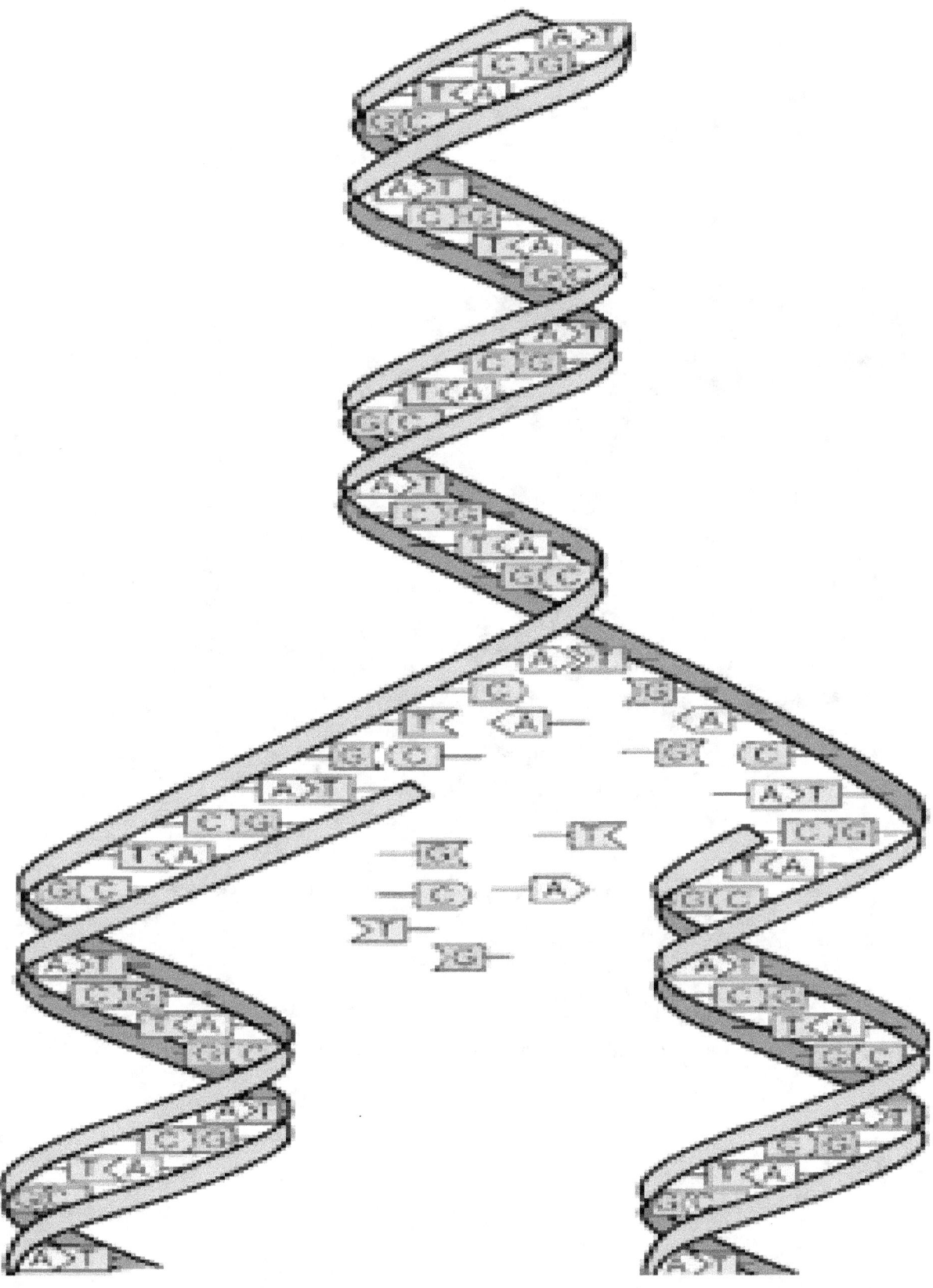

DNA replication. The two base-pair complementary chains of the DNA molecule allow for replication of the genetic instructions.

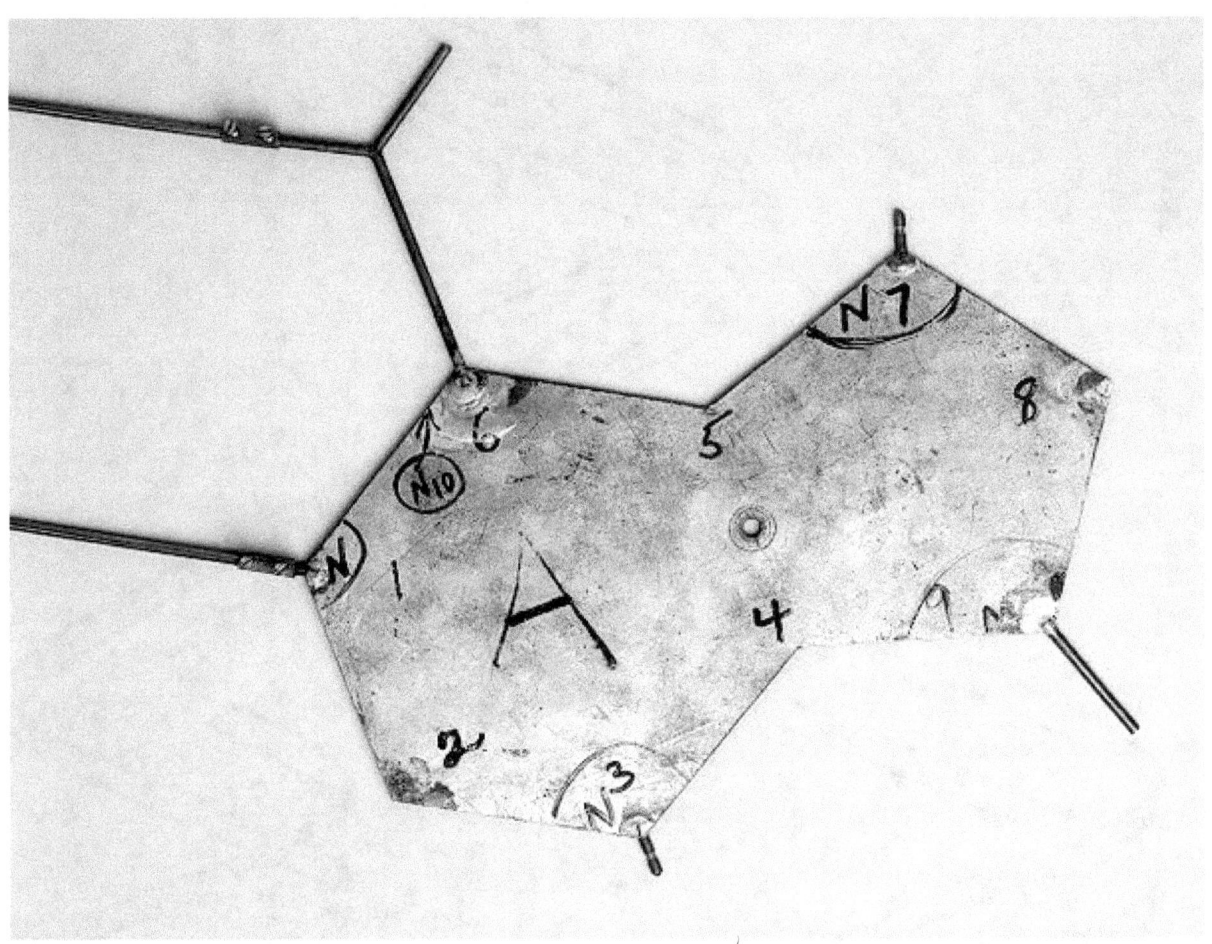

Watson and Crick used many aluminium templates like this one, which is the single base Adenine (A), to build a physical model of DNA in 1953.

Chapter 11

Genetic code

"Codon" redirects here. For the plant genus, see Codon (genus).

 The **genetic code** is the set of rules by which information encoded within genetic material (DNA or mRNA sequences) is translated into proteins by living cells. Biological decoding is accomplished by the ribosome, which links amino acids in an order specified by mRNA, using transfer RNA (tRNA) molecules to carry amino acids and to read the mRNA three nucleotides at a time. The genetic code is highly similar among all organisms and can be expressed in a simple table with 64 entries.

The code defines how sequences of nucleotide triplets, called *codons*, specify which amino acid will be added next during protein synthesis. With some exceptions,[1] a three-nucleotide codon in a nucleic acid sequence specifies a single amino acid. Because the vast majority of genes are encoded with exactly the same code (see the RNA codon table), this particular code is often referred to as the canonical or standard genetic code, or simply *the* genetic code, though in fact some variant codes have evolved. For example, protein synthesis in human mitochondria relies on a genetic code that differs from the standard genetic code.

While the genetic code determines the protein sequence for a given coding region, other genomic regions can influence when and where these proteins are produced.

11.1   Discovery

Serious efforts to understand how proteins are encoded began after the structure of DNA was discovered in 1953. George Gamow postulated that sets of three bases must be employed to encode the 20 standard amino acids used by living cells to build proteins. With four different nucleotides, a code of 2 nucleotides would allow for only a maximum of $4^2 = 16$ amino acids. A code of 3 nucleotides could code for a maximum of $4^3 = 64$ amino acids.[2]

The Crick, Brenner et al. experiment first demonstrated that codons consist of three DNA bases; Marshall Nirenberg and Heinrich J. Matthaei were the first to elucidate the nature of a codon in 1961 at the National Institutes of Health. They used a cell-free system to translate a poly-uracil RNA sequence (i.e., UUUUU...) and discovered that the polypeptide that they had synthesized consisted of only the amino acid phenylalanine.[3] They thereby deduced that the codon UUU specified the amino acid phenylalanine. This was followed by experiments in Severo Ochoa's laboratory that demonstrated that the poly-adenine RNA sequence (AAAAA...) coded for the polypeptide poly-lysine[4] and that the poly-cytosine RNA sequence (CCCCC...) coded for the polypeptide poly-proline.[5] Therefore, the codon AAA specified the amino acid lysine, and the codon CCC specified the amino acid proline. Using different copolymers most of the remaining codons were then determined. Subsequent work by Har Gobind Khorana identified the rest of the genetic code. Shortly thereafter, Robert W. Holley determined the structure of transfer RNA (tRNA), the adapter molecule that facilitates the process of translating RNA into protein. This work was based upon earlier studies by Severo Ochoa, who received the Nobel Prize in Physiology or Medicine in 1959 for his work on the enzymology of RNA synthesis.[6]

Extending this work, Nirenberg and Philip Leder revealed the triplet nature of the genetic code and deciphered the codons of the standard genetic code. In these experiments, various combinations of mRNA were passed through a filter that contained ribosomes, the components of cells that translate RNA into protein. Unique triplets promoted the binding of specific tRNAs to the ribosome. Leder and Nirenberg were able to determine the sequences of 54 out of 64 codons in their experiments.[7] In 1968, Khorana, Holley and Nirenberg received the Nobel Prize in Physiology or Medicine for their work.[8]

11.2   Salient features

11.2.1   Sequence reading frame

A codon is defined by the initial nucleotide from which translation starts. For example, the string GGGAAACCC, if read from the first position, contains the codons GGG, AAA, and CCC; and, if read from the second position, it contains the codons GGA and AAC; if read starting from the third position, GAA and ACC. Every sequence can, thus, be read in three reading frames, each of which will produce a different amino acid sequence (in the given example, Gly-Lys-Pro, Gly-Asn, or Glu-Thr, respectively). With double-stranded DNA, there are six possible reading frames, three in the forward orientation on one strand and three reverse on the opposite strand.[9]:330 The actual frame in which a protein sequence is translated is defined by a start codon, usually the first AUG codon in the mRNA sequence.

11.2.2   Start/stop codons

Translation starts with a chain initiation codon or start codon. Unlike stop codons, the codon alone is not sufficient to begin the process. Nearby sequences such as the Shine-Dalgarno sequence in *E. coli* and initiation factors are also required to start translation. The most common start codon is AUG, which is read as methionine or, in bacteria, as formylmethionine. Alternative start codons depending on the organism include "GUG" or "UUG"; these codons normally represent valine and leucine, respectively, but as start codons they are translated as methionine or formylmethionine.[10]

The three stop codons have been given names: UAG is *amber*, UGA is *opal* (sometimes also called *umber*), and UAA is *ochre*. "Amber" was named by discoverers Richard Epstein and Charles Steinberg after their friend Harris Bernstein, whose last name means "amber" in German.[11] The other two stop codons were named "ochre" and "opal" in order to keep the "color names" theme. Stop codons are also called "termination" or "nonsense" codons. They signal release of the nascent polypeptide from the ribosome because there is no cognate tRNA that has anticodons complementary to these stop signals, and so a release factor binds to the ribosome instead.[12]

11.2.3   Effect of mutations

During the process of DNA replication, errors occasionally occur in the polymerization of the second strand. These errors, called mutations, can have an impact on the phenotype of an organism, especially if they occur within the protein coding sequence of a gene. Error rates are usually very low—1 error in every 10–100 million bases—due to the "proofreading" ability of DNA polymerases.[14][15]

Missense mutations and nonsense mutations are examples of point mutations, which can cause genetic diseases such as sickle-cell disease and thalassemia respectively.[16][17][18] Clinically important missense mutations generally change the properties of the coded amino acid residue between being basic, acidic, polar or non-polar, whereas nonsense mutations result in a stop codon.[9]:266

Mutations that disrupt the reading frame sequence by indels (insertions or deletions) of a non-multiple of 3 nucleotide bases are known as frameshift mutations. These mutations usually result in a completely different translation from the original, and are also very likely to cause a stop codon to be read, which truncates the creation of the protein.[19] These mutations may impair the function of the resulting protein, and are thus rare in *in vivo* protein-coding sequences. One reason inheritance of frameshift mutations is rare is that, if the protein being translated is essential for growth under the selective pressures the organism faces, absence of a functional protein may cause death before the organism is viable.[20] Frameshift mutations may result in severe genetic diseases such as Tay-Sachs disease.[21]

Although most mutations that change protein sequences are harmful or neutral, some mutations have a beneficial effect on an organism.[22] These mutations may enable the mutant organism to withstand particular environmental stresses better than wild-type organisms, or reproduce more quickly. In these cases a mutation will tend to become more common in a population through natural selection.[23] Viruses that use RNA as their genetic material have rapid mutation rates,[24] which can be an advantage, since these viruses will evolve constantly and rapidly, and thus evade the defensive responses of e.g. the human immune system.[25] In large populations of asexually reproducing organisms, for example, *E. coli*, multiple beneficial mutations may co-occur. This phenomenon is called clonal interference and causes competition among the mutations.[26]

11.2.4 Degeneracy

Main article: Codon degeneracy

Degeneracy is the redundancy of the genetic code. The genetic code has redundancy but no ambiguity (see the codon tables below for the full correlation). For example, although codons GAA and GAG both specify glutamic acid (redundancy), neither of them specifies any other amino acid (no ambiguity). The codons encoding one amino acid may differ in any of their three positions. For example, the amino acid leucine is specified by **YUR** or CUN (UUA, UUG, CUU, CUC, CUA, or CUG) codons (difference in the first or third position indicated using IUPAC notation), while the amino acid serine is specified by UCN or AGY (UCA, UCG, UCC, UCU, AGU, or AGC) codons (difference in the first, second, or third position).[27]:521–522 A practical consequence of redundancy is that errors in the third position of the triplet codon cause only a silent mutation or an error that would not affect the protein because the hydrophilicity or hydrophobicity is maintained by equivalent substitution of amino acids; for example, a codon of NUN (where N = any nucleotide) tends to code for hydrophobic amino acids. NCN yields amino acid residues that are small in size and moderate in hydropathy; NAN encodes average size hydrophilic residues. The genetic code is so well-structured for hydropathy that a mathematical analysis (Singular Value Decomposition) of 12 variables (4 nucleotides x 3 positions) yields a remarkable correlation (C = 0.95) for predicting the hydropathy of the encoded amino acid directly from the triplet nucleotide sequence, *without translation*.[28][29] Note in the table, below, eight amino acids are not affected at all by mutations at the third position of the codon, whereas in the figure above, a mutation at the second position is likely to cause a radical change in the physicochemical properties of the encoded amino acid.

11.3 Transfer of information via the genetic code

The genome of an organism is inscribed in DNA, or, in the case of some viruses, RNA. The portion of the genome that codes for a protein or an RNA is called a gene. Those genes that code for proteins are composed of tri-nucleotide units called **codons**, each coding for a single amino acid. Each nucleotide sub-unit consists of a phosphate, a deoxyribose sugar, and one of the four nitrogenous nucleobases. The purine bases adenine (A) and guanine (G) are larger and consist of two aromatic rings. The pyrimidine bases cytosine (C) and thymine (T) are smaller and consist of only one aromatic ring. In the double-helix configuration, two strands of DNA are joined to each other by hydrogen bonds in an arrangement known as base pairing. These bonds almost always form between an adenine base on one strand and a thymine base on the other strand, or between a cytosine base on one strand and a guanine base on the other. This means that the number of A and T bases will be the same in a given double helix, as will the number of G and C bases.[27]:102–117 In RNA, thymine (T) is replaced by uracil (U), and the deoxyribose is substituted by ribose.[27]:127

Each protein-coding gene is transcribed into a molecule of the related RNA polymer. In prokaryotes, this RNA functions as messenger RNA or mRNA; in eukaryotes, the transcript needs to be processed to produce a mature mRNA. The mRNA is, in turn, translated on a ribosome into a chain of amino acids otherwise known as a polypeptide.[27]:Chp 12 The process of translation requires transfer RNAs which are covalently attached to a specific amino acid, guanosine triphosphate as an energy source, and a number of translation factors. tRNAs have anticodons complementary to the codons in an mRNA and can be covalently "charged" with specific amino acids at their 3' terminal CCA ends by enzymes known as aminoacyl tRNA synthetases, which have high specificity for both their cognate amino acid and tRNA. The high specificity of these enzymes is a major reason why the fidelity of protein translation is maintained.[27]:464–469

There are 4^3 = 64 different codon combinations possible with a triplet codon of three nucleotides; all 64 codons are assigned to either an amino acid or a stop signal. If, for example, an RNA sequence UUUAAACCC is considered and the reading frame starts with the first U (by convention, 5' to 3'), there are three codons, namely, UUU, AAA, and CCC, each of which specifies one amino acid. Therefore, this 9 base RNA sequence will be translated into an amino acid sequence that is three amino acids long.[27]:521–539 A given amino acid may be encoded by between one and six different codon sequences. A comparison may be made using bioinformatics tools wherein the codon is similar to a word, which is the standard data "chunk" and a nucleotide is similar to a bit, in that it is the smallest unit. This allows for powerful comparisons across species as well as within organisms.

The standard genetic code is shown in the following tables. Table 1 shows which amino acid each of the 64 codons specifies. Table 2 shows which codons specify each of the 20 standard amino acids involved in translation. These are called forward and reverse codon tables, respectively. For example, the codon "AAU" represents the amino acid asparagine, and "UGU" and "UGC" represent cysteine (standard three-letter designations, Asn and Cys, respectively).[27]:522

11.4   RNA codon table

^A The codon AUG both codes for methionine and serves as an initiation site: the first AUG in an mRNA's coding region is where translation into protein begins.[30]

11.5   DNA codon table

Main article: DNA codon table

The DNA codon table is essentially identical to that for RNA, but with U replaced by T.

11.6   Variations to the standard genetic code

See also: List of genetic codes

While slight variations on the standard code had been predicted earlier,[31] none were discovered until 1979, when researchers studying human mitochondrial genes discovered they used an alternative code. Many slight variants have been discovered since then,[32] including various alternative mitochondrial codes,[33] and small variants such as translation of the codon UGA as tryptophan in *Mycoplasma* species, and translation of CUG as a serine rather than a leucine in yeasts of the "CTG clade" (*Candida albicans* is member of this group).[34][35][36] Because viruses must use the same genetic code as their hosts, modifications to the standard genetic code could interfere with the synthesis or functioning of viral proteins. However, some viruses (such as totiviruses) have adapted to the genetic code modification of the host.[37] In bacteria and archaea, GUG and UUG are common start codons, but in rare cases, certain proteins may use alternative start codons not normally used by that species.[32]

In certain proteins, non-standard amino acids are substituted for standard stop codons, depending on associated signal sequences in the messenger RNA. For example, UGA can code for selenocysteine and UAG can code for pyrrolysine. Selenocysteine is now viewed as the 21st amino acid, and pyrrolysine is viewed as the 22nd.[32] Unlike selenocysteine, pyrrolysine encoded UAG is translated with the participation of a dedicated aminoacyl-tRNA synthetase.[38] Both selenocysteine and pyrrolysine may be present in the same organism.[39] Although the genetic code is normally fixed in an organism the achaeal prokaryote *Acetohalobium arabaticum* can expand its genetic code from 20 to 21 amino acids (by including pyrrolysine) under different conditions of growth.[40]

Despite these differences, all known naturally-occurring codes are very similar to each other, and the coding mechanism is the same for all organisms: three-base codons, tRNA, ribosomes, reading the code in the same direction and translating the code three letters at a time into sequences of amino acids.

11.6.1 Predicting the genetic code

The genetic code used by a genome can be predicted by identifying the genes encoded on that genome, and comparing the codons on the DNA to the amino acids in homologous proteins in other genomes. The evolutionary conservation of protein sequences makes it possible to predict the amino acid translation for each codon as the one that is most often aligned to that codon. The program FACIL[41] allows the automated prediction of the genetic code, searching which amino acids in homologous protein domains are most often aligned to every codon. The resulting amino acid probabilities for each codon are displayed in a genetic code logo, that also shows the support for a stop codon.

11.6.2 Expanded genetic code

Main article: Expanded genetic code
See also: Nucleic acid analogues

Since 2001, 40 non-natural amino acids have been added into protein by creating a unique codon (recoding) and a corresponding transfer-RNA:aminoacyl – tRNA-synthetase pair to encode it with diverse physicochemical and biological properties in order to be used as a tool to exploring protein structure and function or to create novel or enhanced proteins.[42] [43]

H. Murakami and M. Sisido have extended some codons to have four and five bases. Steven A. Benner constructed a functional 65th (*in vivo*) codon.[44]

11.7 Origin

If amino acids were randomly assigned to triplet codons, then there would be 1.5×10^{84} possible genetic codes to choose from.[45]:163 This number is found by calculating how many ways there are to place 21 items (20 amino acids plus one stop) in 64 bins, wherein each item is used at least once. The genetic code used by all known forms of life is nearly universal with few minor variations. One could ask: Has all life on Earth descended from a single ancestor that mutated to make the final optimization in the genetic code? Many hypotheses on the evolutionary origins of the genetic code have been proposed.

Four themes run through the many hypotheses about the evolution of the genetic code:[46]

- **Chemical principles** govern specific RNA interaction with amino acids. Experiments with aptamers showed that some amino acids have a selective chemical affinity for the base triplets that code for them.[47] Recent experiments show that of the 8 amino acids tested, 6 show some RNA triplet-amino acid association.[45]:170[48]

- **Biosynthetic expansion**. The standard modern genetic code grew from a simpler earlier code through a process of "biosynthetic expansion". Here the idea is that primordial life "discovered" new amino acids (for example, as by-products of metabolism) and later incorporated some of these into the machinery of genetic coding.[49] Although much circumstantial evidence has been found to suggest that fewer different amino acids were used in the past than today,[50] precise and detailed hypotheses about which amino acids entered the code in what order have proved far more controversial.[51][52]

- **Natural selection** has led to codon assignments of the genetic code that minimize the effects of mutations.[53] A recent hypothesis[54] suggests that the triplet code was derived from codes that used longer than triplet codons (such as quadruplet codons). Longer than triplet decoding would have higher degree of codon redundancy and would be more error resistant than the triplet decoding. This feature could allow accurate decoding in the absence of highly complex translational machinery such as the ribosome and before cells began making ribosomes.

- **Information channels:** Information-theoretic approaches model the process of translating the genetic code into corresponding amino acids as an error-prone information channel.[55] The inherent noise (that is, the error) in the channel poses the organism with a fundamental question: how can a genetic code be constructed to withstand the impact of noise[56] while accurately and efficiently translating information? These "rate-distortion" models[57] suggest that the genetic code originated as a result of the interplay of the three conflicting evolutionary forces: the needs for diverse amino-acids,[58] for error-tolerance[53] and for minimal cost of resources. The code emerges at a coding transition when the mapping of codons to amino-acids becomes nonrandom. The emergence of the code is governed by the topology defined by the probable errors and is related to the map coloring problem.[59]

Transfer RNA molecules appear to have evolved before modern aminoacyl-tRNA synthetases, so the latter cannot be part of the explanation of its patterns.[60]

Models encompassing aspects of two or more of the above themes have also been explored. For example, models based on signaling games combine elements of game theory, natural selection and information channels. Such models have been used to suggest that the first polypeptides were likely short and had some use other than enzymatic function. Game theoretic models have also suggested that the organization of RNA strings into cells may have been necessary to prevent "deceptive" use of the genetic code, i.e. preventing the ancient equivalent of viruses from overwhelming the RNA world.[61]

The distribution of codon assignments in the genetic code is nonrandom.[62] For example, the genetic code clusters certain amino acid assignments. Amino acids that share the same biosynthetic pathway tend to have the same first base in their codons.[63] Amino acids with similar physical properties tend to have similar codons,[64][65] reducing the problems caused by point mutations and mistranslations.[62] A robust hypothesis for the origin of genetic code should also address or predict the following gross features of the codon table:[66]

1. absence of codons for D-amino acids

2. secondary codon patterns for some amino acids

3. confinement of synonymous positions to third position

4. limitation to 20 amino acids instead of a number closer to 64

5. relation of stop codon patterns to amino acid coding patterns

11.8   See also

11.9   References

[1] Turanov AA, Lobanov AV, Fomenko DE, Morrison HG, Sogin ML, Klobutcher LA, Hatfield DL, Gladyshev VN (Jan 2009). "Genetic code supports targeted insertion of two amino acids by one codon". *Science* **323** (5911): 259–61. doi:10.1126/science. 1164748.PMC3088105. PMID 19131629.

[2] Crick F (1988). "Chapter 8: The genetic code". *What mad pursuit: a personal view of scientific discovery*. New York: Basic Books. pp. 89–101. ISBN 0-465-09138-5.

[3] Nirenberg MW, Matthaei JH (Oct 1961). "The dependence of cell-free protein synthesis in E. coli upon naturally occurring or synthetic polyribonucleotides". *Proceedings of the National Academy of Sciences of the United States of America* **47** (10): 1588–602. Bibcode:1961PNAS...47.1588N. doi:10.1073/pnas.47.10.1588. PMC 223178. PMID 14479932.

[4] Gardner RS, Wahba AJ, Basilio C, Miller RS, Lengyel P, Speyer JF (Dec 1962). "Synthetic polynucleotides and the amino acid code. VII". *Proceedings of the National Academy of Sciences of the United States of America* **48** (12): 2087–94. Bibcode:1962PN AS...48.2087G.doi:10.1073/pnas.48.12.2087. PMC221128. PMID 13946552.

[5] Wahba AJ, Gardner RS, Basilio C, Miller RS, Speyer JF, Lengyel P (Jan 1963). "Synthetic polynucleotides and the amino acid code. VIII". *Proceedings of the National Academy of Sciences of the United States of America* **49** (1): 116–22. Bibcode:1963PN AS...49..116W.doi:10.1073/pnas.49.1.116. PMC300638. PMID 13998282.

[6] "The Nobel Prize in Physiology or Medicine 1959" (Press release). The Royal Swedish Academy of Science. 1959. Retrieved 2010-02-27. The Nobel Prize in Physiology or Medicine 1959 was awarded jointly to Severo Ochoa and Arthur Kornberg 'for their discovery of the mechanisms in the biological synthesis of ribonucleic acid and deoxyribonucleic acid'.

[7] Nirenberg M, Leder P, Bernfield M, Brimacombe R, Trupin J, Rottman F, O'Neal C (May 1965). "RNA codewords and protein synthesis, VII. On the general nature of the RNA code". *Proceedings of the National Academy of Sciences of the United States of America* **53** (5): 1161–8. Bibcode:1965PNAS...53.1161N. doi:10.1073/pnas.53.5.1161. PMC 301388. PMID 5330357.

[8] "The Nobel Prize in Physiology or Medicine 1968" (Press release). The Royal Swedish Academy of Science. 1968. Retrieved 2010-02-27. The Nobel Prize in Physiology or Medicine 1968 was awarded jointly to Robert W. Holley, Har Gobind Khorana and Marshall W. Nirenberg 'for their interpretation of the genetic code and its function in protein synthesis'.

[9] Mulligan PK, King RC, Stansfield WD (2006). *A dictionary of genetics*. Oxford [Oxfordshire]: Oxford University Press. p. 608. ISBN 0-19-530761-5.

[10] Touriol C, Bornes S, Bonnal S, Audigier S, Prats H, Prats AC, Vagner S (2003). "Generation of protein isoform diversity by alternative initiation of translation at non-AUG codons". *Biology of the Cell / Under the Auspices of the European Cell Biology Organization* **95** (3-4): 169–78. doi:10.1016/S0248-4900(03)00033-9. PMID 12867081.

[11] Edgar B (Oct 2004). "The genome of bacteriophage T4: an archeological dig". *Genetics* **168** (2): 575–82. PMC 1448817. PMID 15514035.

[12] Maloy S (2003-11-29). "How nonsense mutations got their names". *Microbial Genetics Course*. San Diego State University. Retrieved 2010-03-10.

[13] References for the image are found in Wikimedia Commons page at: Commons:File:Notable mutations.svg#References.

[14] Griffiths AJ, Miller JH, Suzuki DT, Lewontin RC, et al., eds. (2000). "Spontaneous mutations". *An Introduction to Genetic Analysis* (7th ed.). New York: W. H. Freeman. ISBN 0-7167-3520-2.

[15] Freisinger E, Grollman AP, Miller H, Kisker C (Apr 2004). "Lesion (in)tolerance reveals insights into DNA replication fidelity". *The EMBO Journal* **23** (7): 1494–505. doi:10.1038/sj.emboj.7600158. PMC 391067. PMID 15057282.

[16] (Boillée 2006, p. 39)

[17] Chang JC, Kan YW (Jun 1979). "beta 0 thalassemia, a nonsense mutation in man". *Proceedings of the National Academy of Sciences of the United States of America* **76** (6): 2886–9. Bibcode:1979PNAS...76.2886C. doi:10.1073/pnas.76.6.2886. PMC 383714. PMID 88735.

[18] Boillée S, Vande Velde C, Cleveland DW (Oct 2006). "ALS: a disease of motor neurons and their nonneuronal neighbors". *Neuron* **52** (1): 39–59. doi:10.1016/j.neuron.2006.09.018. PMID 17015226.

[19] Isbrandt D, Hopwood JJ, von Figura K, Peters C (1996). "Two novel frameshift mutations causing premature stop codons in a patient with the severe form of Maroteaux-Lamy syndrome". *Human Mutation* **7** (4): 361–3. doi:10.1002/(SICI)1098-1004(1996)7:4<361::AID-HUMU12>3.0.CO;2-0. PMID 8723688.

[20] Crow JF (1993). "How much do we know about spontaneous human mutation rates?". *Environmental and Molecular Mutagenesis* **21** (2): 122–9. doi:10.1002/em.2850210205. PMID 8444142.

[21] Lewis R (2005). *Human Genetics: Concepts and Applications* (6th ed.). Boston, Mass: McGraw Hill. pp. 227–228. ISBN 0-07-111156-5.

[22] Sawyer SA, Parsch J, Zhang Z, Hartl DL (Apr 2007). "Prevalence of positive selection among nearly neutral amino acid replacements in Drosophila". *Proceedings of the National Academy of Sciences of the United States of America* **104** (16): 6504–10. Bibcode:2007PNAS..104.6504S. doi:10.1073/pnas.0701572104. PMC 1871816. PMID 17409186.

[23] Bridges KR (2002). "Malaria and the Red Cell". *Harvard*.

[24] Drake JW, Holland JJ (Nov 1999). "Mutation rates among RNA viruses". *Proceedings of the National Academy of Sciences of the United States of America* **96** (24): 13910–3. Bibcode:1999PNAS...9613910D. doi:10.1073/pnas.96.24.13910. PMC 24164. PMID 10570172.

[25] Holland J, Spindler K, Horodyski F, Grabau E, Nichol S, VandePol S (Mar 1982). "Rapid evolution of RNA genomes". *Science* **215** (4540): 1577–85. Bibcode:1982Sci...215.1577H. doi:10.1126/science.7041255. PMID 7041255.

[26] de Visser JA, Rozen DE (Apr 2006). "Clonal interference and the periodic selection of new beneficial mutations in Escherichia coli". *Genetics* **172** (4): 2093–100. doi:10.1534/genetics.105.052373. PMC 1456385. PMID 16489229.

[27] Watson JD, Baker TA, Bell SP, Gann A, Levine M, Oosick R (2008). *Molecular Biology of the Gene*. San Francisco: Pearson/Benjamin Cummings. ISBN 0-8053-9592-X.

[28] Yang et al. (1990) in Michel-Beyerle, M. E., ed. Reaction centers of photosynthetic bacteria: Feldafing-II-Meeting 6. Berlin: Springer-Verlag. pp. 209–18. ISBN 3-540-53420-2.

[29] Füllen G, Youvan DC (1994). "Genetic Algorithms and Recursive Ensemble Mutagenesis in Protein Engineering". Complexity International 1.

[30] Nakamoto T (March 2009). "Evolution and the universality of the mechanism of initiation of protein synthesis". *Gene* **432** (1–2): 1–6. doi:10.1016/j.gene.2008.11.001. PMID 19056476.

[31] Crick FH, Orgel LE (1973). "Directed panspermia". *Icarus* **19** (3): 341–6, 344. Bibcode:1973Icar...19..341C. doi:10.1016/0019-1035(73)90110-3. It is a little surprising that organisms with somewhat different codes do not coexist. (Further discussion)

[32] Elzanowski A, Ostell J (2008-04-07). "The Genetic Codes". National Center for Biotechnology Information (NCBI). Retrieved 2010-03-10.

[33] Jukes TH, Osawa S (Dec 1990). "The genetic code in mitochondria and chloroplasts". *Experientia* **46** (11-12): 1117–26. doi:10.1007/BF01936921. PMID 2253709.

[34] Fitzpatrick DA, Logue ME, Stajich JE, Butler G (1 January 2006). "A fungal phylogeny based on 42 complete genomes derived from supertree and combined gene analysis". *BMC Evolutionary Biology* **6**: 99. doi:10.1186/1471-2148-6-99. PMC 1679813. PMID 17121679.

[35] Santos MA, Tuite MF (May 1995). "The CUG codon is decoded in vivo as serine and not leucine in Candida albicans". *Nucleic Acids Research* **23** (9): 1481–6. doi:10.1093/nar/23.9.1481. PMC 306886. PMID 7784200.

[36] Butler G, Rasmussen MD, Lin MF, Santos MA, Sakthikumar S, Munro CA, Rheinbay E, Grabherr M, Forche A, Reedy JL, Agrafioti I, Arnaud MB, Bates S, Brown AJ, Brunke S, Costanzo MC, Fitzpatrick DA, de Groot PW, Harris D, Hoyer LL, Hube B, Klis FM, Kodira C, Lennard N, Logue ME, Martin R, Neiman AM, Nikolaou E, Quail MA, Quinn J, Santos MC, Schmitzberger FF, Sherlock G, Shah P, Silverstein KA, Skrzypek MS, Soll D, Staggs R, Stansfield I, Stumpf MP, Sudbery PE, Srikantha T, Zeng Q, Berman J, Berriman M, Heitman J, Gow NA, Lorenz MC, Birren BW, Kellis M, Cuomo CA (Jun 2009). "Evolution of pathogenicity and sexual reproduction in eight Candida genomes". *Nature* **459** (7247): 657–62. Bibcode:2009Natur.459..657B. doi:10.1038/nature08064. PMC 2834264. PMID 19465905.

[37] Taylor DJ, Ballinger MJ, Bowman SM, Bruenn JA (2013). "Virus-host co-evolution under a modified nuclear genetic code". *PeerJ* **1**: e50. doi:10.7717/peerj.50. PMC 3628385. PMID 23638388.

[38] Krzycki JA (Dec 2005). "The direct genetic encoding of pyrrolysine". *Current Opinion in Microbiology* **8** (6): 706–12. doi:10.1016/j.mib.2005.10.009. PMID 16256420.

[39] Zhang Y, Baranov PV, Atkins JF, Gladyshev VN (May 2005). "Pyrrolysine and selenocysteine use dissimilar decoding strategies". *The Journal of Biological Chemistry* **280** (21): 20740–51. doi:10.1074/jbc.M501458200. PMID 15788401.

[40] Prat L, Heinemann IU, Aerni HR, Rinehart J, O'Donoghue P, Söll D (Dec 2012). "Carbon source-dependent expansion of the genetic code in bacteria". *Proceedings of the National Academy of Sciences of the United States of America* **109** (51): 21070–5. Bibcode:2012PNAS..10921070P. doi:10.1073/pnas.1218613110. PMC 3529041. PMID 23185002.

[41] Dutilh BE, Jurgelenaite R, Szklarczyk R, van Hijum SA, Harhangi HR, Schmid M, de Wild B, Françoijs KJ, Stunnenberg HG, Strous M, Jetten MS, Op den Camp HJ, Huynen MA (Jul 2011). "FACIL: Fast and Accurate Genetic Code Inference and Logo". *Bioinformatics* **27** (14): 1929–33. doi:10.1093/bioinformatics/btr316. PMID 21653513.

[42] Xie J, Schultz PG (Dec 2005). "Adding amino acids to the genetic repertoire". *Current Opinion in Chemical Biology* **9** (6): 548–54. doi:10.1016/j.cbpa.2005.10.011. PMID 16260173.

[43] Wang Q, Parrish AR, Wang L (Mar 2009). "Expanding the genetic code for biological studies". *Chemistry & Biology* **16** (3): 323–36. doi:10.1016/j.chembiol.2009.03.001. PMC 2696486. PMID 19318213.

[44] Simon M (2005). *Emergent computation: emphasizing bioinformatics*. New York: AIP Press/Springer Science+Business Media. pp. 105–106. ISBN 0-387-22046-1.

[45] Yarus M (2010). *Life from an RNA World: The Ancestor Within*. Cambridge: Harvard University Press. p. 163. ISBN 0-674-05075-4.

[46] Knight RD, Freeland SJ, Landweber LF (Jun 1999). "Selection, history and chemistry: the three faces of the genetic code". *Trends in Biochemical Sciences* **24** (6): 241–7. doi:10.1016/S0968-0004(99)01392-4. PMID 10366854.

[47] Knight RD, Landweber LF (Sep 1998). "Rhyme or reason: RNA-arginine interactions and the genetic code". *Chemistry & Biology* **5** (9): R215–20. doi:10.1016/S1074-5521(98)90001-1. PMID 9751648.

[48] Yarus M, Widmann JJ, Knight R (Nov 2009). "RNA-amino acid binding: a stereochemical era for the genetic code". *Journal of Molecular Evolution* **69** (5): 406–29. doi:10.1007/s00239-009-9270-1. PMID 19795157.

[49] Sengupta, S., and P. G. Higgs (2015). Pathways of genetic code evolution in ancient and modern organisms. Journal of Molecular Evolution 80:229-243.

[50] Brooks DJ, Fresco JR, Lesk AM, Singh M (Oct 2002). "Evolution of amino acid frequencies in proteins over deep time: inferred order of introduction of amino acids into the genetic code". *Molecular Biology and Evolution* **19** (10): 1645–55. doi:10.1093/oxfordjournals.molbev.a003988. PMID 12270892.

[51] Amirnovin R (May 1997). "An analysis of the metabolic theory of the origin of the genetic code". *Journal of Molecular Evolution* **44** (5): 473–6. doi:10.1007/PL00006170. PMID 9115171.

[52] Ronneberg TA, Landweber LF, Freeland SJ (Dec 2000). "Testing a biosynthetic theory of the genetic code: fact or artifact?". *Proceedings of the National Academy of Sciences of the United States of America* **97** (25): 13690–5. Bibcode:2000PNAS...97136 doi:10.1073/pnas.250403097. PMC17637. PMID11087835.

[53] Freeland SJ, Wu T, Keulmann N (Oct 2003). "The case for an error minimizing standard genetic code". *Origins of Life and Evolution of the Biosphere* **33** (4-5): 457–77. doi:10.1023/A:1025771327614. PMID 14604186.

[54] Baranov PV, Venin M, Provan G (2009). Gemmell NJ, ed. "Codon size reduction as the origin of the triplet genetic code". *PloS One* **4** (5): e5708. Bibcode:2009PLoSO...4.5708B. doi:10.1371/journal.pone.0005708. PMC 2682656. PMID 19479032.

[55] Tlusty T (Nov 2007). "A model for the emergence of the genetic code as a transition in a noisy information channel". *Journal of Theoretical Biology* **249** (2): 331–42. doi:10.1016/j.jtbi.2007.07.029. PMID 17826800.

[56] Sonneborn TM (1965). Bryson V, Vogel H, eds. *Evolving genes and proteins*. New York: Academic Press. pp. 377–397.

[57] Tlusty T (Feb 2008). "Rate-distortion scenario for the emergence and evolution of noisy molecular codes". *Physical Review Letters* **100** (4): 048101. arXiv:1007.4149. Bibcode:2008PhRvL.100d8101T. doi:10.1103/PhysRevLett.100.048101. PMID 18352335.

[58] Sella G, Ardell DH (Sep 2006). "The coevolution of genes and genetic codes: Crick's frozen accident revisited". *Journal of Molecular Evolution* **63** (3): 297–313. doi:10.1007/s00239-004-0176-7. PMID 16838217.

[59] Tlusty T (Sep 2010). "A colorful origin for the genetic code: information theory, statistical mechanics and the emergence of molecular codes". *Physics of Life Reviews* **7** (3): 362–76. arXiv:1007.3906. Bibcode:2010PhLRv...7..362T. doi:10.1016/j.plrev.2010.06.002.PMID20558115.

[60] Ribas de Pouplana L, Turner RJ, Steer BA, Schimmel P (Sep 1998). "Genetic code origins: tRNAs older than their synthetases?". *Proceedings of the National Academy of Sciences of the United States of America* **95** (19): 11295–300. Bibcode:1998PNAS...9511295D.doi:10.1073/pnas.95.19.11295. PMC21636. PMID 9736730.

[61] Jee J, Sundstrom A, Massey SE, Mishra B (Nov 2013). "What can information-asymmetric games tell us about the context of Crick's 'frozen accident'?". *Journal of the Royal Society, Interface / the Royal Society* **10** (88): 20130614. doi:10.1098/rsif.2013.0 614.PMC3785830. PMID 23985735.

[62] Freeland SJ, Hurst LD (Sep 1998). "The genetic code is one in a million". *Journal of Molecular Evolution* **47** (3): 238–48. doi:10.1007/PL00006381. PMID 9732450.

[63] Taylor FJ, Coates D (1989). "The code within the codons". *Bio Systems* **22** (3): 177–87. doi:10.1016/0303-2647(89)90059-2. PMID 2650752.

[64] Di Giulio M (Oct 1989). "The extension reached by the minimization of the polarity distances during the evolution of the genetic code". *Journal of Molecular Evolution* **29** (4): 288–93. doi:10.1007/BF02103616. PMID 2514270.

[65] Wong JT (Feb 1980). "Role of minimization of chemical distances between amino acids in the evolution of the genetic code". *Proceedings of the National Academy of Sciences of the United States of America* **77** (2): 1083–6. Bibcode:1980PNAS...77.1083W. doi:10.1073/pnas.77.2.1083. PMC 348428. PMID 6928661.

[66] Erives A (Aug 2011). "A model of proto-anti-codon RNA enzymes requiring L-amino acid homochirality". *Journal of Molecular Evolution* **73** (1-2): 10–22. doi:10.1007/s00239-011-9453-4. PMC 3223571. PMID 21779963.

11.10   Further reading

- Griffiths AJ, Miller JH, Suzuki DT, Lewontin RC, Gilbert WM (1999). *An Introduction to genetic analysis* (7th ed.). San Francisco: W.H. Freeman. ISBN 0-7167-3771-X.

- Alberts B, Johnson A, Lewis J, Raff M, Roberts K, Walter P (2002). *Molecular biology of the cell* (4th ed.). New York: Garland Science. ISBN 0-8153-3218-1.

- Lodish HF, Berk A, Zipursky SL, Matsudaira P, Baltimore D, Darnell JE (2000). *Molecular cell biology* (4th ed.). San Francisco: W.H. Freeman. ISBN 0-7167-3706-X.

- Caskey CT, Leder P (Apr 2014). "The RNA code: nature's Rosetta Stone". *Proceedings of the National Academy of Sciences of the United States of America* **111** (16): 5758–9. Bibcode:2014PNAS..111.5758C. doi:10.1073/pnas.1404819111.PMID24756939.

11.11   External links

- The Genetic Codes → Genetic Code Tables

- The Codon Usage Database → Codon frequency tables for many organisms

- History of deciphering the genetic code

- American Scientist: Ode to the code (Origin)

- Alphabet of Life (Origin)

- Symmetries in the genetic code

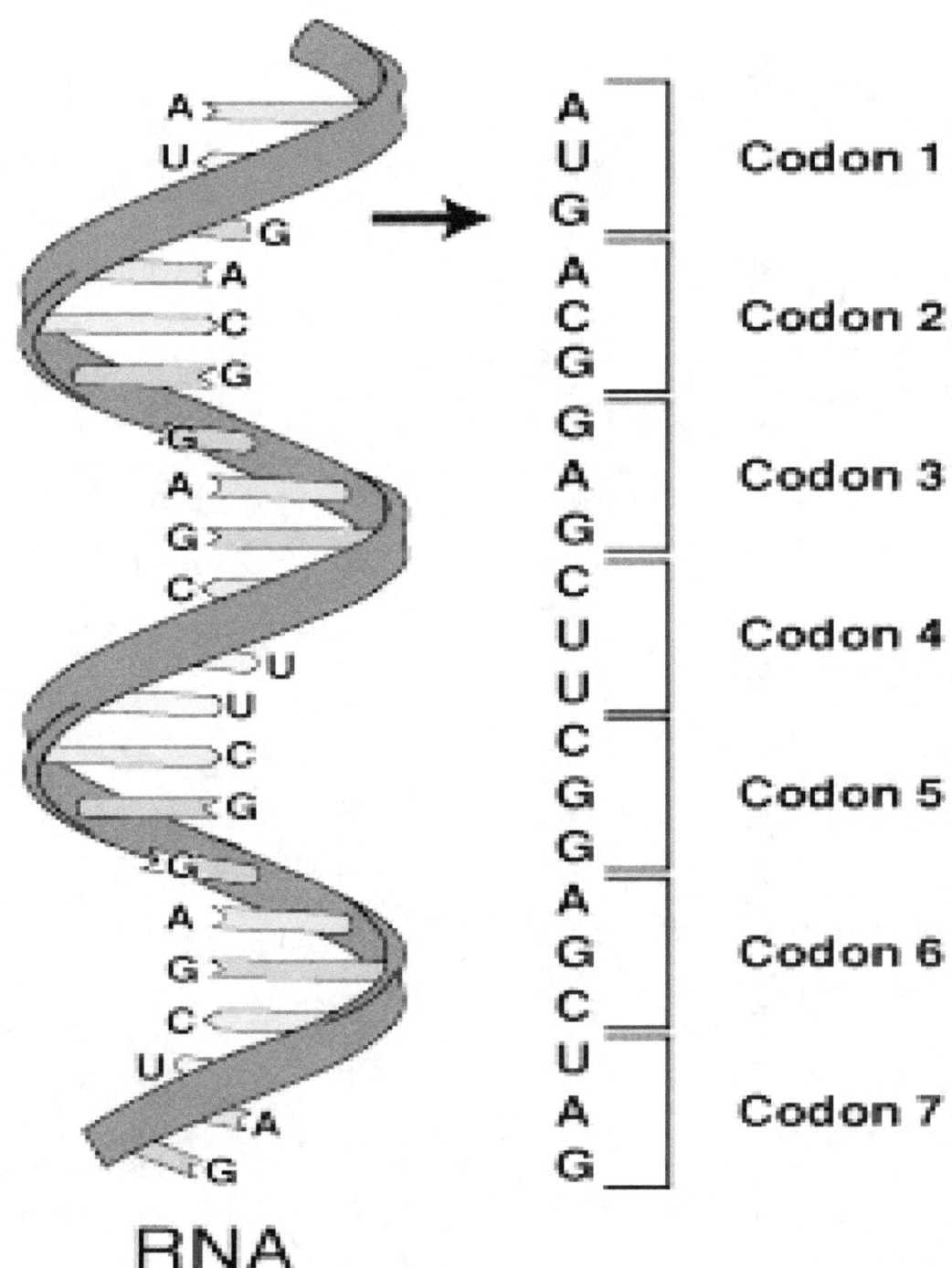

RNA

Ribonucleic acid

A series of codons in part of a messenger RNA (mRNA) molecule. Each codon consists of three nucleotides, usually corresponding to a single amino acid. The nucleotides are abbreviated with the letters A, U, G and C. This is mRNA, which uses U (uracil). DNA uses T (thymine) instead. This mRNA molecule will instruct a ribosome to synthesize a protein according to this code.

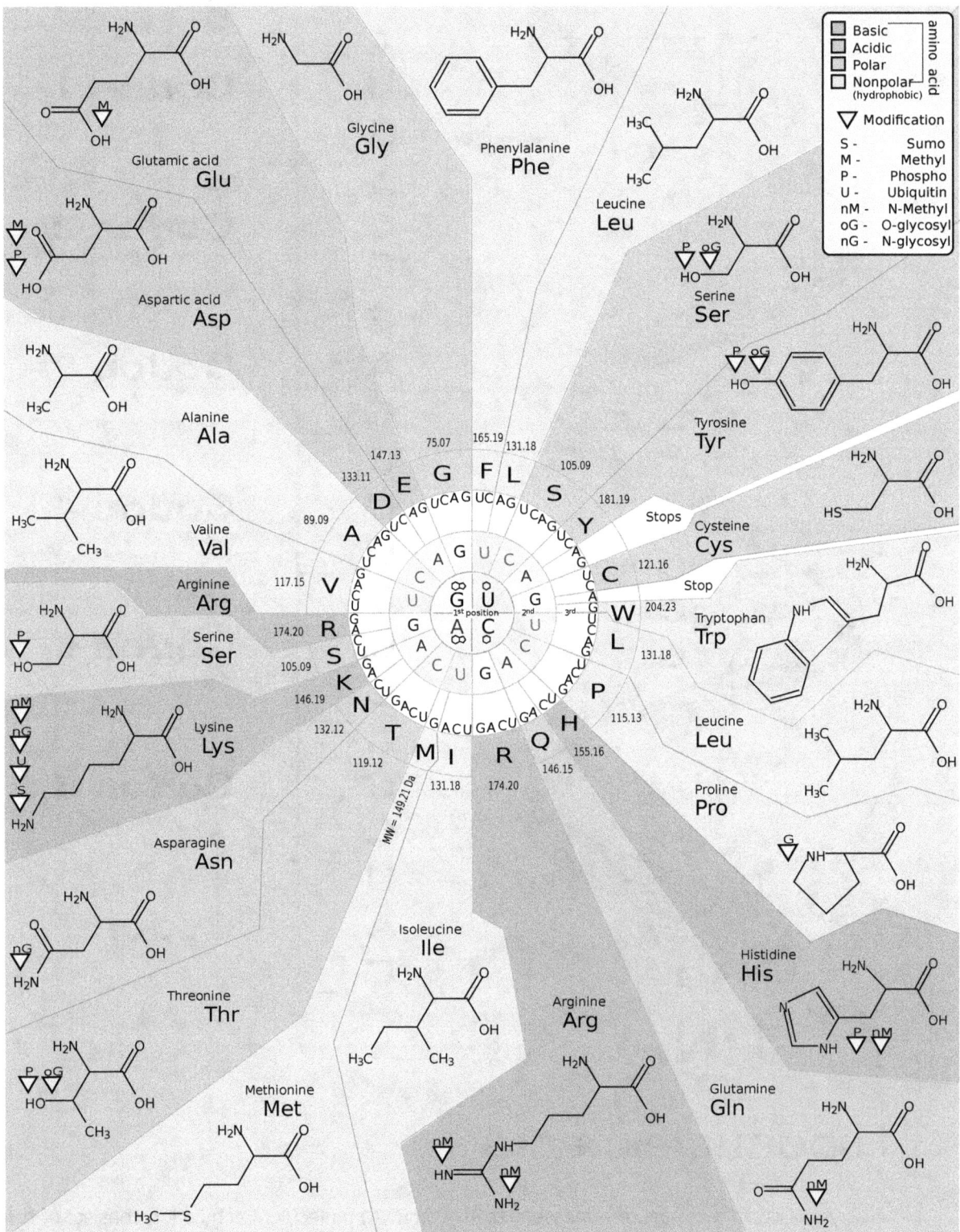

The genetic code

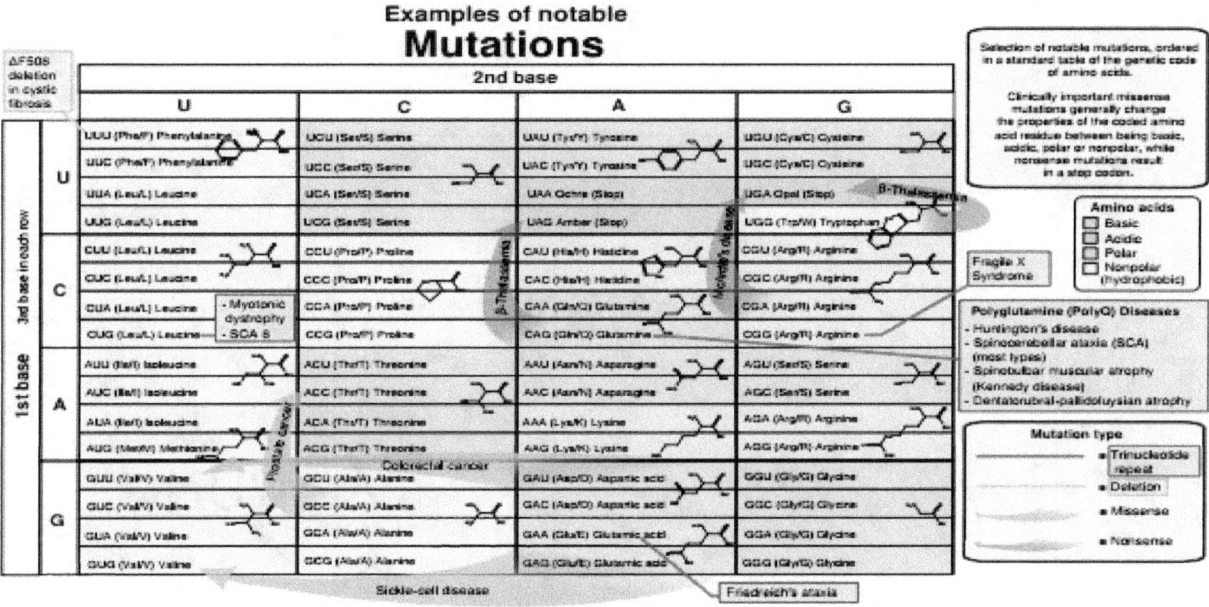

Examples of notable mutations that can occur in humans.

| nonpolar | polar | basic | acidic | (stop codon) |

Standard genetic code

1st base	2nd base								3rd base
	U		**C**		**A**		**G**		
U	UUU	(Phe/F) Phenylalanine	UCU	(Ser/S) Serine	UAU	(Tyr/Y) Tyrosine	UGU	(Cys/C) Cysteine	U
	UUC		UCC		UAC		UGC		C
	UUA		UCA		UAA	Stop (*Ochre*)	UGA	Stop (*Opal*)	A
	UUG		UCG		UAG	Stop (*Amber*)	UGG	(Trp/W) Tryptophan	G
C	CUU	(Leu/L) Leucine	CCU	(Pro/P) Proline	CAU	(His/H) Histidine	CGU	(Arg/R) Arginine	U
	CUC		CCC		CAC		CGC		C
	CUA		CCA		CAA	(Gln/Q) Glutamine	CGA		A
	CUG		CCG		CAG		CGG		G
A	AUU	(Ile/I) Isoleucine	ACU	(Thr/T) Threonine	AAU	(Asn/N) Asparagine	AGU	(Ser/S) Serine	U
	AUC		ACC		AAC		AGC		C
	AUA		ACA		AAA	(Lys/K) Lysine	AGA	(Arg/R) Arginine	A
	AUG[A]	(Met/M) Methionine	ACG		AAG		AGG		G
G	GUU	(Val/V) Valine	GCU	(Ala/A) Alanine	GAU	(Asp/D) Aspartic acid	GGU	(Gly/G) Glycine	U
	GUC		GCC		GAC		GGC		C
	GUA		GCA		GAA	(Glu/E) Glutamic acid	GGA		A
	GUG		GCG		GAG		GGG		G

The codon AUG both codes for methionine and serves as an initiation site: the first AUG in an mRNA's coding region is where translation into protein begins.

Inverse table (compressed using IUPAC notation)

Amino acid	Codons	Compressed	Amino acid	Codons	Compressed
Ala/A	GCU, GCC, GCA, GCG	GCN	**Leu/L**	UUA, UUG, CUU, CUC, CUA, CUG	YUR, CUN
Arg/R	CGU, CGC, CGA, CGG, AGA, AGG	CGN, MGR	**Lys/K**	AAA, AAG	AAR
Asn/N	AAU, AAC	AAY	**Met/M**	AUG	
Asp/D	GAU, GAC	GAY	**Phe/F**	UUU, UUC	UUY
Cys/C	UGU, UGC	UGY	**Pro/P**	CCU, CCC, CCA, CCG	CCN
Gln/Q	CAA, CAG	CAR	**Ser/S**	UCU, UCC, UCA, UCG, AGU, AGC	UCN, AGY
Glu/E	GAA, GAG	GAR	**Thr/T**	ACU, ACC, ACA, ACG	ACN
Gly/G	GGU, GGC, GGA, GGG	GGN	**Trp/W**	UGG	
His/H	CAU, CAC	CAY	**Tyr/Y**	UAU, UAC	UAY
Ile/I	AUU, AUC, AUA	AUH	**Val/V**	GUU, GUC, GUA, GUG	GUN
START	AUG		**STOP**	UAA, UGA, UAG	UAR, URA

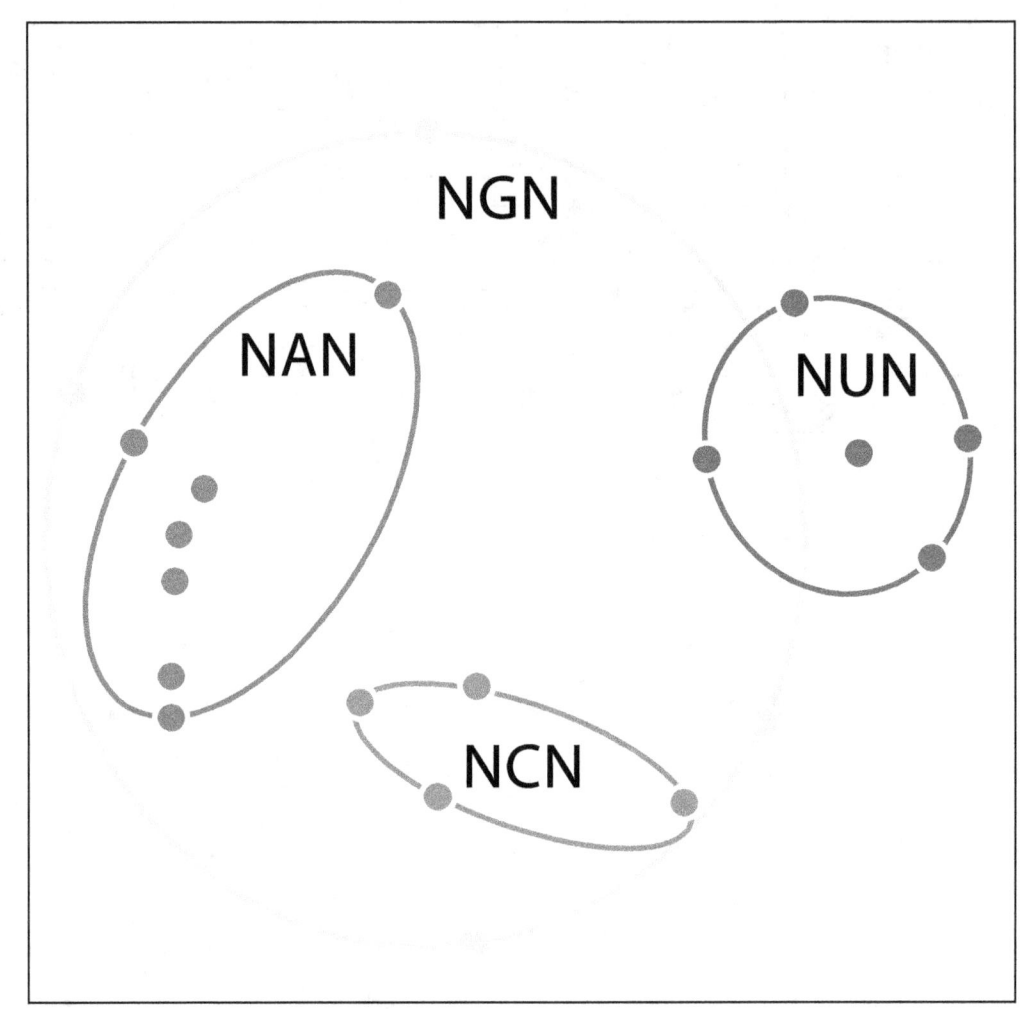

Grouping of codons by amino acid residue molar volume and hydropathy. A more detailed version is available.

Genetic code logo of the Globobulimina pseudospinescens *mitochondrial genome. The logo shows the 64 codons from left to right, predicted alternatives in red (relative to the standard genetic code). Red line: stop codons. The height of each amino acid in the stack shows how often it is aligned to the codon in homologous protein domains. The stack height indicates the support for the prediction.*

Chapter 12

Conformational isomerism

Rotation about single bond of butane to interconvert one conformer to another. Above: Newman projection; below: depiction of spatial orientation.

In chemistry, **conformational isomerism** is a form of stereoisomerism in which the isomers can be interconverted exclusively by rotations about formally single bonds (refer to figure on single bond rotation).[1] Such isomers are generally referred to as **conformational isomers** or **conformers** and, specifically, as **rotamers**.[2] Rotations about single bonds are restricted by a rotational energy barrier which must be overcome to interconvert one conformer to another. Conformational isomerism arises when the rotation about a single bond is relatively unhindered. That is, the energy barrier must be small enough for the interconversion to occur.

Conformational isomers are thus distinct from the other classes of stereoisomers (i. e. configurational isomers) where interconversion necessarily involves breaking and reforming of chemical bonds.[3] For example, L- & D- and R- & S-configurations of organic molecules have different handedness and optical activities, and can only be interconverted by breaking one or more bonds connected to the chiral atom and reforming a similar bond in a different direction or spatial orientation.

The study of the energetics between different rotamers is referred to as **conformational analysis**.[4] It is useful for understanding the stability of different isomers, for example, by taking into account the spatial orientation and through-space interactions of substituents. In addition, conformational analysis can be used to predict and explain product(s) selectivity, mechanisms, and rates of reactions.[5]

12.1   Types of conformational isomerism

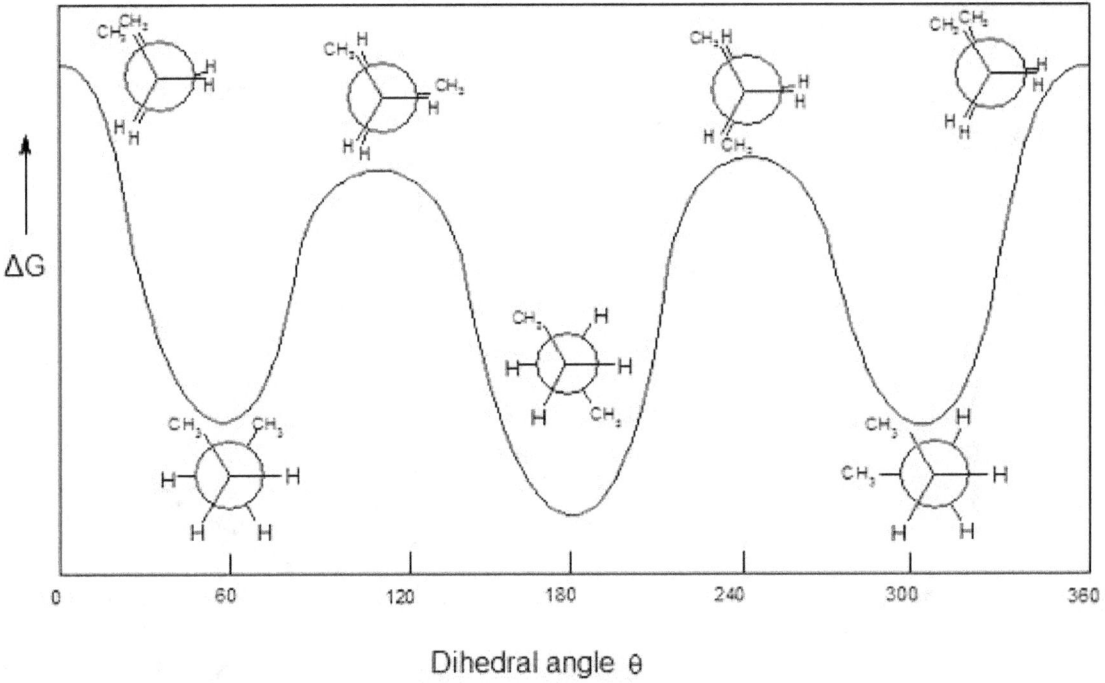

ΔG

Dihedral angle θ

Free energy diagram of butane as a function of dihedral angle.

The types of conformational isomers are related to the spatial orientations of the substituents between two vicinal atoms. These are eclipsed and staggered. The staggered conformation includes the gauche (±60°) and anti (180°) conformations, depending on the spatial orientations of the two substituents.

For example, butane has three rotamers relating to its two methyl (CH_3) groups: two gauche conformers, which have the methyls ±60° apart and are enantiomeric, and an anti conformer, where the four carbon centres are coplanar and the substituents are 180° apart (refer to free energy diagram of butane). The energy difference between gauche and anti is 0.9 kcal/mol associated with the strain energy of the gauche conformer.[4] The anti conformer is, therefore, the most stable (~ 0 kcal/mol). The three eclipsed conformations with dihedral angles of 0°, 120°, and 240° are not considered to be rotamers, but are instead transition states of higher energy.[4] Note that the two eclipsed conformations have different energies: at 0° the two methyl groups are eclipsed, resulting in higher energy (~ 5 kcal/mol) than at 120°, where the methyl groups are eclipsed with hydrogens (~ 3.5 kcal/mol).[6]

While simple molecules can be described by these types of conformations, more complex molecules require the use of the Klyne–Prelog system to describe the different conformers.[4]

More specific examples of conformational isomerism are detailed elsewhere:

1. Ring conformation

   - Cyclohexane conformations with chair and boat conformers.
   - Carbohydrate conformation.

2. Allylic strain – energetics related to rotation about the single bond between sp^2 and sp^3 carbons.

3. Atropisomerism – due to restricted rotation about a bond, a molecule can become chiral.

4. Folding of molecules, where some shapes are stable and functional, but others are not.

12.2 Free energy and equilibria of conformational isomers

12.2.1 Equilibrium of conformers

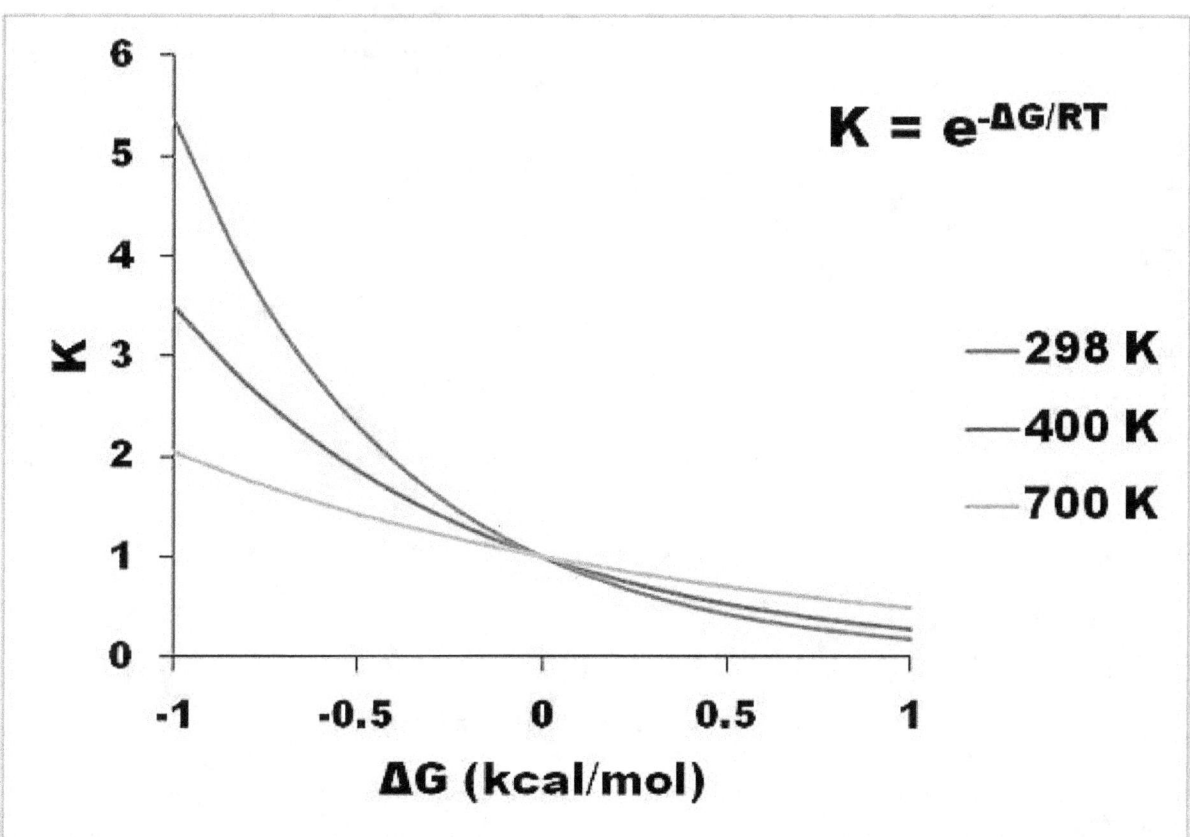

Equilibrium distribution of two conformers at different temperatures given the free energy of their interconversion.

Conformational isomers exist in a dynamic equilibrium, where the relative free energies of isomers determines the population of each isomer and the energy barrier of rotation determines the rate of interconversion between isomers:[7]

$$K = e^{-\Delta G/RT}$$

where K is the equilibrium constant, ΔG is the difference in free energy between the two conformers in kcal/mol, R is the universal gas constant (0.002 kcal/mol K), and T is the system's temperature in Kelvin (K).

Three isotherms are given in the diagram depicting the equilibrium distribution of two conformers at different temperatures. At a free energy difference of 0 kcal/mol, this gives an equilibrium constant of 1, meaning that two conformers exist in a 1:1 ratio. The two have equal free energy; neither is more stable, so neither predominates compared to the other. A negative difference in free energy means that a conformer interconverts to a thermodynamically more stable conformation, thus the equilibrium constant will always be greater than 1. For example, the ΔG of butane from gauche to anti is −0.9 kcal/mol, therefore the equilibrium constant is 4.5, favoring the anti conformation. Conversely, a positive difference in free energy means the conformer already is the more stable one, so the interconversion is an unfavorable equilibrium (K is less than one). Even for highly unfavorable changes (large positive ΔG), the equilibrium constant between two conformers can be increased by increasing temperature, meaning the amount of the less stable conformer present at equilibrium does increase slightly.

12.2.2   Population distribution of conformers

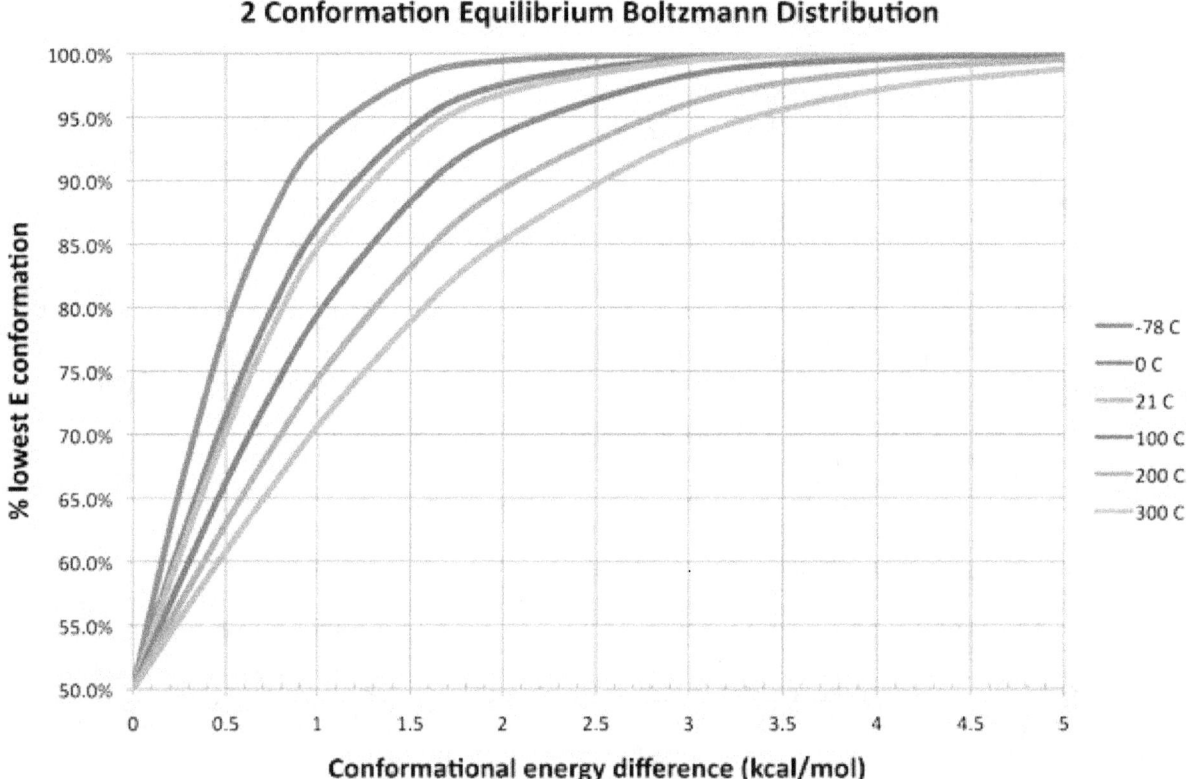

Boltzmann distribution % of lowest energy conformation in a two component equilibrating system at various temperatures (degrees Celsius, color) and energy difference in kcal/mol (x-axis)

The fractional population distribution of different conformers follows a Boltzmann distribution:[8]

$$\frac{N_i}{N_{\text{total}}} = \frac{e^{-E_{\text{rel}}/RT}}{\sum_{k=1}^{N_{\text{total}}} e^{-E_k/RT}}$$

The left hand side is the equilibrium ratio of conformer i to the total. E_{rel} is the relative energy of the i-th conformer from the minimum energy conformer. E_k is the relative energy of the k-th conformer from the minimum energy conformer. R is the molar ideal gas constant equal to 8.31 J/(mol·K) and T is the temperature in kelvins (K). The denominator of the right side is the partition function.

12.2.3 Factors contributing to the free energy of conformers

The effects of electrostatic and steric interactions of the substituents as well as orbital interactions such as hyperconjugation are responsible for the relative stability of conformers and their transition states. The contributions of these factors vary depending on the nature of the substituents and may either contribute positively or negatively to the energy barrier. Computational studies of small molecules such as ethane suggest that electrostatic effects make the greatest contribution to the energy barrier; however, the barrier is traditionally attributed primarily to steric interactions.[9][10]

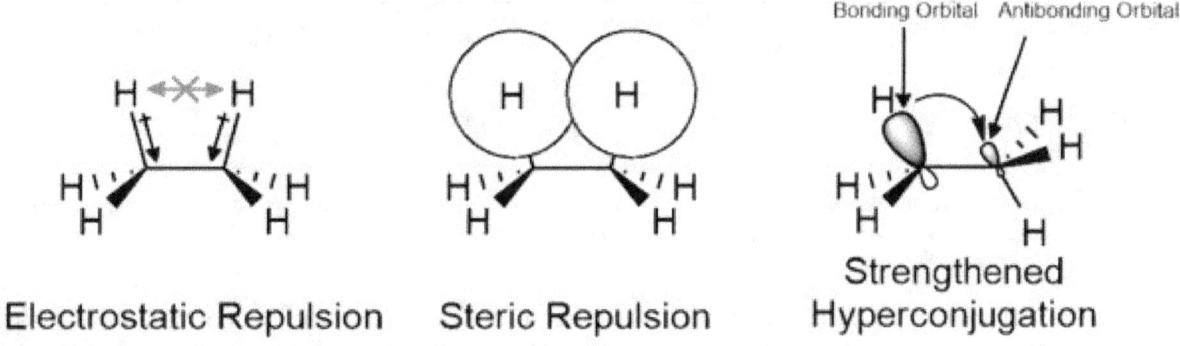

Contributions to rotational energy barrier

In the case of cyclic systems, the steric effect and contribution to the free energy can be approximated by A values, which measure the energy difference when a substituent is in the axial or equatorial position.

12.3 Isolation or observation of the conformational isomers

The short timescale of interconversion precludes the separation of conformational isomers in many cases. Atropisomers are conformational isomers which can be separated due to restricted rotation.[11]

Protein folding also generates stable conformational isomers which can be observed. The Karplus equation relates the dihedral angle of vicinal protons to their J-coupling constants as measured by NMR. The equation aids in the elucidation of protein folding as well as the conformations of other rigid aliphatic molecules.[12]

The equilibrium between conformational isomers may also be observed using spectroscopic techniques:

In cyclohexane derivatives, the two chair conformers interconvert with rates on the order of 10^5 ring-flips/sec, which precludes their separation.[13] The conformer in which the substituent is equatorial crystallizes selectively, and when these crystals are dissolved at very low temperatures, one can directly monitor the approach to equilibrium by NMR spectroscopy.[14]

The *dynamics* of conformational (and other kinds of) isomerism can be monitored by NMR spectroscopy at varying temperatures. The technique applies to barriers of 8–14 kcal/mol, and species exhibiting such dynamics are often called "fluxional".

IR spectroscopy is ordinarily used to measure conformer ratios. For the axial and equatorial conformer of bromocyclohexane, vCB_r differs by almost 50 cm^{-1}.[13]

12.4 Conformation-dependent reactions

Reaction rates are highly dependent on the conformation of the reactants. This theme is especially well elucidated in organic chemistry. One example is provided by the elimination reactions, which involve the simultaneous removal of a proton and a leaving group from vicinal positions under the influence of a base.

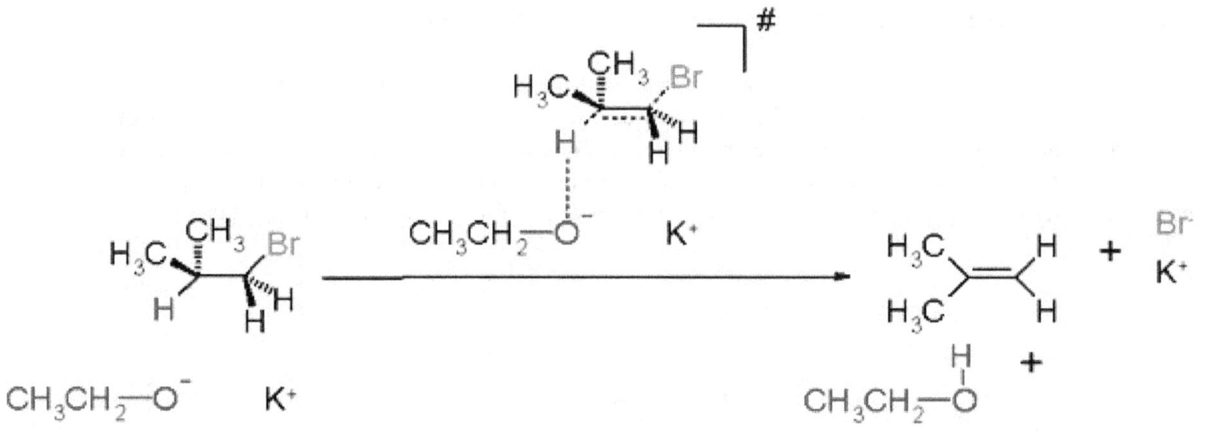

Base-induced bimolecular dehydrohalogenation (an E2 type reaction mechanism). The optimum geometry for the transition state requires the breaking bonds to be antiperiplanar, as they are in the appropriate staggered conformation

The mechanism requires that the departing atoms or groups follow antiparallel trajectories. For open chain substrates this geometric prerequisite is met by at least one of the three staggered conformers. For some cyclic substrates such as cyclohexane, however, an antiparallel arrangement may not be attainable depending on the substituents which might set a conformational lock.[15] Adjacent substituents on a cyclohexane ring can achieve antiperiplanarity only when they occupy trans diaxial positions.

One consequence of this analysis is that *trans*−4-tert-butylcyclohexyl chloride cannot easily eliminate but instead undergoes substitution (see diagram below) because the most stable conformation has the bulky *t*Bu group in the equatorial position, therefore the chloride group is not antiperiplanar with any vicinal hydrogen. The thermodynamically unfavored conformation has the *t*Bu group in the axial position, which exhibits the high energetic 7-atoms interactions (see A value) of 4.7–4.9 kcal/mol.[16] As a result, the *t*Bu group "locks" the ring in the conformation where it is in the equatorial position and substitution reaction is observed. On the other hand, *cis*−4-tert-butylcyclohexyl chloride undergoes elimination because antiperiplanarity of Cl and H can be achieved when the *t*Bu group is in the favorable equatorial position.

12.5   See also

- Isomer

- Steric effects

- Molecular configuration

- Molecular modelling

- Macrocyclic stereocontrol

- Klyne–Prelog system

- Anomeric effect

12.6   References

[1] IUPAC definition of a conformer.

[2] IUPAC, *Compendium of Chemical Terminology*, 2nd ed. (the "Gold Book") (1997). Online corrected version: (1996) "Rotamer".

[3] Hunt, Ian. "Stereochemistry". *University of Calgary*. Retrieved 28 October 2013.

[4] Anslyn, Eric; Dennis Dougherty (2006). *Modern Physical Organic Chemistry*. University Science. p. 95. ISBN 978-1891389313.

[5] Barton, Derek. "The Principles of Conformational Analysis.". *Nobel Media AB 2013*. Elsevier Publishing Co. Retrieved 10 November 2013.

[6] Bauld, Nathan. "Butane Conformational Analysis". *University of Texas*. Retrieved 28 October 2013.

[7] Bruzik, Karol. "Chapter 6: Conformation". *University of Illinois at Chicago*. Retrieved 10 November 2013.

[8] Rzepa, Henry. "Conformational Analysis". *Imperial College London*. Retrieved 11 November 2013.

[9] Liu, Shubin (7 February 2013). "Origin and Nature of Bond Rotation Barriers: A Unified View". *The Journal of Physical Chemistry A* **117** (5): 962–965. doi:10.1021/jp312521z.

[10] Carey, Francis A. (2011). *Organic chemistry* (8th ed.). New York: McGraw-Hill. p. 105. ISBN 0-07-340261-3.

[11] McNaught (1997). *IUPAC Compendium of Chemical Terminology*. Oxford: Blackwell Scientific Publications. ISBN 0967855098.

[12] Dalton, Louisa. "Karplus Equation". *Chemical and Engineering News*. American Chemical Society. Retrieved 2013-10-27.

[13] Eliel, E. L.; Wilen, S. H.; Mander, L. N. "Stereochemistry Of Organic Compounds", J. Wiley and Sons, 1994. ISBN 0-471-01670-5.

[14] Dunbrack, R. (2002). "Rotamer Libraries in the 21st Century". *Current Opinion in Structural Biology* **12** (4): 431–440. doi:10.1016/S0959-440X(02)00344-5. PMID 12163064.

[15] "Cycloalkanes". Imperial College London. Retrieved 28 October 2013.

[16] Dougherty, Eric V. Anslyn ; Dennis A. (2006). *Modern physical organic chemistry* (Dodr. ed.). Sausalito, Calif.: University Science Books. p. 104. ISBN 978-1-891389-31-3.

Chapter 13

A-DNA

A-DNA is one of the possible double helical structures which DNA can adopt. A-DNA is thought to be one of three biologically active double helical structures along with B-DNA and Z-DNA. It is a right-handed double helix fairly similar to the more common B-DNA form, but with a shorter, more compact helical structure whose base pairs are not perpendicular to the helix-axis as in B-DNA. It was discovered by Rosalind Franklin, who also named the A and B forms. She showed that DNA is driven into the A form when under dehydrating conditions. Such conditions are commonly used in to form crystals, and many DNA crystal structures are in the A form. The same helical conformation occurs in double-stranded RNAs, and in DNA-RNA hybrid double helices.

13.1   Structure

A-DNA is fairly similar to B-DNA given that it is a right-handed double helix with major and minor grooves. However, as shown in the comparison table below, there is a slight increase in the number of base pairs (bp) per turn (resulting in a smaller twist angle), and smaller rise per per base pair (making A-DNA 20-25% shorter than B-DNA). The major groove of A-DNA is deep and narrow, while the minor groove is wide and shallow.

13.2   Comparison geometries of the most common DNA forms

13.3   Biological Functions

Dehydration of DNA drives it into the A form, and this apparently protects DNA under conditions such as the extreme desiccation of bacteria.[1] Protein binding can also strip solvent off of DNA and convert it to the A form, as revealed by the structure of a rod-shaped virus.[2]

It has been proposed that the motors that package double-stranded DNA in bacteriophages exploit the fact that A-DNA is shorter than B-DNA, and that conformational changes in the DNA itself are the source of the large forces generated by these motors.[3] In this model, ATP hydrolysis is used to drive protein conformational changes that alternatively dehydrate and rehydrate the DNA, and the DNA shortening/lengthening cycle is coupled to a protein-DNA grip/release cycle to generate the forward motion that moves DNA into the capsid.

13.4   See also

- Mechanical properties of DNA
- DNA

- B-DNA

- Z-DNA

- C-DNA

13.5 References

[1] Whelan DR, et al. (2014). "Detection of an en masse and reversible B- to A-DNA conformational transition in prokaryotes in response to desiccation". *J R Soc Interface* **11**: 20140454. doi:10.1098/rsif.2014.0454. PMID 24898023.

[2] Di Maio F, Egelman EH, et al. (2015). "A virus that infects a hyperthermophile encapsidates A-form DNA". *Science* **348**: 914–917. doi:10.1126/science.aaa4181. PMID 25999507.

[3] Harvey, SC (2015). "The scrunchworm hypothesis: Transitions between A-DNA and B-DNA provide the driving force for genome packaging in double-stranded DNA bacteriophages". *Journal of Structural Biology* **189**: 1–8. doi:10.1016/j.jsb.2014.11.012. PMID 25486612.

13.6 External links

- Cornell Comparison of DNA structures

- Nucleic Acid Nomenclature

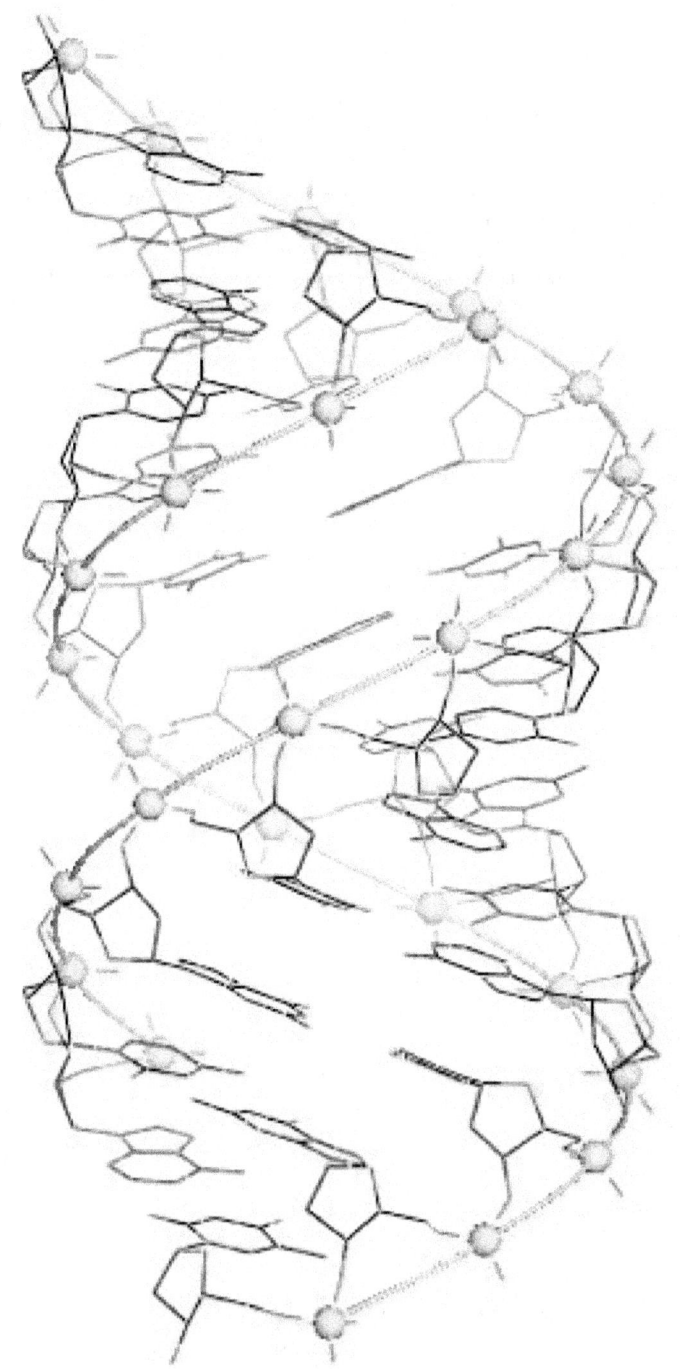

The A-DNA structure.

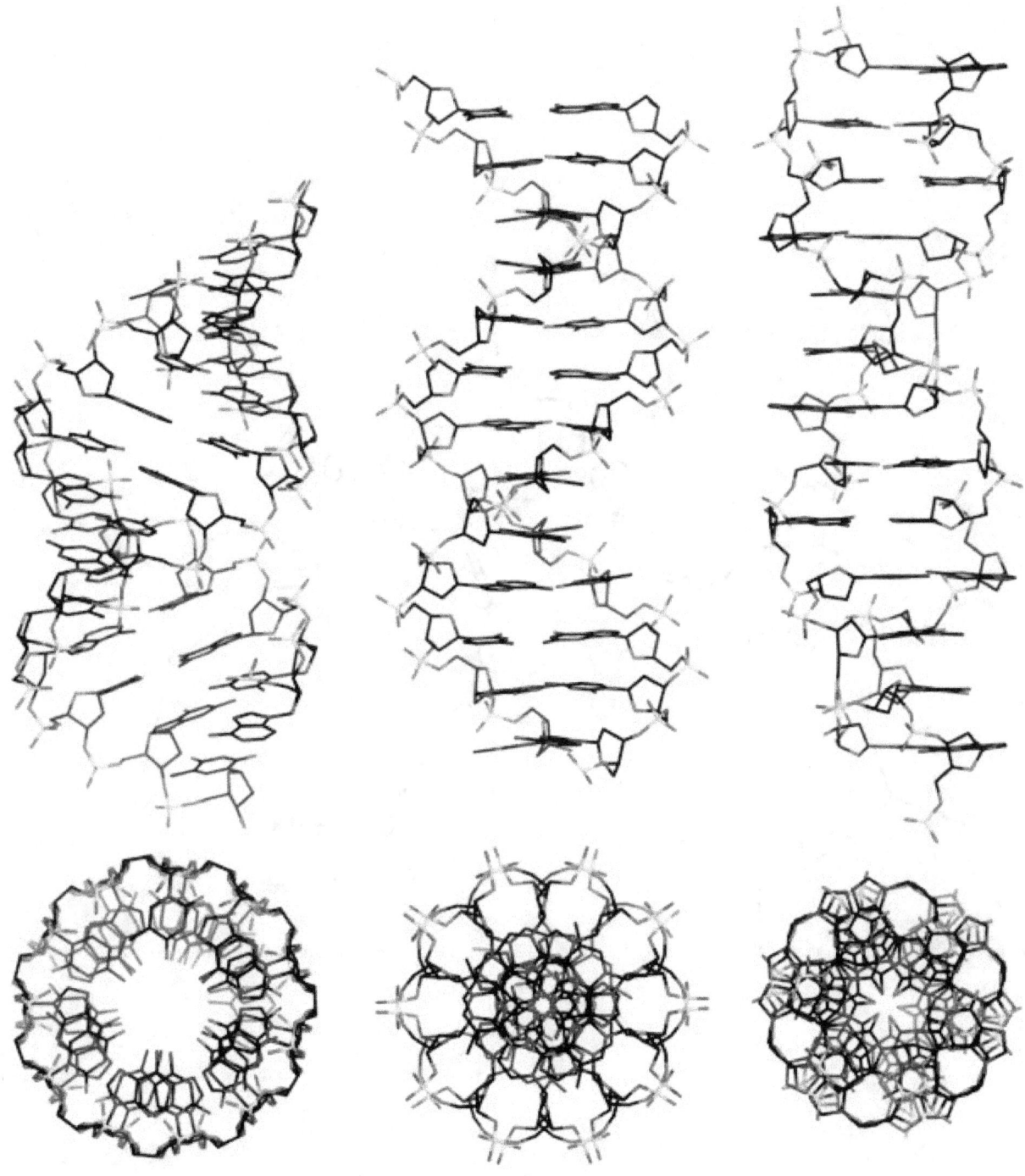

Side and top view of A-, B-, and Z-DNA conformations.

Yellow dots represent the location of the helical axis of A-, B-, and Z-DNA with respect to a Guanine-Cytosine base pair.

Chapter 14

Z-DNA

Z-DNA is one of the many possible double helical structures of DNA. It is a left-handed double helical structure in which the double helix winds to the left in a zig-zag pattern (instead of to the right, like the more common B-DNA form). Z-DNA is thought to be one of three biologically active double helical structures along with A- and B-DNA.

14.1 History

Left handed DNA was first discovered by Robert Wells and colleagues, during their studies of a repeating polymer of inosine-cytosine.[1] They observed a "reverse" circular dichroism spectrum for such DNAs, and interpreted this (correctly) to mean that the strands wrapped around one another in a left handed fashion. The relationship between Z-DNA and the more familiar B-DNA was indicated by the work of Pohl and Jovin,[2] who showed that the ultraviolet circular dichroism of poly(dG-dC) was nearly inverted in 4 M sodium chloride solution. The suspicion that this was the result of a conversion from B-DNA to Z-DNA was confirmed by examining the raman spectra of these solutions and the Z-DNA crystals.[3] Subsequently, a crystal structure of "Z-DNA" was published which turned out to be the first single-crystal X-ray structure of a DNA fragment (a self-complementary DNA hexamer $d(CG)_3$). It was resolved as a left-handed double helix with two anti-parallel chains that were held together by Watson-Crick base pairs (see: x-ray crystallography). It was solved by Andrew Wang, Alexander Rich, and co-workers in 1979 at MIT.[4] The crystallisation of a B- to Z-DNA junction in 2005[5] provided a better understanding of the potential role Z-DNA plays in cells. Whenever a segment of Z-DNA forms, there must be B-Z junctions at its two ends, interfacing it to the B-form of DNA found in the rest of the genome.

In 2007, the RNA version of Z-DNA, Z-RNA, was described as a transformed version of an A-RNA double helix into a left-handed helix.[6] The transition from A-RNA to Z-RNA, however, was already described in 1984.[7]

14.2 Structure

Z-DNA is quite different from the right-handed forms. In fact, Z-DNA is often compared against B-DNA in order to illustrate the major differences. The Z-DNA helix is left-handed and has a structure that repeats every 2 base pairs. The major and minor grooves, unlike A- and B-DNA, show little difference in width. Formation of this structure is generally unfavourable, although certain conditions can promote it; such as alternating purine-pyrimidine sequence (especially poly(dGC)$_2$), negative DNA supercoiling or high salt and some cations (all at physiological temperature, 37 °C, and pH 7.3-7.4). Z-DNA can form a junction with B-DNA (called a "B-to-Z junction box") in a structure which involves the extrusion of a base pair.[8] The Z-DNA conformation has been difficult to study because it does not exist as a stable feature of the double helix. Instead, it is a transient structure that is occasionally induced by biological activity and then quickly disappears.[9]

14.2.1 Predicting Z-DNA structure

It is possible to predict the likelihood of a DNA sequence forming a Z-DNA structure. An algorithm for predicting the propensity of DNA to flip from the B-form to the Z-form, *ZHunt*, was written by Dr. P. Shing Ho in 1984 (at MIT).[10] This algorithm was later developed by Tracy Camp, P. Christoph Champ, Sandor Maurice, and Jeffrey M. Vargason for genome-wide mapping of Z-DNA (with P. Shing Ho as the principal investigator).[11]

Z-Hunt is available at Z-Hunt online.

14.3 Biological significance

While no definitive biological significance of Z-DNA has been found, it is commonly believed to provide torsional strain relief (supercoiling) while DNA transcription occurs.[5][12] The potential to form a Z-DNA structure also correlates with regions of active transcription. A comparison of regions with a high sequence-dependent, predicted propensity to form Z-DNA in human chromosome 22 with a selected set of known gene transcription sites suggests there is a correlation.[11]

Toxic effect of ethidium bromide on trypanosomas is caused by shift of their kinetoplastid DNA to Z-form. The shift is caused by intercalation of EtBr and subsequent loosening of DNA structure that leads to unwinding of DNA, shift to Z-form and inhibition of DNA replication.[13]

14.3.1 Z-DNA formed after transcription initiation

The first domain to bind Z-DNA with high affinity was discovered in ADAR1 using an approach developed by Alan Herbert.[14][15] Crystallographic and NMR studies confirmed the biochemical findings that this domain bound Z-DNA in a non-sequence-specific manner.[16][17][18] Related domains were identified in a number of other proteins through sequence homology.[15] The identification of the Z-alpha domain provided a tool for other crystallographic stuides that lead to the characterization of Z-RNA and the B-Z junction. Biological studies suggested that the Z-DNA binding domain of ADAR1 may localize this enzyme that modifies the sequence of the newly formed RNA to sites of active transcription.[19][20]

In 2003, Alex Rich noticed that a poxvirus virulence factor, called E3L that has a Z-alpha related domain, mimicked a mammalian protein that binds Z-DNA.[21][22] In 2005, Rich and his colleagues pinned down what E3L does for the poxvirus. When expressed in human cells, E3L increases by five- to 10-fold the production of several genes that block a cell's ability to self-destruct in response to infection.

Rich speculates that the Z-DNA is necessary for transcription and that E3L stabilizes the Z-DNA, thus prolonging expression of the anti-apoptotic genes. He suggests that a small molecule that interferes with the E3L binding to Z-DNA could thwart the activation of these genes and help protect people from pox infections.

14.4 Comparison geometries of some DNA forms

14.5 See also

- ADAR1

- A-DNA

- B-DNA

- DNA

- DNA supercoil

- E3L

- Mechanical properties of DNA

- *Proteopedia Z-DNA*

- Z-DNA binding protein 1 (ZBP1)

- Zuotin

14.6 References

[1] Mitsui; et al. (1970). "Physical and enzymatic studies on poly d(I-C)-poly d(I-C), an unusual double-helical DNA". *Nature (London)* **228** (5277): 1166–1169. doi:10.1038/2281166a0. PMID 4321098.

[2] Pohl FM, Jovin TM (1972). "Salt-induced co-operative conformational change of a synthetic DNA: equilibrium and kinetic studies with poly(dG-dC)". *J. Mol. Biol.* **67** (3): 375–396. doi:10.1016/0022-2836(72)90457-3. PMID 5045303.

[3] Thamann TJ, Lord RC, Wang AHJ, Rich A (1981). "High salt form of poly(dG-dC)•poly(dG-dC) is left handed Z-DNA: raman spectra of crystals and solutions". *Nucl. Acids Res.* **9** (20): 5443–5457. doi:10.1093/nar/9.20.5443. PMC 327531. PMID 7301594.

[4] Wang AHJ, Quigley GJ, Kolpak FJ, Crawford JL, van Boom JH, Van der Marel G, Rich A (1979). "Molecular structure of a left-handed double helical DNA fragment at atomic resolution". *Nature (London)* **282** (5740): 680–686. Bibcode:1979Natur.282..680W.doi:10.1038/282680a0. PMID514347.

[5] Ha SC, Lowenhaupt K, Rich A, Kim YG, Kim KK (2005). "Crystal structure of a junction between B-DNA and Z-DNA reveals two extruded bases". *Nature* **437** (7062): 1183–1186. Bibcode:2005Natur.437.1183H. doi:10.1038/nature04088. PMID 16237447.

[6] Placido D, Brown BA 2nd, Lowenhaupt K, Rich A, Athanasiadis A (2007). "A left-handed RNA double helix bound by the Zalpha domain of the RNA-editing enzyme ADAR1". *Structure* **15** (4): 395–404. doi:10.1016/j.str.2007.03.001. PMC 2082211. PMID 17437712.

[7] Hall K, Cruz P, Tinoco I Jr, Jovin TM, van de Sande JH (October 1984). "'Z-RNA'--a left-handed RNA double helix". *Nature* **311** (5986): 584–586. Bibcode:1984Natur.311..584H. doi:10.1038/311584a0. PMID 6482970.

[8] de Rosa M, de Sanctis D, Rosario AL, Archer M, Rich A, Athanasiadis A, Carrondo MA (2010-05-18). "Crystal structure of a junction between two Z-DNA helices". *Proc Natl Acad Sci USA* **107** (20): 9088–9092. Bibcode:2010PNAS..107.9088D. doi:10.1073/pnas.1003182107. PMC 2889044. PMID 20439751.

[9] Zhang H, Yu H, Ren J, Qu X (2006). "Reversible B/Z-DNA transition under the low salt condition and non-B-form poly-dApolydT selectivity by a cubane-like europium-L-aspartic acid complex". *Biophysical Journal* **90** (9): 3203–3207. Bibcode:2006BpJ....90.3203Z.doi:10.1529/biophysj.105.078402. PMC1432110. PMID 16473901.

[10] Ho PS, Ellison MJ, Quigley GJ, Rich A (1986). "A computer aided thermodynamic approach for predicting the formation of Z-DNA in naturally occurring sequences". *EMBO Journal* **5** (10): 2737–2744. PMC 1167176. PMID 3780676.

[11] Champ PC, Maurice S, Vargason JM, Camp T, Ho PS (2004). "Distributions of Z-DNA and nuclear factor I in human chromosome 22: a model for coupled transcriptional regulation". *Nucleic Acids Res.* **32** (22): 6501–6510. doi:10.1093/nar/gkh988. PMC 545456. PMID 15598822.

[12] Rich A, Zhang S (2003). "Timeline: Z-DNA: the long road to biological function". *Nature Reviews Genetics* **4** (7): 566–572. doi:10.1038/nrg1115. PMID 12838348.

[13] Roy Chowdhury, Arnab; Bakshi, Rahul; Wang, Jianyang; Yildirir, Gokben; Liu, Beiyu; Pappas-Brown, Valeria; Tolun, Gökhan; Griffith, Jack D.; Shapiro, Theresa A.; Jensen, Robert E.; Englund, Paul T.; Ullu, Elisabetta (16 December 2010). "The Killing of African Trypanosomes by Ethidium Bromide". *PLoS Pathogens* **6** (12): e1001226. doi:10.1371/journal.ppat.1001226.

[14] Herbert A, Rich A (1993). "A method to identify and characterize Z-DNA binding proteins using a linear oligodeoxynucleotide". *Nucleic Acids Res* **21** (11): 2669–72. doi:10.1093/nar/21.11.2669. PMC 309597. PMID 8332463.

[15] Herbert A, Alfken J, Kim YG, Mian IS, Nishikura K, Rich A (1997). "A Z-DNA binding domain present in the human editing enzyme, double-stranded RNA adenosine deaminase.". *Proc Natl Acad Sci USA* **94** (16): 8421–6. Bibcode:1997PNAS...94.8421H.doi:10.1073/pnas.94.16.8421. PMC22942. PMID9237992.

[16] Herbert A, Schade M, Lowenhaupt K, Alfken J, Schwartz T, Shlyakhtenko LS, Lyubchenko YL, Rich A (1998). "The Zalpha domain from human ADAR1 binds to the Z-DNA conformer of many different sequences". *Nucleic Acids Res* **26** (15): 2669–72. doi:10.1093/nar/26.15.3486. PMC 147729. PMID 9671809.

[17] Schwartz T, Rould MA, Lowenhaupt K, Herbert A, Rich A (1999). "Crystal structure of the Zalpha domain of the human editing enzyme ADAR1 bound to left-handed Z-DNA". *Science* **284** (5421): 1841–5. doi:10.1126/science.284.5421.1841. PMID 10364558.

[18] Schade M, Turner CJ, Kühne R, Schmieder P, Lowenhaupt K, Herbert A, Rich A, Oschkinat H (1999). "The solution structure of the Zalpha domain of the human RNA editing enzyme ADAR1 reveals a prepositioned binding surface for Z-DNA". *Proc Natl Acad Sci USA* **96** (22): 2465–70. Bibcode:1999PNAS...9612465S. doi:10.1073/pnas.96.22.12465. PMC 22950. PMID 10535945.

[19] Herbert A, Rich A (2001). "The role of binding domains for dsRNA and Z-DNA in the in vivo editing of minimal substrates by ADAR1". *Proc Natl Acad Sci USA* **98** (21): 12132–7. Bibcode:2001PNAS...9812132H. doi:10.1073/pnas.211419898. PMC 59780. PMID 11593027.

[20] Halber D (1999-09-11). "Scientists observe biological activities of 'left-handed' DNA". MIT News Office. Retrieved 2008-09-29.

[21] Kim YG, Muralinath M, Brandt T, Pearcy M, Hauns K, Lowenhaupt K, Jacobs BL, Rich A (2003). "A role for Z-DNA binding in vaccinia virus pathogenesis". *Proc Natl Acad Sci USA* **100** (12): 6974–6979. Bibcode:2003PNAS..100.6974K. doi:10.1073/pnas.0431131100. PMC 165815. PMID 12777633.

[22] Kim YG, Lowenhaupt K, Oh DB, Kim KK, Rich A (2004). "Evidence that vaccinia virulence factor E3L binds to Z-DNA in vivo: Implications for development of a therapy for poxvirus infection". *Proc Natl Acad Sci USA* **101** (6): 1514–1518. Bibcode:2004PNAS..101.1514K. doi:10.1073/pnas.0308260100. PMC 341766. PMID 14757814.

[23] Sinden, Richard R (1994-01-15). *DNA structure and function* (1st ed.). Academic Press. p. 398. ISBN 0-126-45750-6.

[24] Rich A, Norheim A, Wang AHJ (1984). "The chemistry and biology of left-handed Z-DNA". *Annual Review of Biochemistry* **53** (1): 791–846. doi:10.1146/annurev.bi.53.070184.004043. PMID 6383204.

[25] Ho PS (1994-09-27). "The non-B-DNA structure of d(CA/TG)n does not differ from that of Z-DNA". *Proc Natl Acad Sci USA* **91** (20): 9549–9553. Bibcode:1994PNAS...91.9549H. doi:10.1073/pnas.91.20.9549. PMC 44850. PMID 7937803.

14.7   External links

- ZHunt Online Server

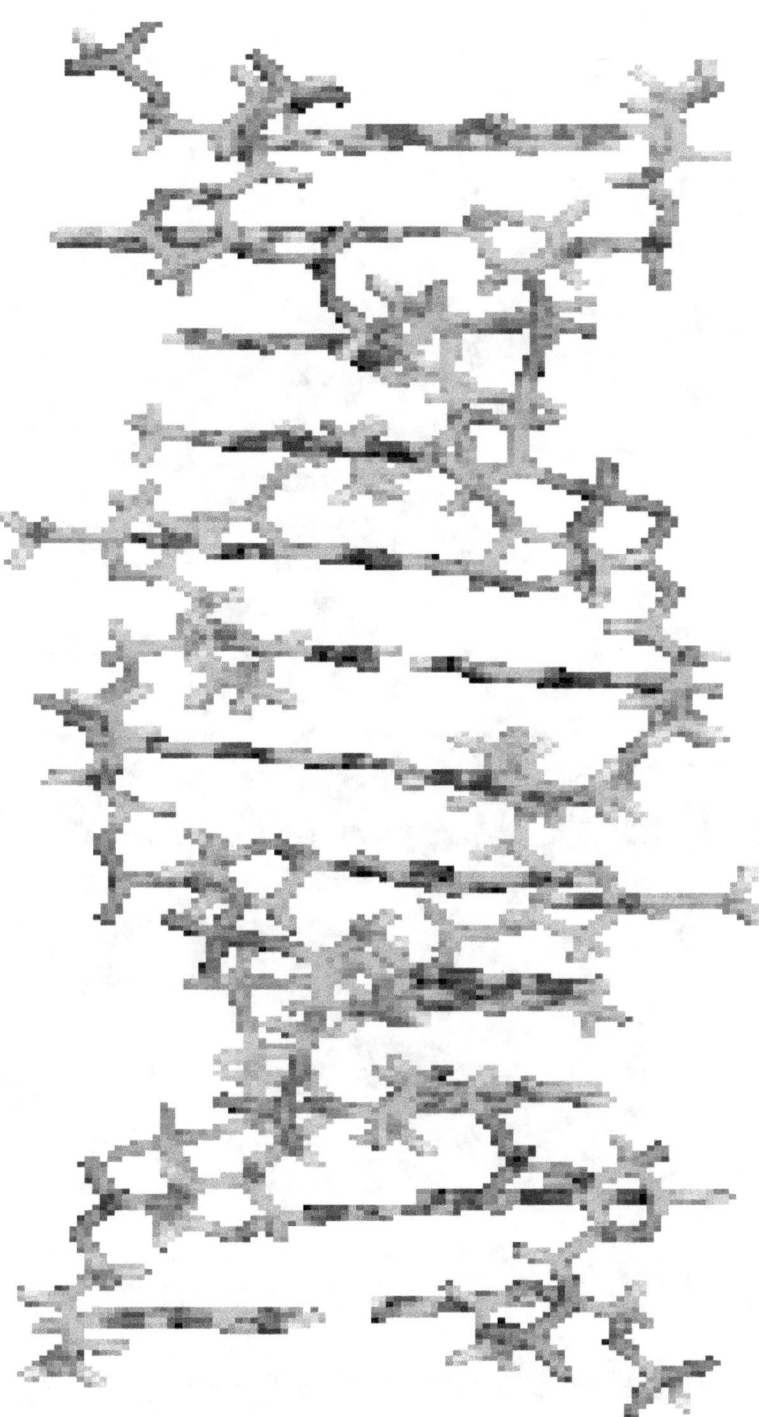

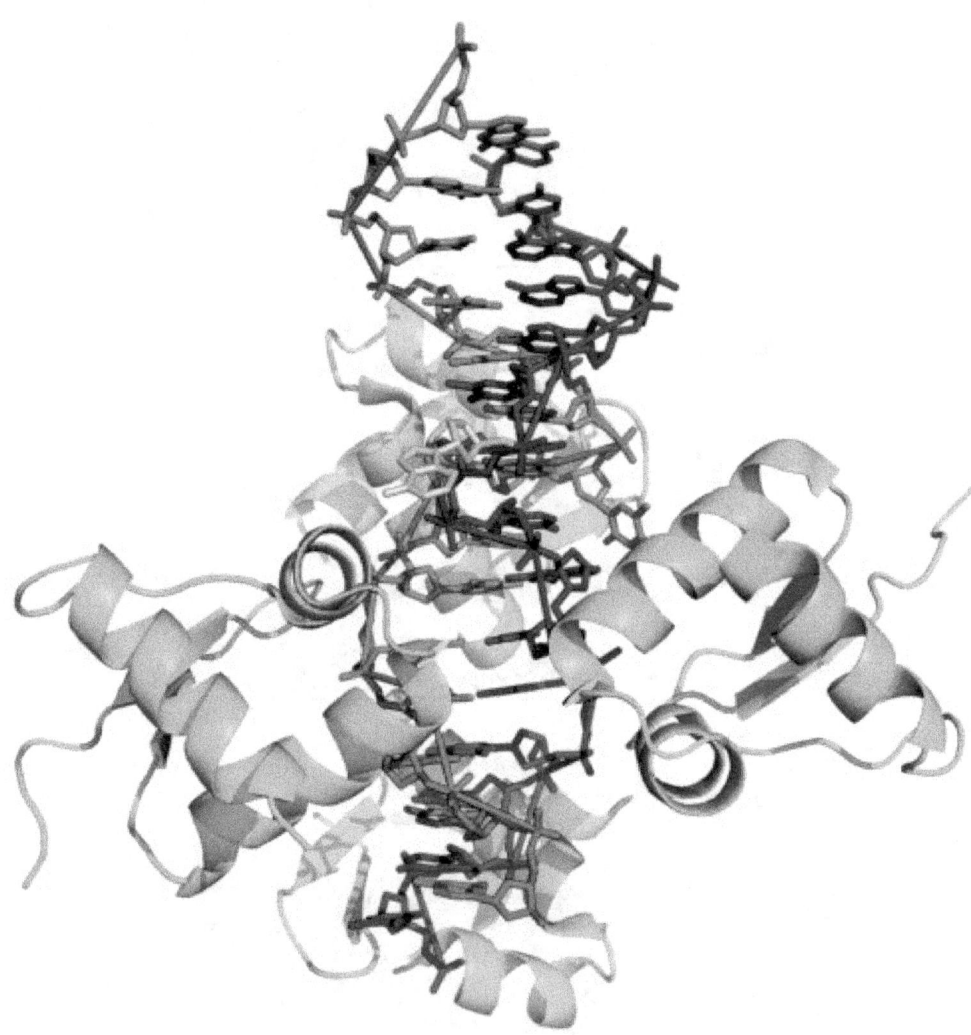

B-/Z-DNA junction bound to a Z-DNA binding domain. Note the two highlighted extruded bases. From PDB: 2ACJ.

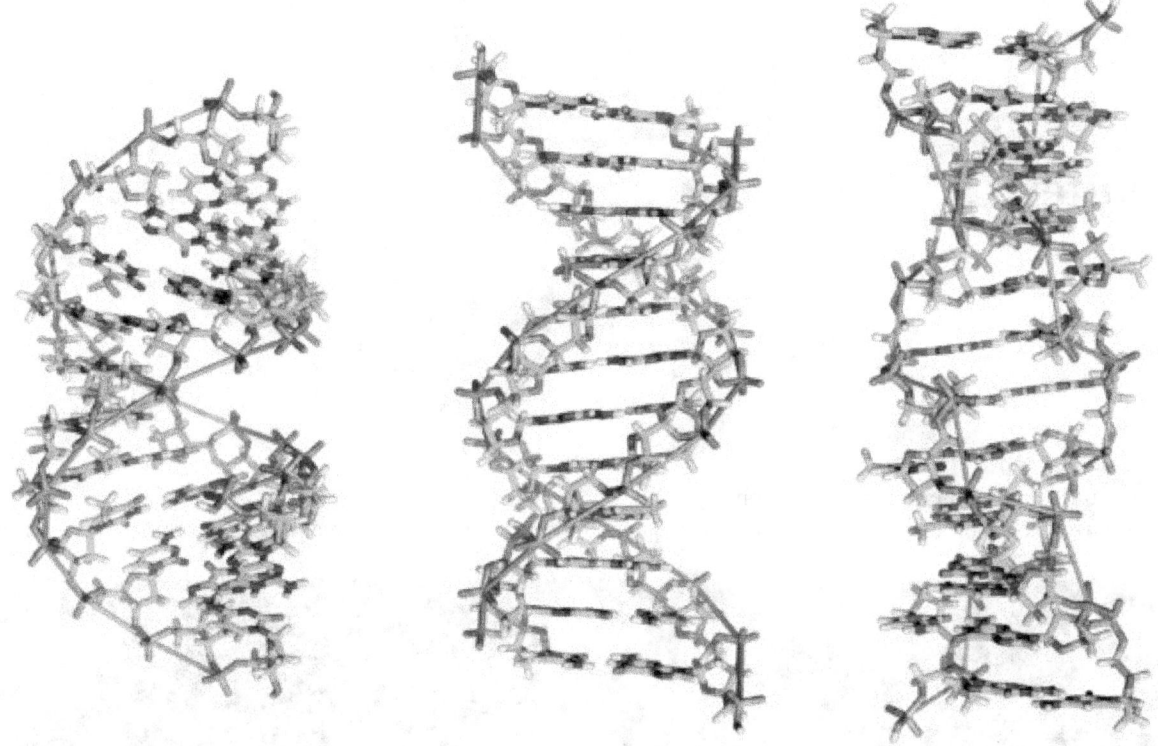

Side view of A-, B-, and Z-DNA.

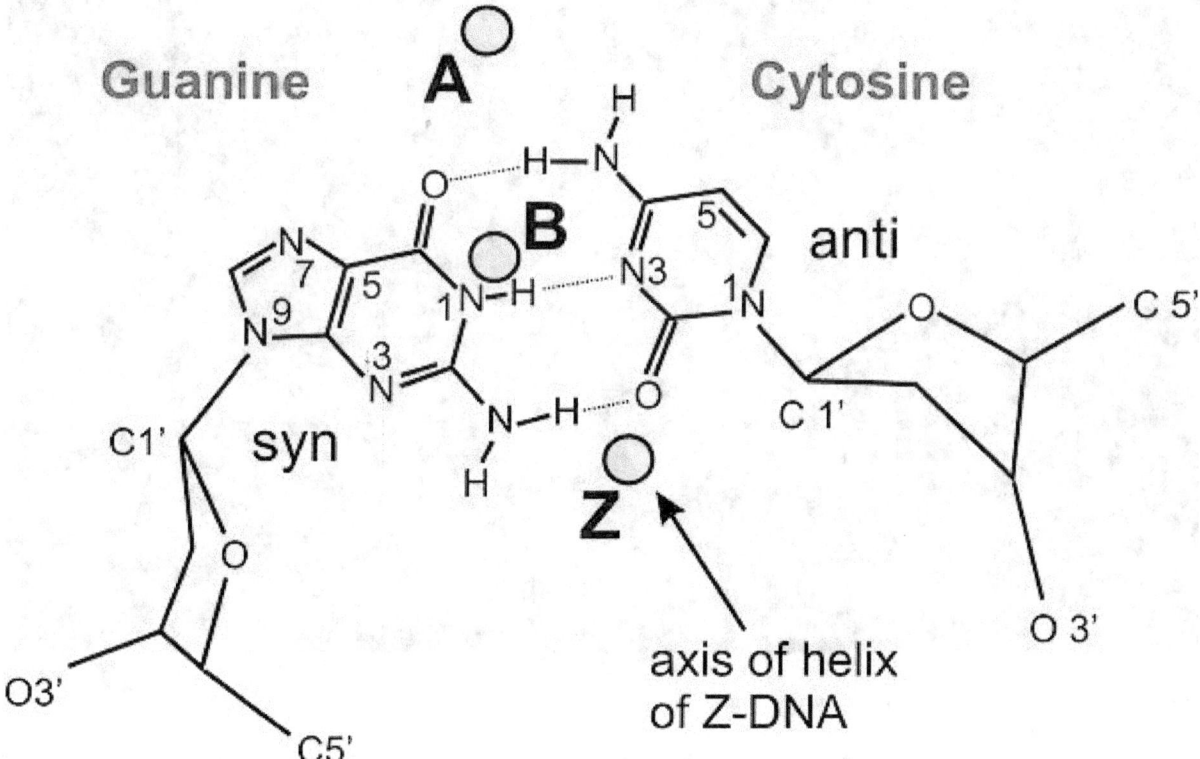

The helix axis of A-, B-, and Z-DNA.

Chapter 15

Telomere

For other uses, see Telomere (disambiguation).

A **telomere** is a region of repetitive nucleotide sequences at each end of a chromatid, which protects the end of the

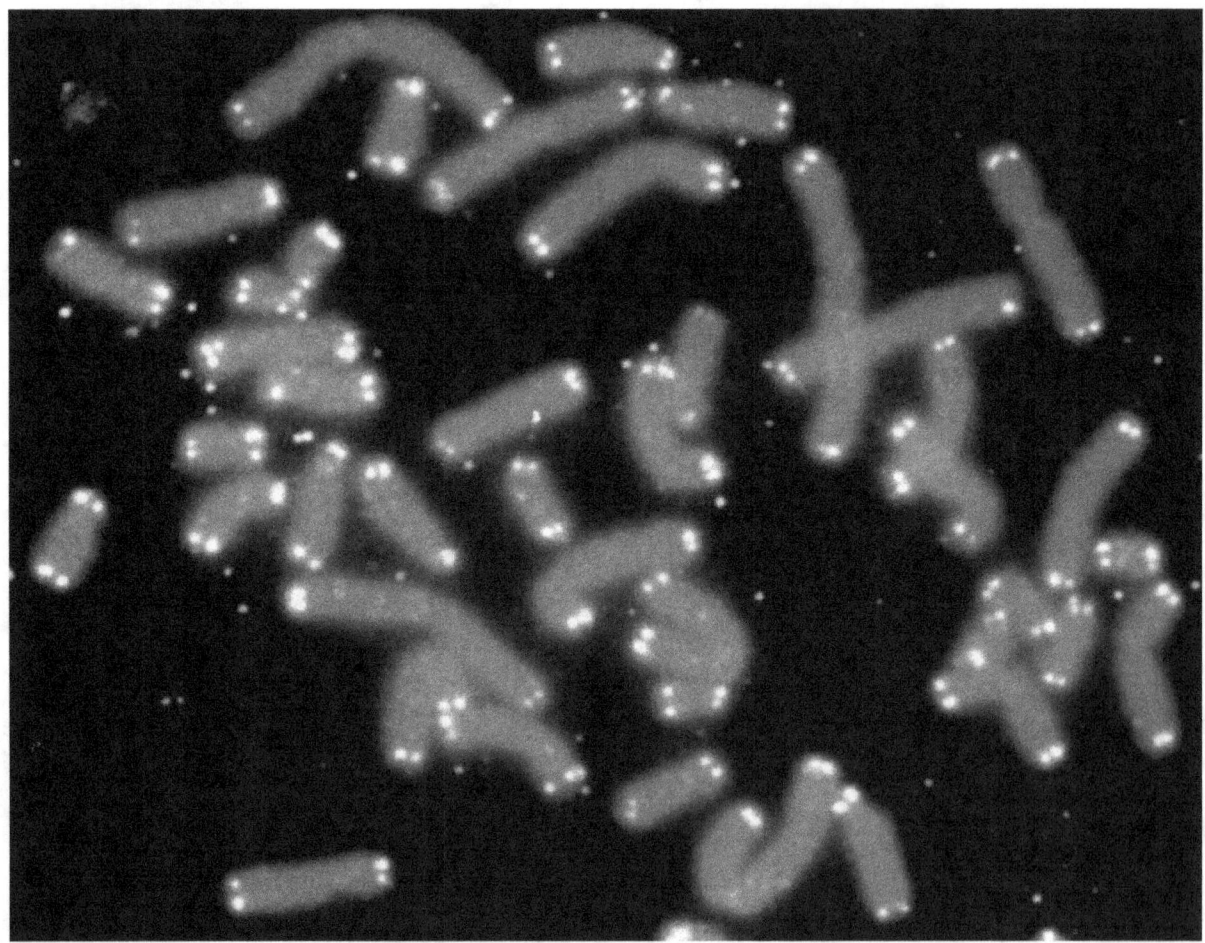

Human chromosomes (grey) capped by telomeres (white)

chromosome from deterioration or from fusion with neighboring chromosomes. Its name is derived from the Greek nouns telos ($\tau\acute{\varepsilon}\lambda o\varsigma$) 'end' and meros ($\mu\acute{\varepsilon}\rho o\varsigma$, root: $\mu\varepsilon\rho$-) 'part.' For vertebrates, the sequence of nucleotides in telomeres is TTAGGG. This sequence of TTAGGG is repeated approximately 2,500 times in humans. [1]

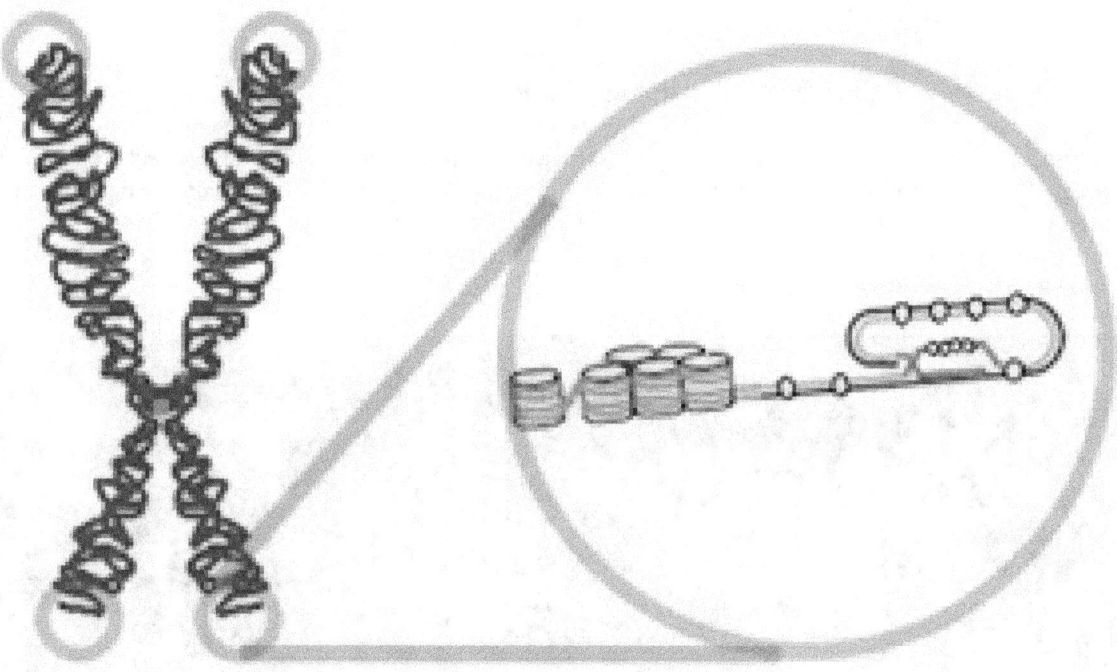

Telomere

During chromosome replication, the enzymes that duplicate DNA cannot continue their duplication all the way to the end of a chromosome, so in each duplication the end of the chromosome is shortened[2] (this is because the synthesis of Okazaki fragments requires RNA primers attaching ahead on the lagging strand). The telomeres are disposable buffers at the ends of chromosomes which are truncated during cell division; their presence protects the genes before them on the chromosome from being truncated instead.

Over time, due to each cell division, the telomere ends become shorter.[3] They are replenished by an enzyme, telomerase reverse transcriptase.

15.1 Discovery

In the early 1970s, Russian theorist Alexei Olovnikov first recognized that chromosomes could not completely replicate their ends. Building on this, and to accommodate Leonard Hayflick's idea of limited somatic cell division, Olovnikov suggested that DNA sequences are lost every time a cell/DNA replicates until the loss reaches a critical level, at which point cell division ends.[4][5] However, Olovnikov's prediction was not widely known except by a handful of researchers studying cellular aging and immortalization.[6]

In 1975–1977, Elizabeth Blackburn, working as a postdoctoral fellow at Yale University with Joseph Gall, discovered the unusual nature of telomeres, with their simple repeated DNA sequences composing chromosome ends.[7] Blackburn, Carol Greider, and Jack Szostak were awarded the 2009 Nobel Prize in Physiology or Medicine for the discovery of how chromosomes are protected by telomeres and the enzyme telomerase.[8]

Nevertheless, in the 1970s there was no recognition that the telomere-shortening mechanism normally limits cells to a fixed number of divisions, and no animal study suggesting that this could be responsible for aging on the cellular level and sets a limit on lifespans.[9][10]

It remained for a privately funded collaboration from biotechnology company Geron to isolate the genes for the RNA and protein component of human telomerase in order to establish the role of telomere shortening in cellular aging and telomerase reactivation in cell immortalization.[11]

15.2   Nature and function

15.2.1   Structure, function and evolutionary biology

Telomeres are repetitive nucleotide sequences located at the termini of linear chromosomes of most eukaryotic organisms. For vertebrates, the sequence of nucleotides in telomeres is TTAGGG. Most prokaryotes, lacking this linear arrangement, do not have telomeres. Telomeres compensate for incomplete semi-conservative DNA replication at chromosomal ends. A protein complex known as shelterin serves as protection against double-strand break (DSB) repair by homologous recombination (HR) and non-homologous end joining (NHEJ).[12][13]

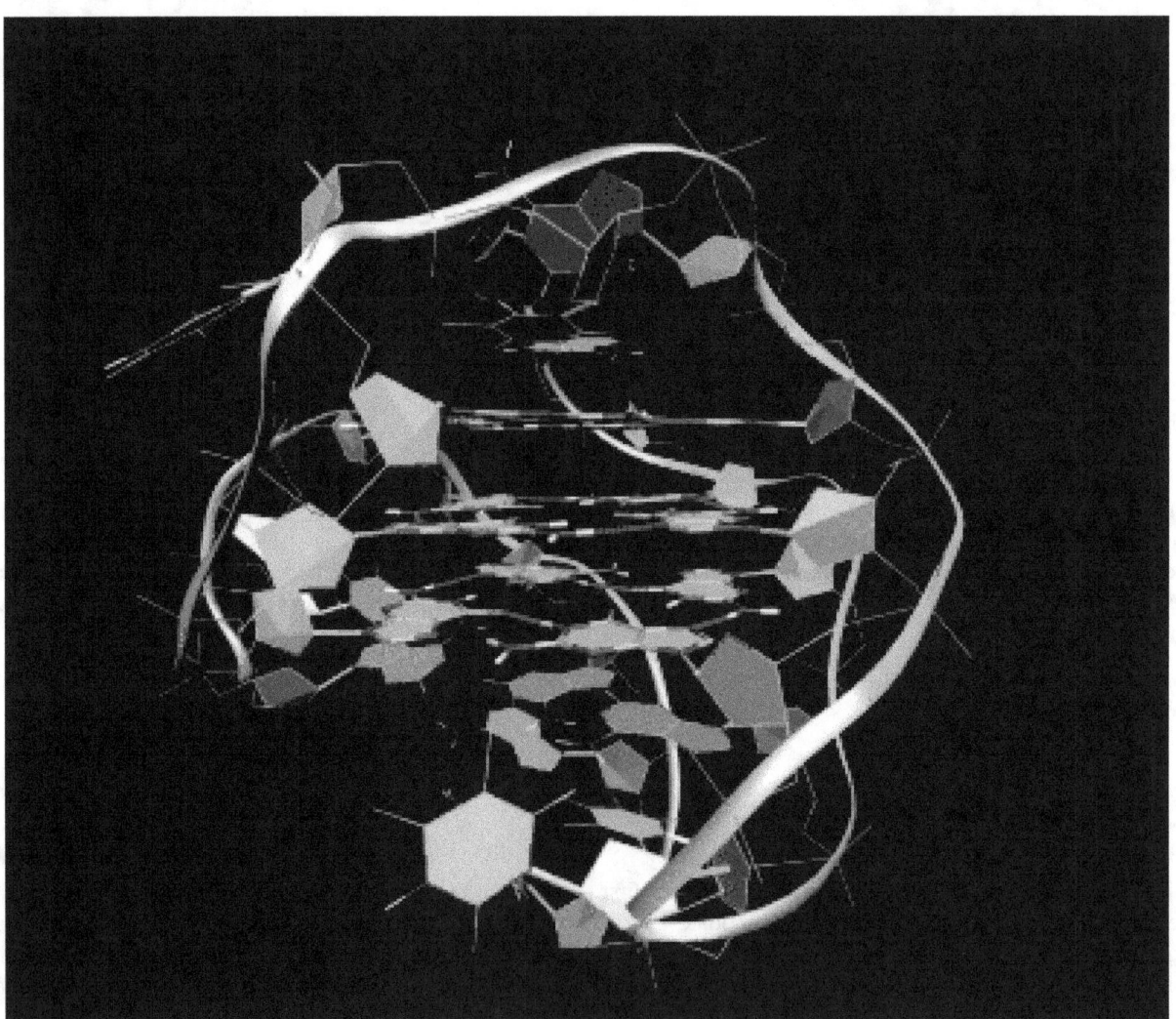

Three-dimensional representation of the molecular structure of a telomere (G-quadruplex)

In most prokaryotes, chromosomes are circular and, thus, do not have ends to suffer premature replication termination. A small fraction of bacterial chromosomes (such as those in *Streptomyces*, *Agrobacterium*, and *Borrelia*) are linear and possess telomeres, which are very different from those of the eukaryotic chromosomes in structure and functions. The known structures of bacterial telomeres take the form of proteins bound to the ends of linear chromosomes, or hairpin loops of single-stranded DNA at the ends of the linear chromosomes.[14]

While replicating DNA, the eukaryotic DNA replication enzymes (the DNA polymerase protein complex) cannot replicate the sequences present at the ends of the chromosomes (or more precisely the chromatid fibres). Hence, these sequences and the information they carry may get lost. This is the reason telomeres are so important in context of successful cell division: They "cap" the end-sequences and themselves get lost in the process of DNA replication. But the cell has an enzyme called telomerase, which carries out the task of adding repetitive nucleotide sequences to the ends of the DNA. Telomerase, thus, "replenishes" the telomere "cap" of the DNA. In most multicellular eukaryotic organisms, telomerase is active only in germ cells, some types of stem cells such as embryonic stem cells, and certain white blood cells. Telomerase can be re activated and telomeres reset back to an embryonic state by somatic cell nuclear transfer.[15] There are theories that claim that the steady shortening of telomeres with each replication in somatic (body) cells may have a role in senescence and in the prevention of cancer. This is because the telomeres act as a sort of time-delay "fuse", eventually running out after a certain number of cell divisions and resulting in the eventual loss of vital genetic information from the cell's chromosome with future divisions.

Telomere length varies greatly between species, from approximately 300 base pairs in yeast[16] to many kilobases in humans, and usually is composed of arrays of guanine-rich, six- to eight-base-pair-long repeats. Eukaryotic telomeres normally terminate with 3′ single-stranded-DNA overhang, which is essential for telomere maintenance and capping. Multiple proteins binding single- and double-stranded telomere DNA have been identified.[17] These function in both telomere maintenance and capping. Telomeres form large loop structures called telomere loops, or T-loops. Here, the single-stranded DNA curls around in a long circle stabilized by telomere-binding proteins.[18] At the very end of the T-loop, the single-stranded telomere DNA is held onto a region of double-stranded DNA by the telomere strand disrupting the double-helical DNA and base pairing to one of the two strands. This triple-stranded structure is called a displacement loop or D-loop.[19]

Telomere shortening in humans can induce replicative senescence, which blocks cell division. This mechanism appears to prevent genomic instability and development of cancer in human aged cells by limiting the number of cell divisions. However, shortened telomeres impair immune function that might also increase cancer susceptibility.[20] If telomeres become too short, they have the potential to unfold from their presumed closed structure. The cell may detect this uncapping as DNA damage and then either stop growing, enter cellular old age (senescence), or begin programmed cell self-destruction (apoptosis) depending on the cell's genetic background (p53 status). Uncapped telomeres also result in chromosomal fusions. Since this damage cannot be repaired in normal somatic cells, the cell may even go into apoptosis. Many aging-related diseases are linked to shortened telomeres. Organs deteriorate as more and more of their cells die off or enter cellular senescence.

At the very distal end of the telomere is a 300 bp single-stranded portion, which forms the T-Loop. This loop is analogous to a *knot*, which stabilizes the telomere, preventing the telomere ends from being recognized as break points by the DNA repair machinery. Should non-homologous end joining occur at the telomeric ends, chromosomal fusion will result. The T-loop is held together by several proteins, the most notable ones being TRF1, TRF2, POT1, TIN1, and TIN2, collectively referred to as the shelterin complex. In humans, the shelterin complex consists of six proteins identified as TRF1, TRF2, TIN2, POT1, TPP1, and RAP1.[12]

Main article: Shelterin

15.2.2   Cancer, telomerase and ALT (alternative lengthening of telomeres)

Malignant cells that bypass this arrest become immortalized by telomere extension due mostly to the activation of telomerase (the reverse transcriptase enzyme responsible for synthesis of telomeres). Telomerase is a "ribonucleoprotein complex" composed of a protein component and an RNA primer sequence that acts to protect the terminal ends of chromosomes from being broken down by enzymes. The telomeres (and the actions of telomerase) are necessary because, during replication, DNA polymerase can synthesize DNA in only a 5′ to 3′ direction (each DNA strand having a polarity that is determined by the precise manner in which sugar molecules of the strand's "backbone" are linked together) and can do so only by adding nucleotides to RNA primers (that have already been placed at various points along the length of the DNA). The RNA strands are replaced with newly synthesized DNA, but DNA polymerase can only "backfill" deoxyribonucleotides if there is already DNA "upstream" from (i.e., located 5′ to) the RNA primer. At the chromosome terminal, however, there is no nucleotide sequence in the 5′ direction (and therefore no upstream RNA primer or DNA), so DNA polymerase cannot function and genetic sequence might be lost through chromosomal fraying. Chromosomal ends might also be processed as breaks in double-strand DNA with chromosome-to-chromosome telomere fusions resulting.

Telomeres at the end of DNA prevent the chromosome from growing shorter during replications (with loss of genetic information) by employing "telomerases" to synthesize DNA at the chromosome terminal. These include a protein subgroup of specialized reverse transcriptase enzymes known as TERT (**TE**lomerase **R**everse **T**ranscriptases) and are involved in synthesis of telomeres in humans and many other, but not all, organisms. Because DNA replication mechanisms are affected by oxidative stress and because TERT expression is very low in most types of human cell, telomeres shrink a little bit every time a cell divides. Among cell types characterized by extensive cell division (such as stem cells and certain white blood cells), however, TERT is expressed at higher levels and telomere shortening is partially or fully prevented.

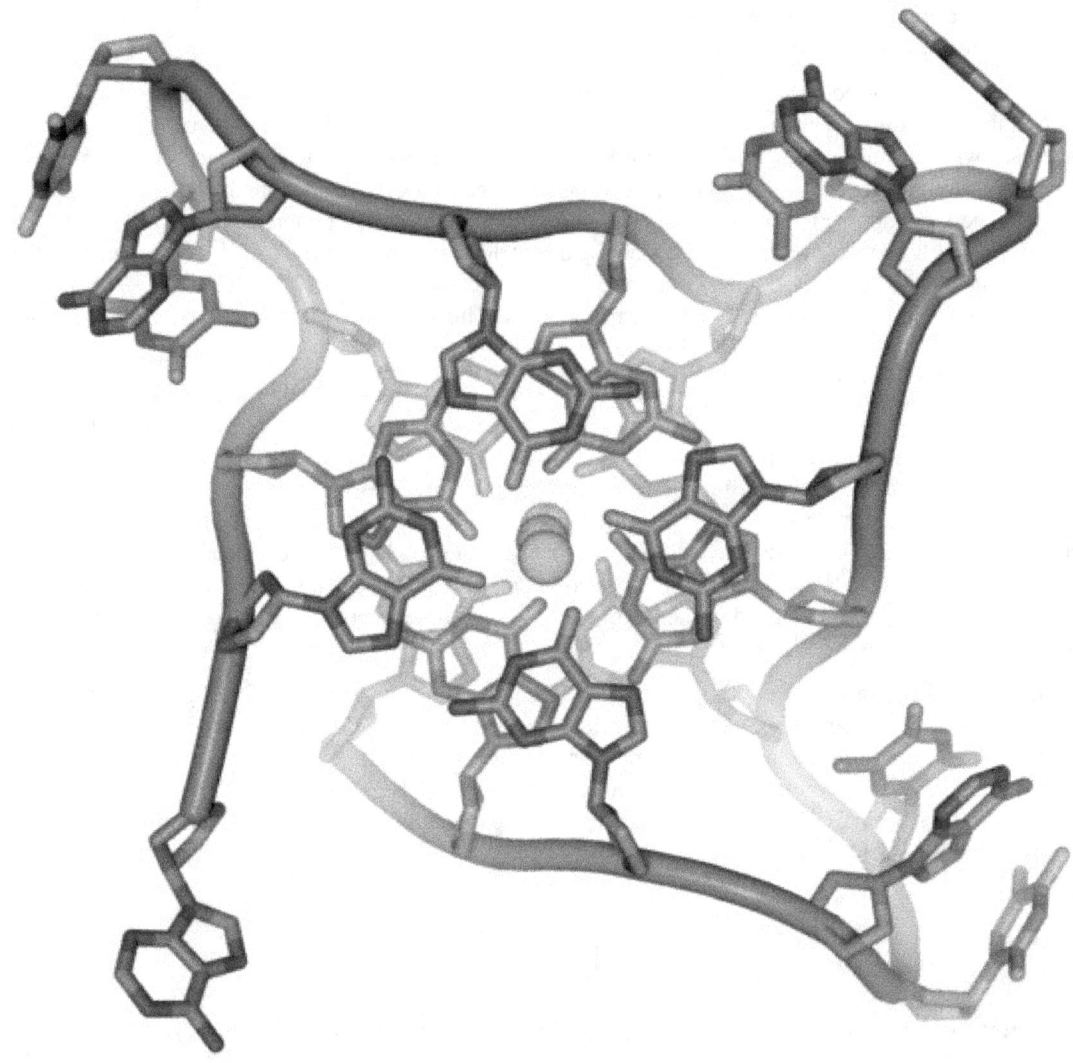

Structure of parallel quadruplexes that can be formed by human telomeric DNA. Image created from NDB UD0017.

In addition to its TERT protein component, telomerase also contains a piece of template RNA known as the TERC (**TE**lomerase **R**NA **C**omponent) or TR (**T**elomerase **R**NA). In humans, this TERC telomere sequence is a repeating string of TTAGGG, between 3 and 20 kilobases in length. There are an additional 100-300 kilobases of telomere-associated repeats between the telomere and the rest of the chromosome. Telomere sequences vary from species to species, but, in general, one strand is rich in G with fewer Cs. These G-rich sequences can form four-stranded structures (G-quadruplexes), with sets of four bases held in plane and then stacked on top of each other, with either a sodium or a potassium ion between the planar quadruplexes.

Mammalian (and other) somatic cells without telomerase gradually lose telomeric sequences as a result of incomplete replication (Counter *et al.*, 1992). As mammalian telomeres shorten, eventually cells reach their replicative limit and progress into senescence or old age. Senescence involves p53 and pRb pathways and leads to the halting of cell proliferation

(Campisi, 2005). Senescence may play an important role in suppression of cancer emergence, although inheriting shorter telomeres probably does not protect against cancer.[20] With critically shortened telomeres, further cell proliferation can be achieved by inactivation of p53 and pRb pathways. Cells entering proliferation after inactivation of p53 and pRb pathways undergo crisis. Crisis is characterized by gross chromosomal rearrangements and genome instability, and almost all cells die.

Alternative Lengthening of Telomeres

However, 5–10% of human cancers activate the Alternative Lengthening of Telomeres (ALT) pathway, which relies on recombination-mediated elongation.[21] Rarely, cells emerge from crisis immortalized through telomere lengthening by either activated telomerase or ALT (Colgina and Reddel, 1999; Reddel and Bryan, 2003). The first description of an ALT cell line demonstrated that their telomeres are highly heterogeneous in length and predicted a mechanism involving recombination (Murnane et al., 1994). Subsequent studies have confirmed a role for recombination in telomere maintenance by ALT (Dunham et al., 2000), however the exact mechanism of this pathway is yet to be determined. ALT cells produce abundant t-circles, possible products of intratelomeric recombination and t-loop resolution (Tomaska *et al.*, 2000; 2009; Cesare and Griffith, 2004; Wang *et al.*, 2004).

Evolutionary aspects

Since shorter telomeres are thought by some to be a cause of aging, this raises the question of why longer telomeres are not selected for to ameliorate these effects. A prominent explanation suggests that inheriting longer telomeres would cause increased cancer rates (e.g. Weinstein and Ciszek, 2002). However, a recent literature review and analysis [20] suggests this is unlikely, because shorter telomeres and telomerase inactivation is more often associated with increased cancer rates, and the mortality from cancer occurs late in life when the force of natural selection is very low. An alternative explanation to the hypothesis that long telomeres are selected against due to their cancer promoting effects is the "thrifty telomere" hypothesis, which suggests that the cellular proliferation effects of longer telomeres causes increased energy expenditures.[20] In environments of energetic limitation, shorter telomeres might be an energy sparing mechanism.

Relation to breast cancer

In healthy female breast, a proportion of cells called luminal progenitors that line the milk ducts have proliferative and differentiation potential and most of them contain critically short telomeres with DNA damage foci. These cells are believed to be the possible common cellular loci where cancers of the breast involving telomere dysregulation may arise.[22] The telomere shortening in these progenitors is not age dependent but is speculated to be basal to luminal epithelial differentiation program-dependent. Also, the telomerase activity are unusually high in these cells when isolated from younger women but decline with age.[23]

15.3 Shortening

Telomeres shorten in part because of the *end replication problem* that is exhibited during DNA replication in eukaryotes only. Because DNA replication does not begin at either end of the DNA strand, but starts in the center, and considering that all known DNA polymerases move in the 5' to 3' direction, one finds a leading and a lagging strand on the DNA molecule being replicated.

On the leading strand, DNA polymerase can make a complementary DNA strand without any difficulty because it goes from 5' to 3'. However, there is a problem going in the other direction on the lagging strand. To counter this, short sequences of RNA acting as primers attach to the lagging strand a short distance ahead of where the initiation site was. The DNA polymerase can start replication at that point and go to the end of the initiation site. This causes the formation of Okazaki fragments. More RNA primers attach further on the DNA strand and DNA polymerase comes along and continues to make a new DNA strand.

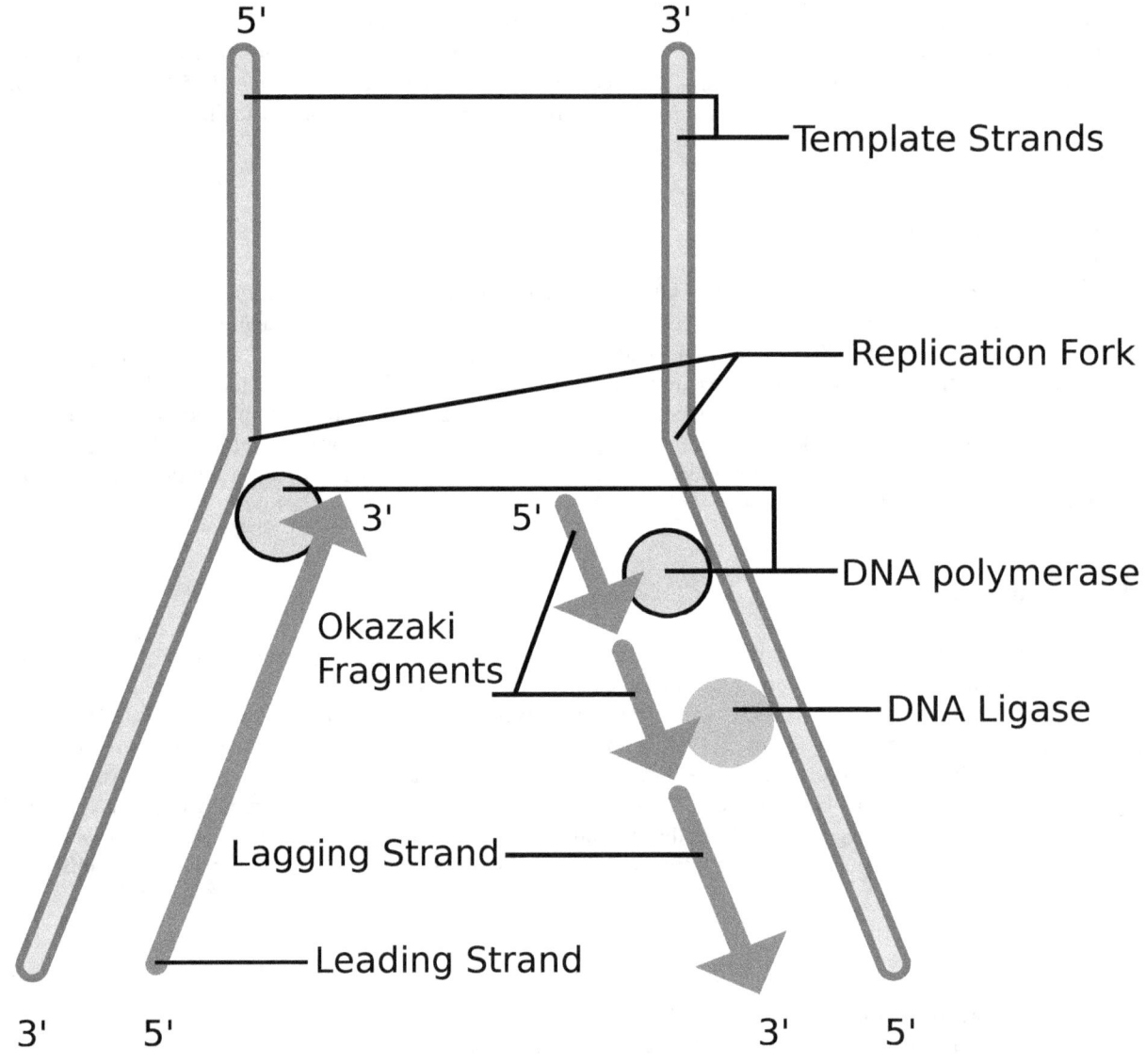

Lagging strand during DNA replication

Eventually, the last RNA primer attaches, and DNA polymerase, RNA nuclease, and DNA ligase come along to convert the RNA (of the primers) to DNA and to seal the gaps in between the Okazaki fragments. But, in order to change RNA to DNA, there must be another DNA strand in front of the RNA primer. This happens at all the sites of the lagging strand, but it does not happen at the end where the last RNA primer is attached. Ultimately, that RNA is destroyed by enzymes that degrade any RNA left on the DNA. Thus, a section of the telomere is lost during each cycle of replication at the 5' end of the lagging strand's daughter.

However, *in vitro* studies have shown that telomeres are highly susceptible to oxidative stress, and Richter and Zglinicki presented evidence that oxidative stress-mediated DNA damage is an important determinant of telomere shortening.[24] Telomere shortening due to free radicals explains the difference between the estimated loss per division because of the end-replication problem (c. 20 bp) and actual telomere shortening rates (50–100 bp), and has a greater absolute impact on telomere length than shortening caused by the end-replication problem. Population-based studies have also indicated an interaction between anti-oxidant intake and telomere length. In the Long Island Breast Cancer Study Project (LIBCSP), authors found a moderate increase in breast cancer risk among women with the shortest telomeres and lower dietary intake of beta carotene, vitamin C or E.[25] These results suggest that cancer risk due to telomere shortening may interact with other mechanisms of DNA damage, specifically oxidative stress.

Telomere shortening is associated with ageing, mortality and ageing-related diseases. In 2003, Richard Cawthon discovered that those with longer telomeres lead longer lives than those with short telomeres.[26] However, it is not known whether short telomeres are just a sign of cellular age or actually contribute to the aging process.

15.4 Lengthening

The phenomenon of limited cellular division was first observed by Leonard Hayflick, and is now referred to as the Hayflick limit.[27][28] Significant discoveries were subsequently made by a group of scientists organized at Geron Corporation by Geron's founder Michael D. West that tied telomere shortening with the Hayflick limit.[29] The cloning of the catalytic component of telomerase enabled experiments to test whether the expression of telomerase at levels sufficient to prevent telomere shortening was capable of immortalizing human cells. Telomerase was demonstrated in a 1998 publication in Science to be capable of extending cell lifespan, and now is well-recognized as capable of immortalizing human somatic cells.[30]

It is becoming apparent that reversing shortening of telomeres through temporary activation of telomerase may be a potent means to slow aging. They reason that this would extend human life because it would extend the Hayflick limit. Three routes have been proposed to reverse telomere shortening: drugs, gene therapy, or metabolic suppression, so-called, torpor/hibernation. So far these ideas have not been proven in humans, but it has been demonstrated that telomere shortening is reversed in hibernation and aging is slowed (Turbill, et al. 2012 & 2013) and that hibernation prolongs life-span (Lyman et al. 1981). It has also been demonstrated that telomere extension has successfully reversed some signs of aging in laboratory mice [31][32] and the nematode worm species *Caenorhabditis elegans*.[33] It has been hypothesized that longer telomeres and especially telomerase activation might cause increased cancer (e.g. Weinstein and Ciszek, 2002). However, longer telomeres might also protect against cancer, because short telomeres are associated with cancer. It has also been suggested that longer telomeres might cause increased energy consumption.[20]

Techniques to extend telomeres could be useful for tissue engineering, because they might permit healthy, noncancerous mammalian cells to be cultured in amounts large enough to be engineering materials for biomedical repairs.

Two recent studies on long-lived seabirds demonstrate that the role of telomeres is far from being understood . In 2003, scientists observed that the telomeres of Leach's storm-petrel (*Oceanodroma leucorhoa*) seem to lengthen with chronological age, the first observed instance of such behaviour of telomeres.[34] In 2006, Juola *et al.*[35] reported that in another unrelated, long-lived seabird species, the great frigatebird (*Fregata minor*), telomere length did decrease until at least c.40 years of age (i.e. probably over the entire lifespan), but the speed of decrease slowed down massively with increasing ages, and that rates of telomere length decrease varied strongly between individual birds. They concluded that in this species (and probably in frigatebirds and their relatives in general), telomere length could not be used to determine a bird's age sufficiently well. Thus, it seems that there is much more variation in the behavior of telomere length than initially believed.

Furthermore, Gomes et al. found, in a study of the comparative biology of mammalian telomeres, that telomere length of different mammalian species correlates inversely, rather than directly, with lifespan, and they concluded that the contribution of telomere length to lifespan remains controversial.[36] Harris et al. found little evidence that, in humans, telomere length is a significant biomarker of normal aging with respect to important cognitive and physical abilities.[37] Gilley and Blackburn tested whether cellular senescence in paramecium is caused by telomere shortening, and found that telomeres were not shortened during senescence.[38]

15.4.1 Exercise-induced lengthening

A 2013 pilot study from UCSF took 35 men with localized early-stage prostate cancer and had 10 of them begin "lifestyle changes that included: a plant-based diet (high in fruits, vegetables and unrefined grains, and low in fat and refined carbohydrates); moderate exercise (walking 30 minutes a day, six days a week); stress reduction (gentle yoga-based stretching, breathing, meditation)" and also "weekly group support". When compared to the other 25 study participants, "The group that made the lifestyle changes experienced a 'significant' increase in telomere length of approximately 10 percent. Further, the more people changed their behavior by adhering to the recommended lifestyle program, the more dramatic their improvements in telomere length."[39] A 2014 study entitled "Stand up for health--avoiding sedentary behaviour might lengthen your telomeres: secondary outcomes from a physical activity RCT in older people" indicated somewhat contradictory results, stating, "In the intervention group, there was a negative correlation between changes in time spent exercising and changes in telomere length (rho=−0.39, p=0.07). On the other hand, in the intervention group, telomere lengthening was significantly associated with reduced sitting time (rho=−0.68, p=0.02).[40]

15.5 Sequences

Known, up-to-date telomere nucleotide sequences are listed in Telomerase Database website.

15.6 Cancer

Telomeres are critical for maintaining genomic integrity and studies show that telomere dysfunction or shortening is commonly acquired during the process of tumor development.[42] Short telomeres can lead to genomic instability, chromosome loss and the formation of non-reciprocal translocations; and telomeres in tumor cells and their precursor lesions are significantly shorter than surrounding normal tissue.[43][44]

Observational studies have found shortened telomeres in many cancers: including pancreatic, bone, prostate, bladder, lung, kidney, and head and neck. In addition, people with many types of cancer have been found to possess shorter leukocyte telomeres than healthy controls.[45] Recent meta-analyses suggest 1.4 to 3.0 fold increased risk of cancer for those with the shortest vs. longest telomeres.[46][47] However the increase risk varies by age, sex, tumor type and differences in lifestyle factors.

Some of the same lifestyle factors which increase risk of developing cancer have also been associated with shortened telomeres: including stress, smoking, physical inactivity and diet high in refined sugars [47] Diet and physical activity influence inflammation and oxidative stress. These factors are thought to influence telomere maintenance.[48] Psychologic stress has also been linked to accelerated cell aging, as reflected by decreased telomerase activity and short telomeres.[49] It has been suggested that a combination of lifestyle modifications, including healthy diet, exercise and stress reduction, have the potential to increase telomere length, reverse cellular aging, and reduce the risk for aging-related diseases. In a recent clinical trial for early-stage prostate cancer patients, comprehensive lifestyle changes resulted in a short-term increase in telomerase activity and long-term modification in telomere length.[50][51] Lifestyle modifications have the potential to naturally regulate telomere maintenance without promoting tumorgenesis, as traditional mechanisms of telomere lengthening involve the use of telomerase activating agents.

Cancer cells require a mechanism to maintain their telomeric DNA in order to continue dividing indefinitely (immortalization). A mechanism for telomere elongation or maintenance is one of the key steps in cellular immortalization and can be used as a diagnostic marker in the clinic. Telomerase, the enzyme complex responsible for elongating telomeres through the addition of telomere repeats to the ends of chromosomes, is activated in approximately 80% of tumors.[52] However, a sizeable fraction of cancerous cells employ alternative lengthening of telomeres (ALT),[53] a non-conservative telomere lengthening pathway involving the transfer of telomere tandem repeats between sister-chromatids.[54]

Telomerase is the natural enzyme that promotes telomere repair. It is active in stem cells, germ cells, hair follicles, and 90 percent of cancer cells, but its expression is low or absent in somatic cells. Telomerase functions by adding bases to the ends of the telomeres. Cells with sufficient telomerase activity are considered immortal in the sense that they can divide past the Hayflick limit without entering senescence or apoptosis. For this reason, telomerase is viewed as a potential target for anti-cancer drugs (such as Geron's Imetelstat currently in human clinical trials and telomestatin).[55]

Studies using knockout mice have demonstrated that the role of telomeres in cancer can both be limiting to tumor growth, as well as promote tumorigenesis, depending on the cell type and genomic context.[56][57]

15.7 Measurement

Several techniques are currently employed to assess average telomere length in eukaryotic cells. The most widely used method is the Terminal Restriction Fragment (TRF) southern blot,[58] which involves hybridization of a radioactive 32P-(TTAGGG)n oligonucleotide probe to Hinf / Rsa I digested genomic DNA embedded on a nylon membrane and subsequently exposed to autoradiographic film or phosphoimager screen. Another histochemical method, termed Q-FISH, involves fluorescent in situ hybridization (FISH).[59] Q-FISH, however, requires significant amounts of genomic DNA (2-20 micrograms) and labor that renders its use limited in large epidemiological studies. Some of these impediments have been overcome with a Real-Time PCR assay for telomere length and Flow-FISH. Real-time PCR assay involves determining the Telomere-to-Single Copy Gene (T/S)ratio,[60] which is demonstrated to be proportional to the average telomere length in a cell.

Another technique, referred to as single telomere elongation length analysis (STELA), was developed in 2003 by Duncan Baird. This technique allows investigations that can target specific telomere ends, which is not possible with TRF analysis. However, due to this technique's being PCR-based, telomeres larger than 25Kb cannot be amplified and there is a bias towards shorter telomeres.

Telomere length is associated with the general health of an individual as well as certain diseases, beyond cancer.[61][62] While multiple companies offer telomere length measurement services,[63][64][65] the utility of these measurements for widespread clinical or personal use has been questioned by prominent scientists without financial interests in these companies.[66][67] Nobel Prize winner Elizabeth Blackburn, who was the co-founder of one of these companies and has prominently promoted the clinical utility of telomere length measures,[68] resigned from the company in June 2013 "owing to an impending change in the control of Telome Health".[69]

15.8 See also

- Biological clock and epigenetic clock
- Centromere
- DNA damage theory of aging
- Immortality
- Maximum life span
- Rejuvenation (aging)
- Senescence, biological aging

15.9 References

[1] Sadava, D., Hillis, D., Heller, C., & Berenbaum, M. (2011). *Life: The science of biology.* (9th ed.) Sunderland, MA: Sinauer Associates Inc.

[2] AtGoogleTalks, August 20, 2008 Molecular biologist Elizabeth Blackburn

[3] Passarge, Eberhard. Color atlas of genetics, 2007.

[4] Olovnikov, Alexei M. (1971). Принцип маргинотомии в матричном синтезе полинуклеотидов [Principle of marginotomy in template synthesis of polynucleotides]. *Doklady Akademii Nauk SSSR* (in Russian) **201** (6): 1496–9. PMID 5158754.

[5] Olovnikov AM (September 1973). "A theory of marginotomy. The incomplete copying of template margin in enzymic synthesis of polynucleotides and biological significance of the phenomenon". *J. Theor. Biol.* **41** (1): 181–90. doi:10.1016/0022-5193(73)90198-7. PMID 4754905.

[6] "No Nobel physiology and medicine award for Russian gerontologist Aleksey Olovnikov". *Telegraph.* October 21, 2009.

[7] Blackburn AM; Gall, Joseph G. (March 1978). "A tandemly repeated sequence at the termini of the extrachromosomal ribosomal RNA genes in Tetrahymena". *J. Mol. Biol.* **120** (1): 33–53. doi:10.1016/0022-2836(78)90294-2. PMID 642006.

[8] "The 2009 Nobel Prize in Physiology or Medicine - Press Release". Nobelprize.org. 2009-10-05. Retrieved 2012-06-12.

[9] Harrison's Principles of Internal Medicine, Ch. 69, Cancer cell biology and angiogenesis, Robert G. Fenton and Dan L. Longo, p. 454.

[10] "Portfolio".

[11] "Unravelling the secret of ageing". *COSMOS: The Science of Everything.* October 5, 2009. Archived from the original on January 14, 2015.

[12] Blasco, Maria; Paula Martínez (21 Jun 2010). "Role of shelterin in cancer and aging". *Aging Cell* **9**(5): 653–666. doi:10.1111/j.9726.2010.00596.x. PMID 20569239.

[13] Lundblad, 2000; Ferreira *et al.*, 2004

[14] Maloy, Stanley (July 12, 2002). "Bacterial Chromosome Structure". Retrieved 2008-06-22.

[15] Robert P. Lanza, Jose B. Cibelli, Catherine Blackwell, Vincent J. Cristofalo, Mary Kay Francis, Gabriela M. Baerlocher, Jennifer Mak, Michael Schertzer, Elizabeth A. Chavez, Nancy Sawyer, Peter M. Lansdorp, Michael D. West1 (28 April 2000). "Extension of Cell Life-Span and Telomere Length in Animals Cloned from Senescent Somatic Cells" (PDF). Science.

[16] Shampay , Szostak J.W., Blackburn E.H.; Szostak; Blackburn (1984). "DNA sequences of telomeres maintained in yeast". *Nature* **310** (5973): 154–157. doi:10.1038/310154a0. PMID 6330571.

[17] Blackburn, 2001; Smogorzewska and de Lange, 2004; Cech, 2004; De Lange *et al.*, 2005; Kota and Runge, 1999

[18] Griffith J, Comeau L, Rosenfield S, Stansel R, Bianchi A, Moss H, de Lange T; Comeau; Rosenfield; Stansel; Bianchi; Moss; De Lange (1999). "Mammalian telomeres end in a large duplex loop". *Cell* **97** (4): 503–14. doi:10.1016/S0092-8674(00)80760-6. PMID 10338214.

[19] Burge S, Parkinson G, Hazel P, Todd A, Neidle S; Parkinson; Hazel; Todd; Neidle (2006). "Quadruplex DNA: sequence, topology and structure". *Nucleic Acids Res* **34** (19): 5402–15. doi:10.1093/nar/gkl655. PMC 1636468. PMID 17012276.

[20] Eisenberg DTA (2011). "An evolutionary review of human telomere biology: The thrifty telomere hypothesis and notes on potential adaptive paternal effects". *American Journal of Human Biology* **23** (2): 149–167. doi:10.1002/ajhb.21127. PMID 21319244.

[21] Henson, JD; Neumann, AA; Yeager, TR; Reddel, RR (2002). "Alternative lengthening of telomeres in mammalian cells". *Oncogene* **21** (4): 598–610. doi:10.1038/sj.onc.1205058. PMID 11850785.

[22] BBC, World/Mundo. "Resuelven misterio sobre el origen del cáncer de mama".

[23] Kannan, Nagarajan; Nazmul Huda, LiRen Tu, Radina Droumeva, Geraldine Aubert, Elizabeth Chavez, Ryan R. Brinkman, Peter Lansdorp, Joanne Emerman, Satoshi Abe, Connie Eaves, David Gilley (4 June 2013). "The Luminal Progenitor Compartment of the Normal Human Mammary Gland Constitutes a Unique Site of Telomere Dysfunction". *Stem Cell Reports* **1** (1): 28–31. doi:10.1016/j.stemcr.2013.04.003. PMID 24052939.

[24] Richter, T; von Zglinicki, T (2007). "A continuous correlation between oxidative stress and telomere shortening in fibroblasts". *Exp Gerontol* **42** (11): 1039–1042. doi:10.1016/j.exger.2007.08.005. PMID 17869047.

[25] Shen, J; Gammon, MD; Terry, MB; Wang, Q; Bradshaw, P; Teitelbaum, SL; Neugut, AI; Santella, RM (Apr 2009). "Telomere length, oxidative damage, antioxidants and breast cancer risk". *Int J Cancer* **124** (7): 1637–43. doi:10.1002/ijc.24105.

[26] Cawthon, RM; Smith, KR; O'Brien, E; Sivatchenko, A; Kerber, RA (2003). "Association between telomere length in blood and mortality in people aged 60 years or older". *Lancet* **361** (9355): 393–395. doi:10.1016/s0140-6736(03)12384-7.

[27] Hayflick L, Moorhead PS; Moorhead (1961). "The serial cultivation of human diploid cell strains". *Exp Cell Res* **25** (3): 585–621. doi:10.1016/0014-4827(61)90192-6. PMID 13905658.

[28] Hayflick L. (1965). "The limited in vitro lifetime of human diploid cell strains". *Exp. Cell Res.* **37**(3): 614–636. doi:10.1016/004827(65)90211-9. PMID 14315085.

[29] Feng J, Funk WD, Wang SS, Weinrich SL, Avilion AA, Chiu CP, Adams RR, Chang E, Allsopp RC, Yu J; Funk; Wang; Weinrich; Avilion; Chiu; Adams; Chang; Allsopp; Yu (September 1995). "The RNA component of human telomerase". *Science* **269** (5228): 1236–41. doi:10.1126/science.7544491. PMID 7544491.

[30] Bodnar, A.G.; Ouellette, M.; Frolkis, M.; Holt, S.E.; Chiu, C.P.; Morin, G.B.; Harley, C.B.; Shay, J.W.; Lichtsteiner, S.; Wright, W.E. (1998). "Extension of life-span by introduction of telomerase into normal human cells". *Science* **279** (5349): 349–352. doi:10.1126/science.279.5349.349.

[31] Sample, Ian (November 28, 2010). "Harvard scientists reverse the ageing process in mice – now for humans". *The Guardian* (London).

[32] http://www.nature.com/nature/journal/vaop/ncurrent/full/nature09603.html

[33] Joeng KS, Song EJ, Lee KJ, Lee J; Song; Lee; Lee (2004). "Long lifespan in worms with long telomeric DNA". *Nature Genetics* **36** (6): 607–11. doi:10.1038/ng1356. PMID 15122256.

[34] Nakagawa S, Gemmell NJ, Burke T; Gemmell; Burke (September 2004). "Measuring vertebrate telomeres: applications and limitations". *Mol. Ecol.* **13** (9): 2523–33. doi:10.1111/j.1365-294X.2004.02291.x. PMID 15315667.

[35] Juola, Frans A; Haussmann, Mark F; Dearborn, Donald C; Vleck, Carol M (2006). "Telomere shortening in a long-lived marine bird: Cross-sectional analysis and test of an aging tool". *The Auk* **123** (3): 775. doi:10.1642/0004-8038(2006)123[775:TSIALM]2.0.CO;2. ISSN 0004-8038.

[36] Gomes, NM; Ryder, OA; Houck, ML; Charter, SJ; Walker, W; Forsyth, NR; Austad, SN; Venditti, C; Pagel, M; Shay, JW; Wright, WE (2011). "Comparative biology of mammalian telomeres: hypotheses on ancestral states and the roles of telomeres in longevity determination". *Aging Cell* **10** (5): 761–768. doi:10.1111/j.1474-9726.2011.00718.x. PMC 3387546. PMID 21518243.

[37] Harris, SE; Martin-Ruiz, C; von Zglinicki, T; Starr, JM; Deary, IJ (2010). "Telomere length and aging biomarkers in 70-year-olds: the Lothian Birth Cohort 1936". *Neurobiol Aging* **33** (7): 1486.e3–1486.e8. doi:10.1016/j.neurobiolaging.2010.11.013. PMID 21194798.

[38] Gilley, D; Blackburn, EH (1994). "Lack of telomere shortening during senescence in Paramecium". *Proc Natl Acad Sci U S A* **91** (5): 1955–1958. doi:10.1073/pnas.91.5.1955. PMID 8127914.

[39] Fernandez, Elizabeth (2013-09-16). "Lifestyle Changes May Lengthen Telomeres, A Measure of Cell Aging". *http://www.ucsf.edu/*. University of California, San Francisco. Retrieved 2015-03-16.

[40] Sjögren, P; Fisher, R; Kallings, L; Svenson, U; Roos, G; Hellénius, M (2014-09-03). "Stand up for health--avoiding sedentary behaviour might lengthen your telomeres: secondary outcomes from a physical activity RCT in older people.". *http://www.ncbi.nlm.nih.gov/*. National Center for Biotechnology Information. Retrieved 2014-03-16.

[41] Peška, Vratislav; Fajkus, Petr; Fojtová, Miloslava; Dvořáčková, Martina; Hapala, Jan; Dvořáček, Vojtěch; Polanská, Pavla; Leitch, Andrew R.; Sýkorová, Eva; Fajkus, Jiří (May 2015). "Characterisation of an unusual telomere motif (TTTTTTAGGG) in the plant (Solanaceae), a species with a large genome". *The Plant Journal* **82** (4): 644–654. doi:10.1111/tpj.12839.

[42] Raynaud, CM; Sabatier, L; Philipot, O; Olaussen, KA; Soria, JC (2008). "Telomere length, telomeric proteins and genomic instability during the multistep carcinogenic process". *Crit Rev Oncol Hematol* **66**: 99–117. doi:10.1016/j.critrevonc.2007.11.006.

[43] Blasco, MA; Lee, HW; Hande, MP; Samper, E; Lansdorp, PM; et al. (1997). "Telomere shortening and tumor formation by mouse cells lacking telomerase RNA". *Cell* **91** (1): 25–34. doi:10.1016/s0092-8674(01)80006-4.

[44] Artandi, SE; Chang, S; Lee, SL; Alson, S; Gottlieb, GJ; et al. (2000). "Telomere dysfunction promotes non-reciprocal translocations and epithelial cancers in mice". *Nature* **406**: 641–645.

[45] Willeit Peter, Willeit Johann, Mayr Anita, Weger Siegfried, Oberhollenzer Friedrich, Brandstätter Anita, Kronenberg Florian, Kiechl Stefan; Willeit; Mayr; Weger; Oberhollenzer; Brandstätter; Kronenberg; Kiechl (2010). "Telomere length and risk of incident cancer and cancer mortality". *JAMA* **304** (1): 69–75. doi:10.1001/jama.2010.897. PMID 20606151.

[46] Ma, H; Zhou, Z; Wei, S; et al. (2011). "Shortened telomere length is associated with increased risk of cancer: a meta-analysis". *PLOS ONE* **6** (6): e20466. doi:10.1371/journal.pone.0020466.

[47] Wentzensen, IM; Mirabello, L; Pfeiffer, RM; Savage, SA (2011). "The association of telomere length and cancer: a meta-analysis". *Cancer Epidemiol Biomarkers Prev.* **20** (6): 1238–1250. doi:10.1158/1055-9965.epi-11-0005.

[48] Paul, L (Oct 2011). "Diet, nutrition and telomere length". *J Nurt Biochem* **22** (10): 895–901. doi:10.1016/j.jnutbio.2010.12.001.

[49] Epel, ES; Lin, J; Wilhelm, FH; Wolkowitz, OM; Cawthon, R; Adler, NE; Dolbier, C; Mendes, WB; Blackburn, EH (April 2006). "Cell aging in relation to stress arousal and cardiovascular disease risk factors". *Psychoneuroendocrinology* **31** (3): 277–87. doi:10.1016/j.psyneuen.2005.08.011.

[50] Ornish, D; Lin, J; Chan, JM; Epel, E; Kemp, C; Weidner, G; Marlin, R; Frenda, SJ; Magbanua, MJ; Daubenmier, J; Estay, I; Hills, NK; Chainani-Wu, N; Carroll, PR; Blackburn, EH (Oct 2013). "Effect of comprehensive lifestyle changes on telomerase activity and telomerelength in men with biopsy-proven low-risk prostate cancer: 5-year follow-up of a descriptive pilot study". *Lancet Oncol* **14** (11): 1112–20. doi:10.1016/S1470-2045(13)70366-8.

[51] Ornish, D; Lin, J; Daubenmier, J; Weidner, G; Epel, E; Kemp, C; Magbanua, MJ; Marlin, R; Yglecias, L; Carroll, PR; Blackburn, EH (Nov 2008). "Increased telomerase activity and comprehensive lifestyle changes: a pilot study". *Lancet Oncol* **9** (11): 1048–57. doi:10.1016/S1470-2045(08)70234-1.

[52] Aschacher; Wolf; Enzmann; Kienzl (2015). "ALINE-1 induces hTERT and ensures telomere maintenance in tumour cell lines". doi:10.1038/onc.2015.65. PMID 11850785.

[53] Henson JD, Neumann AA, Yeager TR, Reddel RR; Neumann; Yeager; Reddel (2002). "Alternative lengthening of telomeres in mammalian cells". *Oncogene* **21** (4): 598–610. doi:10.1038/sj.onc.1205058. PMID 11850785.

[54] Chris Molenaar, Karien Wiesmeijer, Nico P. Verwoerd, Shadi Khazen, Roland Eils, Hans J. Tanke, and Roeland W. Dirks (2003-12-15). "Visualizing telomere dynamics in living mammalian cells using PNA probes". *The EMBO Journal* (The European Molecular Biology Organization) **22** (24): 6631–6641. doi:10.1093/emboj/cdg633. PMC 291828. PMID 14657034.

[55] Philippi C, Loretz B, Schaefer UF, Lehr CM.; Loretz; Schaefer; Lehr (April 2010). "Telomerase as an emerging target to fight cancer - Opportunities and challenges for nanomedicine". *Journal of Controlled Releases* **146** (2): 228–40. doi:10.1016/j.jconrel .2010.03.025.PMID20381558.

[56] Chin L, Artandi SE, Shen Q; et al. (May 1999). "p53 deficiency rescues the adverse effects of telomere loss and cooperates with telomere dysfunction to accelerate carcinogenesis". *Cell* **97** (4): 527–38. doi:10.1016/S0092-8674(00)80762-X. PMID 10338216.

[57] Greenberg RA, Chin L, Femino A; et al. (May 1999). "Short dysfunctional telomeres impair tumorigenesis in the INK4a(delta2/3) cancer-prone mouse". *Cell* **97** (4): 515–25. doi:10.1016/S0092-8674(00)80761-8. PMID 10338215.

[58] Allshire RC; et al. (1989). "Human telomeres contain at least three types of G-rich repeat distributed non-randomly". *Nucleic Acids Res.* **17** (12): 4611–4627. doi:10.1093/nar/17.12.4611. PMC 318019. PMID 2664709.

[59] Rufer N; et al. (1998). "Telomere length dynamics in human lymphocyte subpopulations measured by flow cytometry". *Nat Biotechnol* **16** (8): 743–747. doi:10.1038/nbt0898-743. PMID 9702772.

[60] Cawthon, RM (2002). "Telomere measurement by quantitative PCR". *Nucleic Acids Research* **30**(10): e47. doi:10.1093/nar/30.1 PMC 115301. D   12000852.

[61] Armanios, M; Blackburn, EH (2012). "The telomere syndromes". *Nature Reviews Genetics* **13**(10): 693–704. doi:10.1038/nrg32 PMC 3548426. PMID 22965356.

[62] Codd, V; Nelson, CP; Albrecht, E; Mangino, M; Deelen, J; Buxton, JL; Hottenga, JJ; Fischer, K; Esko, T; Surakka, Ida; Broer, Linda; Nyholt, Dale R; Leach, Irene Mateo; Salo, Perttu; Hägg, Sara; Matthews, Mary K; Palmen, Jutta; Norata, Giuseppe D; O'Reilly, Paul F; Saleheen, Danish; Amin, Najaf; Balmforth, Anthony J; Beekman, Marian; De Boer, Rudolf A; Böhringer, Stefan; Braund, Peter S; Burton, Paul R; Craen, Anton J Mde; Denniff, Matthew; Dong, Yanbin (2013). "Identification of seven loci affecting mean telomere length and their association with disease". *Nature Genetics* **45** (4): 422–7, 427e1–2. doi:10.1038/ng.2528. PMID 23535734.

[63] "Titanovo, Inc". Titanovo.com. Retrieved 2015-04-15.

[64] "Telome Health, Inc". Telomehealth.com. Retrieved 2013-07-13.

[65] "TeloMe Home". Telome.com. Retrieved 2013-07-13.

[66] "A Blood Test Offers Clues to Longevity".

[67] Zglinicki, T. v. (13 March 2012). "Will your telomeres tell your future?" (PDF). *BMJ* **344** (mar13 1): e1727–e1727. doi:10.1136/bmj.e1727.

[68] Jo Marchant. "Spit test offers guide to health : Nature News". Nature.com. Retrieved 2013-07-13.

[69] "Elizabeth Blackburn calls time on 'fountain of youth' firm Telome Health".

15.10   Further reading

- Aubert G., Lansdorp P.M. (April 2008). "Telomeres and Aging". *Physiological Reviews* **88**(2): 557–579. doi:10.107. PMID 18391173.

- Cong YS, Wright WE, Shay JW (September 2002). "Human telomerase and its regulation". *Microbiol. Mol. Biol. Rev.* **66** (3): 407–25, table of contents. doi:10.1128/MMBR.66.3.407-425.2002. PMC 120798. PMID 12208997.

- Eisenberg DTA (2011). "An evolutionary review of human telomere biology: The thrifty telomere hypothesis and notes on potential adaptive paternal effects". *American Journal of Human Biology* **23** (2): n/a–n/a. doi:10.1002/ajhb 21319244.

- Tomaska L., Nosek J., Kramara J., Griffith J.D. (2009). "Telomeric circles: universal players in telomere maintenance". *Nature Structural & Molecular Biology* **16** (10): 1010–1015. doi:10.1038/nsmb.1660. PMID 19809492.

- Weinstein BS, Ciszek D; Lansdorp (May 2002). "The reserve-capacity hypothesis: evolutionary origins and modern implications of the trade-off between tumor-suppression and tissue-repair". *Exp. Gerontol.* **37** (5): 615–27. doi:10.1016/S0531-5565(02)00012-8. PMID 11909679. — A paper detailing the evolutionary origins and medical implications of the vertebrate telomere system, including the pervasive trade-off between cancer prevention and damage repair. Also addresses the probable danger posed by the elongation of telomeres in lab mice.

15.11   External links

- Elizabeth Blackburn's seminars: "Telomeres and Telomerase"

- Telomeres and Telomerase: The Means to the End Nobel Lecture by Elizabeth Blackburn, which includes a reference to the impact of stress, and pessimism on telomere length

- Telomerase and the Consequences of Telomere Dysfunction Nobel Lecture by Carol Greider

- DNA Ends: Just the Beginning Nobel Lecture by Jack Szostak

Chapter 16

Branched DNA assay

In biology, a **branched DNA assay** is a signal amplification assay (as opposed to a target amplification assay) that is used to detect nucleic acid molecules.

From the base up, a branched DNA assay begins with a dish or some other solid support (e.g., a plastic dipstick). The dish is peppered with small, single stranded DNA molecules (or chains) that 'stick up' into the air. These are known as capture probe DNA molecules. Next, an extender DNA molecule is added. Each extender has two domains, one that hybridizes to the capture DNA molecule and one that "hangs out" in the air. The purpose of the extender is two-fold. First, it creates more available surface area for target DNA molecules to bind, and second, it allows the assay to be easily adapted to detect a variety of target DNA molecules.

Once the capture and extender molecules are in place and they have hybridized, the sample can be added. Target molecules in the sample will bind to the extender molecule. So we have a base peppered with capture probes, which are hybridized to extender probes, which in turn are hybridized to target molecules.

At this point, signal amplification takes place. A label extender DNA molecule is added that has two domains (similar to the first extender). The label extender hybridizes to the target and to a pre-amplified molecule. The preamplifier molecule has two domains. First, it binds to the label extender and second, it binds to the amplifier molecule. An example amplifier molecule is an oligonucleotide chain bound to the enzyme alkaline phosphatase.

Diagrammatically, we have Base -> Capture Probe -> Extender -> Target -> label extender -> pre-amplifier -> amplifier

The assay can be used to detect and quantify many types of RNA or DNA target. In the assay, branched DNA is mixed with a sample to be tested. The detection is done using a non-radioactive method and does not require preamplification of the nucleic acid to be detected. The assay entirely relies on hybridization. Enzymes are used to indicate the extent of hybridization but are not used to manipulate the nucleic acids. Thus, small amounts of a nucleic acid can be detected and quantified without a reverse transcription step (in the case of RNA) and/or PCR. The assay can be run as a high throughput assay, unlike quantitative Northern-blotting or the RNAse-protection assay, which are labor-intensive and thus difficult to perform on a large number of samples. The other major high throughput technique employed in the quantification of specific RNA molecules is quantitative PCR, after reverse transcription of the RNA to cDNA.

Several different short single-stranded DNA molecules (oligonucleotides) are used in a branched DNA-assay. The capture and capture-extender oligonucleotide bind to the target nucleic acid and immobilize it on a solid support. The label oligonucleotide and the branched DNA then detects the immobilized target nucleic acid. The immobilization of the target on a solid support makes extensive washing easier, which reduces false positive results. After binding of the target to the solid support it can be detected by branched DNA which is coupled to an enzyme (e.g. alkaline phosphatase). The branched DNA binds to the sample nucleic acid by specific hybridization in areas which are not occupied by capture hybrids. The branching of the DNA allows for very dense decorating of the DNA with the enzyme, which is important for the high sensitivity of the assay. The enzyme catalyzes a reaction of a substrate which generates light (detectable in a luminometer). The amount of light emitted increases with the amount of the specific nucleic acid present in the sample. The design of the branched DNA and the way it is hybridized to the nucleic acid to be investigated differs between different generations of the bDNA assay.[1] Despite the fact that the starting material is not preamplified, bDNA assays can detect less than 100 copies of HIV-RNA per mL of blood.[1]

16.1   See also

- Dendrimer

16.2   Notes and references

[1] Collins, M. L.; Irvine, B.; Tyner, D.; Fine, E.; Zayati, C.; Chang, C.; Horn, T.; Ahle, D.; Detmer, J.; Shen, L. P.; Kolberg, J.; Bushnell, S.; Urdea, M. S.; Ho, D. D. (1997). "A branched DNA signal amplification assay for quantification of nucleic acid targets below 100 molecules/ml". *Nucleic Acids Research* **25** (15): 2979–2984. doi:10.1093/nar/25.15.2979. PMC 146852. PMID 9224596.

16.3   External links

- Branched DNA Assay at the US National Library of Medicine Medical Subject Headings (MeSH)

Chapter 17

DNA nanotechnology

DNA nanotechnology is the design and manufacture of artificial nucleic acid structures for technological uses. In this field, nucleic acids are used as non-biological engineering materials for nanotechnology rather than as the carriers of genetic information in living cells. Researchers in the field have created static structures such as two- and three-dimensional crystal lattices, nanotubes, polyhedra, and arbitrary shapes, as well as functional devices such as molecular machines and DNA computers. The field is beginning to be used as a tool to solve basic science problems in structural biology and biophysics, including applications in crystallography and spectroscopy for protein structure determination. Potential applications in molecular scale electronics and nanomedicine are also being investigated.

The conceptual foundation for DNA nanotechnology was first laid out by Nadrian Seeman in the early 1980s, and the field began to attract widespread interest in the mid-2000s. This use of nucleic acids is enabled by their strict base pairing rules, which cause only portions of strands with complementary base sequences to bind together to form strong, rigid double helix structures. This allows for the rational design of base sequences that will selectively assemble to form complex target structures with precisely controlled nanoscale features. A number of assembly methods are used to make these structures, including tile-based structures that assemble from smaller structures, folding structures using the DNA origami method, and dynamically reconfigurable structures using strand displacement techniques. While the field's name specifically references DNA, the same principles have been used with other types of nucleic acids as well, leading to the occasional use of the alternative name **nucleic acid nanotechnology**.

17.1 Fundamental concepts

17.1.1 Properties of nucleic acids

Nanotechnology is often defined as the study of materials and devices with features on a scale below 100 nanometers. DNA nanotechnology, specifically, is an example of bottom-up molecular self-assembly, in which molecular components spontaneously organize into stable structures; the particular form of these structures is induced by the physical and chemical properties of the components selected by the designers.[4] In DNA nanotechnology, the component materials are strands of nucleic acids such as DNA; these strands are often synthetic and are almost always used outside the context of a living cell. DNA is well-suited to nanoscale construction because the binding between two nucleic acid strands depends on simple base pairing rules which are well understood, and form the specific nanoscale structure of the nucleic acid double helix. These qualities make the assembly of nucleic acid structures easy to control through nucleic acid design. This property is absent in other materials used in nanotechnology, including proteins, for which protein design is very difficult, and nanoparticles, which lack the capability for specific assembly on their own.[5]

The structure of a nucleic acid molecule consists of a sequence of nucleotides distinguished by which nucleobase they contain. In DNA, the four bases present are adenine (A), cytosine (C), guanine (G), and thymine (T). Nucleic acids have the property that two molecules will only bind to each other to form a double helix if the two sequences are complementary, meaning that they form matching sequences of base pairs, with A only binding to T, and C only to G.[5][6] Because the

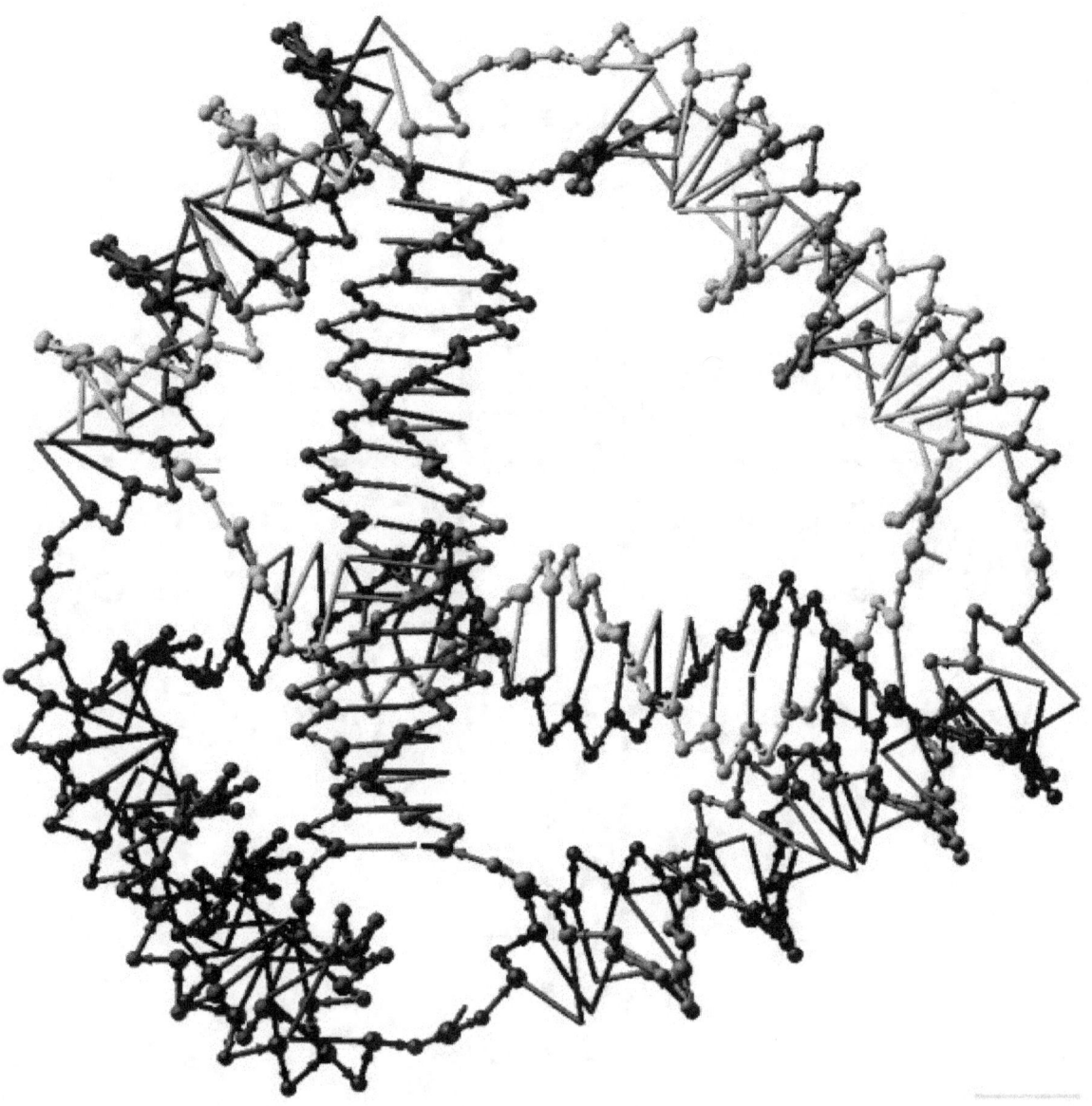

DNA nanotechnology involves the creation of artificial, designed nanostructures out of nucleic acids, such as this DNA tetrahedron.[1] Each edge of the tetrahedron is a 20 base pair DNA double helix, and each vertex is a three-arm junction. The 4 DNA strands that form the 4 tetrahedral faces are color-coded.

formation of correctly matched base pairs is energetically favorable, nucleic acid strands are expected in most cases to bind to each other in the conformation that maximizes the number of correctly paired bases. The sequences of bases in a system of strands thus determine the pattern of binding and the overall structure in an easily controllable way. In DNA nanotechnology, the base sequences of strands are rationally designed by researchers so that the base pairing interactions cause the strands to assemble in the desired conformation.[3][5] While DNA is the dominant material used, structures incorporating other nucleic acids such as RNA and peptide nucleic acid (PNA) have also been constructed.[7][8]

17.1.2 Subfields

DNA nanotechnology is sometimes divided into two overlapping subfields: structural DNA nanotechnology and dynamic DNA nanotechnology. Structural DNA nanotechnology, sometimes abbreviated as SDN, focuses on synthesizing and

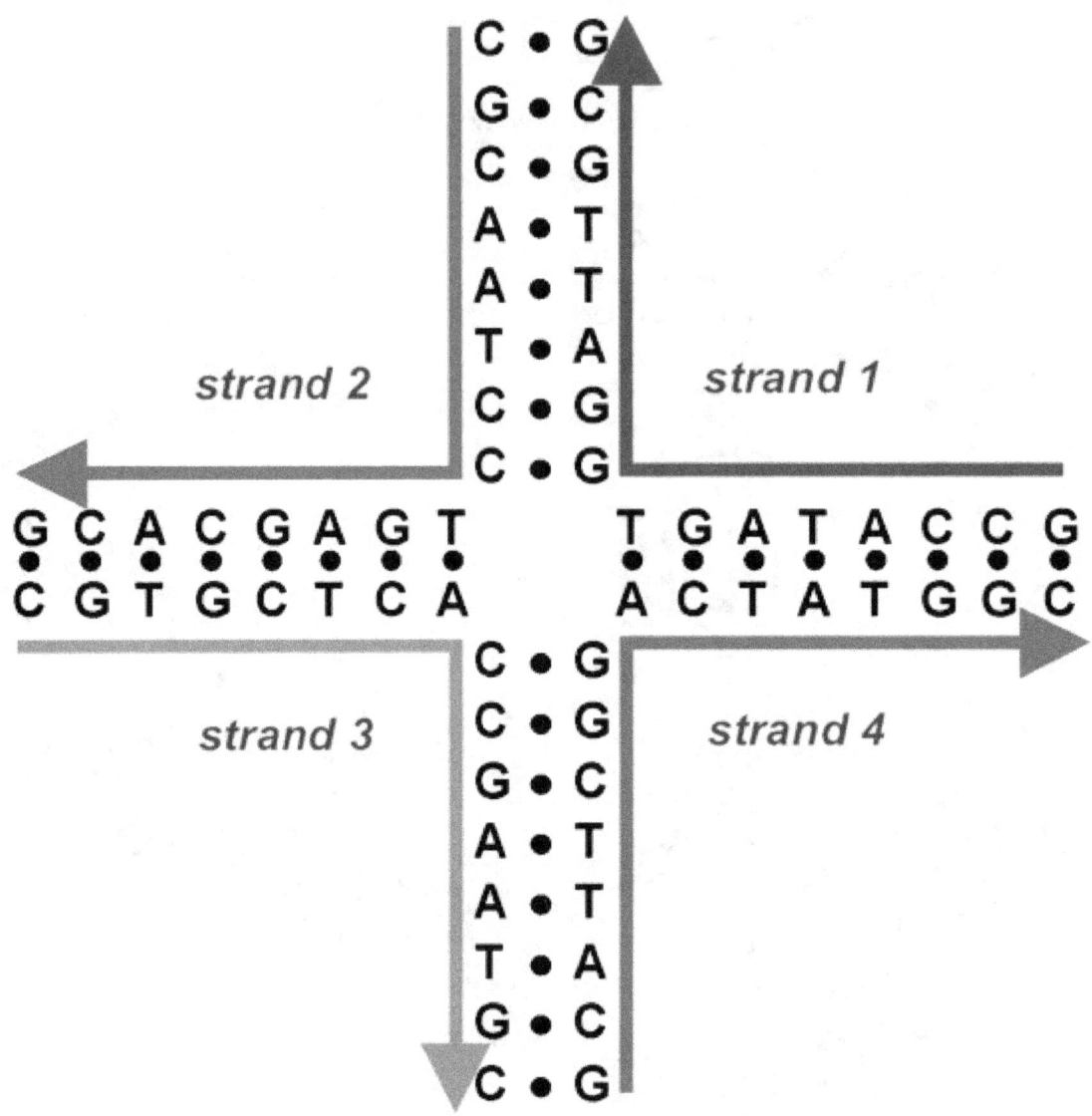

These four strands associate into a DNA four-arm junction because this structure maximizes the number of correct base pairs, with A matched to T and C matched to G.[2][3] See this image for a more realistic model of the four-arm junction showing its tertiary structure.

characterizing nucleic acid complexes and materials that assemble into a static, equilibrium end state. On the other hand, dynamic DNA nanotechnology focuses on complexes with useful non-equilibrium behavior such as the ability to reconfigure based on a chemical or physical stimulus. Some complexes, such as nucleic acid nanomechanical devices, combine features of both the structural and dynamic subfields.[9][10]

The complexes constructed in structural DNA nanotechnology use topologically branched nucleic acid structures containing junctions. (In contrast, most biological DNA exists as an unbranched double helix.) One of the simplest branched structures is a four-arm junction that consists of four individual DNA strands, portions of which are complementary in a specific pattern. Unlike in natural Holliday junctions, each arm in the artificial immobile four-arm junction has a different base sequence, causing the junction point to be fixed at a certain position. Multiple junctions can be combined in the same complex, such as in the widely used double-crossover (DX) motif, which contains two parallel double helical domains with individual strands crossing between the domains at two crossover points. Each crossover point is itself topologically a four-arm junction, but is constrained to a single orientation, as opposed to the flexible single four-arm junction, providing

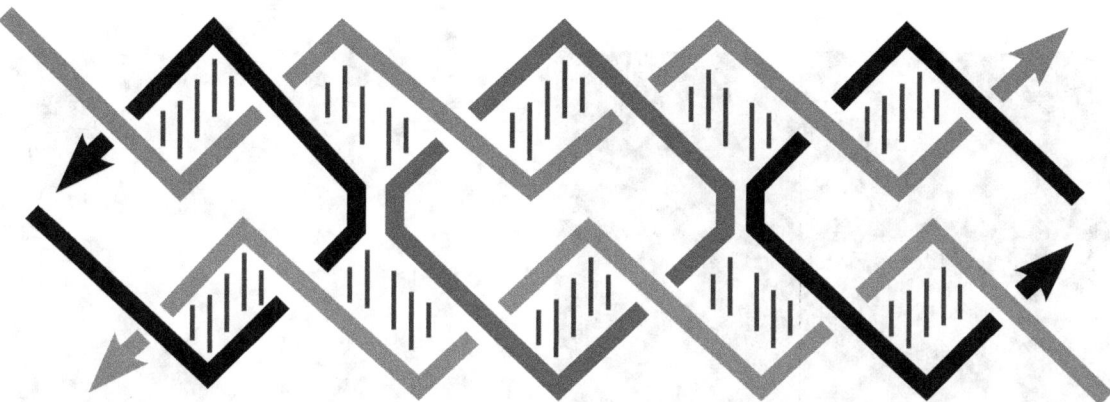

This double-crossover (DX) supramolecular complex consists of five DNA single strands that form two double-helical domains, on the top and the bottom in this image. There are two crossover points where the strands cross from one domain into the other.[2]

a rigidity that makes the DX motif suitable as a structural building block for larger DNA complexes.[3][5]

Dynamic DNA nanotechnology uses a mechanism called toehold-mediated strand displacement to allow the nucleic acid complexes to reconfigure in response to the addition of a new nucleic acid strand. In this reaction, the incoming strand binds to a single-stranded toehold region of a double-stranded complex, and then displaces one of the strands bound in the original complex through a branch migration process. The overall effect is that one of the strands in the complex is replaced with another one.[9] In addition, reconfigurable structures and devices can be made using functional nucleic acids such as deoxyribozymes and ribozymes, which are capable of performing chemical reactions, and aptamers, which can bind to specific proteins or small molecules.[11]

17.2 Structural DNA nanotechnology

Structural DNA nanotechnology, sometimes abbreviated as SDN, focuses on synthesizing and characterizing nucleic acid complexes and materials where the assembly has a static, equilibrium endpoint. The nucleic acid double helix has a robust, defined three-dimensional geometry that makes it possible to predict and design the structures of more complicated nucleic acid complexes. Many such structures have been created, including two- and three-dimensional structures, and periodic, aperiodic, and discrete structures.[10]

17.2.1 Extended lattices

Small nucleic acid complexes can be equipped with sticky ends and combined into larger two-dimensional periodic lattices containing a specific tessellated pattern of the individual molecular tiles.[10] The earliest example of this used double-crossover (DX) complexes as the basic tiles, each containing four sticky ends designed with sequences that caused the DX units to combine into periodic two-dimensional flat sheets that are essentially rigid two-dimensional crystals of DNA.[15][16] Two-dimensional arrays have been made from other motifs as well, including the Holliday junction rhombus lattice,[17] and various DX-based arrays making use of a double-cohesion scheme.[18][19] The top two images at right show examples of tile-based periodic lattices.

Two-dimensional arrays can be made to exhibit aperiodic structures whose assembly implements a specific algorithm, exhibiting one form of DNA computing.[20] The DX tiles can have their sticky end sequences chosen so that they act as Wang tiles, allowing them to perform computation. A DX array whose assembly encodes an XOR operation has been demonstrated; this allows the DNA array to implement a cellular automaton that generates a fractal known as the Sierpinski gasket. The third image at right shows this type of array.[14] Another system has the function of a binary counter, displaying a representation of increasing binary numbers as it grows. These results show that computation can

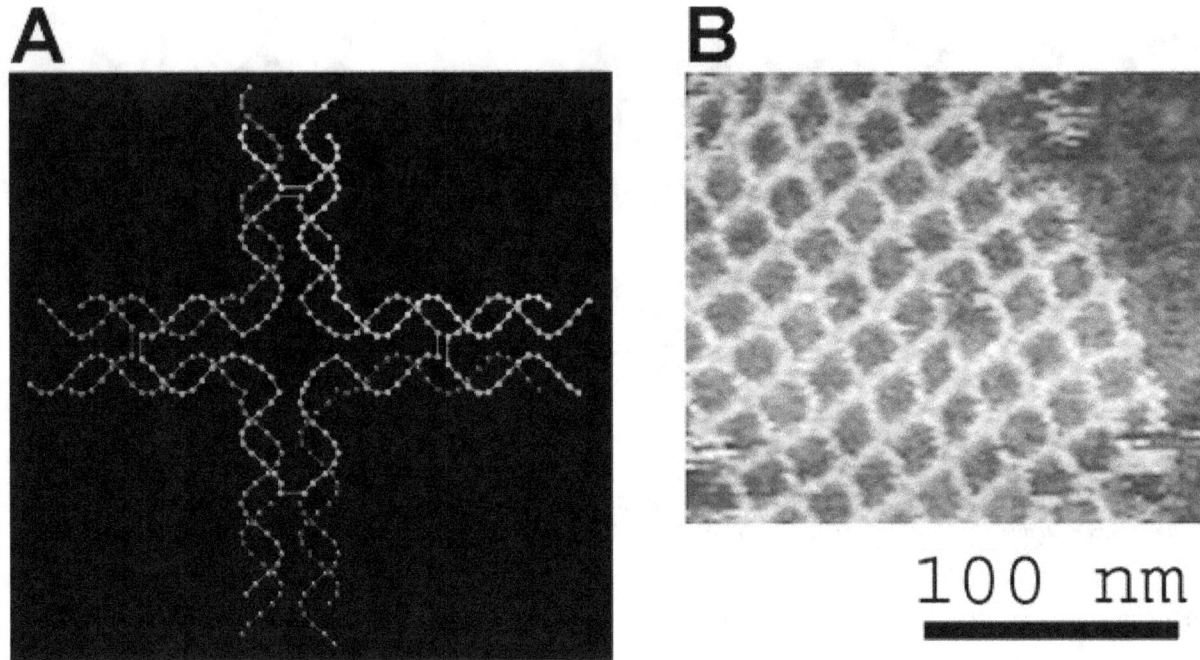

Left, *a model of a DNA tile used to make another two-dimensional periodic lattice.* Right, *an atomic force micrograph of the assembled lattice.*[12][13]

be incorporated into the assembly of DNA arrays.[21]

DX arrays have been made to form hollow nanotubes 4–20 nm in diameter, essentially two-dimensional lattices which curve back upon themselves.[22] These DNA nanotubes are somewhat similar in size and shape to carbon nanotubes, and while they lack the electrical conductance of carbon nanotubes, DNA nanotubes are more easily modified and connected to other structures. One of many schemes for constructing DNA nanotubes uses a lattice of curved DX tiles that curls around itself and closes into a tube.[23] In an alternative method that allows the circumference to be specified in a simple, modular fashion using single-stranded tiles, the rigidity of the tube is an emergent property.[24]

The creation of three-dimensional lattices out of DNA was the earliest goal of DNA nanotechnology, but this proved to be one of the most difficult to realize. Success using a motif based on the concept of tensegrity, a balance between tension and compression forces, was finally reported in 2009.[20][25]

17.2.2   Discrete structures

Researchers have synthesized a number of three-dimensional DNA complexes that each have the connectivity of a polyhedron, such as a cube or octahedron, meaning that the DNA duplexes trace the edges of a polyhedron with a DNA junction at each vertex.[26] The earliest demonstrations of DNA polyhedra were very work-intensive, requiring multiple ligations and solid-phase synthesis steps to create catenated polyhedra.[27] Subsequent work yielded polyhedra whose synthesis was much easier. These include a DNA octahedron made from a long single strand designed to fold into the correct conformation,[28] and a tetrahedron that can be produced from four DNA strands in a single step, pictured at the top of this article.[1]

Nanostructures of arbitrary, non-regular shapes are usually made using the DNA origami method. These structures consist of a long, natural virus strand as a "scaffold", which is made to fold into the desired shape by computationally designed short "staple" strands. This method has the advantages of being easy to design, as the base sequence is predetermined by the scaffold strand sequence, and not requiring high strand purity and accurate stoichiometry, as most other DNA nanotechnology methods do. DNA origami was first demonstrated for two-dimensional shapes, such as a smiley face and a coarse map of the Western Hemisphere.[26][29] Solid three-dimensional structures can be made by using parallel DNA helices arranged in a honeycomb pattern,[30] and structures with two-dimensional faces can be made to fold into a hollow overall three-dimensional shape, akin to a cardboard box. These can be programmed to open and reveal or release a molecular cargo in response to a stimulus, making them potentially useful as programmable molecular cages.[31][32]

17.2.3 Templated assembly

Nucleic acid structures can be made to incorporate molecules other than nucleic acids, sometimes called heteroelements, including proteins, metallic nanoparticles, quantum dots, and fullerenes. This allows the construction of materials and devices with a range of functionalities much greater than is possible with nucleic acids alone. The goal is to use the self-assembly of the nucleic acid structures to template the assembly of the nanoparticles hosted on them, controlling their position and in some cases orientation.[26][33] Many of these schemes use a covalent attachment scheme, using oligonucleotides with amide or thiol functional groups as a chemical handle to bind the heteroelements. This covalent binding scheme has been used to arrange gold nanoparticles on a DX-based array,[34] and to arrange streptavidin protein molecules into specific patterns on a DX array.[35] A non-covalent hosting scheme using Dervan polyamides on a DX array was used to arrange streptavidin proteins in a specific pattern on a DX array.[36] Carbon nanotubes have been hosted on DNA arrays in a pattern allowing the assembly to act as a molecular electronic device, a carbon nanotube field-effect transistor.[37] In addition, there are nucleic acid metallization methods, in which the nucleic acid is replaced by a metal which assumes the general shape of the original nucleic acid structure,[38] and schemes for using nucleic acid nanostructures as lithography masks, transferring their pattern into a solid surface.[39]

17.3 Dynamic DNA nanotechnology

Dynamic DNA nanotechnology focuses on creating nucleic acid systems with designed dynamic functionalities related to their overall structures, such as computation and mechanical motion. There is some overlap between structural and dynamic DNA nanotechnology, as structures can be formed through annealing and then reconfigured dynamically, or can be made to form dynamically in the first place.[26][40]

17.3.1 Nanomechanical devices

Main article: DNA machine

DNA complexes have been made that change their conformation upon some stimulus, making them one form of nanorobotics. These structures are initially formed in the same way as the static structures made in structural DNA nanotechnology, but are designed so that dynamic reconfiguration is possible after the initial assembly.[9][40] The earliest such device made use of the transition between the B-DNA and Z-DNA forms to respond to a change in buffer conditions by undergoing a twisting motion.[41] This reliance on buffer conditions, however, caused all devices to change state at the same time. Subsequent systems could change states based upon the presence of control strands, allowing multiple devices to be independently operated in solution. Some examples of such systems are a "molecular tweezers" design that has an open and a closed state,[42] a device that could switch from a paranemic-crossover (PX) conformation to a double-junction (JX2) conformation, undergoing rotational motion in the process,[43] and a two-dimensional array that could dynamically expand and contract in response to control strands.[44] Structures have also been made that dynamically open or close, potentially acting as a molecular cage to release or reveal a functional cargo upon opening.[31][45][46]

DNA walkers are a class of nucleic acid nanomachines that exhibit directional motion along a linear track. A large number of schemes have been demonstrated.[40] One strategy is to control the motion of the walker along the track using control strands that need to be manually added in sequence.[47][48] Another approach is to make use of restriction enzymes or deoxyribozymes to cleave the strands and cause the walker to move forward, which has the advantage of running autonomously.[49][50] A later system could walk upon a two-dimensional surface rather than a linear track, and demonstrated the ability to selectively pick up and move molecular cargo.[51] Additionally, a linear walker has been demonstrated that performs DNA-templated synthesis as the walker advances along the track, allowing autonomous multistep chemical synthesis directed by the walker.[52] The synthetic DNA walkers' function is similar to that of the proteins dynein and kinesin.[53]

17.3.2   Strand displacement cascades

Cascades of strand displacement reactions can be used for either computational or structural purposes. An individual strand displacement reaction involves revealing a new sequence in response to the presence of some initiator strand. Many such reactions can be linked into a cascade where the newly revealed output sequence of one reaction can initiate another strand displacement reaction elsewhere. This in turn allows for the construction of chemical reaction networks with many components, exhibiting complex computational and information processing abilities. These cascades are made energetically favorable through the formation of new base pairs, and the entropy gain from disassembly reactions. Strand displacement cascades allow for isothermal operation of the assembly or computational process, as opposed to traditional nucleic acid assembly's requirement for a thermal annealing step, where the temperature is raised and then slowly lowered to ensure proper formation of the desired structure. They can also support catalytic functionality of the initiator species, where less than one equivalent of the initiator can cause the reaction to go to completion.[9][54]

Strand displacement complexes can be used to make molecular logic gates capable of complex computation.[55] Unlike traditional electronic computers, which use electric current as inputs and outputs, molecular computers use the concentrations of specific chemical species as signals. In the case of nucleic acid strand displacement circuits, the signal is the presence of nucleic acid strands that are released or consumed by binding and unbinding events to other strands in displacement complexes. This approach has been used to make logic gates such as AND, OR, and NOT gates.[56] More recently, a four-bit circuit was demonstrated that can compute the square root of the integers 0–15, using a system of gates containing 130 DNA strands.[57]

Another use of strand displacement cascades is to make dynamically assembled structures. These use a hairpin structure for the reactants, so that when the input strand binds, the newly revealed sequence is on the same molecule rather than disassembling. This allows new opened hairpins to be added to a growing complex. This approach has been used to make simple structures such as three- and four-arm junctions and dendrimers.[54]

17.4   Applications

DNA nanotechnology provides one of the few ways to form designed, complex structures with precise control over nanoscale features. The field is beginning to see application to solve basic science problems in structural biology and biophysics. The earliest such application envisaged for the field, and one still in development, is in crystallography, where molecules that are difficult to crystallize in isolation could be arranged within a three-dimensional nucleic acid lattice, allowing determination of their structure. Another application is the use of DNA origami rods to replace liquid crystals in residual dipolar coupling experiments in protein NMR spectroscopy; using DNA origami is advantageous because, unlike liquid crystals, they are tolerant of the detergents needed to suspend membrane proteins in solution. DNA walkers have been used as nanoscale assembly lines to move nanoparticles and direct chemical synthesis. Furthermore, DNA origami structures have aided in the biophysical studies of enzyme function and protein folding.[10][58]

DNA nanotechnology is moving towards potential real-world applications. The ability of nucleic acid arrays to arrange other molecules indicates its potential applications in molecular scale electronics. The assembly of a nucleic acid structure could be used to template the assembly of a molecular electronic elements such as molecular wires, providing a method for nanometer-scale control of the placement and overall architecture of the device analogous to a molecular breadboard.[10][26] DNA nanotechnology has been compared to the concept of programmable matter because of the coupling of computation to its material properties.[59]

In a study conducted by a group of scientists from iNANO center and CDNA Center in Aarhus university (Aarhus), researchers were able to construct a small multi-switchable 3D DNA Box Origami. The proposed nanoparticle was characterized by AFM, TEM and FRET. The constructed box was shown to have a unique reclosing mechanism, which enabled it to repeatedly open and close in response to a unique set of DNA or RNA keys. The authors proposed that this "DNA device can potentially be used for a broad range of applications such as controlling the function of single molecules, controlled drug delivery, and molecular computing.".[60]

There are potential applications for DNA nanotechnology in nanomedicine, making use of its ability to perform computation in a biocompatible format to make "smart drugs" for targeted drug delivery. One such system being investigated uses a hollow DNA box containing proteins that induce apoptosis, or cell death, that will only open when in proximity to a cancer cell.[58][61] There has additionally been interest in expressing these artificial structures in engineered living bacterial cells, most likely using the transcribed RNA for the assembly, although it is unknown whether these complex structures are able to efficiently fold or assemble in the cell's cytoplasm. If successful, this could enable directed evolution of nucleic acid nanostructures.[26] Scientists at Oxford University reported the self-assembly of four short strands of synthetic DNA into a cage which is capable of entering cells and surviving for at least 48 hours. The fluorescently labeled DNA tetrahedra were found to remain intact in the laboratory cultured human kidney cells despite the attack by cellular enzymes after two days. This experiment showed the potential of drug delivery inside the living cells using the DNA 'cage'.[62][63] A DNA tetrahedron was used to deliver RNA Interference (RNAi) in a mouse model, reported a team of researchers in MIT. Delivery of the interfering RNA for treatment has showed some success using polymer or lipid, but there are limitations of safety and imprecise targeting, in addition to short shelf life in the blood stream. The DNA nanostructure created by the team consists of six strands of DNA to form a tetrahedron, with a single strand of RNA affixed to each of the six edges. The tetrahedron is further equipped with targeting protein, three folate molecules, which lead the DNA nanoparticles to the abundant folate receptors found on some tumors. The result showed that the gene expression targeted by the RNAi, luciferase, dropped by more than half. This study shows promise in using DNA nanotechnology as an effective tool to deliver treatment using the emerging RNA Interference technology.[64][65]

17.5 Design

DNA nanostructures must be rationally designed so that the individual nucleic acid strands will assemble into the desired structures. This process usually begins with the specification of a desired target structure or functionality. Then, the overall secondary structure of the target complex is determined, specifying the arrangement of nucleic acid strands within the structure, and which portions of those strands should be bound to each other. The last step is the primary structure design, which is the specification of the actual base sequences of each nucleic acid strand.[22][66]

17.5.1 Structural design

The first step in designing a nucleic acid nanostructure is to decide how a given structure should be represented by a specific arrangement of nucleic acid strands. This design step determines the secondary structure, or the positions of the base pairs that hold the individual strands together in the desired shape.[22] Several approaches have been demonstrated:

- **Tile-based structures.** This approach breaks the target structure into smaller units with strong binding between the strands contained in each unit, and weaker interactions between the units. It is often used to make periodic lattices, but can also be used to implement algorithmic self-assembly, making them a platform for DNA computing. This was the dominant design strategy used from the mid-1990s until the mid-2000s, when the DNA origami methodology was developed.[22][67]

- **Folding structures.** An alternative to the tile-based approach, folding approaches make the nanostructure from a single long strand. This long strand can either have a designed sequence that folds due to its interactions with itself, or it can be folded into the desired shape by using shorter, "staple" strands. This latter method is called DNA origami, which allows the creation of nanoscale two- and three-dimensional shapes (see Discrete structures below).[26][29]

- **Dynamic assembly.** This approach directly controls the kinetics of DNA self-assembly, specifying all of the intermediate steps in the reaction mechanism in addition to the final product. This is done using starting materials which adopt a hairpin structure; these then assemble into the final conformation in a cascade reaction, in a specific order (see Strand displacement cascades below). This approach has the advantage of proceeding isothermally, at a constant temperature. This is in contrast to the thermodynamic approaches, which require a thermal annealing step where a temperature change is required to trigger the assembly and favor proper formation of the desired structure.[26][54]

17.5.2   Sequence design

Main article: Nucleic acid design

After any of the above approaches are used to design the secondary structure of a target complex, an actual sequence of nucleotides that will form into the desired structure must be devised. Nucleic acid design is the process of assigning a specific nucleic acid base sequence to each of a structure's constituent strands so that they will associate into a desired conformation. Most methods have the goal of designing sequences so that the target structure has the lowest energy, and is thus the most thermodynamically favorable, while incorrectly assembled structures have higher energies and are thus disfavored. This is done either through simple, faster heuristic methods such as sequence symmetry minimization, or by using a full nearest-neighbor thermodynamic model, which is more accurate but slower and more computationally intensive. Geometric models are used to examine tertiary structure of the nanostructures and to ensure that the complexes are not overly strained.[66][68]

Nucleic acid design has similar goals to protein design. In both, the sequence of monomers is designed to favor the desired target structure and to disfavor other structures. Nucleic acid design has the advantage of being much computationally easier than protein design, because the simple base pairing rules are sufficient to predict a structure's energetic favorability, and detailed information about the overall three-dimensional folding of the structure is not required. This allows the use of simple heuristic methods that yield experimentally robust designs. However, nucleic acid structures are less versatile than proteins in their functionality because of proteins' increased ability to fold into complex structures, as well as the limited chemical diversity of the four nucleotides as compared to the twenty proteinogenic amino acids.[68]

17.6   Materials and methods

The sequences of the DNA strands making up a target structure are designed computationally, using molecular modeling and thermodynamic modeling software.[66][68] The nucleic acids themselves are then synthesized using standard oligonucleotide synthesis methods, usually automated in an oligonucleotide synthesizer, and strands of custom sequences are commercially available.[69] Strands can be purified by denaturing gel electrophoresis if needed,[70] and precise concentrations determined via any of several nucleic acid quantitation methods using ultraviolet absorbance spectroscopy.[71]

The fully formed target structures can be verified using native gel electrophoresis, which gives size and shape information for the nucleic acid complexes. An electrophoretic mobility shift assay can assess whether a structure incorporates all desired strands.[72] Fluorescent labeling and Förster resonance energy transfer (FRET) are sometimes used to characterize the structure of the complexes.[73]

Nucleic acid structures can be directly imaged by atomic force microscopy, which is well suited to extended two-dimensional structures, but less useful for discrete three-dimensional structures because of the microscope tip's interaction with the fragile nucleic acid structure; transmission electron microscopy and cryo-electron microscopy are often used in this case. Extended three-dimensional lattices are analyzed by X-ray crystallography.[74][75]

17.7   History

The conceptual foundation for DNA nanotechnology was first laid out by Nadrian Seeman in the early 1980s.[76] Seeman's original motivation was to create a three-dimensional DNA lattice for orienting other large molecules, which would simplify their crystallographic study by eliminating the difficult process of obtaining pure crystals. This idea had reportedly come to him in late 1980, after realizing the similarity between the woodcut *Depth* by M. C. Escher and an array of DNA six-arm junctions.[3][77] A number of natural branched DNA structures were known at the time, including the DNA replication fork and the mobile Holliday junction, but Seeman's insight was that immobile nucleic acid junctions could be created by properly designing the strand sequences to remove symmetry in the assembled molecule, and that these immobile junctions could in principle be combined into rigid crystalline lattices. The first theoretical paper proposing this scheme was published in 1982, and the first experimental demonstration of an immobile DNA junction was published the following year.[5][26]

In 1991, Seeman's laboratory published a report on the synthesis of a cube made of DNA, the first synthetic three-dimensional nucleic acid nanostructure, for which he received the 1995 Feynman Prize in Nanotechnology. This was followed by a DNA truncated octahedron. However, it soon became clear that these structures, polygonal shapes with flexible junctions as their vertices, were not rigid enough to form extended three-dimensional lattices. Seeman developed the more rigid double-crossover (DX) motif, and in 1998, in collaboration with Erik Winfree, published the creation of two-dimensional lattices of DX tiles.[3][76][78] These tile-based structures had the advantage that they provided the capability to implement DNA computing, which was demonstrated by Winfree and Paul Rothemund in their 2004 paper on the algorithmic self-assembly of a Sierpinski gasket structure, and for which they shared the 2006 Feynman Prize in Nanotechnology. Winfree's key insight was that the DX tiles could be used as Wang tiles, meaning that their assembly was capable of performing computation.[76] The synthesis of a three-dimensional lattice was finally published by Seeman in 2009, nearly thirty years after he had set out to achieve it.[58]

New capabilities continued to be discovered for designed DNA structures throughout the 2000s. The first DNA nanomachine—a motif that changes its structure in response to an input—was demonstrated in 1999 by Seeman. An improved system, which was the first nucleic acid device to make use of toehold-mediated strand displacement, was demonstrated by Bernard Yurke the following year. The next advance was to translate this into mechanical motion, and in 2004 and 2005, a number of DNA walker systems were demonstrated by the groups of Seeman, Niles Pierce, Andrew Turberfield, and Chengde Mao.[40] The idea of using DNA arrays to template the assembly of other molecules such as nanoparticles and proteins, first suggested by Bruche Robinson and Seeman in 1987,[79] was demonstrated in 2002 by Seeman, Kiehl et al.[80] and subsequently by numerous other groups.

In 2006, Rothemund first demonstrated the DNA origami technique for easily and robustly creating folded DNA structures of arbitrary shape. Rothemund had conceived of this method as being conceptually intermediate between Seeman's DX lattices, which used many short strands, and William Shih's DNA octahedron, which consisted mostly of one very long strand. Rothemund's DNA origami contains a long strand whose folding is assisted by a number of short strands. This method allowed the creation of much larger structures than were previously possible, and which are less technically demanding to design and synthesize.[78] DNA origami was the cover story of *Nature* on March 15, 2006.[29] Rothemund's research demonstrating two-dimensional DNA origami structures was followed by the demonstration of solid three-dimensional DNA origami by Douglas *et al.* in 2009,[30] while the labs of Jørgen Kjems and Yan demonstrated hollow three-dimensional structures made out of two-dimensional faces.[58]

DNA nanotechnology was initially met with some skepticism due to the unusual non-biological use of nucleic acids as materials for building structures and doing computation, and the preponderance of proof of principle experiments that extended the capabilities of the field but were far from actual applications. Seeman's 1991 paper on the synthesis of the DNA cube was rejected by the journal *Science* after one reviewer praised its originality while another criticized it for its lack of biological relevance. By the early 2010s, however, the field was considered to have increased its capabilities to the point that applications for basic science research were beginning to be realized, and practical applications in medicine and other fields were beginning to be considered feasible.[58][81] The field had grown from very few active laboratories in 2001 to at least 60 in 2010, which increased the talent pool and thus the number of scientific advances in the field during that decade.[20]

17.8 See also

- International Society for Nanoscale Science, Computation, and Engineering
- Nanobiotechnology
- Molecular models of DNA
- List of nucleic acid simulation software

17.9 References

[1] **DNA polyhedra:** Goodman, Russel P.; Schaap, Iwan A. T.; Tardin, C. F.; Erben, Christof M.; Berry, Richard M.; Schmidt, C.F.; Turberfield, Andrew J. (9 December 2005). "Rapid chiral assembly of rigid DNA building blocks for molecular nanofab-

rication". *Science* **310** (5754): 1661–1665. Bibcode:2005Sci...310.1661G. doi:10.1126/science.1120367. PMID 16339440.

[2] **Overview:** Mao, Chengde (December 2004). "The emergence of complexity: lessons from DNA". *PLoS Biology* **2** (12): 2036–2038. doi:10.1371/journal.pbio.0020431. PMC 535573. PMID 15597116.

[3] **Overview:** Seeman, Nadrian C. (June 2004). "Nanotechnology and the double helix". *Scientific American* **290** (6): 64–75. doi:10.1038/scientificamerican0604-64. PMID 15195395.

[4] **Background:** Pelesko, John A. (2007). *Self-assembly: the science of things that put themselves together.* New York: Chapman & Hall/CRC. pp. 5, 7. ISBN 978-1-58488-687-7.

[5] **Overview:** Seeman, Nadrian C. (2010). "Nanomaterials based on DNA". *Annual Review of Biochemistry***79**: 65–87. doi:10.11 biochem-060308-102244. PMC 3454582. PMID 20222824.

[6] **Background:** Long, Eric C. (1996). "Fundamentals of nucleic acids". In Hecht, Sidney M. *Bioorganic chemistry: nucleic acids.* New York: Oxford University Press. pp. 4–10. ISBN 0-19-508467-5.

[7] **RNA nanotechnology:** Chworos, Arkadiusz; Severcan, Isil; Koyfman, Alexey Y.; Weinkam, Patrick; Oroudjev, Emin; Hansma, Helen G.; Jaeger, Luc (2004). "Building Programmable Jigsaw Puzzles with RNA". *Science* **306** (5704): 2068–2072. Bibcode:2 doi:10.1126/science.1104686. PMID 15604402.

[8] **RNA nanotechnology:** Guo, Peixuan (2010). "The Emerging Field of RNA Nanotechnology". *Nature Nanotechnology* **5** (12): 833–842. Bibcode:2010NatNa...5..833G. doi:10.1038/nnano.2010.231. PMC 3149862. PMID 21102465.

[9] **Dynamic DNA nanotechnology:** Zhang, D. Y.; Seelig, G. (February 2011). "Dynamic DNA nanotechnology using strand-displacement reactions". *Nature Chemistry* **3** (2): 103–113. Bibcode:2011NatCh...3..103Z. doi:10.1038/nchem.957. PMID 21258382.

[10] **Structural DNA nanotechnology:** Seeman, Nadrian C. (November 2007). "An overview of structural DNA nanotechnology". *Molecular Biotechnology* **37** (3): 246–257. doi:10.1007/s12033-007-0059-4. PMC 3479651. PMID 17952671.

[11] **Dynamic DNA nanotechnology:** Lu, Y.; Liu, J. (December 2006). "Functional DNA nanotechnology: Emerging applications of DNAzymes and aptamers". *Current Opinion in Biotechnology* **17** (6): 580–588. doi:10.1016/j.copbio.2006.10.004. PMID 17056247.

[12] **Other arrays:** Strong, Michael (March 2004). "Protein Nanomachines". *PLoS Biology***2**(3): e73. doi:10.1371/journal.pbio.00 PMC 368168. PMID 15024422.

[13] Yan, H.; Park, S. H.; Finkelstein, G.; Reif, J. H.; Labean, T. H. (26 September 2003). "DNA-templated self-assembly of protein arrays and highly conductive nanowires". *Science* **301** (5641): 1882–1884. Bibcode:2003Sci...301.1882Y. doi:10.1126/science. PMID14512621.

[14] **Algorithmic self-assembly:** Rothemund, Paul W. K.; Papadakis, Nick; Winfree, Erik (December 2004). "Algorithmic self-assembly of DNA Sierpinski triangles". *PLoS Biology* **2** (12): 2041–2053. doi:10.1371/journal.pbio.0020424. PMC 534809. PMID 15583715.

[15] **DX arrays:** Winfree, Erik; Liu, Furong; Wenzler, Lisa A.; Seeman, Nadrian C. (6 August 1998). "Design and self-assembly of two-dimensional DNA crystals". *Nature* **394** (6693): 529–544. Bibcode:1998Natur.394..539W. doi:10.1038/28998. PMID 9707114.

[16] **DX arrays:** Liu, Furong; Sha, Ruojie; Seeman, Nadrian C. (10 February 1999). "Modifying the surface features of two-dimensional DNA crystals". *Journal of the American Chemical Society* **121** (5): 917–922. doi:10.1021/ja982824a.

[17] **Other arrays:** Mao, Chengde; Sun, Weiqiong; Seeman, Nadrian C. (16 June 1999). "Designed two-dimensional DNA Holliday junction arrays visualized by atomic force microscopy". *Journal of the American Chemical Society* **121** (23): 5437–5443. doi:10.1021/ja9900398.

[18] **Other arrays:** Constantinou, Pamela E.; Wang, Tong; Kopatsch, Jens; Israel, Lisa B.; Zhang, Xiaoping; Ding, Baoquan; Sherman, William B.; Wang, Xing; Zheng, Jianping; Sha, Ruojie; Seeman, Nadrian C. (21 September 2006). "Double cohesion in structural DNA nanotechnology". *Organic and Biomolecular Chemistry* **4** (18): 3414–3419. doi:10.1039/b605212f. PMC 3491902. PMID 17036134.

[19] **Other arrays:** Mathieu, Frederick; Liao, Shiping; Kopatsch, Jens; Wang, Tong; Mao, Chengde; Seeman, Nadrian C. (April 2005). "Six-helix bundles designed from DNA". *Nano Letters* **5** (4): 661–665. Bibcode:2005NanoL...5..661M. doi:10.1021/nl0 50084f.PMC3464188. PMID 15826105.

[20] **History**: Seeman, Nadrian (9 June 2010). "Structural DNA nanotechnology: growing along with Nano Letters". *Nano Letters* **10** (6): 1971–1978. Bibcode:2010NanoL..10.1971S. doi:10.1021/nl101262u. PMC 2901229. PMID 20486672.

[21] **Algorithmic self-assembly:** Barish, Robert D.; Rothemund, Paul W. K.; Winfree, Erik (December 2005). "Two computational primitives for algorithmic self-assembly: copying and counting". *Nano Letters* **5** (12): 2586–2592. Bibcode:2005NanoL...5 .2586B.doi:10.1021/nl052038l. PMID 16351220.

[22] **Design:** Feldkamp, U.; Niemeyer, C. M. (13 March 2006). "Rational design of DNA nanoarchitectures". *Angewandte Chemie International Edition* **45** (12): 1856–1876. doi:10.1002/anie.200502358. PMID 16470892.

[23] **DNA nanotubes:** Rothemund, Paul W. K.; Ekani-Nkodo, Axel; Papadakis, Nick; Kumar, Ashish; Fygenson, Deborah Kuchnir & Winfree, Erik (22 December 2004). "Design and Characterization of Programmable DNA Nanotubes". *Journal of the American Chemical Society* **126** (50): 16344–16352. doi:10.1021/ja0443191. PMID 15600335.

[24] **DNA nanotubes:** Yin, P.; Hariadi, R. F.; Sahu, S.; Choi, H. M. T.; Park, S. H.; Labean, T. H.; Reif, J. H. (8 August 2008). "Programming DNA Tube Circumferences". *Science* **321** (5890): 824–826. Bibcode:2008Sci...321..824Y. doi:10.1126/science.115 7312.PMID18687961.

[25] **Three-dimensional arrays:** Zheng, Jianping; Birktoft, Jens J.; Chen, Yi; Wang, Tong; Sha, Ruojie; Constantinou, Pamela E.; Ginell, Stephan L.; Mao, Chengde; Seeman, Nadrian C. (3 September 2009). "From molecular to macroscopic via the rational design of a self-assembled 3D DNA crystal". *Nature* **461** (7260): 74–77. Bibcode:2009Natur.461...74Z. doi:10.1038/nature08274. PMC 2764300. PMID 19727196.

[26] **Overview:** Pinheiro, A. V.; Han, D.; Shih, W. M.; Yan, H. (December 2011). "Challenges and opportunities for structural DNA nanotechnology". *Nature Nanotechnology* **6** (12): 763–772. Bibcode:2011NatNa...6..763P. doi:10.1038/nnano.2011.187. PMC 3334823. PMID 22056726.

[27] **DNA polyhedra:** Zhang, Yuwen; Seeman, Nadrian C. (1 March 1994). "Construction of a DNA-truncated octahedron". *Journal of the American Chemical Society* **116** (5): 1661–1669. doi:10.1021/ja00084a006.

[28] **DNA polyhedra:** Shih, William M.; Quispe, Joel D.; Joyce, Gerald F. (12 February 2004). "A 1.7-kilobase single-stranded DNA that folds into a nanoscale octahedron". *Nature* **427** (6975): 618–621. Bibcode:2004Natur.427..618S. doi:10.1038/natur e02307.PMID14961116.

[29] **DNA origami:** Rothemund, Paul W. K. (16 March 2006). "Folding DNA to create nanoscale shapes and patterns". *Nature* **440** (7082): 297–302. Bibcode:2006Natur.440..297R. doi:10.1038/nature04586. PMID 16541064.

[30] **DNA origami:** Douglas, Shawn M.; Dietz, Hendrik; Liedl, Tim; Högberg, Björn; Graf, Franziska; Shih, William M. (21 May 2009). "Self-assembly of DNA into nanoscale three-dimensional shapes". *Nature* **459** (7245): 414–418. Bibcode:2009Natur.45 9..414D.doi:10.1038/nature08016. PMC2688462. PMID 19458720.

[31] **DNA boxes:** Andersen, Ebbe S.; Dong, Mingdong; Nielsen, Morten M.; Jahn, Kasper; Subramani, Ramesh; Mamdouh, Wael; Golas, Monika M.; Sander, Bjoern; et al. (7 May 2009). "Self-assembly of a nanoscale DNA box with a controllable lid". *Nature* **459** (7243): 73–76. Bibcode:2009Natur.459...73A. doi:10.1038/nature07971. PMID 19424153.

[32] **DNA boxes:** Ke, Yonggang; Sharma, Jaswinder; Liu, Minghui; Jahn, Kasper; Liu, Yan; Yan, Hao (10 June 2009). "Scaffolded DNA origami of a DNA tetrahedron molecular container". *Nano Letters* **9** (6): 2445–2447. Bibcode:2009NanoL...9.2445K. doi:10.1021/nl901165f. PMID 19419184.

[33] **Overview:** Endo, M.; Sugiyama, H. (12 October 2009). "Chemical approaches to DNA nanotechnology". *ChemBioChem* **10** (15): 2420–2443. doi:10.1002/cbic.200900286. PMID 19714700.

[34] **Nanoarchitecture:** Zheng, Jiwen; Constantinou, Pamela E.; Micheel, Christine; Alivisatos, A. Paul; Kiehl, Richard A.; Seeman Nadrian C. (July 2006). "2D Nanoparticle Arrays Show the Organizational Power of Robust DNA Motifs". *Nano Letters* **6** (7): 1502–1504. Bibcode:2006NanoL...6.1502Z. doi:10.1021/nl060994c. PMC 3465979. PMID 16834438.

[35] **Nanoarchitecture:** Park, Sung Ha; Pistol, Constantin; Ahn, Sang Jung; Reif, John H.; Lebeck, Alvin R.; Dwyer, Chris; LaBean, Thomas H. (October 2006). "Finite-size, fully addressable DNA tile lattices formed by hierarchical assembly procedures". *Angewandte Chemie* **118** (40): 749–753. doi:10.1002/ange.200690141.

[36] **Nanoarchitecture:** Cohen, Justin D.; Sadowski, John P.; Dervan, Peter B. (22 October 2007). "Addressing single molecules on DNA nanostructures". *Angewandte Chemie International Edition* **46** (42): 7956–7959. doi:10.1002/anie.200702767. PMID 17763481.

[37] **Nanoarchitecture:** Maune, Hareem T.; Han, Si-Ping; Barish, Robert D.; Bockrath, Marc; Goddard III, William A.; Rothemund, Paul W. K.; Winfree, Erik (January 2009). "Self-assembly of carbon nanotubes into two-dimensional geometries using DNA origami templates". *Nature Nanotechnology* **5** (1): 61–66. Bibcode:2010NatNa...5...61M. doi:10.1038/nnano.2009.311. PMID 19898497.

[38] **Nanoarchitecture:** Liu, J.; Geng, Y.; Pound, E.; Gyawali, S.; Ashton, J. R.; Hickey, J.; Woolley, A. T.; Harb, J. N. (22 March 2011). "Metallization of branched DNA origami for nanoelectronic circuit fabrication". *ACS Nano* **5** (3): 2240–2247. doi:10.1021/nn1035075. PMID 21323323.

[39] **Nanoarchitecture:** Deng, Z.; Mao, C. (6 August 2004). "Molecular lithography with DNA nanostructures". *Angewandte Chemie International Edition* **43** (31): 4068. doi:10.1002/anie.200460257.

[40] **DNA machines:** Bath, Jonathan; Turberfield, Andrew J. (May 2007). "DNA nanomachines". *Nature Nanotechnology* **2** (5): 275–284. Bibcode:2007NatNa...2..275B. doi:10.1038/nnano.2007.104. PMID 18654284.

[41] **DNA machines:** Mao, Chengde; Sun, Weiqiong; Shen, Zhiyong; Seeman, Nadrian C. (14 January 1999). "A DNA nanomechanical device based on the B-Z transition". *Nature* **397** (6715): 144–146. Bibcode:1999Natur.397..144M. doi:10.1038/16437. PMID 9923675.

[42] **DNA machines:** Yurke, Bernard; Turberfield, Andrew J.; Mills, Allen P., Jr; Simmel, Friedrich C.; Neumann, Jennifer L. (10 August 2000). "A DNA-fuelled molecular machine made of DNA". *Nature* **406** (6796): 605–609. Bibcode:2000Natur.406..605Y. doi:10.1038/35020524. PMID 10949296.

[43] **DNA machines:** Yan, Hao; Zhang, Xiaoping; Shen, Zhiyong; Seeman, Nadrian C. (3 January 2002). "A robust DNA mechanical device controlled by hybridization topology". *Nature* **415** (6867): 62–65. Bibcode:2002Natur.415...62Y. doi:10.1038/415062a. PMID 11780115.

[44] **DNA machines:** Feng, L.; Park, S. H.; Reif, J. H.; Yan, H. (22 September 2003). "A two-state DNA lattice switched by DNA nanoactuator". *Angewandte Chemie* **115** (36): 4478. doi:10.1002/ange.200351818.

[45] **DNA machines:** Goodman, R. P.; Heilemann, M.; Doose, S. R.; Erben, C. M.; Kapanidis, A. N.; Turberfield, A. J. (February 2008). "Reconfigurable, braced, three-dimensional DNA nanostructures". *Nature Nanotechnology* **3** (2): 93–96. Bibcode:2008NatNa...3...93G.doi:10.1038/nnano.2008.3. PMID 18654468.

[46] **Applications:** Douglas, Shawn M.; Bachelet, Ido; Church, George M. (17 February 2012). "A logic-gated nanorobot for targeted transport of molecular payloads". *Science* **335**(6070): 831–834. Bibcode:2012Sci...335..831D.doi:10.1126/science.1214

[47] **DNA walkers:** Shin, Jong-Shik; Pierce, Niles A. (8 September 2004). "A synthetic DNA walker for molecular transport". *Journal of the American Chemical Society* **126** (35): 10834–10835. doi:10.1021/ja047543j. PMID 15339155.

[48] **DNA walkers:** Sherman, William B.; Seeman, Nadrian C. (July 2004). "A precisely controlled DNA biped walking device". *Nano Letters* **4** (7): 1203–1207. Bibcode:2004NanoL...4.1203S. doi:10.1021/nl049527q.

[49] **DNA walkers:** Tian, Ye; He, Yu; Chen, Yi; Yin, Peng; Mao, Chengde (11 July 2005). "A DNAzyme that walks processively and autonomously along a one-dimensional track". *Angewandte Chemie* **117** (28): 4429–4432. doi:10.1002/ange.200500703.

[50] **DNA walkers:** Bath, Jonathan; Green, Simon J.; Turberfield, Andrew J. (11 July 2005). "A free-running DNA motor powered by a nicking enzyme". *Angewandte Chemie International Edition* **44** (28): 4358–4361. doi:10.1002/anie.200501262.

[51] **Functional DNA walkers:** Lund, Kyle; Manzo, Anthony J.; Dabby, Nadine; Michelotti, Nicole; Johnson-Buck, Alexander; Nangreave, Jeanette; Taylor, Steven; Pei, Renjun; Stojanovic, Milan N.; Walter, Nils G.; Winfree, Erik; Yan, Hao (13 May 2010). "Molecular robots guided by prescriptive landscapes". *Nature* **465** (7295): 206–210. Bibcode:2010Natur.465..206L. doi:10.1038/nature09012. PMC 2907518. PMID 20463735.

[52] **Functional DNA walkers:** He, Yu; Liu, David R. (November 2010). "Autonomous multistep organic synthesis in a single isothermal solution mediated by a DNA walker". *Nature Nanotechnology* **5** (11): 778–782. Bibcode:2010NatNa...5..778H. doi:10.1038/nnano.2010.190. PMC 2974042. PMID 20935654.

[53] "Recent progress on DNA based walkers". *www.sciencedirect.com.proxy1.lib.uwo.ca*. Retrieved 2015-09-28.

[54] **Kinetic assembly:** Yin, Peng; Choi, Harry M. T.; Calvert, Colby R.; Pierce, Niles A. (17 January 2008). "Programming biomolecular self-assembly pathways". *Nature* **451** (7176): 318–322. Bibcode:2008Natur.451..318Y. doi:10.1038/nature06451. PMID 18202654.

[55] **Fuzzy and Boolean logic gates based on DNA:** Zadegan, R. M.; Jepsen, M. D. E.; Hildebrandt, L. L.; Birkedal, V.; Kjems, J. R. (2015). "Construction of a Fuzzy and Boolean Logic Gates Based on DNA". *Small* **11** (15): 1811. doi:10.1002/smll.201402755. PMID 25565140.

[56] **Strand displacement cascades:** Seelig, G.; Soloveichik, D.; Zhang, D. Y.; Winfree, E. (8 December 2006). "Enzyme-free nucleic acid logic circuits". *Science* **314** (5805): 1585–1588. Bibcode:2006Sci...314.1585S. doi:10.1126/science.1132493. PMID 17158324.

[57] **Strand displacement cascades:** Qian, Lulu; Winfree, Erik (3 June 2011). "Scaling up digital circuit computation with DNA strand displacement cascades". *Science* **332** (6034): 1196–1201. Bibcode:2011Sci...332.1196Q. doi:10.1126/science.1200520. PMID 21636773.

[58] **History/applications:** Service, Robert F. (3 June 2011). "DNA nanotechnology grows up". *Science* **332** (6034): 1140–1143. doi:10.1126/science.332.6034.1140.

[59] **Applications:** Rietman, Edward A. (2001). *Molecular engineering of nanosystems.* Springer. pp. 209–212. ISBN 978-0-387-98988-4. Retrieved 17 April 2011.

[60] M. Zadegan, Reza; et, al. (2012). "Construction of a 4 Zeptoliters Switchable 3D DNA Box Origami". *ACS Nano* **6** (11): 10050–10053. doi:10.1021/nn303767b.

[61] **Applications:** Jungmann, Ralf; Renner, Stephan; Simmel, Friedrich C. (March 2008). "From DNA nanotechnology to synthetic biology". *HFSP journal* **2** (2): 99–109. doi:10.2976/1.2896331. PMC 2645571. PMID 19404476.

[62] Lovy, Howard (5 July 2011). "DNA cages can unleash meds inside cells". fiercedrugdelivery.com. Retrieved 22 September 2013.

[63] Walsh, Anthony; Yin, Hai; Erben, Christoph; Wood, Matthew; Turberfield, Andrew (2011). "DNA Cage Delivery to Mammalian Cells". *ACS Nano* (ACS Publications) **5** (7): 5427–5432. doi:10.1021/nn2005574. PMID 21696187.

[64] Trafton, Anne (4 June 2012). "Researchers achieve RNA interference, in a lighter package". MIT News. Retrieved 22 September 2013.

[65] Lee, Hyukjin; Lytton-Jean, Abigail; Chen, Yi; Love, Kevin; Park, Angela; Karagiannis, Emmanouil; Sehgal, Alfica; Querbes, William; et al. (2012). "Molecularly self-assembled nucleic acid nanoparticles for targeted in vivo siRNA delivery" (PDF). *Nature Nanotechnology* (Nature) **7** (6): 389–393. Bibcode:2012NatNa...7..389L. doi:10.1038/NNANO.2012.73.

[66] **Design:** Brenneman, Arwen; Condon, Anne (25 September 2002). "Strand design for biomolecular computation". *Theoretical Computer Science* **287**: 39–58. doi:10.1016/S0304-3975(02)00135-4.

[67] **Overview:** Lin, Chenxiang; Liu, Yan; Rinker, Sherri; Yan, Hao (11 August 2006). "DNA tile based self-assembly: building complex nanoarchitectures". *ChemPhysChem* **7** (8): 1641–1647. doi:10.1002/cphc.200600260. PMID 16832805.

[68] **Design:** Dirks, Robert M.; Lin, Milo; Winfree, Erik; Pierce, Niles A. (15 February 2004). "Paradigms for computational nucleic acid design". *Nucleic Acids Research* **32** (4): 1392–1403. doi:10.1093/nar/gkh291. PMC 390280. PMID 14990744.

[69] **Methods:** Ellington, A.; Pollard, J. D. (1 May 2001). "Synthesis and purification of oligonucleotides". *Current Protocols in Molecular Biology*. doi:10.1002/0471142727.mb0211s42. ISBN 0471142727.

[70] **Methods:** Ellington, A.; Pollard, J. D. (1 May 2001). "Purification of oligonucleotides using denaturing polyacrylamide gel electrophoresis". *Current Protocols in Molecular Biology*. doi:10.1002/0471142727.mb0212s42. ISBN 0471142727.

[71] **Methods:** Gallagher, S. R.; Desjardins, P. (1 July 2011). "Quantitation of nucleic acids and proteins". *Current Protocols Essential Laboratory Techniques*. doi:10.1002/9780470089941.et0202s5. ISBN 0470089938.

[72] **Methods:** Chory, J.; Pollard, J. D. (1 May 2001). "Separation of small DNA fragments by conventional gel electrophoresis". *Current Protocols in Molecular Biology*. doi:10.1002/0471142727.mb0207s47. ISBN 0471142727.

[73] **Methods:** Walter, N. G. (1 February 2003). "Probing RNA structural dynamics and function by fluorescence resonance energy transfer (FRET)". *Current Protocols in Nucleic Acid Chemistry*. doi:10.1002/0471142700.nc1110s11. ISBN 0471142700.

[74] **Methods:** Lin, C.; Ke, Y.; Chhabra, R.; Sharma, J.; Liu, Y.; Yan, H. (2011). "Synthesis and Characterization of Self-Assembled DNA Nanostructures". In Zuccheri, G. and Samorì, B. *DNA Nanotechnology: Methods and Protocols.* Methods in Molecular Biology **749**. pp. 1–11. doi:10.1007/978-1-61779-142-0_1. ISBN 978-1-61779-141-3.

[75] **Methods:** Bloomfield, Victor A.; Crothers, Donald M.; Tinoco, Jr., Ignacio (2000). *Nucleic acids: structures, properties, and functions*. Sausalito, Calif: University Science Books. pp. 84–86, 396–407. ISBN 0-935702-49-0.

[76] **History:** Pelesko, John A. (2007). *Self-assembly: the science of things that put themselves together*. New York: Chapman & Hall/CRC. pp. 201, 242, 259. ISBN 978-1-58488-687-7.

[77] **History:** See "Current crystallization protocol". Nadrian Seeman Lab. for a statement of the problem, and "DNA cages containing oriented guests". Nadrian Seeman Laboratory. for the proposed solution.

[78] **DNA origami:** Rothemund, Paul W. K. (2006). "Scaffolded DNA origami: from generalized multicrossovers to polygonal networks". In Chen, Junghuei; Jonoska, Natasha; Rozenberg, Grzegorz. *Nanotechnology: science and computation*. Natural Computing Series. New York: Springer. pp. 3–21. doi:10.1007/3-540-30296-4_1. ISBN 978-3-540-30295-7.

[79] **Nanoarchitecture:** Robinson, Bruche H.; Seeman, Nadrian C. (August 1987). "The design of a biochip: a self-assembling molecular-scale memory device". *Protein Engineering* **1** (4): 295–300. doi:10.1093/protein/1.4.295. PMID 3508280.

[80] **Nanoarchitecture:** Xiao, Shoujun; Liu, Furong; Rosen, Abbey E.; Hainfeld, James F.; Seeman, Nadrian C.; Musier-Forsyth, Karin; Kiehl, Richard A. (August 2002). "Selfassembly of metallic nanoparticle arrays by DNA scaffolding". *Journal of Nanoparticle Research* **4** (4): 313–317. doi:10.1023/A:1021145208328.

[81] **History:** Hopkin, Karen (August 2011). "Profile: 3-D seer". *The Scientist*. Retrieved 8 August 2011.

17.10   Further reading

General:

- Seeman, Nadrian C. (June 2004). "Nanotechnology and the double helix". *Scientific American* **290** (6): 64–75. doi:10.1038/scientificamerican0604-64. PMID 15195395.—An article written for laypeople by the founder of the field

- Seeman, Nadrian C. (9 June 2010). "Structural DNA nanotechnology: growing along with Nano Letters". *Nano Letters* **10** (6): 1971–1978. Bibcode:2010NanoL..10.1971S. doi:10.1021/nl101262u. PMC 2901229. PMID 20486672.—A review of results in the period 2001–2010

- Seeman, Nadrian C. (2010). "Nanomaterials based on DNA". *Annual Review of Biochemistry* **79**: 65–87. doi:10.114 biochem-060308-102244. PMC 3454582. PMID 20222824.—A more comprehensive review including both old and new results in the field

- Service, Robert F. (3 June 2011). "DNA nanotechnology grows up". *Science* **332** (6034): 1140–1143. doi:10.1126/s0. and doi:10.1126/science.332.6034.1142.—A news article focusing on the history of the field and development of new applications

- Zadegan, Reza M.; Norton, Michael L. (June 2012). "Structural DNA Nanotechnology: From Design to Applications". *Int. J. Mol. Sci.* **13** (6): 7149–7162. doi:10.3390/ijms13067149. PMC 3397516. PMID 22837684.—A very recent and comprehensive review in the field

Specific subfields:

- Bath, Jonathan; Turberfield, Andrew J. (5 May 2007). "DNA nanomachines". *Nature Nanotechnology* **2** (5): 275–284. Bibcode:2007NatNa...2..275B. doi:10.1038/nnano.2007.104. PMID 18654284.—A review of nucleic acid nanomechanical devices

- Feldkamp, Udo; Niemeyer, Christof M. (13 March 2006). "Rational design of DNA nanoarchitectures". *Angewandte Chemie International Edition* **45** (12): 1856–76. doi:10.1002/anie.200502358. PMID 16470892.—A review coming from the viewpoint of secondary structure design

- Lin, Chenxiang; Liu, Yan; Rinker, Sherri; Yan, Hao (11 August 2006). "DNA tile based self-assembly: building complex nanoarchitectures". *ChemPhysChem* **7** (8): 1641–1647. doi:10.1002/cphc.200600260. PMID 16832805.—A minireview specifically focusing on tile-based assembly

- Zhang, David Yu; Seelig, Georg (February 2011). "Dynamic DNA nanotechnology using strand-displacement reactions". *Nature Chemistry* **3** (2): 103–113. Bibcode:2011NatCh...3..103Z. doi:10.1038/nchem.957. PMID 21258382.—A review of DNA systems making use of strand displacement mechanisms

17.11 External links

- International Society for Nanoscale Science, Computation and Engineering

- What is Bionanotechnology?—a video introduction to DNA nanotechnology

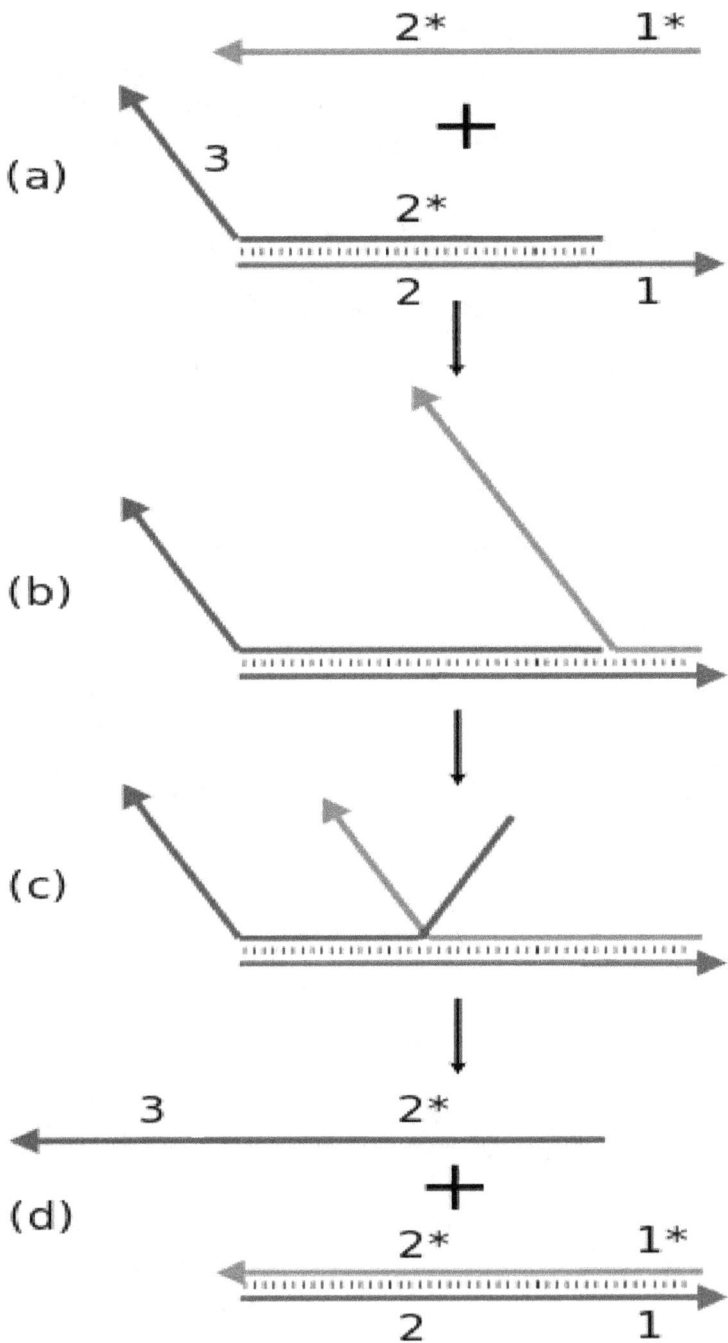

Dynamic DNA nanotechnology often makes use of toehold-mediated strand displacement reactions. In this example, the red strand binds to the single stranded toehold region on the green strand (region 1), and then in a branch migration process across region 2, the blue strand is displaced and freed from the complex. Reactions like these are used to dynamically reconfigure or assemble nucleic acid nanostructures. In addition, the red and blue strands can be used as signals in a molecular logic gate.

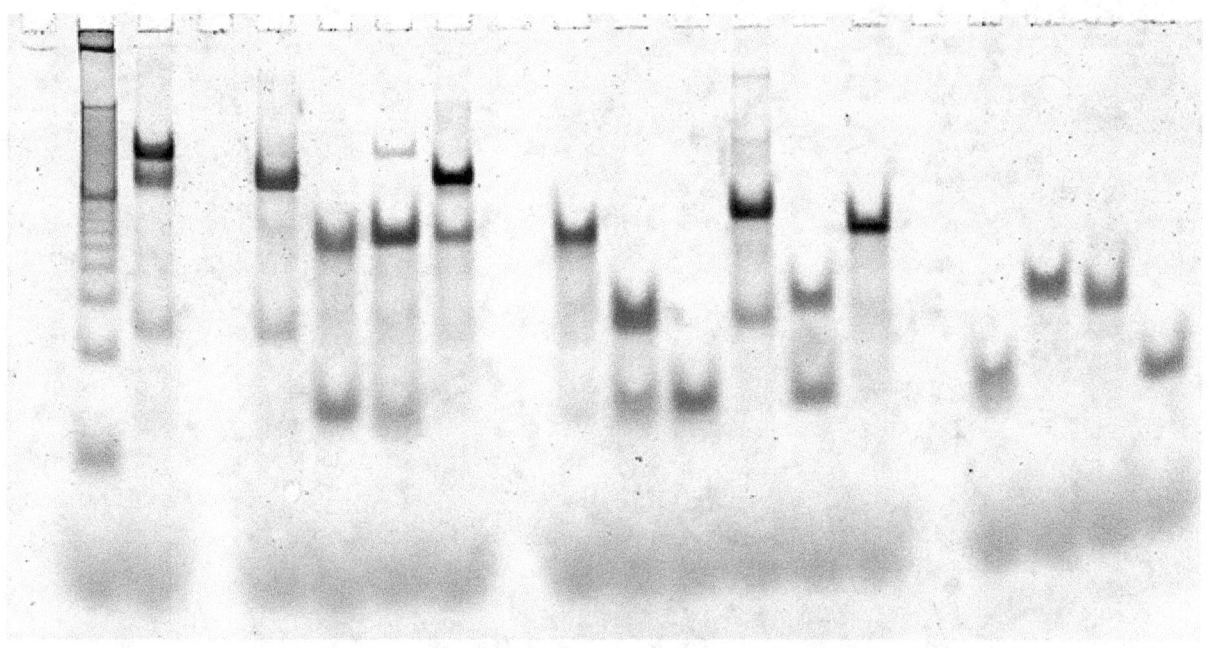

Gel electrophoresis methods, such as this formation assay on a DX complex, are used to ascertain whether the desired structures are forming properly. Each vertical lane contains a series of bands, where each band is characteristic of a particular reaction intermediate.

The woodcut Depth *(pictured) by M. C. Escher reportedly inspired Nadrian Seeman to consider using three-dimensional lattices of DNA to orient hard-to-crystallize molecules. This led to the beginning of the field of DNA nanotechnology.*

Chapter 18

DNA methylation

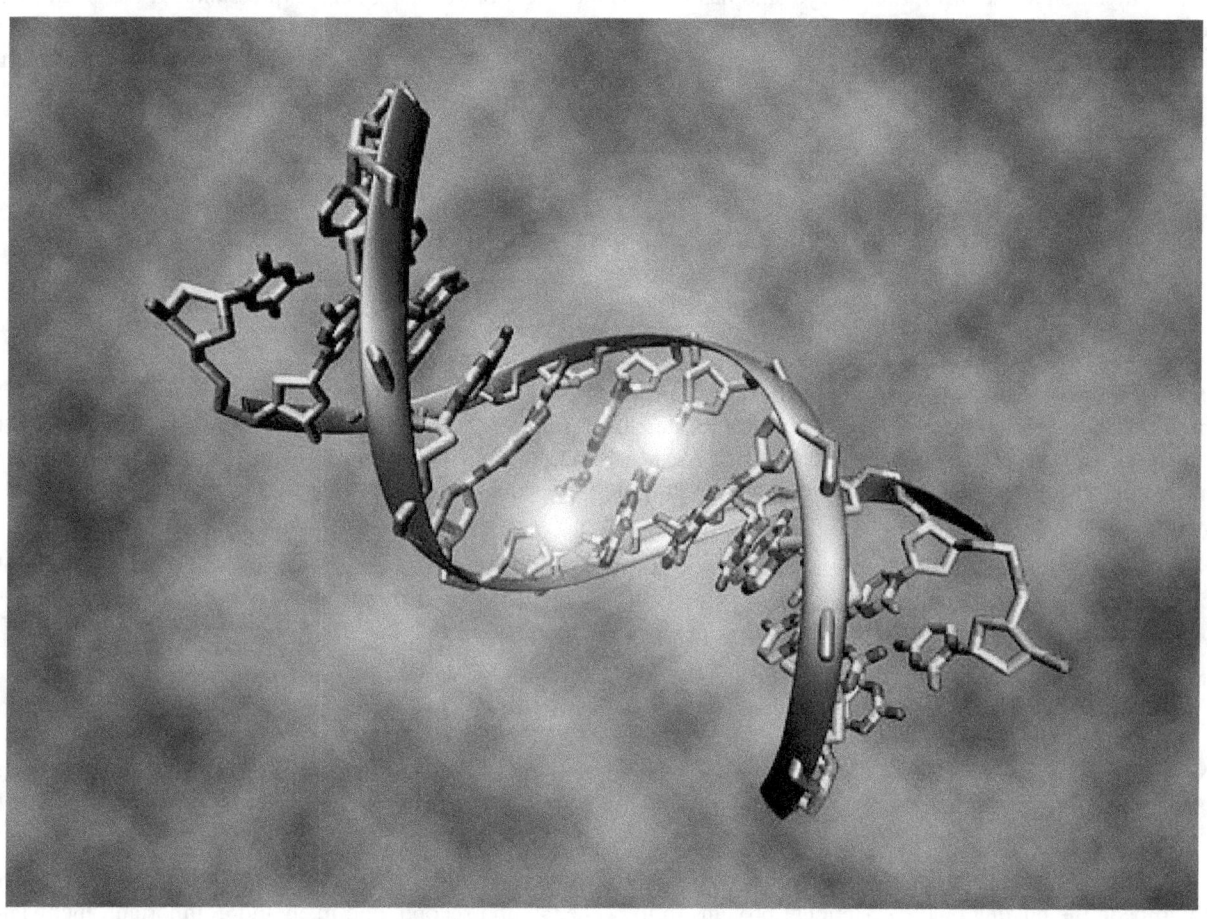

Illustration of a DNA molecule that is methylated at the two center cytosines. DNA methylation plays an important role for epigenetic gene regulation in development and disease.

DNA methylation is a process by which methyl groups are added to DNA. Methylation modifies the function of the DNA. When located in a gene promoter, DNA methylation typically acts to repress gene transcription. DNA methylation is essential for normal development and is associated with a number of key processes including genomic imprinting, X-chromosome inactivation, repression of repetitive elements, and carcinogenesis.

Two of DNA's four nucleotides, cytosine and adenine, can be methylated. Adenine methylation is restricted to prokaryotes. The rate of cytosine DNA methylation differs strongly between species: 14% of cytosines are methylated in *Arabidopsis thaliana*, 4% in *Mus musculus*, 2.3% in *Escherichia coli*, 0.03% in *Drosophila*, and virtually none (< 0.0002%) in yeast species.[1]

DNA methylation can stably alter the expression of genes in cells as cells divide and differentiate from embryonic stem cells into specific tissues. The resulting change is normally permanent and unidirectional, preventing a cell from reverting to a stem cell or converting into a different cell type. However, DNA methylation can be removed either passively, by dilution as cells divide, or by a faster, active, process. The latter process occurs via hydroxylation of the methyl groups that are to be removed, rather than by complete removal of methyl groups.[2][3] DNA methylation is typically removed during zygote formation and re-established through successive cell divisions during development. Methylation modifications that regulate gene expression are usually heritable through mitotic cell division; some methylation is also heritable through the specialized meiotic cell division that creates egg and sperm cells, resulting in genomic imprinting. DNA methylation suppresses the expression of endogenous retroviral genes and other harmful stretches of DNA that have been incorporated into the host genome over time. DNA methylation also forms the basis of chromatin structure, which enables a single cell to grow into multiple organs or perform multiple functions. DNA methylation also plays a crucial role in the development of nearly all types of cancer.[4]

DNA methylation at the 5 position of cytosine has the specific effect of reducing gene expression and has been found in every vertebrate examined. In adult somatic cells (cells in the body, not used for reproduction), DNA methylation typically occurs in a CpG dinucleotide context; non-CpG methylation is prevalent in embryonic stem cells,[5][6][7] and has also been indicated in neural development.[8]

18.1   In mammals

For significant overlapping coverage, see also Methylation.

Between 60% and 90% of all CpGs are methylated in mammals.[9][10] Methylated C residues spontaneously deaminate to form T residues over time; hence CpG dinucleotides steadily deaminate to TpG dinucleotides, which is evidenced by the under-representation of CpG dinucleotides in the human genome (they occur at only 21% of the expected frequency).[11] (On the other hand, spontaneous deamination of unmethylated C residues gives rise to U residues, a change that is quickly recognized and repaired by the cell.)

DNA methylation has been noted to target different organisms in different areas, and have varying functions. For example, the pattern of DNA methylation in mammals is generally evenly distributed throughout CpG sites (with exceptions). However, invertebrates experience the opposite: CpG methylation patterns that are grouped in clusters. [12]

Unmethylated CpGs are often grouped in clusters called *CpG islands*, which are present in the 5' regulatory regions of many genes. In many disease processes, such as cancer, gene promoter CpG islands acquire abnormal hypermethylation, which results in transcriptional silencing that can be inherited by daughter cells following cell division. Alterations of DNA methylation have been recognized as an important component of cancer development. Hypomethylation, in general, arises earlier and is linked to chromosomal instability and loss of imprinting, whereas hypermethylation is associated with promoters and can arise secondary to gene (oncogene suppressor) silencing, but might be a target for epigenetic therapy.[13]

DNA methylation may affect the transcription of genes in two ways. First, the methylation of DNA itself may physically impede the binding of transcriptional proteins to the gene,[14] and second, and likely more important, methylated DNA may be bound by proteins known as methyl-CpG-binding domain proteins (MBDs). MBD proteins then recruit additional proteins to the locus, such as histone deacetylases and other chromatin remodeling proteins that can modify histones, thereby forming compact, inactive chromatin, termed heterochromatin. This link between DNA methylation and chromatin structure is very important. In particular, loss of methyl-CpG-binding protein 2 (MeCP2) has been implicated in Rett syndrome; and methyl-CpG-binding domain protein 2 (MBD2) mediates the transcriptional silencing of hypermethylated genes in cancer.

Research has suggested that long-term memory storage in humans may be regulated by DNA methylation.[15][16]

DNA methylation levels can be used to estimate age, forming an accurate biological clock in humans and chimpanzees.[17]

18.1.1 In cancer

DNA methylation is an important regulator of gene transcription and a large body of evidence has demonstrated that genes with high levels of 5-methylcytosine in their promoter region are transcriptionally silent, and that DNA methylation gradually accumulates upon long-term gene silencing. DNA methylation is essential during embryonic development, and in somatic cells, patterns of DNA methylation are generally transmitted to daughter cells with a high fidelity. Aberrant DNA methylation patterns – hypermethylation and hypomethylation compared to normal tissue – have been associated with a large number of human malignancies. Hypermethylation typically occurs at CpG islands in the promoter region and is associated with gene inactivation. A lower level of leukocyte DNA methylation is associated with many types of cancer.[18] Global hypomethylation has also been implicated in the development and progression of cancer through different mechanisms.[19] Typically, there is hypermethylation of tumor suppressor genes and hypomethylation of oncogenes.[20]

18.1.2 In atherosclerosis

Epigenetic modifications such as DNA methylation have been implicated in cardiovascular disease, including atherosclerosis. In animal models of atherosclerosis, vascular tissue as well as blood cells such as mononuclear blood cells exhibit global hypomethylation with gene-specific areas of hypermethylation. DNA methylation polymorphisms may be used as an early biomarker of atherosclerosis since they are present before lesions are observed, which may provide an early tool for detection and risk prevention.[21]

Two of the cell types targeted for DNA methylation polymorphisms are monocytes and lymphocytes, which experience an overall hypomethylation. One proposed mechanism behind this global hypomethylation is elevated homocysteine levels causing hyperhomocysteinemia, a known risk factor for cardiovascular disease. High plasma levels of homocysteine inhibit DNA methyltransferases, which causes hypomethylation. Hypomethylation of DNA affects gene that alter smooth muscle cell proliferation, cause endothelial cell dysfunction, and increase inflammatory mediators, all of which are critical in forming atherosclerotic lesions.[22] High levels of homocysteine also result in hypermethylation of CpG islands in the promoter region of the estrogen receptor alpha (ERα) gene, causing its down regulation.[23] ERα protects against atherosclerosis due to its action as a growth suppressor, causing the smooth muscle cells to remain in a quiescent state.[24] Hypermethylation of the ERα promoter thus allows intimal smooth muscle cells to proliferate excessively and contribute to the development of the atherosclerotic lesion.[25]

Another gene that experiences a change in methylation status in atherosclerosis is the monocarboxylate transporter (MCT3), which produces a protein responsible for the transport of lactate and other ketone bodies out of many cell types, including vascular smooth muscle cells. In atherosclerosis patients, there is an increase in methylation of the CpG islands in exon 2, which decreases MCT3 protein expression. The down regulation of MCT3 impairs lactate transport, and significantly increases smooth muscle cell proliferation, which further contributes to the atherosclerotic lesion. An ex vivo experiment using the demethylating agent Decitabine (5-aza-2 -deoxycytidine) was shown to induce MCT3 expression in a dose dependant manner, as all hypermethylated sites in the exon 2 CpG island became demethylated after treatment. This may serve as a novel therapeutic agent to treat atherosclerosis, although no human studies have been conducted thus far.[26]

18.1.3 In aging

A longitudinal study of twin children showed that, between the ages of 5 and 10, there was divergence of methylation patterns due to environmental rather than genetic influences.[27] There is a global loss of DNA methylation during aging.[20] However, some genes become hypermethylated with age, including genes for the estrogen receptor, p16, and insulin-like growth factor 2.[20] Biological clocks, such as an epigenetic clock, are promising biomarkers of aging.[28][29]

18.1.4 In exercise

High intensity exercise has been shown to result in reduced DNA methylation in skeletal muscle.[30] Promoter methylation of PGC-1α and PDK4 were immediately reduced after high intensity exercise, whereas PPAR-γ methylation was not reduced until three hours after exercise.[30] By contrast, six months of exercise in previously sedentary middle-age men resulted in increased methylation in adipose tissue.[31] One study showed a possible increase in global genomic DNA methylation of white blood cells with more physical activity in non-Hispanics.[18]

18.2 DNA methyltransferases

In mammalian cells, DNA methylation occurs mainly at the C5 position of CpG dinucleotides and is carried out by two general classes of enzymatic activities – maintenance methylation and *de novo* methylation.[32]

Maintenance methylation activity is necessary to preserve DNA methylation after every cellular DNA replication cycle. Without the DNA methyltransferase (DNMT), the replication machinery itself would produce daughter strands that are unmethylated and, over time, would lead to passive demethylation. DNMT1 is the proposed maintenance methyltransferase that is responsible for copying DNA methylation patterns to the daughter strands during DNA replication. Mouse models with both copies of DNMT1 deleted are embryonic lethal at approximately day 9, due to the requirement of DNMT1 activity for development in mammalian cells.

It is thought that DNMT3a and DNMT3b are the *de novo* methyltransferases that set up DNA methylation patterns early in development. DNMT3L is a protein that is homologous to the other DNMT3s but has no catalytic activity. Instead, DNMT3L assists the *de novo* methyltransferases by increasing their ability to bind to DNA and stimulating their activity. Finally, DNMT2 (TRDMT1) has been identified as a DNA methyltransferase homolog, containing all 10 sequence motifs common to all DNA methyltransferases; however, DNMT2 (TRDMT1) does not methylate DNA but instead methylates cytosine-38 in the anticodon loop of aspartic acid transfer RNA.[33]

Since many tumor suppressor genes are silenced by DNA methylation during carcinogenesis, there have been attempts to re-express these genes by inhibiting the DNMTs. 5-Aza-2'-deoxycytidine (decitabine) is a nucleoside analog that inhibits DNMTs by trapping them in a covalent complex on DNA by preventing the β-elimination step of catalysis, thus resulting in the enzymes' degradation. However, for decitabine to be active, it must be incorporated into the genome of the cell, which can cause mutations in the daughter cells if the cell does not die. In addition, decitabine is toxic to the bone marrow, which limits the size of its therapeutic window. These pitfalls have led to the development of antisense RNA therapies that target the DNMTs by degrading their mRNAs and preventing their translation. However, it is currently unclear whether targeting DNMT1 alone is sufficient to reactivate tumor suppressor genes silenced by DNA methylation.

18.3 In plants

Significant progress has been made in understanding DNA methylation in the model plant *Arabidopsis thaliana*. DNA methylation in plants differs from that of mammals: while DNA methylation in mammals mainly occurs on the cytosine nucleotide in a CpG site, in plants the cytosine can be methylated at CpG, CpHpG, and CpHpH sites, where H represents any nucleotide but guanine. Overall, *Arabidopsis* DNA is highly methylated, mass spectrometry analysis estimated 14% of cytosines to be modified.[1]

The principal *Arabidopsis* DNA methyltransferase enzymes, which transfer and covalently attach methyl groups onto DNA, are DRM2, MET1, and CMT3. Both the DRM2 and MET1 proteins share significant homology to the mammalian methyltransferases DNMT3 and DNMT1, respectively, whereas the CMT3 protein is unique to the plant kingdom. There are currently two classes of DNA methyltransferases: 1) the *de novo* class, or enzymes that create new methylation marks on the DNA; and 2) a maintenance class that recognizes the methylation marks on the parental strand of DNA and transfers new methylation to the daughters strands after DNA replication. DRM2 is the only enzyme that has been implicated as a *de novo* DNA methyltransferase. DRM2 has also been shown, along with MET1 and CMT3 to be involved in maintaining methylation marks through DNA replication.[34] Other DNA methyltransferases are expressed in plants but have no known function (see the Chromatin Database).

It is not clear how the cell determines the locations of *de novo* DNA methylation, but evidence suggests that, for many (though not all) locations, RNA-directed DNA methylation (RdDM) is involved. In RdDM, specific RNA transcripts are produced from a genomic DNA template, and this RNA forms secondary structures called double-stranded RNA molecules.[35] The double-stranded RNAs, through either the small interfering RNA (siRNA) or microRNA (miRNA) pathways direct de-novo DNA methylation of the original genomic location that produced the RNA.[35] This sort of mechanism is thought to be important in cellular defense against RNA viruses and/or transposons, both of which often form a double-stranded RNA that can be mutagenic to the host genome. By methylating their genomic locations, through an as yet poorly understood mechanism, they are shut off and are no longer active in the cell, protecting the genome from their mutagenic effect.

18.4   In fungi

Many fungi have low levels (0.1 to 0.5%) of cytosine methylation, whereas other fungi have as much as 5% of the genome methylated.[36] This value seems to vary both among species and among isolates of the same species.[37] There is also evidence that DNA methylation may be involved in state-specific control of gene expression in fungi. However, at a detection limit of 250 attomoles by using ultra-high sensitive mass spectrometry DNA methylation was not confirmed in single cellular yeast species such as *Saccharomyces cerevisiae* or *Schizosaccharomyces pombe*, indicating that yeasts do not possess this DNA modification.[1]

Although brewers' yeast (*Saccharomyces*) and fission yeast (*Schizosaccharomyces*) have no detectable DNA methylation, the model filamentous fungus *Neurospora crassa* has a well-characterized methylation system.[38] Several genes control methylation in *Neurospora* and mutation of the DNA methyl transferase, *dim-2*, eliminates all DNA methylation but does not affect growth or sexual reproduction. While the *Neurospora* genome has very little repeated DNA, half of the methylation occurs in repeated DNA including transposon relics and centromeric DNA. The ability to evaluate other important phenomena in a DNA methylase-deficient genetic background makes *Neurospora* an important system in which to study DNA methylation.

18.5   In insects

Further information: Epigenetics in insects

Functional DNA methylation has been discovered in Honey Bees.[39] DNA methylation marks are mainly on the gene body, and current opinions on the function of DNA methylation is gene regulation via alternative splicing [40]

Drosophila melanogaster possess a very low level of DNA methylation only [1] that is however too low to be studied by methods such as bisulphite sequencing. Takayama et al.[41] developed a sensitive method that allowed to find that the fly genome DNA sequence patterns that associate with methylation are very different from the patterns seen in humans, or in other animal or plant species to date. Genome methylation in D. melanogaster was found at specific short motifs (concentrated in specific 5-base sequence motifs that are CA- and CT-rich but depleted of guanine) and is independent of DNMT2 activity. By using highly sensitive mass spectrometry approaches, Zhang et al.[42] have now demonstrated the presence of low (0.07%) but significant levels of adenine methylation during the earliest stages of Drosophila embryogenesis.

18.6   In bacteria

Adenine or cytosine methylation is part of the restriction modification system of many bacteria, in which specific DNA sequences are methylated periodically throughout the genome. A methylase is the enzyme that recognizes a specific sequence and methylates one of the bases in or near that sequence. Foreign DNAs (which are not methylated in this manner) that are introduced into the cell are degraded by sequence-specific restriction enzymes and cleaved. Bacterial genomic DNA is not recognized by these restriction enzymes. The methylation of native DNA acts as a sort of primitive immune system, allowing the bacteria to protect themselves from infection by bacteriophage.

E. coli DNA adenine methyltransferase (Dam) is an enzyme of ~32 kDa that does not belong to a restriction/modification system. The target recognition sequence for *E. coli* Dam is GATC, as the methylation occurs at the N6 position of the adenine in this sequence (G meATC). The three base pairs flanking each side of this site also influence DNA–Dam binding. Dam plays several key roles in bacterial processes, including mismatch repair, the timing of DNA replication, and gene expression. As a result of DNA replication, the status of GATC sites in the *E. coli* genome changes from fully methylated to hemimethylated. This is because adenine introduced into the new DNA strand is unmethylated. Re-methylation occurs within two to four seconds, during which time replication errors in the new strand are repaired. Methylation, or its absence, is the marker that allows the repair apparatus of the cell to differentiate between the template and nascent strands. It has been shown that altering Dam activity in bacteria results in increased spontaneous mutation rate. Bacterial viability is compromised in dam mutants that also lack certain other DNA repair enzymes, providing further evidence for the role of Dam in DNA repair.

One region of the DNA that keeps its hemimethylated status for longer is the origin of replication, which has an abundance of GATC sites. This is central to the bacterial mechanism for timing DNA replication. SeqA binds to the origin of replication, sequestering it and thus preventing methylation. Because hemimethylated origins of replication are inactive, this mechanism limits DNA replication to once per cell cycle.

Expression of certain genes, for example those coding for pilus expression in *E. coli*, is regulated by the methylation of GATC sites in the promoter region of the gene operon. The cells' environmental conditions just after DNA replication determine whether Dam is blocked from methylating a region proximal to or distal from the promoter region. Once the pattern of methylation has been created, the pilus gene transcription is locked in the on or off position until the DNA is again replicated. In *E. coli*, these pilus operons have important roles in virulence in urinary tract infections. It has been proposed that inhibitors of Dam may function as antibiotics.

On the other hand, DNA cytosine methylase targets CCAGG and CCTGG sites to methylate cytosine at the C5 position (C meC(A/T) GG). The other methylase enzyme, EcoKI, causes methylation of adenines in the sequences $AAC(N_6)GTGC$ and $GCAC(N_6)GTT$.

18.6.1   Molecular cloning

Most strains used by molecular biologists are derivatives of *E. coli* K-12, and possess both Dam and Dcm, but there are commercially available strains that are dam-/dcm- (lack of activity of either methylase). In fact, it is possible to unmethylate the DNA extracted from dam+/dcm+ strains by transforming it into dam-/dcm- strains. This would help digest sequences that are not being recognized by methylation-sensitive restriction enzymes.[43][44]

The restriction enzyme DpnI can recognize 5'-GmeATC-3' sites and digest the methylated DNA. Being such a short motif, it occurs frequently in sequences by chance, and as such its primary use for researchers is to degrade template DNA following PCRs (PCR products lack methylation, as no methylases are present in the reaction). Similarly, some commercially available restriction enzymes are sensitive to methylation at their cognate restriction sites, and must as mentioned previously be used on DNA passed through a dam-/dcm- strain to allow cutting.

18.7   Detection

DNA methylation can be detected by the following assays currently used in scientific research:

- Mass spectrometry is a very sensitive and reliable analytical method to detect DNA methylation. MS in general is however not informative about the sequence context of the methylation, thus limited in studying the function of this DNA modification.

- Methylation-Specific PCR (MSP), which is based on a chemical reaction of sodium bisulfite with DNA that converts unmethylated cytosines of CpG dinucleotides to uracil or UpG, followed by traditional PCR.[45] However, methylated cytosines will not be converted in this process, and primers are designed to overlap the CpG site of interest, which allows one to determine methylation status as methylated or unmethylated.

- Whole genome bisulfite sequencing, also known as BS-Seq, which is a high-throughput genome-wide analysis of DNA methylation. It is based on aforementioned sodium bisulfite conversion of genomic DNA, which is then sequenced on a Next-generation sequencing platform. The sequences obtained are then re-aligned to the reference genome to determine methylation states of CpG dinucleotides based on mismatches resulting from the conversion of unmethylated cytosines into uracil.

- The HELP assay, which is based on restriction enzymes' differential ability to recognize and cleave methylated and unmethylated CpG DNA sites.

- ChIP-on-chip assays, which is based on the ability of commercially prepared antibodies to bind to DNA methylation-associated proteins like MeCP2.

- Restriction landmark genomic scanning, a complicated and now rarely used assay based upon restriction enzymes' differential recognition of methylated and unmethylated CpG sites; the assay is similar in concept to the HELP assay.

- Methylated DNA immunoprecipitation (MeDIP), analogous to chromatin immunoprecipitation, immunoprecipitation is used to isolate methylated DNA fragments for input into DNA detection methods such as DNA microarrays (MeDIP-chip) or DNA sequencing (MeDIP-seq).

- Pyrosequencing of bisulfite treated DNA. This is sequencing of an amplicon made by a normal forward primer but a biotinylated reverse primer to PCR the gene of choice. The Pyrosequencer then analyses the sample by denaturing the DNA and adding one nucleotide at a time to the mix according to a sequence given by the user. If there is a mis-match, it is recorded and the percentage of DNA for which the mis-match is present is noted. This gives the user a percentage methylation per CpG island.

- Molecular break light assay for DNA adenine methyltransferase activity – an assay that relies on the specificity of the restriction enzyme DpnI for fully methylated (adenine methylation) GATC sites in an oligonucleotide labeled with a fluorophore and quencher. The adenine methyltransferase methylates the oligonucleotide making it a substrate for DpnI. Cutting of the oligonucleotide by DpnI gives rise to a fluorescence increase.[46][47]

- Methyl Sensitive Southern Blotting is similar to the HELP assay, although uses Southern blotting techniques to probe gene-specific differences in methylation using restriction digests. This technique is used to evaluate local methylation near the binding site for the probe.

- MethylCpG Binding Proteins (MBPs) and fusion proteins containing just the Methyl Binding Domain (MBD) are used to separate native DNA into methylated and unmethylated fractions. The percentage methylation of individual CpG islands can be determined by quantifying the amount of the target in each fraction.[48] Extremely sensitive detection can be achieved in FFPE tissues with abscription-based detection.

- High Resolution Melt Analysis (HRM or HRMA), is a post-PCR analytical technique. The target DNA is treated with sodium bisulfite, which chemically converts unmethylated cytosines into uracils, while methylated cytosines are preserved. PCR amplification is then carried out with primers designed to amplify both methylated and unmethylated templates. After this amplification, highly methylated DNA sequences contain a higher number of CpG sites compared to unmethylated templates, which results in a different melting temperature that can be used in quantitative methylation detection.[49][50]

- Ancient DNA methylation reconstruction, a method to reconstruct high-resolution DNA methylation from ancient DNA samples. The method is based on the natural degradation processes that occur in ancient DNA: with time, methylated cytosines are degraded into thymines, whereas unmethylated cytosines are degraded into uracils. This asymmetry in degradation signals was used to reconstruct the full methylation maps of the Neanderthal and the Denisovan [51]

18.8   Differentially methylated regions (DMRs)

Differentially methylated regions (DMRs), are genomic regions with different methylation statuses among multiple samples (tissues, cells, individuals or others), are regarded as possible functional regions involved in gene transcriptional regulation. The identification of DMRs among multiple tissues (T-DMRs) provides a comprehensive survey of epigenetic differences among human tissues.[52] DMRs between cancer and normal samples (C-DMRs) demonstrate the aberrant methylation in cancers.[53] It is well known that DNA methylation is associated with cell differentiation and proliferation.[54] Many DMRs have been found in the development stages (D-DMRs) [55] and in the reprogrammed progress (R-DMRs).[56] In addition, there are intra-individual DMRs (Intra-DMRs) with longitudinal changes in global

DNA methylation along with the increase of age in a given individual.[57] There are also inter-individual DMRs (Inter-DMRs) with different methylation patterns among multiple individuals.[58]

QDMR (Quantitative Differentially Methylated Regions) is a quantitative approach to quantify methylation difference and identify DMRs from genome-wide methylation profiles by adapting Shannon entropy (http://bioinfo.hrbmu.edu.cn/qdmr). The platform-free and species-free nature of QDMR makes it potentially applicable to various methylation data. This approach provides an effective tool for the high-throughput identification of the functional regions involved in epigenetic regulation. QDMR can be used as an effective tool for the quantification of methylation difference and identification of DMRs across multiple samples.[59]

Gene-set analysis (a.k.a. pathway analysis; usually performed tools such as DAVID, GoSeq or GSEA) has been shown to be severely biased when applied to high-throughput methylation data (e.g. MeDIP-seq, MeDIP-ChIP, HELP-seq etc.), and a wide range of studies have thus mistakenly reported hyper-methylation of genes related to development and differentiation; it has been suggested that this can be corrected using sample label permutations or using a statistical model to control for differences in the numberes of CpG probes / CpG sites that target each gene.[60]

18.9   Computational prediction

DNA methylation can also be detected by computational models through sophisticated algorithms and methods. Computational models can facilitate the global profiling of DNA methylation across chromosomes, and often such models are faster and cheaper to perform than biological assays. Such up-to-date computational models include Bhasin, *et al.*,[61] Bock, *et al.*,[62] and Zheng, *et al.*[63] [64] Together with biological assay, these methods greatly facilitate the DNA methylation analysis.

18.10   See also

- 5-Hydroxymethylcytosine

- 5-Methylcytosine

- 7-Methylguanosine

- Demethylating agent

- DNA demethylation

- DNA methylation age

- Epigenetics, of which DNA methylation is a significant contributor

- Epigenome

- Genome

- Genomic imprinting, an inherited repression of an allele, relying on DNA methylation

- MethDB DNA Methylation database

- N^6-Methyladenosine

- Reprogramming

18.11 References

[1] Capuano, F; Muelleder, M; Kok, R. M.; Blom, H. J.; Ralser, M (2014). "Cytosine DNA methylation is found in *Drosophila melanogaster* but absent in *Saccharomyces cerevisiae, Schizosaccharomyces pombe* and other yeast species". *Analytical Chemistry* **86** (8): 140318143747008. doi:10.1021/ac500447w. PMC 4006885. PMID 24640988.

[2] Iqbal, K.; Jin, S. -G.; Pfeifer, G. P.; Szabo, P. E. (2011). "Reprogramming of the paternal genome upon fertilization involves genome-wide oxidation of 5-methylcytosine". *Proceedings of the National Academy of Sciences* **108** (9): 3642–3647. doi:10.1073/pnas.1014033108. PMC 3048122. PMID 21321204.

[3] Wossidlo, M.; Nakamura, T.; Lepikhov, K.; Marques, C. J.; Zakhartchenko, V.; Boiani, M.; Arand, J.; Nakano, T.; Reik, W.; Walter, J. R. (2011). "5-Hydroxymethylcytosine in the mammalian zygote is linked with epigenetic reprogramming". *Nature Communications* **2**: 241. doi:10.1038/ncomms1240. PMID 21407207.

[4] Jaenisch, R.; Bird, A. (2003). "Epigenetic regulation of gene expression: how the genome integrates intrinsic and environmental signals". *Nature Genetics*. 33 Suppl (3s): 245–254. doi:10.1038/ng1089. PMID 12610534.

[5] Dodge JE, Ramsahoye BH, Wo ZG, Okano M, Li E (2002). "De novo methylation of MMLV provirus in embryonic stem cells: CpG versus non-CpG methylation". *Gene* **289** (1–2): 41–48. doi:10.1016/S0378-1119(02)00469-9.

[6] Haines TR, Rodenhiser DI, Ainsworth PJ (2001). "Allele-Specific Non-CpG Methylation of the Nf1 Gene during Early Mouse Development". *Developmental Biology* **240** (2): 585–598. doi:10.1006/dbio.2001.0504. PMID 11784085.

[7] Lister R, Pelizzola M, Dowen RH; et al. (October 2009). "Human DNA methylomes at base resolution show widespread epigenomic differences". *Nature* **462** (7271): 315–22. doi:10.1038/nature08514. PMC 2857523. PMID 19829295.

[8] Lister, R.; Mukamel, E. A.; Nery, J. R.; Urich, M.; Puddifoot, C. A.; Johnson, N. D.; Lucero, J.; Huang, Y.; Dwork, A. J.; Schultz, M. D.; Yu, M.; Tonti-Filippini, J.; Heyn, H.; Hu, S.; Wu, J. C.; Rao, A.; Esteller, M.; He, C.; Haghighi, F. G.; Sejnowski, T. J.; Behrens, M. M.; Ecker, J. R. (4 July 2013). "Global Epigenomic Reconfiguration During Mammalian Brain Development". *Science* **341** (6146): 1237905. doi:10.1126/science.1237905.

[9] Ehrlich M, Gama Sosa MA, Huang L-H., Midgett RM, Kuo KC, McCune RA, Gehrke C (April 1982). "Amount and distribution of 5-methylcytosine in human DNA from different types of tissues or cells". *Nucleic Acids Research* **10** (8): 2709–2721. doi:10.1093/nar/10.8.2709. PMC 320645. PMID 7079182.

[10] Tucker KL (June 2001). "Methylated cytosine and the brain: a new base for neuroscience". *Neuron* **30** (3): 649–652. doi:10.1016/S0896-6273(01)00325-7. PMID 11430798.

[11] International Human Genome Sequencing Consortium; et al. (February 2001). "Initial sequencing and analysis of the human genome". *Nature* **409** (6822): 860–921. doi:10.1038/35057062. PMID 11237011.

[12] "The Role of Methylation in Gene Expression | Learn Science at Scitable". *www.nature.com*. Retrieved 2015-09-27.

[13] Daura-Oller E, Cabre M, Montero MA, Paternain JL, Romeu A (2009). "Specific gene hypomethylation and cancer: New insights into coding region feature trends". *Bioinformation* **3** (8): 340–343. doi:10.6026/97320630003340. PMC 2720671. PMID 19707296.

[14] Choy MK, Movassagh M, Goh HG, Bennett M, Down T, Foo R (2010). "Genome-wide conserved consensus transcription factor binding motifs are hyper-methylated". *BMC Genomics* **11** (1): 519. doi:10.1186/1471-2164-11-519. PMC 2997012. PMID 20875111.

[15] Miller C, Sweatt J (2007-03-15). "Covalent modification of DNA regulates memory formation". *Neuron* **53** (6): 857–869. doi:10.1016/j.neuron.2007.02.022. PMID 17359920.

[16] Powell, Devin (2008-12-02). "Memories may be stored on your DNA". New Scientist. Retrieved 2008-12-02.

[17] Horvath S (2013). "DNA methylation age of human tissues and cell types". *Genome Biology* **14** (R115): R115. doi:10.1186/gb-2013-14-10-r115. PMC 4015143. PMID 24138928.

[18] Zhang FF1, Cardarelli R, Carroll J, Zhang S, Fulda KG, Gonzalez K, Vishwanatha JK, Morabia A, Santella RM (2011). "Physical activity and global genomic DNA methylation in a cancer-free population". *EPIGENETICS* **6** (3): 293–299. doi:10.4161/epi.6.3.14378.PMC3092677. PMID 21178401.

[19] Craig, JM; Wong, NC (editor) (2011). *Epigenetics: A Reference Manual*. Caister Academic Press. ISBN 978-1-904455-88-2.

[20] Gonzalo S (2010). "Epigenetic alterations in aging". *Journal of Applied Physiology* **109** (2): 586–597. doi:10.1152/japplphysiol PMC 2928596. PMID 20448029. .00238.2010.

[21] Lund, G.L.; Andersson, L.; Lauria, M.; Lindholm, M.; Fraga, M.F.; Villar-Garea, A.; Ballestar, E.; Esteller, M.; Zaina, S. (2004). "DNA methylation polymorphisms precede any histological sign of atherosclerosis in mice lacking Apolipoprotein E.". *J Biol Chem* **279** (28): 29147–29154. doi:10.1074/jbc.m403618200.

[22] Castro, R.; Rivera, I.; Struys, E.A.; Jansen, E.E.; Ravasco, P.; Camilo, M.E.; Blom, H.J.; Jakobs, C.; Tavares; de Almeida, T. (2003). "Increased homocysteine concentrations and S-adenosylhomocysteine concentrations and DNA hypomethylation in vascular disease". *Clin Chem* **49** (8): 1292–1296. doi:10.1373/49.8.1292.

[23] Huang, Y.S.; Zhi, Y.F.; Wang, S.R. (2009). "Hypermethylation of estrogen receptor-α gene in atheromatosis patients and its correlation with homocysteine". *Pathophysiology* **16** (4): 259–265. doi:10.1016/j.pathophys.2009.02.010.

[24] Dong, C.D.; Yoon, W.; Goldschmidt-Clermont, P.J. (2002). "DNA methylation and atherosclerosis". *J Nutr* **132** (8): 2406S–2409S.

[25] Ying, A.K.; Hassanain, H.H.; Roos, C.M.; Smiraglia, D.J.; Issa, J.J.; Michler, R.E.; Caligiuri, M.; Plass, C.; Goldschmidt-Clermont, P.J. (2000). "Methylation of the estrogen receptor- α gene promoter is selectively increased in proliferating human aortic smooth muscle cells". *Cardiovas Res* **46** (1): 172–179. doi:10.1016/s0008-6363(00)00004-3.

[26] Zhu, S.; Goldschmidt-Clermont, P.J.; Dong, C. (2005). "Inactivation of Monocarboxylate Transporter MCT3 by DNA methylation in atherosclerosis". *Circulation* **112** (9): 1353–1361. doi:10.1161/circulationaha.104.519025.

[27] Wong CC1, Caspi A, Williams B, Craig IW, Houts R, Ambler A, Moffitt TE, Mill J (2010). "A longitudinal study of epigenetic variation in twins". *EPIGENETICS* **5** (6): 516–526. doi:10.4161/epi.5.6.12226. PMC 3322496. PMID 20505345.

[28] Horvath S (2013). "DNA methylation age of human tissues and cell types". *Genome Biology* **14** (10): R115. doi:10.1186/gb-2013-14-10-r115. PMC 4015143. PMID 24138928.

[29] Jones, M. J., Goodman, S. J. and Kobor, M. S. (2015), DNA methylation and healthy human aging. Aging Cell. doi:10.1111/acel.

[30] Barrès R1, Yan J, Egan B, Treebak JT, Rasmussen M, Fritz T, Caidahl K, Krook A, O'Gorman DJ, Zierath JR (2012). "Acute exercise remodels promoter methylation in human skeletal muscle". *Cell Metabolism* **15** (3): 405–411. doi:10.1016/j.cmet.2012. 01.001.PMID22405075.

[31] Rönn T1, Volkov P, Davegårdh C, Dayeh T, Hall E, Olsson AH, Nilsson E, Tornberg A, Dekker Nitert M, Eriksson KF, Jones HA, Groop L, Ling C (2013). "A six months exercise intervention influences the genome-wide DNA methylation pattern in human adipose tissue". *PLOS Genetics* **9** (6): e1003572. doi:10.1371/journal.pgen.1003572. PMC 3694844. PMID 23825961.

[32] Gratchev, Alexei. Review on DNA Methylation. (n.d.) Retrieved from http://www.methods.info/Methods/DNA_methylation/ Methylation_review.html

[33] Goll MG, Kirpekar F, Maggert KA, Yoder JA, Hsieh CL, Zhang X, Golic KG, Jacobsen SE, Bestor TH (January 2006). "Methylation of tRNAAsp by the DNA methyltransferase homolog Dnmt2". *Science* **311** (5759): 395–398. doi:10.1126/science.1120976. PMID 16424344.

[34] Cao X and Jacobsen SE (December 2002). "Locus-specific control of asymmetric and CpNpG methylation by the DRM and CMT3 methyltransferase genes". *PNAS* **99** (Suppl 4): 16491–16498. doi:10.1073/pnas.162371599. PMC 139913. PMID 12151602.

[35] Aufsatz W, Mette MF, van der Winden J, Matzke AJM, Matzke M (2002). "RNA-directed DNA methylation in Arabidopsis". *PNAS* **99** (90004): 16499–16506. doi:10.1073/pnas.162371499. PMC 139914. PMID 12169664.

[36] Antequera F, Tamame M, Villanueva JR, Santos T (July 1984). "DNA methylation in the fungi". *J. Biol. Chem.* **259** (13): 8033–8036. PMID 6330093.

[37] Binz T, D'Mello N, Horgen PA (1998). "A comparison of DNA methylation levels in selected isolates of higher fungi". *Mycologia* (Mycological Society of America) **90** (5): 785–790. doi:10.2307/3761319. JSTOR 3761319.

[38] Selker EU, Tountas NA, Cross SH, Margolin BS, Murphy JG, Bird AP, Freitag M (2003). "The methylated component of the Neurospora crassa genome". *Nature* **422** (6934): 893–897. doi:10.1038/nature01564. PMID 12712205.

[39] "Functional CpG Methylation System in a Social Insect". *Science*. doi:10.1126/science.1135213.

[40] Li-Byarlay H.; et al. "RNA interference knockdown of DNA methyl-transferase 3 affects gene alternative splicing in the honey bee". *P.N.A.S.* doi:10.1073/pnas.1310735110.

[41] S. Takayama, J. Dhahbi, A. Roberts, G. Mao, S.-J. Heo, L. Pachter, D. I. K. Martin, D. Boffelli (2014). Genome methylation in D. melanogaster is found at specific short motifs and is independent of DNMT2 activity. Genome Research,doi:10.1101/gr.1624

[42] . doi:10.1016/j.cell.2015.04.018. Missing or empty |title= (help)

[43] Palmer BR and Marinus MG (1994). "The dam and dcm strains of Escherichia coli—a review". *Gene* **143** (1): 1–12. doi:10.1016/0378-1119(94)90597-5. PMID 8200522.

[44] "Making unmethylated (dam-/dcm-) DNA".

[45] Hernández, H. G.; Tse, M. Y.; Pang, S. C.; Arboleda, H.; Forero, D. A. (2013). "Optimizing methodologies for PCR-based DNA methylation analysis". *BioTechniques* **55** (4): 181–197. doi:10.2144/000114087. PMID 24107250.

[46] Wood RJ, Maynard-Smith MD, Robinson VL, Oyston PC, Titball RW, Roach PL (2007). Fugmann, Sebastian, ed. "Kinetic analysis of Yersinia pestis DNA adenine methyltransferase activity using a hemimethylated molecular break light oligonucleotide". *PLoS ONE* **2** (8): e801. doi:10.1371/journal.pone.0000801. PMC 1949145. PMID 17726531.

[47] Li J, Yan H, Wang K, Tan W, Zhou X (February 2007). "Hairpin fluorescence DNA probe for real-time monitoring of DNA methylation". *Anal. Chem.* **79** (3): 1050–1056. doi:10.1021/ac061694i. PMID 17263334.

[48] ^ David R. McCarthy, Philip D. Cotter, and Michelle M. Hanna (2012). MethylMeter(r): A Quantitative, Sensitive, and Bisulfite-Free Method for Analysis of DNA Methylation, DNA Methylation - From Genomics to Technology, Dr. Tatiana Tatarinova (Ed.), ISBN 978-953-51-0320-2, InTech, DOI: 10.5772/36090. Available from: http://www.intechopen.com/books/dna-methylation-from-genomics-to-technology/methylmeter-a-quantitative-sensititive-and-bisulfite-free-method-for-ation

[49] Wojdacz, TK; Dobrovic, A (2007). "Methylation-sensitive high resolution melting (MS-HRM): a new approach for sensitive and high-throughput assessment of methylation". *Nucleic Acids Res.* **35** (6): e41. doi:10.1093/nar/gkm013. PMC 1874596. PMID 17289753.

[50] Malentacchi, F; Forni, G; Vinci, S; Orlando, C (2009). "Quantitative evaluation of DNA methylation by optimization of a differential-high resolution melt analysis protocol". *Nucleic Acids Res.* **37** (12): e86. doi:10.1093/nar/gkp383. PMC 2709587. PMID 19454604.

[51] Gokhman D1, Lavi E, Prüfer K, Fraga MF, Riancho JA, Kelso J, Pääbo S, Meshorer E, Carmel L. (2014). "Reconstructing the DNA methylation maps of the Neandertal and the Denisovan.". *Science* **344** (6183): 523–7. doi:10.1126/science.1250368. PMID 24786081.

[52] Rakyan, VK; Down, TA; Thorne, NP; Flicek, P; Kulesha, E; Gräf, S; Tomazou, EM; Bäckdahl, L; Johnson, N; Herberth, M; Howe, KL; Jackson, DK; Miretti, MM; Fiegler, H; Marioni, JC; Birney, E; Hubbard, TJ; Carter, NP; Tavaré, S; Beck, S (September 2008). "An integrated resource for genome-wide identification and analysis of human tissue-specific differentially methylated regions (tDMRs).". *Genome Research* **18** (9): 1518–29. doi:10.1101/gr.077479.108. PMC 2527707. PMID 18577705.

[53] Irizarry, RA; Ladd-Acosta, C; Wen, B; Wu, Z; Montano, C; Onyango, P; Cui, H; Gabo, K; Rongione, M; Webster, M; Ji, H; Potash, JB; Sabunciyan, S; Feinberg, AP (February 2009). "The human colon cancer methylome shows similar hypo- and hypermethylation at conserved tissue-specific CpG island shores.". *Nature Genetics* **41** (2): 178–86. doi:10.1038/ng.298. PMC 2729128. PMID 19151715.

[54] Reik, W; Dean, W; Walter, J (Aug 10, 2001). "Epigenetic reprogramming in mammalian development.". *Science* **293** (5532): 1089–93. doi:10.1126/science.1063443. PMID 11498579.

[55] Meissner, A; Mikkelsen, TS; Gu, H; Wernig, M; Hanna, J; Sivachenko, A; Zhang, X; Bernstein, BE; Nusbaum, C; Jaffe, DB; Gnirke, A; Jaenisch, R; Lander, ES (Aug 7, 2008). "Genome-scale DNA methylation maps of pluripotent and differentiated cells.". *Nature* **454** (7205): 766–70. doi:10.1038/nature07107. PMC 2896277. PMID 18600261.

[56] Doi, A; Park, IH; Wen, B; Murakami, P; Aryee, MJ; Irizarry, R; Herb, B; Ladd-Acosta, C; Rho, J; Loewer, S; Miller, J; Schlaeger, T; Daley, GQ; Feinberg, AP (December 2009). "Differential methylation of tissue- and cancer-specific CpG island shores distinguishes human induced pluripotent stem cells, embryonic stem cells and fibroblasts.". *Nature Genetics* **41** (12): 1350–3. doi:10.1038/ng.471. PMC 2958040. PMID 19881528.

[57] Bjornsson, HT; Sigurdsson, MI; Fallin, MD; Irizarry, RA; Aspelund, T; Cui, H; Yu, W; Rongione, MA; Ekström, TJ; Harris, TB; Launer, LJ; Eiriksdottir, G; Leppert, MF; Sapienza, C; Gudnason, V; Feinberg, AP (Jun 25, 2008). "Intra-individual change over time in DNA methylation with familial clustering.". *JAMA: the Journal of the American Medical Association* **299** (24): 2877–83. doi:10.1001/jama.299.24.2877. PMC 2581898. PMID 18577732.

[58] Bock, C; Walter, J; Paulsen, M; Lengauer, T (June 2008). "Inter-individual variation of DNA methylation and its implications for large-scale epigenome mapping.". *Nucleic Acids Research* **36** (10): e55. doi:10.1093/nar/gkn122. PMC 2425484. PMID 18413340.

[59] Zhang, Y; Liu, H; Lv, J; Xiao, X; Zhu, J; Liu, X; Su, J; Li, X; Wu, Q; Wang, F; Cui, Y (May 2011). "QDMR: a quantitative method for identification of differentially methylated regions by entropy.". *Nucleic Acids Research* **39** (9): e58. doi:10.1093/nar/gkr053.PMC3089487. PMID 21306990.

[60] Geeleher P, Hartnett L, Egan LJ, Golden A, Raja Ali RA, Seoighe C (June 2013). "Gene-Set Analysis is Severely Biased When Applied to Genome-wide Methylation Data". *Bioinformatics* **29** (15): 1851–7. doi:10.1093/bioinformatics/btt311. PMID 23732277.

[61] Bhasin M, Zhang H, Reinherz EL, Reche PA. (Aug 2005). "Prediction of methylated CpGs in DNA sequences using a support vector machine". *FEBS Lett.* **579** (20): 4302–8. doi:10.1016/j.febslet.2005.07.002. PMID 16051225.

[62] Bock C, Paulsen M, Tierling S, Mikeska T, Lengauer T, Walter J. (Mar 2006). "CpG island methylation in human lymphocytes is highly correlated with DNA sequence, repeats, and predicted DNA structure". *PLoS Genet.* **2** (3): e26. doi:10.1371/journal.pgen .0020026.PMC1386721. PMID 16520826.

[63] Zheng H, Jiang SW, Wu H (2011). "Enhancement on the predictive power of the prediction model for human genomic DNA methylation". *International Conference on Bioinformatics and Computational Biology (BIOCOMP'11)*.

[64] Zheng H, Jiang SW, Li J, Wu H (2013). "CpGIMethPred: computational model for predicting methylation status of CpG islands in human genome". *BMC Medical Genomics)*.

18.12   Further reading

- Law J, Jacobsen SE (2010). "Establishing, maintaining and modifying DNA methylation patterns in plants and animals". *Nat. Rev. Genet.* **11** (3): 204–220. doi:10.1038/nrg2719. PMC 3034103. PMID 20142834.

- Straussman R, Nejman D, Roberts D; et al. (2009). "Developmental programming of CpG island methylation profiles in the human genome". *Nat. Struct. Mol. Biol.* **16** (5): 564–571. doi:10.1038/nsmb.1594. PMID 19377480.

- Patra SK (2008). "Ras regulation of DNA-methylation and cancer". *Exp Cell Res***314**(6): 1193–1201. doi:10.1016 PMID 18282569.

- Patra SK, Patra A, Ghosh TC; et al. (2008). "Demethylation of (cytosine-5-C-methyl) DNA and regulation of transcription in the epigenetic pathways of cancer development". *Cancer Metast. Rev.* **27** (2): 315–334. doi:10.1007/s10555-008-9118-y. PMID 18246412.

18.13   External links

- DNA Methylation at the US National Library of Medicine Medical Subject Headings (MeSH)

- ENCODE threads explorer Non-coding RNA characterization. Nature (journal)

- PCMdb Pancreatic Cancer Methylation Database. Nature Scientific Report 4:4197

Chapter 19

DNA damage (naturally occurring)

DNA damage is an alteration in the chemical structure of DNA, such as a break in a strand of DNA, a base missing from the backbone of DNA, or a chemically changed base such as 8-OHdG. Damage to DNA that occurs naturally can result from metabolic or hydrolytic processes. Metabolism releases compounds that damage DNA including reactive oxygen species, reactive nitrogen species, reactive carbonyl species, lipid peroxidation products and alkylating agents, among others, while hydrolysis cleaves chemical bonds in DNA.[1] Naturally occurring oxidative DNA damages arise at least 10,000 times per cell per day in humans and 50,000 times or more per cell per day in rats,[2] as documented below.

DNA damage is distinctly different from mutation, although both are types of error in DNA. DNA damage is an abnormal chemical structure in DNA, while a mutation is a change in the sequence of standard base pairs.

DNA damage and mutation have different biological consequences. While most DNA damages can undergo DNA repair, such repair is not 100% efficient. Un-repaired DNA damages accumulate in non-replicating cells, such as cells in the brains or muscles of adult mammals and can cause aging.[3][4][5] (Also see DNA damage theory of aging.) In replicating cells, such as cells lining the colon, errors occur upon replication of past damages in the template strand of DNA or during repair of DNA damages. These errors can give rise to mutations or epigenetic alterations.[6] Both of these types of alteration can be replicated and passed on to subsequent cell generations. These alterations can change gene function or regulation of gene expression and possibly contribute to progression to cancer.

19.1 DNA damages are a major problem for life

One indication that DNA damages are a major problem for life is that DNA repair processes, to cope with ubiquitously occurring DNA damages, have been found in all cellular organisms in which DNA repair has been investigated. For example, in bacteria, a regulatory network aimed at repairing DNA damages (called the SOS response in *Escherichia coli*) has been found in many bacterial species. *E. coli* RecA, a key enzyme in the SOS response pathway, is the defining member of a ubiquitous class of DNA strand-exchange proteins that are essential for homologous recombination, a pathway that maintains genomic integrity by repairing broken DNA.[7] Genes homologous to *RecA* and to other central genes in the SOS response pathway are found in almost all the bacterial genomes sequenced to date, covering a large number of phyla, suggesting both an ancient origin and a widespread occurrence of recombinational repair of DNA damage.[8] Eukaryotic recombinases that are homologues of RecA are also widespread in eukaryotic organisms. For example, in fission yeast and humans, RecA homologues promote duplex-duplex DNA-strand exchange needed for repair of many types of DNA lesions.[9][10]

Another indication that DNA damages are a major problem for life is that cells make large investments in DNA repair processes. As pointed out by Hoeijmakers,[4] repairing just one double-strand break could require more than 10,000 ATP molecules, as used in signaling the presence of the damage, the generation of repair foci, and the formation (in humans) of the RAD51 nucleofilament (an intermediate in homologous recombinational repair). (RAD51 is a homologue of bacterial RecA.)

19.2   Frequencies of endogenous DNA damages

The list below shows some frequencies with which new naturally occurring DNA damages arise per day, due to endogenous cellular processes.

- Oxidative damages

  - Humans, per cell per day

    - 10,000[11]
      11,500[12]
      2,800[13] specific damages 8-oxoGua, 8-oxodG plus 5-HMUra
      2,800[14] specific damages 8-oxoGua, 8-oxodG plus 5-HMUra

  - Rats, per cell per day

    - 74,000[12]
      86,000[15]
      100,000[11]

  - Mice, per cell per day

    - 34,000[13] specific damages 8-oxoGua, 8-oxodG plus 5-HMUra
      47,000[16] specific damages oxo8dG in mouse liver
      28,000[14] specific damages 8-oxoGua, 8-oxodG, 5-HMUra

- Depurinations

  - Mammalian cells, per cell per day

    - 2,000 to 10,000[17][18]
      9,000[19]
      12,000[20]
      13,920[21]

- Depyrimidinations

  - Mammalian cells, per cell per day

    - 600[20]
      696[21]

- Single-strand breaks

  - Mammalian cells, per cell per day

    - 55,200[21]

- Double-strand breaks

  - Human cells, per cell cycle

    - 10[22]
      50[23]

- O6-methylguanines

  - Mammalian cells, per cell per day

    - 3,120[21]

- Cytosine deamination

  - Mammalian cells, per cell per day

    - 192[21]

Another important endogenous DNA damage is M1dG, short for (3-(2'-deoxy-beta-D-erythro-pentofuranosyl)-pyrimidoa]-purin-10(3H)-one). The excretion in urine(likely reflecting rate of occurrence)of M1dG may be as much as1,000-fold lower than that of8-oxodG.[24]However,a more important measure may be the steady-state level in DNA,reflecting both rate of occurrence and rate of DNA repair. The steady-state level of M1dG is higher than that of8-oxodG.[25]This pointsout that some DNA damages produced at a low rate may be difficult to repair and remain in DNA at a high steady-statelevel. Both M1dG[26]and8-oxodG[27]   are mutagenic.

19.3   Steady-state levels of DNA damages

Steady-state levels of DNA damages represent the balance between formation and repair. More than 100 types of ox-idative DNA damage have been characterized, and 8-oxodG constitutes about 5% of the steady state oxidative damages in DNA.[28] Helbock et al.[29] estimated that there were 24,000 steady state oxidative DNA adducts per cell in young rats and 66,000 adducts per cell in old rats. This reflects the accumulation of DNA damage with age. DNA damage accumulation with age is further described in DNA damage theory of aging.

Swenberg et al.[30] measured average amounts of selected steady state endogenous DNA damages in mammalian cells. The seven most common damages they evaluated are shown in Table 1.

Evaluating steady-state damages in specific tissues of the rat, Nakamura and Swenberg[31] indicated that the number of abasic sites varied from about 50,000 per cell in liver, kidney and lung to about 200,000 per cell in the brain.

19.4   Consequences of naturally occurring DNA damages

Differentiated somatic cells of adult mammals generally replicate infrequently or not at all. Such cells, including, for example, brain neurons and muscle myocytes, have little or no cell turnover. Non-replicating cells do not generally generate mutations due to DNA damage-induced errors of replication. These non-replicating cells do not commonly give rise to cancer, but they do accumulate DNA damages with time that likely contribute to aging (see DNA damage theory of aging). In a non-replicating cell, a single-strand break or other type of damage in the transcribed strand of DNA can block RNA polymerase II catalysed transcription.[32] This would interfere with the synthesis of the protein coded for by the gene in which the blockage occurred.

Brasnjevic et al.[33] summarized the evidence showing that single-strand breaks accumulate with age in the brain (though accumulation differed in different regions of the brain) and that single-strand breaks are the most frequent steady-state DNA damages in the brain. As discussed above, these accumulated single-strand breaks would be expected to block transcription of genes. Consistent with this, as reviewed by Hetman et al.,[34] 182 genes were identified and shown to have reduced transcription in the brains of individuals older than 72 years, compared to transcription in the brains of those less than 43 years old. When 40 particular proteins were evaluated in a muscle of rats, the majority of the proteins showed significant decreases during aging from 18 months (mature rat) to 30 months (aged rat) of age.[35]

Another type of DNA damage, the double strand break, was shown to cause cell death (loss of cells) through apoptosis.[36] This type of DNA damage would not accumulate with age, since once a cell was lost through apoptosis, its double strand damage would be lost with it.

19.5   See also

- Ageing

- Aging brain

- AP site

- Direct DNA damage

- DNA

- DNA adduct

- DNA damage theory of aging

- DNA repair

- DNA replication

- Free Radical damage to DNA

- Homologous recombination

- Meiosis

- Mutation

- Natural competence

- Origin and function of meiosis

- Reactive oxygen species

19.6 References

[1] De Bont R, van Larebeke N. (2004) Endogenous DNA damage in humans: a review of quantitative data. Mutagenesis 19(3):169-185. Review. PMID 15123782

[2] Bernstein C, Prasad AR, Nfonsam V, Bernstein H (2013). "DNA Damage, DNA Repair and Cancer". In Chen C. *New Research Directions in DNA Repair*. Rijeka, Croatia: InTech. doi:10.5772/53919. ISBN 978-953-51-1114-6.

[3] Bernstein H, Payne CM, Bernstein C, Garewal H, Dvorak K (2008). Cancer and aging as consequences of un-repaired DNA damage. In: New Research on DNA Damages (Editors: Honoka Kimura and Aoi Suzuki) Nova Science Publishers, Inc., New York, Chapter 1, pp. 1-47. open access, but read only https://www.novapublishers.com/catalog/product_info.php?products_id=43247 ISBN 978-1604565812

[4] Hoeijmakers JH. (2009) DNA damage, aging, and cancer. *N Engl J Med*. 361(15):1475-1485. Review. PMID 19812404

[5] Freitas AA, de Magalhães JP. (2011) A review and appraisal of the DNA damage theory of ageing. Mutat Res. 728(1-2):12-22. Review. doi:10.1016/j.mrrev.2011.05.001 PMID 21600302

[6] O'Hagan HM, Mohammad HP, Baylin SB. (2008) Double strand breaks can initiate gene silencing and SIRT1-dependent onset of DNA methylation in an exogenous promoter CpG island. *PLoS Genet*. 4(8):e1000155. doi:10.1371/journal.pgen.1000155 PMID 18704159

[7] Bell JC, Plank JL, Dombrowski CC, Kowalczykowski SC. (2012) Direct imaging of RecA nucleation and growth on single molecules of SSB-coated ssDNA. *Nature* 491(7423):274-278. doi:10.1038/nature11598. PMID 23103864

[8] Erill I, Campoy S, Barbé J. (2007) Aeons of distress: an evolutionary perspective on the bacterial SOS response. FEMS Microbiol Rev. 31(6):637-656. Review. doi:10.1111/j.1574-6976.2007.00082.x PMID 17883408

[9] Murayama Y, Kurokawa Y, Mayanagi K, Iwasaki H. (2008) Formation and branch migration of Holliday junctions mediated by eukaryotic recombinases. *Nature* 451(7181):1018-1021. PMID 18256600

[10] Holthausen JT, Wyman C, Kanaar R. (2010) Regulation of DNA strand exchange in homologous recombination. DNA Repair (Amst) 9(12):1264-1272. PMID 20971042

[11] Ames BN, Shigenaga MK, Hagen TM. (1993) Oxidants, antioxidants, and the degenerative diseases of aging. *Proc Natl Acad Sci U S A*. 90(17):7915-7922. Review. PMID 8367443

[12] Helbock HJ, Beckman KB, Shigenaga MK, Walter PB, Woodall AA, Yeo HC, Ames BN. (1998) DNA oxidation matters: the HPLC-electrochemical detection assay of 8-oxo-deoxyguanosine and 8-oxo-guanine. *Proc Natl Acad Sci U S A*. 95(1): 288-293. PMID 9419368

[13] Foksinski M, Rozalski R, Guz J, Ruszkowska B, Sztukowska P, Piwowarski M, Klungland A, Olinski R. (2004) Urinary excretion of DNA repair products correlates with metabolic rates as well as with maximum life spans of different mammalian species. Free Radic Biol Med 37(9) 1449-1454. PMID 15454284

[14] Tudek B, Winczura A, Janik J, Siomek A, Foksinski M, Oliński R. (2010). Involvement of oxidatively damaged DNA and repair in cancer development and aging. Am J Transl Res 2(3):254-284. PMID 20589166

[15] Fraga CG, Shigenaga MK, Park JW, Degan P, Ames BN. Oxidative damage to DNA during aging: 8-hydroxy-2'-deoxyguanosine in rat organ DNA and urine. *Proc Natl Acad Sci U S A* 1990;87(12) 4533-4537. PMID 2352934

[16] Hamilton ML, Guo Z, Fuller CD, Van Remmen H, Ward WF, Austad SN, Troyer DA, Thompson I, Richardson A. (2001). A reliable assessment of 8-oxo-2-deoxyguanosine levels in nuclear and mitochondrial DNA using the sodium iodide method to isolate DNA. *Nucleic Acids Res* 29(10):2117-2126. PMID 11353081

[17] Lindahl T, Nyberg B. (1972) Rate of depurination of native deoxyribonucleic acid. *Biochemistry*11(19) 3610-3618.doi:10.1038 PMID 4626532

[18] Lindahl T. (1993) Instability and decay of the primary structure of DNA. *Nature* 362(6422) 709-715. PMID 8469282

[19] Nakamura J, Walker VE, Upton PB, Chiang SY, Kow YW, Swenberg JA. Highly sensitive apurinic/apyrimidinic site assay can detect spontaneous and chemically induced depurination under physiological conditions. *Cancer Res* 1998;58(2) 222-225. PMID 9443396

[20] Lindahl T. (1977) DNA repair enzymes acting on spontaneous lesions in DNA. In: Nichols WW and Murphy DG (eds.) DNA Repair Processes. Symposia Specialists, Miami p225-240. ISBN 088372099X ISBN 978-0883720998

[21] Tice, R.R., and Setlow, R.B. (1985) DNA repair and replication in aging organisms and cells. In: Finch EE and Schneider EL (eds.) Handbook of the Biology of Aging. Van Nostrand Reinhold, New York. Pages 173-224. ISBN 0442225296 ISBN 978-0442225292

[22] Haber JE. (1999) DNA recombination: the replication connection. *Trends Biochem Sci* 24(7) 271-275. PMID 10390616

[23] Vilenchik MM, Knudson AG. (2003) Endogenous DNA double-strand breaks: production, fidelity of repair, and induction of cancer. *Proc Natl Acad Sci U S A* 100(22) 12871-12876. PMID 14566050

[24] Chan SW, Dedon PC. (2010) The biological and metabolic fates of endogenous DNA damage products. J Nucleic Acids 2010:929047. PMID 21209721

[25] Kadlubar FF, Anderson KE, Häussermann S, Lang NP, Barone GW, Thompson PA, MacLeod SL, Chou MW, Mikhailova M, Plastaras J, Marnett LJ, Nair J, Velic I, Bartsch H. (1998) Comparison of DNA adduct levels associated with oxidative stress in human pancreas. Mutat Res. 405(2):125-33. PMID 9748537

[26] VanderVeen LA, Hashim MF, Shyr Y, Marnett LJ. Induction of frameshift and base pair substitution mutations by the major DNA adduct of the endogenous carcinogen malondialdehyde. (2003) Proc Natl Acad Sci U S A 100(24):14247-14252. PMID 14603032

[27] Tan X, Grollman AP, Shibutani S. (1999) Comparison of the mutagenic properties of 8-oxo-7,8-dihydro-2'-deoxyadenosine and 8-oxo-7,8-dihydro-2'-deoxyguanosine DNA lesions in mammalian cells. *Carcinogenesis* 20(12):2287-2292. PMID 10590221

[28] Hamilton ML, Guo Z, Fuller CD, Van Remmen H, Ward WF, Austad SN, Troyer DA, Thompson I, Richardson A. (2001) A reliable assessment of 8-oxo-2-deoxyguanosine levels in nuclear and mitochondrial DNA using the sodium iodide method to isolate DNA. *Nucleic Acids Res.* 29(10):2117-26. PMID 11353081

[29] Helbock HJ, Beckman KB, Shigenaga MK, Walter PB, Woodall AA, Yeo HC, Ames BN. (1998) DNA oxidation matters: the HPLC-electrochemical detection assay of 8-oxo-deoxyguanosine and 8-oxo-guanine. *Proc Natl Acad Sci U S A* 95(1):288-293. PMID 9419368

[30] Swenberg JA, Lu K, Moeller BC, Gao L, Upton PB, Nakamura J, Starr TB. (2011) Endogenous versus exogenous DNA adducts: their role in carcinogenesis, epidemiology, and risk assessment. Toxicol Sci. 120(Suppl 1):S130-45. PMID 21163908

[31] Nakamura J, Swenberg JA. (1999) Endogenous apurinic/apyrimidinic sites in genomic DNA of mammalian tissues. *Cancer Res.* 59(11):2522-2526. PMID 10363965

[32] Kathe SD, Shen GP, Wallace SS. (2004) Single-stranded breaks in DNA but not oxidative DNA base damages block transcriptional elongation by RNA polymerase II in HeLa cell nuclear extracts. *J Biol Chem.* 279(18):18511-18520. PMID 14978042

[33] Brasnjevic I, Hof PR, Steinbusch HW, Schmitz C. (2008) Accumulation of nuclear DNA damage or neuron loss: molecular basis for a new approach to understanding selective neuronal vulnerability in neurodegenerative diseases. DNA Repair (Amst). 7(7):1087-1097. PMID 18458001

[34] Hetman M, Vashishta A, Rempala G. (2010) Neurotoxic mechanisms of DNA damage: focus on transcriptional inhibition. J Neurochem. 114(6):1537-1549. doi: 10.1111/j.1471-4159.2010.06859.x. Review. PMID 20557419

[35] Piec I, Listrat A, Alliot J, Chambon C, Taylor RG, Bechet D. (2005) Differential proteome analysis of aging in rat skeletal muscle. *FASEB J.* 19(9):1143-5. PMID 15831715

[36] Carnevale J, Palander O, Seifried LA, Dick FA. (2012) DNA damage signals through differentially modified E2F1 molecules to induce apoptosis. *Mol Cell Biol.* 32(5):900-912. PMID 22184068

Chapter 20

Mutation

For other uses, see Mutation (disambiguation).

In biology, a **mutation** is a permanent change of the nucleotide sequence of the genome of an organism, virus, or extrachromosomal DNA or other genetic elements. Mutations result from damage to DNA which is not repaired, errors in the process of replication, or from the insertion or deletion of segments of DNA by mobile genetic elements.[1][2][3] Mutations may or may not produce discernible changes in the observable characteristics (phenotype) of an organism. Mutations play a part in both normal and abnormal biological processes including: evolution, cancer, and the development of the immune system, including junctional diversity.

Mutation can result in several different types of change in sequences. Mutations in genes can either have no effect, alter the product of a gene, or prevent the gene from functioning properly or completely. Mutations can also occur in nongenic regions. One study on genetic variations between different species of *Drosophila* suggests that, if a mutation changes a protein produced by a gene, the result is likely to be harmful, with an estimated 70 percent of amino acid polymorphisms that have damaging effects, and the remainder being either neutral or weakly beneficial.[4] Due to the damaging effects that mutations can have on genes, organisms have mechanisms such as DNA repair to prevent or correct mutations by reverting the mutated sequence back to its original state.[1]

20.1 Description

Mutations can involve the duplication of large sections of DNA, usually through genetic recombination.[5] These duplications are a major source of raw material for evolving new genes, with tens to hundreds of genes duplicated in animal genomes every million years.[6] Most genes belong to larger gene families of shared ancestry, known as homology.[7] Novel genes are produced by several methods, commonly through the duplication and mutation of an ancestral gene, or by recombining parts of different genes to form new combinations with new functions.[8][9]

Here, protein domains act as modules, each with a particular and independent function, that can be mixed together to produce genes encoding new proteins with novel properties.[10] For example, the human eye uses four genes to make structures that sense light: three for cone cell or color vision and one for rod cell or night vision; all four arose from a single ancestral gene.[11] Another advantage of duplicating a gene (or even an entire genome) is that this increases engineering redundancy; this allows one gene in the pair to acquire a new function while the other copy performs the original function.[12][13] Other types of mutation occasionally create new genes from previously noncoding DNA.[14][15]

Changes in chromosome number may involve even larger mutations, where segments of the DNA within chromosomes break and then rearrange. For example, in the Homininae, two chromosomes fused to produce human chromosome 2; this fusion did not occur in the lineage of the other apes, and they retain these separate chromosomes.[16] In evolution, the most important role of such chromosomal rearrangements may be to accelerate the divergence of a population into new species by making populations less likely to interbreed, thereby preserving genetic differences between these populations.[17]

211

Sequences of DNA that can move about the genome, such as transposons, make up a major fraction of the genetic material of plants and animals, and may have been important in the evolution of genomes.[18] For example, more than a million copies of the Alu sequence are present in the human genome, and these sequences have now been recruited to perform functions such as regulating gene expression.[19] Another effect of these mobile DNA sequences is that when they move within a genome, they can mutate or delete existing genes and thereby produce genetic diversity.[2]

Nonlethal mutations accumulate within the gene pool and increase the amount of genetic variation.[20] The abundance of some genetic changes within the gene pool can be reduced by natural selection, while other "more favorable" mutations may accumulate and result in adaptive changes.

Prodryas persephone, *a Late Eocene butterfly*

For example, a butterfly may produce offspring with new mutations. The majority of these mutations will have no effect; but one might change the color of one of the butterfly's offspring, making it harder (or easier) for predators to see. If this color change is advantageous, the chance of this butterfly's surviving and producing its own offspring are a little better, and over time the number of butterflies with this mutation may form a larger percentage of the population.

Neutral mutations are defined as mutations whose effects do not influence the fitness of an individual. These can accumulate over time due to genetic drift. It is believed that the overwhelming majority of mutations have no significant effect on an organism's fitness. Also, DNA repair mechanisms are able to mend most changes before they become permanent mutations, and many organisms have mechanisms for eliminating otherwise-permanently mutated somatic cells.

Beneficial mutations can improve reproductive success.

20.2 Causes

Main article: Mutagenesis

Four classes of mutations are (1) spontaneous mutations (molecular decay), (2) mutations due to error-prone replication bypass of naturally occurring DNA damage (also called error-prone translesion synthesis), (3) errors introduced during DNA repair, and (4) induced mutations caused by mutagens. Scientists may also deliberately introduce mutant sequences through DNA manipulation for the sake of scientific experimentation.

20.2.1 Spontaneous mutation

Spontaneous mutations on the molecular level can be caused by:[21]

- Tautomerism — A base is changed by the repositioning of a hydrogen atom, altering the hydrogen bonding pattern of that base, resulting in incorrect base pairing during replication.

- Depurination — Loss of a purine base (A or G) to form an apurinic site (AP site).

- Deamination — Hydrolysis changes a normal base to an atypical base containing a keto group in place of the original amine group. Examples include C → U and A → HX (hypoxanthine), which can be corrected by DNA repair mechanisms; and 5MeC (5-methylcytosine) → T, which is less likely to be detected as a mutation because thymine is a normal DNA base.

- Slipped strand mispairing — Denaturation of the new strand from the template during replication, followed by renaturation in a different spot ("slipping"). This can lead to insertions or deletions.

20.2.2 Error-prone replication bypass

There is increasing evidence that the majority of spontaneously arising mutations are due to error-prone replication (translesion synthesis) past a DNA damage in the template strand. Naturally occurring oxidative DNA damages arise at least 10,000 times per cell per day in humans and 50,000 times or more per cell per day in rats.[22] In mice, the majority of mutations are caused by translesion synthesis.[23] Likewise, in yeast, Kunz et al.[24] found that more than 60% of the spontaneous single base pair substitutions and deletions were caused by translesion synthesis.

20.2.3 Errors introduced during DNA repair

See also: DNA damage (naturally occurring)

Although naturally occurring double-strand breaks occur at a relatively low frequency in DNA, their repair often causes mutation. Non-homologous end joining (NHEJ) is a major pathway for repairing double-strand breaks. NHEJ involves removal of a few nucleotides to allow somewhat inaccurate alignment of the two ends for rejoining followed by addition of nucleotides to fill in gaps. As a consequence, NHEJ often introduces mutations.[25]

20.2.4 Induced mutation

Induced mutations on the molecular level can be caused by:-

- Chemicals
  - Hydroxylamine
  - Base analogs (e.g., Bromodeoxyuridine (BrdU))

- Alkylating agents (e.g., *N*-ethyl-*N*-nitrosourea (ENU)). These agents can mutate both replicating and non-replicating DNA. In contrast, a base analog can mutate the DNA only when the analog is incorporated in replicating the DNA. Each of these classes of chemical mutagens has certain effects that then lead to transitions, transversions, or deletions.

- Agents that form DNA adducts (e.g., ochratoxin A)[27]

- DNA intercalating agents (e.g., ethidium bromide)

- DNA crosslinkers

- Oxidative damage

- Nitrous acid converts amine groups on A and C to diazo groups, altering their hydrogen bonding patterns, which leads to incorrect base pairing during replication.

- Radiation

  - Ultraviolet light (UV) (non-ionizing radiation). Two nucleotide bases in DNA—cytosine and thymine—are most vulnerable to radiation that can change their properties. UV light can induce adjacent pyrimidine bases in a DNA strand to become covalently joined as a pyrimidine dimer. UV radiation, in particular longer-wave UVA, can also cause oxidative damage to DNA.[28]

20.3   Classification of mutation types

See also: Chromosome abnormality

20.3.1   By effect on structure

The sequence of a gene can be altered in a number of ways. Gene mutations have varying effects on health depending on where they occur and whether they alter the function of essential proteins. Mutations in the structure of genes can be classified as:

- Small-scale mutations, such as those affecting a small gene in one or a few nucleotides, including:

  - **Point mutations**, often caused by chemicals or malfunction of DNA replication, exchange a single nucleotide for another.[30] These changes are classified as transitions or transversions.[31] Most common is the transition that exchanges a purine for a purine (A ↔ G) or a pyrimidine for a pyrimidine, (C ↔ T). A transition can be caused by nitrous acid, base mis-pairing, or mutagenic base analogs such as BrdU. Less common is a transversion, which exchanges a purine for a pyrimidine or a pyrimidine for a purine (C/T ↔ A/G). An example of a transversion is the conversion of adenine (A) into a cytosine (C). A point mutation can be reversed by another point mutation, in which the nucleotide is changed back to its original state (true reversion) or by second-site reversion (a complementary mutation elsewhere that results in regained gene functionality). Point mutations that occur within the protein coding region of a gene may be classified into three kinds, depending upon what the erroneous codon codes for:

    - Silent mutations, which code for the same (or a sufficiently similar) amino acid.
    - Missense mutations, which code for a different amino acid.
    - Nonsense mutations, which code for a stop codon and can truncate the protein.

  - **Insertions** add one or more extra nucleotides into the DNA. They are usually caused by transposable elements, or errors during replication of repeating elements. Insertions in the coding region of a gene may alter splicing of the mRNA (splice site mutation), or cause a shift in the reading frame (frameshift), both of which can significantly alter the gene product. Insertions can be reversed by excision of the transposable element.

- **Deletions** remove one or more nucleotides from the DNA. Like insertions, these mutations can alter the reading frame of the gene. In general, they are irreversible: Though exactly the same sequence might in theory be restored by an insertion, transposable elements able to revert a very short deletion (say 1–2 bases) in *any* location either are highly unlikely to exist or do not exist at all.

- Large-scale mutations in chromosomal structure, including:

  - **Amplifications** (or gene duplications) leading to multiple copies of all chromosomal regions, increasing the dosage of the genes located within them.

  - **Deletions** of large chromosomal regions, leading to loss of the genes within those regions.

  - Mutations whose effect is to juxtapose previously separate pieces of DNA, potentially bringing together separate genes to form functionally distinct fusion genes (e.g., bcr-abl). These include:

    - **Chromosomal translocations**: interchange of genetic parts from nonhomologous chromosomes.
    - **Interstitial deletions**: an intra-chromosomal deletion that removes a segment of DNA from a single chromosome, thereby apposing previously distant genes. For example, cells isolated from a human astrocytoma, a type of brain tumor, were found to have a chromosomal deletion removing sequences between the Fused in Glioblastoma (FIG) gene and the receptor tyrosine kinase (ROS), producing a fusion protein (FIG-ROS). The abnormal FIG-ROS fusion protein has constitutively active kinase activity that causes oncogenic transformation (a transformation from normal cells to cancer cells).
    - **Chromosomal inversions**: reversing the orientation of a chromosomal segment.

  - **Loss of heterozygosity**: loss of one allele, either by a deletion or a genetic recombination event, in an organism that previously had two different alleles.

20.3.2   By effect on function

See also: Behavior mutation

- **Loss-of-function mutations**, also called **inactivating mutations**, result in the gene product having less or no function (being partially or wholly inactivated). When the allele has a complete loss of function (null allele), it is often called an **amorphic mutation** in the Muller's morphs schema. Phenotypes associated with such mutations are most often recessive. Exceptions are when the organism is haploid, or when the reduced dosage of a normal gene product is not enough for a normal phenotype (this is called haploinsufficiency).

- **Gain-of-function mutations**, also called **activating mutations**, change the gene product such that its effect gets stronger (enhanced activation) or even is superseded by a different and abnormal function. When the new allele is created, a heterozygote containing the newly created allele as well as the original will express the new allele; genetically this defines the mutations as dominant phenotypes. Often called a neomorphic mutation.[32]

- **Dominant negative mutations** (also called **antimorphic mutations**) have an altered gene product that acts antagonistically to the wild-type allele. These mutations usually result in an altered molecular function (often inactive) and are characterized by a dominant or semi-dominant phenotype. In humans, dominant negative mutations have been implicated in cancer (e.g., mutations in genes p53,[33] ATM,[34] CEBPA[35] and PPARgamma[36]). Marfan syndrome is caused by mutations in the *FBN1* gene, located on chromosome 15, which encodes fibrillin-1, a glycoprotein component of the extracellular matrix.[37] Marfan syndrome is also an example of dominant negative mutation and haploinsufficiency.[38][39]

- **Lethal mutations** are mutations that lead to the death of the organisms that carry the mutations.

- A **back mutation** or **reversion** is a point mutation that restores the original sequence and hence the original phenotype.[40]

20.3.3   By effect on fitness

See also: Fitness (biology)

In applied genetics, it is usual to speak of mutations as either harmful or beneficial.

- A **harmful**, or **deleterious**, mutation decreases the fitness of the organism.

- A **beneficial**, or **advantageous** mutation increases the fitness of the organism. Mutations that promotes traits that are desirable, are also called beneficial. In theoretical population genetics, it is more usual to speak of mutations as deleterious or advantageous than harmful or beneficial.

- A **neutral mutation** has no harmful or beneficial effect on the organism. Such mutations occur at a steady rate, forming the basis for the molecular clock. In the neutral theory of molecular evolution, neutral mutations provide genetic drift as the basis for most variation at the molecular level.

- A **nearly neutral mutation** is a mutation that may be slightly deleterious or advantageous, although most nearly neutral mutations are slightly deleterious.

Distribution of fitness effects

Attempts have been made to infer the distribution of fitness effects (DFE) using mutagenesis experiments and theoretical models applied to molecular sequence data. DFE, as used to determine the relative abundance of different types of mutations (i.e., strongly deleterious, nearly neutral or advantageous), is relevant to many evolutionary questions, such as the maintenance of genetic variation,[41] the rate of genomic decay,[42] the maintenance of outcrossing sexual reproduction as opposed to inbreeding[43] and the evolution of sex and genetic recombination.[44] In summary, the DFE plays an important role in predicting evolutionary dynamics.[45][46] A variety of approaches have been used to study the DFE, including theoretical, experimental and analytical methods.

- **Mutagenesis experiment**: The direct method to investigate the DFE is to induce mutations and then measure the mutational fitness effects, which has already been done in viruses, bacteria, yeast, and *Drosophila*. For example, most studies of the DFE in viruses used site-directed mutagenesis to create point mutations and measure relative fitness of each mutant.[47][48][49][50] In *Escherichia coli*, one study used transposon mutagenesis to directly measure the fitness of a random insertion of a derivative of Tn10.[51] In yeast, a combined mutagenesis and deep sequencing approach has been developed to generate high-quality systematic mutant libraries and measure fitness in high throughput.[52] However, given that many mutations have effects too small to be detected[53] and that mutagenesis experiments can detect only mutations of moderately large effect; DNA sequence data analysis can provide valuable information about these mutations.

- **Molecular sequence analysis**: With rapid development of DNA sequencing technology, an enormous amount of DNA sequence data is available and even more is forthcoming in the future. Various methods have been developed to infer the DFE from DNA sequence data.[54][55][56][57] By examining DNA sequence differences within and between species, we are able to infer various characteristics of the DFE for neutral, deleterious and advantageous mutations.[20] To be specific, the DNA sequence analysis approach allows us to estimate the effects of mutations with very small effects, which are hardly detectable through mutagenesis experiments.

One of the earliest theoretical studies of the distribution of fitness effects was done by Motoo Kimura, an influential theoretical population geneticist. His neutral theory of molecular evolution proposes that most novel mutations will be highly deleterious, with a small fraction being neutral.[58][59] Hiroshi Akashi more recently proposed a bimodal model for the DFE, with modes centered around highly deleterious and neutral mutations.[60] Both theories agree that the vast majority of novel mutations are neutral or deleterious and that advantageous mutations are rare, which has been supported by experimental results. One example is a study done on the DFE of random mutations in vesicular stomatitis virus.[47] Out of all mutations, 39.6% were lethal, 31.2% were non-lethal deleterious, and 27.1% were neutral. Another example comes from a high throughput mutagenesis experiment with yeast.[52] In this experiment it was shown that the overall DFE is bimodal, with a cluster of neutral mutations, and a broad distribution of deleterious mutations.

Though relatively few mutations are advantageous, those that are play an important role in evolutionary changes.[61] Like neutral mutations, weakly selected advantageous mutations can be lost due to random genetic drift, but strongly selected advantageous mutations are more likely to be fixed. Knowing the DFE of advantageous mutations may lead to increased ability to predict the evolutionary dynamics. Theoretical work on the DFE for advantageous mutations has been done by John H. Gillespie[62] and H. Allen Orr.[63] They proposed that the distribution for advantageous mutations should be exponential under a wide range of conditions, which, in general, has been supported by experimental studies, at least for strongly selected advantageous mutations.[64][65][66]

In general, it is accepted that the majority of mutations are neutral or deleterious, with rare mutations being advantageous; however, the proportion of types of mutations varies between species. This indicates two important points: first, the proportion of effectively neutral mutations is likely to vary between species, resulting from dependence on effective population size; second, the average effect of deleterious mutations varies dramatically between species.[20] In addition, the DFE also differs between coding regions and noncoding regions, with the DFE of noncoding DNA containing more weakly selected mutations.[20]

20.3.4   By impact on protein sequence

- A **frameshift mutation** is a mutation caused by insertion or deletion of a number of nucleotides that is not evenly divisible by three from a DNA sequence. Due to the triplet nature of gene expression by codons, the insertion or deletion can disrupt the reading frame, or the grouping of the codons, resulting in a completely different translation from the original.[67] The earlier in the sequence the deletion or insertion occurs, the more altered the protein produced is.

  In contrast, any insertion or deletion that is evenly divisible by three is termed an *in-frame mutation*

- A **nonsense mutation** is a point mutation in a sequence of DNA that results in a premature stop codon, or a *nonsense codon* in the transcribed mRNA, and possibly a truncated, and often nonfunctional protein product. (See Stop codon.)

- **Missense mutations** or *nonsynonymous mutations* are types of point mutations where a single nucleotide is changed to cause substitution of a different amino acid. This in turn can render the resulting protein nonfunctional. Such mutations are responsible for diseases such as Epidermolysis bullosa, sickle-cell disease, and SOD1-mediated ALS.[68]

- A **neutral mutation** is a mutation that occurs in an amino acid codon that results in the use of a different, but chemically similar, amino acid. The similarity between the two is enough that little or no change is often rendered in the protein. For example, a change from AAA to AGA will encode arginine, a chemically similar molecule to the intended lysine.

- **Silent mutations** are mutations that do not result in a change to the amino acid sequence of a protein, unless the changed amino acid is sufficiently similar to the original. They may occur in a region that does not code for a protein, or they may occur within a codon in a manner that does not alter the final amino acid sequence. The phrase *silent mutation* is often used interchangeably with the phrase *synonymous mutation*; however, synonymous mutations are a subcategory of the former, occurring only within exons (and necessarily exactly preserving the amino acid sequence of the protein). Synonymous mutations occur due to the degenerate nature of the genetic code.

20.3.5   By inheritance

In multicellular organisms with dedicated reproductive cells, mutations can be subdivided into germline mutations, which can be passed on to descendants through their reproductive cells, and somatic mutations (also called acquired mutations),[69] which involve cells outside the dedicated reproductive group and which are not usually transmitted to descendants.

A germline mutation gives rise to a *constitutional mutation* in the offspring, that is, a mutation that is present in every cell. A constitutional mutation can also occur very soon after fertilisation, or continue from a previous constitutional mutation in a parent.[70]

The distinction between germline and somatic mutations is important in animals that have a dedicated germline to produce reproductive cells. However, it is of little value in understanding the effects of mutations in plants, which lack dedicated germline. The distinction is also blurred in those animals that reproduce asexually through mechanisms such as budding, because the cells that give rise to the daughter organisms also give rise to that organism's germline. A new mutation that was not inherited from either parent is called a *de novo* mutation.

Diploid organisms (e.g., humans) contain two copies of each gene—a paternal and a maternal allele. Based on the occurrence of mutation on each chromosome, we may classify mutations into three types.

- A **heterozygous mutation** is a mutation of only one allele.

- A **homozygous mutation** is an identical mutation of both the paternal and maternal alleles.

- **Compound heterozygous** mutations or a **genetic compound** comprises two different mutations in the paternal and maternal alleles.[71]

A **wild type** or **homozygous non-mutated** organism is one in which neither allele is mutated.

20.3.6   Special classes

- **Conditional mutation** is a mutation that has wild-type (or less severe) phenotype under certain "permissive" environmental conditions and a mutant phenotype under certain "restrictive" conditions. For example, a temperature-sensitive mutation can cause cell death at high temperature (restrictive condition), but might have no deleterious consequences at a lower temperature (permissive condition).

- **Replication timing quantitative trait loci affects DNA replication.**

20.3.7   Nomenclature

In order to categorize a mutation as such, the "normal" sequence must be obtained from the DNA of a "normal" or "healthy" organism (as opposed to a "mutant" or "sick" one), it should be identified and reported; ideally, it should be made publicly available for a straightforward nucleotide-by-nucleotide comparison, and agreed upon by the scientific community or by a group of expert geneticists and biologists, who have the responsibility of establishing the *standard* or so-called "consensus" sequence. This step requires a tremendous scientific effort. (See DNA sequencing.) Once the consensus sequence is known, the mutations in a genome can be pinpointed, described, and classified. The committee of the Human Genome Variation Society (HGVS) has developed the standard human sequence variant nomenclature,[72] which should be used by researchers and DNA diagnostic centers to generate unambiguous mutation descriptions. In principle, this nomenclature can also be used to describe mutations in other organisms. The nomenclature specifies the type of mutation and base or amino acid changes.

- Nucleotide substitution (e.g., 76A>T) — The number is the position of the nucleotide from the 5′ end; the first letter represents the wild-type nucleotide, and the second letter represents the nucleotide that replaced the wild type. In the given example, the adenine at the 76th position was replaced by a thymine.

  - If it becomes necessary to differentiate between mutations in genomic DNA, mitochondrial DNA, and RNA, a simple convention is used. For example, if the 100th base of a nucleotide sequence mutated from G to C, then it would be written as g.100G>C if the mutation occurred in genomic DNA, m.100G>C if the mutation occurred in mitochondrial DNA, or r.100g>c if the mutation occurred in RNA. Note that, for mutations in RNA, the nucleotide code is written in lower case.

- Amino acid substitution (e.g., D111E) — The first letter is the one letter code of the wild-type amino acid, the number is the position of the amino acid from the N-terminus, and the second letter is the one letter code of the amino acid present in the mutation. Nonsense mutations are represented with an X for the second amino acid (e.g. D111X).

- Amino acid deletion (e.g., ΔF508) — The Greek letter Δ (delta) indicates a deletion. The letter refers to the amino acid present in the wild type and the number is the position from the N terminus of the amino acid were it to be present as in the wild type.

20.4 Mutation rates

Further information: Mutation rate

Mutation rates vary across species. Evolutionary biologists have theorized that higher mutation rates are beneficial in some situations, because they allow organisms to evolve and therefore adapt more quickly to their environments. For example, repeated exposure of bacteria to antibiotics, and selection of resistant mutants, can result in the selection of bacteria that have a much higher mutation rate than the original population (mutator strains).

According to one study, two children of different parents had 35 and 49 new mutations. Of them, in one case 92% were from the paternal germline, in another case, 64% were from the maternal germline.[73]

20.5 Harmful mutations

Changes in DNA caused by mutation can cause errors in protein sequence, creating partially or completely non-functional proteins. Each cell, in order to function correctly, depends on thousands of proteins to function in the right places at the right times. When a mutation alters a protein that plays a critical role in the body, a medical condition can result. A condition caused by mutations in one or more genes is called a genetic disorder. Some mutations alter a gene's DNA base sequence but do not change the function of the protein made by the gene. One study on the comparison of genes between different species of *Drosophila* suggests that if a mutation does change a protein, this will probably be harmful, with an estimated 70 percent of amino acid polymorphisms having damaging effects, and the remainder being either neutral or weakly beneficial.[4] Studies have shown that only 7% of point mutations in noncoding DNA of yeast are deleterious and 12% in coding DNA are deleterious. The rest of the mutations are either neutral or slightly beneficial.[74]

If a mutation is present in a germ cell, it can give rise to offspring that carries the mutation in all of its cells. This is the case in hereditary diseases. In particular, if there is a mutation in a DNA repair gene within a germ cell, humans carrying such germline mutations may have an increased risk of cancer. A list of 34 such germline mutations is given in the article DNA repair-deficiency disorder. An example of one is albinism. A mutation that occurs in the OCA1 or OCA2 gene. Individuals with this disorder are more prone to many types of cancers, other disorders and have impaired vision. On the other hand, a mutation may occur in a somatic cell of an organism. Such mutations will be present in all descendants of this cell within the same organism, and certain mutations can cause the cell to become malignant, and, thus, cause cancer.[75]

A DNA damage can cause an error when the DNA is replicated, and this error of replication can cause a gene mutation that, in turn, could cause a genetic disorder. DNA damages are repaired by the DNA repair system of the cell. Each cell has a number of pathways through which enzymes recognize and repair damages in DNA. Because DNA can be damaged in many ways, the process of DNA repair is an important way in which the body protects itself from disease. Once DNA damage has given rise to a mutation, the mutation cannot be repaired. DNA repair pathways can only recognize and act on "abnormal" structures in the DNA. Once a mutation occurs in a gene sequence it then has normal DNA structure and cannot be repaired.

20.6   Beneficial mutations

Although mutations that cause changes in protein sequences can be harmful to an organism, on occasions the effect may be positive in a given environment. In this case, the mutation may enable the mutant organism to withstand particular environmental stresses better than wild-type organisms, or reproduce more quickly. In these cases a mutation will tend to become more common in a population through natural selection.

For example, a specific 32 base pair deletion in human CCR5 (CCR5-Δ32) confers HIV resistance to homozygotes and delays AIDS onset in heterozygotes.[76] One possible explanation of the etiology of the relatively high frequency of CCR5-Δ32 in the European population is that it conferred resistance to the bubonic plague in mid-14th century Europe. People with this mutation were more likely to survive infection; thus its frequency in the population increased.[77] This theory could explain why this mutation is not found in Southern Africa, which remained untouched by bubonic plague. A newer theory suggests that the selective pressure on the CCR5 Delta 32 mutation was caused by smallpox instead of the bubonic plague.[78]

Another example is sickle-cell disease, a blood disorder in which the body produces an abnormal type of the oxygen-carrying substance hemoglobin in the red blood cells. One-third of all indigenous inhabitants of Sub-Saharan Africa carry the gene, because, in areas where malaria is common, there is a survival value in carrying only a single sickle-cell gene (sickle cell trait).[79] Those with only one of the two alleles of the sickle-cell disease are more resistant to malaria, since the infestation of the malaria *Plasmodium* is halted by the sickling of the cells that it infests.

20.7   Prion mutations

Prions are proteins and do not contain genetic material. However, prion replication has been shown to be subject to mutation and natural selection just like other forms of replication.[80]

20.8   Somatic mutations

Main article: Loss of heterozygosity
See also: Carcinogenesis

A change in the genetic structure that is not inherited from a parent, and also not passed to offspring, is called a *somatic cell genetic mutation* or *acquired mutation*.[69]

Cells with heterozygous mutations (one good copy of gene and one mutated copy) may function normally with the unmutated copy until the good copy has been spontaneously somatically mutated. This kind of mutation happens all the time in living organisms, but it is difficult to measure the rate. Measuring this rate is important in predicting the rate at which people may develop cancer.[81]

Point mutations may arise from spontaneous mutations that occur during DNA replication. The rate of mutation may be increased by mutagens. Mutagens can be physical, such as radiation from UV rays, X-rays or extreme heat, or chemical (molecules that misplace base pairs or disrupt the helical shape of DNA). Mutagens associated with cancers are often studied to learn about cancer and its prevention.

20.9   See also

- Aneuploidy

- Antioxidant

- Budgerigar colour genetics

- Carcinogenesis

- DNA sequencing

- Ecogenetics

- Embryology

- Frameshift mutation

- Homeobox

- Muller's morphs

- Mutagenesis

- Mutant

- Mutation rate

- Polyploidy

- Robertsonian translocation

- Saltation (biology)

- Signature-tagged mutagenesis

- Site-directed mutagenesis

- TILLING (molecular biology)

- Trinucleotide repeat expansion

20.10 References

[1] Bertram, John S. (December 2000). "The molecular biology of cancer". *Molecular Aspects of Medicine* (Amsterdam, the Netherlands: Elsevier) **21** (6): 167–223. doi:10.1016/S0098-2997(00)00007-8. ISSN 0098-2997. PMID 11173079.

[2] Aminetzach, Yael T.; Macpherson, J. Michael; Petrov, Dmitri A. (July 29, 2005). "Pesticide Resistance via Transposition-Mediated Adaptive Gene Truncation in *Drosophila*". *Science* (Washington, D.C.: American Association for the Advancement of Science) **309** (5735): 764–767. Bibcode:2005Sci...309..764A. doi:10.1126/science.1112699. ISSN 0036-8075. PMID 16051794.

[3] Burrus, Vincent; Waldor, Matthew K. (June 2004). "Shaping bacterial genomes with integrative and conjugative elements". *Research in Microbiology* (Amsterdam, the Netherlands: Elsevier) **155** (5): 376–386. doi:10.1016/j.resmic.2004.01.012. ISSN 0923-2508. PMID 15207870.

[4] Sawyer, Stanley A.; Parsch, John; Zhi Zhang; et al. (April 17, 2007). "Prevalence of positive selection among nearly neutral amino acid replacements in *Drosophila*". *Proc. Natl. Acad. Sci. U.S.A.* (Washington, D.C.: National Academy of Sciences) **104** (16): 6504–6510. Bibcode:2007PNAS..104.6504S. doi:10.1073/pnas.0701572104. ISSN 0027-8424. PMC 1871816. PMID 17409186.

[5] Hastings, P. J.; Lupski, James R.; Rosenberg, Susan M.; et al. (August 2009). "Mechanisms of change in gene copy number". *Nature Reviews Genetics* (London: Nature Publishing Group) **10** (8): 551–564. doi:10.1038/nrg2593. ISSN 1471-0056. PMC 2864001. PMID 19597530.

[6] Carroll, Grenier & Weatherbee 2005

[7] Harrison, Paul M.; Gerstein, Mark (May 17, 2002). "Studying Genomes Through the Aeons: Protein Families, Pseudogenes and Proteome Evolution". *Journal of Molecular Biology* (Amsterdam, the Netherlands: Elsevier) **318** (5): 1155–1174. doi:10.1016/S0022-2836(02)00109-2. ISSN 0022-2836. PMID 12083509.

[8] Orengo, Christine A.; Thornton, Janet M. (July 2005). "Protein families and their evolution—a structural perspective". *Annual Review of Biochemistry* (Palo Alto, CA: Annual Reviews) **74**: 867–900. doi:10.1146/annurev.biochem.74.082803.133029. ISSN 0066-4154. PMID 15954844.

[9] Manyuan Long; Betrán, Esther; Thornton, Kevin; et al. (November 2003). "The origin of new genes: glimpses from the young and old". *Nature Reviews Genetics* (London: Nature Publishing Group) **4** (11): 865–875. doi:10.1038/nrg1204. ISSN 1471-0056. PMID 14634634.

[10] Minglei Wang; Caetano-Anollés, Gustavo (January 14, 2009). "The Evolutionary Mechanics of Domain Organization in Proteomes and the Rise of Modularity in the Protein World". *Structure* (Cambridge, MA: Cell Press) **17** (1): 66–78. doi:10.1016/j.str.2008.11.008.ISSN 0969-2126. PMID19141283.

[11] Bowmaker, James K. (May 1998). "Evolution of colour vision in vertebrates". *Eye* (London: Nature Publishing Group) **12** (Pt 3b): 541–547. doi:10.1038/eye.1998.143. ISSN 0950-222X. PMID 9775215.

[12] Gregory, T. Ryan; Hebert, Paul D. N. (April 1999). "The Modulation of DNA Content: Proximate Causes and Ultimate Consequences". *Genome Research* (Cold Spring Harbor, NY: Cold Spring Harbor Laboratory Press) **9** (4): 317–324. doi:10.1101/gr.9.4.317(inactive2015-01-09). ISSN 1088-9051. PMID 10207154. Retrieved2015-10-01.

[13] Hurles, Matthew (July 13, 2004). "Gene Duplication: The Genomic Trade in Spare Parts". *PLOS Biology* (San Francisco, CA: Public Library of Science) **2** (7): E206. doi:10.1371/journal.pbio.0020206. ISSN 1545-7885. PMC 449868. PMID 15252449.

[14] Na Liu; Okamura, Katsutomo; Tyler, David M.; et al. (October 2008). "The evolution and functional diversification of animal microRNA genes". *Cell Research* (London: Nature Publishing Group on behalf of the Shanghai Institutes for Biological Sciences) **18** (10): 985–996. doi:10.1038/cr.2008.278. ISSN 1001-0602. PMC 2712117. PMID 18711447.

[15] Siepel, Adam (October 2009). "Darwinian alchemy: Human genes from noncoding DNA". *Genome Research* (Cold Spring Harbor, NY: Cold Spring Harbor Laboratory Press) **19** (10): 1693–1695. doi:10.1101/gr.098376.109. ISSN 1088-9051. PMC 2765273. PMID 19797681.

[16] Jianzhi Zhang; Xiaoxia Wang; Podlaha, Ondrej (May 2004). "Testing the Chromosomal Speciation Hypothesis for Humans and Chimpanzees". *Genome Research* (Cold Spring Harbor, NY: Cold Spring Harbor Laboratory Press) **14** (5): 845–851. doi:10.1101/gr.1891104. ISSN 1088-9051. PMC 479111. PMID 15123584.

[17] Ayala, Francisco J.; Coluzzi, Mario (May 3, 2005). "Chromosome speciation: Humans, *Drosophila*, and mosquitoes". *Proc. Natl. Acad. Sci. U.S.A.* (Washington, D.C.: National Academy of Sciences) **102** (Suppl 1): 6535–6542. Bibcode:2005PNAS..102.6535A.doi:10.1073/pnas.0501847102. ISSN 0027-8424. PMC 1131864. PMID15851677.

[18] Hurst, Gregory D. D.; Werren, John H. (August 2001). "The role of selfish genetic elements in eukaryotic evolution". *Nature Reviews Genetics* (London: Nature Publishing Group) **2** (8): 597–606. doi:10.1038/35084545. ISSN 1471-0056. PMID 11483984.

[19] Häsler, Julien; Strub, Katharina (November 2006). "*Alu* elements as regulators of gene expression". *Nucleic Acids Research* (Oxford, UK: Oxford University Press) **34** (19): 5491–5497. doi:10.1093/nar/gkl706. ISSN 0305-1048. PMC 1636486. PMID 17020921.

[20] Eyre-Walker, Adam; Keightley, Peter D. (August 2007). "The distribution of fitness effects of new mutations" (PDF). *Nature Reviews Genetics* (London: Nature Publishing Group) **8** (8): 610–618. doi:10.1038/nrg2146. ISSN 1471-0056. PMID 17637733. Retrieved 2015-10-02.

[21] Montelone, Beth A. (1998). "Mutation, Mutagens, and DNA Repair". *www-personal.ksu.edu*. Retrieved 2015-10-02.

[22] Bernstein et al. 2013

[23] Stuart, Gregory R.; Oda, Yoshimitsu; de Boer, Johan G.; Glickman, Barry W. (March 2000). "Mutation Frequency and Specificity With Age in Liver, Bladder and Brain of *lacI* Transgenic Mice". *Genetics* (Bethesda, MD: Genetics Society of America) **154** (3): 1291–1300. ISSN 0016-6731. PMC 1460990. PMID 10757770. Retrieved 2015-10-03.

[24] Kunz, Bernard A.; Ramachandran, Karthikeyan; Vonarx, Edward J. (April 1998). "DNA Sequence Analysis of Spontaneous Mutagenesis in *Saccharomyces cerevisiae*". *Genetics* (Bethesda, MD: Genetics Society of America) **148** (4): 1491–1505. ISSN 0016-6731. PMC 1460101. PMID 9560369. Retrieved 2015-10-03.

[25] Lieber, Michael R. (July 2010). "The Mechanism of Double-Strand DNA Break Repair by the Nonhomologous DNA End-Joining Pathway". *Annual Review of Biochemistry* (Palo Alto, CA: Annual Reviews) **79**: 181–211. doi:10.1146/annurev.bioche m.052308.093131.ISSN 0066-4154. PMC3079308. PMID 20192759.

[26] Created from PDB 1JDG

[27] Pfohl-Leszkowicz, Annie; Manderville, Richard A. (January 2007). "Ochratoxin A: An overview on toxicity and carcino-genicity in animals and humans". *Molecular Nutrition & Food Research* (Hoboken, NJ: Wiley-Blackwell) **51** (1): 61–99. doi:10.1002/mnfr.200600137. ISSN 1613-4125. PMID 17195275.

[28] Kozmin, Stanislav; Slezak, Guenaelle; Reynaud-Angelin, Anne; et al. (September 20, 2005). "UVA radiation is highly mutagenic in cells that are unable to repair 7,8-dihydro-8-oxoguanine in Saccharomyces cerevisiae". *Proc. Natl. Acad. Sci. U.S.A.* (Washington, D.C.: National Academy of Sciences) **102** (38): 13538–13543. Bibcode:2005PNAS..10213538K. doi:10.1073/pnas.0504497102. ISSN 0027-8424. PMC 1224634. PMID 16157879.

[29] References for the image are found in Wikimedia Commons page at: Commons:File:Notable mutations.svg#References.

[30] Freese, Ernst (April 15, 1959). "The difference between Spontaneous and Base-Analogue Induced Mutations of Phage T4". *Proc. Natl. Acad. Sci. U.S.A.* (Washington, D.C.: National Academy of Sciences) **45** (4): 622–633. doi:10.1073/pnas.45.4.622. ISSN 0027-8424. PMC 222607. PMID 16590424.

[31] Freese, Ernst (June 1959). "The specific mutagenic effect of base analogues on Phage T4". *Journal of Molecular Biology* (Amsterdam, the Netherlands: Elsevier) **1** (2): 87–105. doi:10.1016/S0022-2836(59)80038-3. ISSN 0022-2836.

[32] McClean, Phillip (1999). "Types of Mutations". *Genes and Mutations*. Course material from PLSC 431/631 - Intermediate Genetics.

[33] Goh, Amanda M.; Coffill, Cynthia R; Lane, David P. (January 2011). "The role of mutant p53 in human cancer". *The Journal of Pathology* (Hoboken, NJ: John Wiley & Sons) **223** (2): 116–126. doi:10.1002/path.2784. ISSN 0022-3417. PMID 21125670.

[34] Chenevix-Trench G, Spurdle AB, Gatei M, et al. (February 6, 2002). "Dominant Negative ATM Mutations in Breast Cancer Families". *Journal of the National Cancer Institute* (Oxford, UK: Oxford University Press) **94** (3): 205–215. doi:10.1093/jnci/94 .3.205.ISSN 0027-8874. PMID11830610.

[35] Paz-Priel, Ido; Friedman, Alan (2011). "C/EBPα dysregulation in AML and ALL". *Critical Reviews in Oncogenesis* (Redding, CT: Begell House) **16** (1–2): 93–102. doi:10.1615/critrevoncog.v16.i1-2.90. ISSN 0893-9675. PMC 3243939. PMID 22150310.

[36] Capaccio D, Ciccodicola A, Sabatino L, et al. (June 2010). "A novel germline mutation in Peroxisome Proliferator-Activated Receptor γ gene associated with large intestine polyp formation and dyslipidemia". *Biochimica et Biophysica Acta (BBA) - Molecular Basis of Disease* (Amsterdam, the Netherlands: Elsevier) **1802** (6): 572–581. doi:10.1016/j.bbadis.2010.01.012. ISSN 0925-4439. PMID 20123124.

[37] McKusick, Victor A. (July 25, 1991). "The defect in Marfan syndrome". *Nature* (London: Nature Publishing Group) **352** (6333): 279–281. Bibcode:1991Natur.352..279M. doi:10.1038/352279a0. ISSN 0028-0836. PMID 1852198.

[38] Judge DP, Biery NJ, Keene DR, et al. (July 15, 2004). "Evidence for a critical contribution of haploinsufficiency in the complex pathogenesis of Marfan syndrome". *Journal of Clinical Investigation* (Ann Arbor, MI: American Society for Clinical Investigation) **114** (2): 172–181. doi:10.1172/JCI20641. ISSN 0021-9738. PMC 449744. PMID 15254584.

[39] Judge, Daniel P.; Dietz, Harry C. (December 3, 2005). "Marfan's syndrome". *The Lancet* (Amsterdam, the Netherlands: Elsevier) **366** (9501): 1965–1976. doi:10.1016/S0140-6736(05)67789-6. ISSN 0140-6736. PMC 1513064. PMID 16325700.

[40] Ellis, Nathan A.; Ciocci, Susan; German, James (February 2001). "Back mutation can produce phenotype reversion in Bloom syndrome somatic cells". *Human Genetics* (Berlin; New York: Springer-Verlag) **108** (2): 167–173. doi:10.1007/s004390000447. ISSN 0340-6717. PMID 11281456.

[41] Charlesworth, Deborah; Charlesworth, Brian; Morgan, Martin T. (December 1995). "The Pattern of Neutral Molecular Variation Under the Background Selection Model". *Genetics* (Bethesda, MD: Genetics Society of America) **141** (4): 1619–1632. ISSN 0016-6731. PMC 1206892. PMID 8601499.

[42] Loewe, Laurence (April 2006). "Quantifying the genomic decay paradox due to Muller's ratchet in human mitochondrial DNA". *Genetical Research* (London; New York: Cambridge University Press) **87** (2): 133–159. doi:10.1017/S0016672306008123. ISSN 0016-6723. PMID 16709275.

[43] Bernstein, Hopf & Michod 1987, pp. 323–370

[44] Peck, Joel R.; Barreau, Guillaume; Heath, Simon C. (April 1997). "Imperfect Genes, Fisherian Mutation and the Evolution of Sex". *Genetics* (Bethesda, MD: Genetics Society of America) **145** (4): 1171–1199. ISSN 0016-6731. PMC 1207886. PMID 9093868.

[45] Keightley, Peter D.; Lynch, Michael (March 2003). "Toward a Realistic Model of Mutations Affecting Fitness". *Evolution* (Hoboken, NJ: John Wiley & Sons for the Society for the Study of Evolution) **57** (3): 683–689. doi:10.1554/0014-3820(2003)057[0683:tarmom]2.0.co;2.ISSN 0014-3820. JSTOR3094781. PMID 12703958.

[46] Barton, Nicholas H.; Keightley, Peter D. (January 2002). "Understanding quantitative genetic variation". *Nature Reviews Genetics* (London: Nature Publishing Group) **3** (1): 11–21. doi:10.1038/nrg700. ISSN 1471-0056. PMID 11823787.

[47] Sanjuán, Rafael; Moya, Andrés; Elena, Santiago F. (June 1, 2004). "The distribution of fitness effects caused by single-nucleotide substitutions in an RNA virus". *Proc. Natl. Acad. Sci. U.S.A.* (Washington, D.C.: National Academy of Sciences) **101** (22): 8396–8401. doi:10.1073/pnas.0400146101. ISSN 0027-8424. PMC 420405. PMID 15159545.

[48] Carrasco, Purificación; de la Iglesia, Francisca; Elena, Santiago F. (December 2007). "Distribution of Fitness and Virulence Effects Caused by Single-Nucleotide Substitutions in *Tobacco Etch Virus*". *Journal of Virology* (Washington, D.C.: American Society for Microbiology) **81** (23): 12979–12984. doi:10.1128/JVI.00524-07. ISSN 0022-538X. PMC 2169111. PMID 17898073.

[49] Sanjuán, Rafael (June 27, 2010). "Mutational fitness effects in RNA and single-stranded DNA viruses: common patterns revealed by site-directed mutagenesis studies". *Philosophical Transactions of the Royal Society B* (London: Royal Society) **365** (1548): 1975–1982. doi:10.1098/rstb.2010.0063. ISSN 0962-8436. PMID 20478892.

[50] Peris, Joan B.; Davis, Paulina; Cuevas, José M.; et al. (June 2010). "Distribution of Fitness Effects Caused by Single-Nucleotide Substitutions in Bacteriophage f1". *Genetics* (Bethesda, MD: Genetics Society of America) **185** (2): 603–609. doi:10.1534/genetics.110.115162. ISSN 0016-6731. PMC 2881140. PMID 20382832.

[51] Elena, Santiago F.; Ekunwe, Lynette; Hajela, Neerja; et al. (March 1998). "Distribution of fitness effects caused by random insertion mutations in Escherichia coli". *Genetica* (Kluwer Academic Publishers). **102–103** (1–6): 349–358. doi:10.1023/A:1017031008316.ISSN 0016-6707. PMID9720287.

[52] Hietpas, Ryan T.; Jensen, Jeffrey D.; Bolon, Daniel N. A. (May 10, 2011). "Experimental illumination of a fitness landscape". *Proc. Natl. Acad. Sci. U.S.A.* (Washington, D.C.: National Academy of Sciences) **108** (19): 7896–7901. doi:10.1073/pnas.1016024108.ISSN 0027-8424. PMC3093508. PMID 21464309.

[53] Davies, Esther K.; Peters, Andrew D.; Keightley, Peter D. (September 10, 1999). "High Frequency of Cryptic Deleterious Mutations in *Caenorhabditis elegans*". *Science* (Washington, D.C.: American Association for the Advancement of Science) **285** (5434): 1748–1751. doi:10.1126/science.285.5434.1748. ISSN 0036-8075. PMID 10481013.

[54] Loewe, Laurence; Charlesworth, Brian (September 22, 2006). "Inferring the distribution of mutational effects on fitness in *Drosophila*". *Biology Letters* (London: Royal Society) **2** (3): 426–430. doi:10.1098/rsbl.2006.0481. ISSN 1744-9561. PMC 1686194. PMID 17148422.

[55] Eyre-Walker, Adam; Woolfit, Megan; Phelps, Ted (June 2006). "The Distribution of Fitness Effects of New Deleterious Amino Acid Mutations in Humans". *Genetics* (Bethesda, MD: Genetics Society of America) **173** (2): 891–900. doi:10.1534/genetics.106.057570.ISSN 0016-6731. PMC1526495. PMID 16547091.

[56] Sawyer, Stanley A.; Kulathinal, Rob J.; Bustamante, Carlos D.; et al. (August 2003). "Bayesian Analysis Suggests that Most Amino Acid Replacements in *Drosophila* Are Driven by Positive Selection". *Journal of Molecular Evolution* (New York: Springer-Verlag) **57** (1): S154–S164. doi:10.1007/s00239-003-0022-3. ISSN 0022-2844. PMID 15008412.

[57] Piganeau, Gwenaël; Eyre-Walker, Adam (September 2, 2003). "Estimating the distribution of fitness effects from DNA sequence data: implications for the molecular clock". *Proc. Natl. Acad. Sci. U.S.A.* (Washington, D.C.: National Academy of Sciences) **100** (18): 10335–10340. doi:10.1073/pnas.1833064100. ISSN 0027-8424. PMC 193562. PMID 12925735.

[58] Kimura, Motoo (February 17, 1968). "Evolutionary Rate at the Molecular Level". *Nature* (London: Nature Publishing Group) **217** (5129): 624–626. doi:10.1038/217624a0. ISSN 0028-0836. PMID 5637732.

[59] Kimura 1983

[60] Akashi, Hiroshi (September 30, 1999). "Within- and between-species DNA sequence variation and the 'footprint' of natural selection". *Gene* (Amsterdam, the Netherlands: Elsevier) **238** (1): 39–51. doi:10.1016/S0378-1119(99)00294-2. ISSN 0378-1119. PMID 10570982.

[61] Eyre-Walker, Adam (October 2006). "The genomic rate of adaptive evolution". *Trends in Ecology & Evolution* (Cambridge, MA: Cell Press) **21** (10): 569–575. doi:10.1016/j.tree.2006.06.015. ISSN 0169-5347. PMID 16820244.

[62] Gillespie, John H. (September 1984). "Molecular Evolution Over the Mutational Landscape". *Evolution* (Hoboken, NJ: John Wiley & Sons for the Society for the Study of Evolution) **38** (5): 1116–1129. doi:10.2307/2408444. ISSN 0014-3820. JSTOR 2408444.

[63] Orr, H. Allen (April 2003). "The Distribution of Fitness Effects Among Beneficial Mutations". *Genetics* (Bethesda, MD: Genetics Society of America) **163** (4): 1519–1526. ISSN 0016-6731. PMC 1462510. PMID 12702694. Retrieved 2015-10-07.

[64] Kassen, Rees; Bataillon, Thomas (April 2006). "Distribution of fitness effects among beneficial mutations before selection in experimental populations of bacteria". *Nature Genetics* (London: Nature Publishing Group) **38** (4): 484–488. doi:10.1038/ng1751. ISSN 1061-4036. PMID 16550173.

[65] Rokyta, Darin R.; Joyce, Paul; Caudle, S. Brian; et al. (April 2005). "An empirical test of the mutational landscape model of adaptation using a single-stranded DNA virus". *Nature Genetics* (London: Nature Publishing Group) **37** (4): 441–444. doi:10.1038/ng1535. ISSN 1061-4036. PMID 15778707.

[66] Imhof, Marianne; Schlotterer, Christian (January 30, 2001). "Fitness effects of advantageous mutations in evolving *Escherichia coli* populations". *Proc. Natl. Acad. Sci. U.S.A.* (Washington, D.C.: National Academy of Sciences) **98** (3): 1113–1137. doi:10.1073/pnas.98.3.1113. ISSN 0027-8424. PMC 14717. PMID 11158603.

[67] Hogan, C. Michael (October 12, 2010). "Mutation". In Monosson, Emily. *Encyclopedia of Earth*. Washington, D.C.: Environmental Information Coalition, National Council for Science and the Environment. OCLC 72808636. Retrieved 2015-10-08.

[68] Boillée, Séverine; Vande Velde, Christine; Cleveland, Don W. (October 5, 2006). "ALS: A Disease of Motor Neurons and Their Nonneuronal Neighbors". *Neuron* (Cambridge, MA: Cell Press) **52** (1): 39–59. doi:10.1016/j.neuron.2006.09.018. ISSN 0896-6273. PMID 17015226.

[69] "Somatic cell genetic mutation". *Genome Dictionary*. Athens, Greece: Information Technology Associates. June 30, 2007. Retrieved 2010-06-06.

[70] "*RB1* Genetics". *Daisy's Eye Cancer Fund*. Oxford, UK. Archived from the original on 2011-11-26. Retrieved 2015-10-09.

[71] "Compound heterozygote". *MedTerms*. New York: WebMD. June 14, 2012. Retrieved 2015-10-09.

[72] den Dunnen, Johan T.; Antonarakis, Stylianos E. (January 2000). "Mutation Nomenclature Extensions and Suggestions to Describe Complex Mutations: A Discussion". *Human Mutation* (Hoboken, NJ: Wiley-Liss, Inc.) **15** (1): 7–12. doi:10.1002/(SICI)1098-1004(200001)15:1<7::AID-HUMU4>3.0.CO;2-N.ISSN 1059-7794. PMID 10612815.

[73] Conrad DF, Keebler JE, DePristo MA, et al. (July 2011). "Variation in genome-wide mutation rates within and between human families". *Nature Genetics* (London: Nature Publishing Group) **43** (7): 712–714. doi:10.1038/ng.862. ISSN 1061-4036. PMC 3322360. PMID 21666693. Retrieved 2015-10-09.

[74] Doniger, Scott W.; Hyun Seok Kim; Swain, Devjanee; et al. (August 29, 2008). Pritchard, Jonathan K., ed. "A Catalog of Neutral and Deleterious Polymorphism in Yeast". *PLOS Genetics* (San Francisco, CA: Public Library of Science) **4** (8): e1000183. doi:10.1371/journal.pgen.1000183. ISSN 1553-7404. PMC 2515631. PMID 18769710.

[75] Ionov, Yurij; Peinado, Miguel A.; Malkhosyan, Sergei; et al. (June 10, 1993). "Ubiquitous somatic mutations in simple repeated sequences reveal a new mechanism for colonic carcinogenesis". *Nature* (London: Nature Publishing Group) **363** (6429): 558–561. Bibcode:1993Natur.363..558I. doi:10.1038/363558a0. ISSN 0028-0836. PMID 8505985.

[76] Sullivan, Amy D.; Wigginton, Janis; Kirschner, Denise (August 28, 2001). "The coreceptor mutation CCR5Δ32 influences the dynamics of HIV epidemics and is selected for by HIV". *Proc. Natl. Acad. Sci. U.S.A.* (Washington, D.C.: National Academy of Sciences) **95** (18): 10214–10219. Bibcode:2001PNAS...9810214S. doi:10.1073/pnas.181325198. ISSN 0027-8424. PMC 56941. PMID 11517319.

[77] "Mystery of the Black Death". *Secrets of the Dead*. Season 3. Episode 2. October 30, 2002. PBS. Retrieved 2015-10-10. Episode background.

[78] Galvani, Alison P.; Slatkin, Montgomery (December 9, 2003). "Evaluating plague and smallpox as historical selective pressures for the CCR5-Δ32 HIV-resistance allele". *Proc. Natl. Acad. Sci. U.S.A.* (Washington, D.C.: National Academy of Sciences) **100** (25): 15276–15279. Bibcode:2003PNAS..10015276G. doi:10.1073/pnas.2435085100. ISSN 0027-8424. PMC 299980. PMID 14645720.

[79] Konotey-Ahulu, Felix. "Frequently Asked Questions [FAQ's]". *sicklecell.md*.

[80] "'Lifeless' prion proteins are 'capable of evolution'". Health. *BBC News Online* (London). January 1, 2010. Retrieved 2015-10-10.

[81] Araten, David J.; Golde, David W.; Rong H. Zhang; et al. (September 15, 2005). "A Quantitative Measurement of the Human Somatic Mutation Rate". *Cancer Research* (Philadelphia, PA: American Association for Cancer Research) **65** (18): 8111–8117. doi:10.1158/0008-5472.CAN-04-1198. ISSN 0008-5472. PMID 16166284.

20.11   Bibliography

- Bernstein, Carol; Prasad, Anil R.; Nfonsam, Valentine; Bernstein, Harris (2013). "DNA Damage, DNA Repair and Cancer". In Chen, Clark. *New Research Directions in DNA Repair*. Rijeka, Croatia: InTech. doi:10.5772/53919. ISBN 978-953-51-1114-6.

- Bernstein, Harris; Hopf, Frederic A.; Michod, Richard E. (1987). "The Molecular Basis of the Evolution of Sex". In Scandalios, John G. *Molecular Genetics of Development*. Advances in Genetics **24**. San Diego, CA: Academic Press. doi:10.1016/S0065-2660(08)60012-7. ISBN 0-12-017624-6. ISSN 0065-2660. LCCN 47030313. OCLC 18561279. PMID 3324702.

- Carroll, Sean B.; Grenier, Jennifer K.; Weatherbee, Scott D. (2005). *From DNA to Diversity: Molecular Genetics and the Evolution of Animal Design* (2nd ed.). Malden, MA: Blackwell Publishing. ISBN 1-4051-1950-0. LCCN 2003027991. OCLC 53972564.

- Kimura, Motoo (1983). *The Neutral Theory of Molecular Evolution*. Cambridge, UK; New York: Cambridge University Press. ISBN 0-521-23109-4. LCCN 82022225. OCLC 9081989.

20.12   External links

- den Dunnen, Johan T. "Nomenclature for the description of sequence variants". Melbourne, Australia: Human Genome Variation Society. Retrieved 2015-10-18.

- Jones, Steve; Woolfson, Adrian; Partridge, Linda (December 6, 2007). "Genetic Mutation". *In Our Time*. BBC Radio 4. Retrieved 2015-10-18.

- Liou, Stephanie (February 5, 2011). "All About Mutations". *HOPES*. Standford, CA: Huntington's Disease Outreach Project for Education at Stanford. Retrieved 2015-10-18.

- "Locus Specific Mutation Databases". Leiden, the Netherlands: Leiden University Medical Center. Retrieved 2015-10-18.

- "Welcome to the Mutalyzer website". Leiden, the Netherlands: Leiden University Medical Center. Retrieved 2015-10-18. — The Mutalyzer website.

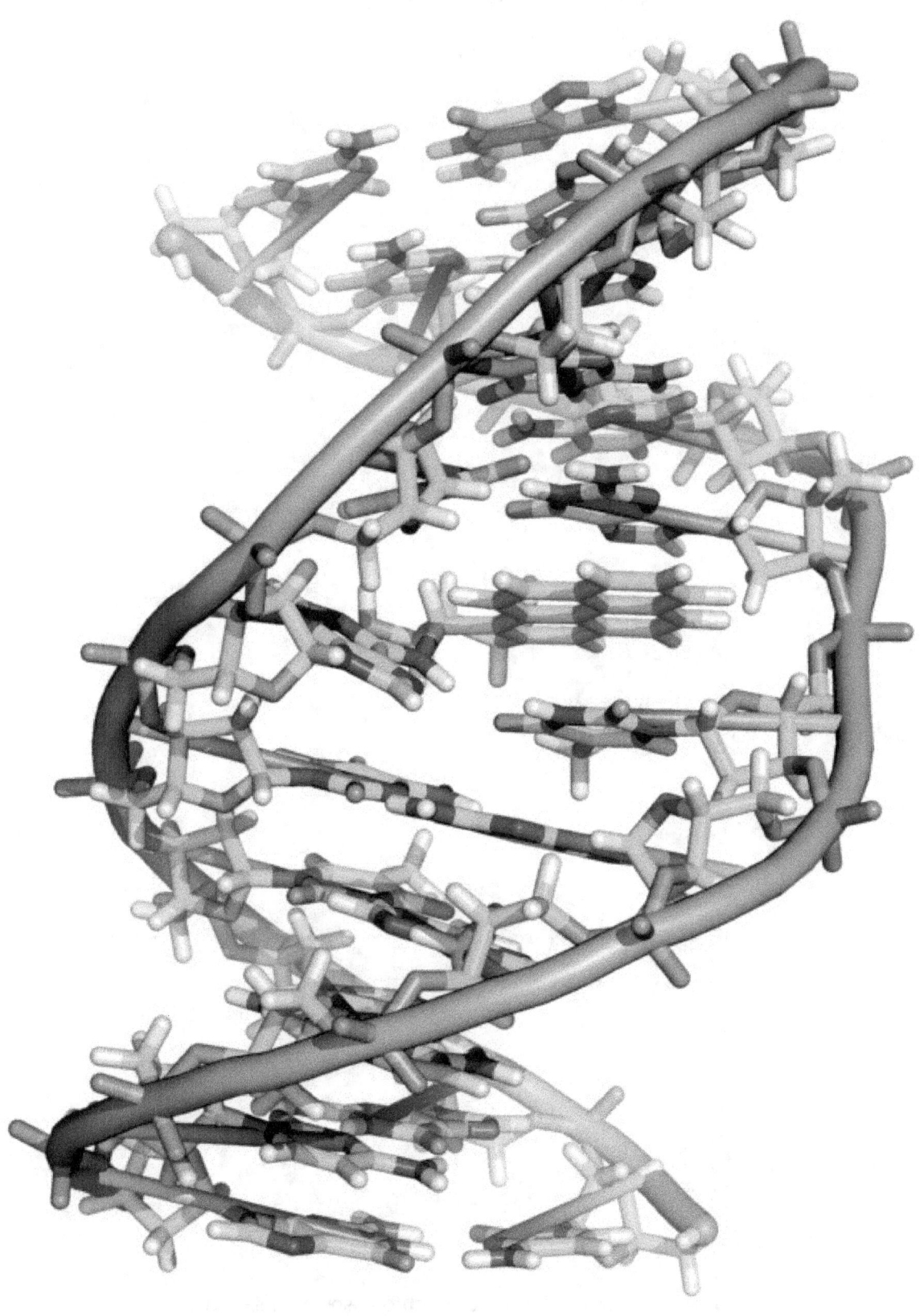

A covalent adduct between benzo[a]pyrene, the major mutagen in tobacco smoke, and DNA[26]

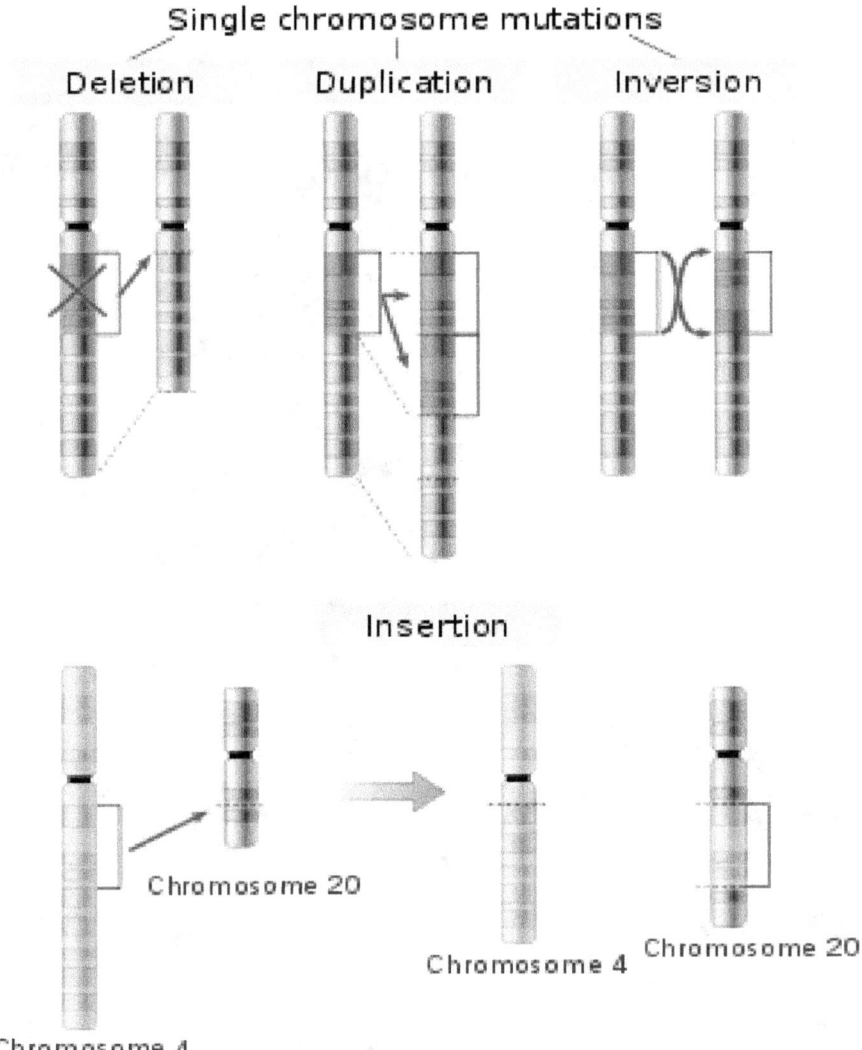

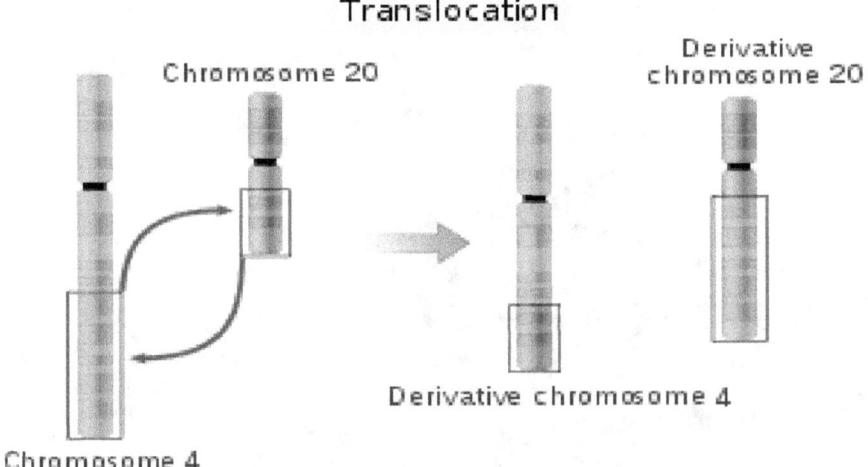

Illustrations of five types of chromosomal mutations.

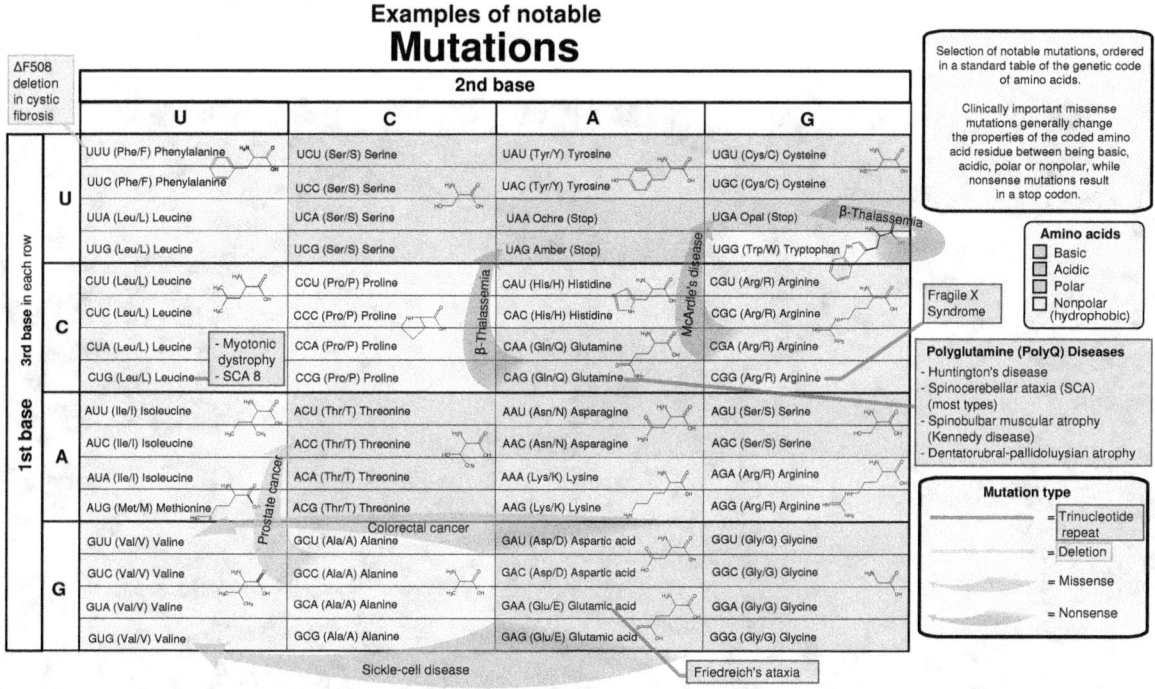

Selection of disease-causing mutations, in a standard table of the genetic code of amino acids.[29]

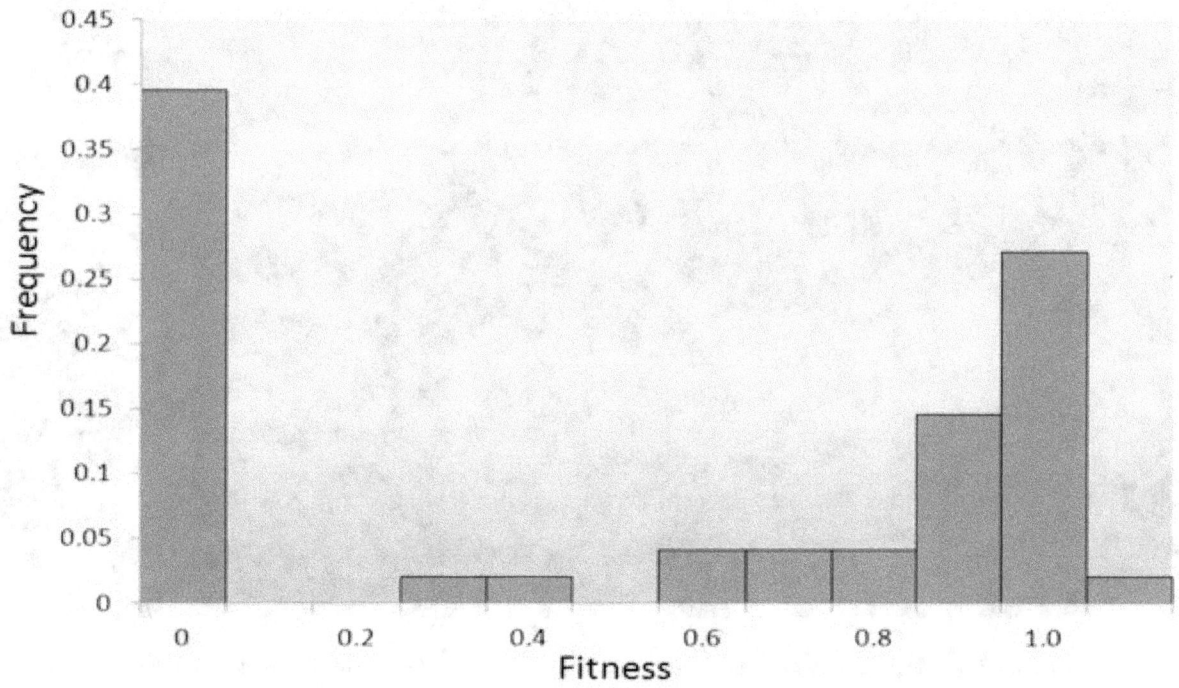

The distribution of fitness effects (DFE) of mutations in vesicular stomatitis virus. In this experiment, random mutations were introduced into the virus by site-directed mutagenesis, and the fitness of each mutant was compared with the ancestral type. A fitness of zero, less than one, one, more than one, respectively, indicates that mutations are lethal, deleterious, neutral, and advantageous.[47]

A mutation has caused this garden moss rose to produce flowers of different colors. This is a somatic mutation that may also be passed on in the germline.

Chapter 21

DNA damage theory of aging

The **DNA damage theory of aging** proposes that aging is a consequence of unrepaired accumulation of naturally occurring DNA damages. Damage in this context is a DNA alteration that has an abnormal structure. Although both mitochondrial and nuclear DNA damage can contribute to aging, nuclear DNA is the main subject of this analysis. Nuclear DNA damage can contribute to aging either indirectly (by increasing apoptosis or cellular senescence) or directly (by increasing cell dysfunction).[1][2]

In humans and other mammals, DNA damage occurs frequently and DNA repair processes have evolved to compensate. In estimates made for mice, on average approximately 1,500 to 7,000 DNA lesions occur per hour in each mouse cell, or about 36,000 to 160,000 per cell per day.[3] In any cell some DNA damage may remain despite the action of repair processes. The accumulation of unrepaired DNA damage is more prevalent in certain types of cells, particularly in non-replicating or slowly replicating cells, such as cells in the brain, skeletal and cardiac muscle.

21.1 DNA damage and mutation

Further information: DNA repair, DNA damage (naturally occurring), Mutation

To understand the DNA damage theory of aging it is important to distinguish between DNA damage and mutation, the two major types of errors that occur in DNA. Damages and mutation are fundamentally different. DNA damages are physical abnormalities in the DNA, such as single and double strand breaks, 8-hydroxydeoxyguanosine residues and polycyclic aromatic hydrocarbon adducts. DNA damages can be recognized by enzymes, and thus they can be correctly repaired if redundant information, such as the undamaged sequence in the complementary DNA strand or in a homologous chromosome, is available for copying. If a cell retains DNA damage, transcription of a gene can be prevented and thus translation into a protein will also be blocked. Replication may also be blocked and/or the cell may die. Descriptions of decrements in function, characteristic of aging, associated with accumulation of DNA damages, are given later in this article.

In contrast to DNA damage, a mutation is a change in the base sequence of the DNA. A mutation cannot be recognized by enzymes once the base change is present in both DNA strands, and thus a mutation cannot be repaired. At the cellular level, mutations can cause alterations in protein function and regulation. Mutations are replicated when the cell replicates. In a population of cells, mutant cells will increase or decrease in frequency according to the effects of the mutation on the ability of the cell to survive and reproduce. Although distinctly different from each other, DNA damages and mutations are related because DNA damages often cause errors of DNA synthesis during replication or repair and these errors are a major source of mutation.

Given these properties of DNA damage and mutation, it can be seen that DNA damages are a special problem in non-dividing or slowly dividing cells, where unrepaired damages will tend to accumulate over time. On the other hand, in rapidly dividing cells, unrepaired DNA damages that do not kill the cell by blocking replication will tend to cause replication errors and thus mutation. The great majority of mutations that are not neutral in their effect are deleterious to a cell's survival. Thus, in a population of cells comprising a tissue with replicating cells, mutant cells will tend to be lost. However, infrequent mutations that provide a survival advantage will tend to clonally expand at the expense of neighboring cells in the tissue. This advantage to the cell is disadvantageous to the whole organism, because such mutant cells can give rise to cancer. Thus DNA damages in frequently dividing cells, because they give rise to mutations, are a prominent cause of cancer. In contrast, DNA damages in infrequently dividing cells are likely a prominent cause of aging.

The first person to suggest that DNA damage, as distinct from mutation, is the primary cause of aging was Alexander in 1967.[4] By the early 1980s there was significant experimental support for this idea in the literature.[5] By the early 1990s experimental support for this idea was substantial, and furthermore it had become increasingly evident that oxidative DNA damage, in particular, is a major cause of aging.[6][7][8][9][10]

In a series of articles from 1970 to 1977, PV Narasimh Acharya, Phd. (1924–1993) theorized and scientifically proved that cells undergo "irreparable DNA damage," whereby DNA crosslinks occur when both normal cellular repair processes fail and cellular apoptosis does not occur. Specifically, PVN Acharya noted that double-strand breaks and a "cross-linkage joining both strands at the same point is irreparable because neither strand can then serve as a template for repair. The cell will die in the next mitosis or in some rare instances, mutate."[11][12][13][14][15] Acharya's research also showed how irreparable DNA damage is caused by environmental pollutants, low dose ionizing radiation and food additives, particularly nitrites and nitrates and such damage to the DNA is a causal factor for pre-mature aging and cancer.

21.2   Age-associated accumulation of DNA damage and decline in gene expression

Further information: DNA damage (naturally occurring)

In tissues composed of non- or infrequently replicating cells, DNA damage can accumulate with age and lead either to loss of cells, or, in surviving cells, loss of gene expression. Accumulated DNA damage is usually measured directly. Numerous studies of this type have indicated that oxidative damage to DNA is particularly important.[16] The loss of expression of specific genes can be detected at both the mRNA level and protein level.

21.2.1   Brain

Further information: Aging brain

The adult brain is composed in large part of terminally differentiated non-dividing neurons. Many of the conspicuous features of aging reflect a decline in neuronal function. Accumulation of DNA damage with age in the mammalian brain has been reported during the period 1971 to the present in at least 29 studies. A review (/book published) of the role of DNA damage in aging, including a comprehensive summary of the studies showing DNA damage accumulation with age in brain, muscle, liver and kidney, was presented by Bernstein et al.[17] Here, we mention only some recent studies involving rodents plus one human study. Rutten et al.[18] showed that single-strand breaks accumulate in the mouse brain with age. Sen et al.[19] showed that DNA damages which block the polymerase chain reaction in rat brain accumulate with age. Swain and Rao observed marked increases in several types of DNA damages in aging rat brain, including single-strand breaks, double-strand breaks and modified bases (8-OHdG and uracil).[20] Wolf et al.[21] also showed that the oxidative DNA damage 8-OHdG accumulates in rat brain with age. Similarly, it was shown that as humans age from 48–97 years, 8-OHdG accumulates in the brain.[22]

Decrements in function were noted in aging human brain, where transcription of a set of evaluated genes declines with age from 40 to 106 years.[23] These genes play central roles in synaptic plasticity, vesicular transport and mitochondrial function. In the brain, promoters of genes with reduced expression have markedly increased DNA damage.[23] In cultured human neurons, these gene promoters are selectively damaged by oxidative stress. Thus Lu et al.[23] concluded that DNA damage may reduce the expression of selectively vulnerable genes involved in learning, memory and neuronal survival, initiating a program of brain aging that starts early in adult life.

21.2.2 Muscle

Further information: Muscle

Muscle strength, and stamina for sustained physical effort, have decrements in function with age in humans and other species. Skeletal muscle is a tissue composed largely of multinucleated myofibers, elements that arise from the fusion of mononucleated myoblasts. Accumulation of DNA damage with age in mammalian muscle has been reported in at least 18 studies since 1971.[17] We will mention here only two of the more recent studies in rodents plus one in humans. Hamilton et al.[24] reported that the oxidative DNA damage 8-OHdG accumulates in heart and skeletal muscle (as well as in brain, kidney and liver) of both mouse and rat with age. In humans, increases in 8-OHdG with age were reported for skeletal muscle.[25] Catalase is an enzyme that removes hydrogen peroxide, a reactive oxygen species, and thus limits oxidative DNA damage. In mice, when catalase expression is increased specifically in mitochondria, oxidative DNA damage (8-OHdG) in skeletal muscle is decreased and lifespan is increased by about 20%.[26][27] These findings suggest that mitochondria are a significant source of the oxidative damages contributing to aging.

Protein synthesis and protein degradation decline with age in skeletal and heart muscle, as would be expected, since DNA damage blocks gene transcription. In a recent study Piec et al.[28] found numerous changes in protein expression in rat skeletal muscle with age, including lower levels of several proteins related to myosin and actin. Force is generated in striated muscle by the interactions between myosin thick filaments and actin thin filaments.

21.2.3 Liver

Further information: Liver

Liver hepatocytes do not ordinarily divide and appear to be terminally differentiated, but they retain the ability to proliferate when injured. With age, the mass of the liver decreases, blood flow is reduced, metabolism is impaired, and alterations in microcirculation occur. At least 21 studies have reported an increase in DNA damage with age in liver.[17] For instance, Helbock et al.[29] estimated that the steady state level of oxidative DNA base alterations increased from 24,000 per cell in the liver of young rats to 66,000 per cell in the liver of old rats.

21.2.4 Kidney

Further information: Kidney

In kidney, changes with age include reduction in both renal blood flow and glomerular filtration rate, and impairment in the ability to concentrate urine and to conserve sodium and water. DNA damages, particularly oxidative DNA damages, increase with age (at least 8 studies).[17] For instance Hashimoto et al.[30] showed that 8-OHdG accumulates in rat kidney DNA with age.

21.2.5 Long-lived stem cells

Further information: Stem cell, Stem cell theory of aging

Tissue-specific stem cells produce differentiated cells through a series of increasingly more committed progenitor intermediates. In hematopoiesis (blood cell formation), the process begins with long-term hematopoietic stem cells that self-renew and also produce progeny cells that upon further replication go through a series of stages leading to differentiated cells without self-renewal capacity. In mice, deficiencies in DNA repair appear to limit the capacity of hematopoietic stem cells to proliferate and self-renew with age.[31] Sharpless and Depinho reviewed evidence that hematopoietic stem cells, as well as stem cells in other tissues, undergo intrinsic aging.[32] They speculated that stem cells grow old, in part, as a result of DNA damage. DNA damage may trigger signalling pathways, such as apoptosis, that contribute to depletion of stem cell stocks. This has been observed in several cases of accelerated aging and may occur in normal aging too.[33]

21.3   Glucose theory of aging

It is proposed that high "blood glucose" (a.k.a. "blood sugar") is a cause of DNA damage. This, in turn, causes cells to divide, and the telomere to shorten, which in theory advances aging.

Healthy average non-diabetic fasting blood glucose, according to Dr. Richard Bernstein, is about 83 mg/dl, ideally between 60 and 100 mg/dl. It is believed that higher levels of blood sugar above about 100 mg/dl are a principle cause of aging, as seen in diabetics.

21.4   Mutation theories of aging

Further information: Evolution of ageing

A popular idea, that has failed to gain significant experimental support, is the idea that mutation, as distinct from DNA damage, is the primary cause of aging. As discussed above, mutations tend to arise in frequently replicating cells as a result of errors of DNA synthesis when template DNA is damaged, and can give rise to cancer. However, in mice there is no increase in mutation in the brain with aging.[34][35][36] Mice defective in a gene (Pms2) that ordinarily corrects base mispairs in DNA have about a 100-fold elevated mutation frequency in all tissues, but do not appear to age more rapidly.[37] On the other hand, mice defective in one particular DNA repair pathway show clear premature aging, but do not have elevated mutation.[38]

One variation of the idea that mutation is the basis of aging, that has received much attention, is that mutations specifically in mitochondrial DNA are the cause of aging. Several studies have shown that mutations accumulate in mitochondrial DNA in infrequently replicating cells with age. DNA polymerase gamma is the enzyme that replicates mitochondrial DNA. A mouse mutant with a defect in this DNA polymerase is only able to replicate its mitochondrial DNA inaccurately, so that the mutation rate is 500-fold higher than in normal mice. Yet these mice showed no obvious features of rapidly accelerated aging.[39] The probable explanation for the apparent lack of effect of the additional mutations in mitochondrial DNA is that, within a typical cell, there are large numbers of mitochondria and each mitochondrion can have multiple copies of mitochondrial DNA. Since most mutations are recessive, any particular deleterious mutation would not be expected to have a pronounced effect when many copies of the correct DNA sequence are present in the same and in other mitochondria in the cell. Overall, the observations discussed in this section indicate that mutations are not the primary cause of aging.

21.5   Dietary restriction

Further information: Calorie restriction

In rodents, caloric restriction slows aging and extends lifespan. At least 4 studies have shown that caloric restriction reduces 8-OHdG damages in various organs of rodents. One of these studies showed that caloric restriction reduced accumulation of 8-OHdG with age in rat brain, heart and skeletal muscle, and in mouse brain, heart, kidney and liver.[24] More recently, Wolf et al.[21] showed that dietary restriction reduced accumulation of 8-OHdG with age in rat brain, heart, skeletal muscle, and liver. Thus reduction of oxidative DNA damage is associated with a slower rate of aging and increased lifespan.

21.6   Inherited defects that cause premature aging

Further information: DNA repair-deficiency disorder

If DNA damage is the underlying cause of aging, it would be expected that humans with inherited defects in the ability to repair DNA damages should age at a faster pace than persons without such a defect. Numerous examples of rare inherited conditions with DNA repair defects are known. Several of these show multiple striking features of premature aging, and others have fewer such features. Perhaps the most striking premature aging conditions are Werner syndrome (mean lifespan 47 years), Huchinson-Gilford Progeria (mean lifespan 13 years), and Cockayne syndrome (mean lifespan 13 years). Werner syndrome is due to an inherited defect in an enzyme (a helicase and exonuclease) that acts in base excision repair of DNA (e.g. see Harrigan et al.[40]). Hutchinson-Guilford Progeria is due to a defect in Lamin A protein which forms a scaffolding within the cell nucleus to organize chromatin and is needed for repair of double-strand breaks in DNA.[41] Cockayne Syndrome is due to a defect in a protein necessary for the repair process, transcription coupled nucleotide excision repair, which can remove damages, particularly oxidative DNA damages, that block transcription.[42] In addition to these three conditions, several other human syndromes, that also have defective DNA repair, show several features of premature aging. These include ataxia telangiectasia, Nijmegen breakage syndrome, some subgroups of xeroderma pigmentosum, trichothiodystrophy, Fanconi anemia, Bloom syndrome and Rothmund-Thomson syndrome.

In addition to human inherited syndromes, experimental mouse models with genetic defects in DNA repair show features of premature aging and reduced lifespan.(e.g. refs.[43][44][45]) In particular, mutant mice defective in Ku70, or Ku80, or double mutant mice deficient in both Ku70 and Ku80 exhibit early aging.[46] The mean lifespans of the three mutant mouse strains were similar to each other, at about 37 weeks, compared to 108 weeks for the wild-type control. Six specific signs of aging were examined, and the three mutant mice were found to display the same aging signs as the control mice, but at a much earlier age. Cancer incidence was not increased in the mutant mice. Ku70 and Ku80 form the heterodimer Ku protein essential for the non-homologous end joining (NHEJ) pathway of DNA repair, active in repairing DNA double-strand breaks. This suggests an important role of NHEJ in longevity assurance.

21.7   Lifespan in different mammalian species

Further information: Maximum life span

Studies comparing DNA repair capacity in different mammalian species have shown that repair capacity correlates with lifespan. The initial study of this type, by Hart and Setlow,[47] showed that the ability of skin fibroblasts of seven mammalian species to perform DNA repair after exposure to a DNA damaging agent correlated with lifespan of the species. The species studied were shrew, mouse, rat, hamster, cow, elephant and human. This initial study stimulated many additional studies involving a wide variety of mammalian species, and the correlation between repair capacity and lifespan generally held up. In one of the more recent studies, Burkle et al.[48] studied the level of a particular enzyme, poly(ADP-ribose) polymerase, which is involved in repair of single-strand breaks in DNA. They found that the lifespan of 13 mammalian species correlated with the activity of this enzyme. In addition, they found that humans who lived past 100 years had a significantly higher activity of this enzyme than younger individuals.

21.8   Conclusions

Numerous studies have shown that DNA damage accumulates in brain, muscle, liver, kidney, and in long-lived stem cell. These accumulated DNA damages are the likely cause of the decline in gene expression and loss of functional capacity observed with increasing age. On the other hand, accumulation of mutations, as distinct from DNA damages, is not a plausible candidate as the primary cause of aging. A calorie-restricted diet in mammals improves lifespan, and this improvement is associated with a decrease in oxidative DNA damage. Several inherited genetic defects in ability to repair DNA damage give rise to premature aging suggesting a causal relationship between DNA damage and aging. In comparisons of different mammalian species that differ in lifespan, DNA repair capacity is found to correlate with lifespan. The principal source of the DNA damages leading to normal aging appears to be reactive oxygen species, produced as byproducts of normal cellular metabolism.

21.9   See also

- Aging brain

- Biological immortality

- DNA damage (naturally occurring)

- DNA repair

- Life expectancy

- Longevity

- Maximum life span

- Rejuvenation (aging)

- Senescence

- Telomere

21.10   References

[1] Best,BP (2009). "Nuclear DNA damage as a direct cause of aging"(PDF).*Rejuvenation Research***12**(3): 199–208. doi:10.1089/r PMID 19594328.

[2] Freitas AA, de Magalhães JP (2011). "A review and appraisal of the DNA damage theory of ageing". *Mutation Research (journal)* **728** (1-2): 12–22. doi:10.1016/j.mrrev.2011.05.001. PMID 19594328.

[3] Vilenchik, MM; Knudson, AG (May 2000). "Inverse radiation dose-rate effects on somatic and germ-line mutations and DNA damage rates". *Proc Natl Acad Sci U S A.* **97** (10): 5381–6. doi:10.1073/pnas.090099497. PMID 10792040.

[4] "The role of DNA lesions in the processes leading to aging in mice." **21**. 1967. pp. 29–50. PMID 4860956.

[5] Gensler, HL; Bernstein, H (Sep 1981). "DNA damage as the primary cause of aging". *Q Rev Biol.* **56** (3): 279–303. doi:10.1086/412317. PMID 7031747.

[6] Bernstein C, Bernstein H. (1991) Aging, Sex, and DNA Repair. Academic Press, San Diego. ISBN 978-0120928606 partly available at http://books.google.com/books?id=BaXYYUXy71cC&pg=PA3&lpg=PA3&dq=Aging,+Sex,+and+DNA+Repair& source=bl&ots=9E6VrRl7fJ&sig=kqUROJfBM6EZZeIrkuEFygsVVpo&hl=en&sa=X&ei=z8BqUpi7D4KQiALC54Ew&ved= 0CFUQ6AEwBg#v=onepage&q=Aging%2C%20Sex%2C%20and%20DNA%20Repair&f=false

[7] "Endogenous mutagens and the causes of aging and cancer". *Mutation Research/Fundamental and Molecular Mechanisms of Mutagenesis* **250** (1-2): 3–16. 1991. doi:10.1016/0027-5107(91)90157-j. PMID 1944345.

[8] Holmes, GE; Bernstein, C; Bernstein, H (1992). "Oxidative and other DNA damages as the basis of aging: a review". *Mutat Res* **275** (3-6): 305–315. doi:10.1016/0921-8734(92)90034-M. PMID 1383772.

[9] "DNA damage and repair in brain: relationship to aging". *Mutation Research/DNAging* **275** (3-6): 317–29. September 1992. doi:10.1016/0921-8734(92)90035-N. PMID 1383773.

[10] "Oxidants, antioxidants, and the degenerative diseases of aging". *Proceedings of the National Academy of Sciences* **90** (17): 7915–22. September 1993. doi:10.1073/pnas.90.17.7915. PMC 47258. PMID 8367443.

[11] Acharya PV (1972). "The isolation and partial characterization of age-correlated oligo-deoxyribo-ribonucleotides with covalently linked aspartyl-glutamyl polypeptides". *Johns Hopkins Med. J. Suppl.* (1): 254–60. PMID 5055816.

[12] Acharya, PV; Ashman, SM; Bjorksten, J; The isolation and partial characterization of age-correlated oligo-deoxyribo-ribo nucleo peptides. Finska Kemists Medd. 81 No. 3 (1972) Suomen Kemists. Tied. Chemical Abstacts, Vol 78, No. 19. May 14, 1973. Abs. N. 122001 g.

[13] Acharya, PVN. Isolation and Partial Characterization of Age-Correlated Oligo-nucleotides with Covalently Bound Peptides. 14th Nordic Congress, Umea, Sweden, June 19, 1971.

[14] Acharya, PVN. DNA-damage: The Cause of Aging. Ninth International Congress of Biochemistry: Stockholm. July 1–7, 1973 (Abs.3 m 12).

[15] Acharya, PVN (1977). "Irreparable DNA-damage by Industrial Pollutants in Pre-mature Aging, Chemical Carcinogenesis and Cardiac Hypertrophy: Experiments and Theory". *Israel Journal of Medical Sciences* **13**: 441.

[16] Sinha, Jitendra Kumar; Ghosh, Shampa; Swain, Umakanta; Giridharan, Nappan Veethil; Raghunath, Manchala (2014). "Increased macromolecular damage due to oxidative stress in the neocortex and hippocampus of WNIN/Ob, a novel rat model of premature aging". *Neuroscience* **269**: 256–64. doi:10.1016/j.neuroscience.2014.03.040. PMID 24709042.

[17] Bernstein H, Payne CM, Bernstein C, Garewal H, Dvorak K (2008). Cancer and aging as consequences of un-repaired DNA damage. In: New Research on DNA Damages (Editors: Honoka Kimura and Aoi Suzuki) Nova Science Publishers, Inc., New York, Chapter 1, pp. 1–47. open access, but read only https://www.novapublishers.com/catalog/product_info.php?products_id=43247 ISBN 1604565810 ISBN 978-1604565812

[18] Rutten, BP; Schmitz, C; Gerlach, OH; Oyen, HM; de Mesquita, EB; Steinbusch, HW; Korr, H (Jan 2007). "The aging brain: accumulation of DNA damage or neuron loss?". *Neurobiol Aging* **28** (1): 91–8. doi:10.1016/j.neurobiolaging.2005.10.019. PMID 16338029.

[19] Sen, T; Jana, S; Srcctama, S; Chatterjee, U; Chakrabarti, S (Mar 2007). "Gene-specific oxidative lesions in aged rat brain detected by polymerase chain reaction inhibition assay". *Free Radic Res.* **41** (3): 288–94. doi:10.1080/10715760601083722. PMID 17364957.

[20] Swain, U; Subba Rao, K (Aug 2011). "Study of DNA damage via the comet assay and base excision repair activities in rat brain neurons and astrocytes during aging". *Mech Ageing Dev* **132** (8-9): 374–81. doi:10.1016/j.mad.2011.04.012. PMID 21600238.

[21] Wolf, FI; Fasanella, S; Tedesco, B; Cavallini, G; Donati, A; Bergamini, E; Cittadini, A (Mar 2005). "Peripheral lymphocyte 8-OHdG levels correlate with age-associated increase of tissue oxidative DNA damage in Sprague-Dawley rats. Protective effects of caloric restriction". *Exp Gerontol* **40** (3): 181–8. doi:10.1016/j.exger.2004.11.002. PMID 15763395.

[22] Mecocci, P; MacGarvey, U; Kaufman, AE; Koontz, D; Shoffner, JM; Wallace, DC; Beal, MF (Oct 1993). "Oxidative damage to mitochondrial DNA shows marked age-dependent increases in human brain". *Ann Neurol* **34** (4): 609–16. doi:10.1002/ana.410340416.PMID8215249.

[23] Lu, T; Pan, Y; Kao, SY; Li, C; Kohane, I; Chan, J; Yankner, BA (Jun 2004). "Gene regulation and DNA damage in the ageing human brain". *Nature* **429** (6994): 883–91. doi:10.1038/nature02661. PMID 15190254.

[24] Hamilton, ML; Van Remmen, H; Drake, JA; Yang, H; Guo, ZM; Kewitt, K; Walter, CA; Richardson, A (Aug 2001). "Does oxidative damage to DNA increase with age?". *Proc Natl Acad Sci U S A.* **98** (18): 10469–74. doi:10.1073/pnas.171202698. PMID 11517304.

[25] "Age-dependent increases in oxidative damage to DNA, lipids, and proteins in human skeletal muscle.". *Free Radic Biol Med* **26** (3-4): 303–8. Feb 1999. doi:10.1016/s0891-5849(98)00208-1. PMID 9895220.

[26] Schriner SE, Linford NJ, Martin GM, Treuting P, Ogburn CE, Emond M, Coskun PE, Ladiges W, Wolf N, Van Remmen H, Wallace DC, Rabinovitch PS. "Extension of murine life span by overexpression of catalase targeted to mitochondria. Science. 2005 Jun 24;308(5730):1909-11. doi:10.1126/science.1106653 PMID 15879174

[27] Linford NJ, Schriner SE, Rabinovitch PS. "Oxidative damage and aging: spotlight on mitochondria. Cancer Res. 2006 Mar 1;66(5):2497-9. doi:10.1158/0008-5472.CAN-05-3163 PMID 16510562

[28] Piec, I; Listrat, A; Alliot, J; Chambon, C; Taylor, RG; Bechet, D (Jul 2005). "Differential proteome analysis of aging in rat skeletal muscle". *FASEB J* **19** (9): 1143–5. doi:10.1096/fj.04-3084fje. PMID 15831715.

[29] "DNA oxidation matters: the HPLC-electrochemical detection assay of 8-oxo-deoxyguanosine and 8-oxo-guanine" **95** (1). January 1998. pp. 288–93. PMC 18204. PMID 9419368.

[30] "DNA damage measured by comet assay and 8-OH-dG formation related to blood chemical analyses in aged rats.". *J Toxicol Sci* **32** (3): 249–59. Aug 2007. doi:10.2131/jts.32.249. PMID 17785942.

[31] Rossi DJ, Bryder D, Seita J, Nussenzweig A, Hoeijmakers J, Weissman IL. Deficiencies in DNA damage repair limit the function of haematopoietic stem cells with age. Nature. 2007 Jun 7;447(7145):725-9. doi:10.1038/nature05862 PMID 17554309

[32] Sharpless NE, DePinho RA. How stem cells age and why this makes us grow old. Nat Rev Mol Cell Biol. 2007 Sep;8(9):703-13. Review. doi:10.1038/nrm2241 PMID 17717515

[33] Freitas AA1, de Magalhães JP. A review and appraisal of the DNA damage theory of ageing. Mutat Res. 2011 Jul-Oct;728(1-2):12-22. doi: 10.1016/j.mrrev.2011.05.001. PMID 21600302

[34] Dollé ME, Giese H, Hopkins CL, Martus HJ, Hausdorff JM, Vijg J. Rapid accumulation of genome rearrangements in liver but not in brain of old mice" *Nat Genet* 1997 Dec;17(4):431-4. doi:10.1038/ng1297-431 PMID 9398844

[35] "Mutation frequency and specificity with age in liver, bladder and brain of lacI transgenic mice" **154** (3). March 2000. pp. 1291–300. PMC 1460990. PMID 10757770.

[36] Hill KA, Halangoda A, Heinmoeller PW, Gonzalez K, Chitaphan C, Longmate J, Scaringe WA, Wang JC, Sommer SS. Tissue-specific time courses of spontaneous mutation frequency and deviations in mutation pattern are observed in middle to late adulthood in Big Blue mice. Environ Mol Mutagen. 2005 Jun;45(5):442-54. doi:10.1002/em.20119 PMID 15690342

[37] "Elevated levels of mutation in multiple tissues of mice deficient in the DNA mismatch repair gene Pms2.". *Proceedings of the National Academy of Sciences* **94** (7): 3122–7. Apr 1997. doi:10.1073/pnas.94.7.3122. PMC 20332. PMID 9096356.

[38] Dollé ME, Busuttil RA, Garcia AM, Wijnhoven S, van Drunen E, Niedernhofer LJ, van der Horst G, Hoeijmakers JH, van Steeg H, Vijg J. Increased genomic instability is not a prerequisite for shortened lifespan in DNA repair deficient mice. Mutat Res. 2006 Apr 11;596(1-2):22-35. doi:10.1016/j.mrfmmm.2005.11.008 PMID 16472827

[39] Vermulst M, Bielas JH, Kujoth GC, Ladiges WC, Rabinovitch PS, Prolla TA, Loeb LA. Mitochondrial point mutations do not limit the natural lifespan of mice" *Nat Genet* 2007 Apr;39(4):540-3. doi:10.1038/ng1988 PMID 17334366

[40] Harrigan JA, Wilson DM 3rd, Prasad R, Opresko PL, Beck G, May A, Wilson SH, Bohr VA. The Werner syndrome protein operates in base excision repair and cooperates with DNA polymerase beta. Nucleic Acids Res. 2006 Jan 30;34(2):745-54. doi:10.1093/nar/gkj475 PMID 16449207

[41] Liu Y, Wang Y, Rusinol AE, Sinensky MS, Liu J, Shell SM, Zou Y. Involvement of xeroderma pigmentosum group A (XPA) in progeria arising from defective maturation of prelamin A" *FASEB J* 2008 Feb;22(2):603-11. doi:10.1096/fj.07-8598com PMID 17848622

[42] D'Errico M, Parlanti E, Teson M, Degan P, Lemma T, Calcagnile A, Iavarone I, Jaruga P, Ropolo M, Pedrini AM, Orioli D, Frosina G, Zambruno G, Dizdaroglu M, Stefanini M, Dogliotti E. The role of CSA in the response to oxidative DNA damage in human cells. Oncogene. 2007 Jun 28;26(30):4336-43. doi:10.1038/sj.onc.1210232 PMID 17297471

[43] Vogel H, Lim DS, Karsenty G, Finegold M, Hasty P (1999). "Deletion of Ku86 causes early onset of senescence in mice". *Proc. Natl. Acad. Sci. U.S.A.* **96** (19): 10770–5. PMC 17958. PMID 10485901.

[44] Niedernhofer LJ, Garinis GA, Raams A, Lalai AS, Robinson AR, Appeldoorn E, Odijk H, Oostendorp R, Ahmad A, van Leeuwen W, Theil AF, Vermeulen W, van der Horst GT, Meinecke P, Kleijer WJ, Vijg J, Jaspers NG, Hoeijmakers JH. A new progeroid syndrome reveals that genotoxic stress suppresses the somatotroph axis. Nature. 2006 Dec 21;444(7122):1038-43. doi:10.1038/nature05456 PMID 17183314

[45] Mostoslavsky R, Chua KF, Lombard DB, Pang WW, Fischer MR, Gellon L, Liu P, Mostoslavsky G, Franco S, Murphy MM, Mills KD, Patel P, Hsu JT, Hong AL, Ford E, Cheng HL, Kennedy C, Nunez N, Bronson R, Frendewey D, Auerbach W, Valenzuela D, Karow M, Hottiger MO, Hursting S, Barrett JC, Guarente L, Mulligan R, Demple B, Yancopoulos GD, Alt FW. Genomic instability and aging-like phenotype in the absence of mammalian SIRT6. Cell. 2006 Jan 27;124(2):315-29. doi:10.1016/j.cell.2005.11.044 PMID 16439206

[46] Li H, Vogel H, Holcomb VB, Gu Y, Hasty P (2007). "Deletion of Ku70, Ku80, or both causes early aging without substantially increased cancer". *Mol. Cell. Biol.* **27** (23): 8205–14. doi:10.1128/MCB.00785-07. PMC 2169178. PMID 17875923.

[47] "Correlation between deoxyribonucleic acid excision-repair and life-span in a number of mammalian species.". *Proceedings of the National Academy of Sciences* **71** (6): 2169–73. Jun 1974. doi:10.1073/pnas.71.6.2169. PMID 4526202.

[48] Bürkle A, Brabeck C, Diefenbach J, Beneke S. "The emerging role of poly(ADP-ribose) polymerase-1 in longevity. Int J Biochem Cell Biol. 2005 May;37(5):1043-53. Review. doi:10.1016/j.biocel.2004.10.006 PMID 15743677

Chapter 22

Cell nucleus

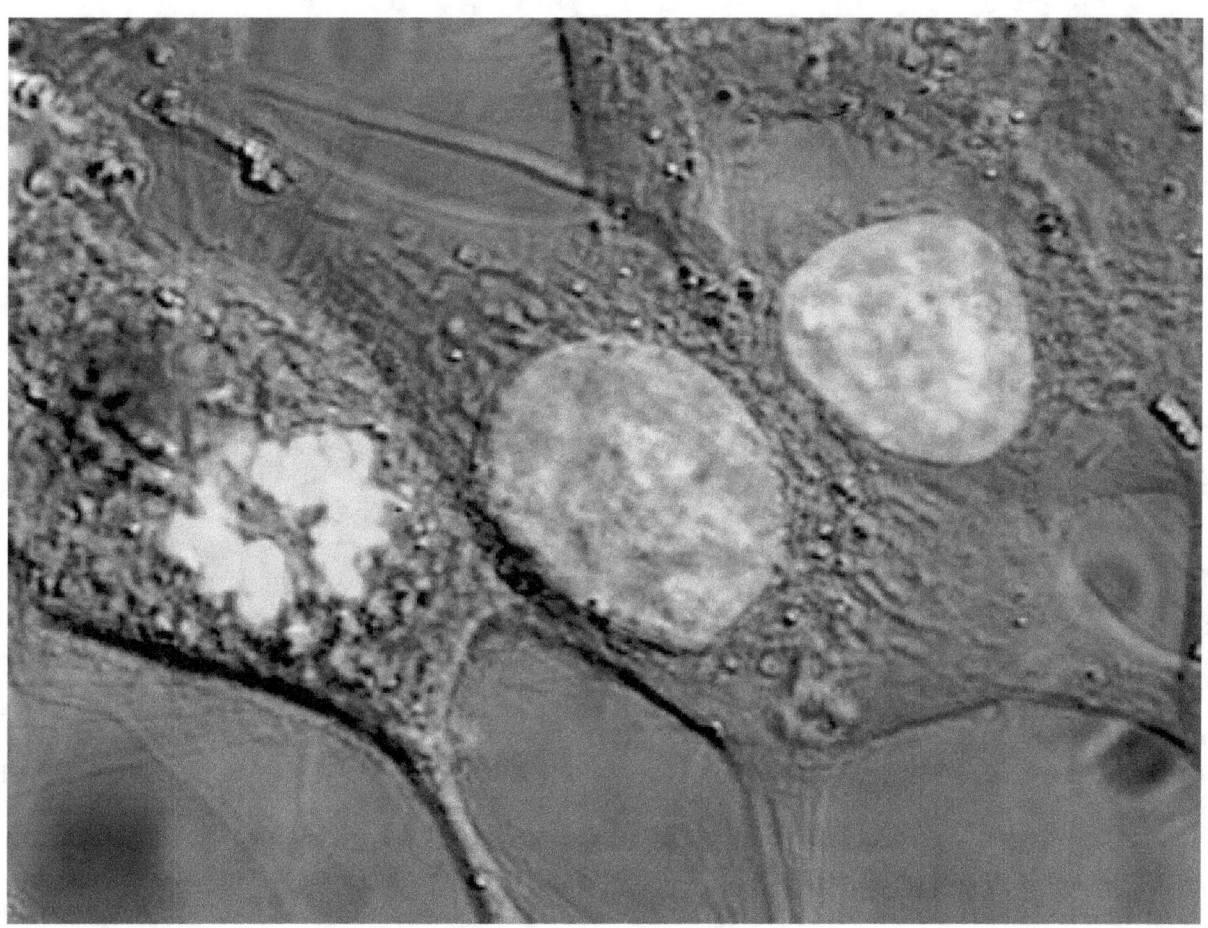

HeLa cells stained for nuclear DNA with the Blue Hoechst dye. The central and rightmost cell are in interphase, thus their entire nuclei are labeled. On the left, a cell is going through mitosis and its DNA has condensed.

In cell biology, the **nucleus** (pl. *nuclei*; from Latin *nucleus* or *nuculeus*, meaning kernel) is a membrane-enclosed organelle found in eukaryotic cells. Eukaryotes usually have a single nucleus, but a few cell types have no nuclei, and a few others have many.

Cell nuclei contain most of the cell's genetic material, organized as multiple long linear DNA molecules in complex with a large variety of proteins, such as histones, to form chromosomes. The genes within these chromosomes are the cell's nuclear genome. The function of the nucleus is to maintain the integrity of these genes and to control the activities of the cell by regulating gene expression—the nucleus is, therefore, the control center of the cell. The main structures making up the nucleus are the nuclear envelope, a double membrane that encloses the entire organelle and isolates its contents from the cellular cytoplasm, and the nucleoskeleton (which includes nuclear lamina), a network within the nucleus that adds mechanical support, much like the cytoskeleton, which supports the cell as a whole.

Because the nuclear membrane is impermeable to large molecules, nuclear pores are required that regulate nuclear transport of molecules across the envelope. The pores cross both nuclear membranes, providing a channel through which larger molecules must be actively transported by carrier proteins while allowing free movement of small molecules and ions. Movement of large molecules such as proteins and RNA through the pores is required for both gene expression and the maintenance of chromosomes. The interior of the nucleus does not contain any membrane-bound sub compartments, its contents are not uniform, and a number of *sub-nuclear bodies* exist, made up of unique proteins, RNA molecules, and particular parts of the chromosomes. The best-known of these is the nucleolus, which is mainly involved in the assembly of ribosomes. After being produced in the nucleolus, ribosomes are exported to the cytoplasm where they translate mRNA.

22.1   History

Oldest known depiction of cells and their nuclei by Antonie van Leeuwenhoek, 1719

The nucleus was the first organelle to be discovered. What is most likely the oldest preserved drawing dates back to the early microscopist Antonie van Leeuwenhoek (1632–1723). He observed a "Lumen", the nucleus, in the red blood cells of salmon.[1] Unlike mammalian red blood cells, those of other vertebrates still contain nuclei.

The nucleus was also described by Franz Bauer in 1804[2] and in more detail in 1831 by Scottish botanist Robert Brown in a talk at the Linnean Society of London. Brown was studying orchids under microscope when he observed an opaque area, which he called the areola or nucleus, in the cells of the flower's outer layer.[3]

He did not suggest a potential function. In 1838, Matthias Schleiden proposed that the nucleus plays a role in generating cells, thus he introduced the name "Cytoblast" (cell builder). He believed that he had observed new cells assembling around "cytoblasts". Franz Meyen was a strong opponent of this view, having already described cells multiplying by division and believing that many cells would have no nuclei. The idea that cells can be generated de novo, by the "cytoblast" or otherwise, contradicted work by Robert Remak (1852) and Rudolf Virchow (1855) who decisively propagated the new paradigm that cells are generated solely by cells ("Omnis cellula e cellula"). The function of the nucleus remained unclear.[4]

Between 1877 and 1878, Oscar Hertwig published several studies on the fertilization of sea urchin eggs, showing that the nucleus of the sperm enters the oocyte and fuses with its nucleus. This was the first time it was suggested that an individual develops from a (single) nucleated cell. This was in contradiction to Ernst Haeckel's theory that the complete phylogeny of a species would be repeated during embryonic development, including generation of the first nucleated cell from a "Monerula", a structureless mass of primordial mucus ("Urschleim"). Therefore, the necessity of the sperm nucleus for

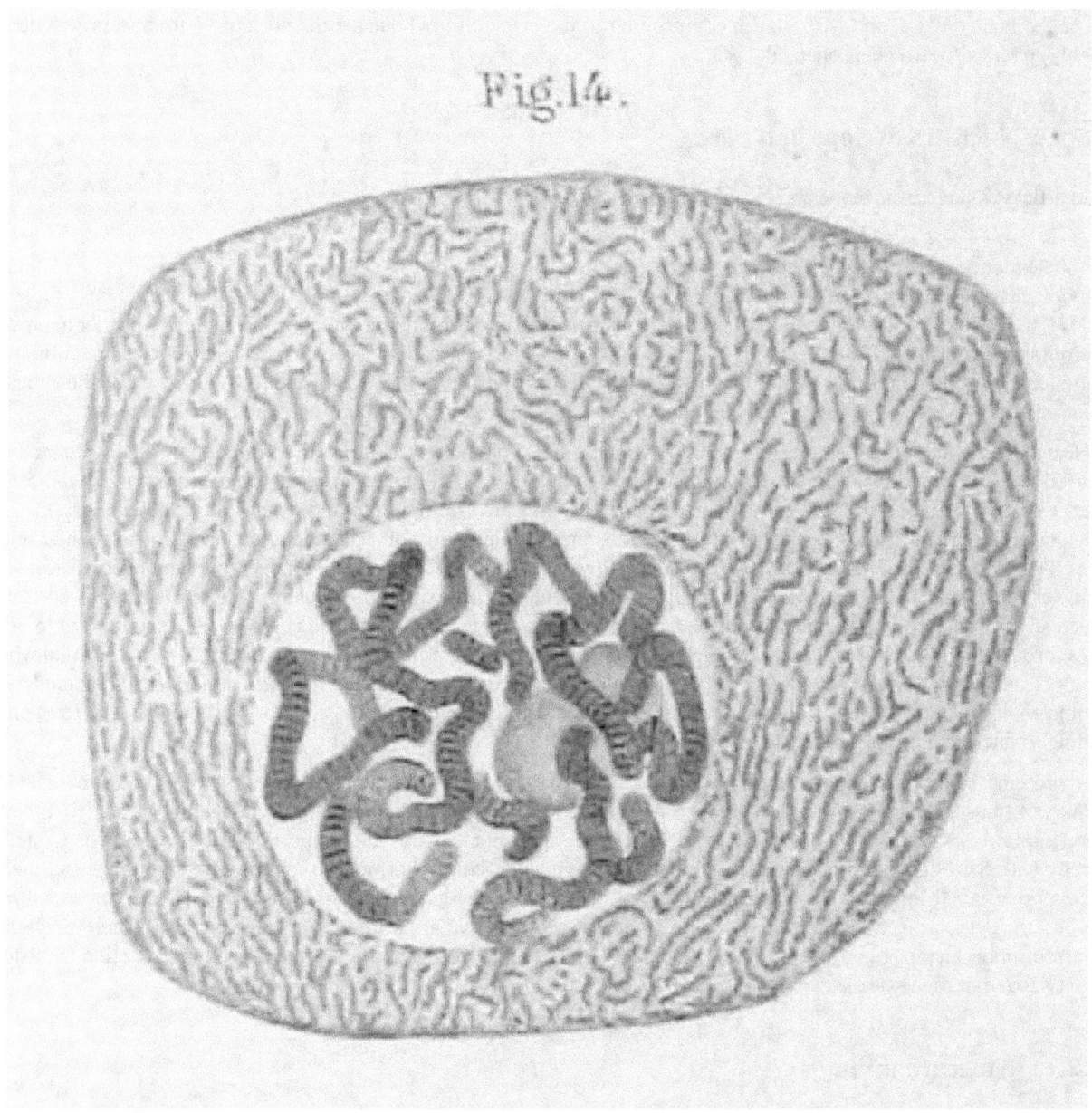

Fig.14.

Drawing of a Chironomus *salivary gland cell published by Walther Flemming in 1882. The nucleus contains Polytene chromosomes.*

fertilization was discussed for quite some time. However, Hertwig confirmed his observation in other animal groups, including amphibians and molluscs. Eduard Strasburger produced the same results for plants in 1884. This paved the way to assign the nucleus an important role in heredity. In 1873, August Weismann postulated the equivalence of the maternal and paternal germ *cells* for heredity. The function of the nucleus as carrier of genetic information became clear only later, after mitosis was discovered and the Mendelian rules were rediscovered at the beginning of the 20th century; the chromosome theory of heredity was therefore developed.[4]

22.2 Structures

The nucleus is the largest cellular organelle in animal cells.[5] In mammalian cells, the average diameter of the nucleus is approximately 6 micrometres (μm), which occupies about 10% of the total cell volume.[6] The viscous liquid within it is called nucleoplasm, and is similar in composition to the cytosol found outside the nucleus.[7] It appears as a dense, roughly spherical or irregular organelle.

22.2.1   Nuclear envelope and pores

Main articles: Nuclear envelope and Nuclear pores

The nuclear envelope, otherwise known as nuclear membrane, consists of two cellular membranes, an inner and an outer membrane, arranged parallel to one another and separated by 10 to 50 nanometres (nm). The nuclear envelope completely encloses the nucleus and separates the cell's genetic material from the surrounding cytoplasm, serving as a barrier to prevent macromolecules from diffusing freely between the nucleoplasm and the cytoplasm.[8] The outer nuclear membrane is continuous with the membrane of the rough endoplasmic reticulum (RER), and is similarly studded with ribosomes.[8] The space between the membranes is called the perinuclear space and is continuous with the RER lumen.

Nuclear pores, which provide aqueous channels through the envelope, are composed of multiple proteins, collectively referred to as nucleoporins. The pores are about 125 million daltons in molecular weight and consist of around 50 (in yeast) to several hundred proteins (in vertebrates).[5] The pores are 100 nm in total diameter; however, the gap through which molecules freely diffuse is only about 9 nm wide, due to the presence of regulatory systems within the center of the pore. This size selectively allows the passage of small water-soluble molecules while preventing larger molecules, such as nucleic acids and larger proteins, from inappropriately entering or exiting the nucleus. These large molecules must be actively transported into the nucleus instead. The nucleus of a typical mammalian cell will have about 3000 to 4000 pores throughout its envelope,[9] each of which contains an eightfold-symmetric ring-shaped structure at a position where the inner and outer membranes fuse.[10] Attached to the ring is a structure called the *nuclear basket* that extends into the nucleoplasm, and a series of filamentous extensions that reach into the cytoplasm. Both structures serve to mediate binding to nuclear transport proteins.[5]

Most proteins, ribosomal subunits, and some DNAs are transported through the pore complexes in a process mediated by a family of transport factors known as karyopherins. Those karyopherins that mediate movement into the nucleus are also called importins, whereas those that mediate movement out of the nucleus are called exportins. Most karyopherins interact directly with their cargo, although some use adaptor proteins.[11] Steroid hormones such as cortisol and aldosterone, as well as other small lipid-soluble molecules involved in intercellular signaling, can diffuse through the cell membrane and into the cytoplasm, where they bind nuclear receptor proteins that are trafficked into the nucleus. There they serve as transcription factors when bound to their ligand; in the absence of ligand, many such receptors function as histone deacetylases that repress gene expression.[5]

22.2.2   Nuclear lamina

Main article: Nuclear lamina

In animal cells, two networks of intermediate filaments provide the nucleus with mechanical support: The nuclear lamina forms an organized meshwork on the internal face of the envelope, while less organized support is provided on the cytosolic face of the envelope. Both systems provide structural support for the nuclear envelope and anchoring sites for chromosomes and nuclear pores.[6]

The nuclear lamina is composed mostly of lamin proteins. Like all proteins, lamins are synthesized in the cytoplasm and later transported to the nucleus interior, where they are assembled before being incorporated into the existing network of nuclear lamina.[12][13] Lamins found on the cytosolic face of the membrane, such as emerin and nesprin, bind to the cytoskeleton to provide structural support. Lamins are also found inside the nucleoplasm where they form another regular structure, known as the *nucleoplasmic veil*,[14] that is visible using fluorescence microscopy. The actual function of the veil is not clear, although it is excluded from the nucleolus and is present during interphase.[15] Lamin structures that make up the veil, such as LEM3, bind chromatin and disrupting their structure inhibits transcription of protein-coding genes.[16]

Like the components of other intermediate filaments, the lamin monomer contains an alpha-helical domain used by two monomers to coil around each other, forming a dimer structure called a coiled coil. Two of these dimer structures then join side by side, in an antiparallel arrangement, to form a tetramer called a *protofilament*. Eight of these protofilaments form a lateral arrangement that is twisted to form a ropelike *filament*. These filaments can be assembled or disassembled in a dynamic manner, meaning that changes in the length of the filament depend on the competing rates of filament addition and removal.[6]

Mutations in lamin genes leading to defects in filament assembly cause a group of rare genetic disorders known as *laminopathies*. The most notable laminopathy is the family of diseases known as progeria, which causes the appearance of premature aging in its sufferers. The exact mechanism by which the associated biochemical changes give rise to the aged phenotype is not well understood.[17]

22.2.3 Chromosomes

Main article: Chromosome

 The cell nucleus contains the majority of the cell's genetic material in the form of multiple linear DNA molecules organized into structures called chromosomes. Each human cell contains roughly two meters of DNA. During most of the cell cycle these are organized in a DNA-protein complex known as chromatin, and during cell division the chromatin can be seen to form the well-defined chromosomes familiar from a karyotype. A small fraction of the cell's genes are located instead in the mitochondria.

There are two types of chromatin. Euchromatin is the less compact DNA form, and contains genes that are frequently expressed by the cell.[18] The other type, heterochromatin, is the more compact form, and contains DNA that is infrequently transcribed. This structure is further categorized into *facultative* heterochromatin, consisting of genes that are organized as heterochromatin only in certain cell types or at certain stages of development, and *constitutive* heterochromatin that consists of chromosome structural components such as telomeres and centromeres.[19] During interphase the chromatin organizes itself into discrete individual patches,[20] called *chromosome territories*.[21] Active genes, which are generally found in the euchromatic region of the chromosome, tend to be located towards the chromosome's territory boundary.[22]

Antibodies to certain types of chromatin organization, in particular, nucleosomes, have been associated with a number of autoimmune diseases, such as systemic lupus erythematosus.[23] These are known as anti-nuclear antibodies (ANA) and have also been observed in concert with multiple sclerosis as part of general immune system dysfunction.[24] As in the case of progeria, the role played by the antibodies in inducing the symptoms of autoimmune diseases is not obvious.

22.2.4 Nucleolus

Main article: Nucleolus

 The nucleolus is a discrete densely stained structure found in the nucleus. It is not surrounded by a membrane, and is sometimes called a *suborganelle*. It forms around tandem repeats of rDNA, DNA coding for ribosomal RNA (rRNA). These regions are called nucleolar organizer regions (NOR). The main roles of the nucleolus are to synthesize rRNA and assemble ribosomes. The structural cohesion of the nucleolus depends on its activity, as ribosomal assembly in the nucleolus results in the transient association of nucleolar components, facilitating further ribosomal assembly, and hence further association. This model is supported by observations that inactivation of rDNA results in intermingling of nucleolar structures.[25]

In the first step of ribosome assembly, a protein called RNA polymerase I transcribes rDNA, which forms a large pre-rRNA precursor. This is cleaved into the subunits 5.8S, 18S, and 28S rRNA.[26] The transcription, post-transcriptional processing, and assembly of rRNA occurs in the nucleolus, aided by small nucleolar RNA (snoRNA) molecules, some of which are derived from spliced introns from messenger RNAs encoding genes related to ribosomal function. The assembled ribosomal subunits are the largest structures passed through the nuclear pores.[5]

When observed under the electron microscope, the nucleolus can be seen to consist of three distinguishable regions: the innermost *fibrillar centers* (FCs), surrounded by the *dense fibrillar component* (DFC), which in turn is bordered by the *granular component* (GC). Transcription of the rDNA occurs either in the FC or at the FC-DFC boundary, and, therefore, when rDNA transcription in the cell is increased, more FCs are detected. Most of the cleavage and modification of rRNAs occurs in the DFC, while the latter steps involving protein assembly onto the ribosomal subunits occur in the GC.[26]

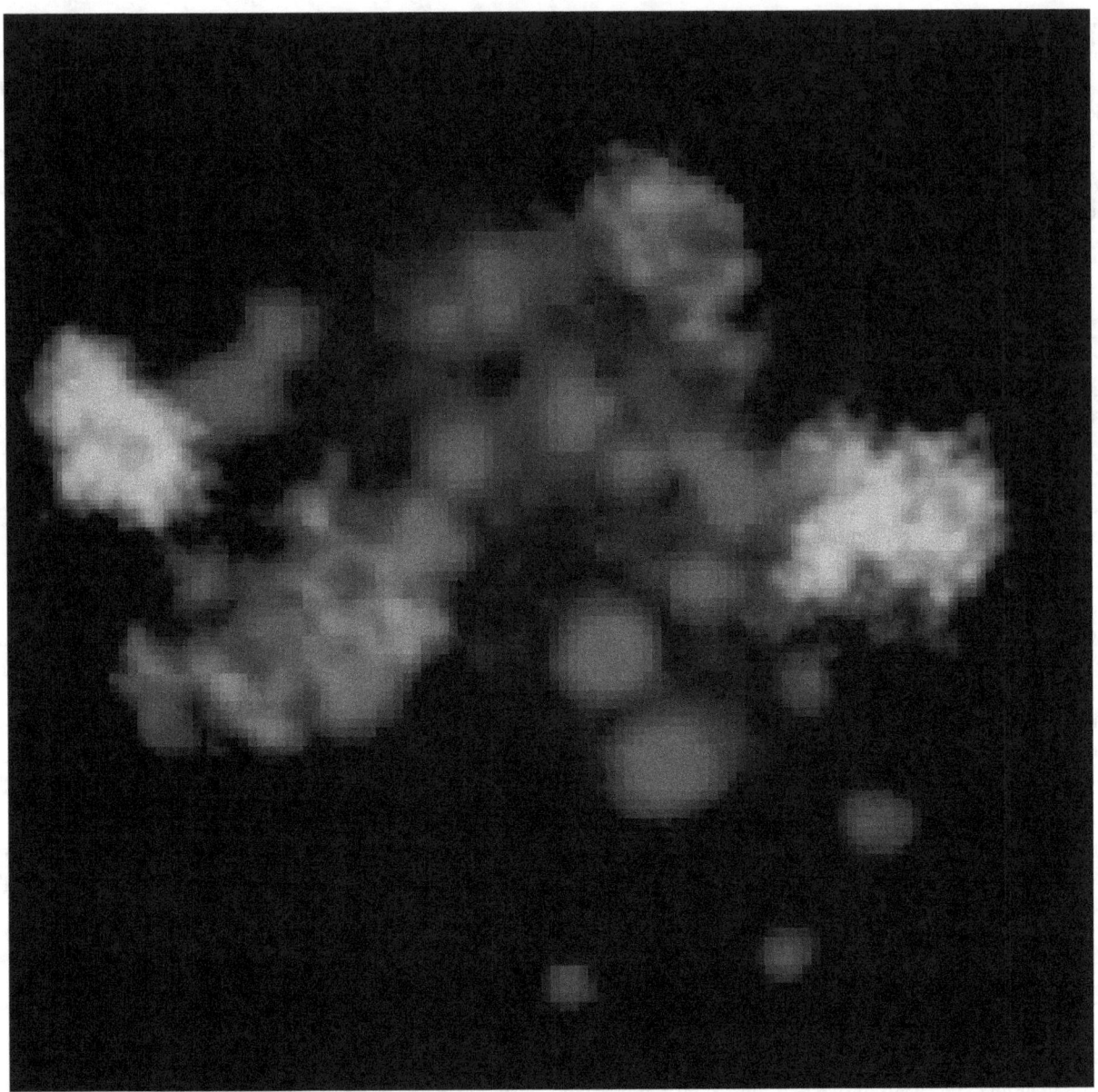

A mouse fibroblast nucleus in which DNA is stained blue. The distinct chromosome territories of chromosome 2 (red) and chromosome 9 (green) are stained with fluorescent in situ hybridization.

22.2.5   Other subnuclear bodies

Besides the nucleolus, the nucleus contains a number of other non-membrane-delineated bodies. These include Cajal bodies, Gemini of coiled bodies, polymorphic interphase karyosomal association (PIKA), promyelocytic leukaemia (PML) bodies, paraspeckles, and splicing speckles. Although little is known about a number of these domains, they are significant in that they show that the nucleoplasm is not uniform mixture, but rather contains organized functional subdomains.[29]

Other subnuclear structures appear as part of abnormal disease processes. For example, the presence of small intranuclear rods has been reported in some cases of nemaline myopathy. This condition typically results from mutations in actin, and the rods themselves consist of mutant actin as well as other cytoskeletal proteins.[31]

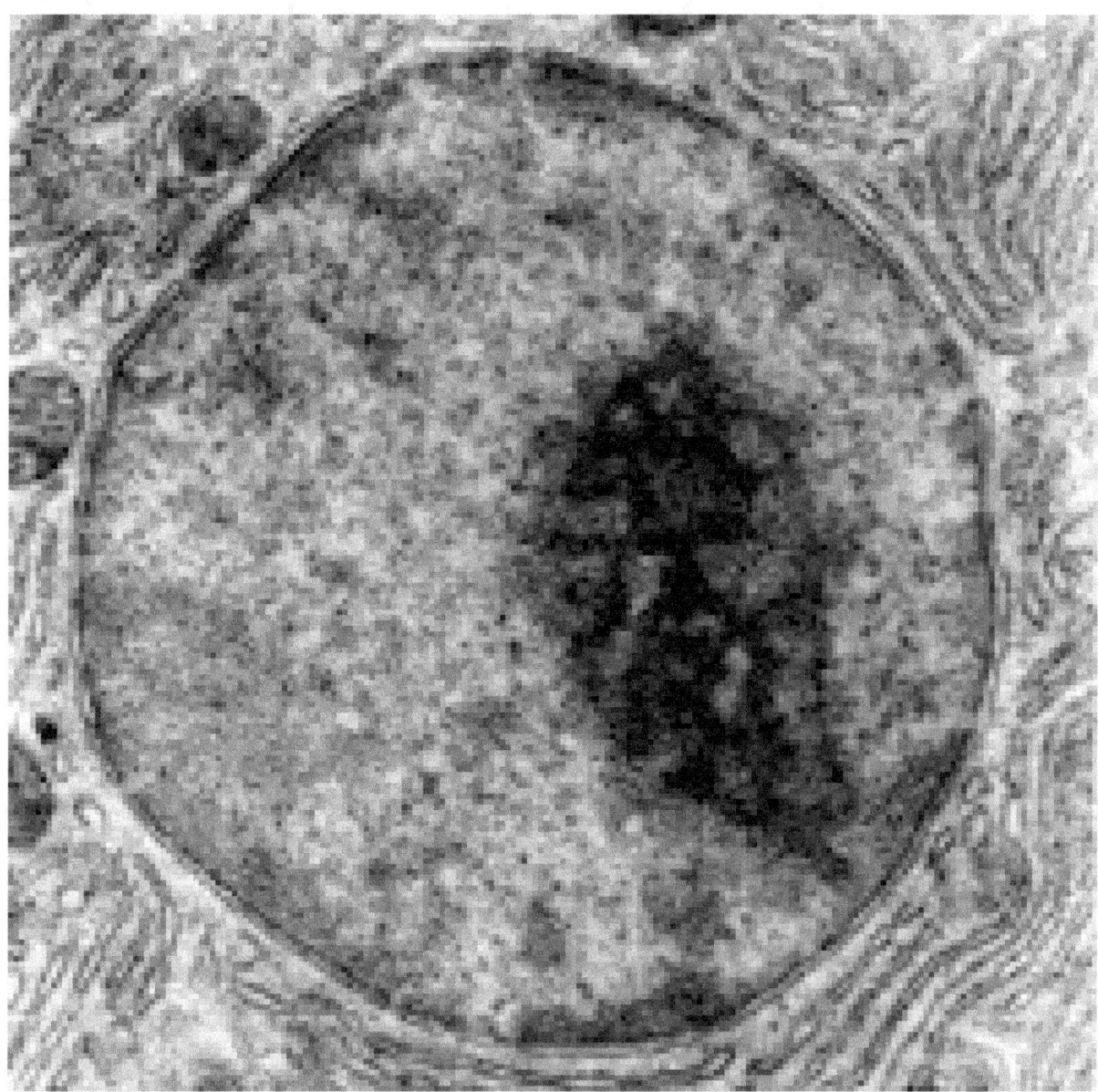

An electron micrograph of a cell nucleus, showing the darkly stained nucleolus

Cajal bodies and gems

A nucleus typically contains between 1 and 10 compact structures called Cajal bodies or coiled bodies (CB), whose diameter measures between 0.2 μm and 2.0 μm depending on the cell type and species.[27] When seen under an electron microscope, they resemble balls of tangled thread[28] and are dense foci of distribution for the protein coilin.[32] CBs are involved in a number of different roles relating to RNA processing, specifically small nucleolar RNA (snoRNA) and small nuclear RNA (snRNA) maturation, and histone mRNA modification.[27]

Similar to Cajal bodies are Gemini of coiled bodies, or gems, whose name is derived from the Gemini constellation in reference to their close "twin" relationship with CBs. Gems are similar in size and shape to CBs, and in fact are virtually indistinguishable under the microscope.[32] Unlike CBs, gems do not contain small nuclear ribonucleoproteins (snRNPs), but do contain a protein called survival of motor neuron (SMN) whose function relates to snRNP biogenesis. Gems are believed to assist CBs in snRNP biogenesis,[33] though it has also been suggested from microscopy evidence that CBs and gems are different manifestations of the same structure.[32]

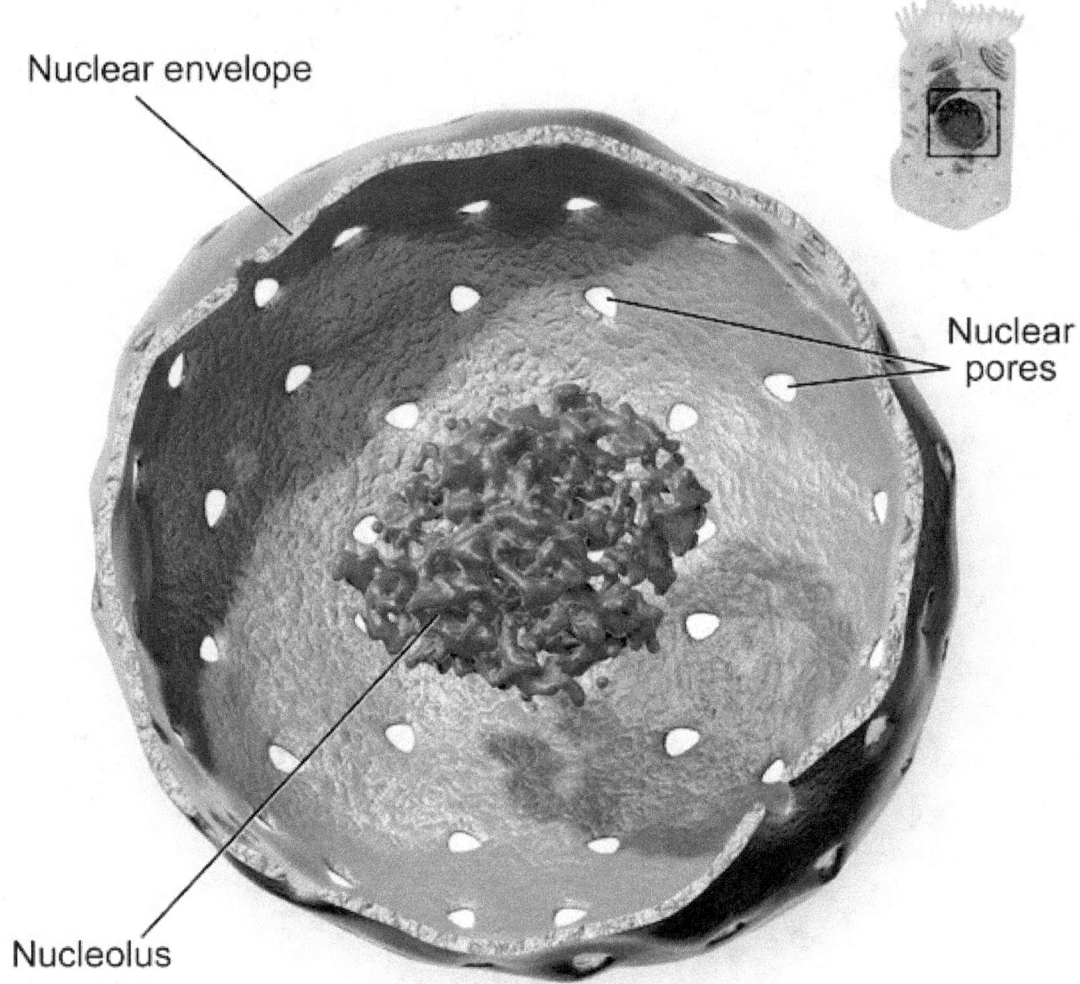

Nuclear envelope

Nuclear pores

Nucleolus

Nucleus

3D rendering of nucleus with location of nucleolus

RAFA and PTF domains

RAFA domains, or polymorphic interphase karyosomal associations, were first described in microscopy studies in 1991. Their function was and remains unclear, though they were not thought to be associated with active DNA replication, transcription, or RNA processing.[34] They have been found to often associate with discrete domains defined by dense localization of the transcription factor PTF, which promotes transcription of small nuclear RNA (snRNA).[35]

PML bodies

Promyelocytic leukaemia bodies (PML bodies) are spherical bodies found scattered throughout the nucleoplasm, measuring around 0.1–1.0 μm. They are known by a number of other names, including nuclear domain 10 (ND10), Kremer bodies, and PML oncogenic domains. PML bodies are named after one of their major components, the promyelocytic leukemia protein (PML). They are often seen in the nucleus in association with Cajal bodies and cleavage bodies.[29] PML bodies belong to the nuclear matrix, an ill-defined super-structure of the nucleus proposed to anchor and regulate many nuclear functions, including DNA replication, transcription, or epigenetic silencing.[36] The PML protein is the key organizer of these domains that recruits an ever-growing number of proteins, whose only common known feature to date is their ability to be SUMOylated. Yet, pml-/- mice (which have their PML gene deleted) cannot assemble nuclear bodies, develop normally and live well, demonstrating that PML bodies are dispensable for most basic biological functions.[36]

Splicing speckles

Speckles are subnuclear structures that are enriched in pre-messenger RNA splicing factors and are located in the interchromatin regions of the nucleoplasm of mammalian cells. At the fluorescence-microscope level they appear as irregular, punctate structures, which vary in size and shape, and when examined by electron microscopy they are seen as clusters of interchromatin granules. Speckles are dynamic structures, and both their protein and RNA-protein components can cycle continuously between speckles and other nuclear locations, including active transcription sites. Studies on the composition, structure and behaviour of speckles have provided a model for understanding the functional compartmentalization of the nucleus and the organization of the gene-expression machinery[37] splicing snRNPs[38][39] and other splicing proteins necessary for pre-mRNA processing.[40] Because of a cell's changing requirements, the composition and location of these bodies changes according to mRNA transcription and regulation via phosphorylation of specific proteins.[41] The splicing speckles are also known as nuclear speckles (nuclear specks), splicing factor compartments (SF compartments), interchromatin granule clusters (IGCs), B snurposomes.[42] B snurposomes are found in the amphibian oocyte nuclei and in *Drosophila melanogaster* embryos. B snurposomes appear alone or attached to the Cajal bodies in the electron micrographs of the amphibian nuclei.[43] IGCs function as storage sites for the splicing factors.[44]

Paraspeckles

Main article: Paraspeckle

Discovered by Fox et al. in 2002, paraspeckles are irregularly shaped compartments in the nucleus' interchromatin space.[45] First documented in HeLa cells, where there are generally 10–30 per nucleus,[46] paraspeckles are now known to also exist in all human primary cells, transformed cell lines, and tissue sections.[47] Their name is derived from their distribution in the nucleus; the "para" is short for parallel and the "speckles" refers to the splicing speckles to which they are always in close proximity.[46]

Paraspeckles are dynamic structures that are altered in response to changes in cellular metabolic activity. They are transcription dependent[45] and in the absence of RNA Pol II transcription, the paraspeckle disappears and all of its associated protein components (PSP1, p54nrb, PSP2, CFI(m)68, and PSF) form a crescent shaped perinucleolar cap in the nucleolus. This phenomenon is demonstrated during the cell cycle. In the cell cycle, paraspeckles are present during interphase and during all of mitosis except for telophase. During telophase, when the two daughter nuclei are formed, there is no RNA Pol II transcription so the protein components instead form a perinucleolar cap.[47]

Perichromatin fibrils

Perichromatin fibrils are visible only under electron microscope. They are located next to the transcriptionally active chromatin and is hypothesized to be the site of active pre-mRNA processing.[44]

22.3   Function

The nucleus provides a site for genetic transcription that is segregated from the location of translation in the cytoplasm, allowing levels of gene regulation that are not available to prokaryotes. The main function of the cell nucleus is to control gene expression and mediate the replication of DNA during the cell cycle.

22.3.1   Cell compartmentalization

The nuclear envelope allows the nucleus to control its contents, and separate them from the rest of the cytoplasm where necessary. This is important for controlling processes on either side of the nuclear membrane. In most cases where a cytoplasmic process needs to be restricted, a key participant is removed to the nucleus, where it interacts with transcription factors to downregulate the production of certain enzymes in the pathway. This regulatory mechanism occurs in the case of glycolysis, a cellular pathway for breaking down glucose to produce energy. Hexokinase is an enzyme responsible for the first the step of glycolysis, forming glucose-6-phosphate from glucose. At high concentrations of fructose-6-phosphate, a molecule made later from glucose-6-phosphate, a regulator protein removes hexokinase to the nucleus,[48] where it forms a transcriptional repressor complex with nuclear proteins to reduce the expression of genes involved in glycolysis.[49]

In order to control which genes are being transcribed, the cell separates some transcription factor proteins responsible for regulating gene expression from physical access to the DNA until they are activated by other signaling pathways. This prevents even low levels of inappropriate gene expression. For example, in the case of NF-κB-controlled genes, which are involved in most inflammatory responses, transcription is induced in response to a signal pathway such as that initiated by the signaling molecule TNF-α, binds to a cell membrane receptor, resulting in the recruitment of signalling proteins, and eventually activating the transcription factor NF-κB. A nuclear localisation signal on the NF-κB protein allows it to be transported through the nuclear pore and into the nucleus, where it stimulates the transcription of the target genes.[6]

The compartmentalization allows the cell to prevent translation of unspliced mRNA.[50] Eukaryotic mRNA contains introns that must be removed before being translated to produce functional proteins. The splicing is done inside the nucleus before the mRNA can be accessed by ribosomes for translation. Without the nucleus, ribosomes would translate newly transcribed (unprocessed) mRNA, resulting in malformed and nonfunctional proteins.

22.3.2   Gene expression

Main article: Gene expression

Gene expression first involves transcription, in which DNA is used as a template to produce RNA. In the case of genes encoding proteins, that RNA produced from this process is messenger RNA (mRNA), which then needs to be translated by ribosomes to form a protein. As ribosomes are located outside the nucleus, mRNA produced needs to be exported.[51]

Since the nucleus is the site of transcription, it also contains a variety of proteins that either directly mediate transcription or are involved in regulating the process. These proteins include helicases, which unwind the double-stranded DNA molecule to facilitate access to it, RNA polymerases, which synthesize the growing RNA molecule, topoisomerases, which change the amount of supercoiling in DNA, helping it wind and unwind, as well as a large variety of transcription factors that regulate expression.[52]

22.3.3   Processing of pre-mRNA

Main article: Post-transcriptional modification

Newly synthesized mRNA molecules are known as primary transcripts or pre-mRNA. They must undergo post-transcriptional modification in the nucleus before being exported to the cytoplasm; mRNA that appears in the cytoplasm without these modifications is degraded rather than used for protein translation. The three main modifications are 5' capping, 3' polyadenylation, and RNA splicing. While in the nucleus, pre-mRNA is associated with a variety of proteins in complexes known as heterogeneous ribonucleoprotein particles (hnRNPs). Addition of the 5' cap occurs co-transcriptionally and is the first step in post-transcriptional modification. The 3' poly-adenine tail is only added after transcription is complete.

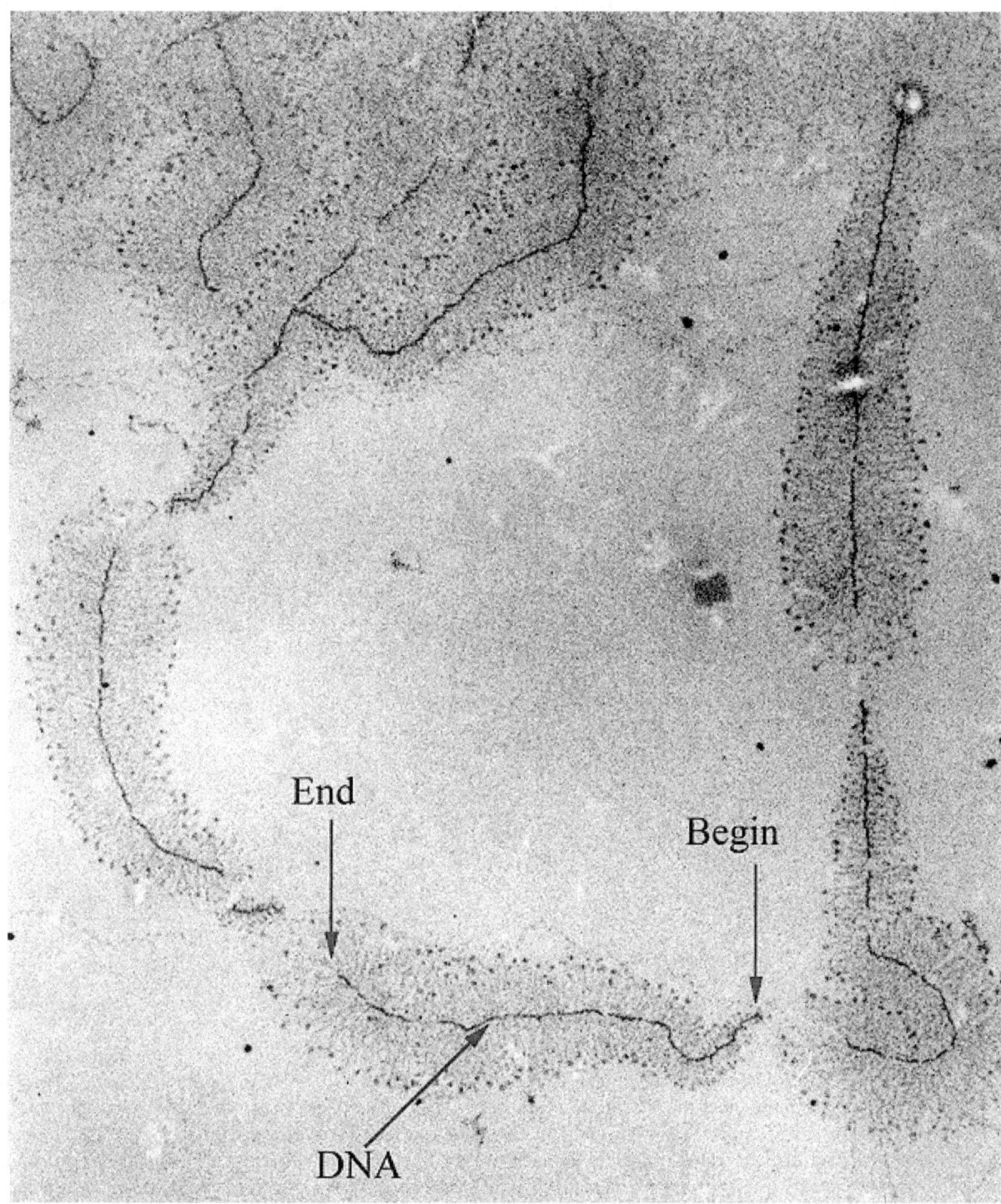

A micrograph of ongoing gene transcription of ribosomal RNA illustrating the growing primary transcripts. "Begin" indicates the 5' end of the DNA, where new RNA synthesis begins; "end" indicates the 3' end, where the primary transcripts are almost complete.

RNA splicing, carried out by a complex called the spliceosome, is the process by which introns, or regions of DNA that do not code for protein, are removed from the pre-mRNA and the remaining exons connected to re-form a single continuous molecule. This process normally occurs after 5' capping and 3' polyadenylation but can begin before synthesis is complete in transcripts with many exons.[5] Many pre-mRNAs, including those encoding antibodies, can be spliced in multiple ways to produce different mature mRNAs that encode different protein sequences. This process is known as alternative splicing, and allows production of a large variety of proteins from a limited amount of DNA.

22.4   Dynamics and regulation

22.4.1   Nuclear transport

Main article: Nuclear transport
The entry and exit of large molecules from the nucleus is tightly controlled by the nuclear pore complexes. Although

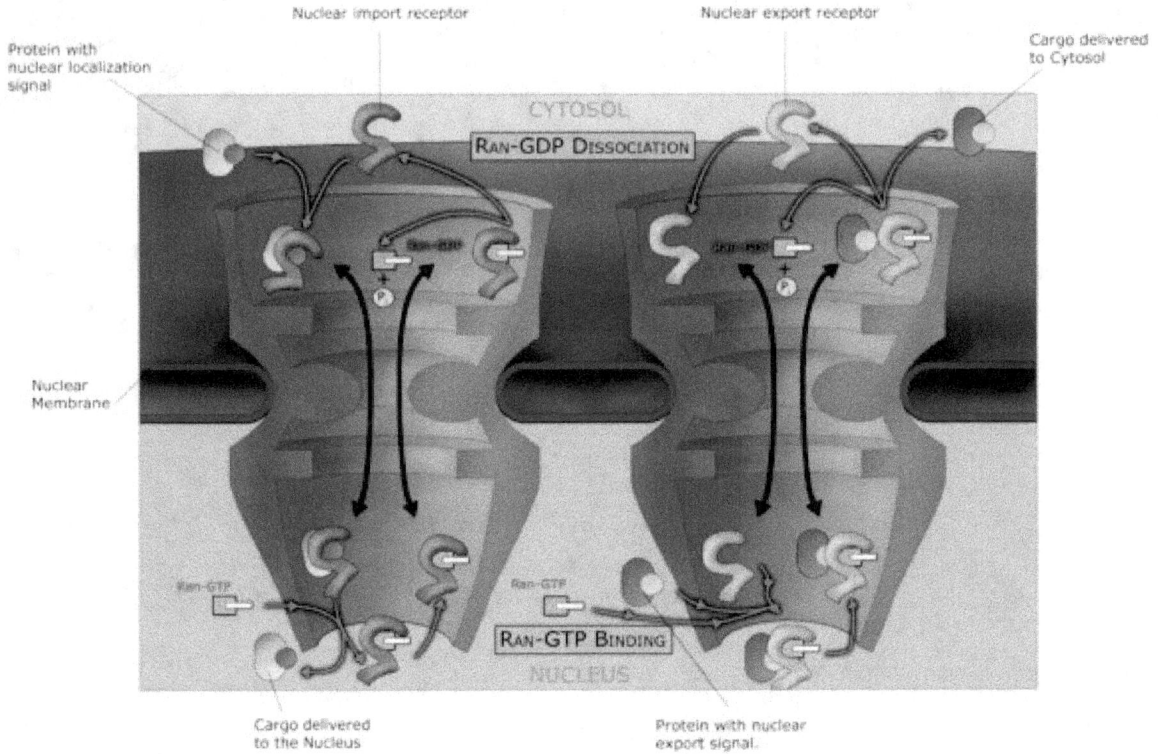

Macromolecules, such as RNA and proteins, are actively transported across the nuclear membrane in a process called the Ran-GTP nuclear transport cycle.

small molecules can enter the nucleus without regulation,[53] macromolecules such as RNA and proteins require association karyopherins called importins to enter the nucleus and exportins to exit. "Cargo" proteins that must be translocated from the cytoplasm to the nucleus contain short amino acid sequences known as nuclear localization signals, which are bound by importins, while those transported from the nucleus to the cytoplasm carry nuclear export signals bound by exportins. The ability of importins and exportins to transport their cargo is regulated by GTPases, enzymes that hydrolyze the molecule guanosine triphosphate to release energy. The key GTPase in nuclear transport is Ran, which can bind either GTP or GDP (guanosine diphosphate), depending on whether it is located in the nucleus or the cytoplasm. Whereas importins depend on RanGTP to dissociate from their cargo, exportins require RanGTP in order to bind to their cargo.[11]

Nuclear import depends on the importin binding its cargo in the cytoplasm and carrying it through the nuclear pore into the nucleus. Inside the nucleus, RanGTP acts to separate the cargo from the importin, allowing the importin to exit the nucleus and be reused. Nuclear export is similar, as the exportin binds the cargo inside the nucleus in a process facilitated by RanGTP, exits through the nuclear pore, and separates from its cargo in the cytoplasm.

Specialized export proteins exist for translocation of mature mRNA and tRNA to the cytoplasm after post-transcriptional modification is complete. This quality-control mechanism is important due to these molecules' central role in protein translation; mis-expression of a protein due to incomplete excision of exons or mis-incorporation of amino acids could have negative consequences for the cell; thus, incompletely modified RNA that reaches the cytoplasm is degraded rather than used in translation.[5]

22.4.2 Assembly and disassembly

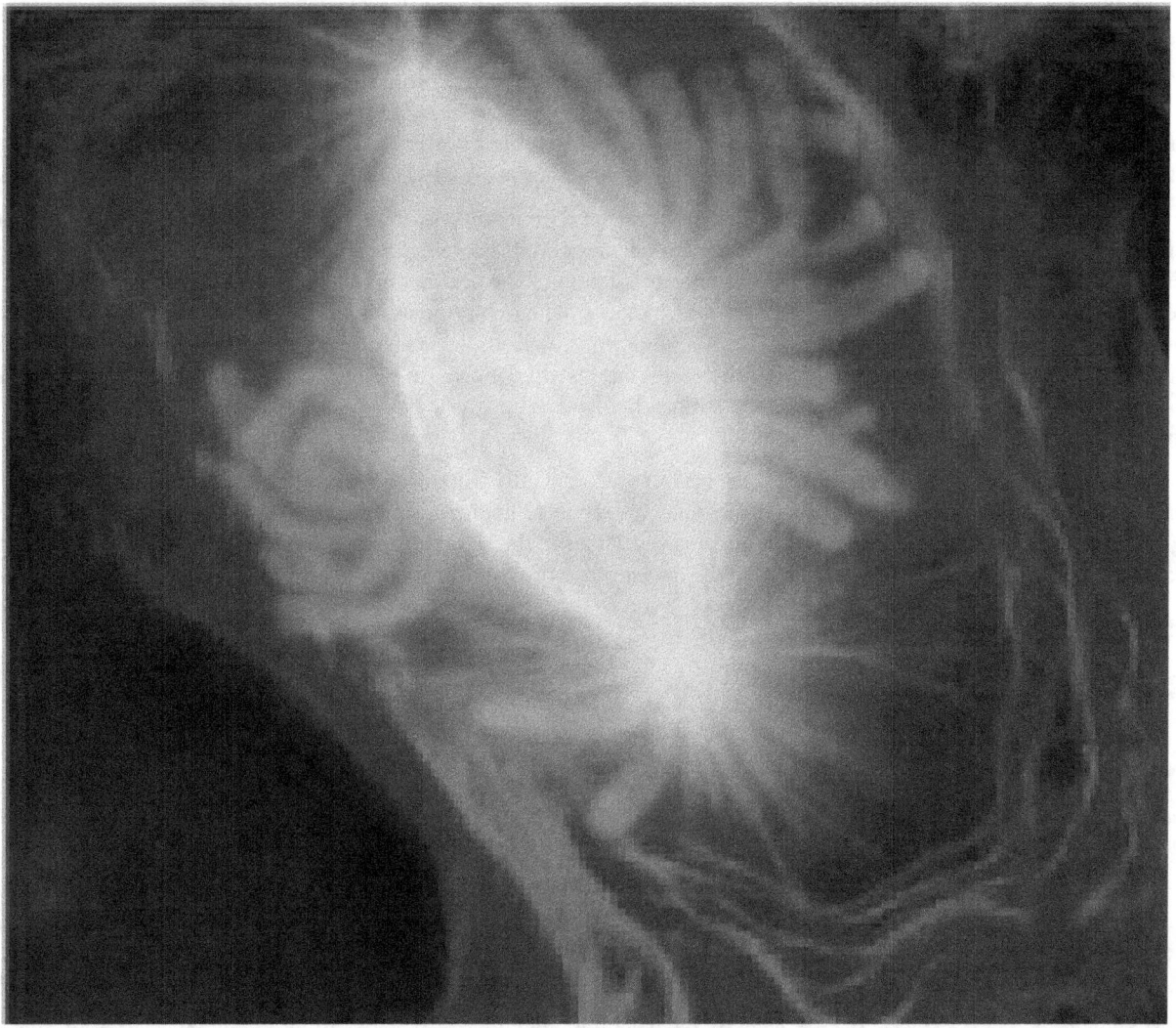

An image of a newt lung cell stained with fluorescent dyes during metaphase. The mitotic spindle can be seen, stained green, attached to the two sets of chromosomes, stained light blue. All chromosomes but one are already at the metaphase plate.

During its lifetime, a nucleus may be broken down or destroyed, either in the process of cell division or as a consequence of apoptosis (the process of programmed cell death). During these events, the structural components of the nucleus — the envelope and lamina — can be systematically degraded. In most cells, the disassembly of the nuclear envelope marks the end of the prophase of mitosis. However, this disassembly of the nucleus is not a universal feature of mitosis and does not occur in all cells. Some unicellular eukaryotes (e.g., yeasts) undergo so-called closed mitosis, in which the nuclear envelope remains intact. In closed mitosis, the daughter chromosomes migrate to opposite poles of the nucleus, which then divides in two. The cells of higher eukaryotes, however, usually undergo open mitosis, which is characterized by breakdown of the nuclear envelope. The daughter chromosomes then migrate to opposite poles of the mitotic spindle, and new nuclei reassemble around them.

At a certain point during the cell cycle in open mitosis, the cell divides to form two cells. In order for this process to be possible, each of the new daughter cells must have a full set of genes, a process requiring replication of the chromosomes as well as segregation of the separate sets. This occurs by the replicated chromosomes, the sister chromatids, attaching to microtubules, which in turn are attached to different centrosomes. The sister chromatids can then be pulled to separate locations in the cell. In many cells, the centrosome is located in the cytoplasm, outside the nucleus; the microtubules would be unable to attach to the chromatids in the presence of the nuclear envelope.[54] Therefore, the early stages in the cell cycle, beginning in prophase and until around prometaphase, the nuclear membrane is dismantled.[14] Likewise, during the same period, the nuclear lamina is also disassembled, a process regulated by phosphorylation of the lamins by protein kinases such as the CDC2 protein kinase.[55] Towards the end of the cell cycle, the nuclear membrane is reformed, and around the same time, the nuclear lamina are reassembled by dephosphorylating the lamins.[55]

However, in dinoflagellates, the nuclear envelope remains intact, the centrosomes are located in the cytoplasm, and the microtubules come in contact with chromosomes, whose centromeric regions are incorporated into the nuclear envelope (the so-called closed mitosis with extranuclear spindle). In many other protists (e.g., ciliates, sporozoans) and fungi, the centrosomes are intranuclear, and their nuclear envelope also does not disassemle during cell division.

Apoptosis is a controlled process in which the cell's structural components are destroyed, resulting in death of the cell. Changes associated with apoptosis directly affect the nucleus and its contents, for example, in the condensation of chromatin and the disintegration of the nuclear envelope and lamina. The destruction of the lamin networks is controlled by specialized apoptotic proteases called caspases, which cleave the lamin proteins and, thus, degrade the nucleus' structural integrity. Lamin cleavage is sometimes used as a laboratory indicator of caspase activity in assays for early apoptotic activity.[14] Cells that express mutant caspase-resistant lamins are deficient in nuclear changes related to apoptosis, suggesting that lamins play a role in initiating the events that lead to apoptotic degradation of the nucleus.[14] Inhibition of lamin assembly itself is an inducer of apoptosis.[56]

The nuclear envelope acts as a barrier that prevents both DNA and RNA viruses from entering the nucleus. Some viruses require access to proteins inside the nucleus in order to replicate and/or assemble. DNA viruses, such as herpesvirus replicate and assemble in the cell nucleus, and exit by budding through the inner nuclear membrane. This process is accompanied by disassembly of the lamina on the nuclear face of the inner membrane.[14]

22.4.3 Disease-related dynamics

Initially, it has been suspected that immunoglobulins in general and autoantibodies in particular do not enter the nucleus. Now there is a body of evidence that under pathological conditions (e.g. lupus erythematosus) IgG can enter the nucleus.[57]

22.5 Nuclei per cell

Most eukaryotic cell types usually have a single nucleus, but some have no nuclei, while others have several. This can result from normal development, as in the maturation of mammalian red blood cells, or from faulty cell division.

22.5.1 Anucleated cells

Anucleated cells contain no nucleus and are, therefore, incapable of dividing to produce daughter cells. The best-known anucleated cell is the mammalian red blood cell, or erythrocyte, which also lacks other organelles such as mitochondria, and serves primarily as a transport vessel to ferry oxygen from the lungs to the body's tissues. Erythrocytes mature through erythropoiesis in the bone marrow, where they lose their nuclei, organelles, and ribosomes. The nucleus is expelled during the process of differentiation from an erythroblast to a reticulocyte, which is the immediate precursor of the mature erythrocyte.[58] The presence of mutagens may induce the release of some immature "micronucleated" erythrocytes into the bloodstream.[59][60] Anucleated cells can also arise from flawed cell division in which one daughter lacks a nucleus and the other has two nuclei.

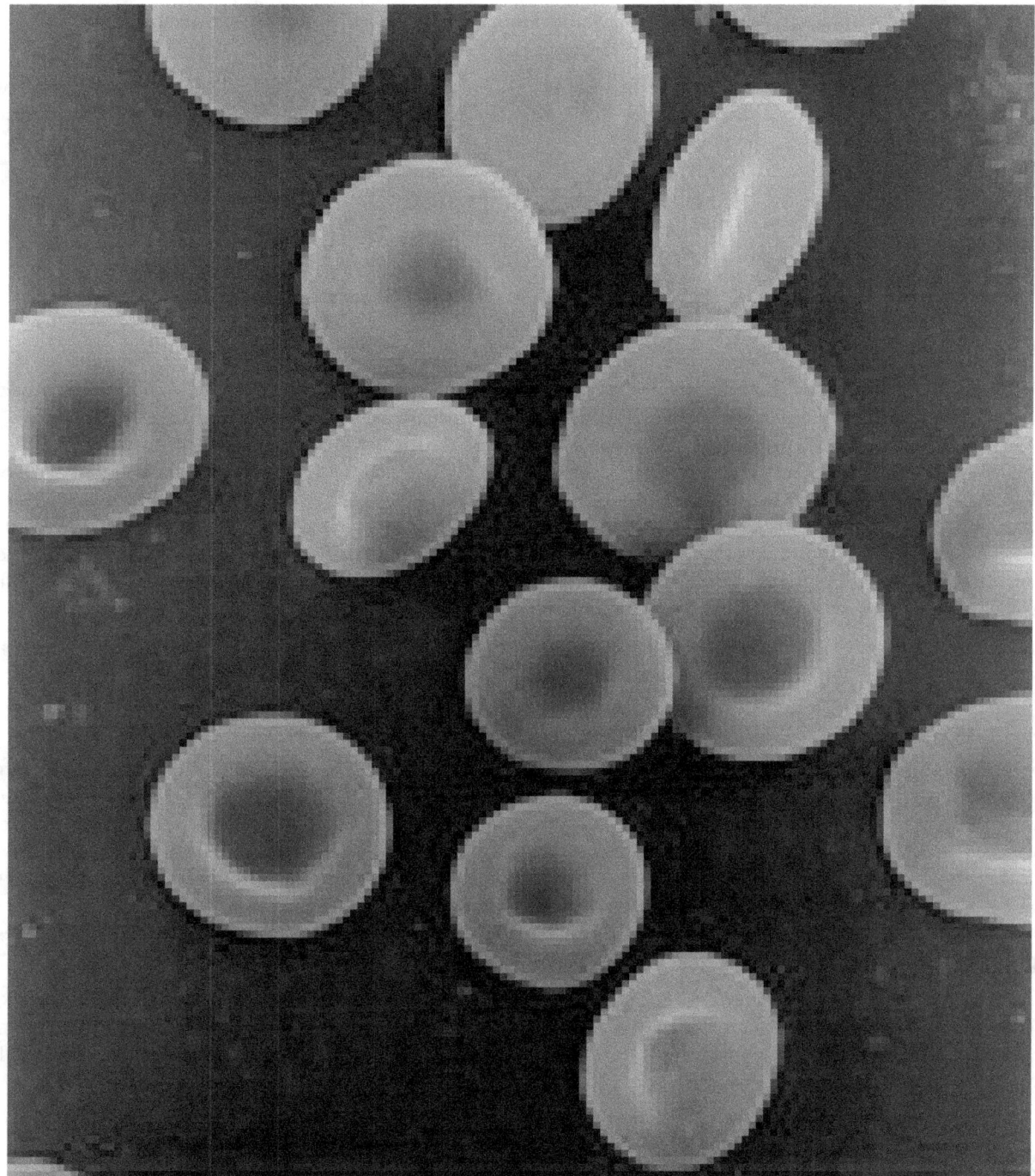

Human red blood cells, like those of other mammals, lack nuclei. This occurs as a normal part of the cells' development.

22.5.2   Multinucleated cells

Multinucleated cells contain multiple nuclei. Most acantharean species of protozoa[61] and some fungi in mycorrhizae[62] have naturally multinucleated cells. Other examples include the intestinal parasites in the genus *Giardia*, which have two nuclei per cell.[63] In humans, skeletal muscle cells, called myocytes and syncytium, become multinucleated during development; the resulting arrangement of nuclei near the periphery of the cells allows maximal intracellular space for myofibrils.[5] Multinucleated and binucleated cells can also be abnormal in humans; for example, cells arising from the fusion of monocytes and macrophages, known as giant multinucleated cells, sometimes accompany inflammation[64] and are also implicated in tumor formation.[65]

A number of dinoflagelates are known to have two nuclei.[66] Unlike other multinucleated cells these nuclei contain two distinct lineages of DNA: one from the dinoflagelate and the other from a symbiotic diatom. Curiously the mitochondrion and the plastid of the diatom remain functional.

22.6   Evolution

As the major defining characteristic of the eukaryotic cell, the nucleus' evolutionary origin has been the subject of much speculation. Four major hypotheses have been proposed to explain the existence of the nucleus, although none have yet earned widespread support.[67]

The first model known as the "syntrophic model" proposes that a symbiotic relationship between the archaea and bacteria created the nucleus-containing eukaryotic cell. (Organisms of the Archaea and Bacteria domain have no cell nucleus.[68]) It is hypothesized that the symbiosis originated when ancient archaea, similar to modern methanogenic archaea, invaded and lived within bacteria similar to modern myxobacteria, eventually forming the early nucleus. This theory is analogous to the accepted theory for the origin of eukaryotic mitochondria and chloroplasts, which are thought to have developed from a similar endosymbiotic relationship between proto-eukaryotes and aerobic bacteria.[69] The archaeal origin of the nucleus is supported by observations that archaea and eukarya have similar genes for certain proteins, including histones. Observations that myxobacteria are motile, can form multicellular complexes, and possess kinases and G proteins similar to eukarya, support a bacterial origin for the eukaryotic cell.[70]

A second model proposes that proto-eukaryotic cells evolved from bacteria without an endosymbiotic stage. This model is based on the existence of modern planctomycetes bacteria that possess a nuclear structure with primitive pores and other compartmentalized membrane structures.[71] A similar proposal states that a eukaryote-like cell, the chronocyte, evolved first and phagocytosed archaea and bacteria to generate the nucleus and the eukaryotic cell.[72]

The most controversial model, known as *viral eukaryogenesis*, posits that the membrane-bound nucleus, along with other eukaryotic features, originated from the infection of a prokaryote by a virus. The suggestion is based on similarities between eukaryotes and viruses such as linear DNA strands, mRNA capping, and tight binding to proteins (analogizing histones to viral envelopes). One version of the proposal suggests that the nucleus evolved in concert with phagocytosis to form an early cellular "predator".[73] Another variant proposes that eukaryotes originated from early archaea infected by poxviruses, on the basis of observed similarity between the DNA polymerases in modern poxviruses and eukaryotes.[74][75] It has been suggested that the unresolved question of the evolution of sex could be related to the viral eukaryogenesis hypothesis.[76]

A more recent proposal, the *exomembrane hypothesis*, suggests that the nucleus instead originated from a single ancestral cell that evolved a second exterior cell membrane; the interior membrane enclosing the original cell then became the nuclear membrane and evolved increasingly elaborate pore structures for passage of internally synthesized cellular components such as ribosomal subunits.[77]

22.7   See also

- Nucleus (neuroanatomy)

22.8   Gallery

- Comparison of human and chimpanzee chromosomes.

- Mouse chromosome territories in different cell types.

- 24 chromosome territories in human cells.

22.9 References

[1] Leeuwenhoek, A. van: Opera Omnia, seu Arcana Naturae ope exactissimorum Microscopiorum detecta, experimentis variis comprobata, Epistolis ad varios illustres viros. J. Arnold et Delphis, A. Beman, Lugdinum Batavorum 1719–1730. Cited after: Dieter Gerlach, Geschichte der Mikroskopie. Verlag Harry Deutsch, Frankfurt am Main, Germany, 2009. ISBN 978-3-8171-1781-9.

[2] Harris, H (1999). *The Birth of the Cell*. New Haven: Yale University Press. ISBN 0-300-07384-4.

[3] Brown, Robert (1866). "On the Organs and Mode of Fecundation of Orchidex and Asclepiadea". *Miscellaneous Botanical Works I*: 511–514.

[4] Cremer, Thomas (1985). *Von der Zellenlehre zur Chromosomentheorie*. Berlin, Heidelberg, New York, Tokyo: Springer Verlag. ISBN 3-540-13987-7. Online Version here

[5] Lodish, H; Berk A; Matsudaira P; Kaiser CA; Krieger M; Scott MP; Zipursky SL; Darnell J. (2004). *Molecular Cell Biology* (5th ed.). New York: WH Freeman. ISBN 0-7167-2672-6.

[6] Bruce Alberts, Alexander Johnson, Julian Lewis, Martin Raff, Keith Roberts, Peter Walter, ed. (2002). *Molecular Biology of the Cell, Chapter 4, pages 191–234* (4th ed.). Garland Science.

[7] Clegg JS (February 1984). "Properties and metabolism of the aqueous cytoplasm and its boundaries". *Am. J. Physiol.* **246** (2 Pt 2): R133–51. PMID 6364846.

[8] Paine P, Moore L, Horowitz S (1975). "Nuclear envelope permeability". *Nature* **254** (5496): 109–114. doi:10.1038/254109a0. PMID 1117994.

[9] Rodney Rhoades, Richard Pflanzer, ed. (1996). "Ch3". *Human Physiology* (3rd ed.). Saunders College Publishing.

[10] Shulga N, Mosammaparast N, Wozniak R, Goldfarb D (2000). "Yeast nucleoporins involved in passive nuclear envelope permeability". *J Cell Biol* **149** (5): 1027–1038. doi:10.1083/jcb.149.5.1027. PMC 2174828. PMID 10831607.

[11] Pemberton L, Paschal B (2005). "Mechanisms of receptor-mediated nuclear import and nuclear export". *Traffic* **6** (3): 187–198. doi:10.1111/j.1600-0854.2005.00270.x. PMID 15702987.

[12] Stuurman N, Heins S, Aebi U (1998). "Nuclear lamins: their structure, assembly, and interactions". *J Struct Biol* **122** (1–2): 42–66. doi:10.1006/jsbi.1998.3987. PMID 9724605.

[13] Goldman A, Moir R, Montag-Lowy M, Stewart M, Goldman R (1992). "Pathway of incorporation of microinjected lamin A into the nuclear envelope". *J Cell Biol* **119** (4): 725–735. doi:10.1083/jcb.119.4.725. PMC 2289687. PMID 1429833.

[14] Goldman R, Gruenbaum Y, Moir R, Shumaker D, Spann T (2002). "Nuclear lamins: building blocks of nuclear architecture". *Genes Dev* **16** (5): 533–547. doi:10.1101/gad.960502. PMID 11877373.

[15] Moir RD, Yoona M, Khuona S, Goldman RD. (2000). "Nuclear Lamins A and B1: Different Pathways of Assembly during Nuclear Envelope Formation in Living Cells". *Journal of Cell Biology* **151** (6): 1155–1168. doi:10.1083/jcb.151.6.1155. PMC 2190592. PMID 11121432.

[16] Spann TP, Goldman AE, Wang C, Huang S, Goldman RD. (2002). "Alteration of nuclear lamin organization inhibits RNA polymerase II–dependent transcription". *Journal of Cell Biology* **156** (4): 603–608. doi:10.1083/jcb.200112047. PMC 2174089. PMID 11854306.

[17] Mounkes LC, Stewart CL (2004). "Aging and nuclear organization: lamins and progeria". *Current Opinion in Cell Biology* **16** (3): 322–327. doi:10.1016/j.ceb.2004.03.009. PMID 15145358.

[18] Ehrenhofer-Murray A (2004). "Chromatin dynamics at DNA replication, transcription and repair". *Eur J Biochem* **271** (12): 2335–2349. doi:10.1111/j.1432-1033.2004.04162.x. PMID 15182349.

[19] Grigoryev S, Bulynko Y, Popova E (2006). "The end adjusts the means: heterochromatin remodelling during terminal cell differentiation". *Chromosome Res* **14** (1): 53–69. doi:10.1007/s10577-005-1021-6. PMID 16506096.

[20] Schardin, Margit; Cremer, T; Hager, HD; Lang, M (December 1985). "Specific staining of human chromosomes in Chinese hamster x man hybrid cell lines demonstrates interphase chromosome territories". *Human Genetics* (Springer Berlin / Heidelberg) **71** (4): 281–287. doi:10.1007/BF00388452. PMID 2416668.

[21] Lamond, Angus I.; William C. Earnshaw (1998-04-24). "Structure and Function in the Nucleus". *Science* **280** (5363): 547–553. doi:10.1126/science.280.5363.547. PMID 9554838.

[22] Kurz, A; Lampel, S; Nickolenko, JE; Bradl, J; Benner, A; Zirbel, RM; Cremer, T; Lichter, P (1996). "Active and inactive genes localize preferentially in the periphery of chromosome territories". *The Journal of Cell Biology* (The Rockefeller University Press) **135** (5): 1195–1205. doi:10.1083/jcb.135.5.1195. PMC 2121085. PMID 8947544.

[23] NF Rothfield, BD Stollar (1967). "The Relation of Immunoglobulin Class, Pattern of Antinuclear Antibody, and Complement-Fixing Antibodies to DNA in Sera from Patients with Systemic Lupus Erythematosus". *J Clin Invest* **46** (11): 1785–1794. doi:10.1172/JCI105669. PMC 292929. PMID 4168731.

[24] S Barned, AD Goodman, DH Mattson (1995). "Frequency of anti-nuclear antibodies in multiple sclerosis". *Neurology* **45** (2): 384–385. doi:10.1212/WNL.45.2.384. PMID 7854544.

[25] Hernandez-Verdun, Daniele (2006). "Nucleolus: from structure to dynamics". *Histochem. Cell. Biol* **125** (1–2): 127–137. doi:10.1007/s00418-005-0046-4. PMID 16328431.

[26] Lamond, Angus I.; Judith E. Sleeman (October 2003). "Nuclear substructure and dynamics". *current biology* **13** (21): R825–828. doi:10.1016/j.cub.2003.10.012. PMID 14588256.

[27] Cioce M, Lamond A (2005). "Cajal bodies: a long history of discovery". *Annu Rev Cell Dev Biol* **21**: 105–131. doi:10.1146/an PMID 16212489.

[28] Pollard, Thomas D.; William C. Earnshaw (2004). *Cell Biology*. Philadelphia: Saunders. ISBN 0-7216-3360-9.

[29] Dundr, Miroslav; Tom Misteli (2001). "Functional architecture in the cell nucleus". *Biochem. J.* **356** (Pt 2): 297–310. doi:10.1042/0264-6021:3560297. PMC 1221839. PMID 11368755.

[30] Fox, Archa (2007-03-07). *Paraspeckle Size*. Interview with R. Sundby. E-mail Correspondence.

[31] Goebel, H.H.; I Warlow (January 1997). "Nemaline myopathy with intranuclear rods—intranuclear rod myopathy". *Neuromuscular Disorders* **7** (1): 13–19. doi:10.1016/S0960-8966(96)00404-X. PMID 9132135.

[32] Matera AG, Frey MA. (1998). "Coiled Bodies and Gems: Janus or Gemini?". *American Journal of Human Genetics* **63** (2): 317–321. doi:10.1086/301992. PMC 1377332. PMID 9683623.

[33] Matera, A. Gregory (1998). "Of Coiled Bodies, Gems, and Salmon". *Journal of Cellular Biochemistry* **70** (2): 181–192. doi:10.1002/(sici)1097-4644(19980801)70:2<181::aid-jcb4>3.0.co;2-k. PMID 9671224.

[34] Saunders WS, Cooke CA, Earnshaw WC (1991). "Compartmentalization within the nucleus: discovery of a novel subnuclear region.". *Journal of Cellular Biology* **115** (4): 919–931. doi:10.1083/jcb.115.4.919. PMID 1955462

[35] Pombo A, Cuello P, Schul W, Yoon J, Roeder R, Cook P, Murphy S (1998). "Regional and temporal specialization in the nucleus: a transcriptionally active nuclear domain rich in PTF, Oct1 and PIKA antigens associates with specific chromosomes early in the cell cycle". *The EMBO Journal* **17** (6): 1768–1778. doi:10.1093/emboj/17.6.1768. PMC 1170524. PMID 9501098.

[36] Lallemand-Breitenbach, V.; De The, H. (2010). "PML Nuclear Bodies". *Cold Spring Harbor Perspectives in Biology* **2** (5): a000661. doi:10.1101/cshperspect.a000661. PMC 2857171. PMID 20452955.

[37] Lamond AI, Spector DL (August 2003). "Nuclear speckles: a model for nuclear organelles". *Nature Reviews Molecular Cell Biology* **4** (8): 605–12. doi:10.1038/nrm1172. PMID 12923522.

[38] Tripathi K, Parnaik VK (September 2008). "Differential dynamics of splicing factor SC35 during the cell cycle" (PDF). *J. Biosci.* **33** (3): 345–54. doi:10.1007/s12038-008-0054-3. PMID 19005234.

[39] Tripathi, K.; Parnaik, V. K. (2008). "Differential dynamics of splicing factor SC35 during the cell cycle". *Journal of biosciences* **33** (3): 345–354. doi:10.1007/s12038-008-0054-3. PMID 19005234.

[40] Lamond AI, Spector DL (August 2003). "Nuclear speckles: a model for nuclear organelles". *Nature Reviews Molecular Cell Biology* **4** (8): 605–12. doi:10.1038/nrm1172. PMID 12923522.

[41] Handwerger, Korie E.; Joseph G. Gall (January 2006). "Subnuclear organelles: new insights into form and function". *TRENDS in Cell Biology* **16** (1): 19–26. doi:10.1016/j.tcb.2005.11.005. PMID 16325406.

[42] "Cellular component Nucleus speckle". UniProt: UniProtKB. Retrieved 2013-08-30.

[43] Gall, Joseph G.; Bellini, Michel; Wu, Zheng'an; Murphy, Christine (December 1999). "Assembly of the Nuclear Transcription and Processing Machinery: Cajal Bodies (Coiled Bodies) and Transcriptosomes". *Molecular Biology of the Cell* **10** (12): 4385–4402. doi:10.1091/mbc.10.12.4385. ISSN 1059-1524. PMC 25765. PMID 10588665.

[44] Matera, A. Gregory; Rebecca M. Terns; Michael P. Terns (March 2007). "Non-coding RNAs: lessons from the small nuclear and small nucleolar RNAs". *Nature Reviews Molecular Cell Biology* **8** (3): 209–220. doi:10.1038/nrm2124. ISSN 1471-0072. PMID 17318225. Retrieved 2013-08-09.

[45] Fox, Archa; Lam, YW; Leung, AK; Lyon, CE; Andersen, J; Mann, M; Lamond, AI (2002). "Paraspeckles:A Novel Nuclear Domain". *Current Biology* **12** (1): 13–25. doi:10.1016/S0960-9822(01)00632-7. PMID 11790299.

[46] Fox, Archa; Wendy Bickmore (2004). "Nuclear Compartments: Paraspeckles". Nuclear Protein Database. Archived from the original on May 2, 2006. Retrieved 2007-03-06.

[47] Fox, A.; et al. (2005). "P54nrb Forms a Heterodimer with PSP1 That Localizes to Paraspeckles in an RNA-dependent Manner". *Molecular Biology of the Cell* **16** (11): 5304–5315. doi:10.1091/mbc.E05-06-0587. PMC 1266428. PMID 16148043.

[48] Lehninger, Albert L.; Nelson, David L.; Cox, Michael M. (2000). *Lehninger principles of biochemistry* (3rd ed.). New York: Worth Publishers. ISBN 1-57259-931-6.

[49] Moreno F, Ahuatzi D, Riera A, Palomino CA, Herrero P. (2005). "Glucose sensing through the Hxk2-dependent signalling pathway.". *Biochem Soc Trans* **33** (1): 265–268. doi:10.1042/BST0330265. PMID 15667322. PMID 15667322

[50] Görlich, Dirk; Ulrike Kutay (1999). "Transport between the cell nucleus and the cytoplasm". *Ann. Rev. Cell Dev. Biol.* **15** (1): 607–660. doi:10.1146/annurev.cellbio.15.1.607. PMID 10611974.

[51] Nierhaus, Knud H.; Daniel N. Wilson (2004). *Protein Synthesis and Ribosome Structure: Translating the Genome*. Wiley-VCH. ISBN 3-527-30638-2.

[52] Nicolini, Claudio A. (1997). *Genome Structure and Function: From Chromosomes Characterization to Genes Technology*. Springer. ISBN 0-7923-4565-7.

[53] Watson, JD; Baker TA; Bell SP; Gann A; Levine M; Losick R. (2004). "Ch9–10". *Molecular Biology of the Gene* (5th ed.). Peason Benjamin Cummings; CSHL Press. ISBN 0-8053-9603-9.

[54] Lippincott-Schwartz, Jennifer (2002-03-07). "Cell biology: Ripping up the nuclear envelope". *Nature* **416** (6876): 31–32. doi:10.1038/416031a. PMID 11882878.

[55] Boulikas T (1995). "Phosphorylation of transcription factors and control of the cell cycle". *Crit Rev Eukaryot Gene Expr* **5** (1): 1–77. PMID 7549180.

[56] Steen R, Collas P (2001). "Mistargeting of B-type lamins at the end of mitosis: implications on cell survival and regulation of lamins A/C expression". *J Cell Biol* **153** (3): 621–626. doi:10.1083/jcb.153.3.621. PMC 2190567. PMID 11331311.

[57] Böhm I. IgG deposits can be detected in cell nuclei of patients with both lupus erythematosus and malignancy. *Clin Rheumatol* 2007;26(11) 1877-1882

[58] Skutelsky, E.; Danon D. (June 1970). "Comparative study of nuclear expulsion from the late erythroblast and cytokinesis". *J Cell Biol* **60** (60(3)): 625–635. doi:10.1016/0014-4827(70)90536-7. PMID 5422968.

[59] Torous, DK; Dertinger SD; Hall NE; Tometsko CR. (2000). "Enumeration of micronucleated reticulocytes in rat peripheral blood: a flow cytometric study". *Mutat Res* **465** (465(1–2)): 91–99. doi:10.1016/S1383-5718(99)00216-8. PMID 10708974.

[60] Hutter, KJ; Stohr M. (1982). "Rapid detection of mutagen induced micronucleated erythrocytes by flow cytometry". *Histochemistry* **75** (3): 353–362. doi:10.1007/bf00496738. PMID 7141888.

[61] Zettler, LA; Sogin ML; Caron DA (1997). "Phylogenetic relationships between the Acantharea and the Polycystinea: A molecular perspective on Haeckel's Radiolaria". *Proc Natl Acad Sci USA* **94** (21): 11411–11416. doi:10.1073/pnas.94.21.11411. PMC 23483. PMID 9326623.

[62] Horton, TR (2006). "The number of nuclei in basidiospores of 63 species of ectomycorrhizal Homobasidiomycetes". *Mycologia* **98** (2): 233–238. doi:10.3852/mycologia.98.2.233. PMID 16894968.

[63] Adam RD (December 1991). "The biology of Giardia spp". *Microbiol. Rev.* **55** (4): 706–32. PMC 372844. PMID 1779932.

[64] McInnes, A; Rennick DM (1988). "Interleukin 4 induces cultured monocytes/macrophages to form giant multinucleated cells". *J Exp Med* **167** (2): 598–611. doi:10.1084/jem.167.2.598. PMC 2188835. PMID 3258008.

[65] Goldring, SR; Roelke MS; Petrison KK; Bhan AK (1987). "Human giant cell tumors of bone identification and characterization of cell types". *J Clin Invest* **79** (2): 483–491. doi:10.1172/JCI112838. PMC 424109. PMID 3027126.

[66] Imanian, B; Pombert, JF; Dorrell, RG; Burki, F; Keeling, PJ (2012). "Tertiary endosymbiosis in two dinotoms has generated little change in the mitochondrial genomes of their dinoflagellate hosts and diatom endosymbionts". *PLOS ONE* **7** (8): e43763. doi:10.1371/journal.pone.0043763.

[67] Pennisi E. (2004). "Evolutionary biology. The birth of the nucleus". *Science***305**(5685): 766–768. doi:10.1126/science.305.56 PMID 15297641.

[68] C.Michael Hogan. 2010. *Archaea*. eds. E.Monosson & C.Cleveland, Encyclopedia of Earth. National Council for Science and the Environment, Washington DC.

[69] Margulis, Lynn (1981). *Symbiosis in Cell Evolution*. San Francisco: W. H. Freeman and Company. pp. 206–227. ISBN 0-7167-1256-3.

[70] Lopez-Garcia P, Moreira D. (2006). "Selective forces for the origin of the eukaryotic nucleus". *BioEssays* **28** (5): 525–533. doi:10.1002/bies.20413. PMID 16615090.

[71] Fuerst JA. (2005). "Intracellular compartmentation in planctomycetes". *Annu Rev Microbiol.***59**: 299–328. doi:10.1146/annurev PMID 15910279.

[72] Hartman H, Fedorov A. (2002). "The origin of the eukaryotic cell: a genomic investigation". *Proc Natl Acad Sci U S A*. **99** (3): 1420–1425. doi:10.1073/pnas.032658599. PMC 122206. PMID 11805300.

[73] Bell PJ (September 2001). "Viral eukaryogenesis: was the ancestor of the nucleus a complex DNA virus?". *J. Mol. Evol.* **53** (3): 251–6. doi:10.1007/s002390010215. PMID 11523012.

[74] Takemura M (2001). "Poxviruses and the origin of the eukaryotic nucleus". *J Mol Evol***52**(5): 419–425. doi:10.1007/s00239001 PMID 11443345.

[75] Villarreal L, DeFilippis V (2000). "A hypothesis for DNA viruses as the origin of eukaryotic replication proteins". *J Virol* **74** (15): 7079–7084. doi:10.1128/JVI.74.15.7079-7084.2000. PMC 112226. PMID 10888648.

[76] Bell PJ (November 2006). "Sex and the eukaryotic cell cycle is consistent with a viral ancestry for the eukaryotic nucleus". *J. Theor. Biol.* **243** (1): 54–63. doi:10.1016/j.jtbi.2006.05.015. PMID 16846615.

[77] de Roos AD (2006). "The origin of the eukaryotic cell based on conservation of existing interfaces". *Artif Life* **12** (4): 513–523. doi:10.1162/artl.2006.12.4.513. PMID 16953783.

22.10   Further reading

- Goldman, Robert D.; Gruenbaum, Y; Moir, RD; Shumaker, DK; Spann, TP (2002). "Nuclear lamins: building blocks of nuclear architecture". *Genes & Dev.* **16** (5): 533–547. doi:10.1101/gad.960502. PMID 11877373.

  A review article about nuclear lamins, explaining their structure and various roles

- Görlich, Dirk; Kutay, U (1999). "Transport between the cell nucleus and the cytoplasm". *Ann. Rev. Cell Dev. Biol.* **15**: 607–660. doi:10.1146/annurev.cellbio.15.1.607. PMID 10611974.

  A review article about nuclear transport, explains the principles of the mechanism, and the various transport pathways

- Lamond, Angus I.; Earnshaw, WC (1998-04-24). "Structure and Function in the Nucleus". *Science* **280** (5363): 547–553. doi:10.1126/science.280.5363.547. PMID 9554838.

A review article about the nucleus, explaining the structure of chromosomes within the organelle, and describing the nucleolus and other subnuclear bodies

- Pennisi E. (2004). "Evolutionary biology. The birth of the nucleus". *Science***305**(5685): 766–768. doi:10.1126 PMID 15297641.

A review article about the evolution of the nucleus, explaining a number of different theories

- Pollard, Thomas D.; William C. Earnshaw (2004). *Cell Biology*. Philadelphia: Saunders. ISBN 0-7216-3360-9.

A university level textbook focusing on cell biology. Contains information on nucleus structure and function, including nuclear transport, and subnuclear domains

22.11 External links

- MBInfo - The Nucleus
- cellnucleus.com Website covering structure and function of the nucleus from the Department of Oncology at the University of Alberta.
- http://npd.hgu.mrc.ac.uk/user/?page=compartment The Nuclear Protein Database] Information on nuclear components.
- The Nucleus Collection in the Image & Video Library of The American Society for Cell Biology contains peer-reviewed still images and video clips that illustrate the nucleus.
- Nuclear Envelope and Nuclear Import Section from *Landmark Papers in Cell Biology, Joseph G. Gall, J. Richard McIntosh, eds., contains digitized commentaries and links to seminal research papers on the nucleus. Published online in the Image & Video Library of The American Society for Cell Biology*
- Cytoplasmic patterns generated by human antibodies

Chapter 23

Chromatin

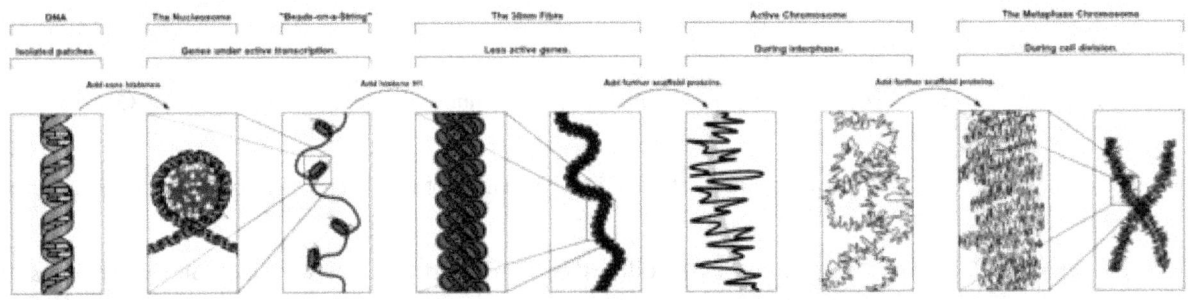

The major structures in DNA compaction: DNA, the nucleosome, the 10 nm "beads-on-a-string" fibre, the 30 nm chromatin fibre and the metaphase chromosome.

Chromatin is a complex of macromolecules found in cells, consisting of DNA, protein and RNA. The primary functions of chromatin are 1) to package DNA into a smaller volume to fit in the cell, 2) to reinforce the DNA macromolecule to allow mitosis, 3) to prevent DNA damage, and 4) to control gene expression and DNA replication. The primary protein components of chromatin are histones that compact the DNA. Chromatin is only found in eukaryotic cells (cells with defined nuclei). Prokaryotic cells have a different organization of their DNA (the prokaryotic chromosome equivalent is called genophore and is localized within the nucleoid region).

The structure of chromatin depends on several factors. The overall structure depends on the stage of the cell cycle. During interphase, the chromatin is structurally loose to allow access to RNA and DNA polymerases that transcribe and replicate the DNA. The local structure of chromatin during interphase depends on the genes present on the DNA: DNA coding genes that are actively transcribed ("turned on") are more loosely packaged and are found associated with RNA polymerases (referred to as euchromatin) while DNA coding inactive genes ("turned off") are found associated with structural proteins and are more tightly packaged (heterochromatin).[1][2] Epigenetic chemical modification of the structural proteins in chromatin also alters the local chromatin structure, in particular chemical modifications of histone proteins by methylation and acetylation. As the cell prepares to divide, i.e. enters mitosis or meiosis, the chromatin packages more tightly to facilitate segregation of the chromosomes during anaphase. During this stage of the cell cycle this makes the individual chromosomes in many cells visible by optical microscope.

In general terms, there are three levels of chromatin organization:

1. DNA wraps around histone proteins forming nucleosomes; the "beads on a string" structure (euchromatin).

2. Multiple histones wrap into a 30 nm fibre consisting of nucleosome arrays in their most compact form (heterochromatin). (Definitively established to exist in vitro, the 30-nanometer fibre was not seen in recent X-ray studies of human mitotic chromosomes.[3])

3. Higher-level DNA packaging of the 30 nm fibre into the metaphase chromosome (during mitosis and meiosis).

There are, however, many cells that do not follow this organisation. For example, spermatozoa and avian red blood cells have more tightly packed chromatin than most eukaryotic cells, and trypanosomatid protozoa do not condense their chromatin into visible chromosomes for mitosis.

23.1 During interphase

The structure of chromatin during interphase of mitosis is optimized to allow simple access of transcription and DNA repair factors to the DNA while compacting the DNA into the nucleus. The structure varies depending on the access required to the DNA. Genes that require regular access by RNA polymerase require the looser structure provided by euchromatin.

23.2 Dynamic chromatin structure and hierarchy

Chromatin undergoes various structural changes during a cell cycle. Histone proteins are the basic packer and arranger of chromatin and can be modified by various post-translational modifications to alter chromatin packing (Histone modification). Most of the modifications occur on the histone tail. The consequences in term of chromatin accessibility and compaction depend both on the amino-acid that is modified and the type of modification. For example, Histone acetylation results in loosening and increased accessibility of chromatin for replication and transcription. Lysine tri-methylation can either be correlated with transcriptional activity (tri-methylation of histone H3 Lysine 4) or transcriptional repression and chromatin compaction (tri-methylation of histone H3 Lysine 9 or 27). Several studies suggested that different modifications could occur simultaneously. For example, it was proposed that a bivalent structure (with tri-methylation of both Lysine 4 and 27 on histone H3) was involved in mammalian early development.[4]

Polycomb-group proteins play a role in regulating genes through modulation of chromatin structure.[5]

For additional information, see Histone modifications in chromatin regulation and RNA polymerase control by chromatin structure.

23.2.1 DNA structure

Main articles: Mechanical properties of DNA and Z-DNA

In nature, DNA can form three structures, A-, B-, and Z-DNA. A- and B-DNA are very similar, forming right-handed helices, whereas Z-DNA is a left-handed helix with a zig-zag phosphate backbone. Z-DNA is thought to play a specific role in chromatin structure and transcription because of the properties of the junction between B- and Z-DNA.

At the junction of B- and Z-DNA, one pair of bases is flipped out from normal bonding. These play a dual role of a site of recognition by many proteins and as a sink for torsional stress from RNA polymerase or nucleosome binding.

23.2.2 Nucleosomes and beads-on-a-string

Main articles: Nucleosome, Chromatosome and Histone

The basic repeat element of chromatin is the nucleosome, interconnected by sections of linker DNA, a far shorter arrangement than pure DNA in solution.

In addition to the core histones, there is the linker histone, H1, which contacts the exit/entry of the DNA strand on the nucleosome. The nucleosome core particle, together with histone H1, is known as a chromatosome. Nucleosomes,

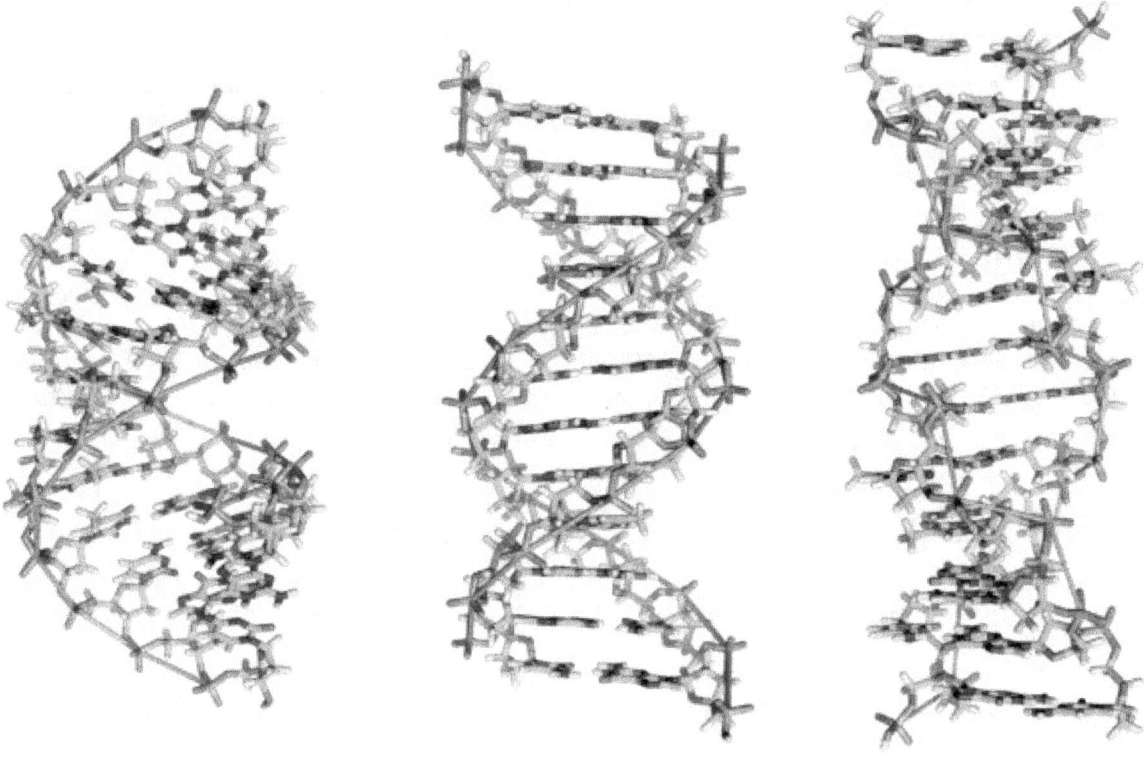

The structures of A-, B-, and Z-DNA.

with about 20 to 60 base pairs of linker DNA, can form, under non-physiological conditions, an approximately 10 nm "beads-on-a-string" fibre. (Fig. 1-2). .

The nucleosomes bind DNA non-specifically, as required by their function in general DNA packaging. There are, however, large DNA sequence preferences that govern nucleosome positioning. This is due primarily to the varying physical properties of different DNA sequences: For instance, adenine and thymine are more favorably compressed into the inner minor grooves. This means nucleosomes can bind preferentially at one position approximately every 10 base pairs (the helical repeat of DNA)- where the DNA is rotated to maximise the number of A and T bases that will lie in the inner minor groove. (See mechanical properties of DNA.)

23.2.3   30 nanometer chromatin fibre

With addition of H1, the beads-on-a-string structure in turn coils into a 30 nm diameter helical structure known as the 30 nm fibre or filament. The precise structure of the chromatin fibre in the cell is not known in detail, and there is still some debate over this.[6]

This level of chromatin structure is thought to be the form of euchromatin, which contains actively transcribed genes. EM studies have demonstrated that the 30 nm fibre is highly dynamic such that it unfolds into a 10 nm fiber ("beads-on-a-string") structure when transversed by an RNA polymerase engaged in transcription.

The existing models commonly accept that the nucleosomes lie perpendicular to the axis of the fibre, with linker histones arranged internally. A stable 30 nm fibre relies on the regular positioning of nucleosomes along DNA. Linker DNA is relatively resistant to bending and rotation. This makes the length of linker DNA critical to the stability of the fibre, requiring nucleosomes to be separated by lengths that permit rotation and folding into the required orientation without excessive stress to the DNA. In this view, different length of the linker DNA should produce different folding topologies of the chromatin fiber. Recent theoretical work, based on electron-microscopy images[7] of reconstituted fibers supports this view.[8]

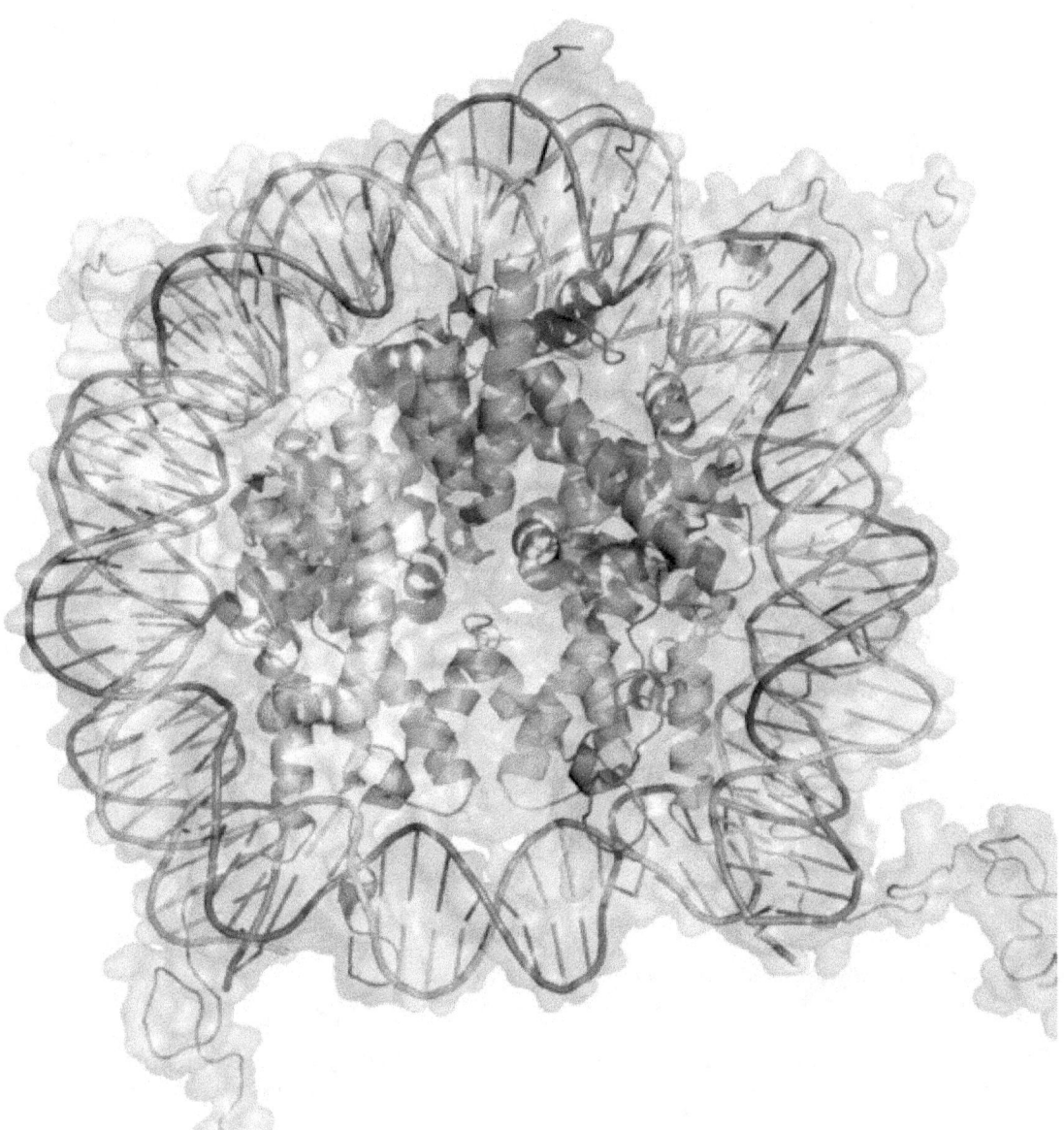

A cartoon representation of the nucleosome structure. From PDB: 1KX5.

23.2.4   Spatial organization of chromatin in the cell nucleus

The spatial arrangement of the chromatin within the nucleus is not random - specific regions of the chromatin can be found in certain territories. Territories are, for example, the lamina-associated domains (LADs), and the topological association domains (TADs), which are bound together by protein complexes.[9] Currently, polymer models such as the Strings & Binders Switch (SBS) model[10] and the Dynamic Loop (DL) model[11] are used to describe the folding of chromatin within the nucleus.

23.3   Chromatin and bursts of transcription

Chromatin and its interaction with enzymes has been researched, and a conclusion being made is that it is relevant and an important factor in gene expression. Vincent G. Allfrey, a professor at Rockefeller University, stated that RNA synthesis

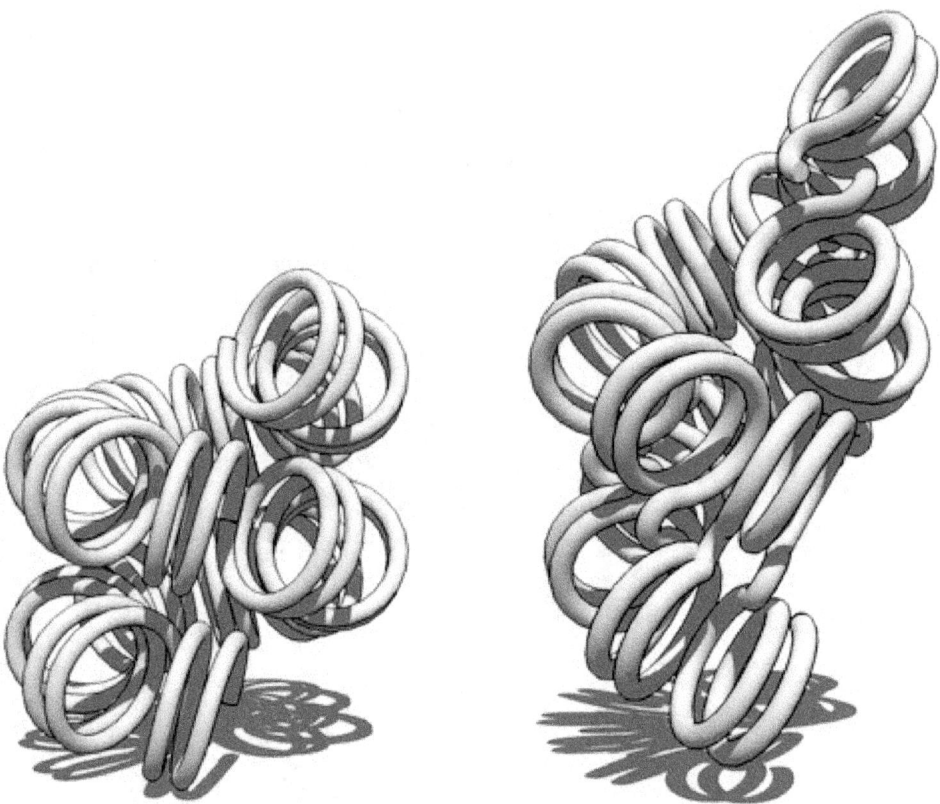

Two proposed structures of the 30nm chromatin filament.
Left: 1 start helix "solenoid" structure.
Right: 2 start loose helix structure.
Note: the histones are omitted in this diagram - only the DNA is shown.

is related to histone acetylation.[12] The lysine amino acid attached to the end of the histones is positively charged. The acetylation of these tails would make the chromatin ends neutral, allowing for DNA access.

When the chromatin decondenses, the DNA is open to entry of molecular machinery. Fluctuations between open and closed chromatin may contribute to the discontinuity of transcription, or transcriptional bursting. Other factors are probably involved, such as the association and dissociation of transcription factor complexes with chromatin. The phenomenon, as opposed to simple probabilistic models of transcription, can account for the high variability in gene expression occurring between cells in isogenic populations[13]

23.4   Metaphase chromatin (chromosomes)

The metaphase structure of chromatin differs vastly to that of interphase. It is optimised for physical strength and manageability, forming the classic chromosome structure seen in karyotypes. The structure of the condensed chromatin is thought to be loops of 30 nm fibre to a central scaffold of proteins. It is, however, not well-characterised.

The physical strength of chromatin is vital for this stage of division to prevent shear damage to the DNA as the daughter chromosomes are separated. To maximise strength the composition of the chromatin changes as it approaches the centromere, primarily through alternative histone H1 anologues.

It should also be noted that, during mitosis, while most of the chromatin is tightly compacted, there are small regions that

Four proposed structures of the 30 nm chromatin filament for DNA repeat length per nucleosomes ranging from 177 to 207 bp. Linker DNA in yellow and nucleosomal DNA in pink.

are not as tightly compacted. These regions often correspond to promoter regions of genes that were active in that cell type prior to entry into chromitosis. The lack of compaction of these regions is called bookmarking, which is an epigenetic mechanism believed to be important for transmitting to daughter cells the "memory" of which genes were active prior to entry into mitosis.[14] This bookmarking mechanism is needed to help transmit this memory because transcription ceases during mitosis.

23.5 Chromatin: alternative definitions

1. **Simple and concise definition:** Chromatin is a macromolecular complex of a DNA macromolecule and protein macromolecules (and RNA). The proteins package and arrange the DNA and control its functions within the cell nucleus.

2. **A biochemists' operational definition:** Chromatin is the DNA/protein/RNA complex extracted from eukaryotic lysed interphase nuclei. Just which of the multitudinous substances present in a nucleus will constitute a part of the extracted material partly depends on the technique each researcher uses. Furthermore, the composition and properties of chromatin vary from one cell type to the another, during development of a specific cell type, and at different stages in the cell cycle.

3. **The *DNA + histone = chromatin* definition:** The DNA double helix in the cell nucleus is packaged by special proteins termed histones. The formed protein/DNA complex is called chromatin. The basic structural unit of chromatin is the nucleosome.

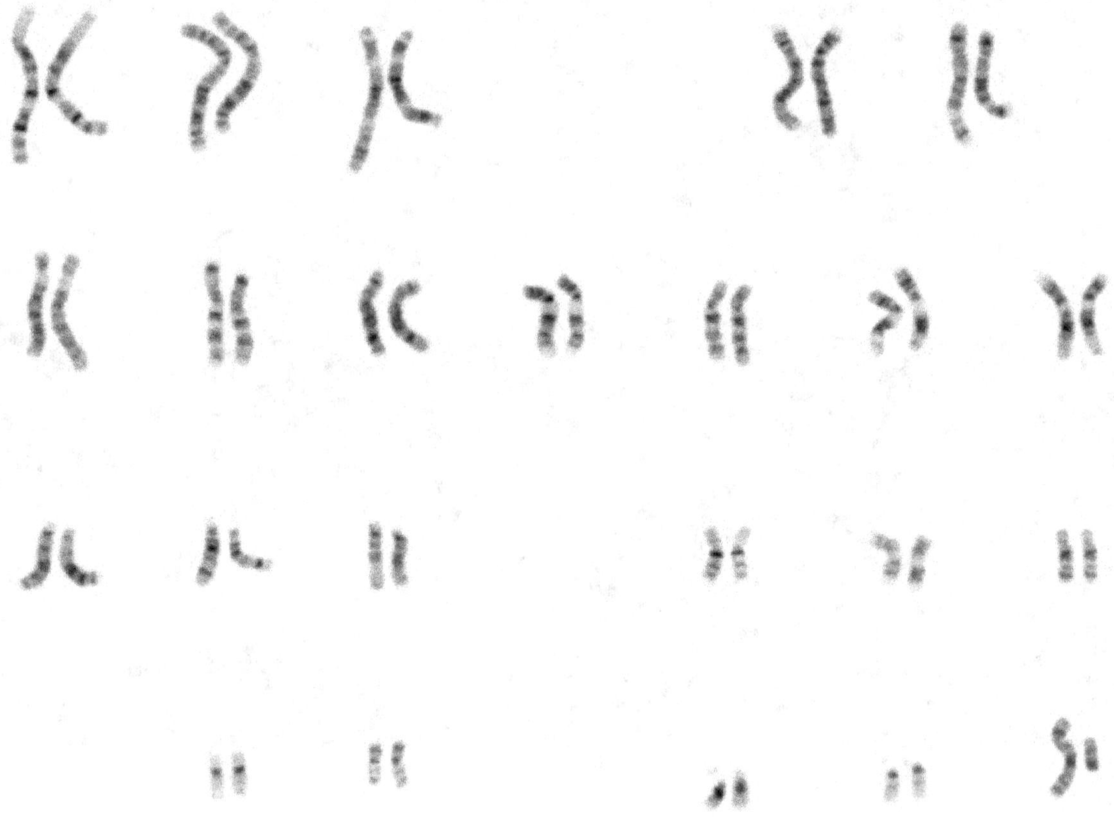

Karyogram of human male using Giemsa staining, showing the classic metaphase chromatin structure.

23.6   Alternative chromatin organizations

During metazoan spermiogenesis, the spermatid's chromatin is remodelled into a more spaced-packaged, widened, almost crystal-like structure. This process is associated with the cessation of transcription and involves nuclear protein exchange. The histones are mostly displaced, and replaced by protamines (small, arginine-rich proteins).[15]

23.7   Nobel Prizes

The following scientists were recognized for their contributions to chromatin research with Nobel Prizes:

23.8   See also

- Chromatid

- Epigenetics

- Histone-Modifying Enzymes

- Position-effect variegation

- Salt-and-pepper chromatin

- Transcriptional bursting

23.9 References

[1] "Chromatin Network Home Page.". Retrieved 2008-11-18.

[2] Dame, R.T. (May 2005). "The role of nucleoid-associated proteins in the organization and compaction of bacterial chromatin". *Molecular Microbiology* **56** (4): 858–870. doi:10.1111/j.1365-2958.2005.04598.x. PMID 15853876.

[3] Hansen, Jeffrey (March 2012). "Human mitotic chromosome structure: what happened to the 30-nm fibre?". *The EMBO Journal* **31** (7): 1621–1623. doi:10.1038/emboj.2012.66. PMC 3321215. PMID 22415369.

[4] Bernstein, B.E., T.S. Mikkelsen, X. Xie, M. Kamal, D.J. Huebert, J. Cuff, B. Fry, A. Meissner, M. Wernig, K. Plath, R. Jaenisch, A. Wagschal, R. Feil, S.L. Schreiber & E.S. Lander (April 2006). "A bivalent chromatin structure marks key developmental genes in embryonic stem cells". *Cell* **125** (2): 315–26. doi:10.1016/j.cell.2006.02.041. ISSN 0092-8674. PMID 16630819.

[5] Portoso M and Cavalli G (2008). "The Role of RNAi and Noncoding RNAs in Polycomb Mediated Control of Gene Expression and Genomic Programming". *RNA and the Regulation of Gene Expression: A Hidden Layer of Complexity*. Caister Academic Press. isbn=978-1-904455-25-7.

[6] Annunziato, Anthony T. "DNA Packaging: Nucleosomes and Chromatin". *Scitable*. Nature Education. Retrieved 2015-10-29.

[7] Robinson DJ, Fairall L, Huynh VA, Rhodes D. (April 2006). "EM measurements define the dimensions of the "30-nm" chromatin fiber: Evidence for a compact, interdigitated structure". *PNAS* **103** (17): 6506–11. doi:10.1073/pnas.0601212103. PMC 1436021. PMID 16617109.

[8] Wong H, Victor JM, Mozziconacci J. (September 2007). Chen, Pu, ed. "An All-Atom Model of the Chromatin Fiber Containing Linker Histones Reveals a Versatile Structure Tuned by the Nucleosomal Repeat Length". *PLoS ONE* **2** (9): e877. doi:10.1371/journal.pone.0000877. PMC 1963316. PMID 17849006.

[9] Nicodemi M, Pombo A (June 2014). "Models of chromosome structure". *Curr. Opin. Cell Biol.* **28**: 90–5. doi:10.1016/j.ceb.2014.04.004. PMID 24804566.

[10] Nicodemi M, Panning B, Prisco A (May 2008). "A thermodynamic switch for chromosome colocalization". *Genetics* **179** (1): 717–21. doi:10.1534/genetics.107.083154. PMC 2390650. PMID 18493085.

[11] Bohn M, Heermann DW (2010). "Diffusion-driven looping provides a consistent framework for chromatin organization". *PLoS ONE* **5** (8): e12218. doi:10.1371/journal.pone.0012218. PMC 2928267. PMID 20811620.

[12] ALLFREY VG, FAULKNER R, MIRSKY AE (May 1964). "ACETYLATION AND METHYLATION OF HISTONES AND THEIR POSSIBLE ROLE IN THE REGULATION OF RNA SYNTHESIS". *Proc. Natl. Acad. Sci. U.S.A.* **51** (5): 786–94. doi:10.1073/pnas.51.5.786. PMC 300163. PMID 14172992.

[13] Kaochar S, Tu BP (November 2012). "Gatekeepers of chromatin: Small metabolites elicit big changes in gene expression". *Trends Biochem. Sci.* **37** (11): 477–83. doi:10.1016/j.tibs.2012.07.008. PMC 3482309. PMID 22944281.

[14] Xing H, Vanderford NL, Sarge KD (November 2008). "The TBP-PP2A mitotic complex bookmarks genes by preventing condensin action". *Nat. Cell Biol.* **10** (11): 1318–23. doi:10.1038/ncb1790. PMC 2577711. PMID 18931662.

[15] De Vries M, Ramos L, Housein Z, De Boer P (May 2012). "Chromatin remodelling initiation during human spermiogenesis". *Biol Open* **1** (5): 446–57. doi:10.1242/bio.2012844. PMC 3507207. PMID 23213436.

[16] "Thomas Hunt Morgan and His Legacy". Nobelprize.org. 7 Sep 2012

23.10   Other references

- Cooper, Geoffrey M. 2000. The Cell, 2nd edition, A Molecular Approach. Chapter 4.2 Chromosomes and Chromatin.

- Corces, V. G. (1995). "Chromatin insulators. Keeping enhancers under control". *Nature* **376** (6540): 462–463. doi:10.1038/376462a0.

- Cremer, T. 1985. Von der Zellenlehre zur Chromosomentheorie: Naturwissenschaftliche Erkenntnis und Theorienwechsel in der frühen Zell- und Vererbungsforschung, Veröffentlichungen aus der Forschungsstelle für Theoretische Pathologie der Heidelberger Akademie der Wissenschaften. Springer-Vlg., Berlin, Heidelberg.

- Elgin, S. C. R. (ed.). 1995. Chromatin Structure and Gene Expression, vol. 9. IRL Press, Oxford, New York, Tokyo.

- Gerasimova, T. I.; Corces, V. G. (1996). "Boundary and insulator elements in chromosomes". *Current Op. Genet. and Dev.* **6**: 185–192. doi:10.1016/s0959-437x(96)80049-9.

- Gerasimova, T. I.; Corces, V. G. (1998). "Polycomb and Trithorax group proteins mediate the function of a chromatin insulator". *Cell* **92**: 511–521. doi:10.1016/s0092-8674(00)80944-7.

- Gerasimova, T. I.; Corces, V. G. (2001). "CHROMATIN INSULATORS AND BOUNDARIES: Effects on Transcription and Nuclear Organization". *Annu Rev Genet* **35**: 193–208.

- Gerasimova, T. I.; Byrd, K.; Corces, V. G. (2000). "A chromatin insulator determines the nuclear localization of DNA [In Process Citation]". *Mol Cell* **6**: 1025–35. doi:10.1016/s1097-2765(00)00101-5.

- Ha, S. C.; Lowenhaupt, K.; Rich, A.; Kim, Y. G.; Kim, K. K. (2005). "Crystal structure of a junction between B-DNA and Z-DNA reveals two extruded bases". *Nature* **437**: 1183–6. doi:10.1038/nature04088. PMID 16237447.

- Pollard, T., and W. Earnshaw. 2002. Cell Biology. Saunders.

- Saumweber, H. 1987. Arrangement of Chromosomes in Interphase Cell Nuclei, p. 223-234. In W. Hennig (ed.), Structure and Function of Eucaryotic Chromosomes, vol. 14. Springer-Verlag, Berlin, Heidelberg.

- Sinden, R. R. (2005). "Molecular biology: DNA twists and flips". *Nature* **437**: 1097–8. doi:10.1038/4371097a.

- Van Holde KE. 1989. Chromatin. New York: Springer-Verlag. ISBN 0-387-96694-3.

- Van Holde, K., J. Zlatanova, G. Arents, and E. Moudrianakis. 1995. Elements of chromatin structure: histones, nucleosomes, and fibres, p. 1-26. In S. C. R. Elgin (ed.), Chromatin structure and gene expression. IRL Press at Oxford University Press, Oxford.

23.11   External links

- Chromatin, Histones & Cathepsin; PMAP The Proteolysis Map-animation
- Recent chromatin publications and news
- Protocol for *in vitro* Chromatin Assembly
- ENCODE threads Explorer Chromatin patterns at transcription factor binding sites. Nature (journal)

Chapter 24

Chromosome

A **chromosome** (*chromo-* + *-some*) is a packaged and organized structure containing most of the DNA of a living organism. It is not usually found on its own, but rather is structured by being wrapped around protein complexes called nucleosomes, which consist of histones. The DNA in chromosomes is also associated with transcription (copying of genetic sequences) factors and several other macromolecules. During most of the duration of the Cell cycle, a chromosome consists of one long double-stranded DNA molecule (with associated proteins). During S phase, the chromosome gets replicated, resulting in an 'X'-shaped structure called a metaphase chromosome. Both the original and the newly copied DNA are now called chromatids. The two "sister" chromatids join together at a protein junction called a centromere. Chromosomes are normally visible under a light microscope only when the cell is undergoing mitosis. Even then, the full chromosome containing both joined sister chromatids becomes visible only during a sequence of mitosis known as metaphase (when chromosomes align together, attached to the mitotic spindle and prepare to divide).[1] This DNA and its associated proteins and macromolecules is collectively known as chromatin, which is further packaged along with its associated molecules into a discrete structure called a nucleosome. Chromatin is present in most cells, with a few exceptions - erythrocytes for example. Occurring only in the nucleus of eukaryotic cells, chromatin composes the vast majority of all DNA, except for a small amount inherited maternally which is found in mitochondria. In prokaryotic cells, chromatin occurs free-floating in cytoplasm, as these cells lack organelles and a defined nucleus. Bacteria also lack histones. The main information-carrying macromolecule is a single piece of coiled double-stranded DNA, containing many genes, regulatory elements and other noncoding DNA.[2] The DNA-bound macromolecules are proteins, which serve to package the DNA and control its functions. Chromosomes vary widely between different organisms. Some species such as certain bacteria also contain plasmids or other extrachromosomal DNA. These are circular structures in the cytoplasm which contain cellular DNA and play a role in horizontal gene transfer.[1]

Compaction of the duplicated chromosomes during cell division (mitosis or meiosis) results either in a four-arm structure (pictured to the right) if the centromere is located in the middle of the chromosome or a two-arm structure if the centromere is located near one of the ends. Chromosomal recombination during meiosis and subsequent sexual reproduction plays a vital role in genetic diversity. If these structures are manipulated incorrectly, through processes known as chromosomal instability and translocation, the cell may undergo mitotic catastrophe and die, or it may unexpectedly evade apoptosis leading to the progression of cancer.

In prokaryotes (see nucleoids) and viruses,[2] the DNA is often densely packed and organized. In the case of archaea by homologs to eukaryotic histones, in the case of bacteria by histone-like proteins. Small circular genomes called plasmids are often found in bacteria and also in mitochondria and chloroplasts, reflecting their bacterial origins.

24.1 History of discovery

The word *chromosome* comes from the Greek χρῶμα (*chroma*, "colour") and σῶμα (*soma*, "body"). Chromatin and chromosomes are both very strongly stained by particular dyes.[3]

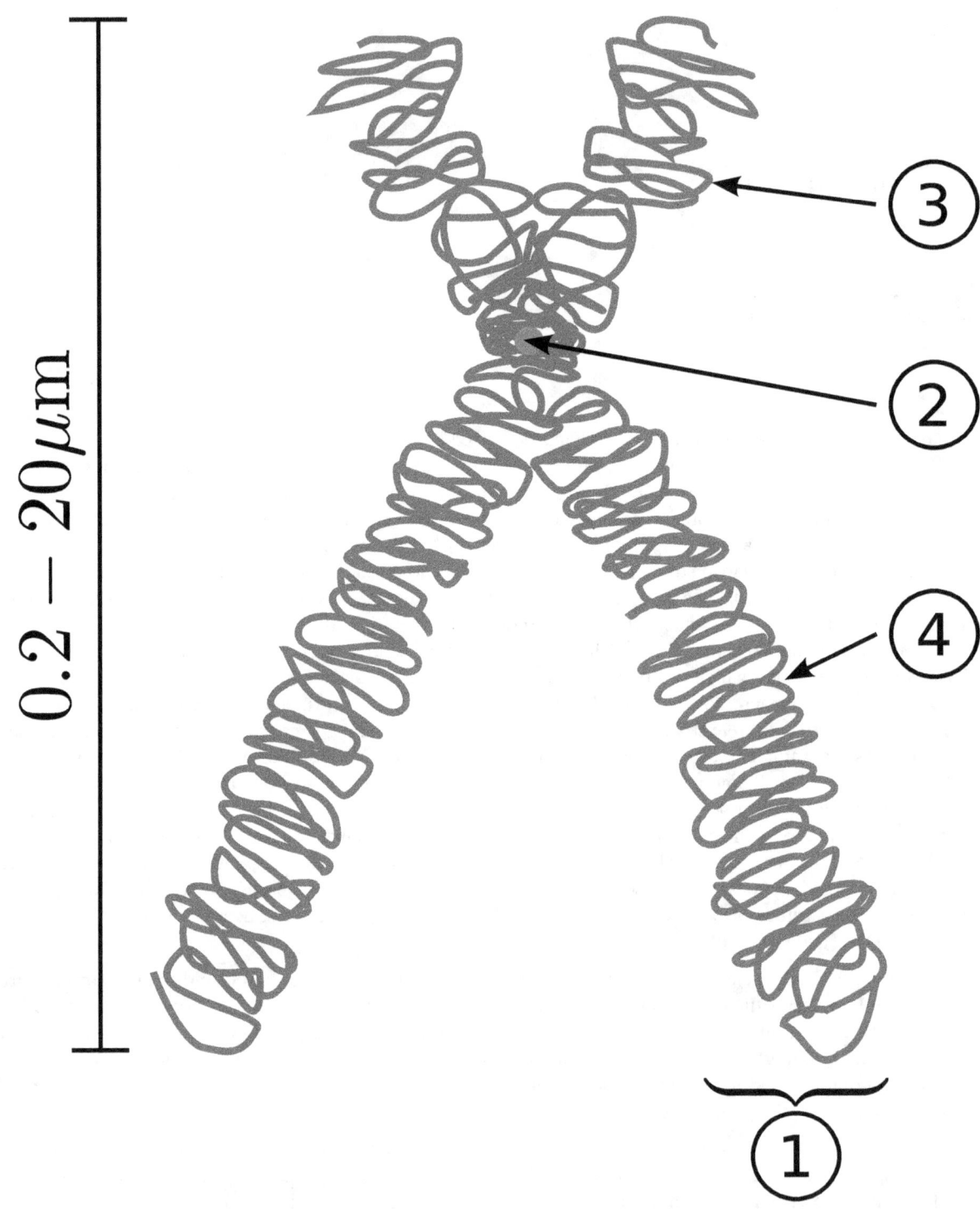

0.2 – 20µm

Diagram of a replicated and condensed metaphase eukaryotic chromosome. (1) Chromatid – one of the two identical parts of the chromosome after S phase. (2) Centromere – the point where the two chromatids touch. (3) Short arm. (4) Long arm.

Schleiden,[1] Virchow and Bütschli were among the first scientists who recognized the structures now so familiar to everyone as chromosomes.[4] The term was coined by von Waldeyer-Hartz,[5] referring to the term chromatin, which was introduced by Walther Flemming.

In a series of experiments beginning in the mid-1880s, Theodor Boveri gave the definitive demonstration that chromosomes are the vectors of heredity. His two principles were the *continuity* of chromosomes and the *individuality* of chromosomes. It is the second of these principles that was so original. Wilhelm Roux suggested that each chromosome carries a different genetic load. Boveri was able to test and confirm this hypothesis. Aided by the rediscovery at the start of the 1900s of Gregor Mendel's earlier work, Boveri was able to point out the connection between the rules of inheritance and the behaviour of the chromosomes. Boveri influenced two generations of American cytologists: Edmund Beecher Wilson, Walter Sutton and Theophilus Painter were all influenced by Boveri (Wilson and Painter actually worked with him).

In his famous textbook *The Cell in Development and Heredity*, Wilson linked together the independent work of Boveri and Sutton (both around 1902) by naming the chromosome theory of inheritance the Boveri–Sutton chromosome theory (the names are sometimes reversed).[6] Ernst Mayr remarks that the theory was hotly contested by some famous geneticists: William Bateson, Wilhelm Johannsen, Richard Goldschmidt and T.H. Morgan, all of a rather dogmatic turn of mind. Eventually, complete proof came from chromosome maps in Morgan's own lab.[7]

The number of human chromosomes was published in 1923 by Theophilus Painter. By inspection through the microscope he counted 24 pairs which would mean 48 chromosomes. His error was copied by others and it was not until 1956 that the true number, 46, was determined by Indonesia-born cytogeneticist Joe Hin Tjio.[8]

24.2 Prokaryotes

The prokaryotes – bacteria and archaea – typically have a single circular chromosome, but many variations exist.[9] The chromosomes of most bacteria can range in size from only 130,000 base pairs in the endosymbiotic bacteria *Candidatus Hodgkinia cicadicola*[10] and *Candidatus Tremblaya princeps*,[11] to over 14,000,000 base pairs in the soil-dwelling bacterium *Sorangium cellulosum*.[12] Spirochaetes of the genus *Borrelia* are a notable exception to this arrangement, with bacteria such as *Borrelia burgdorferi*, the cause of Lyme disease, containing a single *linear* chromosome.[13]

24.2.1 Structure in sequences

Prokaryotic chromosomes have less sequence-based structure than eukaryotes. Bacteria typically have a single point (the origin of replication) from which replication starts, whereas some archaea contain multiple replication origins.[14] The genes in prokaryotes are often organized in operons, and do not usually contain introns, unlike eukaryotes.

24.2.2 DNA packaging

Prokaryotes do not possess nuclei. Instead, their DNA is organized into a structure called the nucleoid.[15] The nucleoid is a distinct structure and occupies a defined region of the bacterial cell. This structure is, however, dynamic and is maintained and remodeled by the actions of a range of histone-like proteins, which associate with the bacterial chromosome.[16] In archaea, the DNA in chromosomes is even more organized, with the DNA packaged within structures similar to eukaryotic nucleosomes.[17][18]

Bacterial chromosomes tend to be tethered to the plasma membrane of the bacteria. In molecular biology application, this allows for its isolation from plasmid DNA by centrifugation of lysed bacteria and pelleting of the membranes (and the attached DNA).

Prokaryotic chromosomes and plasmids are, like eukaryotic DNA, generally supercoiled. The DNA must first be released into its relaxed state for access for transcription, regulation, and replication.

24.3 Eukaryotes

See also: Eukaryotic chromosome fine structure

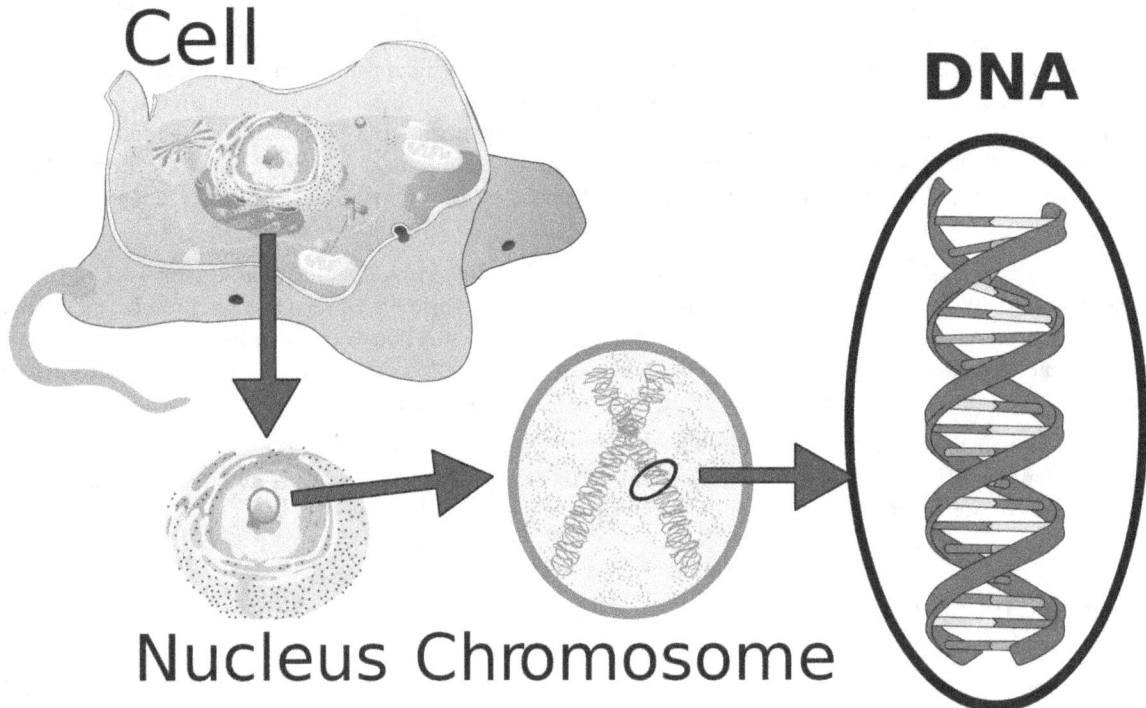

Organization of DNA in a eukaryotic cell.

In eukaryotes, nuclear chromosomes are packaged by proteins into a condensed structure called chromatin. This allows the very long DNA molecules to fit into the cell nucleus. The structure of chromosomes and chromatin varies through the cell cycle. Chromosomes are even more condensed than chromatin and are an essential unit for cellular division. Chromosomes must be replicated, divided, and passed successfully to their daughter cells so as to ensure the genetic diversity and survival of their progeny. Chromosomes may exist as either duplicated or unduplicated. Unduplicated chromosomes are single linear strands, whereas duplicated chromosomes contain two identical copies (called chromatids or sister chromatids) joined by a centromere.

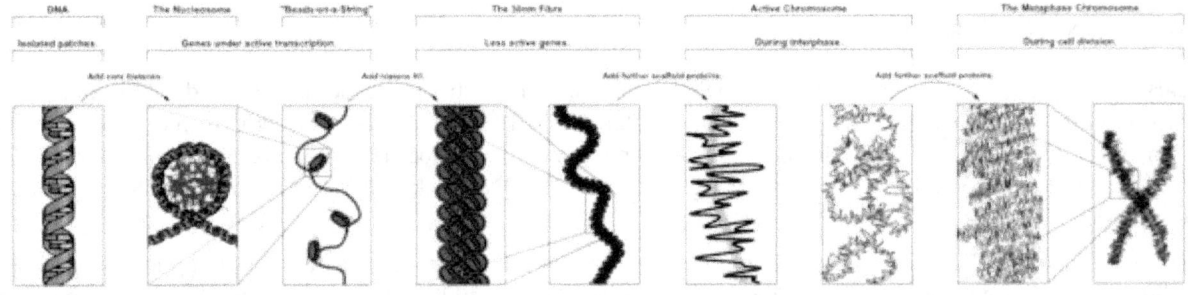

Fig. 2: *The major structures in DNA compaction: DNA, the nucleosome, the 10 nm "beads-on-a-string" fibre, the 30 nm fibre and the metaphase chromosome.*

Eukaryotes (cells with nuclei such as those found in plants, yeast, and animals) possess multiple large linear chromosomes contained in the cell's nucleus. Each chromosome has one centromere, with one or two arms projecting from the centromere, although, under most circumstances, these arms are not visible as such. In addition, most eukaryotes have a small circular mitochondrial genome, and some eukaryotes may have additional small circular or linear cytoplasmic chromosomes.

In the nuclear chromosomes of eukaryotes, the uncondensed DNA exists in a semi-ordered structure, where it is wrapped around histones (structural proteins), forming a composite material called chromatin.

24.3.1 Chromatin

Main article: Chromatin

Chromatin is the complex of DNA and protein found in the eukaryotic nucleus, which packages chromosomes. The structure of chromatin varies significantly between different stages of the cell cycle, according to the requirements of the DNA.

Interphase chromatin

During interphase (the period of the cell cycle where the cell is not dividing), two types of chromatin can be distinguished:

- Euchromatin, which consists of DNA that is active, e.g., being expressed as protein.

- Heterochromatin, which consists of mostly inactive DNA. It seems to serve structural purposes during the chromosomal stages. Heterochromatin can be further distinguished into two types:
  - *Constitutive heterochromatin*, which is never expressed. It is located around the centromere and usually contains repetitive sequences.
  - *Facultative heterochromatin*, which is sometimes expressed.

Metaphase chromatin and division

See also: mitosis and meiosis

In the early stages of mitosis or meiosis (cell division), the chromatin strands become more and more condensed. They cease to function as accessible genetic material (transcription stops) and become a compact transportable form. This compact form makes the individual chromosomes visible, and they form the classic four arm structure, a pair of sister chromatids attached to each other at the centromere. The shorter arms are called *p arms* (from the French *petit*, small) and the longer arms are called *q arms* (q follows p in the Latin alphabet; q-g "grande"; alternatively it is sometimes said q is short for *queue* meaning tail in French[19]). This is the only natural context in which individual chromosomes are visible with an optical microscope.

During mitosis, microtubules grow from centrosomes located at opposite ends of the cell and also attach to the centromere at specialized structures called kinetochores, one of which is present on each sister chromatid. A special DNA base sequence in the region of the kinetochores provides, along with special proteins, longer-lasting attachment in this region. The microtubules then pull the chromatids apart toward the centrosomes, so that each daughter cell inherits one set of chromatids. Once the cells have divided, the chromatids are uncoiled and DNA can again be transcribed. In spite of their appearance, chromosomes are structurally highly condensed, which enables these giant DNA structures to be contained within a cell nucleus (Fig. 2).

24.3.2 Human chromosomes

Chromosomes in humans can be divided into two types: autosomes and sex chromosomes. Certain genetic traits are linked to a person's sex and are passed on through the sex chromosomes. The autosomes contain the rest of the genetic hereditary information. All act in the same way during cell division. Human cells have 23 pairs of chromosomes (22 pairs of autosomes and one pair of sex chromosomes), giving a total of 46 per cell. In addition to these, human cells have many hundreds of copies of the mitochondrial genome. Sequencing of the human genome has provided a great deal of information about each of the chromosomes. Below is a table compiling statistics for the chromosomes, based on the Sanger Institute's human genome information in the Vertebrate Genome Annotation (VEGA) database.[20] Number of genes is an estimate as it is in part based on gene predictions. Total chromosome length is an estimate as well, based on the estimated size of unsequenced heterochromatin regions.

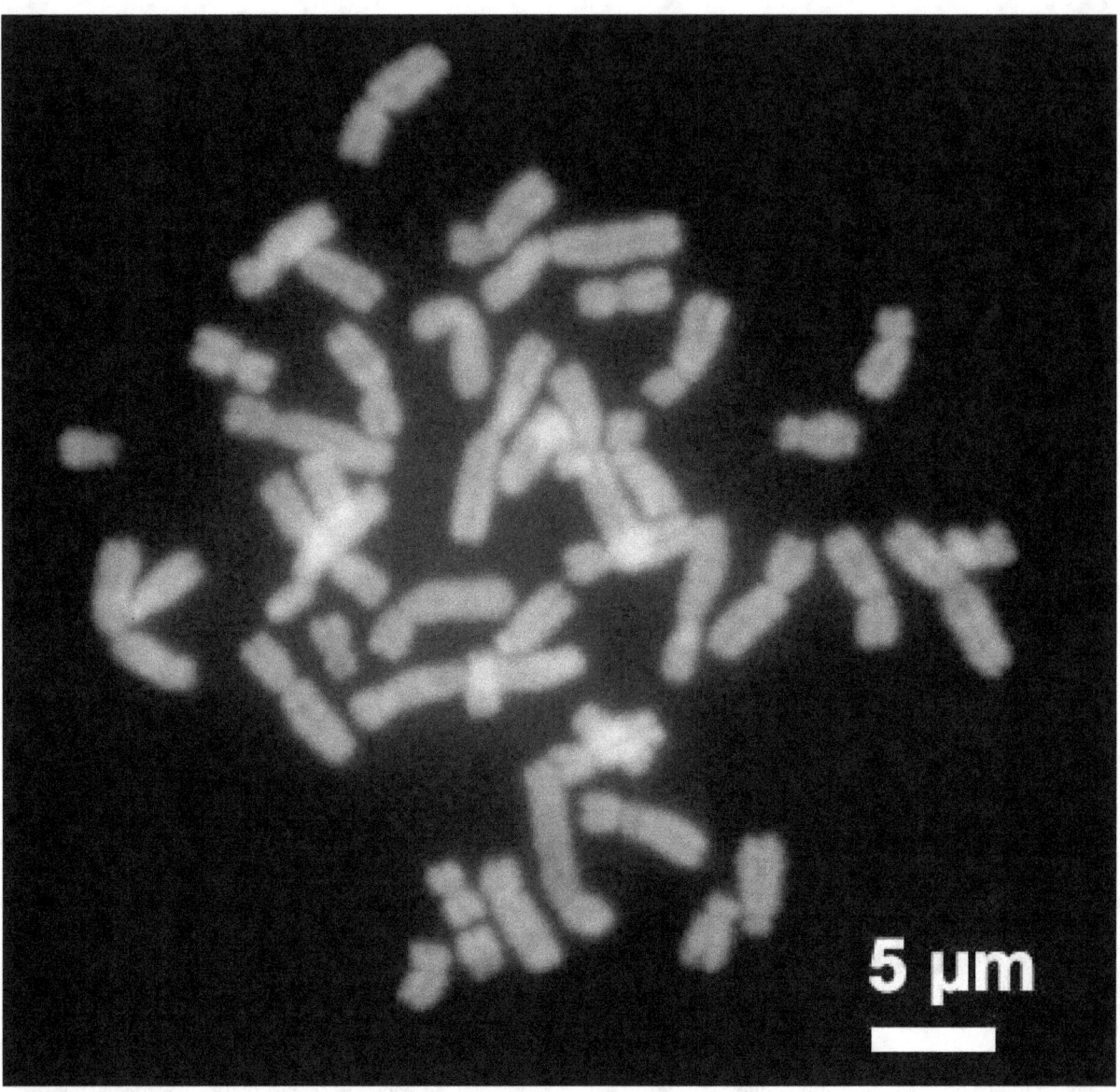

Human chromosomes during metaphase

24.4   Number of chromosomes in various organisms

Main article: List of number of chromosomes of various organisms

24.4.1   In eukaryotes

These tables give the total number of chromosomes (including sex chromosomes) in a cell nucleus. For example, human cells are diploid and have 22 different types of autosome, each present as two copies, and two sex chromosomes. This gives 46 chromosomes in total. Other organisms have more than two copies of their chromosome types, such as bread wheat, which is *hexaploid* and has six copies of seven different chromosome types – 42 chromosomes in total.

Normal members of a particular eukaryotic species all have the same number of nuclear chromosomes (see the table). Other eukaryotic chromosomes, i.e., mitochondrial and plasmid-like small chromosomes, are much more variable in

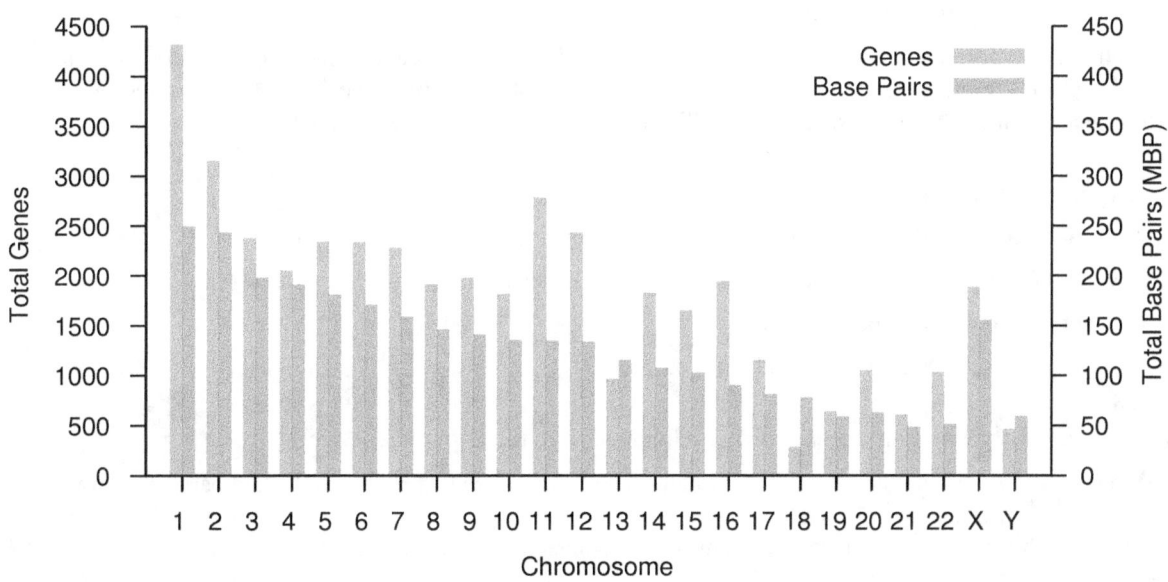

Estimated number of genes and base pairs (in mega base pairs) on each human chromosome

number, and there may be thousands of copies per cell.

Asexually reproducing species have one set of chromosomes, which are the same in all body cells. However, asexual species can be either haploid or diploid.

Sexually reproducing species have somatic cells (body cells), which are diploid [2n] having two sets of chromosomes (23 pairs in humans with one set of 23 chromosomes from each parent), one set from the mother and one from the father. Gametes, reproductive cells, are haploid [n]: They have one set of chromosomes. Gametes are produced by meiosis of a diploid germ line cell. During meiosis, the matching chromosomes of father and mother can exchange small parts of themselves (crossover), and thus create new chromosomes that are not inherited solely from either parent. When a male and a female gamete merge (fertilization), a new diploid organism is formed.

Some animal and plant species are polyploid [Xn]: They have more than two sets of homologous chromosomes. Plants important in agriculture such as tobacco or wheat are often polyploid, compared to their ancestral species. Wheat has a haploid number of seven chromosomes, still seen in some cultivars as well as the wild progenitors. The more-common pasta and bread wheats are polyploid, having 28 (tetraploid) and 42 (hexaploid) chromosomes, compared to the 14 (diploid) chromosomes in the wild wheat.[48]

24.4.2   In prokaryotes

Prokaryote species generally have one copy of each major chromosome, but most cells can easily survive with multiple copies.[49] For example, *Buchnera*, a symbiont of aphids has multiple copies of its chromosome, ranging from 10–400 copies per cell.[50] However, in some large bacteria, such as *Epulopiscium fishelsoni* up to 100,000 copies of the chromosome can be present.[51] Plasmids and plasmid-like small chromosomes are, as in eukaryotes, highly variable in copy number. The number of plasmids in the cell is almost entirely determined by the rate of division of the plasmid – fast division causes high copy number.

24.5   Karyotype

Main article: Karyotype
 In general, the **karyotype** is the characteristic chromosome complement of a eukaryote species.[52] The preparation and

study of karyotypes is part of cytogenetics.

Although the replication and transcription of DNA is highly standardized in eukaryotes, *the same cannot be said for their karyotypes*, which are often highly variable. There may be variation between species in chromosome number and in detailed organization. In some cases, there is significant variation within species. Often there is:

1. variation between the two sexes

2. variation between the germ-line and soma (between gametes and the rest of the body)

3. variation between members of a population, due to balanced genetic polymorphism

4. geographical variation between races

5. mosaics or otherwise abnormal individuals.

Also, variation in karyotype may occur during development from the fertilised egg.

The technique of determining the karyotype is usually called *karyotyping*. Cells can be locked part-way through division (in metaphase) in vitro (in a reaction vial) with colchicine. These cells are then stained, photographed, and arranged into a *karyogram*, with the set of chromosomes arranged, autosomes in order of length, and sex chromosomes (here X/Y) at the end: Fig. 3.

Like many sexually reproducing species, humans have special gonosomes (sex chromosomes, in contrast to autosomes). These are XX in females and XY in males.

24.5.1   Historical note

Investigation into the human karyotype took many years to settle the most basic question: *How many chromosomes does a normal diploid human cell contain?* In 1912, Hans von Winiwarter reported 47 chromosomes in spermatogonia and 48 in oogonia, concluding an XX/XO sex determination mechanism.[53] Painter in 1922 was not certain whether the diploid number of man is 46 or 48, at first favouring 46.[54] He revised his opinion later from 46 to 48, and he correctly insisted on humans having an XX/XY system.[55]

New techniques were needed to definitively solve the problem:

1. Using cells in culture

2. Arresting mitosis in metaphase by a solution of colchicine

3. Pretreating cells in a hypotonic solution 0.075 M KCl, which swells them and spreads the chromosomes

4. Squashing the preparation on the slide forcing the chromosomes into a single plane

5. Cutting up a photomicrograph and arranging the result into an indisputable karyogram.

It took until 1954 before the human diploid number was confirmed as 46.[56][57] Considering the techniques of Winiwarter and Painter, their results were quite remarkable.[58] Chimpanzees (the closest living relatives to modern humans) have 48 chromosomes (as well as the other great apes: in humans two chromosomes fused to form chromosome 2).

24.6   Aberrations

Main articles: Chromosome abnormality and aneuploidy

  Chromosomal aberrations are disruptions in the normal chromosomal content of a cell and are a major cause of genetic conditions in humans, such as Down syndrome, although most aberrations have little to no effect. Some chromosome abnormalities do not cause disease in carriers, such as translocations, or chromosomal inversions, although they may lead to a higher chance of bearing a child with a chromosome disorder. Abnormal numbers of chromosomes or chromosome sets, called aneuploidy, may be lethal or may give rise to genetic disorders. Genetic counseling is offered for families that may carry a chromosome rearrangement.

The gain or loss of DNA from chromosomes can lead to a variety of genetic disorders. Human examples include:

- Cri du chat, which is caused by the deletion of part of the short arm of chromosome 5. "Cri du chat" means "cry of the cat" in French; the condition was so-named because affected babies make high-pitched cries that sound like those of a cat. Affected individuals have wide-set eyes, a small head and jaw, moderate to severe mental health problems, and are very short.

- Down's syndrome, the most common trisomy, usually caused by an extra copy of chromosome 21 (trisomy 21). Characteristics include decreased muscle tone, stockier build, asymmetrical skull, slanting eyes and mild to moderate developmental disability.[59]

- Edwards syndrome, or trisomy-18, the second most common trisomy. Symptoms include motor retardation, developmental disability and numerous congenital anomalies causing serious health problems. Ninety percent of those affected die in infancy. They have characteristic clenched hands and overlapping fingers.

- Isodicentric 15, also called idic(15), partial tetrasomy 15q, or inverted duplication 15 (inv dup 15).

- Jacobsen syndrome, which is very rare. It is also called the terminal 11q deletion disorder.[60] Those affected have normal intelligence or mild developmental disability, with poor expressive language skills. Most have a bleeding disorder called Paris-Trousseau syndrome.

- Klinefelter syndrome (XXY). Men with Klinefelter syndrome are usually sterile, and tend to be taller and have longer arms and legs than their peers. Boys with the syndrome are often shy and quiet, and have a higher incidence of speech delay and dyslexia. Without testosterone treatment, some may develop gynecomastia during puberty.

- Patau Syndrome, also called D-Syndrome or trisomy-13. Symptoms are somewhat similar to those of trisomy-18, without the characteristic folded hand.

- Small supernumerary marker chromosome. This means there is an extra, abnormal chromosome. Features depend on the origin of the extra genetic material. Cat-eye syndrome and isodicentric chromosome 15 syndrome (or Idic15) are both caused by a supernumerary marker chromosome, as is Pallister-Killian syndrome.

- Triple-X syndrome (XXX). XXX girls tend to be tall and thin and have a higher incidence of dyslexia.

- Turner syndrome (X instead of XX or XY). In Turner syndrome, female sexual characteristics are present but underdeveloped. Females with Turner syndrome often have a short stature, low hairline, abnormal eye features and bone development and a "caved-in" appearance to the chest.

- XYY syndrome. XYY boys are usually taller than their siblings. Like XXY boys and XXX girls, they are more likely to have learning difficulties.

- Wolf-Hirschhorn syndrome, which is caused by partial deletion of the short arm of chromosome 4. It is characterized by growth retardation, delayed motor skills development, "Greek Helmet" facial features, and mild to profound mental health problems.

24.7 Etymology and pronunciation

The word *chromosome* (/ˈkroʊsməˌzoʊm/) uses combining forms of *chromo-* and *-some*, yielding "colored body", which describes a chromosome's appearance on microscopy.

24.8   See also

- Genetic deletion

- DNA

- For information about chromosomes in genetic algorithms, see chromosome (genetic algorithm)

- Genetic genealogy

    - Genealogical DNA test

- Lampbrush chromosome

- List of number of chromosomes of various organisms

- Locus (explains gene location nomenclature)

- Maternal influence on sex determination

- Sex-determination system

    - XY sex-determination system
        - X-chromosome
            - X-inactivation
        - Y-chromosome
            - Y-chromosomal Aaron
            - Y-chromosomal Adam

- Polytene chromosome

- Neochromosome

24.9   Notes and references

[1] Schleyden, M.J. (1847). *Microscopical researches into the accordance in the structure and growth of animals and plants.*

[2] Johnson, J; Chiu, W (1 April 2000). "Structures of virus and virus-like particles". *Current Opinion in Structural Biology* **10** (2): 229–235. doi:10.1016/S0959-440X(00)00073-7. PMID 10753814.

[3] Coxx, H. J. (1925). *Biological Stains - A Handbook on the Nature and Uses of the Dyes Employed in the Biological Laboratory.* Commission on Standardization of Biological Stains.

[4] Fokin, S.I. (2013). "Otto Bütschli (1848–1920): Where we will genuflect?" *Protistology*, 8 (1), 22–35, .

[5] Waldeyer-Hartz, "Über Karyokinese und ihre Beziehungen zu den Befruchtungsvorgängen," *Archiv für mikroskopische Anatomie und Entwicklungsmechanik*, 1888, 32: 27.

[6] Wilson, E.B. (1925). *The Cell in Development and Heredity*, Ed. 3. Macmillan, New York. p. 923.

[7] Mayr, E. (1982). *The growth of biological thought.* Harvard. p. 749.

[8] Matthews, Robert. "The bizarre case of the chromosome that never was" (PDF). Retrieved 13 July 2013.

[9] Thanbichler M; Shapiro L (2006). "Chromosome organization and segregation in bacteria". *J. Struct. Biol.* **156** (2): 292–303. doi:10.1016/j.jsb.2006.05.007. PMID 16860572.

[10] Van Leuven, JT; Meister, RC; Simon, C; McCutcheon, JP (11 September 2014). "Sympatric speciation in a bacterial endosymbiont results in two genomes with the functionality of one.". *Cell* **158** (6): 1270–80. PMID 25175626.

[11] McCutcheon, JP; von Dohlen, CD (23 August 2011). "An interdependent metabolic patchwork in the nested symbiosis of mealybugs.". *Current biology : CB* **21** (16): 1366–72. PMID 21835622.

[12] Han, K; Li, ZF; Peng, R; Zhu, LP; Zhou, T; Wang, LG; Li, SG; Zhang, XB; Hu, W; Wu, ZH; Qin, N; Li, YZ (2013). "Extraordinary expansion of a Sorangium cellulosum genome from an alkaline milieu.". *Scientific reports* **3**: 2101. PMID 23812535.

[13] Hinnebusch J; Tilly K (1993). "Linear plasmids and chromosomes in bacteria". *Mol Microbiol* **10** (5): 917–22. doi:10.1111/j.1365-2958.1993.tb00963.x. PMID 7934868.

[14] Kelman LM; Kelman Z (2004). "Multiple origins of replication in archaea". *Trends Microbiol.***12**(9): 399–401. doi:10.1016/j PMID 15337158.

[15] Thanbichler M; Wang SC; Shapiro L (2005). "The bacterial nucleoid: a highly organized and dynamic structure". *J. Cell. Biochem.* **96** (3): 506–21. doi:10.1002/jcb.20519. PMID 15988757.

[16] Sandman K; Pereira SL; Reeve JN (1998). "Diversity of prokaryotic chromosomal proteins and the origin of the nucleosome". *Cell. Mol. Life Sci.* **54** (12): 1350–64. doi:10.1007/s000180050259. PMID 9893710.

[17] Sandman K; Reeve JN (2000). "Structure and functional relationships of archaeal and eukaryal histones and nucleosomes". *Arch. Microbiol.* **173** (3): 165–9. doi:10.1007/s002039900122. PMID 10763747.

[18] Pereira SL; Grayling RA; Lurz R; Reeve JN (1997). "Archaeal nucleosomes". *Proc. Natl. Acad. Sci. U.S.A.* **94** (23): 12633–7. Bibcode:1997PNAS...9412633P. doi:10.1073/pnas.94.23.12633. PMC 25063. PMID 9356501.

[19] "Chromosome Mapping: Idiograms" *Nature Education* - August 13, 2013

[20] Vega.sanger.ad.uk, all data in this table was derived from this database, November 11, 2008.

[21] Sequenced percentages are based on fraction of euchromatin portion, as the Human Genome Project goals called for determination of only the euchromatic portion of the genome. Telomeres, centromeres, and other heterochromatic regions have been left undetermined, as have a small number of unclonable gaps. See http://www.ncbi.nlm.nih.gov/genome/seq/ for more information on the Human Genome Project.

[22] Chromosomes - Genetics Home Reference

[23] Armstrong SJ; Jones GH (January 2003). "Meiotic cytology and chromosome behaviour in wild-type Arabidopsis thaliana". *J. Exp. Bot.* **54** (380): 1–10. doi:10.1093/jxb/54.380.1. PMID 12456750.

[24] Gill BS; Kimber G (April 1974). "The Giemsa C-Banded Karyotype of Rye". *Proc. Natl. Acad. Sci. U.S.A.* **71** (4): 1247–9. Bibcode:1974PNAS...71.1247G. doi:10.1073/pnas.71.4.1247. PMC 388202. PMID 4133848.

[25] Kato A; Lamb JC; Birchler JA (September 2004). "Chromosome painting using repetitive DNA sequences as probes for somatic chromosome identification in maize". *Proc. Natl. Acad. Sci. U.S.A.* **101** (37): 13554–9. Bibcode:2004PNAS..10113554K. doi:10.1073/pnas.0403659101. PMC 518793. PMID 15342909.

[26] Dubcovsky J; Luo MC; Zhong GY; et al. (1996). "Genetic Map of Diploid Wheat, Triticum Monococcum L., and Its Comparison with Maps of Hordeum Vulgare L". *Genetics* **143** (2): 983–99. PMC 1207354. PMID 8725244.

[27] Kenton A; Parokonny AS; Gleba YY; Bennett MD (August 1993). "Characterization of the Nicotiana tabacum L. genome by molecular cytogenetics". *Mol. Gen. Genet.* **240** (2): 159–69. doi:10.1007/BF00277053. PMID 8355650.

[28] Leitch IJ; Soltis DE; Soltis PS; Bennett MD (2005). "Evolution of DNA amounts across land plants (embryophyta)". *Ann. Bot.* **95** (1): 207–17. doi:10.1093/aob/mci014. PMID 15596468.

[29] Umeko Semba; Yasuko Umeda; Yoko Shibuya; Hiroaki Okabe; Sumio Tanase & Tetsuro Yamamoto (2004). "Primary structures of guinea pig high- and low-molecular-weight kininogens". *International Immunopharmacology* **4** (10–11): 1391–1400. doi:10.1016/j.intimp.2004.06.003. PMID 15313436.

[30] "The Genetics of the Popular Aquarium Pet - Guppy Fish". Retrieved 2009-12-06.

[31] Vitturi R; Libertini A; Sineo L; et al. (2005). "Cytogenetics of the land snails Cantareus aspersus and C. mazzullii (Mollusca: Gastropoda: Pulmonata)". *Micron* **36** (4): 351–7. doi:10.1016/j.micron.2004.12.010. PMID 15857774.

[32] Vitturi R; Colomba MS; Pirrone AM; Mandrioli M (2002). "rDNA (18S-28S and 5S) colocalization and linkage between ribosomal genes and (TTAGGG)(n) telomeric sequence in the earthworm, Octodrilus complanatus (Annelida: Oligochaeta: Lumbricidae), revealed by single- and double-color FISH". *J. Hered* **93** (4): 279–82. doi:10.1093/jhered/93.4.279. PMID 12407215.

[33] Ambarish, C.N. Sridhar, K.R. (2014). "Cytological and karyological observations of two endemic pill-millipedes Arthrosphaera (Pocock, 1895) (Diplopoda: Sphaerotheriida) of the Western Ghats of India". *Caryologia* **66** (1). doi:10.1080/00087114.

[34] Nie W; Wang J; O'Brien PC; et al. (2002). "The genome phylogeny of domestic cat, red panda and five mustelid species revealed by comparative chromosome painting and G-banding". *Chromosome Res.* **10** (3): 209–22. doi:10.1023/A:1015292005631. PMID 12067210.

[35] Romanenko, Svetlana A.; Perelman, Polina L.; Serdukova, Natalya A.; Trifonov, Vladimir A.; Biltueva, Larisa S.; Wang, Jinhuan; Li, Tangliang; Nie, Wenhui; O'Brien, Patricia C.M.; Volobouev, Vitaly T.; Stanyon, Roscoe; Ferguson-Smith, Malcolm A.; Yang, Fengtang; Graphodatsky, Alexander S. (2006). "Reciprocal chromosome painting between three laboratory rodent species". *Mammalian Genome* **17** (12): 1183–92. doi:10.1007/s00335-006-0081-z. PMID 17143584.

[36] Painter, TS (1928). "A Comparison of the Chromosomes of the Rat and Mouse with Reference to the Question of Chromosome Homology in Mammals". *Genetics* **13** (2): 180–9. PMC 1200977. PMID 17246549.

[37] Hayes, H.; Rogel-Gaillard, C.; Zijlstra, C.; De Haan, N.A.; Urien, C.; Bourgeaux, N.; Bertaud, M.; Bosma, A.A. (2002). "Establishment of an R-banded rabbit karyotype nomenclature by FISH localization of 23 chromosome-specific genes on both G- and R-banded chromosomes". *Cytogenetic and Genome Research* **98** (2–3): 199–205. doi:10.1159/000069807. PMID 12698004.

[38] T.J. Robinson; F. Yang; W.R. Harrison (2002). "Chromosome painting refines the history of genome evolution in hares and rabbits (order Lagomorpha)". *Cytogenic and Genetic Research* **96** (1–4): 223–227. doi:10.1159/000063034. PMID 12438803.

[39] "section 4.W4", *Rabbits, Hares and Pikas. Status Survey and Conservation Action Plan*, pp. 61–94

[40] De Grouchy J (1987). "Chromosome phylogenies of man, great apes, and Old World monkeys". *Genetica* **73** (1–2): 37–52. doi:10.1007/bf00057436. PMID 3333352.

[41] Houck, M.L.; Kumamoto, A.T.; Gallagher, D.S.; Benirschke, K. (2001). "Comparative cytogenetics of the African elephant *(Loxodonta africana)* and Asiatic elephant *(Elephas maximus)*". *Cytogenetic and Genome Research* **93** (3–4): 249–52. doi:10.1159/000056992. PMID 11528120.

[42] Wayne RK; Ostrander EA (1999). "Origin, genetic diversity, and genome structure of the domestic dog". *BioEssays* **21** (3): 247–57. doi:10.1002/(SICI)1521-1878(199903)21:3<247::AID-BIES9>3.0.CO;2-Z. PMID 10333734.

[43] Burt DW (2002). "Origin and evolution of avian microchromosomes". *Cytogenet. Genome Res.* **96**(1–4): 97–112. doi:10.1159/ PMID 12438785.

[44] Ciudad J; Cid E; Velasco A; Lara JM; Aijón J; Orfao A (2002). "Flow cytometry measurement of the DNA contents of G0/G1 diploid cells from three different teleost fish species". *Cytometry* **48** (1): 20–5. doi:10.1002/cyto.10100. PMID 12116377.

[45] Yasukochi Y; Ashakumary LA; Baba K; Yoshido A; Sahara K (2006). "A Second-Generation Integrated Map of the Silkworm Reveals Synteny and Conserved Gene Order Between Lepidopteran Insects". *Genetics* **173** (3): 1319–28. doi:10.1534/genetics.106.055541.PMC1526672. PMID 16547103.

[46] Itoh, Masahiro; Ikeuchi, Tatsuro; Shimba, Hachiro; Mori, Michiko; Sasaki, Motomichi; Makino, Sajiro (1969). "A Comparative Karyotype Study in Fourteen Species of Birds". *The Japanese journal of genetics* **44** (3): 163–170. doi:10.1266/jjg.44.163.

[47] Smith J; Burt DW (1998). "Parameters of the chicken genome (Gallus gallus)". *Anim. Genet.* **29** (4): 290–4. doi:10.1046/j.1365-2052.1998.00334.x. PMID 9745667.

[48] Sakamura, Tetsu (1918). "Kurze Mitteilung über die Chromosomenzahlen und die Verwandtschaftsverhältnisse der Triticum-Arten". *Shokubutsugaku Zasshi* **32** (379): 150–3. doi:10.15281/jplantres1887.32.379_150.

[49] Charlebois R.L. (ed) 1999. *Organization of the prokaryote genome*. ASM Press, Washington DC.

[50] Komaki K; Ishikawa H (March 2000). "Genomic copy number of intracellular bacterial symbionts of aphids varies in response to developmental stage and morph of their host". *Insect Biochem. Mol. Biol.* **30** (3): 253–8. doi:10.1016/S0965-1748(99)00125-3. PMID 10732993.

[51] Mendell JE; Clements KD; Choat JH; Angert ER (May 2008). "Extreme polyploidy in a large bacterium". *Proc. Natl. Acad. Sci. U.S.A.* **105** (18): 6730–4. Bibcode:2008PNAS..105.6730M. doi:10.1073/pnas.0707522105. PMC 2373351. PMID 18445653.

[52] White, M. J. D. (1973). *The chromosomes* (6th ed.). London: Chapman and Hall, distributed by Halsted Press, New York. p. 28. ISBN 0-412-11930-7.

[53] von Winiwarter H (1912). "Études sur la spermatogenese humaine". *Arch. Biologie* **27** (93): 147–9.

[54] Painter TS (1922). "The spermatogenesis of man". *Anat. Res.* **23**: 129.

[55] Painter TS (1923). "Studies in mammalian spermatogenesis II. The spermatogenesis of man". *J. Exp. Zoology* **37** (3): 291–336. doi:10.1002/jez.1400370303.

[56] Tjio JH; Levan A (1956). "The chromosome number of man". *Hereditas* **42** (1-2): 1–6. doi:10.1111/j.1601-5223.1956.tb03010.x.

[57] Ford C.E; Hamerton J.L (1956). "The Chromosomes of Man". *Nature* **178** (4541): 1020–1023. Bibcode:1956Natur.178.1020F. doi:10.1038/1781020a0. PMID 13378517.

[58] Hsu T.C. *Human and mammalian cytogenetics: a historical perspective.* Springer-Verlag, N.Y. p10: "It's amazing that he [Painter] even came close!"

[59] Miller, Kenneth R. (2000). "Chapter 9-3". *Biology* (5th ed.). Upper Saddle River, New Jersey: Prentice Hall. pp. 194–5. ISBN 0-13-436265-9.

[60] European Chromosome 11 Network

24.10 External links

- An Introduction to DNA and Chromosomes from HOPES: Huntington's Outreach Project for Education at Stanford
- Chromosome Abnormalities at AtlasGeneticsOncology
- On-line exhibition on chromosomes and genome (SIB)
- What Can Our Chromosomes Tell Us?, from the University of Utah's Genetic Science Learning Center
- Try making a karyotype yourself, from the University of Utah's Genetic Science Learning Center
- Kimballs Chromosome pages
- Chromosome News from Genome News Network
- Eurochromnet, European network for Rare Chromosome Disorders on the Internet
- Ensembl.org, Ensembl project, presenting chromosomes, their genes and syntenic loci graphically via the web
- Genographic Project
- Home reference on Chromosomes from the U.S. National Library of Medicine
- Visualisation of human chromosomes and comparison to other species
- Unique - The Rare Chromosome Disorder Support Group Support for people with rare chromosome disorders

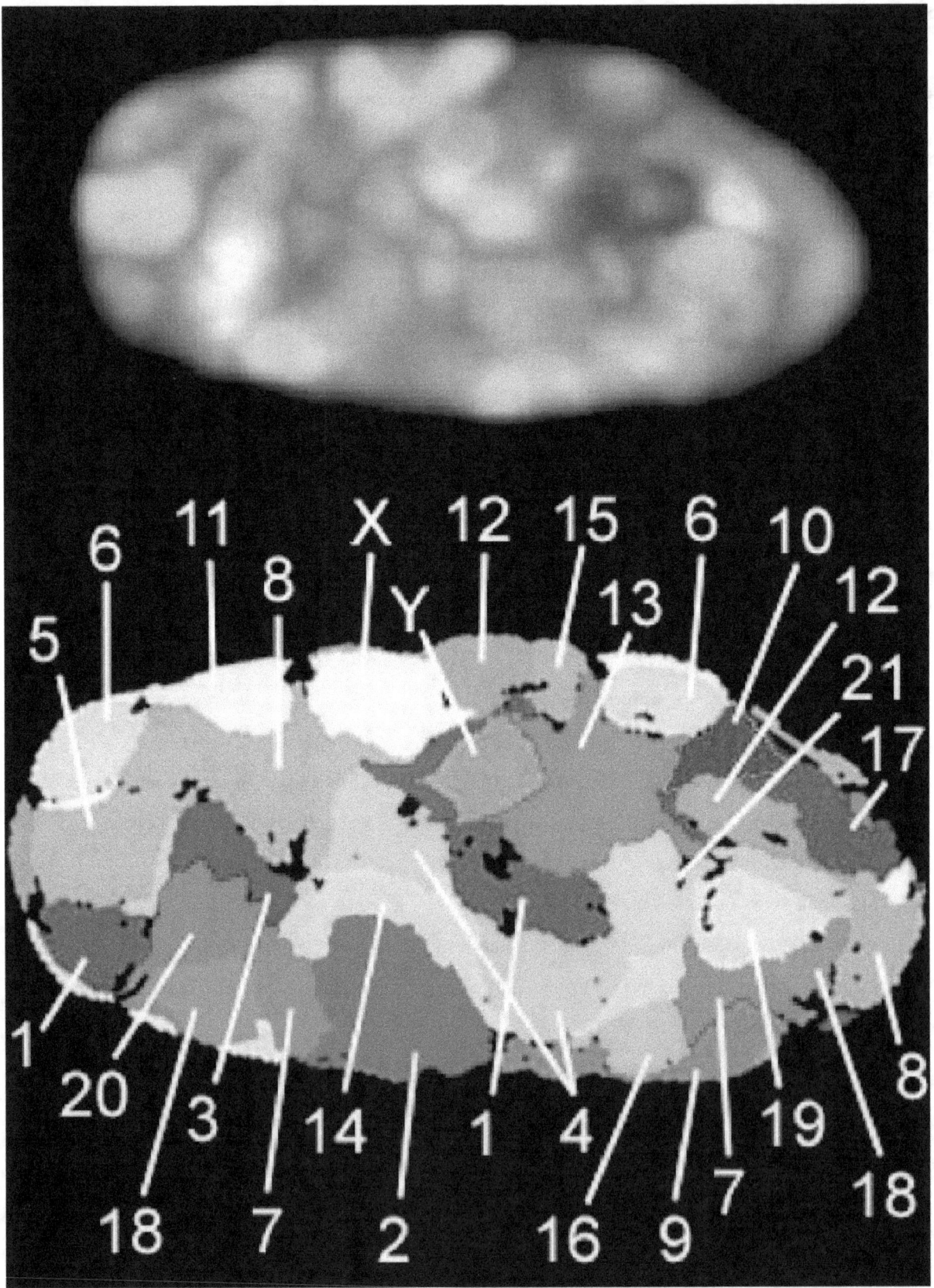

The 23 human chromosome territories during prometaphase in fibroblast cells.

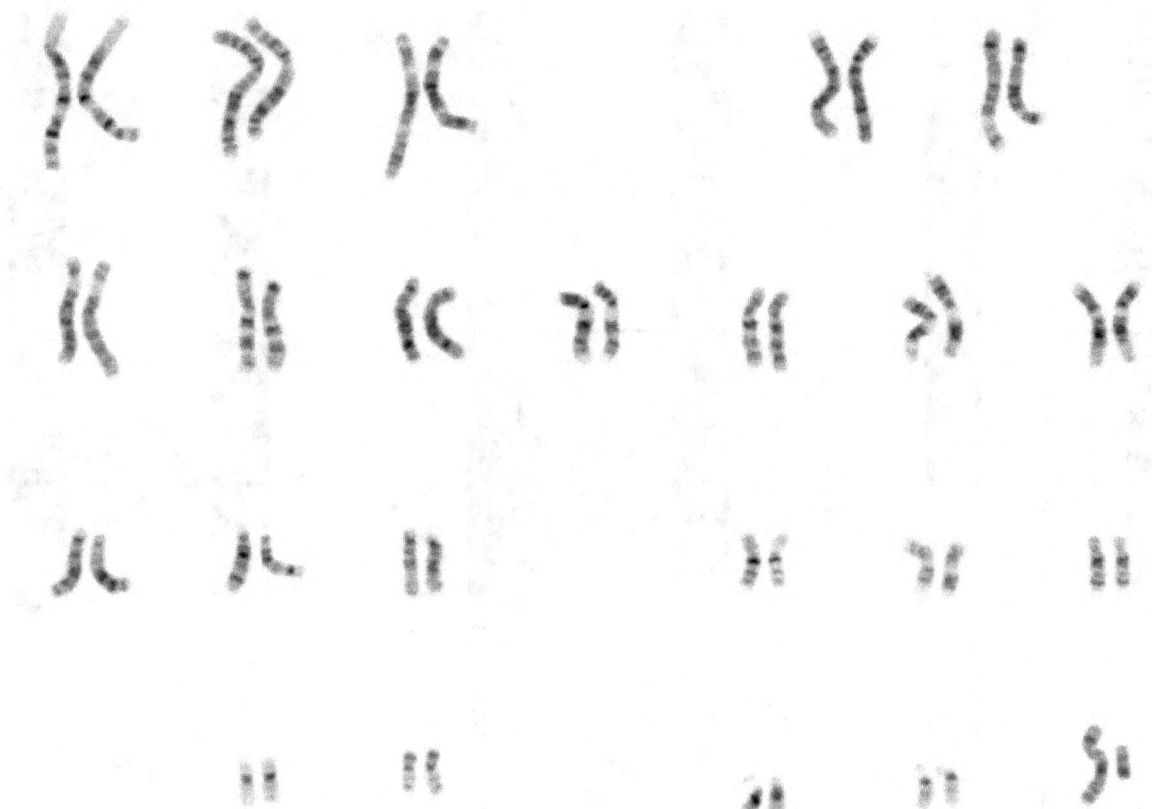

Figure 3: *Karyogram of a human male*

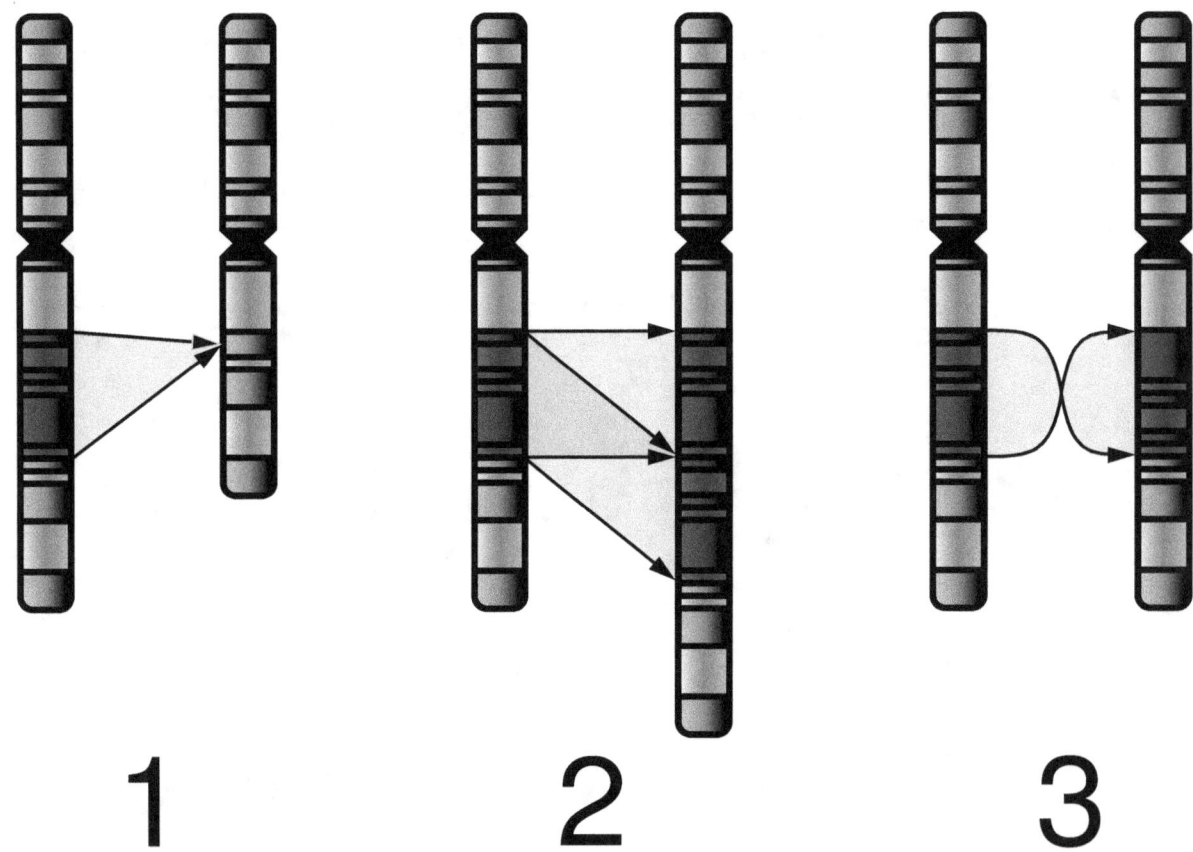

The three major single chromosome mutations; deletion (1), duplication (2) and inversion (3).

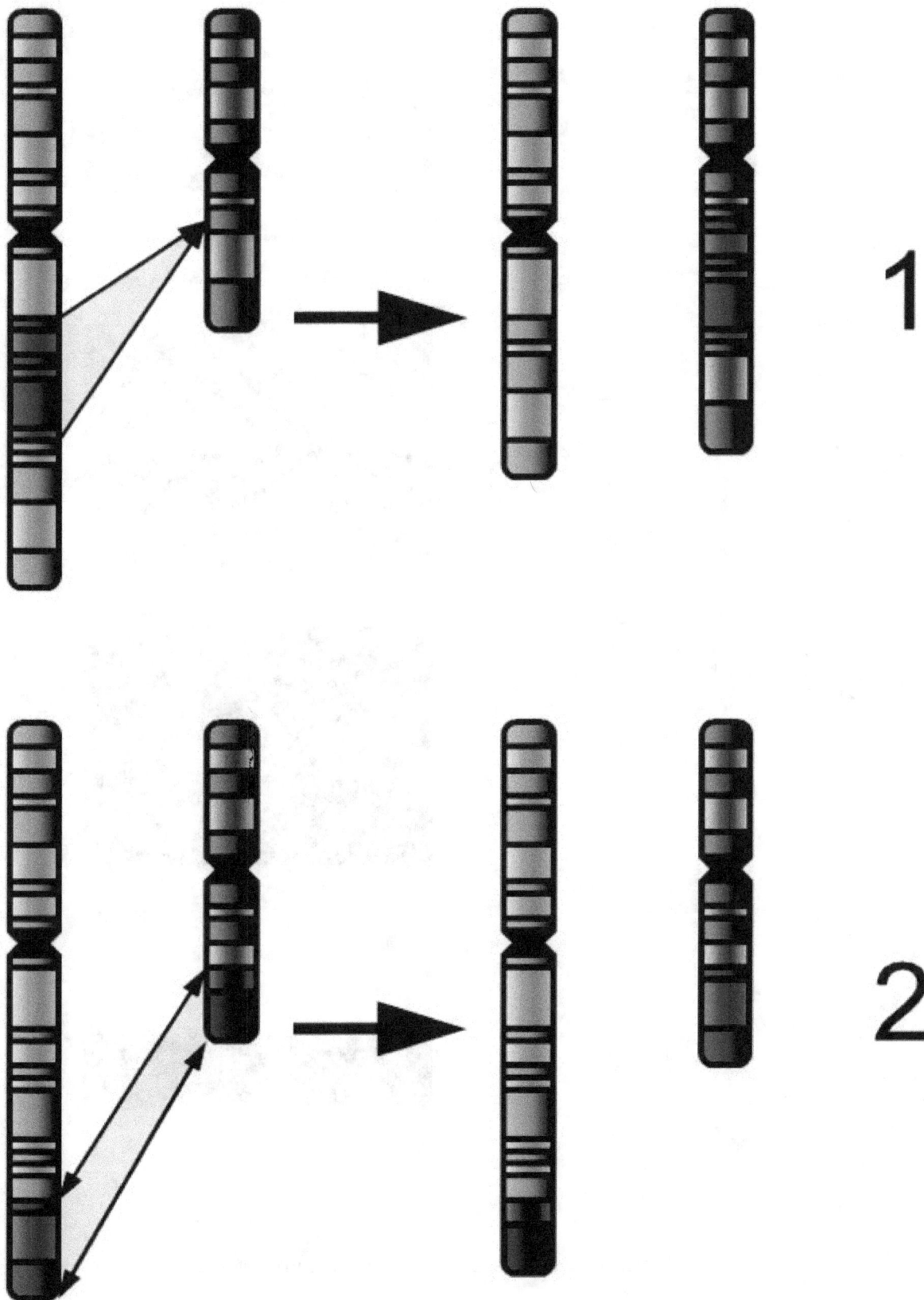

The two major two-chromosome mutations; insertion (1) and translocation (2).

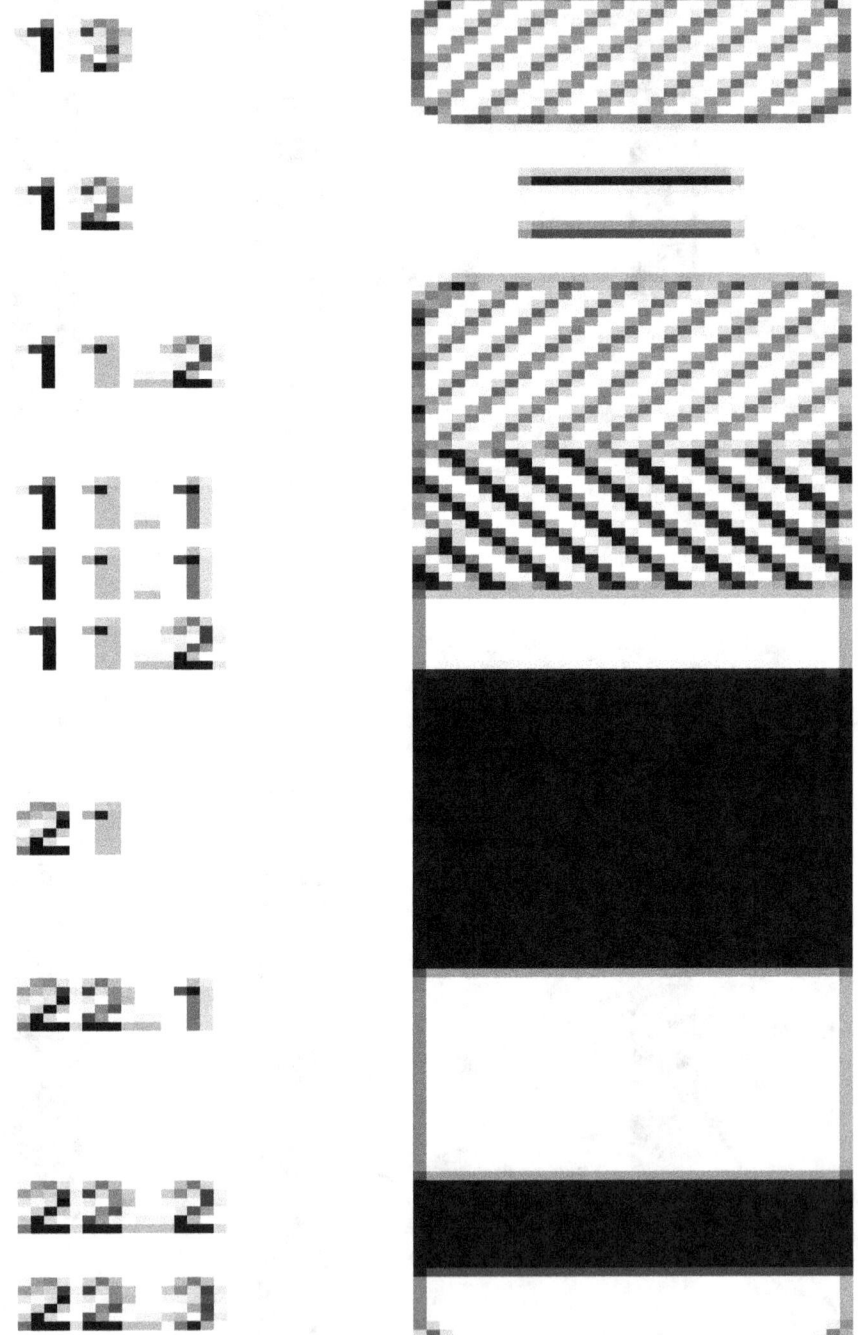

In Down syndrome, there are three copies of chromosome 21

Chapter 25

Gene

This article is about the heritable unit for transmission of biological traits. For the name, see Eugene.

A **gene** is a locus (or region) of DNA that encodes a functional RNA or protein product, and is the molecular unit of heredity.[1][2]:Glossary The transmission of genes to an organism's offspring is the basis of the inheritance of phenotypic traits. Most biological traits are under the influence of polygenes (many different genes) as well as the gene–environment interactions. Some genetic traits are instantly visible, such as eye colour or number of limbs, and some are not, such as blood type, risk for specific diseases, or the thousands of basic biochemical processes that comprise life.

Genes can acquire mutations in their sequence, leading to different variants, known as alleles, in the population. These alleles encode slightly different versions of a protein, which cause different phenotype traits. Colloquial usage of the term "having a gene" (e.g., "good genes," "hair colour gene") typically refers to having a different allele of the gene. Genes evolve due to natural selection or survival of the fittest of the alleles.

The concept of a gene continues to be refined as new phenomena are discovered.[3] For example, regulatory regions of a gene can be far removed from its coding regions, and coding regions can be split into several exons. Some viruses store their genome in RNA instead of DNA and some gene products are functional non-coding RNAs. Therefore, a broad, modern working definition of a gene is any discrete locus of heritable, genomic sequence which affect an organism's traits by being expressed as a functional product or by regulation of gene expression.[4][5]

25.1 History

Main article: History of genetics

25.1.1 Discovery of discrete inherited units

The existence of discrete inheritable units was first suggested by Gregor Mendel (1822–1884).[6] From 1857 to 1864, he studied inheritance patterns in 8000 common edible pea plants, tracking distinct traits from parent to offspring. He described these mathematically as 2^n combinations where n is the number of differing characteristics in the original peas. Although he did not use the term *gene*, he explained his results in terms of discrete inherited units that give rise to observable physical characteristics. This description prefigured the distinction between genotype (the genetic material of an organism) and phenotype (the visible traits of that organism). Mendel was also the first to demonstrate independent assortment, the distinction between dominant and recessive traits, the distinction between a heterozygote and homozygote, and the phenomenon of discontinuous inheritance.

Prior to Mendel's work, the dominant theory of heredity was one of blending inheritance, which suggested that each parent contributed fluids to the fertilisation process and that the traits of the parents blended and mixed to produce the

Gregor Mendel

offspring. Charles Darwin developed a theory of inheritance he termed pangenesis,[7] which used the term *gemmule* to describe hypothetical particles that would mix during reproduction. Although Mendel's work was largely unrecognized after its first publication in 1866, it was 'rediscovered' in 1900 by three European scientists, Hugo de Vries, Carl Correns, and Erich von Tschermak, who claimed to have reached similar conclusions in their own research.[8]

The word *gene* is derived (via *pangene*) from the Ancient Greek word γένος (*génos*) meaning "race, offspring".[9] *Gene* was coined in 1909 by Danish botanist Wilhelm Johannsen to describe the fundamental physical and functional unit of heredity,[10] while the related word *genetics* was first used by William Bateson in 1905.[11]

25.1.2 Discovery of DNA

Advances in understanding genes and inheritance continued throughout the 20th century. Deoxyribonucleic acid (DNA) was shown to be the molecular repository of genetic information by experiments in the 1940s to 1950s.[12][13] The structure of DNA was studied by Rosalind Franklin using X-ray crystallography, which led James D. Watson and Francis Crick to publish a model of the double-stranded DNA molecule whose paired nucleotide bases indicated a compelling hypothesis for the mechanism of genetic replication.[14][15] Collectively, this body of research established the central dogma of molecular biology, which states that proteins are translated from RNA, which is transcribed from DNA. This dogma has since been shown to have exceptions, such as reverse transcription in retroviruses. The modern study of genetics at the level of DNA is known as molecular genetics.

In 1972, Walter Fiers and his team at the University of Ghent were the first to determine the sequence of a gene: the gene for Bacteriophage MS2 coat protein.[16] The subsequent development of chain-termination DNA sequencing in 1977 by Frederick Sanger improved the efficiency of sequencing and turned it into a routine laboratory tool.[17] An automated version of the Sanger method was used in early phases of the Human Genome Project.[18]

25.1.3 Modern evolutionary synthesis

Main article: Modern evolutionary synthesis

The theories developed in the 1930s and 1940s to integrate molecular genetics with Darwinian evolution are called the modern evolutionary synthesis, a term introduced by Julian Huxley.[19] Evolutionary biologists subsequently refined this concept, such as George C. Williams' gene-centric view of evolution. He proposed an evolutionary concept of the gene as a unit of natural selection with the definition: "that which segregates and recombines with appreciable frequency."[20]:24 In this view, the molecular gene *transcribes* as a unit, and the evolutionary gene *inherits* as a unit. Related ideas emphasizing the centrality of genes in evolution were popularized by Richard Dawkins.[21][22]

25.2 Molecular basis

Main article: DNA

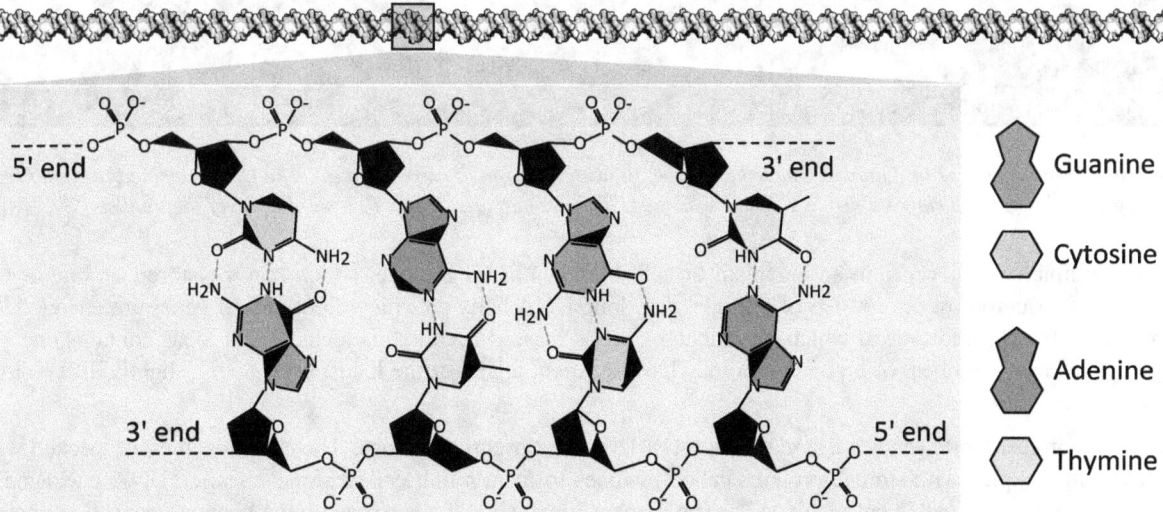

The chemical structure of a four base pair fragment of a DNA double helix. The sugar-phosphate backbone chains run in opposite directions with the bases pointing inwards, base-pairing A to T and C to G with hydrogen bonds.

25.2.1 DNA

The vast majority of living organisms encode their genes in long strands of DNA (deoxyribonucleic acid). DNA consists of a chain made from four types of nucleotide subunits, each composed of: a five-carbon sugar (2'-deoxyribose), a phosphate group, and one of the four bases adenine, cytosine, guanine, and thymine.[2]:2.1

Two chains of DNA twist around each other to form a DNA double helix with the phosphate-sugar backbone spiralling around the outside, and the bases pointing inwards with adenine base pairing to thymine and guanine to cytosine. The specificity of base pairing occurs because adenine and thymine align form two hydrogen bonds, whereas cytosine and guanine form three hydrogen bonds. The two strands in a double helix must therefore be complementary, with their sequence of bases matching such that the adenines of one strand are paired with the thymines of the other strand, and so on.[2]:4.1

Due to the chemical composition of the pentose residues of the bases, DNA strands have directionality. One end of a DNA polymer contains an exposed hydroxyl group on the deoxyribose; this is known as the 3' end of the molecule. The other end contains an exposed phosphate group; this is the 5' end. The two strands of a double-helix run in opposite directions. Nucleic acid synthesis, including DNA replication and transcription occurs in the 5'→3' direction, because new nucleotides are added via a dehydration reaction that uses the exposed 3' hydroxyl as a nucleophile.[23]:27.2

The expression of genes encoded in DNA begins by transcribing the gene into RNA, a second type of nucleic acid that is very similar to DNA, but whose monomers contain the sugar ribose rather than deoxyribose. RNA also contains the base uracil in place of thymine. RNA molecules are less stable than DNA and are typically single-stranded. Genes that encode proteins are composed of a series of three-nucleotide sequences called codons, which serve as the "words" in the genetic "language". The genetic code specifies the correspondence during protein translation between codons and amino acids. The genetic code is nearly the same for all known organisms.[2]:4.1

25.2.2 Chromosomes

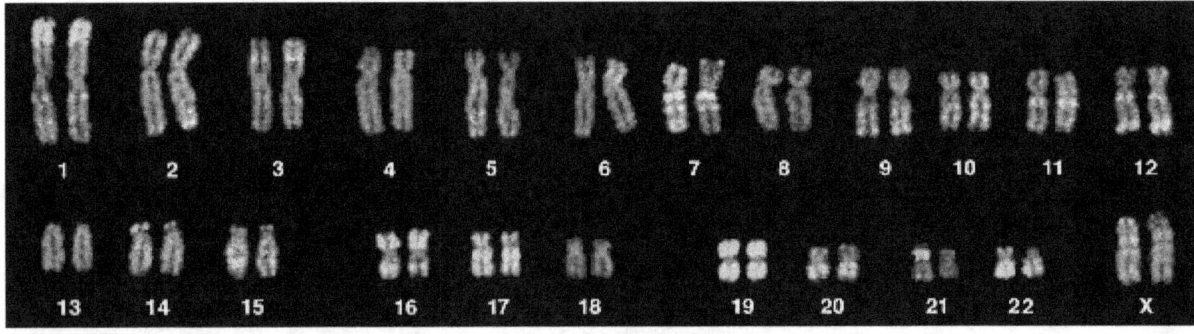

Fluorescent microscopy image of a human female karyotype, showing 23 pairs of chromosomes . The DNA is stained red, with regions rich in housekeeping genes further stained in green. The largest chromosomes are around 10 times the size of the smallest.[24]

The total complement of genes in an organism or cell is known as its genome, which may be stored on one or more chromosomes. A chromosome consists of a single, very long DNA helix on which thousands of genes are encoded.[2]:4.2 The region of the chromosome at which a particular gene is located is called its locus. Each locus contains one allele of a gene; however, members of a population may have different alleles at the locus, each with a slightly different gene sequence.

The majority of eukaryotic genes are stored on a set of large, linear chromosomes. The chromosomes are packed within the nucleus in complex with storage proteins called histones to form a unit called a nucleosome. DNA packaged and condensed in this way is called chromatin.[2]:4.2 The manner in which DNA is stored on the histones, as well as chemical modifications of the histone itself, regulate whether a particular region of DNA is accessible for gene expression. In addition to genes, eukaryotic chromosomes contain sequences involved in ensuring that the DNA is copied without degradation of end regions and sorted into daughter cells during cell division: replication origins, telomeres and the centromere.[2]:4.2 Replication origins are the sequence regions where DNA replication is initiated to make two copies of the chromosome.

Telomeres are long stretches of repetitive sequence that cap the ends of the linear chromosomes and prevent degradation of coding and regulatory regions during DNA replication. The length of the telomeres decreases each time the genome is replicated and has been implicated in the aging process.[25] The centromere is required for binding spindle fibres to separate sister chromatids into daughter cells during cell division.[2]:18.2

Prokaryotes (bacteria and archaea) typically store their genomes on a single large, circular chromosome. Similarly, some eukaryotic organelles contain a remnant circular chromosome with a small number of genes.[2]:14.4 Prokaryotes sometimes supplement their chromosome with additional small circles of DNA called plasmids, which usually encode only a few genes and are transferable between individuals. For example, the genes for antibiotic resistance are usually encoded on bacterial plasmids and can be passed between individual cells, even those of different species, via horizontal gene transfer.[26]

Whereas the chromosomes of prokaryotes are relatively gene-dense, those of eukaryotes often contain regions of DNA that serve no obvious function. Simple single-celled eukaryotes have relatively small amounts of such DNA, whereas the genomes of complex multicellular organisms, including humans, contain an absolute majority of DNA without an identified function.[27] This DNA has often been referred to as "junk DNA". However, more recent analyses suggest that, although protein-coding DNA makes up barely 2% of the human genome, about 80% of the bases in the genome may be expressed, so the term "junk DNA" may be a misnomer.[5]

25.3   Structure and function

The structure of a gene consists of many elements of which the actual protein coding sequence is often only a small part. These include DNA regions that are not transcribed as well as untranslated regions of the RNA.

Firstly, flanking the open reading frame, all genes contain a regulatory sequence that is required for their expression. In order to be expressed, genes require a promoter sequence. The promoter is recognized and bound by transcription factors and RNA polymerase to initiate transcription.[2]:7.1 A gene can have more than one promoter, resulting in messenger RNAs (mRNA) that differ in how far they extend in the 5' end.[28] Promoter regions have a consensus sequence, however highly transcribed genes have "strong" promoter sequences that bind the transcription machinery well, whereas others have "weak" promoters that bind poorly and initiate transcription less frequently.[2]:7.2 Eukaryotic promoter regions are much more complex and difficult to identify than prokaryotic promoters.[2]:7.3

Additionally, genes can have regulatory regions many kilobases upstream or downstream of the open reading frame. These act by binding to transcription factors which then cause the DNA to loop so that the regulatory sequence (and bound transcription factor) become close to the RNA polymerase binding site.[29] For example, enhancers increase transcription by binding an activator protein which then helps to recruit the RNA polymerase to the promoter; conversely silencers bind repressor proteins and make the DNA less available for RNA polymerase.[30]

The transcribed pre-mRNA contains untranslated regions at both ends which contain a ribosome binding site, terminator and start and stop codons.[31] In addition, most eukaryotic open reading frames contain untranslated introns which are removed before the exons are translated. The sequences at the ends of the introns, dictate the splice sites to generate the final mature mRNA which encodes the protein or RNA product.[32]

Many prokaryotic genes are organized into operons, with multiple protein-coding sequences that are transcribed as a unit.[33][34] The products of operon genes typically have related functions and are involved in the same regulatory network.[2]:7.3

25.3.1   Functional definitions

Defining exactly what section of a DNA sequence comprises a gene is difficult.[3] Regulatory regions of a gene such as enhancers do not necessarily have to be close to the coding sequence on the linear molecule because the intervening DNA can be looped out to bring the gene and its regulatory region into proximity. Similarly, a gene's introns can be much larger than its exons. Regulatory regions can even be on entirely different chromosomes and operate *in trans* to allow regulatory regions on one chromosome to come in contact with target genes on another chromosome.[35][36]

Early work in molecular genetics suggested the model that one gene makes one protein. This model has been refined since the discovery of genes that can encode multiple proteins by alternative splicing and coding sequences split in short section across the genome whose mRNAs are concatenated by trans-splicing.[5][37][38]

A broad operational definition is sometimes used to encompass the complexity of these diverse phenomena, where a gene is defined as a union of genomic sequences encoding a coherent set of potentially overlapping functional products.[11] This definition categorizes genes by their functional products (proteins or RNA) rather than their specific DNA loci, with regulatory elements classified as *gene-associated* regions.[11]

25.4   Gene expression

Main article: Gene expression

In all organisms, two steps are required to read the information encoded in a gene's DNA and produce the protein it specifies. First, the gene's DNA is *transcribed* to messenger RNA (mRNA).[2]:6.1 Second, that mRNA is *translated* to protein.[2]:6.2 RNA-coding genes must still go through the first step, but are not translated into protein.[39] The process of producing a biologically functional molecule of either RNA or protein is called gene expression, and the resulting molecule is called a gene product.

25.4.1   Genetic code

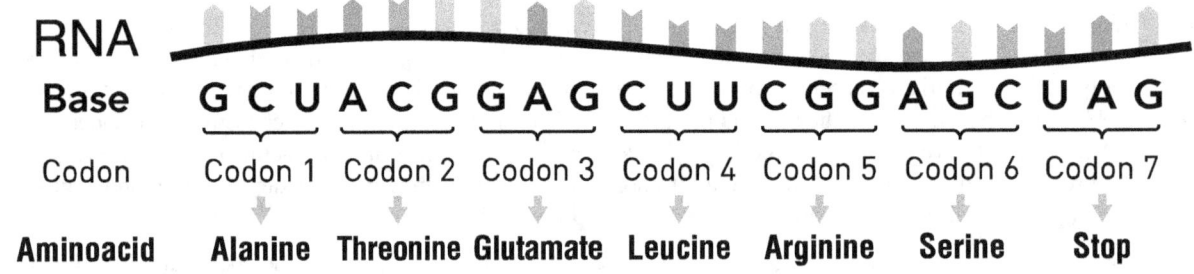

Schematic of a single-stranded RNA molecule illustrating a series of three-base codons. Each three-nucleotide codon corresponds to an amino acid when translated to protein

The nucleotide sequence of a gene's DNA specifies the amino acid sequence of a protein through the genetic code. Sets of three nucleotides, known as codons, each correspond to a specific amino acid.[2]:6 Additionally, a "start codon", and three "stop codons" indicate the beginning and end of the protein coding region. There are 64 possible codons (four possible nucleotides at each of three positions, hence 4^3 possible codons) and only 20 standard amino acids; hence the code is redundant and multiple codons can specify the same amino acid. The correspondence between codons and amino acids is nearly universal among all known living organisms.[40]

25.4.2   Transcription

Transcription produces a single-stranded RNA molecule known as messenger RNA, whose nucleotide sequence is complementary to the DNA from which it was transcribed.[2]:6.1 The mRNA acts as an intermediate between the DNA gene and its final protein product. The gene's DNA is used as a template to generate a complementary mRNA. The mRNA matches the sequence of the gene's DNA coding strand because it is synthesised as the complement of the template strand. Transcription is performed by an enzyme called an RNA polymerase, which reads the template strand in the 3' to 5' direction and synthesizes the RNA from 5' to 3'. To initiate transcription, the polymerase first recognizes and binds a promoter region of the gene. Thus, a major mechanism of gene regulation is the blocking or sequestering the promoter region, either by tight binding by repressor molecules that physically block the polymerase, or by organizing the DNA so that the promoter region is not accessible.[2]:7

In prokaryotes, transcription occurs in the cytoplasm; for very long transcripts, translation may begin at the 5' end of the RNA while the 3' end is still being transcribed. In eukaryotes, transcription occurs in the nucleus, where the cell's DNA is stored. The RNA molecule produced by the polymerase is known as the primary transcript and undergoes post-transcriptional modifications before being exported to the cytoplasm for translation. One of the modifications performed is the splicing of introns which are sequences in the transcribed region that do not encode protein. Alternative splicing mechanisms can result in mature transcripts from the same gene having different sequences and thus coding for different proteins. This is a major form of regulation in eukaryotic cells and also occurs in some prokaryotes.[2]:7.5[41]

25.4.3   Translation

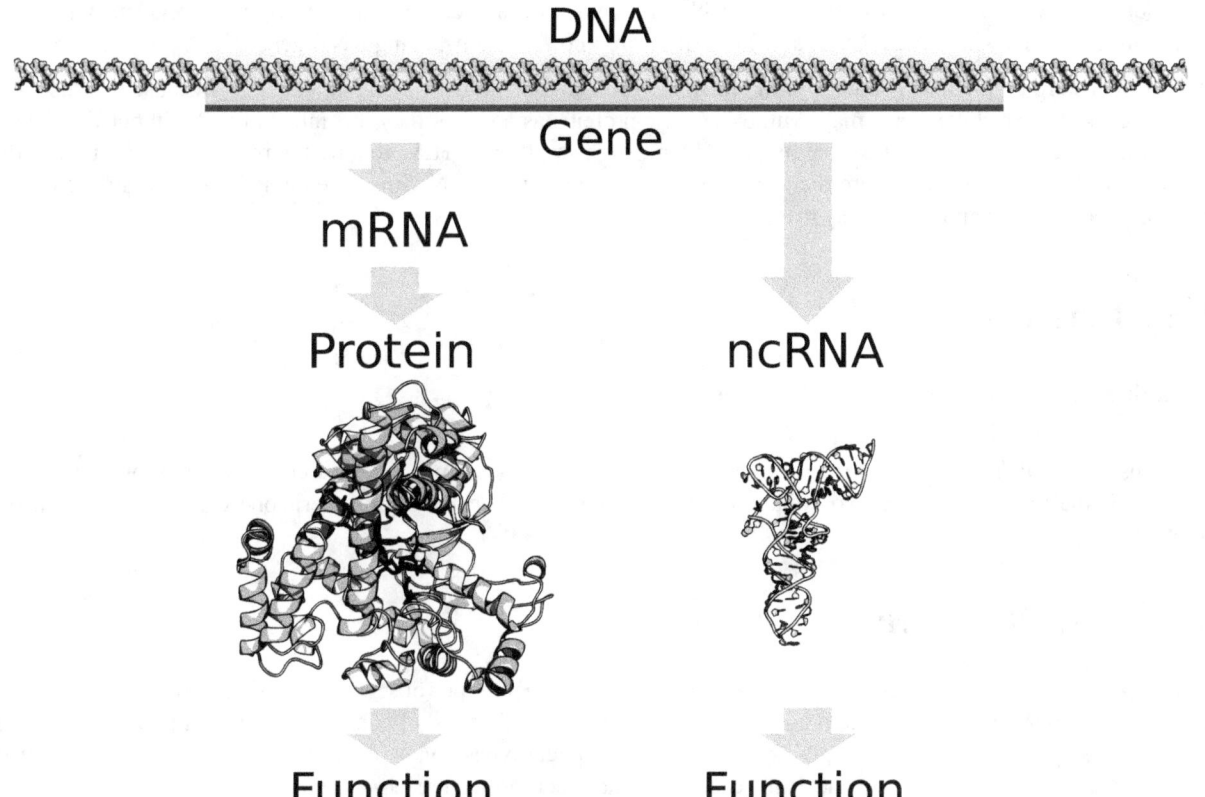

Protein coding genes are transcribed to an mRNA intermediate, then translated to a functional protein. RNA-coding genes are transcribed to a functional non-coding RNA. (PDB: 3BSE, 1OBB, 3TRA)

Translation is the process by which a mature mRNA molecule is used as a template for synthesizing a new protein.[2]:6.2 Translation is carried out by ribosomes, large complexes of RNA and protein responsible for carrying out the chemical reactions to add new amino acids to a growing polypeptide chain by the formation of peptide bonds. The genetic code is read three nucleotides at a time, in units called codons, via interactions with specialized RNA molecules called transfer RNA (tRNA). Each tRNA has three unpaired bases known as the anticodon that are complementary to the codon it reads on the mRNA. The tRNA is also covalently attached to the amino acid specified by the complementary codon. When the tRNA binds to its complementary codon in an mRNA strand, the ribosome attaches its amino acid cargo to the new polypeptide chain, which is synthesized from amino terminus to carboxyl terminus. During and after synthesis, most new proteins must folds to their active three-dimensional structure before they can carry out their cellular functions.[2]:3

25.4.4   Regulation

Genes are regulated so that they are expressed only when the product is needed, since expression draws on limited resources.[2]:7 A cell regulates its gene expression depending on its external environment (e.g. available nutrients, temperature and other stresses), its internal environment (e.g. cell division cycle, metabolism, infection status), and its specific role if in a multicellular organism. Gene expression can be regulated at any step: from transcriptional initiation, to RNA processing, to post-translational modification of the protein. The regulation of lactose metabolism genes in *E. coli* (*lac* operon) was the first such mechanism to be described in 1961.[42]

25.4.5   RNA genes

A typical protein-coding gene is first copied into RNA as an intermediate in the manufacture of the final protein product.[2]:6.1 In other cases, the RNA molecules are the actual functional products, as in the synthesis of ribosomal RNA and transfer RNA. Some RNAs known as ribozymes are capable of enzymatic function, and microRNA has a regulatory role. The DNA sequences from which such RNAs are transcribed are known as non-coding RNA genes.[39]

Some viruses store their entire genomes in the form of RNA, and contain no DNA at all.[43][44] Because they use RNA to store genes, their cellular hosts may synthesize their proteins as soon as they are infected and without the delay in waiting for transcription.[45] On the other hand, RNA retroviruses, such as HIV, require the reverse transcription of their genome from RNA into DNA before their proteins can be synthesized. RNA-mediated epigenetic inheritance has also been observed in plants and very rarely in animals.[46]

25.5   Inheritance

Main articles: Mendelian inheritance and Heredity

Organisms inherit their genes from their parents. Asexual organisms simply inherit a complete copy of their parent's genome. Sexual organisms have two copies of each chromosome because they inherit one complete set from each parent.[2]:1

25.5.1   Mendelian inheritance

According to Mendelian inheritance, variations in an organism's phenotype (observable physical and behavioral characteristics) are due in part to variations in its genotype (particular set of genes). Each gene specifies a particular trait with different sequence of a gene (alleles) giving rise to different phenotypes. Most eukaryotic organisms (such as the pea plants Mendel worked on) have two alleles for each trait, one inherited from each parent.[2]:20

Alleles at a locus may be dominant or recessive; dominant alleles give rise to their corresponding phenotypes when paired with any other allele for the same trait, whereas recessive alleles give rise to their corresponding phenotype only when paired with another copy of the same allele. For example, if the allele specifying tall stems in pea plants is dominant over the allele specifying short stems, then pea plants that inherit one tall allele from one parent and one short allele from the other parent will also have tall stems. Mendel's work demonstrated that alleles assort independently in the production of gametes, or germ cells, ensuring variation in the next generation. Although Mendelian inheritance remains a good model for many traits determined by single genes (including a number of well-known genetic disorders) it does not include the physical processes of DNA replication and cell division.[47][48]

25.5.2   DNA replication and cell division

The growth, development, and reproduction of organisms relies on cell division, or the process by which a single cell divides into two usually identical daughter cells. This requires first making a duplicate copy of every gene in the genome in a process called DNA replication.[2]:5.2 The copies are made by specialized enzymes known as DNA polymerases, which "read" one strand of the double-helical DNA, known as the template strand, and synthesize a new complementary strand. Because the DNA double helix is held together by base pairing, the sequence of one strand completely specifies the sequence of its complement; hence only one strand needs to be read by the enzyme to produce a faithful copy. The

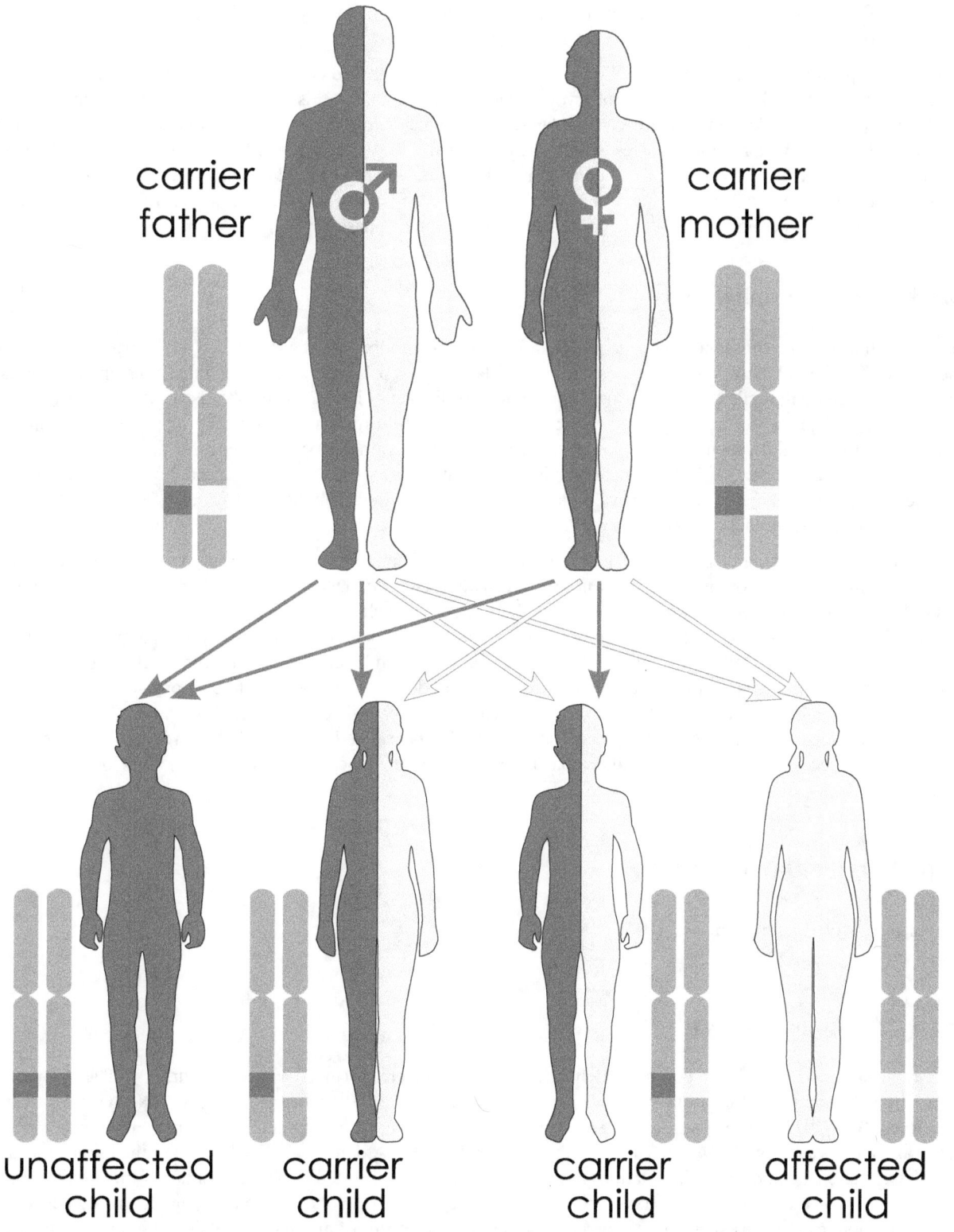

Inheritance of a gene that has two different alleles (blue and white). The gene is located on an autosomal chromosome. The blue allele is recessive to the white allele. The probability of each outcome in the children's generation is one quarter, or 25 percent.

process of DNA replication is semiconservative; that is, the copy of the genome inherited by each daughter cell contains one original and one newly synthesized strand of DNA.[2]:5.2

After DNA replication is complete, the cell must physically separate the two copies of the genome and divide into two distinct membrane-bound cells.[2]:18.2 In prokaryotes (bacteria and archaea) this usually occurs via a relatively simple process called binary fission, in which each circular genome attaches to the cell membrane and is separated into the daughter cells as the membrane invaginates to split the cytoplasm into two membrane-bound portions. Binary fission is extremely fast compared to the rates of cell division in eukaryotes. Eukaryotic cell division is a more complex process known as the cell cycle; DNA replication occurs during a phase of this cycle known as S phase, whereas the process of segregating chromosomes and splitting the cytoplasm occurs during M phase.[2]:18.1

25.5.3   Molecular inheritance

The duplication and transmission of genetic material from one generation of cells to the next is the basis for molecular inheritance, and the link between the classical and molecular pictures of genes. Organisms inherit the characteristics of their parents because the cells of the offspring contain copies of the genes in their parents' cells. In asexually reproducing organisms, the offspring will be a genetic copy or clone of the parent organism. In sexually reproducing organisms, a specialized form of cell division called meiosis produces cells called gametes or germ cells that are haploid, or contain only one copy of each gene.[2]:20.2 The gametes produced by females are called eggs or ova, and those produced by males are called sperm. Two gametes fuse to form a diploid fertilized egg, a single cell that has two sets of genes, with one copy of each gene from the mother and one from the father.[2]:20

During the process of meiotic cell division, an event called genetic recombination or *crossing-over* can sometimes occur, in which a length of DNA on one chromatid is swapped with a length of DNA on the corresponding sister chromatid. This has no effect if the alleles on the chromatids are the same, but results in reassortment of otherwise linked alleles if they are different.[2]:5.5 The Mendelian principle of independent assortment asserts that each of a parent's two genes for each trait will sort independently into gametes; which allele an organism inherits for one trait is unrelated to which allele it inherits for another trait. This is in fact only true for genes that do not reside on the same chromosome, or are located very far from one another on the same chromosome. The closer two genes lie on the same chromosome, the more closely they will be associated in gametes and the more often they will appear together; genes that are very close are essentially never separated because it is extremely unlikely that a crossover point will occur between them. This is known as genetic linkage.[49]

25.6   Molecular evolution

Main article: Molecular evolution

25.6.1   Mutation

DNA replication is for the most part extremely accurate, however errors (mutations) do occur.[2]:7.6 The error rate in eukaryotic cells can be as low as 10^{-8} per nucleotide per replication,[50][51] whereas for some RNA viruses it can be as high as 10^{-3}.[52] This means that each generation, each human genome accumulates 1–2 new mutations.[52] Small mutations can be caused by DNA replication and the aftermath of DNA damage and include point mutations in which a single base is altered and frameshift mutations in which a single base is inserted or deleted. Either of these mutations can change the gene by missense (change a codon to encode a different amino acid) or nonsense (a premature stop codon).[53] Larger mutations can be caused by errors in recombination to cause chromosomal abnormalities including the duplication, deletion, rearrangement or inversion of large sections of a chromosome. Additionally, the DNA repair mechanisms that normally revert mutations can introduce errors when repairing the physical damage to the molecule is more important than restoring an exact copy, for example when repairing double-strand breaks.[2]:5.4

When multiple different alleles for a gene are present in a species's population it is called polymorphic. Most different alleles are functionally equivalent, however some alleles can give rise to different phenotypic traits. A gene's most common allele is called the wild type, and rare alleles are called mutants. The genetic variation in relative frequencies of different alleles in a population is due to both natural selection and genetic drift.[54] The wild-type allele is not necessarily the ancestor of less common alleles, nor is it necessarily fitter.

Most mutations within genes are neutral, having no effect on the organism's phenotype (silent mutations). Some mutations do not change the amino acid sequence because multiple codons encode the same amino acid (synonymous mutations). Other mutations can be neutral if they lead to amino acid sequence changes, but the protein still functions similarly with the new amino acid (e.g. conservative mutations). Many mutations, however, are deleterious or even lethal, and are removed from populations by natural selection. Genetic disorders are the result of deleterious mutations and can be due to spontaneous mutation in the affected individual, or can be inherited. Finally, a small fraction of mutations are beneficial, improving the organism's fitness and are extremely important for evolution, since their directional selection leads to adaptive evolution.[2]:7.6

25.6.2   Sequence homology

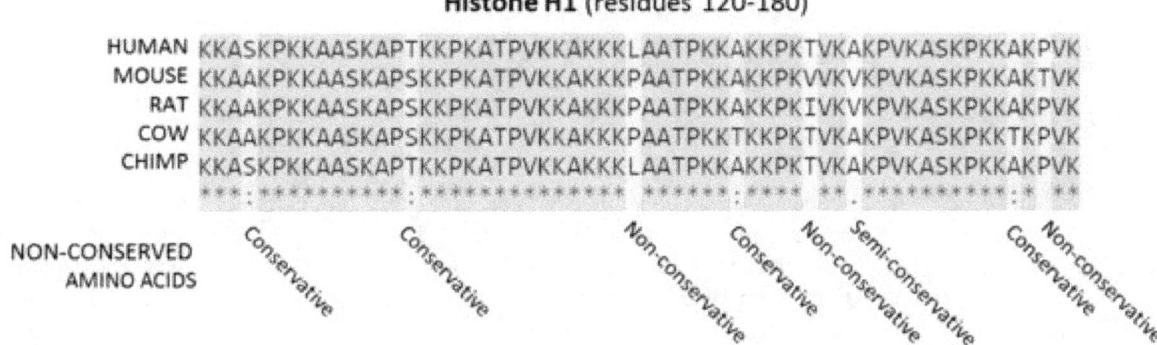

A sequence alignment, produced by ClustalO, of mammalian histone proteins

Genes with a most recent common ancestor, and thus a shared evolutionary ancestry, are known as homologs.[55] These genes appear either from gene duplication within an organism's genome, where they are known as paralogous genes, or are the result of divergence of the genes after a speciation event, where they are known as orthologous genes,[2]:7.6 and often perform the same or similar functions in related organisms. It is often assumed that the functions of orthologous genes are more similar than those of paralogous genes, although the difference is minimal.[56][57]

The relationship between genes can be measured by comparing the sequence alignment of their DNA.[2]:7.6 The degree of sequence similarity between homologous genes is called conserved sequence. Most changes to a gene's sequence do not affect its function and so genes accumulate mutations over time by neutral molecular evolution. Additionally, any selection on a gene will cause its sequence to diverge at a different rate. Genes under stabilizing selection are constrained and so change more slowly whereas genes under directional selection change sequence more rapidly.[58] The sequence differences between genes can be used for phylogenetic analyses to study how those genes have evolved and how the organisms they come from are related.[59][60]

25.6.3   Origins of new genes

The most common source of new genes in eukaryotic lineages is gene duplication, which creates copy number variation of an existing gene in the genome.[61][62] The resulting genes (paralogs) may then diverge in sequence and in function. Sets of genes formed in this way comprise a gene family. Gene duplications and losses within a family are common and represent a major source of evolutionary biodiversity.[63] Sometimes, gene duplication may result in a nonfunctional copy of a gene, or a functional copy may be subject to mutations that result in loss of function; such nonfunctional genes are called pseudogenes.[2]:7.6

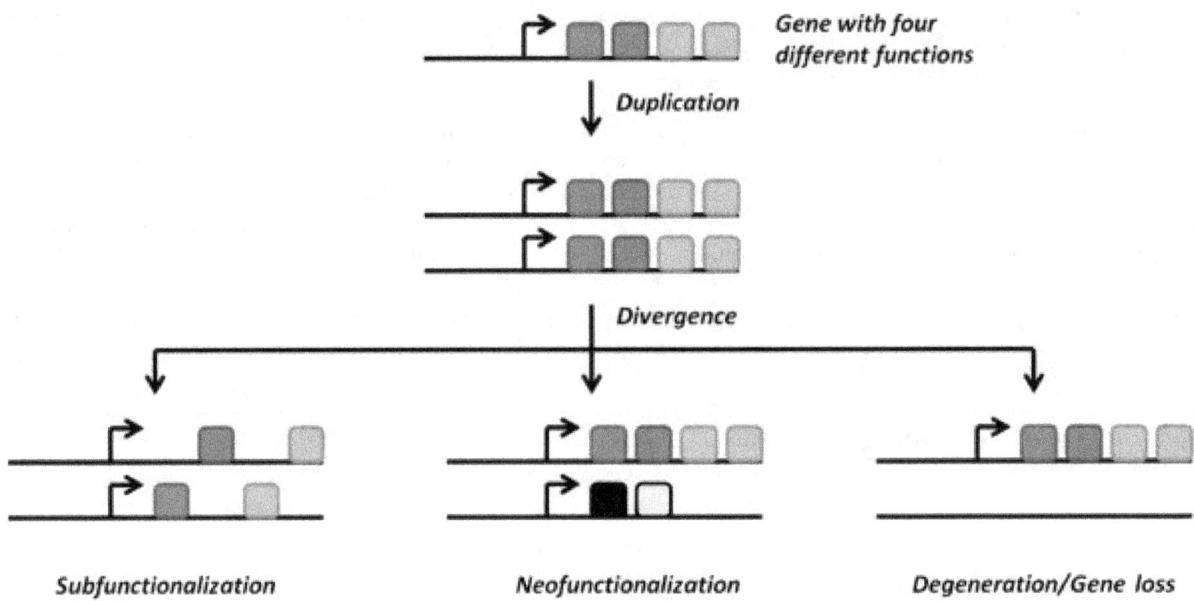

Evolutionary fate of duplicate genes

De novo or "orphan" genes, whose sequence shows no similarity to existing genes, are extremely rare. Estimates of the number of de novo genes in the human genome range from 18[64] to 60.[65] Such genes are typically shorter and simpler in structure than most eukaryotic genes, with few if any introns.[61] Two primary sources of orphan protein-coding genes are gene duplication followed by extremely rapid sequence change, such that the original relationship is undetectable by sequence comparisons, and formation through mutation of "cryptic" transcription start sites that introduce a new open reading frame in a region of the genome that did not previously code for a protein.[66][67]

Horizontal gene transfer refers to the transfer of genetic material through a mechanism other than reproduction. This mechanism is a common source of new genes in prokaryotes, sometimes thought to contribute more to genetic variation than gene duplication.[68] It is a common means of spreading antibiotic resistance, virulence, and adaptive metabolic functions.[26][69] Although horizontal gene transfer is rare in eukaryotes, likely examples have been identified of protist and alga genomes containing genes of bacterial origin.[70][71]

25.7   Genome

The genome is the total genetic material of an organism and includes both the genes and non-coding sequences.[72]

25.7.1   Number of genes

The genome size, and the number of genes it encodes varies widely between organisms. The smallest genomes occur in viruses (which can have as few as 2 protein-coding genes),[81] and viroids (which act as a single non-coding RNA gene).[82] Conversely, plants can have extremely large genomes,[83] with rice containing >46,000 protein-coding genes.[84] The total number of protein-coding genes (the Earth's proteome) is estimated to be 5 million sequences.[85]

Although the number of base-pairs of DNA in the human genome has been known since the 1960s, the estimated number of genes has changed over time as definitions of genes, and methods of detecting them have been refined. Initial theoretical predictions of the number of human genes were as high as 2,000,000.[86] Early experimental measures indicated there to be 50,000–100,000 *transcribed* genes (expressed sequence tags).[87] Subsequently, the sequencing in the Human Genome Project indicated that many of these transcripts were alternative variants of the same genes, and the total number of protein-coding genes was revised down to ~20,000[80] with 13 genes encoded on the mitochondrial genome.[78] Of the

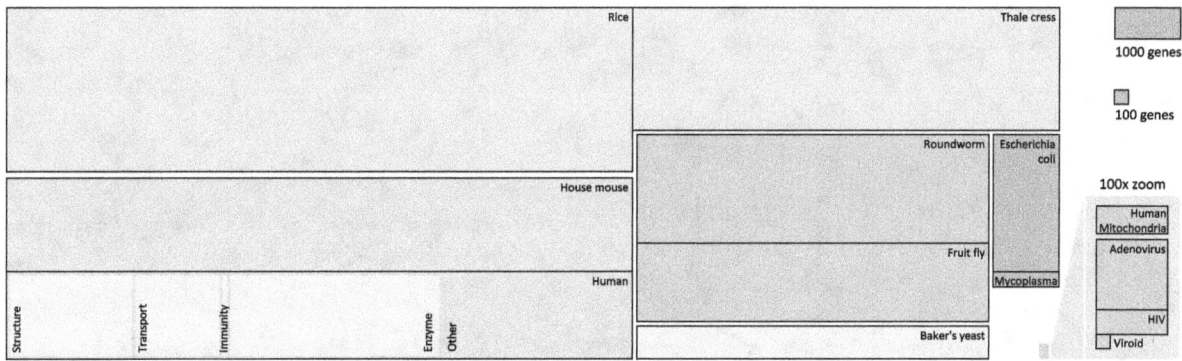

Representative genome sizes for plants (green), vertebrates (blue), invertebrates (red), fungus (yellow), bacteria (purple), and viruses (grey). An inset on the right shows the smaller genomes expanded 100-fold.[73][74][75][76][77][78][79][80]

human genome, only 1–2% consists of protein-coding genes,[88] with the remainder being 'noncoding' DNA such as introns, retrotransposons, and noncoding RNAs.[88][89]

25.7.2   Essential genes

Main article: Essential gene

Essential genes are the set of genes thought to be critical for an organism's survival.[90] This definition assumes the abundant availability of all relevant nutrients and the absence of environmental stress. Only a small portion of an organism's genes are essential. In bacteria, an estimated 250–400 genes are essential for *Escherichia coli* and *Bacillus subtilis*, which is less than 10% of their genes.[91][92][93] Half of these genes are orthologs in both organisms and are largely involved in protein synthesis.[93] In the budding yeast *Saccharomyces cerevisiae* the number of essential genes is slightly higher, at 1000 genes (~20% of their genes).[94] Although the number is more difficult to measure in higher eukaryotes, mice and humans are estimated to have around 2000 essential genes (~10% of their genes).[95]

Housekeeping genes are critical for carrying out basic cell functions and so are expressed at a relatively constant level (constitutively).[96] Since their expression is constant, housekeeping genes are used as experimental controls when analysing gene expression. Not all essential genes are housekeeping genes since some essential genes are developmentally regulated or expressed at certain times during the organism's life cycle.[97]

25.7.3   Genetic and genomic nomenclature

Gene nomenclature has been established by the HUGO Gene Nomenclature Committee (HGNC) for each known human gene in the form of an approved gene name and symbol (short-form abbreviation), which can be accessed through a database maintained by HGNC. Symbols are chosen to be unique, and each gene has only one symbol (although approved symbols sometimes change). Symbols are preferably kept consistent with other members of a gene family and with homologs in other species, particularly the mouse due to its role as a common model organism.[98]

25.8   Genetic engineering

Main article: Genetic engineering

Genetic engineering is the modification of an organism's genome through biotechnology. Since the 1970s, a variety of techniques have been developed to specifically add, remove and edit genes in an organism.[99] Recently developed genome engineering techniques use engineered nuclease enzymes to create targeted DNA repair in a chromosome to either disrupt

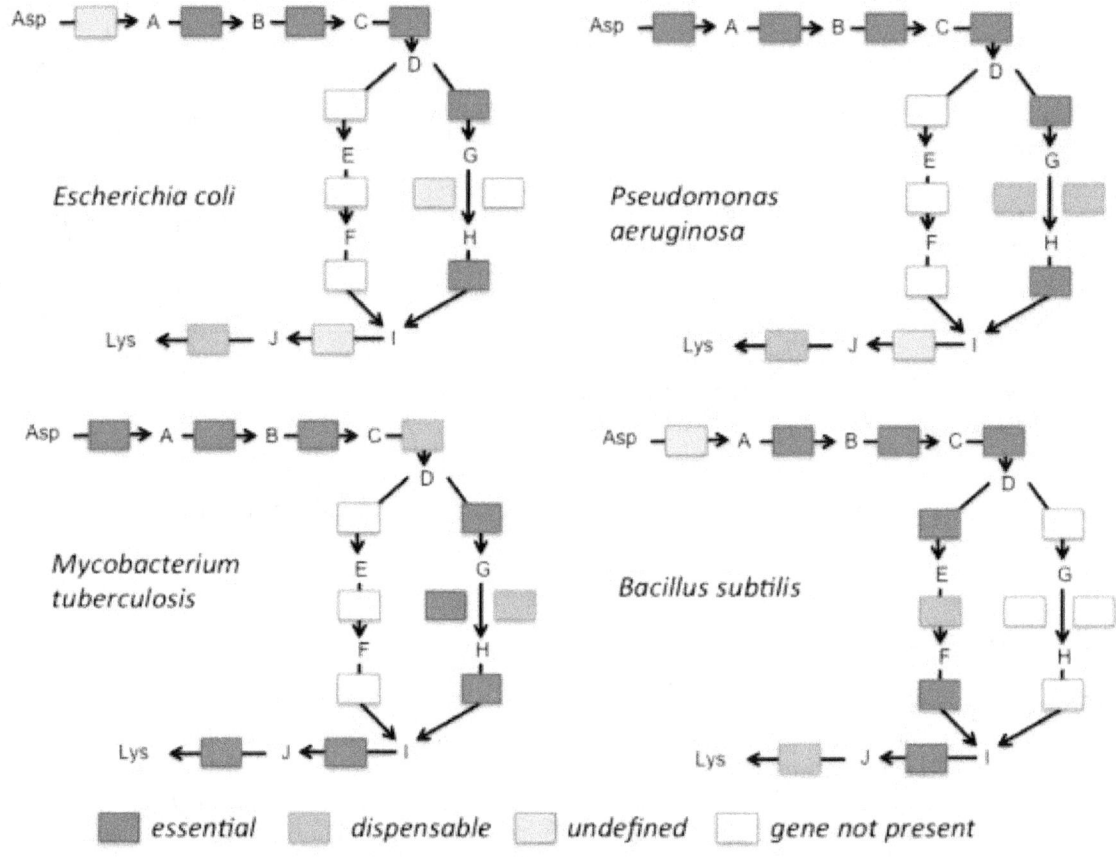

A schematic view of essential genes in lysine biosynthesis of different bacteria. The same protein may be essential in one species but not another.

or edit a gene when the break is repaired.[100][101][102][103] The related term synthetic biology is sometimes used to refer to extensive genetic engineering of an organism.[104]

Genetic engineering is now a routine research tool with model organisms. For example, genes are easily added to bacteria[105] and lineages of knockout mice with a specific gene's function disrupted are used to investigate that gene's function.[106][107] Many organisms have been genetically modified for applications in agriculture, industrial biotechnology, and medicine.

For multicellular organisms, typically the embryo is engineered which grows into the adult genetically modified organism.[108] However, the genomes of cells in an adult organism can be edited using gene therapy techniques to treat genetic diseases.

25.9   See also

- Copy number variation

- Epigenetics

- Full genome sequencing

- Gene-centric view of evolution

- Gene dosage

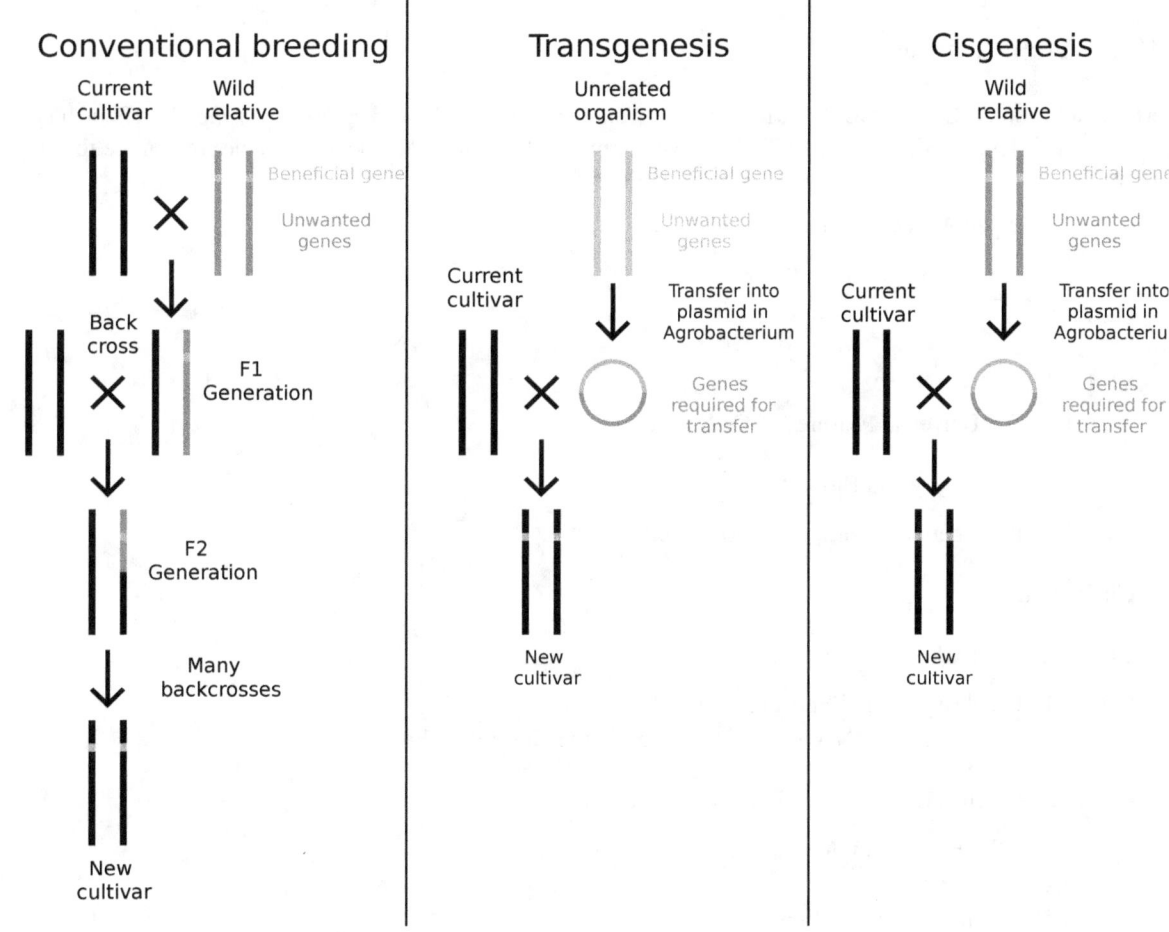

Comparison of conventional plant breeding with transgenic and cisgenic genetic modification.

- Gene expression

- Gene family

- Gene nomenclature

- Gene patent

- Gene pool

- Gene redundancy

- Genetic algorithm

- List of gene prediction software

- List of notable genes

- Predictive medicine

- Pseudogene

- Quantitative trait locus

25.10   References

25.10.1   Main textbook

Alberts B, Johnson A, Lewis J, Raff M, Roberts K, Walter P (2002). *Molecular Biology of the Cell* (Fourth ed.). New York: Garland Science. ISBN 978-0-8153-3218-3. – A molecular biology textbook available free online through NCBI Bookshelf.

Referenced chapters of *Molecular Biology of the Cell*

25.10.2 References

[1] Slack, J.M.W. Genes-A Very Short Introduction. Oxford University Press 2014

[2] Alberts B, Johnson A, Lewis J, Raff M, Roberts K, Walter P (2002). *Molecular Biology of the Cell* (Fourth ed.). New York: Garland Science. ISBN 978-0-8153-3218-3.

[3] Gericke, Niklas Markus; Hagberg, Mariana (5 December 2006). "Definition of historical models of gene function and their relation to students' understanding of genetics". *Science & Education* **16** (7-8): 849–881. Bibcode:2007Sc&Ed..16..849G. doi:10.1007/s11191-006-9064-4.

[4] Pearson H (May 2006). "Genetics: what is a gene?". *Nature* **441** (7092): 398–401. Bibcode:2006Natur.441..398P. doi:10.1038/ PMID 16724031.

[5] Pennisi E (June 2007). "Genomics. DNA study forces rethink of what it means to be a gene". *Science* **316** (5831): 1556–1557. doi:10.1126/science.316.5831.1556. PMID 17569836.

[6] Noble D (September 2008). "Genes and causation" (Free full text). *Philosophical Transactions. Series A, Mathematical, Physical, and Engineering Sciences* **366** (1878): 3001–3015. Bibcode:2008RSPTA.366.3001N. doi:10.1098/rsta.2008.0086. PMID 18559318.

[7] Magner, Lois N. (2002). *A History of the Life Sciences* (Third ed.). Marcel Dekker, CRC Press. p. 371. ISBN 978-0-203-91100-6.

[8] Henig, Robin Marantz (2000). *The Monk in the Garden: The Lost and Found Genius of Gregor Mendel, the Father of Genetics.* Boston: Houghton Mifflin. pp. 1–9. ISBN 978-0395-97765-1.

[9] "gene". *Oxford English Dictionary* (3rd ed.). Oxford University Press. September 2005. (Subscription or UK public library membership required.)

[10] "The Human Genome Project Timeline". Retrieved 13 September 2006.

[11] Gerstein MB, Bruce C, Rozowsky JS, Zheng D, Du J, Korbel JO, Emanuelsson O, Zhang ZD, Weissman S, Snyder M (June 2007). "What is a gene, post-ENCODE? History and updated definition". *Genome Research* **17** (6): 669–681. doi:10.1101/gr.63 PMID17567988.

[12] Avery, OT; MacLeod, CM; McCarty, M (1944). "Studies on the Chemical Nature of the Substance Inducing Transformation of Pneumococcal Types: Induction of Transformation by a Desoxyribonucleic Acid Fraction Isolated from Pneumococcus Type III". *The Journal of experimental medicine* **79** (2): 137–58. doi:10.1084/jem.79.2.137. PMC 2135445. PMID 19871359. Reprint: Avery, OT; MacLeod, CM; McCarty, M (1979). "Studies on the chemical nature of the substance inducing transformation of pneumococcal types. Inductions of transformation by a desoxyribonucleic acid fraction isolated from pneumococcus type III". *The Journal of experimental medicine* **149** (2): 297–326. doi:10.1084/jem.149.2.297. PMC 2184805. PMID 33226.

[13] Hershey, AD; Chase, M (1952). "Independent functions of viral protein and nucleic acid in growth of bacteriophage". *The Journal of General Physiology* **36** (1): 39–56. doi:10.1085/jgp.36.1.39. PMC 2147348. PMID 12981234.

[14] Judson, Horace (1979). *The Eighth Day of Creation: Makers of the Revolution in Biology.* Cold Spring Harbor Laboratory Press. pp. 51–169. ISBN 0-87969-477-7.

[15] Watson, J. D.; Crick, FH (1953). "Molecular Structure of Nucleic Acids: A Structure for Deoxyribose Nucleic Acid" (PDF). *Nature* **171** (4356): 737–8. Bibcode:1953Natur.171..737W. doi:10.1038/171737a0. PMID 13054692.

[16] Min Jou W, Haegeman G, Ysebaert M, Fiers W (May 1972). "Nucleotide sequence of the gene coding for the bacteriophage MS2 coat protein". *Nature* **237** (5350): 82–8. Bibcode:1972Natur.237...82J. doi:10.1038/237082a0. PMID 4555447.

[17] Sanger, F; Nicklen, S; Coulson, AR (1977). "DNA sequencing with chain-terminating inhibitors". *Proceedings of the National Academy of Sciences of the United States of America* **74** (12): 5463–7. Bibcode:1977PNAS...74.5463S. doi:10.1073/pnas.74.12 .5463.PMC431765. PMID271968.

[18] Adams, Jill U. (2008). "DNA Sequencing Technologies". *Nature Education Knowledge.* SciTable (Nature Publishing Group) **1** (1): 193.

[19] Huxley, Julian (1942). *Evolution: the modern synthesis* (Definitive ed.). Cambridge, Mass.: MIT Press. ISBN 978-0262513661.

[20] Williams, George C. (2001). *Adaptation and Natural Selection a Critique of Some Current Evolutionary Thought.* ([Online-Ausg.]. ed.). Princeton: Princeton University Press. ISBN 9781400820108.

[21] Dawkins, Richard (1977). *The selfish gene* (Repr. (with corr.) ed.). London: Oxford Univ. Press. ISBN 0-19-857519-X.

[22] Dawkins, Richard (1989). *The extended phenotype.* (Pbk. ed.). Oxford: Oxford University Press. ISBN 0-19-286088-7.

[23] Stryer L, Berg JM, Tymoczko JL (2002). *Biochemistry* (5th ed.). San Francisco: W.H. Freeman. ISBN 0-7167-4955-6.

[24] Bolzer, Andreas; Kreth, Gregor; Solovei, Irina; Koehler, Daniela; Saracoglu, Kaan; Fauth, Christine; Müller, Stefan; Eils, Roland; Cremer, Christoph; Speicher, Michael R.; Cremer, Thomas (2005). "Three-Dimensional Maps of All Chromosomes in Human Male Fibroblast Nuclei and Prometaphase Rosettes". *PLoS Biology* **3** (5): e157. doi:10.1371/journal.pbio.0030157. PMID 15839726.

[25] Braig M, Schmitt CA (March 2006). "Oncogene-induced senescence: putting the brakes on tumor development". *Cancer Research* **66** (6): 2881–4. doi:10.1158/0008-5472.CAN-05-4006. PMID 16540631.

[26] Bennett, PM (March 2008). "Plasmid encoded antibiotic resistance: acquisition and transfer of antibiotic resistance genes in bacteria.". *British journal of pharmacology.* 153 Suppl 1: S347–57. PMID 18193080.

[27] International Human Genome Sequencing Consortium (October 2004). "Finishing the euchromatic sequence of the human genome". *Nature* **431** (7011): 931–45. Bibcode:2004Natur.431..931H. doi:10.1038/nature03001. PMID 15496913.

[28] Mortazavi A, Williams BA, McCue K, Schaeffer L, Wold B (July 2008). "Mapping and quantifying mammalian transcriptomes by RNA-Seq". *Nature Methods* **5** (7): 621–8. doi:10.1038/nmeth.1226. PMID 18516045.

[29] Pennacchio, L. A.; Bickmore, W.; Dean, A.; Nobrega, M. A.; Bejerano, G. (2013). "Enhancers: Five essential questions". *Nature Reviews Genetics* **14** (4): 288–95. doi:10.1038/nrg3458. PMID 23503198.

[30] Maston, G. A.; Evans, S. K.; Green, M. R. (2006). "Transcriptional Regulatory Elements in the Human Genome". *Annual Review of Genomics and Human Genetics* **7**: 29. doi:10.1146/annurev.genom.7.080505.115623.

[31] Mignone, Flavio; Gissi, Carmela; Liuni, Sabino; Pesole, Graziano (2002-02-28). "Untranslated regions of mRNAs". *Genome Biology* **3** (3): reviews0004. doi:10.1186/gb-2002-3-3-reviews0004. ISSN 1465-6906. PMID 11897027.

[32] Bicknell AA, Cenik C, Chua HN, Roth FP, Moore MJ (December 2012). "Introns in UTRs: why we should stop ignoring them.". *Bioessays* **34** (12): 1025–34. doi:10.1002/bies.201200073. PMID 23108796.

[33] Salgado, H.; Moreno-Hagelsieb, G.; Smith, T.; Collado-Vides, J. (2000). "Operons in Escherichia coli: Genomic analyses and predictions". *Proceedings of the National Academy of Sciences* **97** (12): 6652–6657. Bibcode:2000PNAS...97.6652S. doi:10.1073/pnas.110147297. PMC 18690. PMID 10823905.

[34] Blumenthal, Thomas (November 2004). "Operons in eukaryotes". *Briefings in Functional Genomics & Proteomics* **3** (3): 199–211. doi:10.1093/bfgp/3.3.199. ISSN 2041-2649. PMID 15642184.

[35] Spilianakis CG, Lalioti MD, Town T, Lee GR, Flavell RA (June 2005). "Interchromosomal associations between alternatively expressed loci". *Nature* **435** (7042): 637–45. Bibcode:2005Natur.435..637S. doi:10.1038/nature03574. PMID 15880101.

[36] Williams, A; Spilianakis, CG; Flavell, RA (April 2010). "Interchromosomal association and gene regulation in trans.". *Trends in genetics : TIG* **26** (4): 188–97. PMID 20236724.

[37] Marande W, Burger G (October 2007). "Mitochondrial DNA as a genomic jigsaw puzzle". *Science* (AAAS) **318** (5849): 415. Bibcode:2007Sci...318..415M. doi:10.1126/science.1148033. PMID 17947575.

[38] Parra G, Reymond A, Dabbouseh N, Dermitzakis ET, Castelo R, Thomson TM, Antonarakis SE, Guigó R (January 2006). "Tandem chimerism as a means to increase protein complexity in the human genome". *Genome Research* **16** (1): 37–44. doi:10.1101/gr.4145906. PMC 1356127. PMID 16344564.

[39] Eddy SR (December 2001). "Non-coding RNA genes and the modern RNA world". *Nat. Rev. Genet.* **2** (12): 919–29. doi:10.1038/35103511. PMID 11733745.

[40] Crick, Francis (1962). *The genetic code.* WH Freeman and Company. PMID 13882204.

[41] Woodson SA (May 1998). "Ironing out the kinks: splicing and translation in bacteria". *Genes & Development* **12** (9): 1243–7. doi:10.1101/gad.12.9.1243. PMID 9573040.

[42] Jacob F; Monod J (June 1961). "Genetic regulatory mechanisms in the synthesis of proteins". *J Mol Biol.* **3** (3): 318–56. doi:10.1016/S0022-2836(61)80072-7. PMID 13718526.

[43] Koonin, Eugene V.; Dolja, Valerian V.; Morris, T. Jack (January 1993). "Evolution and Taxonomy of Positive-Strand RNA Viruses: Implications of Comparative Analysis of Amino Acid Sequences". *Critical Reviews in Biochemistry and Molecular Biology* **28** (5): 375–430. doi:10.3109/10409239309078440.

[44] Domingo, Esteban (2001). "RNA Virus Genomes". *eLS.* doi:10.1002/9780470015902.a0001488.pub2.

[45] Domingo, E; Escarmís, C; Sevilla, N; Moya, A; Elena, SF; Quer, J; Novella, IS; Holland, JJ (June 1996). "Basic concepts in RNA virus evolution.". *FASEB journal : official publication of the Federation of American Societies for Experimental Biology* **10** (8): 859–64. PMID 8666162.

[46] Morris, KV; Mattick, JS (June 2014). "The rise of regulatory RNA.". *Nature reviews. Genetics* **15** (6): 423–37. PMID 24776770.

[47] Miko, Ilona (2008). "Gregor Mendel and the Principles of Inheritance". *Nature Education Knowledge.* SciTable (Nature Publishing Group) **1** (1): 134.

[48] Chial, Heidi (2008). "Mendelian Genetics: Patterns of Inheritance and Single-Gene Disorders". *Nature Education Knowledge.* SciTable (Nature Publishing Group) **1** (1): 63.

[49] Lobo, Ingrid; Shaw, Kelly (2008). "Discovery and Types of Genetic Linkage". *Nature Education Knowledge.* SciTable (Nature Publishing Group) **1** (1): 139.

[50] Nachman MW, Crowell SL (September 2000). "Estimate of the mutation rate per nucleotide in humans". *Genetics* **156** (1): 297–304. PMC 1461236. PMID 10978293.

[51] Roach JC, Glusman G, Smit AF, et al. (April 2010). "Analysis of genetic inheritance in a family quartet by whole-genome sequencing". *Science* **328** (5978): 636–9. Bibcode:2010Sci...328..636R. doi:10.1126/science.1186802. PMC 3037280. PMID 20220176.

[52] Drake JW, Charlesworth B, Charlesworth D, Crow JF (April 1998). "Rates of spontaneous mutation". *Genetics* **148** (4): 1667–86. PMC 1460098. PMID 9560386.

[53] "What kinds of gene mutations are possible?". *Genetics Home Reference.* United States National Library of Medicine. 11 May 2015. Retrieved 19 May 2015.

[54] Andrews, Christine A. (2010). "Natural Selection, Genetic Drift, and Gene Flow Do Not Act in Isolation in Natural Populations". *Nature Education Knowledge.* SciTable (Nature Publishing Group) **3** (10): 5.

[55] Patterson, C (November 1988). "Homology in classical and molecular biology.". *Molecular biology and evolution* **5** (6): 603–25. PMID 3065587.

[56] Studer, RA; Robinson-Rechavi, M (May 2009). "How confident can we be that orthologs are similar, but paralogs differ?". *Trends in genetics : TIG* **25** (5): 210–6. PMID 19368988.

[57] Altenhoff, AM; Studer, RA; Robinson-Rechavi, M; Dessimoz, C (2012). "Resolving the ortholog conjecture: orthologs tend to be weakly, but significantly, more similar in function than paralogs.". *PLoS computational biology* **8** (5): e1002514. PMID 22615551.

[58] NOSIL, PATRIK; FUNK, DANIEL J.; ORTIZ-BARRIENTOS, DANIEL (February 2009). "Divergent selection and heterogeneous genomic divergence". *Molecular Ecology* **18** (3): 375–402. doi:10.1111/j.1365-294X.2008.03946.x.

[59] Emery, Laura. "Introduction to Phylogenetics". EMBL-EBI. Retrieved 19 May 2015.

[60] Mitchell, Matthew W.; Gonder, Mary Katherine (2013). "Primate Speciation: A Case Study of African Apes". *Nature Education Knowledge.* SciTable (Nature Publishing Group) **4** (2): 1.

[61] Guerzoni, D; McLysaght, A (November 2011). "De novo origins of human genes.". *PLoS genetics* **7** (11): e1002381. PMID 22102832.

[62] Reams, AB; Roth, JR (2 February 2015). "Mechanisms of gene duplication and amplification.". *Cold Spring Harbor perspectives in biology* **7** (2): a016592. PMID 25646380.

[63] Demuth, JP; De Bie, T; Stajich, JE; Cristianini, N; Hahn, MW (20 December 2006). "The evolution of mammalian gene families.". *PloS one* **1**: e85. Bibcode:2006PLoSO...1...85D. doi:10.1371/journal.pone.0000085. PMID 17183716.

[64] Knowles, DG; McLysaght, A (October 2009). "Recent de novo origin of human protein-coding genes.". *Genome research* **19** (10): 1752–9. PMID 19726446.

[65] Wu, DD; Irwin, DM; Zhang, YP (November 2011). "De novo origin of human protein-coding genes.". *PLoS genetics* **7** (11): e1002379. PMID 22102831.

[66] Tautz, D; Domazet-Lošo, T (31 August 2011). "The evolutionary origin of orphan genes.". *Nature reviews. Genetics* **12** (10): 692–702. PMID 21878963.

[67] Carvunis, AR; Rolland, T; Wapinski, I; Calderwood, MA; Yildirim, MA; Simonis, N; Charloteaux, B; Hidalgo, CA; Barbette, J; Santhanam, B; Brar, GA; Weissman, JS; Regev, A; Thierry-Mieg, N; Cusick, ME; Vidal, M (19 July 2012). "Proto-genes and de novo gene birth.". *Nature* **487** (7407): 370–4. Bibcode:2012Natur.487..370C. doi:10.1038/nature11184. PMID 22722833.

[68] Treangen, TJ; Rocha, EP (27 January 2011). "Horizontal transfer, not duplication, drives the expansion of protein families in prokaryotes.". *PLoS genetics* **7** (1): e1001284. PMID 21298028.

[69] Ochman, H; Lawrence, JG; Groisman, EA (18 May 2000). "Lateral gene transfer and the nature of bacterial innovation.". *Nature* **405** (6784): 299–304. Bibcode:2000Natur.405..299O. PMID 10830951.

[70] Keeling, PJ; Palmer, JD (August 2008). "Horizontal gene transfer in eukaryotic evolution.". *Nature reviews. Genetics* **9** (8): 605–18. PMID 18591983.

[71] Schönknecht, G; Chen, WH; Ternes, CM; Barbier, GG; Shrestha, RP; Stanke, M; Bräutigam, A; Baker, BJ; Banfield, JF; Garavito, RM; Carr, K; Wilkerson, C; Rensing, SA; Gagneul, D; Dickenson, NE; Oesterhelt, C; Lercher, MJ; Weber, AP (8 March 2013). "Gene transfer from bacteria and archaea facilitated evolution of an extremophilic eukaryote.". *Science (New York, N.Y.)* **339** (6124): 1207–10. Bibcode:2013Sci...339.1207S. doi:10.1126/science.1231707. PMID 23471408.

[72] Ridley, M. (2006). *Genome*. New York, NY: Harper Perennial. ISBN 0-06-019497-9

[73] Watson, JD, Baker TA, Bell SP, Gann A, Levine M, Losick R. (2004). "Ch9-10", Molecular Biology of the Gene, 5th ed., Peason Benjamin Cummings; CSHL Press.

[74] "Integr8 – A.thaliana Genome Statistics:".

[75] "Understanding the Basics". *The Human Genome Project*. Retrieved 26 April 2015.

[76] "WS227 Release Letter". WormBase. 10 August 2011. Retrieved 19 November 2013.

[77] Yu, J. (5 April 2002). "A Draft Sequence of the Rice Genome (Oryza sativa L. ssp. indica)". *Science* **296** (5565): 79–92. Bibcode:2002Sci...296...79Y. doi:10.1126/science.1068037.

[78] Anderson, S.; Bankier, A. T.; Barrell, B. G.; de Bruijn, M. H. L.; Coulson, A. R.; Drouin, J.; Eperon, I. C.; Nierlich, D. P.; Roe, B. A.; Sanger, F.; Schreier, P. H.; Smith, A. J. H.; Staden, R.; Young, I. G. (9 April 1981). "Sequence and organization of the human mitochondrial genome". *Nature* **290** (5806): 457–465. Bibcode:1981Natur.290..457A. doi:10.1038/290457a0.

[79] Adams, M. D. (24 March 2000). "The Genome Sequence of Drosophila melanogaster". *Science* **287** (5461): 2185–2195. Bibcode:2000Sci...287.2185.. doi:10.1126/science.287.5461.2185.

[80] Pertea, Mihaela; Salzberg, Steven L (2010). "Between a chicken and a grape: estimating the number of human genes". *Genome Biology* **11** (5): 206. doi:10.1186/gb-2010-11-5-206.

[81] Belyi, V. A.; Levine, A. J.; Skalka, A. M. (22 September 2010). "Sequences from Ancestral Single-Stranded DNA Viruses in Vertebrate Genomes: the Parvoviridae and Circoviridae Are More than 40 to 50 Million Years Old". *Journal of Virology* **84** (23): 12458–12462. doi:10.1128/JVI.01789-10.

[82] Flores, Ricardo; Di Serio, Francesco; Hernández, Carmen (February 1997). "Viroids: The Noncoding Genomes". *Seminars in Virology* **8** (1): 65–73. doi:10.1006/smvy.1997.0107.

[83] Zonneveld, B. J. M. (2010). "New Record Holders for Maximum Genome Size in Eudicots and Monocots". *Journal of Botany* **2010**: 1–4. doi:10.1155/2010/527357.

[84] Yu J, Hu S, Wang J, Wong GK, Li S, Liu B, Deng Y, Dai L, Zhou Y, Zhang X, Cao M, Liu J, Sun J, Tang J, Chen Y, Huang X, Lin W, Ye C, Tong W, Cong L, Geng J, Han Y, Li L, Li W, Hu G, Huang X, Li W, Li J, Liu Z, Li L, Liu J, Qi Q, Liu J, Li L, Li T, Wang X, Lu H, Wu T, Zhu M, Ni P, Han H, Dong W, Ren X, Feng X, Cui P, Li X, Wang H, Xu X, Zhai W, Xu Z, Zhang J, He S, Zhang J, Xu J, Zhang K, Zheng X, Dong J, Zeng W, Tao L, Ye J, Tan J, Ren X, Chen X, He J, Liu D, Tian W, Tian C, Xia H, Bao Q, Li G, Gao H, Cao T, Wang J, Zhao W, Li P, Chen W, Wang X, Zhang Y, Hu J, Wang J, Liu S, Yang J, Zhang G, Xiong Y, Li Z, Mao L, Zhou C, Zhu Z, Chen R, Hao B, Zheng W, Chen S, Guo W, Li G, Liu S, Tao M, Wang J, Zhu L, Yuan L, Yang H (April 2002). "A draft sequence of the rice genome (Oryza sativa L. ssp. indica)". *Science* **296** (5565): 79–92. Bibcode:2002Sci...296...79Y. doi:10.1126/science.1068037. PMID 11935017.

[85] Perez-Iratxeta C, Palidwor G, Andrade-Navarro MA (December 2007). "Towards completion of the Earth's proteome". *EMBO Reports* **8** (12): 1135–1141. doi:10.1038/sj.embor.7401117. PMC 2267224. PMID 18059312.

[86] Kauffman SA (1969). "Metabolic stability and epigenesis in randomly constructed genetic nets". *Journal of Theoretical Biology* (Elsevier) **22** (3): 437–467. doi:10.1016/0022-5193(69)90015-0. PMID 5803332.

[87] Schuler GD, Boguski MS, Stewart EA, Stein LD, Gyapay G, Rice K, White RE, Rodriguez-Tomé P, Aggarwal A, Bajorek E, Bentolila S, Birren BB, Butler A, Castle AB, Chiannilkulchai N, Chu A, Clee C, Cowles S, Day PJ, Dibling T, Drouot N, Dunham I, Duprat S, East C, Edwards C, Fan JB, Fang N, Fizames C, Garrett C, Green L, Hadley D, Harris M, Harrison P, Brady S, Hicks A, Holloway E, Hui L, Hussain S, Louis-Dit-Sully C, Ma J, MacGilvery A, Mader C, Maratukulam A, Matise TC, McKusick KB, Morissette J, Mungall A, Muselet D, Nusbaum HC, Page DC, Peck A, Perkins S, Piercy M, Qin F, Quackenbush J, Ranby S, Reif T, Rozen S, Sanders C, She X, Silva J, Slonim DK, Soderlund C, Sun WL, Tabar P, Thangarajah T, Vega-Czarny N, Vollrath D, Voyticky S, Wilmer T, Wu X, Adams MD, Auffray C, Walter NA, Brandon R, Dehejia A, Goodfellow PN, Houlgatte R, Hudson JR, Ide SE, Iorio KR, Lee WY, Seki N, Nagase T, Ishikawa K, Nomura N, Phillips C, Polymeropoulos MH, Sandusky M, Schmitt K, Berry R, Swanson K, Torres R, Venter JC, Sikela JM, Beckmann JS, Weissenbach J, Myers RM, Cox DR, James MR, Bentley D, Deloukas P, Lander ES, Hudson TJ (October 1996). "A gene map of the human genome". *Science* **274** (5287): 540–6. Bibcode:1996Sci...274..540S. doi:10.1126/science.274.5287.540. PMID 8849440.

[88] Claverie JM (September 2005). "Fewer genes, more noncoding RNA". *Science***309**(5740): 1529–30. Bibcode:2005Sci...309.1 doi:10.1126/science.1116800. PMID 16141064.

[89] Carninci P, Hayashizaki Y (April 2007). "Noncoding RNA transcription beyond annotated genes". *Current Opinion in Genetics & Development* **17** (2): 139–44. doi:10.1016/j.gde.2007.02.008. PMID 17317145.

[90] Glass, J. I.; Assad-Garcia, N.; Alperovich, N.; Yooseph, S.; Lewis, M. R.; Maruf, M.; Hutchison, C. A.; Smith, H. O.; Venter, J. C. (3 January 2006). "Essential genes of a minimal bacterium". *Proceedings of the National Academy of Sciences* **103** (2): 425–430. Bibcode:2006PNAS..103..425G. doi:10.1073/pnas.0510013103.

[91] Gerdes, SY; Scholle, MD; Campbell, JW; Balázsi, G; Ravasz, E; Daugherty, MD; Somera, AL; Kyrpides, NC; Anderson, I; Gelfand, MS; Bhattacharya, A; Kapatral, V; D'Souza, M; Baev, MV; Grechkin, Y; Mseeh, F; Fonstein, MY; Overbeek, R; Barabási, AL; Oltvai, ZN; Osterman, AL (October 2003). "Experimental determination and system level analysis of essential genes in Escherichia coli MG1655.". *Journal of bacteriology* **185** (19): 5673–84. PMID 13129938.

[92] Baba, T; Ara, T; Hasegawa, M; Takai, Y; Okumura, Y; Baba, M; Datsenko, KA; Tomita, M; Wanner, BL; Mori, H (2006). "Construction of Escherichia coli K-12 in-frame, single-gene knockout mutants: the Keio collection.". *Molecular systems biology* **2**: 2006.0008. PMID 16738554.

[93] Juhas, M; Reuß, DR; Zhu, B; Commichau, FM (November 2014). "Bacillus subtilis and Escherichia coli essential genes and minimal cell factories after one decade of genome engineering.". *Microbiology (Reading, England)* **160** (Pt 11): 2341–51. PMID 25092907.

[94] Tu, Z; Wang, L; Xu, M; Zhou, X; Chen, T; Sun, F (21 February 2006). "Further understanding human disease genes by comparing with housekeeping genes and other genes.". *BMC genomics* **7**: 31. PMID 16504025.

[95] Georgi, B; Voight, BF; Bućan, M (May 2013). "From mouse to human: evolutionary genomics analysis of human orthologs of essential genes.". *PLoS genetics* **9** (5): e1003484. PMID 23675308.

[96] Eisenberg, E; Levanon, EY (October 2013). "Human housekeeping genes, revisited.". *Trends in genetics : TIG* **29** (10): 569–74. PMID 23810203.

[97] Amsterdam, A; Hopkins, N (September 2006). "Mutagenesis strategies in zebrafish for identifying genes involved in development and disease.". *Trends in genetics : TIG* **22** (9): 473–8. PMID 16844256.

[98] "About the HGNC". *HGNC Database of Human Gene Names*. HUGO Gene Nomenclature Committee. Retrieved 14 May 2015.

[99] Stanley N. Cohen and Annie C. Y. Chang (1 May 1973). "Recircularization and Autonomous Replication of a Sheared R-Factor DNA Segment in Escherichia coli Transformants — PNAS". Pnas.org. Retrieved 17 July 2010.

[100] Esvelt, KM.; Wang, HH. (2013). "Genome-scale engineering for systems and synthetic biology". *Mol Syst Biol* **9** (1): 641. doi:10.1038/msb.2012.66. PMC 3564264. PMID 23340847.

[101] Tan, WS.; Carlson, DF.; Walton, MW.; Fahrenkrug, SC.; Hackett, PB. (2012). "Precision editing of large animal genomes". *Adv Genet*. Advances in Genetics **80**: 37–97. doi:10.1016/B978-0-12-404742-6.00002-8. ISBN 9780124047426. PMC 3683964. PMID 23084873.

[102] Puchta,H.;Fauser,F. (2013). "Gene targeting in plants: 25years later". *Int. J.Dev. Biol***57**(6–7–8): 629–637. doi:10.1387/ij

[103] Ran FA, Hsu PD, Wright J, Agarwala V, Scott DA, Zhang F (2013). "Genome engineering using the CRISPR-Cas9 system". *Nat Protoc* **8** (11): 2281–308. doi:10.1038/nprot.2013.143. PMC 3969860. PMID 24157548.

[104] Kittleson, Joshua (2012). "Successes and failures in modular genetic engineering". *Current Opinion in Chemical Biology*. doi:10.1016/j.cbpa.2012.06.009.

[105] Berg, P.; Mertz, J. E. (2010). "Personal Reflections on the Origins and Emergence of Recombinant DNA Technology". *Genetics* **184** (1): 9–17. doi:10.1534/genetics.109.112144. PMC 2815933. PMID 20061565.

[106] Austin, Christopher P.; Battey, James F.; Bradley, Allan; Bucan, Maja; Capecchi, Mario; Collins, Francis S.; Dove, William F.; Duyk, Geoffrey; Dymecki, Susan (September 2004). "The Knockout Mouse Project". *Nature Genetics* **36** (9): 921–924. doi:10.1038/ng0904-921. ISSN 1061-4036. PMC 2716027. PMID 15340423.

[107] "A review of current large-scale mouse knockout efforts – Guan – 2010 – genesis – Wiley Online Library". *doi.wiley.com*.

[108] Deng C (2007). "In celebration of Dr. Mario R. Capecchi's Nobel Prize". *International Journal of Biological Sciences* **3** (7): 417–419. doi:10.7150/ijbs.3.417. PMID 17998949.

25.10.3 Further reading

- Watson JD, Baker TA, Bell SP, Gann A, Levine M, Losick R (2013). *Molecular Biology of the Gene* (7th ed.). Benjamin Cummings. ISBN 978-0-321-90537-6.

- Dawkins R (1990). *The Selfish Gene*. Oxford University Press. ISBN 0-19-286092-5. Google Book Search; first published 1976.

- Ridley M (1999). *Genome: The Autobiography of a Species in 23 Chapters*. Fourth Estate. ISBN 0-00-763573-7.

- Brown, T (2002). *Genomes* (2nd ed.). New York: Wiley-Liss. ISBN 0-471-25046-5.

25.11 External links

- Comparative Toxicogenomics Database

- DNA From The Beginning – a primer on genes and DNA

- Genes And DNA – Introduction to genes and DNA aimed at non-biologist

- Entrez Gene – a searchable database of genes

- IDconverter – converts gene IDs between public databases

- iHOP – Information Hyperlinked over Proteins

- TranscriptomeBrowser – Gene expression profile analysis

- The Protein Naming Utility, a database to identify and correct deficient gene names

- *Genes* – an Open Access journal

- IMPC (International Mouse Phenotyping Consortium) – Encyclopedia of mammalian gene function

- Global Genes Project – Leading non-profit organization supporting people living with genetic diseases

- ENCODE threads Explorer Characterization of intergenic regions and gene definition. *Nature*

Chapter 26

Noncoding DNA

In genomics and related disciplines, **noncoding DNA** sequences are components of an organism's DNA that do not encode protein sequences. Some noncoding DNA is transcribed into functional non-coding RNA molecules (e.g. transfer RNA, ribosomal RNA, and regulatory RNAs). Other functions of noncoding DNA include the transcriptional and translational regulation of protein-coding sequences, scaffold attachment regions, origins of DNA replication, centromeres and telomeres.

The amount of noncoding DNA varies greatly among species. For example, over 98% of the human genome is noncoding,[2] while 20% of a typical prokaryote genome is noncoding.[3] When there is much non-coding DNA, a large proportion appears to have no biological function for the organism, as theoretically predicted in the 1960s. Since that time, this non-functional portion has often been referred to as "junk DNA", a term that has elicited strong responses over the years.[4]

The international Encyclopedia of DNA Elements (ENCODE) project uncovered, by direct biochemical approaches, that at least 80% of human genomic DNA has biochemical activity.[5] Though this was not necessarily unexpected due to previous decades of research discovering many functional noncoding regions,[3][6] some scientists criticized the conclusion for conflating biochemical activity with biological function.[7][8][9][10][11] Estimates for the biologically functional fraction of our genome based on comparative genomics range between 8 and 15%.[12][13][14] However, others have argued against relying solely on estimates from comparative genomics due to its limited scope and also because non-coding DNA has been found to be involved in epigenetic activity and making the complexity of species.[6][13][15][16]

26.1 Fraction of noncoding genomic DNA

The amount of total genomic DNA varies widely between organisms, and the proportion of coding and noncoding DNA within these genomes varies greatly as well. More than 98% of the human genome does not encode protein sequences, including most sequences within introns and most intergenic DNA.[2] 20% of a typical prokaryote genome is noncoding.[3]

While overall genome size, and by extension the amount of noncoding DNA, are correlated to organism complexity, there are many exceptions. For example, the genome of the unicellular *Polychaos dubium* (formerly known as *Amoeba dubia*) has been reported to contain more than 200 times the amount of DNA in humans.[17] The pufferfish *Takifugu rubripes* genome is only about one eighth the size of the human genome, yet seems to have a comparable number of genes; approximately 90% of the *Takifugu* genome is noncoding DNA.[2] The extensive variation in nuclear genome size among eukaryotic species is known as the C-value enigma or C-value paradox.[18] Most of the genome size difference appears to lie in the noncoding DNA.

In 2013, a new "record" for the most efficient eukaryotic genome was discovered with *Utricularia gibba*, a bladderwort plant that has only 3% noncoding DNA and 97% of coding DNA. Parts of the noncoding DNA were being deleted by the plant and this suggested that noncoding DNA may not be as critical for plants, even though noncoding DNA is useful for humans.[1] Other studies on plants have discovered crucial functions in portions noncoding DNA that were previously thought to be negligible and have added a new layer to the understanding of gene regulation.[19]

26.2   Types of noncoding DNA sequences

Main article: Conserved non-coding sequence

26.2.1   Noncoding functional RNA

Noncoding RNAs are functional RNA molecules that are not translated into protein. Examples of noncoding RNA include ribosomal RNA, transfer RNA, Piwi-interacting RNA and microRNA.

MicroRNAs are predicted to control the translational activity of approximately 30% of all protein-coding genes in mammals and may be vital components in the progression or treatment of various diseases including cancer, cardiovascular disease, and the immune system response to infection.[20]

26.2.2   *Cis-* and *Trans*-regulatory elements

Cis-regulatory elements are sequences that control the transcription of a nearby gene. Cis-elements may be located in 5' or 3' untranslated regions or within introns. Trans-regulatory elements control the transcription of a distant gene.

Promoters facilitate the transcription of a particular gene and are typically upstream of the coding region. Enhancer sequences may also exert very distant effects on the transcription levels of genes.[21]

26.2.3   Introns

Introns are non-coding sections of a gene, transcribed into the precursor mRNA sequence, but ultimately removed by RNA splicing during the processing to mature messenger RNA. Many introns appear to be mobile genetic elements.[22]

Studies of group I introns from *Tetrahymena* protozoans indicate that some introns appear to be selfish genetic elements, neutral to the host because they remove themselves from flanking exons during RNA processing and do not produce an expression bias between alleles with and without the intron.[22] Some introns appear to have significant biological function, possibly through ribozyme functionality that may regulate tRNA and rRNA activity as well as protein-coding gene expression, evident in hosts that have become dependent on such introns over long periods of time; for example, the *trnL-intron* is found in all green plants and appears to have been vertically inherited for several billions of years, including more than a billion years within chloroplasts and an additional 2–3 billion years prior in the cyanobacterial ancestors of chloroplasts.[22]

26.2.4   Pseudogenes

Pseudogenes are DNA sequences, related to known genes, that have lost their protein-coding ability or are otherwise no longer expressed in the cell. Pseudogenes arise from retrotransposition or genomic duplication of functional genes, and become "genomic fossils" that are nonfunctional due to mutations that prevent the transcription of the gene, such as within the gene promoter region, or fatally alter the translation of the gene, such as premature stop codons or frameshifts.[23] Pseudogenes resulting from the retrotransposition of an RNA intermediate are known as processed pseudogenes; pseudogenes that arise from the genomic remains of duplicated genes or residues of inactivated genes are nonprocessed pseudogenes.[23]

While Dollo's Law suggests that the loss of function in pseudogenes is likely permanent, silenced genes may actually retain function for several million years and can be "reactivated" into protein-coding sequences[24] and a substantial number of pseudogenes are actively transcribed.[23][25] Because pseudogenes are presumed to change without evolutionary constraint, they can serve as a useful model of the type and frequencies of various spontaneous genetic mutations.[26]

26.2.5   Repeat sequences, transposons and viral elements

Transposons and retrotransposons are mobile genetic elements. Retrotransposon repeated sequences, which include long interspersed nuclear elements (LINEs) and short interspersed nuclear elements (SINEs), account for a large proportion of the genomic sequences in many species. Alu sequences, classified as a short interspersed nuclear element, are the most abundant mobile elements in the human genome. Some examples have been found of SINEs exerting transcriptional control of some protein-encoding genes.[27][28][29]

Endogenous retrovirus sequences are the product of reverse transcription of retrovirus genomes into the genomes of germ cells. Mutation within these retro-transcribed sequences can inactivate the viral genome.[30]

Over 8% of the human genome is made up of (mostly decayed) endogenous retrovirus sequences, as part of the over 42% fraction that is recognizably derived of retrotransposons, while another 3% can be identified to be the remains of DNA transposons. Much of the remaining half of the genome that is currently without an explained origin is expected to have found its origin in transposable elements that were active so long ago (> 200 million years) that random mutations have rendered them unrecognizable.[31] Genome size variation in at least two kinds of plants is mostly the result of retrotransposon sequences.[32][33]

26.2.6   Telomeres

Telomeres are regions of repetitive DNA at the end of a chromosome, which provide protection from chromosomal deterioration during DNA replication.

26.3   Junk DNA

The term "junk DNA" became popular in the 1960s.[34][35] According to T. Ryan Gregory, a genomic biologist, the first explicit discussion of the nature of junk DNA was done by David Comings in 1972 and he applied the term to all noncoding DNA.[36] The term was formalized in 1972 by Susumu Ohno,[37] who noted that the mutational load from deleterious mutations placed an upper limit on the number of functional loci that could be expected given a typical mutation rate. Ohno predicted that mammal genomes could not have more than 30,000 loci under selection before the "cost" from the mutational load would cause an inescapable decline in fitness, and eventually extinction. This prediction remains robust, with the human genome containing approximately 20,000 genes. Another source for Ohno's theory was the observation that even closely related species can have widely (orders-of-magnitude) different genome sizes, which had been dubbed the C value paradox in 1971.[8]

Though the fruitfulness of the term "junk DNA" has been questioned on the grounds that it provokes a strong a priori assumption of total non-functionality and though some have recommended using more neutral terminology such as "non-coding DNA" instead;[36] "junk DNA" remains a label for the portions of a genome sequence for which no discernible function has been identified and that through comparative genomics analysis appear under no functional constraint suggesting that the sequence itself has provided no adaptive advantage. Since the late 70s it has become apparent that the majority of non-coding DNA in large genomes finds its origin in the selfish amplification of transposable elements, of which W. Ford Doolittle and Carmen Sapienza in 1980 wrote in the journal *Nature*: "When a given DNA, or class of DNAs, of unproven phenotypic function can be shown to have evolved a strategy (such as transposition) which ensures its genomic survival, then no other explanation for its existence is necessary."[38] The amount of junk DNA can be expected to depend on the rate of amplification of these elements and the rate at which non-functional DNA is lost.[39] In the same issue of *Nature*, Leslie Orgel and Francis Crick wrote that junk DNA has "little specificity and conveys little or no selective advantage to the organism".[40] The term occurs mainly in popular science and in a colloquial way in scientific publications, and it has occasionally been suggested that its connotations may have delayed interest in the biological functions of noncoding DNA.[41]

Several lines of evidence indicate that some "junk DNA" sequences are likely to have unidentified functional activity and that the process of exaptation of fragments of originally selfish or non-functional DNA has been commonplace throughout evolution.[42] In 2012, the ENCODE project, a research program supported by the National Human Genome Research Institute, reported that 76% of the human genome's noncoding DNA sequences were transcribed and that nearly half of the genome was in some way accessible to genetic regulatory proteins such as transcription factors.[4]

However, the suggestion by ENCODE that over 80% of the human genome is biochemically functional has been criticized by other scientists,[7] who argue that neither accessibility of segments of the genome to transcription factors nor their transcription guarantees that those segments have biochemical function and that their transcription is selectively advantageous. Furthermore, the much lower estimates of functionality prior to ENCODE were based on genomic conservation estimates across mammalian lineages.[8][9][10][11]

In response to such views, other scientists argue that the wide spread transcription and splicing that is observed in the human genome directly by biochemical testing is a more accurate indicator of genetic function than genomic conservation because conservation estimates are relative due to incredible variations in genome sizes of even closely related species, it is partially tautological, and these estimates are not based on direct testing for functionality on the genome.[13][15] Conservation estimates may be used to provide clues to identify possible functional elements in the genome, but it does not limit or cap the total amount of functional elements that could possibly exist in the genome since elements that do things at the molecular level can be missed by comparative genomics.[13] Furthermore, much of the apparent junk DNA is involved in epigenetic regulation and appears to be necessary for the development of complex organisms.[6][15][16]

In a 2014 paper, ENCODE researchers tried to address "the question of whether nonconserved but biochemically active regions are truly functional". They noted that in the literature, functional parts of the genome have been identified differently in previous studies depending on the approaches used. There have been three general approaches used to identify functional parts of the human genome: genetic approaches (which rely on changes in phenotype), evolutionary approaches (which rely on conservation) and biochemical approaches (which rely on biochemical testing and was used by ENCODE). All three have limitations: genetic approaches may miss functional elements that do not manifest physically on the organism, evolutionary approaches have difficulties using accurate multispecies sequence alignments since genomes of even closely related species vary considerably, and with biochemical approaches, though having high reproducibility, the biochemical signatures do not always automatically signify a function.[13]

They noted that 70% of the transcription coverage was less than 1 transcript per cell. They noted that this "larger proportion of genome with reproducible but low biochemical signal strength and less evolutionary conservation is challenging to parse between specific functions and biological noise". Furthermore, assay resolution often is much broader than the underlying functional sites so some of the reproducibly "biochemically active but selectively neutral" sequences are unlikely to serve critical functions, especially those with lower-level biochemical signal. To this they added, "However, we also acknowledge substantial limitations in our current detection of constraint, given that some human-specific functions are essential but not conserved and that disease-relevant regions need not be selectively constrained to be functional." On the other hand, they argued that the 12–15% fraction of human DNA under functional constraint, as estimated by a variety of extrapolative evolutionary methods, may still be an underestimate. They concluded that in contrast to evolutionary and genetic evidence, biochemical data offer clues about both the molecular function served by underlying DNA elements and the cell types in which they act. Ultimately genetic, evolutionary, and biochemical approaches can all be used in a complementary way to identify regions that may be functional in human biology and disease.[13]

Some critics have argued that functionality can only be assessed in reference to an appropriate null hypothesis. In this case, the null hypothesis would be that these parts of the genome are non-functional and have properties, be it on the basis of conservation or biochemical activity, that would be expected of such regions based on our general understanding of molecular evolution and biochemistry. According to these critics, until a region in question has been shown to have additional features, beyond what is expected of the null hypothesis, it should provisionally be labelled as non-functional.[43]

26.4   Functions of noncoding DNA

Many noncoding DNA sequences have important biological functions as indicated by comparative genomics studies that report some regions of noncoding DNA that are highly conserved, sometimes on time-scales representing hundreds of millions of years, implying that these noncoding regions are under strong evolutionary pressure and positive selection.[44] For example, in the genomes of humans and mice, which diverged from a common ancestor 65–75 million years ago, protein-coding DNA sequences account for only about 20% of conserved DNA, with the remaining 80% of conserved DNA represented in noncoding regions.[45] Linkage mapping often identifies chromosomal regions associated with a disease with no evidence of functional coding variants of genes within the region, suggesting that disease-causing genetic variants lie in the noncoding DNA.[45] The significance of noncoding DNA mutations in cancer was explored in April 2013.[46]

Noncoding genetic polymorphisms have also been shown to play a role in infectious disease susceptibility, such as hepatitis C.[47]

Some specific sequences of noncoding DNA may be features essential to chromosome structure, centromere function and homolog recognition in meiosis.[48]

According to a comparative study of over 300 prokaryotic and over 30 eukaryotic genomes,[49] eukaryotes appear to require a minimum amount of non-coding DNA. This minimum amount can be predicted using a growth model for regulatory genetic networks, implying that it is required for regulatory purposes. In humans the predicted minimum is about 5% of the total genome.

There is evidence that a significant proportion (over 10%) of 32 mammalian genomes may function through the formation of specific RNA secondary structures.[50] The study used comparative genomics to identify compensatory DNA mutations that maintain RNA base-pairings, a distinctive feature of RNA molecules. Over 80% of the genomic regions presenting evolutionary evidence of RNA structure conservation do not present strong DNA sequence conservation.

26.4.1   Protection of the genome

Main article: Mutation

Noncoding DNA separate genes from each other with long gaps, so mutation in one gene or part of a chromosome, for example deletion or insertion, does not have the "frameshift mutation" on the whole chromosome. When genome complexity is relatively high, like in the case of human genome, not only different genes, but also inside one gene there are gaps of introns to protect the entire coding segment to minimise the changes caused by mutation.

It has been suggested that non-coding DNA may serve to decrease the chance of gene disruption during chromosomal crossover.[51]

26.4.2   Genetic switches

Some noncoding DNA sequences are genetic "switches" that regulate when and where genes are expressed.[52]

26.4.3   Regulation of gene expression

Main article: Regulation of gene expression

Some noncoding DNA sequences determine the expression levels of various genes.[53]

26.4.4   Transcription factor sites

Main article: Transcription factor

Some noncoding DNA sequences determine where transcription factors attach.[53] A transcription factor is a protein that binds to specific non-coding DNA sequences, thereby controlling the flow (or transcription) of genetic information from DNA to mRNA. Transcription factors act at very different locations on the genomes of different people.

Operators

Main article: Operator (biology)

An operator is a segment of DNA to which a repressor binds. A repressor is a DNA-binding protein that regulates the expression of one or more genes by binding to the operator and blocking the attachment of RNA polymerase to the promoter, thus preventing transcription of the genes. This blocking of expression is called repression.

Enhancers

Main article: Enhancer (genetics)

An enhancer is a short region of DNA that can be bound with proteins (trans-acting factors), much like a set of transcription factors, to enhance transcription levels of genes in a gene cluster.

Silencers

Main article: Silencer (DNA)

A silencer is a region of DNA that inactivates gene expression when bound by a regulatory protein. It functions in a very similar way as enhancers, only differing in the inactivation of genes.

Promoters

Main article: Promoter (biology)

A promoter is a region of DNA that facilitates transcription of a particular gene. Promoters are typically located near the genes they regulate.

Insulators

Main article: Insulator (genetics)

A genetic insulator is a boundary element that plays two distinct roles in gene expression, either as an enhancer-blocking code, or rarely as a barrier against condensed chromatin. An insulator in a DNA sequence is comparable to a linguistic word divider such as a comma (,) in a sentence, because the insulator indicates where an enhanced or repressed sequence ends.

26.5   Uses of noncoding DNA

26.5.1   Noncoding DNA and evolution

Shared sequences of apparently non-functional DNA are a major line of evidence of common descent.[54]

Pseudogene sequences appear to accumulate mutations more rapidly than coding sequences due to a loss of selective pressure.[26] This allows for the creation of mutant alleles that incorporate new functions that may be favored by natural selection; thus, pseudogenes can serve as raw material for evolution and can be considered "protogenes".[55]

26.5.2 Long range correlations

A statistical distinction between coding and noncoding DNA sequences has been found. It has been observed that nucleotides in non-coding DNA sequences display long range power law correlations while coding sequences do not.[56][57][58]

26.5.3 Forensic anthropology

Police sometimes gather DNA as evidence for purposes of forensic identification. As described in *Maryland v. King*, a 2013 U.S. Supreme Court decision:[59]

> "The current standard for forensic DNA testing relies on an analysis of the chromosomes located within the nucleus of all human cells. "The DNA material in chromosomes is composed of 'coding' and 'noncoding' regions. The coding regions are known as genes and contain the information necessary for a cell to make proteins. . . . Non-protein coding regions . . . are not related directly to making proteins, [and] have been referred to as 'junk' DNA." The adjective "junk" may mislead the lay person, for in fact this is the DNA region used with near certainty to identify a person.

26.6 See also

- Conserved non-coding sequence
- Eukaryotic chromosome fine structure
- Gene-centered view of evolution
- Gene regulatory network
- Intergenic region
- Intragenomic conflict
- Phylogenetic footprinting
- Transcriptome
- Non-coding RNA

26.7 References

[1] "Worlds Record Breaking Plant: Deletes its Noncoding "Junk" DNA". *Design & Trend*. May 12, 2013. Retrieved 2013-06-04.

[2] Elgar G, Vavouri T; Vavouri (July 2008). "Tuning in to the signals: noncoding sequence conservation in vertebrate genomes". *Trends Genet.* **24** (7): 344–52. doi:10.1016/j.tig.2008.04.005. PMID 18514361.

[3] Costa, Fabrico (2012). "7 Non-coding RNAs, Epigenomics, and Complexity in Human Cells". In Morris, Kevin V. *Non-coding RNAs and Epigenetic Regulation of Gene Expression: Drivers of Natural Selection*. Caister Academic Press. ISBN 1904455948.

[4] Pennisi, E. (6 September 2012). "ENCODE Project Writes Eulogy for Junk DNA". *Science***337**(6099): 1159–1161. doi:10.11

[5] The ENCODE Project Consortium (2012). "An integrated encyclopedia of DNA elements in the human genome". *Nature* **489** (7414): 57–74. Bibcode:2012Natur.489...57T. doi:10.1038/nature11247. PMC 3439153. PMID 22955616..

[6] Carey, Nessa (2015). *Junk DNA: A Journey Through the Dark Matter of the Genome*. Columbia University Press. ISBN 9780231170840.

[7] Robin McKie (24 February 2013). "Scientists attacked over claim that 'junk DNA' is vital to life". *The Observer*.

[8] Sean Eddy (2012) The C-value paradox, junk DNA, and ENCODE, Curr Biol 22(21):R898–R899.

[9] Doolittle, W. Ford (2013). "Is junk DNA bunk? A critique of ENCODE". *Proc Natl Acad Sci USA* **110** (14): 5294–5300. Bibcode:2013PNAS..110.5294D. doi:10.1073/pnas.1221376110. PMC 3619371. PMID 23479647.

[10] Palazzo, Alexander F.; Gregory, T. Ryan (2014). "The Case for Junk DNA". *PLoS Genetics***10**(5): e1004351. doi:10.1371/jou ISSN 1553-7404.

[11] Dan Graur, Yichen Zheng, Nicholas Price, Ricardo B. R. Azevedo1, Rebecca A. Zufall and Eran Elhaik (2013). "On the immortality of television sets: "function" in the human genome according to the evolution-free gospel of ENCODE" (PDF). *Genome Biology and Evolution* **5** (3): 578–90. doi:10.1093/gbe/evt028. PMC 3622293. PMID 23431001.

[12] Ponting, CP; Hardison, RC (2011). "What fraction of the human genome is functional?". *Genome Research* **21**: 1769–1776. doi:10.1101/gr.116814.110. PMC 3205562. PMID 21875934.

[13] Kellis, M.; et al. (2014). "Defining functional DNA elements in the human genome". *PNAS***111**(17): 6131–6138. Bibcode:20 doi:10.1073/pnas.1318948111. PMC 4035993. PMID 24753594.

[14] Chris M. Rands, Stephen Meader, Chris P. Ponting and Gerton Lunter (2014). "8.2% of the Human Genome Is Constrained: Variation in Rates of Turnover across Functional Element Classes in the Human Lineage". *PLoS Genet* **10** (7): e1004525. doi:10.1371/journal.pgen.1004525. PMC 4109858. PMID 25057982.

[15] Mattick JS, Dinger ME (2013). "The extent of functionality in the human genome". *The HUGO Journal***7**(1): 2. doi:10.1186/1 6566-7-2.

[16] Morris, Kevin, ed. (2012). *Non-Coding RNAs and Epigenetic Regulation of Gene Expression: Drivers of Natural Selection.* Norfolk, UK: Caister Academic Press. ISBN 1904455948.

[17] Gregory TR, Hebert PD; Hebert (April 1999). "The modulation of DNA content: proximate causes and ultimate consequences". *Genome Res.* **9** (4): 317–24. doi:10.1101/gr.9.4.317 (inactive 2015-02-01). PMID 10207154.

[18] Wahls, W.P.; et al. (1990). "Hypervariable minisatellite DNA is a hotspot for homologous recombination in human cells". *Cell* **60** (1): 95–103. doi:10.1016/0092-8674(90)90719-U. PMID 2295091.

[19] Waterhouse, Peter M.; Hellens, Roger P. (25 March 2015). "Plant biology: Coding in non-coding RNAs". *Nature* **520** (7545): 41–42. doi:10.1038/nature14378.

[20] Li M, Marin-Muller C, Bharadwaj U, Chow KH, Yao Q, Chen C; Marin-Muller; Bharadwaj; Chow; Yao; Chen (April 2009). "MicroRNAs: Control and Loss of Control in Human Physiology and Disease". *World J Surg* **33** (4): 667–84. doi:10.1007/s002 68-008-9836-x. PMC 2933043. PMID19030926.

[21] Visel A, Rubin EM, Pennacchio LA (September 2009). "Genomic Views of Distant-Acting Enhancers". *Nature* **461** (7261): 199–205. Bibcode:2009Natur.461..199V. doi:10.1038/nature08451. PMC 2923221. PMID 19741700.

[22] Nielsen H, Johansen SD; Johansen (2009). "Group I introns: Moving in new directions". *RNA Biol***6**(4): 375–83. doi:10.4161/ PMID 19667762.

[23] Zheng D, Frankish A, Baertsch R; et al. (June 2007). "Pseudogenes in the ENCODE regions: Consensus annotation, analysis of transcription, and evolution". *Genome Res.* **17** (6): 839–51. doi:10.1101/gr.5586307. PMC 1891343. PMID 17568002.

[24] Marshall CR, Raff EC, Raff RA; Raff; Raff (December 1994). "Dollo's law and the death and resurrection of genes". *Proc. Natl. Acad. Sci. U.S.A.* **91** (25): 12283–7. Bibcode:1994PNAS...9112283M. doi:10.1073/pnas.91.25.12283. PMC 45421. PMID 7991619.

[25] Tutar, Y. (2012). "Pseudogenes". *Comp Funct Genomics* **2012**: 424526. doi:10.1155/2012/424526. PMC 3352212. PMID 22611337.

[26] Petrov DA, Hartl DL; Hartl (2000). "Pseudogene evolution and natural selection for a compact genome". *J. Hered.* **91** (3): 221–7. doi:10.1093/jhered/91.3.221. PMID 10833048.

[27] Ponicsan SL, Kugel JF, Goodrich JA; Kugel; Goodrich (February 2010). "Genomic gems: SINE RNAs regulate mRNA production". *Current Opinion in Genetics & Development* **20** (2): 149–55. doi:10.1016/j.gde.2010.01.004. PMC 2859989. PMID 20176473.

[28] Häsler J, Samuelsson T, Strub K; Samuelsson; Strub (July 2007). "Useful 'junk': Alu RNAs in the human transcriptome". *Cell. Mol. Life Sci.* **64** (14): 1793–800. doi:10.1007/s00018-007-7084-0. PMID 17514354.

[29] Walters RD, Kugel JF, Goodrich JA; Kugel; Goodrich (Aug 2009). "InvAluable junk: the cellular impact and function of Alu and B2 RNAs". *IUBMB Life* **61** (8): 831–7. doi:10.1002/iub.227. PMC 4049031. PMID 19621349.

[30] Nelson, PN.; Hooley, P.; Roden, D.; Davari Ejtehadi, H.; Rylance, P.; Warren, P.; Martin, J.; Murray, PG. (Oct 2004). "Human endogenous retroviruses: transposable elements with potential?". *Clin Exp Immunol* **138** (1): 1–9. doi:10.1111/j.1365-2249.2004.02592.x. PMC 1809191. PMID 15373898.

[31] International Human Genome Sequencing Consortium (February 2001). "Initial sequencing and analysis of the human genome". *Nature* **409** (6822): 879–888. Bibcode:2001Natur.409..860L. doi:10.1038/35057062. PMID 11237011.

[32] Piegu, B.; Guyot, R.; Picault, N.; Roulin, A.; Sanyal, A.; Saniyal, A.; Kim, H.; Collura, K.; et al. (Oct 2006). "Doubling genome size without polyploidization: dynamics of retrotransposition-driven genomic expansions in Oryza australiensis, a wild relative of rice". *Genome Res* **16** (10): 1262–9. doi:10.1101/gr.5290206. PMC 1581435. PMID 16963705.

[33] Hawkins, JS.; Kim, H.; Nason, JD.; Wing, RA.; Wendel, JF. (Oct 2006). "Differential lineage-specific amplification of transposable elements is responsible for genome size variation in Gossypium". *Genome Res* **16** (10): 1252–61. doi:10.1101/gr.5282906. PMC 1581434. PMID 16954538.

[34] Ehret CF, De Haller G; De Haller (1963). "Origin, development, and maturation of organelles and organelle systems of the cell surface in Paramecium". *Journal of Ultrastructure Research*. 9 Supplement 1: 1, 3–42. doi:10.1016/S0022-5320(63)80088-X. PMID 14073743.

[35] Dan Graur, The Origin of Junk DNA: A Historical Whodunnit

[36] Gregory, T. Ryan, ed. (2005). *The Evolution of the Genome*. Elsevier. pp. 29–31. ISBN 0123014638. Comings (1972), on the other hand, gave what must be considered the first explicit discussion of the nature of "junk DNA," and was the first to apply the term to all noncoding DNA."; "For this reason, it is unlikely that any one function for noncoding DNA can account for either its sheer mass or its unequal distribution among taxa. However, dismissing it as no more than "junk" in the pejorative sense of "useless" or "wasteful" does little to advance the understanding of genome evolution. For this reason, the far less loaded term "noncoding DNA" is used throughout this chapter and is recommended in preference to "junk DNA" for future treatments of the subject."

[37] Ohno, Susumu (1972). H. H. Smith, ed. *So Much "junk" DNA in Our Genome*. Gordon and Breach, New York. pp. 366–370. Retrieved 2013-05-15.

[38] Doolittle WF, Sapienza C; Sapienza (1980). "Selfish genes, the phenotype paradigm and genome evolution". *Nature* **284** (5757): 601–603. Bibcode:1980Natur.284..601D. doi:10.1038/284601a0. PMID 6245369.

[39] Another source is genome duplication followed by a loss of function due to redundancy.

[40] Orgel LE, Crick FH; Crick (April 1980). "Selfish DNA: the ultimate parasite". *Nature* **284**(5757): 604–7. Bibcode:1980Natur doi:10.1038/284604a0. PMID 7366731.

[41] Khajavinia A, Makalowski W; Makalowski (May 2007). "What is "junk" DNA, and what is it worth?". *Scientific American* **296** (5): 104. doi:10.1038/scientificamerican0307-104. PMID 17503549. The term "junk DNA" repelled mainstream researchers from studying noncoding genetic material for many years

[42] Biémont, Christian; Vieira, C (2006). "Genetics: Junk DNA as an evolutionary force". *Nature* **443**(7111): 521–4. Bibcode:20 doi:10.1038/443521a. PMID 17024082.

[43] Palazzo, Alexander F.; Lee, Eliza S. (2015). "Non-coding RNA: what is functional and what is junk?". *Frontiers in Genetics* **6**. doi:10.3389/fgene.2015.00002. ISSN 1664-8021.

[44] Ludwig MZ (December 2002). "Functional evolution of noncoding DNA". *Current Opinion in Genetics & Development* **12** (6): 634–9. doi:10.1016/S0959-437X(02)00355-6. PMID 12433575.

[45] Cobb J, Büsst C, Petrou S, Harrap S, Ellis J; Büsst; Petrou; Harrap; Ellis (April 2008). "Searching for functional genetic variants in non-coding DNA". *Clin. Exp. Pharmacol. Physiol.* **35** (4): 372–5. doi:10.1111/j.1440-1681.2008.04880.x. PMID 18307723.

[46] E Khurana; et al. (April 2013). "Integrative annotation of variants from 1092 humans: application to cancer genomics". *Science* **342** (6154): 372–5. doi:10.1126/science.1235587. PMC 3947637. PMID 24092746.

[47] Lu, Yi-Fan; Mauger, David M.; Goldstein, David B.; Urban, Thomas J.; Weeks, Kevin M.; Bradrick, Shelton S. (4 November 2015). "IFNL3 mRNA structure is remodeled by a functional non-coding polymorphism associated with hepatitis C virus clearance". *Scientific Reports* **5**: 16037. doi:10.1038/srep16037.

[48] Subirana JA, Messeguer X; Messeguer (March 2010). "The most frequent short sequences in non-coding DNA". *Nucleic Acids Res.* **38** (4): 1172–81. doi:10.1093/nar/gkp1094. PMC 2831315. PMID 19966278.

[49] S. E. Ahnert; T. M. A. Fink (2008). "How much non-coding DNA do eukaryotes require?" (PDF). *J. Theor. Biol.* **252** (4): 587–592. doi:10.1016/j.jtbi.2008.02.005. PMID 18384817.

[50] Smith MA; et al. (June 2013). "Widespread purifying selection on RNA structure in mammals". *Nucleic Acids Research* **41** (17): 8220–8236. doi:10.1093/nar/gkt596. PMC 3783177. PMID 23847102.

[51] Dileep, V. (2009). The place and function of non-coding DNA in the evolution of variability. *Hypothesis*, **7**(1): e7.

[52] Carroll, Sean B.; et al. (May 2008). "Regulating Evolution". *Scientific American***298**(5): 60–67. doi:10.1038/scientificamerican 60. PMID 18444326.

[53] Callaway, Ewen (March 2010). "Junk DNA gets credit for making us who we are". *New Scientist*.

[54] "Plagiarized Errors and Molecular Genetics", talkorigins, by Edward E. Max, M.D., Ph.D.

[55] Balakirev ES, Ayala FJ; Ayala (2003). "Pseudogenes: are they "junk" or functional DNA?". *Annu. Rev. Genet.* **37**: 123–51. doi:10.1146/annurev.genet.37.040103.103949. PMID 14616058.

[56] C.-K. Peng, S. V. Buldyrev, A. L. Goldberger, S. Havlin, F. Sciortino, M. Simons, H. E. Stanley; Buldyrev, SV; Goldberger, AL; Havlin, S; Sciortino, F; Simons, M; Stanley, HE (1992). "Long-range correlations in nucleotide sequences". *Nature* **356** (6365): 168–70. Bibcode:1992Natur.356..168P. doi:10.1038/356168a0. PMID 1301010.

[57] W. Li and, K. Kaneko; Kaneko, K (1992). "Long-Range Correlation and Partial 1/f^alpha Spectrum in a Non-Coding DNA Sequence" (PDF). *Europhys. Lett* **17** (7): 655–660. Bibcode:1992EL......17..655L. doi:10.1209/0295-5075/17/7/014.

[58] S. V. Buldyrev, A. L. Goldberger, S. Havlin, R. N. Mantegna, M. Matsa, C.-K. Peng, M. Simons, and H. E. Stanley; Goldberger, A.; Havlin, S.; Mantegna, R.; Matsa, M.; Peng, C.-K.; Simons, M.; Stanley, H. (1995). "Long-range correlations properties of coding and noncoding DNA sequences: GenBank analysis". *Phys. Rev. E* **51** (5): 5084–5091. Bibcode:1995PhRvE..51.5084B. doi:10.1103/PhysRevE.51.5084.

[59] Slip opinion for *Maryland v. King* from the U.S. Supreme Court. Retrieved 2013-06-04.

26.8   Further reading

Bennett, Michael D.; Leitch, Ilia J. (2005). "Genome size evolution in plants". In Gregory, T. Ryan. *The Evolution of the Genome*. San Diego: Elsevier. pp. 89–162. ISBN 978-0-08-047052-8.

Gregory, T.R (2005). "Genome size evolution in animals". In T.R. Gregory (ed.). *The Evolution of the Genome*. San Diego: Elsevier. ISBN 0-12-301463-8.

Shabalina SA, Spiridonov NA; Spiridonov (2004). "The mammalian transcriptome and the function of non-coding DNA sequences". *Genome Biol.* **5** (4): 105. doi:10.1186/gb-2004-5-4-105. PMC 395773. PMID 15059247.

Castillo-Davis CI (October 2005). "The evolution of noncoding DNA: how much junk, how much func?". *Trends Genet.* **21** (10): 533–6. doi:10.1016/j.tig.2005.08.001. PMID 16098630.

26.9   External links

- Plant DNA C-values Database at Royal Botanic Gardens, Kew

- Fungal Genome Size Database at Estonian Institute of Zoology and Botany

- ENCODE: The human encyclopaedia at *Nature* ENCODE

Utricularia gibba has 3% noncoding DNA.

Chapter 27

RNA polymerase

Compare RNA-dependent RNA polymerase.

RNA polymerase (**RNAP** or **RNApol**), also known as **DNA-dependent RNA polymerase**, is an enzyme that produces primary transcript RNA. In cells, RNAP is necessary for constructing RNA chains using DNA genes as templates, a process called transcription. RNA polymerase enzymes are essential to life and are found in all organisms and many viruses. In chemical terms, RNAP is a nucleotidyl transferase that polymerizes ribonucleotides at the 3' end of an RNA transcript.

27.1   History

RNAP was discovered independently by Charles Loe, Audrey Stevens, and Jerard Hurwitz in 1960.[1] By this time, one half of the 1959 Nobel Prize in Medicine had been awarded to Severo Ochoa for the discovery of what was believed to be RNAP,[2] but instead turned out to be polynucleotide phosphorylase.

The 2006 Nobel Prize in Chemistry was awarded to Roger D. Kornberg for creating detailed molecular images of RNA polymerase during various stages of the transcription process.[3]

27.2   Control of transcription

Control of the process of gene transcription affects patterns of gene expression and, thereby, allows a cell to adapt to a changing environment, perform specialized roles within an organism, and maintain basic metabolic processes necessary for survival. Therefore, it is hardly surprising that the activity of RNAP is both long and complex and highly regulated. In *Escherichia coli* bacteria, more than 100 transcription factors have been identified, which modify the activity of RNAP.[4]

RNAP can initiate transcription at specific DNA sequences known as promoters. It then produces an RNA chain, which is complementary to the template DNA strand. The process of adding nucleotides to the RNA strand is known as elongation; in eukaryotes, RNAP can build chains as long as 2.4 million nucleotides (the full length of the dystrophin gene). RNAP will preferentially release its RNA transcript at specific DNA sequences encoded at the end of genes, which are known as terminators.

Products of RNAP include:

- Messenger RNA (mRNA)—template for the synthesis of proteins by ribosomes.

- Non-coding RNA or "RNA genes"—a broad class of genes that encode RNA that is not translated into protein. The most prominent examples of RNA genes are transfer RNA (tRNA) and ribosomal RNA (rRNA), both of which

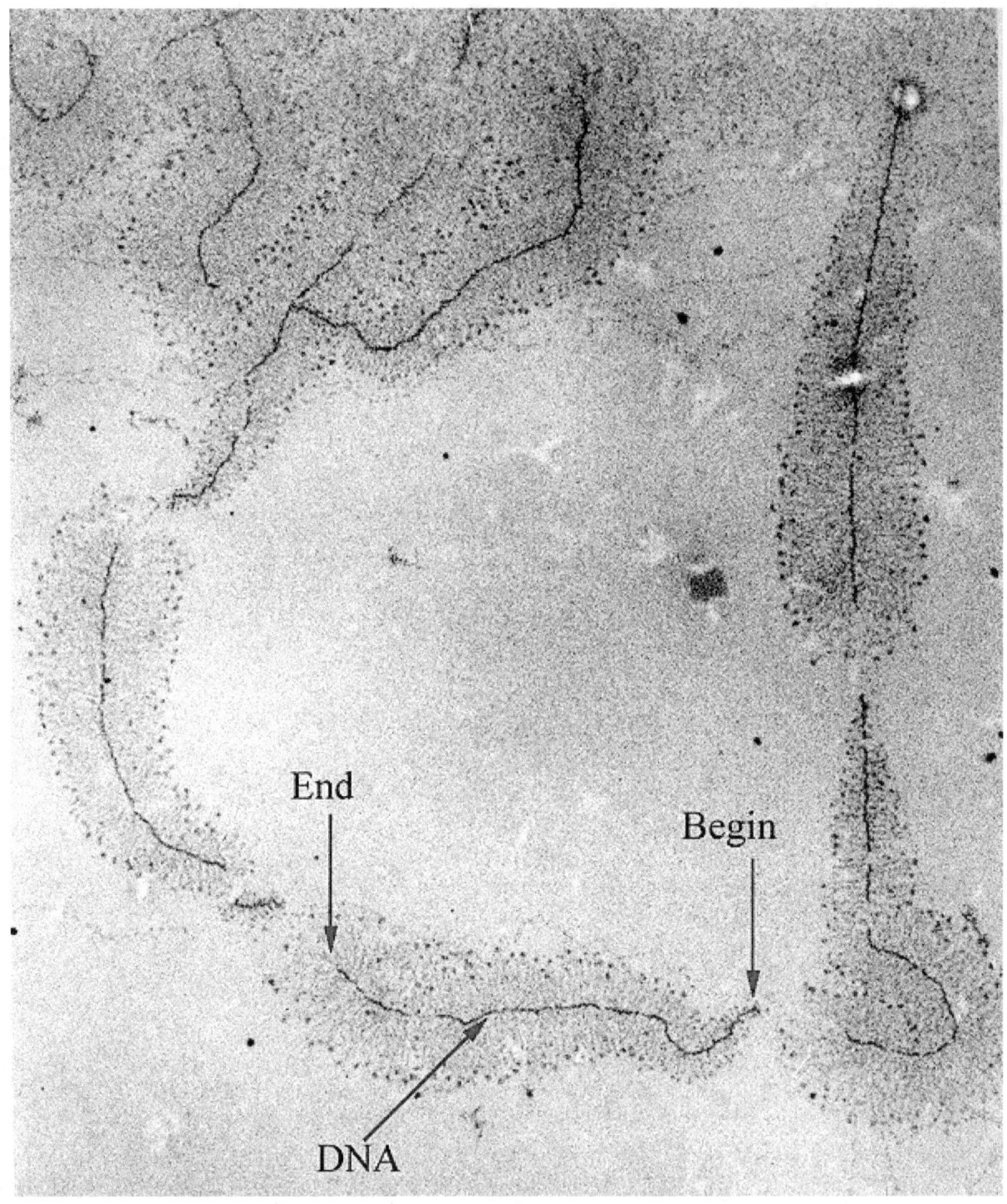

An electron-micrograph of DNA strands decorated by hundreds of RNAP molecules too small to be resolved. Each RNAP is transcribing an RNA strand, which can be seen branching off from the DNA. "Begin" indicates the 3' end of the DNA, where RNAP initiates transcription; "End" indicates the 5' end, where the longer RNA molecules are completely transcribed.

are involved in the process of translation. However, since the late 1990s, many new RNA genes have been found, and thus RNA genes may play a much more significant role than previously thought.

- Transfer RNA (tRNA)—transfers specific amino acids to growing polypeptide chains at the ribosomal site of protein synthesis during translation

- Ribosomal RNA (rRNA)—a component of ribosomes
- Micro RNA—regulates gene activity
- Catalytic RNA (Ribozyme)—enzymatically active RNA molecules

RNAP accomplishes *de novo* synthesis. It is able to do this because specific interactions with the initiating nucleotide hold RNAP rigidly in place, facilitating chemical attack on the incoming nucleotide. Such specific interactions explain why RNAP prefers to start transcripts with ATP (followed by GTP, UTP, and then CTP). In contrast to DNA polymerase, RNAP includes helicase activity, therefore no separate enzyme is needed to unwind DNA.

27.3   Action

RNA polymerase binding in bacteria involves the sigma factor recognizing the core promoter region containing the -35 and -10 elements (located *before* the beginning of sequence to be transcribed) and also, at some promoters, the α subunit C-terminal domain recognizing promoter upstream elements. There are multiple interchangeable sigma factors, each of which recognizes a distinct set of promoters. For example, in *E. coli*, σ^{70} is expressed under normal conditions and recognizes promoters for genes required under normal conditions ("housekeeping genes"), while σ^{32} recognizes promoters for genes required at high temperatures ("heat-shock genes").

After binding to the DNA, the RNA polymerase switches from a closed complex to an open complex. This change involves the separation of the DNA strands to form an unwound section of DNA of approximately 13 bp, referred to as the transcription bubble. Ribonucleotides are base-paired to the template DNA strand, according to Watson-Crick base-pairing interactions. Supercoiling plays an important part in polymerase activity because of the unwinding and rewinding of DNA. Because regions of DNA in front of RNAP are unwound, there are compensatory positive supercoils. Regions behind RNAP are rewound and negative supercoils are present.

As noted above, RNA polymerase makes contacts with the promoter region. However these stabilizing contacts inhibit the enzyme's ability to access DNA further downstream and thus the synthesis of the full-length product. Once the open complex is stabilized, RNA polymerase synthesizes an RNA strand to establish a DNA-RNA heteroduplex (~8-9 bp) at the active center, which stabilizes the elongation complex. In order to accomplish RNA synthesis, RNA polymerase must maintain promoter contacts while unwinding more downstream DNA for synthesis, "scrunching" more downstream DNA into the initiation complex. During the promoter escape transition, RNA polymerase is considered a "stressed intermediate." Thermodynamically the stress accumulates from the DNA-unwinding and DNA-compaction activities. Once the DNA-RNA heteroduplex is long enough, RNA polymerase releases its upstream contacts and effectively achieves the promoter escape transition into the elongation phase. However, promoter escape is not the only outcome. RNA polymerase can also relieve the stress by releasing its downstream contacts, arresting transcription. The paused transcribing complex has two options: (1) release the nascent transcript and begin anew at the promoter or (2) reestablish a new 3'OH on the nascent transcript at the active site via RNA polymerase's catalytic activity and recommence DNA scrunching to achieve promoter escape. Scientists have coined the term "abortive initiation" to explain the unproductive cycling of RNA polymerase before the promoter escape transition. The extent of abortive initiation depends on the presence of transcription factors and the strength of the promoter contacts.

27.3.1   Elongation

Transcription elongation involves the further addition of ribonucleotides and the change of the open complex to the transcriptional complex. RNAP cannot start forming full length transcripts because of its strong binding to the promoter. Transcription at this stage primarily results in short RNA fragments of around 9 bp in a process known as abortive transcription. Once the RNAP starts forming longer transcripts it clears the promoter. At this point, the contacts with the -10 and -35 elements are disrupted, and the σ factor falls off RNAP. This allows the rest of the RNAP complex to move forward, as the σ factor held the RNAP complex in place.

The 17-bp transcriptional complex has an 8-bp DNA-RNA hybrid, that is, 8 base-pairs involve the RNA transcript bound to the DNA template strand. As transcription progresses, ribonucleotides are added to the 3' end of the RNA transcript

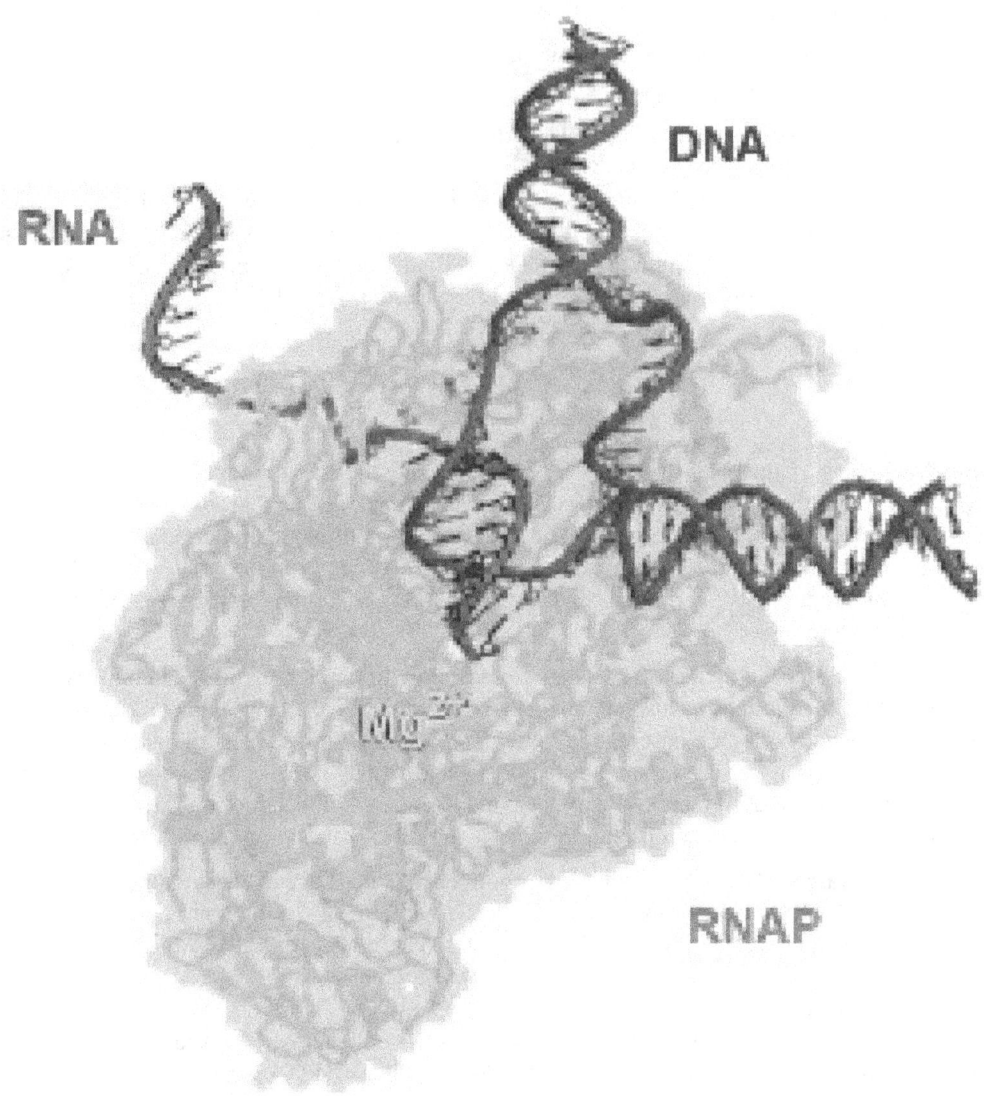

RNAP from T. aquaticus *pictured during elongation. Portions of the enzyme were made transparent so as to make the path of RNA and DNA more clear. The magnesium ion (yellow) is located at the enzyme active site.*

and the RNAP complex moves along the DNA. Although RNAP does not seem to have the 3'exonuclease activity that characterizes the *proofreading* activity found in DNA polymerase, there is evidence of that RNAP will halt at mismatched base-pairs and correct it.

Aspartyl (asp) residues in the RNAP will hold on to Mg^{2+} ions, which will, in turn, coordinate the phosphates of the ribonucleotides. The first Mg^{2+} will hold on to the α-phosphate of the NTP to be added. This allows the nucleophilic attack of the 3'OH from the RNA transcript, adding another NTP to the chain. The second Mg^{2+} will hold on to the pyrophosphate of the NTP. The overall reaction equation is:

$(NMP)_n + NTP \rightarrow (NMP)_{n+1} + PP_i$

27.3.2   Termination

In prokaryotes, termination of RNA transcription can be rho-independent or rho-dependent:

Rho-independent transcription termination is the termination of transcription without the aid of the rho protein. Transcription of a palindromic region of DNA causes the formation of a "hairpin" structure from the RNA transcription looping and binding upon itself. This hairpin structure is often rich in G-C base-pairs, making it more stable than the DNA-RNA hybrid itself. As a result, the 8 bp DNA-RNA hybrid in the transcription complex shifts to a 4 bp hybrid. These last 4 base pairs are weak A-U base pairs, and the entire RNA transcript will fall off the DNA.

27.4   In bacteria

In bacteria, the same enzyme catalyzes the synthesis of mRNA and ncRNA.

RNAP is a large molecule. The core enzyme has five subunits (~400 kDa):[5]

- β': The β' subunit is the largest subunit, and is encoded by the rpoC gene.[6] The β' subunit contains part of the active center responsible for RNA synthesis and contains some of the determinants for non-sequence-specific interactions with DNA and nascent RNA.

- β: The β subunit is the second-largest subunit, and is encoded by the rpoB gene. The β subunit contains the rest of the active center responsible for RNA synthesis and contains the rest of the determinants for non-sequence-specific interactions with DNA and nascent RNA.

- α^I and α^{II}: The α subunit is the third-largest subunit and is present in two copies per molecule of RNAP, α^I and α^{II}. Each α subunit contains two domains: αNTD (N-Terminal domain) and αCTD (C-terminal domain). αNTD contains determinants for assembly of RNAP. αCTD (C-terminal domain) contains determinants for interaction with promoter DNA, making non-sequence-non-specific interactions at most promoters and sequence-specific interactions at upstream-element-containing promoters, and contains determinants for interactions with regulatory factors.

- ω: The ω subunit is the smallest subunit. The ω subunit facilitates assembly of RNAP and stabilizes assembled RNAP.[7]

In order to bind promoters, RNAP core associates with the transcription initiation factor sigma (σ) to form RNA polymerase holoenzyme. Sigma reduces the affinity of RNAP for nonspecific DNA while increasing specificity for promoters, allowing transcription to initiate at correct sites. The complete holoenzyme therefore has 6 subunits: $\beta'\beta\alpha^I$ and $\alpha^{II}\omega\sigma$ (~450 kDa).

27.5   In eukaryotes

Eukaryotes have multiple types of nuclear RNAP, each responsible for synthesis of a distinct subset of RNA. All are structurally and mechanistically related to each other and to bacterial RNAP:

- RNA polymerase I synthesizes a pre-rRNA 45S (35S in yeast), which matures into 28S, 18S and 5.8S rRNAs which will form the major RNA sections of the ribosome.[8]

- RNA polymerase II synthesizes precursors of mRNAs and most snRNA and microRNAs.[9] This is the most studied type, and, due to the high level of control required over transcription, a range of transcription factors are required for its binding to promoters.

- RNA polymerase III synthesizes tRNAs, rRNA 5S and other small RNAs found in the nucleus and cytosol.[10]

- RNA polymerase IV synthesizes siRNA in plants.[11]

- RNA polymerase V synthesizes RNAs involved in siRNA-directed heterochromatin formation in plants.[12]

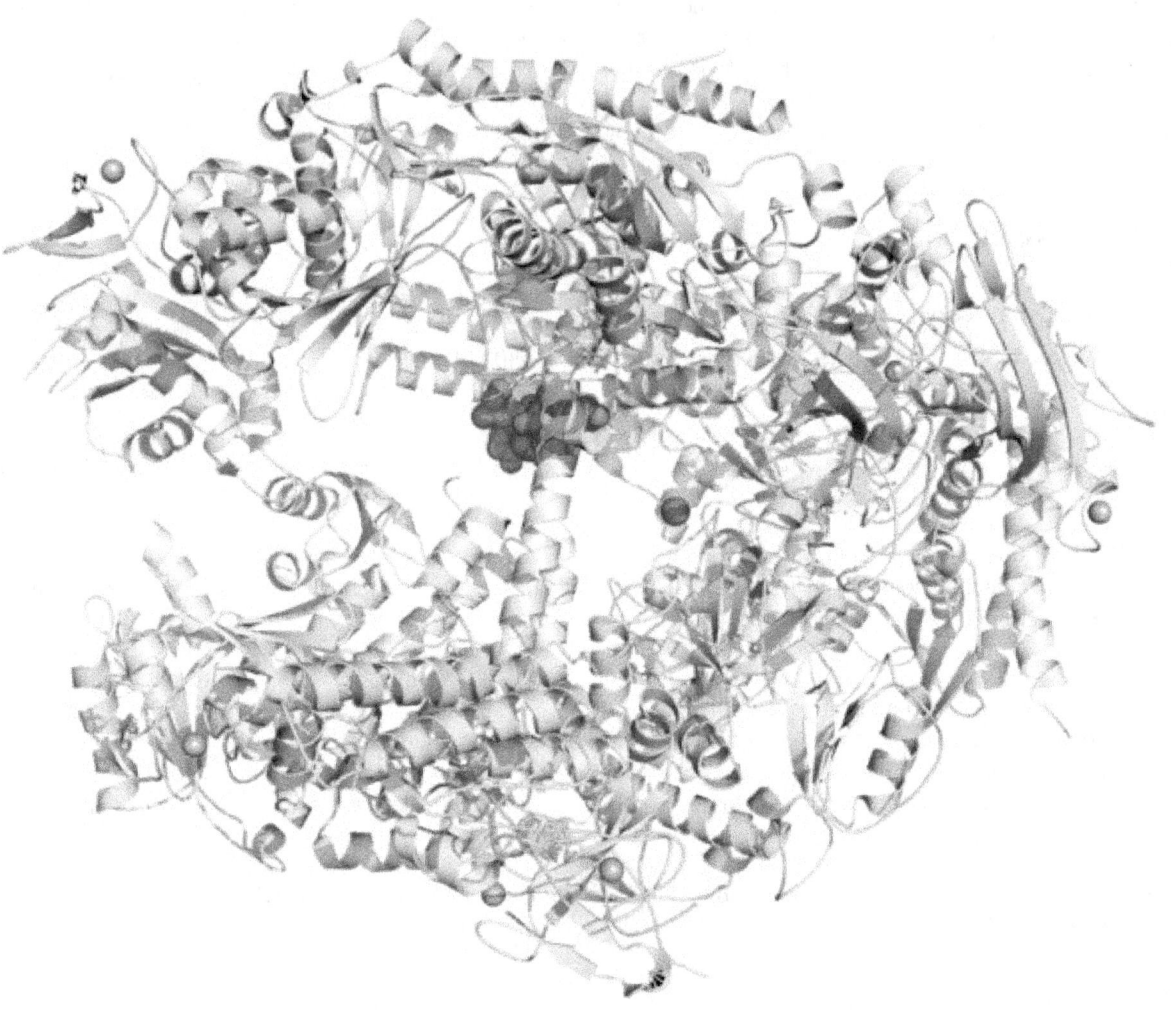

Structure of eukaryotic RNA polymerase II (light blue) in complex with α-amanitin (red), a strong poison found in death cap mushrooms that targets this vital enzyme

Eukaryotic chloroplasts contain an RNAP very highly structurally and mechanistically similar to bacterial RNAP ("plastid-encoded polymerase").

Eukaryotic chloroplasts also contain a second, structurally and mechanistically unrelated, RNAP ("nucleus-encoded polymerase"; member of the "single-subunit RNAP" protein family).

Eukaryotic mitochondria contain a structurally and mechanistically unrelated RNAP (member of the "single-subunit RNAP" protein family).

Given that DNA and RNA polymerases both carry out template-dependent nucleotide polymerization, it might be expected that the two types of enzymes would be structurally related. However, x-ray crystallographic studies of both types of enzymes reveal that, other than containing a critical Mg2+ ion at the catalytic site, they are virtually unrelated to each other; indeed template-dependent nucleotide polymerizing enzymes seem to have arisen independently twice during the early evolution of cells. One lineage led to the modern DNA Polymerases and reverse transcriptases, as well as to a few single-subunit RNA polymerases from viruses. The other lineage formed all of the modern cellular RNA polymerases.

27.6   In archaea

Archaea have a single type of RNAP, responsible for the synthesis of all RNA. Archaeal RNAP is structurally and mechanistically similar to bacterial RNAP and eukaryotic nuclear RNAP I-V, and is especially closely structurally and mechanistically related to eukaryotic nuclear RNAP II.[13][14] The history of the discovery of the archaeal RNA polymerase is quite recent. The first analysis of the RNAP of an archaeon was performed in 1971, when the RNAP from the extreme halophile *Halobacterium cutirubrum* was isolated and purified.[15] Crystal structures of RNAPs from *Sulfolobus solfataricus* and *Sulfolobus shibatae* set the total number of identified archaeal subunits at thirteen.[16][17]

27.7   In viruses

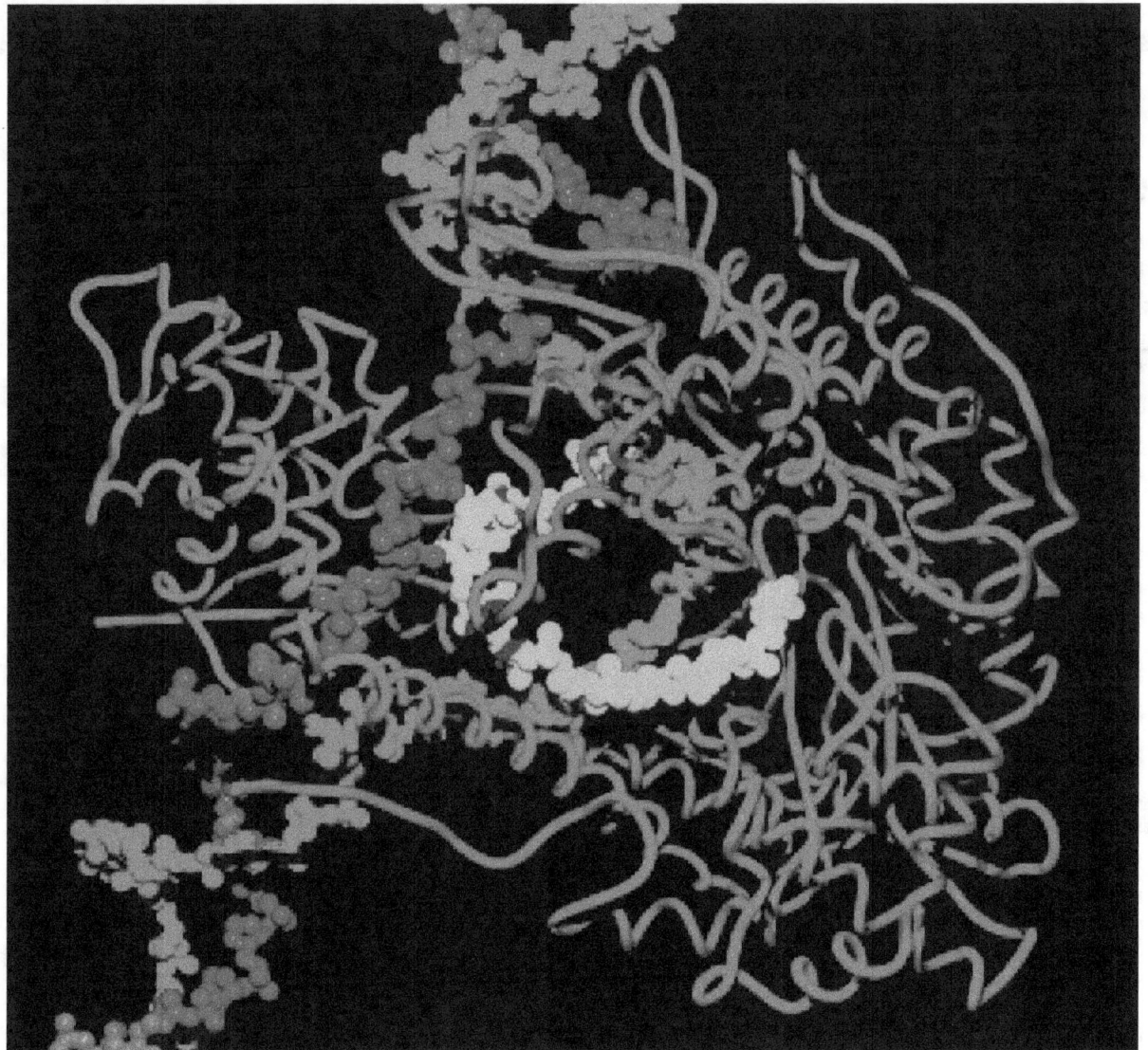

T7 RNA polymerase producing a mRNA (green) from a DNA template. The protein is shown as a purple ribbon. Image derived from PDB 1MSW.

Orthopoxviruses synthesize RNA using a virally encoded RNAP that is structurally and mechanistically related to bacterial RNAP, archaeal RNAP, and eukaryotic nuclear RNAP I-V. Most other viruses that synthesize RNA using a virally encoded RNAP use an RNAP that is not structurally and mechanistically related to bacterial RNAP, archaeal RNAP, and eukaryotic nuclear RNAP I-V. Many viruses use a single-subunit DNA-dependent RNAP that is structurally and mechanistically related to the single-subunit RNAP of eukaryotic chloroplasts and mitochondria and, more distantly, to DNA polymerases and reverse transcriptases. Perhaps the most widely studied such single-subunit RNAP is bacteriophage T7 RNA polymerase. Other viruses use a RNA-dependent RNAP (an RNAP that employs RNA as a template instead of DNA). This occurs in negative strand RNA viruses and dsRNA viruses, both of which exist for a portion of their life cycle as double-stranded RNA. However, some positive strand RNA viruses, such as poliovirus, also contain RNA-dependent RNAP.[18]

27.8   Purification

RNA polymerase can be isolated in the following ways:

- By a phosphocellulose column.[19]

- By glycerol gradient centrifugation.[20]

- By a DNA column.

- By an ion chromatography column.[21]

And also combinations of the above techniques.

27.9   See also

- Alpha-amanitin

- DNA polymerase

- RNA polymerase I

- RNA polymerase II

- RNA polymerase III

- T7 RNA polymerase

- Transcription (genetics)

- Transcription Activators in Eukaryotes

27.10   References

[1] Jerard Hurwitz (December 2005). "The Discovery of RNA Polymerase". *Journal of Biological Chemistry* **280** (52): 42477–85. doi:10.1074/jbc.X500006200. PMID 16230341.

[2] Nobel Prize 1959

[3] Nobel Prize in Chemistry 2006

[4] Akira Ishihama (2000). "Functional modulation of Escherichia coli RNA polymerase". *Annu. Rev. Microbiol.* **54**: 499–518. doi:10.1146/annurev.micro.54.1.499. PMID 11018136.

[5] Ebright RH (2000). "RNA polymerase: structural similarities between bacterial RNA polymerase and eukaryotic RNA polymerase II". *J Mol Biol* **304** (5): 687–98. doi:10.1006/jmbi.2000.4309. PMID 11124018.

[6] Ovchinnikov, Yu; Monastyrskaya, G; Gubanov, V; Guryev, S; Salomatina, I; Shuvaeva, T; Lipkin, V; Sverdlov, E (1982). "The primary structure of E. coli RNA polymerase. Nucleotide sequence of the rpoC gene and amino acid sequence of the β′-subunit". *Nucleic Acids Research* **10** (13): 4035–4044. doi:10.1093/nar/10.13.4035. Retrieved 16 November 2014.

[7] Mathew, Renjith; Chatterji, Dipankar (October 2006). "The evolving story of the omega subunit of bacterial RNA polymerase". *Trends in Microbiology* **14** (10): 450–455. doi:10.1016/j.tim.2006.08.002. Retrieved 17 November 2014.

[8] Grummt I. (1999). "Regulation of mammalian ribosomal gene transcription by RNA polymerase I.". *Prog Nucleic Acid Res Mol Biol.* **62**: 109–54. doi:10.1016/S0079-6603(08)60506-1. PMID 9932453.

[9] Lee Y; Kim M; Han J; Yeom KH; Lee S; Baek SH; Kim VN. (October 2004). "MicroRNA genes are transcribed by RNA polymerase II". *EMBO J.* **23** (20): 4051–60. doi:10.1038/sj.emboj.7600385. PMC 524334. PMID 15372072.

[10] Willis IM. (February 1993). "RNA polymerase III. Genes, factors and transcriptional specificity". *Eur J Biochem.* **212** (1): 1–11. doi:10.1111/j.1432-1033.1993.tb17626.x. PMID 8444147.

[11] Herr AJ, Jensen MB, Dalmay T, Baulcombe DC (2005). "RNA polymerase IV directs silencing of endogenous DNA". *Science* **308** (5718): 118–20. doi:10.1126/science.1106910. PMID 15692015.

[12] Wierzbicki AT, Ream TS, Haag JR, Pikaard CS (May 2009). "RNA Polymerase V transcription guides ARGONAUTE4 to chromatin". *Nat. Genet.* **41** (5): 630–4. doi:10.1038/ng.365. PMC 2674513. PMID 19377477.

[13] Korkhin, Y., U. Unligil, O. Littlefield, P. Nelson, D. Stuart, P. Sigler, S. Bell and N. Abrescia (2009). "Evolution of Complex RNA Polymerases: The Complete Archaeal RNA Polymerase Structure." PLoS biology 7(5): e102.

[14] Werner, F. (2007). "Structure and function of archaeal RNA polymerases." Molecular microbiology 65(6): 1395-1404.

[15] Louis, B. G. and P. S. Fitt (1971). "Nucleic acid enzymology of extremely halophilic bacteria. Halobacterium cutirubrum deoxyribonucleic acid-dependent ribonucleic acid polymerase." The Biochemical journal 121(4): 621-627.

[16] Korkhin, Y., U. Unligil, O. Littlefield, P. Nelson, D. Stuart, P. Sigler, S. Bell and N. Abrescia (2009). "Evolution of Complex RNA Polymerases: The Complete Archaeal RNA Polymerase Structure." PLoS biology 7(5): e102.

[17] Hirata, A.; Klein, B.; Murakami, K. (2008). "The X-ray crystal structure of RNA polymerase from Archaea". *Nature* **451** (7180): 851–854. doi:10.1038/nature06530.

[18] Ahlquist, Paul (2002). "RNA-Dependent RNA Polymerases, Viruses, and RNA Silencing". *Science* **296**: 1270–1273. doi:10.1

[19] Kelly JL; Lehman IR. (August 1986). "Yeast mitochondrial RNA polymerase. Purification and properties of the catalytic subunit". *J Biol Chem.* **261** (22): 10340–7. PMID 3525543.

[20] Honda A, et al. (April 1990). "Purification and molecular structure of RNA polymerase from influenza virus A/PR8". *J Biochem (Tokyo)* **107** (4): 624–8. PMID 2358436.

[21] Hager; et al. (1990). "Use of Mono Q High-Resolution Ion-Exchange Chromatography To Obtain Highly Pure and Active Escherichia coli RNA Polymerase". *Biochemistry* **29** (34): 7890–7894. doi:10.1021/bi00486a016. PMID 2261443.

27.11 External links

- DNAi - DNA Interactive, including information and Flash clips on RNA Polymerase.

- RNA Polymerase at the US National Library of Medicine Medical Subject Headings (MeSH)

- EC 2.7.7.6

- RNA Polymerase - Synthesis RNA from DNA Template

- 3D macromolecular structures of RNA Polymerase from the EM Data Bank(EMDB)

Chapter 28

Ribosome

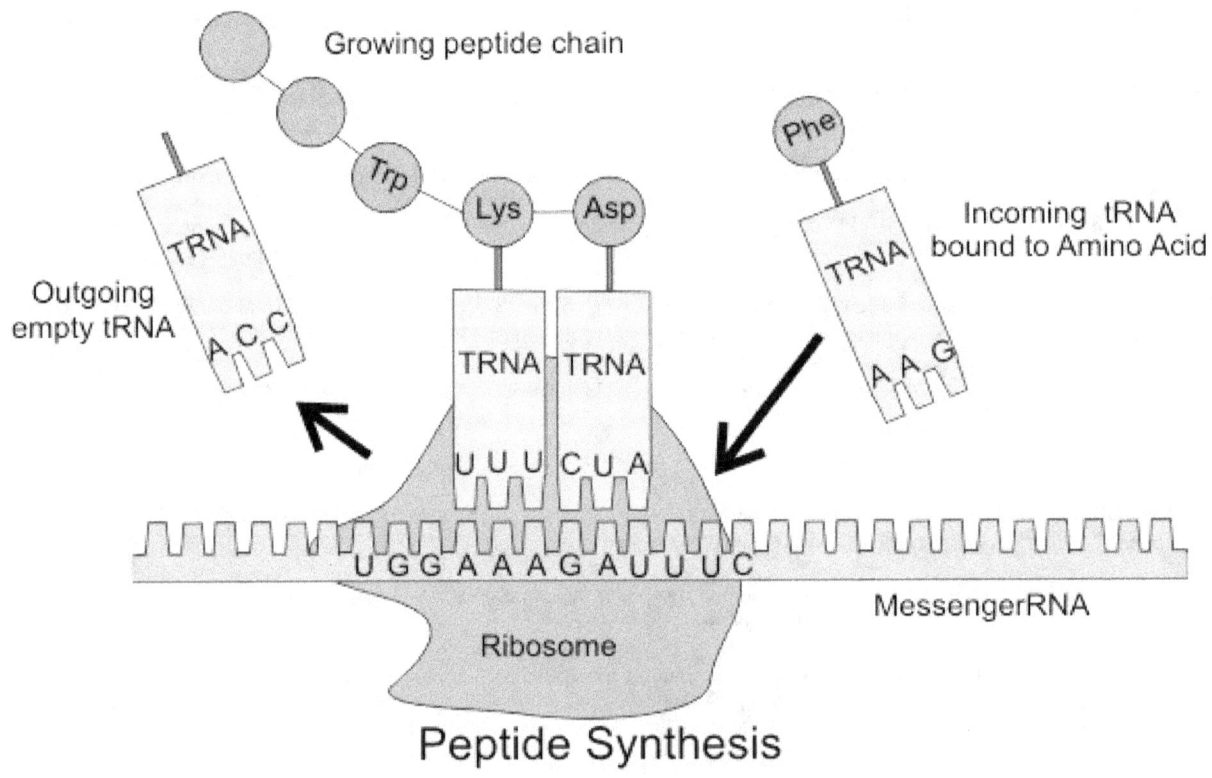

Figure 1 : *The ribosome assembles polymeric protein molecules whose sequence is controlled by the sequence of messenger RNA molecules. This is required by all living cells and associated viruses.*

The **ribosome** (/ˈraɪbəˌzoʊm/) is a large and complex molecular machine, found within all living cells, that serves as the site of biological protein synthesis (translation). Ribosomes link amino acids together in the order specified by messenger RNA (mRNA) molecules. Ribosomes consist of two major components: the small ribosomal subunit, which reads the RNA, and the large subunit, which joins amino acids to form a polypeptide chain. Each subunit is composed of one or more ribosomal RNA (rRNA) molecules and a variety of proteins. The ribosomes and associated molecules are also known as the *translational apparatus*.

The sequence of DNA encoding for a protein may be copied many times into RNA chains of a similar sequence. Ribosomes can bind to an RNA chain and use it as a template for determining the correct sequence of amino acids in a particular protein. Amino acids are selected, collected and carried to the ribosome by transfer RNA (tRNA molecules), which enter one part of the ribosome and bind to the messenger RNA chain. The attached amino acids are then linked together by another part of the ribosome. Once the protein is produced, it can then fold to produce a specific functional three-dimensional structure. It should be noted that during synthesis some proteins already start folding into their correct form.

A ribosome is made from complexes of RNAs and proteins and is therefore a ribonucleoprotein. Each ribosome is divided into two subunits: 1. a smaller subunit which binds to a larger subunit and the mRNA pattern, and 2. a larger subunit which binds to the tRNA, the amino acids, and the smaller subunit. When a ribosome finishes reading an mRNA molecule, these two subunits split apart. Ribosomes are ribozymes, because the catalytic peptidyl transferase activity that links amino acids together is performed by the ribosomal RNA. Ribosomes are often embedded in the intracellular membranes that make up the rough endoplasmic reticulum.

Ribosomes from bacteria, archaea and eukaryotes (the three domains of life on Earth) differ in their size, sequence, structure, and the ratio of protein to RNA. The differences in structure allow some antibiotics to kill bacteria by inhibiting their ribosomes, while leaving human ribosomes unaffected. In bacteria and archaea, more than one ribosome may move along a single mRNA chain at one time, each "reading" its sequence and producing a corresponding protein molecule. The ribosomes in the mitochondria of eukaryotic cells functionally resemble many features of those in bacteria, reflecting the likely evolutionary origin of mitochondria.[1][2]

28.1 Discovery

Ribosomes were first observed in the mid-1950s by Romanian cell biologist George Emil Palade using an electron microscope as dense particles or granules[3] for which, in 1974, he would win a Nobel Prize. The term "ribosome" was proposed by scientist Richard B. Roberts in 1958:

> During the course of the symposium a semantic difficulty became apparent. To some of the participants, "microsomes" mean the ribonucleoprotein particles of the microsome fraction contaminated by other protein and lipid material; to others, the microsomes consist of protein and lipid contaminated by particles. The phrase "microsomal particles" does not seem adequate, and "ribonucleoprotein particles of the microsome fraction" is much too awkward. During the meeting, the word "ribosome" was suggested, which has a very satisfactory name and a pleasant sound. The present confusion would be eliminated if "ribosome" were adopted to designate ribonucleoprotein particles in sizes ranging from 35 to 100S.
> — Roberts, R. B., *Microsomal Particles and Protein Synthesis*[4]

Albert Claude, Christian de Duve, and George Emil Palade were jointly awarded the Nobel Prize in Physiology or Medicine, in 1974, for the discovery of the ribosomes.[5] The Nobel Prize in Chemistry 2009 was awarded to Venkatraman Ramakrishnan, Thomas A. Steitz and Ada E. Yonath for determining the detailed structure and mechanism of the ribosome.[6]

28.2 Structure

The ribosome is responsible for the synthesis of proteins in cells and is found in all cellular organisms. It serves to convert the instructions found in messenger RNA (mRNA, which itself is made from instructions in DNA) into the chains of amino-acids that make up proteins.

The ribosome is a cellular machine which is highly complex. It is made up of dozens of distinct proteins (the exact number varies slightly between species) as well as a few specialized RNA molecules known as ribosomal RNA (rRNA). Note – these rRNAs do not carry instructions to make specific proteins like mRNAs. The ribosomal proteins and rRNAs are arranged into two distinct ribosomal pieces of different size, known generally as the large and small subunit of the ribosome. Ribosomes consist of two subunits that fit together (Figure 2) and work as one to translate the mRNA into a polypeptide chain during protein synthesis (Figure 1). Because they are formed from two subunits of non-equal size, they are slightly longer in the axis than in diameter. Prokaryotic ribosomes are around 20 nm (200 Å) in diameter and

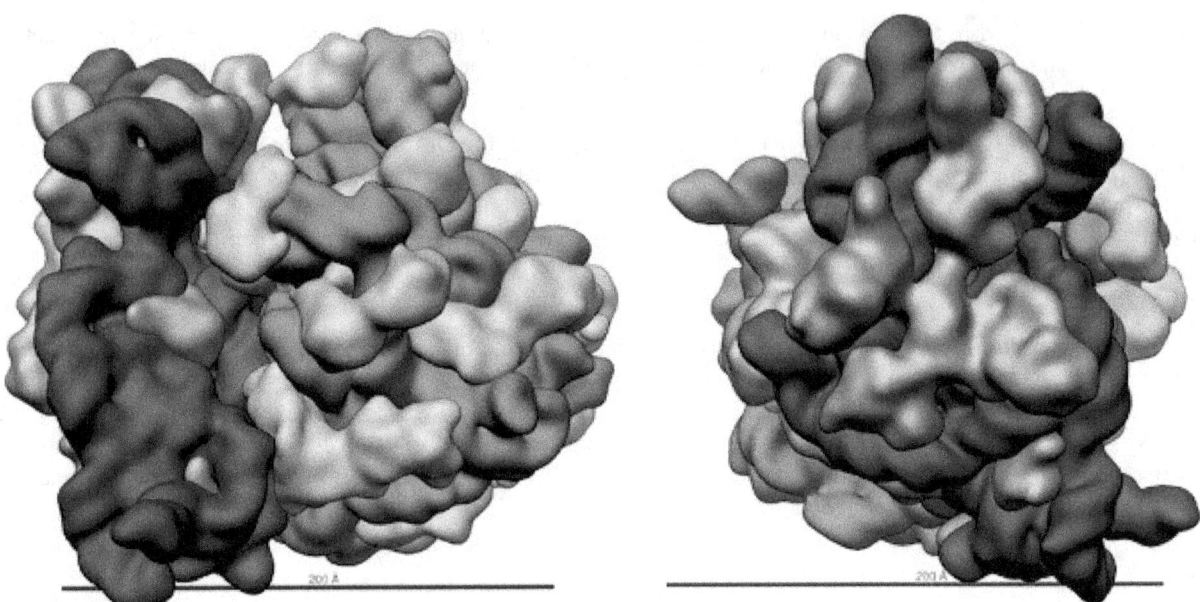

Figure 2 : *Large (red) and small (blue) subunit fit together.*

are composed of 65% rRNA and 35% ribosomal proteins. Eukaryotic ribosomes are between 25 and 30 nm (250–300 Å) in diameter with an rRNA to protein ratio that is close to 1. Bacterial ribosomes are composed of one or two rRNA strands. Eukaryotic ribosomes contain one or three very large rRNA molecules and multiple smaller protein molecules. Crystallographic work has shown that there are no ribosomal proteins close to the reaction site for polypeptide synthesis. This proves that the protein components of ribosomes do not directly participate in peptide bond formation catalysis, but rather suggests that these proteins act as a scaffold that may enhance the ability of rRNA to synthesize protein (See: Ribozyme).

The ribosomal subunits of prokaryotes and eukaryotes are quite similar.[8]

The unit of measurement is the Svedberg unit, a measure of the rate of sedimentation in centrifugation rather than size. This accounts for why fragment names do not add up: for example, prokaryotic 70S ribosomes are made of 50S and 30S subunits.

Prokaryotes have 70S ribosomes, each consisting of a small (30S) and a large (50S) subunit. Their small subunit has a 16S RNA subunit (consisting of 1540 nucleotides) bound to 21 proteins. The large subunit is composed of a 5S RNA subunit (120 nucleotides), a 23S RNA subunit (2900 nucleotides) and 31 proteins.[8] Affinity label for the tRNA binding sites on the E. coli ribosome allowed the identification of A and P site proteins most likely associated with the peptidyltransferase activity; labelled proteins are L27, L14, L15, L16, L2; at least L27 is located at the donor site, as shown by E. Collatz and A.P. Czernilofsky.[9][10] Additional research has demonstrated that the S1 and S21 proteins, in association with the 3'-end of 16S ribosomal RNA, are involved in the initiation of translation.[11]

Eukaryotes have 80S ribosomes, each consisting of a small (40S) and large (60S) subunit. Their 40S subunit has an 18S RNA (1900 nucleotides) and 33 proteins.[12][13] The large subunit is composed of a 5S RNA (120 nucleotides), 28S RNA (4700 nucleotides), a 5.8S RNA (160 nucleotides) subunits and 46 proteins.[8][12][14] During 1977, Czernilofsky published research that used affinity labeling to identify tRNA-binding sites on rat liver ribosomes. Several proteins, including L32/33, L36, L21, L23, L28/29 and L13 were implicated as being at or near the peptidyl transferase center.[15]

The ribosomes found in chloroplasts and mitochondria of eukaryotes also consist of large and small subunits bound together with proteins into one 70S particle.[8] These organelles are believed to be descendants of bacteria (see Endosymbiotic theory) and as such their ribosomes are similar to those of bacteria.[8]

The various ribosomes share a core structure, which is quite similar despite the large differences in size. Much of the RNA is highly organized into various tertiary structural motifs, for example pseudoknots that exhibit coaxial stacking. The extra RNA in the larger ribosomes is in several long continuous insertions, such that they form loops out of the core

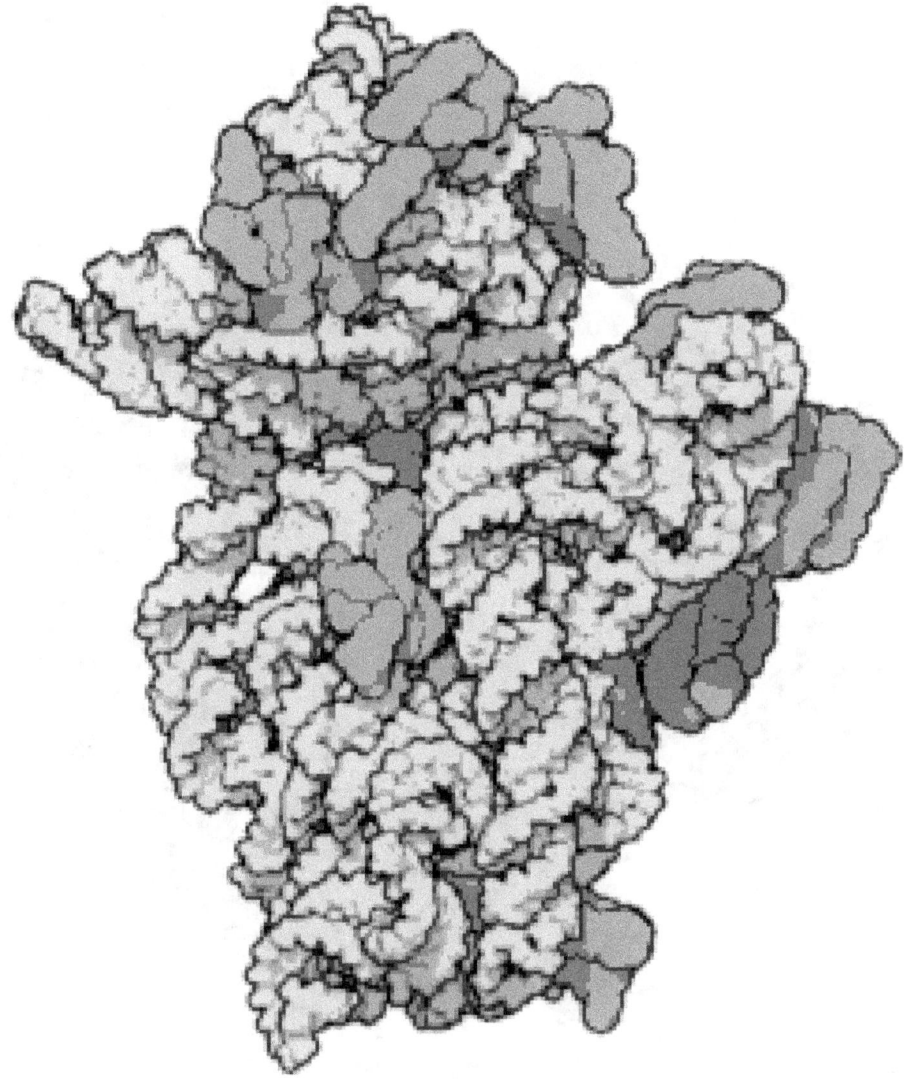

Figure 3 : *Atomic structure of the 30S Subunit from* Thermus thermophilus.[7] *Proteins are shown in blue and the single RNA chain in orange.*

structure without disrupting or changing it.[8] All of the catalytic activity of the ribosome is carried out by the RNA; the proteins reside on the surface and seem to stabilize the structure.[8]

The differences between the bacterial and eukaryotic ribosomes are exploited by pharmaceutical chemists to create antibiotics that can destroy a bacterial infection without harming the cells of the infected person. Due to the differences in their structures, the bacterial 70S ribosomes are vulnerable to these antibiotics while the eukaryotic 80S ribosomes are not.[16] Even though mitochondria possess ribosomes similar to the bacterial ones, mitochondria are not affected by these antibiotics because they are surrounded by a double membrane that does not easily admit these antibiotics into the organelle.[17]

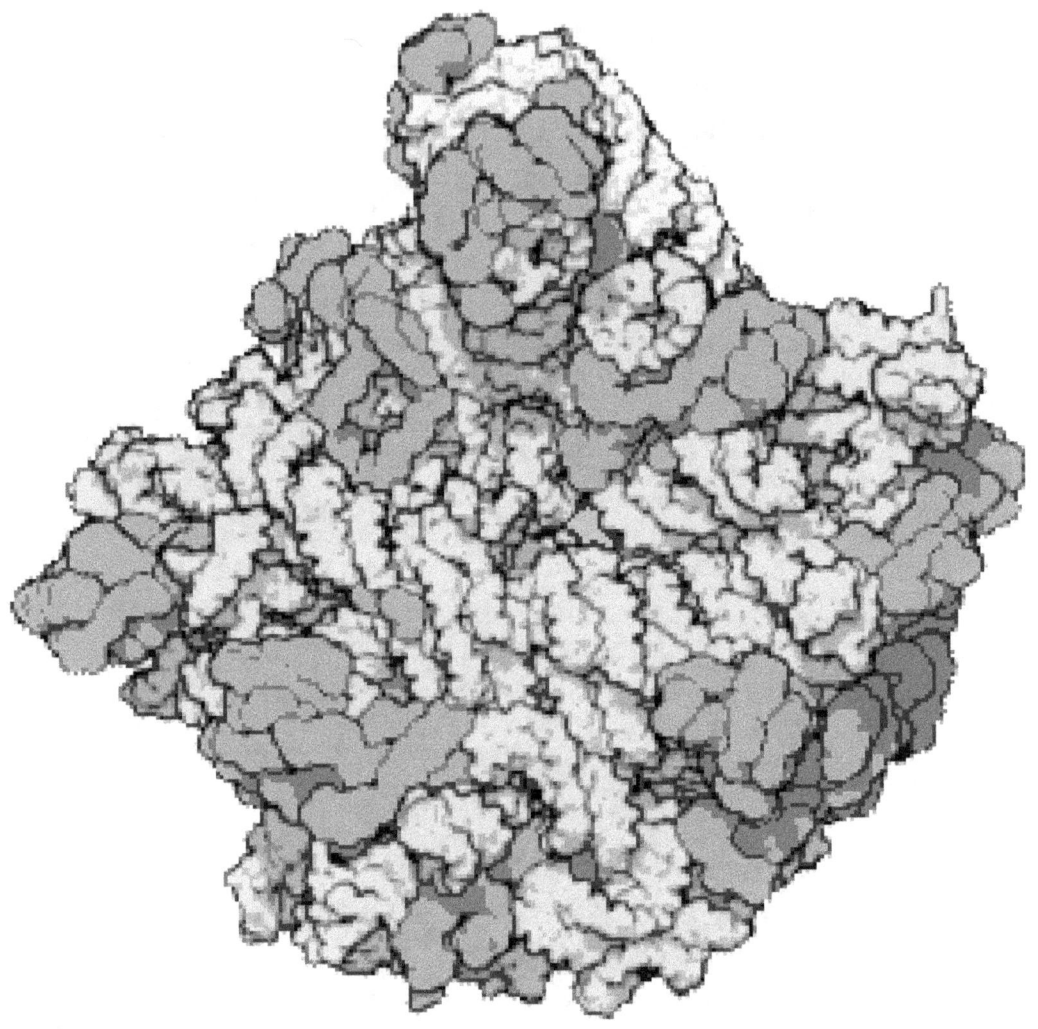

Figure 4 : *Atomic structure of the 50S Subunit from* Haloarcula marismortui. *Proteins are shown in blue and the two RNA chains in orange and yellow.*[18] *The small patch of green in the center of the subunit is the active site.*

28.2.1   High-resolution structure

The general molecular structure of the ribosome has been known since the early 1970s. In the early 2000s the structure has been achieved at high resolutions, of the order of a few Å.

The first papers giving the structure of the ribosome at atomic resolution were published almost simultaneously in late 2000. The 50S (large prokaryotic) subunit was determined from the archaeon *Haloarcula marismortui*[18] and the bacterium *Deinococcus radiodurans*,[19] and the structure of the 30S subunit was determined from *Thermus thermophilus*.[7] These structural studies were awarded the Nobel Prize in Chemistry in 2009. In May 2001 these coordinates were used to reconstruct the entire *T. thermophilus* 70S particle at 5.5 Å resolution.[20]

Two papers were published in November 2005 with structures of the *Escherichia coli* 70S ribosome. The structures of a vacant ribosome were determined at 3.5-Å resolution using x-ray crystallography.[21] Then, two weeks later, a structure based on cryo-electron microscopy was published,[22] which depicts the ribosome at 11–15Å resolution in the act of passing a newly synthesized protein strand into the protein-conducting channel.

The first atomic structures of the ribosome complexed with tRNA and mRNA molecules were solved by using X-ray crystallography by two groups independently, at 2.8 Å[23] and at 3.7 Å.[24] These structures allow one to see the details of interactions of the *Thermus thermophilus* ribosome with mRNA and with tRNAs bound at classical ribosomal sites. Interactions of the ribosome with long mRNAs containing Shine-Dalgarno sequences were visualized soon after that at 4.5- to 5.5-Å resolution.[25]

In 2011, the first complete atomic structure of the eukaryotic 80S ribosome from the yeast *Saccharomyces cerevisiae* was obtained by crystallography.[12] The model reveals the architecture of eukaryote-specific elements and their interaction with the universally conserved core. At the same time, the complete model of a eukaryotic 40S ribosomal structure in *Tetrahymena thermophila* was published and described the structure of the 40S subunit as well as much about the 40S subunit's interaction with eIF1 during translation initiation.[13] Similarly, the eukaryotic 60S subunit structure was also determined from *Tetrahymena thermophila* in complex with eIF6.[14]

28.3   Function

28.3.1   Translation

Main article: Translation (genetics)

Ribosomes are the workplaces of protein biosynthesis, the process of translating mRNA into protein. The mRNA comprises a series of codons that dictate to the ribosome the sequence of the amino acids needed to make the protein. Using the mRNA as a template, the ribosome traverses each codon (3 nucleotides) of the mRNA, pairing it with the appropriate amino acid provided by an aminoacyl-tRNA. aminoacyl-tRNA contains a complementary anticodon on one end and the appropriate amino acid on the other. For fast and accurate recognition of the appropriate tRNA, the ribosome utilizes large conformational changes (conformational proofreading) .[26] The small ribosomal subunit, typically bound to an aminoacyl-tRNA containing the amino acid methionine, binds to an AUG codon on the mRNA and recruits the large ribosomal subunit. The ribosome contains three RNA binding sites, designated A, P and E. The A site binds an aminoacyl-tRNA; the P site binds a peptidyl-tRNA (a tRNA bound to the peptide being synthesized); and the E site binds a free tRNA before it exits the ribosome. Protein synthesis begins at a start codon AUG near the 5' end of the mRNA. mRNA binds to the P site of the ribosome first. The ribosome is able to identify the start codon by use of the Shine-Dalgarno sequence of the mRNA in prokaryotes and Kozak box in eukaryotes.

Although catalysis of the peptide bond involves the C2 hydroxyl of RNA's P-site adenosine in a proton shuttle mechanism, other steps in protein synthesis (such as translocation) are caused by changes in protein conformations. Since their catalytic core is made of RNA, ribosomes are classified as "ribozymes,"[27] and it is thought that they might be remnants of the RNA world.[28]

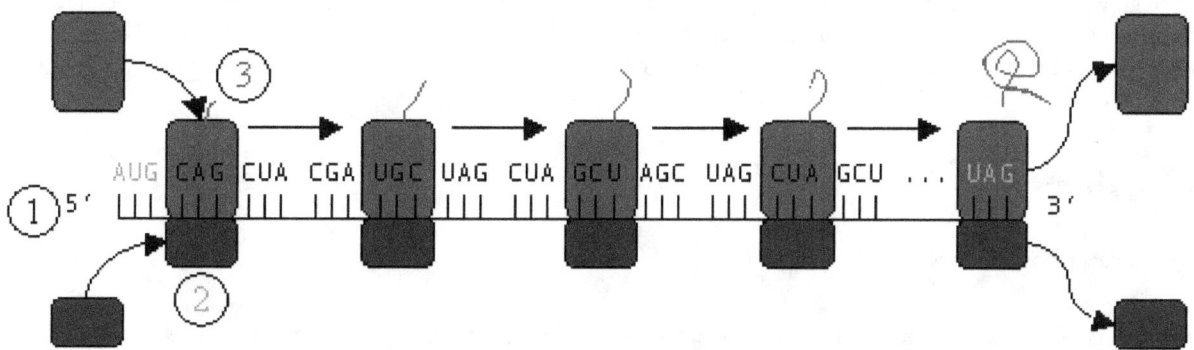

Figure 5 : *Translation of mRNA (1) by a ribosome (2)(shown as small and large subunits) into a polypeptide chain (3). The ribosome begins at the start codon of RNA (AUG) and ends at the stop codon (UAG).*

In Figure 5, both ribosomal subunits (small and large) assemble at the start codon (towards the 5' end of the RNA). The ribosome uses RNA that matches the current codon (triplet) on the mRNA to append an amino acid to the polypeptide chain. This is done for each triplet on the RNA, while the ribosome moves towards the 3' end of the mRNA. Usually in bacterial cells, several ribosomes are working parallel on a single RNA, forming what is called a *polyribosome* or *polysome*.

28.3.2 Addition of translation-independent amino acids

Presence of a ribosome quality control protein Rqc2 is associated with mRNA-independent protein elongation.[29][30] This elongation is a result of ribosomal addition (via tRNAs brought by Rqc2) of *CAT tails*: ribosomes extend the *C*-terminus of a stalled protein with random, translation-independent sequences of *a*lanines and *t*hreonines.[31][32]

28.4 Ribosome locations

Ribosomes are classified as being either "free" or "membrane-bound".

Free and membrane-bound ribosomes differ only in their spatial distribution; they are identical in structure. Whether the ribosome exists in a free or membrane-bound state depends on the presence of an ER-targeting signal sequence on the protein being synthesized, so an individual ribosome might be membrane-bound when it is making one protein, but free in the cytosol when it makes another protein.

Ribosomes are sometimes referred to as organelles, but the use of the term *organelle* is often restricted to describing sub-cellular components that include a phospholipid membrane, which ribosomes, being entirely particulate, do not. For this reason, ribosomes may sometimes be described as "non-membranous organelles".

28.4.1 Free ribosomes

Free ribosomes can move about anywhere in the cytosol, but are excluded from the cell nucleus and other organelles. Proteins that are formed from free ribosomes are released into the cytosol and used within the cell. Since the cytosol contains high concentrations of glutathione and is, therefore, a reducing environment, proteins containing disulfide bonds, which are formed from oxidized cysteine residues, cannot be produced within it.

28.4.2 Membrane-bound ribosomes

When a ribosome begins to synthesize proteins that are needed in some organelles, the ribosome making this protein can become "membrane-bound". In eukaryotic cells this happens in a region of the endoplasmic reticulum (ER) called the "rough ER". The newly produced polypeptide chains are inserted directly into the ER by the ribosome undertaking vectorial synthesis and are then transported to their destinations, through the secretory pathway. Bound ribosomes usually produce proteins that are used within the plasma membrane or are expelled from the cell via *exocytosis*.[33]

28.5 Biogenesis

Main article: Ribosome biogenesis

In bacterial cells, ribosomes are synthesized in the cytoplasm through the transcription of multiple ribosome gene operons. In eukaryotes, the process takes place both in the cell cytoplasm and in the nucleolus, which is a region within the cell nucleus. The assembly process involves the coordinated function of over 200 proteins in the synthesis and processing of the four rRNAs, as well as assembly of those rRNAs with the ribosomal proteins.

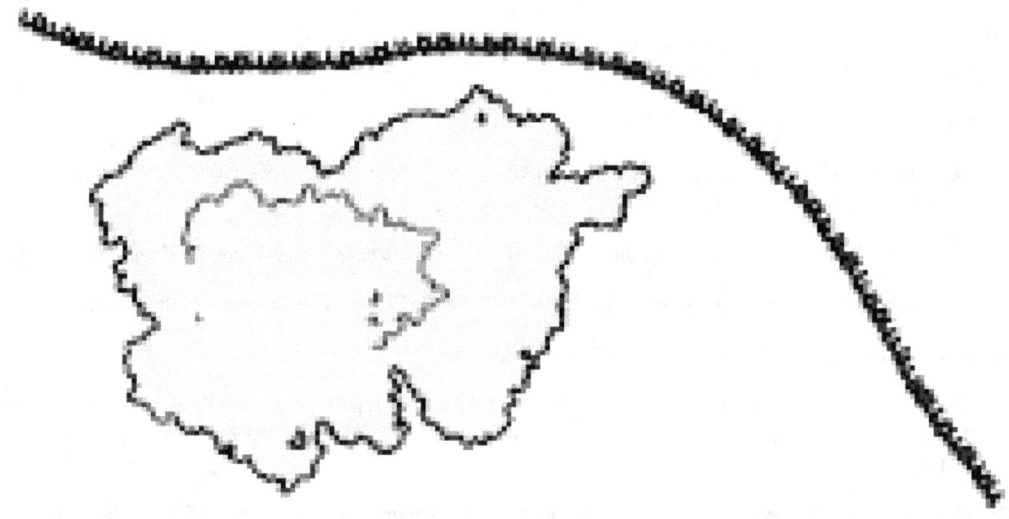

Figure 6 : *A ribosome translating a protein that is secreted into the endoplasmic reticulum.*

28.6   Origin

The ribosome may have first originated in an RNA world, appearing as a self-replicating complex that only later evolved the ability to synthesize proteins when amino acids began to appear.[34] Studies suggest that ancient ribosomes constructed solely of rRNA could have developed the ability to synthesize peptide bonds.[35][36][37] In addition, evidence strongly points to ancient ribosomes as self-replicating complexes, where the rRNA in the ribosomes had informational, structural, and catalytic purposes because it could have coded for tRNAs and proteins needed for ribosomal self-replication.[38] As amino acids gradually appeared in the RNA world under prebiotic conditions,[39][40] their interactions with catalytic RNA would increase both the range and efficiency of function of catalytic RNA molecules.[34] Thus, the driving force for the evolution of the ribosome from an ancient self-replicating machine into its current form as a translational machine may have been the selective pressure to incorporate proteins into the ribosome's self-replicating mechanisms, so as to increase its capacity for self-replication.[38]

28.7 See also

- Aminoglycosides

- Biological machines

- Eukaryotic translation

- Posttranslational modification

- Prokaryotic translation

- Protein dynamics

- RNA tertiary structure

- Translation (genetics)

- Wobble base pair

28.8 References

[1] Benne R, Sloof P (1987). "Evolution of the mitochondrial protein synthetic machinery". *BioSystems***21**(1): 51–68. doi:10.1016 2647(87)90006-2. PMID 2446672.

[2] "Ribosomes". Retrieved 2011-04-28.

[3] PALADE GE (January 1955). "A small particulate component of the cytoplasm". *J Biophys Biochem Cytol* **1** (1): 59–68. doi:10.1083/jcb.1.1.59. PMC 2223592. PMID 14381428.

[4] Roberts, R. B., editor. (1958) "Introduction" in *Microsomal Particles and Protein Synthesis.* New York: Pergamon Press, Inc.

[5] "The Nobel Prize in Physiology or Medicine 1974". *Nobelprize.org*. The Nobel Foundation. Retrieved 10 December 2012.

[6] £€¥20 2009 Nobel Prize in Chemistry, Nobel Foundation.

[7] Wimberly BT, Brodersen DE, Clemons WM Jr, Morgan-Warren RJ, Carter AP, Vonrhein C, Hartsch T, Ramakrishnan V (September 2000). "Structure of the 30S ribosomal subunit". *Nature* **407** (6802): 327–39. doi:10.1038/35030006. PMID 11014182.

[8] The Molecular Biology of the Cell, fourth edition. Bruce Alberts, et al. Garland Science (2002) pg. 342 ISBN 0-8153-3218-1

[9] Czernilofsky, A; Küchler, E; Stöffler, G.; Czernilofsky, P. (1976). "SITE OF REACTION ON RIBOSOMAL-PROTEIN L27 WITH AN AFFINITY LABEL DERIVATIVE OF TRANSFER-RNA-F(MET)". *FEBS Letters* (ELSEVIER SCIENCE BV) **63** (2): 283–286. doi:10.1016/0014-5793(76)80112-3. PMID 770196.

[10] Czernilofsky, A; Collatz, E; Stöffler, G; Küchler, E (1974). "PROTEINS AT TRANSFER-RNA BINDING-SITES OF ESCHERICHIA-COLI RIBOSOMES". *Proceedings of the National Academy of Sciences of the United States of America* (NATL ACAD SCIENCES) **71** (1): 230–234. doi:10.1073/pnas.71.1.230. PMC 387971. PMID 4589893.

[11] Czernilofsky, A; Kurland, C.G.; Stöffler, G. (1975). "30S RIBOSOMAL-PROTEINS ASSOCIATED WITH 3'-TERMINUS OF 16S RNA". *Febs Letters* (ELSEVIER SCIENCE BV) **58** (1): 281–284. doi:10.1016/0014-5793(75)80279-1. PMID 1225593.

[12] Ben-Shem A, Garreau de Loubresse N, Melnikov S, Jenner L, Yusupova G, Yusupov M. (February 2011). "The structure of the eukaryotic ribosome at 3.0 Å resolution". *Science* **334** (6062): 1524–1529. doi:10.1126/science.1212642. PMID 22096102.

[13] Rabl, Leibundgut, Ataide, Haag, Ban (February 2010). "Crystal Structure of the Eukaryotic 40S Ribosomal Subunit in Complex with Initiation Factor 1". *Science* **331** (6018): 730–736. doi:10.1126/science.1198308. PMID 21205638.

[14] Klinge, Voigts-Hoffmann, Leibundgut, Arpagaus, Ban (November 2011). "Crystal Structure of the Eukaryotic 60S Ribosomal Subunit in Complex with Initiation Factor 6". *Science* **334** (6058): 941–948. doi:10.1126/science.1211204. PMID 22052974.

[15] Czernilofsky, A; Collatz, Ekkehard; Gressner, Axel M.; Wool, Ira G.; Küchler, Ernst (1977). "IDENTIFICATION OF TRNA-BINDING SITES ON RAT-LIVER RIBOSOMES BY AFFINITY LABELING". *Molecular and General Genetics* (Springer Verlag) **153** (3): 231–235. doi:10.1007/BF00431588. Retrieved January 2012.

[16] Recht MI, Douthwaite S, Puglisi JD (1999). "Basis for bacterial specificity of action of aminoglycoside antibiotics". *EMBO J* **18** (11): 3133–8. doi:10.1093/emboj/18.11.3133. PMC 1171394. PMID 10357824.

[17] O'Brien, T.W., The General Occurrence of 55S Ribosomes in Mammalian Liver Mitochondria. J. Biol. Chem., 245:3409 (1971).

[18] Ban N, Nissen P, Hansen J, Moore P, Steitz T (2000). "The complete atomic structure of the large ribosomal subunit at 2.4 Å resolution". *Science* **289** (5481): 905–20. doi:10.1126/science.289.5481.905. PMID 10937989.

[19] Schluenzen F, Tocilj A, Zarivach R, Harms J, Gluehmann M, Janell D, Bashan A, Bartels H, Agmon I, Franceschi F, Yonath A (2000). "Structure of functionally activated small ribosomal subunit at 3.3 Å resolution". *Cell* **102** (5): 615–23. doi:10.1016/S0092-8674(00)00084-2. PMID 11007480.

[20] Yusupov MM, Yusupova GZ, Baucom A, et al. (May 2001). "Crystal structure of the ribosome at 5.5 A resolution". *Science* **292** (5518): 883–96. doi:10.1126/science.1060089. PMID 11283358.

[21] Schuwirth BS, Borovinskaya MA, Hau CW, et al. (November 2005). "Structures of the bacterial ribosome at 3.5 A resolution". *Science* **310** (5749): 827–34. doi:10.1126/science.1117230. PMID 16272117.

[22] Mitra K, Schaffitzel C, Shaikh T, et al. (November 2005). "Structure of the E. coli protein-conducting channel bound to a translating ribosome". *Nature* **438** (7066): 318–24. doi:10.1038/nature04133. PMC 1351281. PMID 16292303.

[23] Selmer M, Dunham CM, Murphy FV, et al. (September 2006). "Structure of the 70S ribosome complexed with mRNA and tRNA". *Science* **313** (5795): 1935–42. doi:10.1126/science.1131127. PMID 16959973.

[24] Korostelev A, Trakhanov S, Laurberg M, Noller HF (September 2006). "Crystal structure of a 70S ribosome-tRNA complex reveals functional interactions and rearrangements". *Cell* **126** (6): 1065–77. doi:10.1016/j.cell.2006.08.032. PMID 16962654.

[25] Yusupova G, Jenner L, Rees B, Moras D, Yusupov M (November 2006). "Structural basis for messenger RNA movement on the ribosome". *Nature* **444** (7117): 391–4. doi:10.1038/nature05281. PMID 17051149.

[26] Savir, Y; Tlusty, T (Apr 11, 2013). "The ribosome as an optimal decoder: a lesson in molecular recognition.". *Cell* **153** (2): 471–9. doi:10.1016/j.cell.2013.03.032. PMID 23582332.

[27] Rodnina MV, Beringer M, Wintermeyer W (2007). "How ribosomes make peptide bonds". *Trends Biochem. Sci.* **32** (1): 20–6. doi:10.1016/j.tibs.2006.11.007. PMID 17157507.

[28] Cech T (2000). "Structural biology. The ribosome is a ribozyme". *Science* **289** (5481): 878–9. doi:10.1126/science.289.5481.878. PMID 10960319.

[29] Brandman O, et al. (2012). "A ribosome-bound quality control complex triggers degradation of nascent peptides and signals translation stress". *Cell* **151** (5): 1042–54. doi:10.1016/j.cell.2012.10.044. PMID 23178123.

[30] Defenouillère Q, et al. (2013). "Cdc48-associated complex bound to 60S particles is required for the clearance of aberrant translation products". *Proc Natl Acad Sci U S A.* **110** (13): 5046–51. doi:10.1073/pnas.1221724110. PMID 23479637.

[31] Shen PS, et al. (2015). "A ribosome-bound quality control complex triggers degradation of nascent peptides and signals translation stress". *Science* **347** (6217): 75–8. doi:10.1126/science.1259724. PMID 25554787.

[32] Keeley, Jim; Gutnikoff, Robert (2015-01-02). "Ribosome Studies Turn Up New Mechanism of Protein Synthesis" (Press release). Howard Hughes Medical Institute. Retrieved 2015-01-16.

[33] http://www.ncbi.nlm.nih.gov/books/NBK26841/#A2204

[34] Noller, H. F. (2012). Evolution of protein synthesis from an RNA world. Cold Spring Harbor Perspectives in Biology 4:1-U20.

[35] Dabbs, E.R. (1986). Mutant studies on the prokaryotic ribosome. Springer-Verlag, N.Y.

[36] Noller, H. F., V. Hoffarth, and L. Zimniak (1992). Unusual resistance of peptidyl transferase to protein extraction procedures. Science 256:1416-1419.

[37] Nomura, M., S. Mizushima, M. Ozaki, P. Trau, C. V. Lowry (1969). Structure and function of ribosomes and their molecular components. Cold Spring Harbor Symposium of Quantitative Biology 34:49–61.

[38] Root-Bernstein, M., and R. Root-Bernstein (2015). The ribosome as a missing link in the evolution of life. Journal of Theoretical Biology 367:130-158.

[39] Caetano-Anolles, G., and M. J. Seufferheld (2013). The coevolutionary roots of biochemistry and cellular organization challenge the RNA world paradigm. Journal of Molecular Microbiology and Biotechnology 23:152-177.

[40] Saladino, R., G. Botta, S. Pino, G. Costanzo, and E. Di Mauro (2012). Genetics first or metabolism first? The formamide clue. Chemical Society Reviews 41:5526-5565.

28.9   External links

- Lab computer simulates ribosome in motion
- Role of the Ribosome, Gwen V. Childs, copied here
- Ribosome in *Proteopedia* - The free, collaborative 3D encyclopedia of proteins & other molecules
- Ribosomal proteins families in ExPASy
- Molecule of the Month © RCSB Protein Data Bank:
  - Ribosome
  - Elongation Factors
  - Palade
- 3D electron microscopy structures of ribosomes at the EM Data Bank(EMDB)

This article incorporates public domain material from the NCBI document "Science Primer".

Chapter 29

Transfer RNA

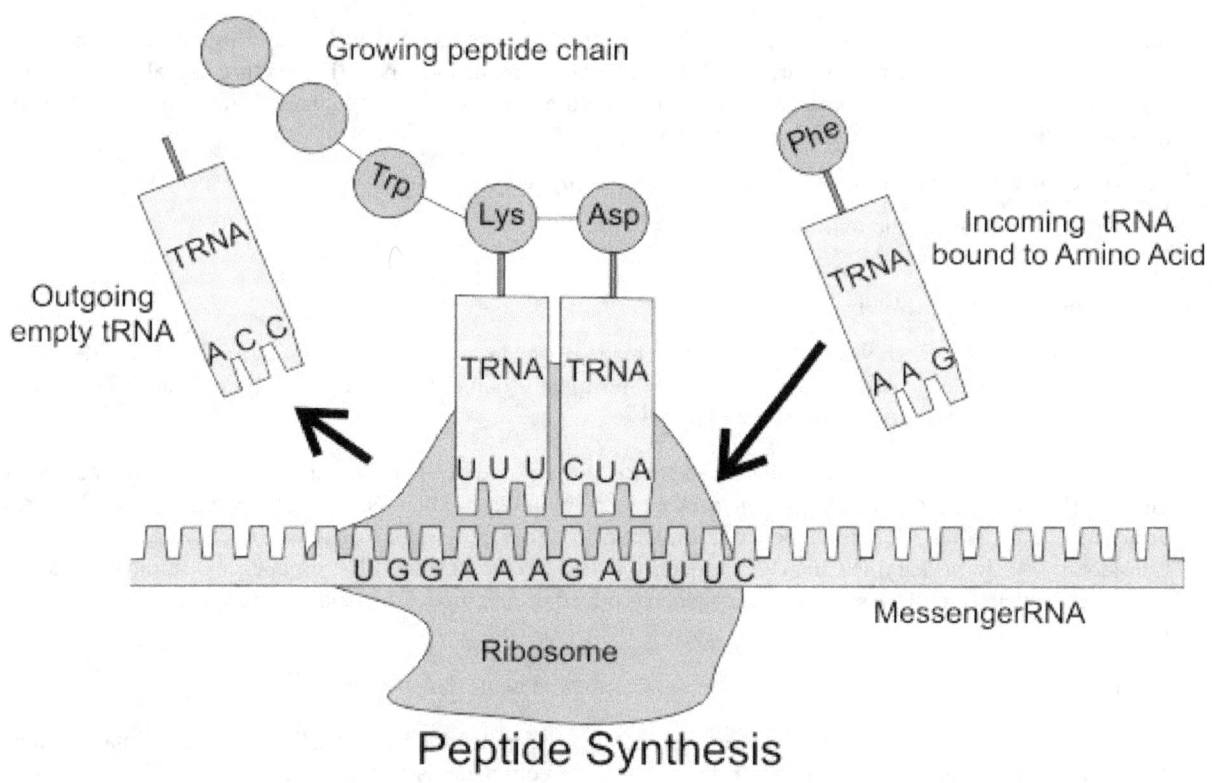

The interaction of tRNA and mRNA in protein synthesis.

A **transfer RNA** (abbreviated **tRNA** and archaically referred to as **sRNA**, for **soluble RNA**[1]) is an adaptor molecule composed of RNA, typically 76 to 90 nucleotides in length,[2] that serves as the physical link between the mRNA and the amino acid sequence of proteins. It does this by carrying an amino acid to the protein synthetic machinery of a cell (ribosome) as directed by a three-nucleotide sequence (codon) in a messenger RNA (mRNA). As such, tRNAs are a necessary component of translation, the biological synthesis of new proteins according to the genetic code.

The specific nucleotide sequence of an mRNA specifies which amino acids are incorporated into the protein product of the gene from which the mRNA is transcribed, and the role of tRNA is to specify which sequence from the genetic code corresponds to which amino acid.[3] One end of the tRNA matches the genetic code in a three-nucleotide sequence called the anticodon. The anticodon forms three base pairs with a codon in mRNA during protein biosynthesis. The mRNA encodes a protein as a series of contiguous codons, each of which is recognized by a particular tRNA. On the other end of the tRNA is a covalent attachment to the amino acid that corresponds to the anticodon sequence. Each type of tRNA molecule can be attached to only one type of amino acid, so each organism has many types of tRNA (in fact, because the genetic code contains multiple codons that specify the same amino acid, there are several tRNA molecules bearing different anticodons which also carry the same amino acid).

The covalent attachment to the tRNA 3' end is catalyzed by enzymes called aminoacyl tRNA synthetases. During protein synthesis, tRNAs with attached amino acids are delivered to the ribosome by proteins called elongation factors (EF-Tu in bacteria, eEF-1 in eukaryotes), which aid in decoding the mRNA codon sequence. If the tRNA's anticodon matches the mRNA, another tRNA already bound to the ribosome transfers the growing polypeptide chain from its 3' end to the amino acid attached to the 3' end of the newly delivered tRNA, a reaction catalyzed by the ribosome.

A large number of the individual nucleotides in a tRNA molecule may be chemically modified, often by methylation or deamidation. These unusual bases sometimes affect the tRNA's interaction with ribosomes and sometimes occur in the anticodon to alter base-pairing properties.[4]:29.1.2

29.1   Structure

The structure of tRNA can be decomposed into its primary structure, its secondary structure (usually visualized as the *cloverleaf structure*), and its tertiary structure[5] (all tRNAs have a similar L-shaped 3D structure that allows them to fit into the P and A sites of the ribosome). The cloverleaf structure becomes the 3D L-shaped structure through coaxial stacking of the helices, which is a common RNA tertiary structure motif.

The lengths of each arm, as well as the loop 'diameter', in a tRNA molecule vary from species to species.[5][6]

The tRNA structure consists of the following:

1. A 5'-terminal phosphate group.

2. The acceptor stem is a 7- to 9-base pair (bp) stem made by the base pairing of the 5'-terminal nucleotide with the 3'-terminal nucleotide (which contains the CCA 3'-terminal group used to attach the amino acid). The acceptor stem may contain non-Watson-Crick base pairs.[5][7]

3. The CCA tail is a cytosine-cytosine-adenine sequence at the 3' end of the tRNA molecule. The amino acid loaded onto the tRNA by aminoacyl tRNA synthetases, to form aminoacyl-tRNA, is covalently bonded to the 3'-hydroxyl group on the CCA tail.[8] This sequence is important for the recognition of tRNA by enzymes and critical in translation.[9][10] In prokaryotes, the CCA sequence is transcribed in some tRNA sequences. In most prokaryotic tRNAs and eukaryotic tRNAs, the CCA sequence is added during processing and therefore does not appear in the tRNA gene.[11]

4. The D arm is a 4- to 6-bp stem ending in a loop that often contains dihydrouridine.[5]

5. The anticodon arm is a 6-bp stem whose loop contains the anticodon.[5] The tRNA 5'-to-3' primary structure contains the anticodon but in reverse order, since 3'-to-5' directionality is required to read the mRNA from 5'-to-3'.

6. The T arm is a 4- to 5- bp stem containing the sequence TΨC where Ψ is pseudouridine, a modified uridine.[5]

7. Bases that have been modified, especially by methylation (e.g. tRNA (guanine-N7-)-methyltransferase), occur in several positions throughout the tRNA. The first anticodon base, or wobble-position, is sometimes modified to inosine (derived from adenine), pseudouridine or lysidine (derived from cytosine).[12]

29.2   Anticodon

An **anticodon**[13] is a unit made up of three nucleotides that correspond to the three bases of the codon on the mRNA. Each tRNA contains a specific anticodon triplet sequence that can base-pair to one or more codons for an amino acid. Some anticodons can pair with more than one codon due to a phenomenon known as wobble base pairing. Frequently, the first nucleotide of the anticodon is one not found on mRNA: inosine, which can hydrogen bond to more than one base

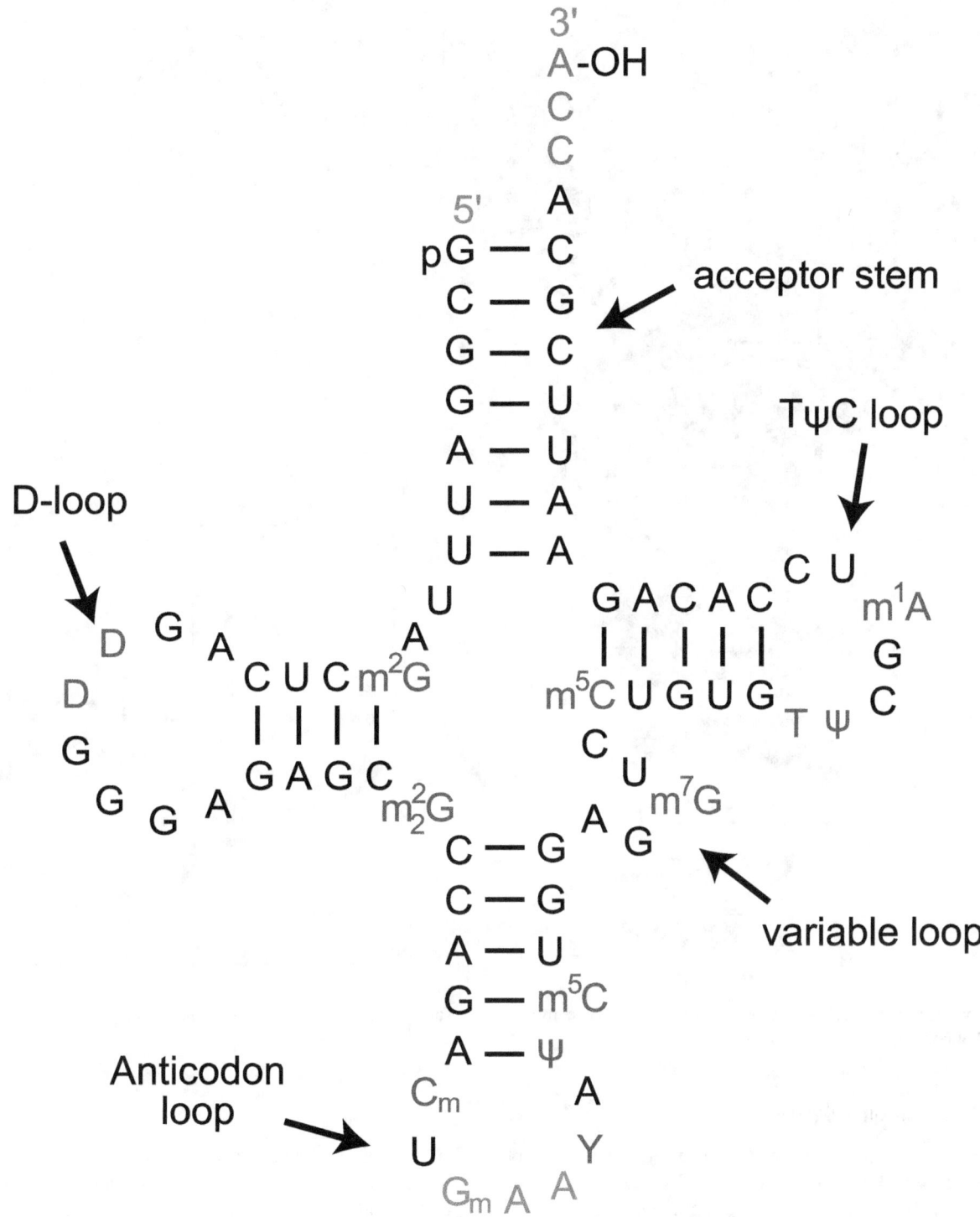

Secondary cloverleaf structure *of tRNA^{Phe} from yeast.*

in the corresponding codon position.[4]:29.3.9 In the genetic code, it is common for a single amino acid to be specified by all four third-position possibilities, or at least by both pyrimidines and purines; for example, the amino acid glycine is coded for by the codon sequences GGU, GGC, GGA, and GGG. Other modified nucleotides may also appear at the first anticodon position - sometimes known as the "wobble position" - resulting in subtle changes to the genetic code, as for

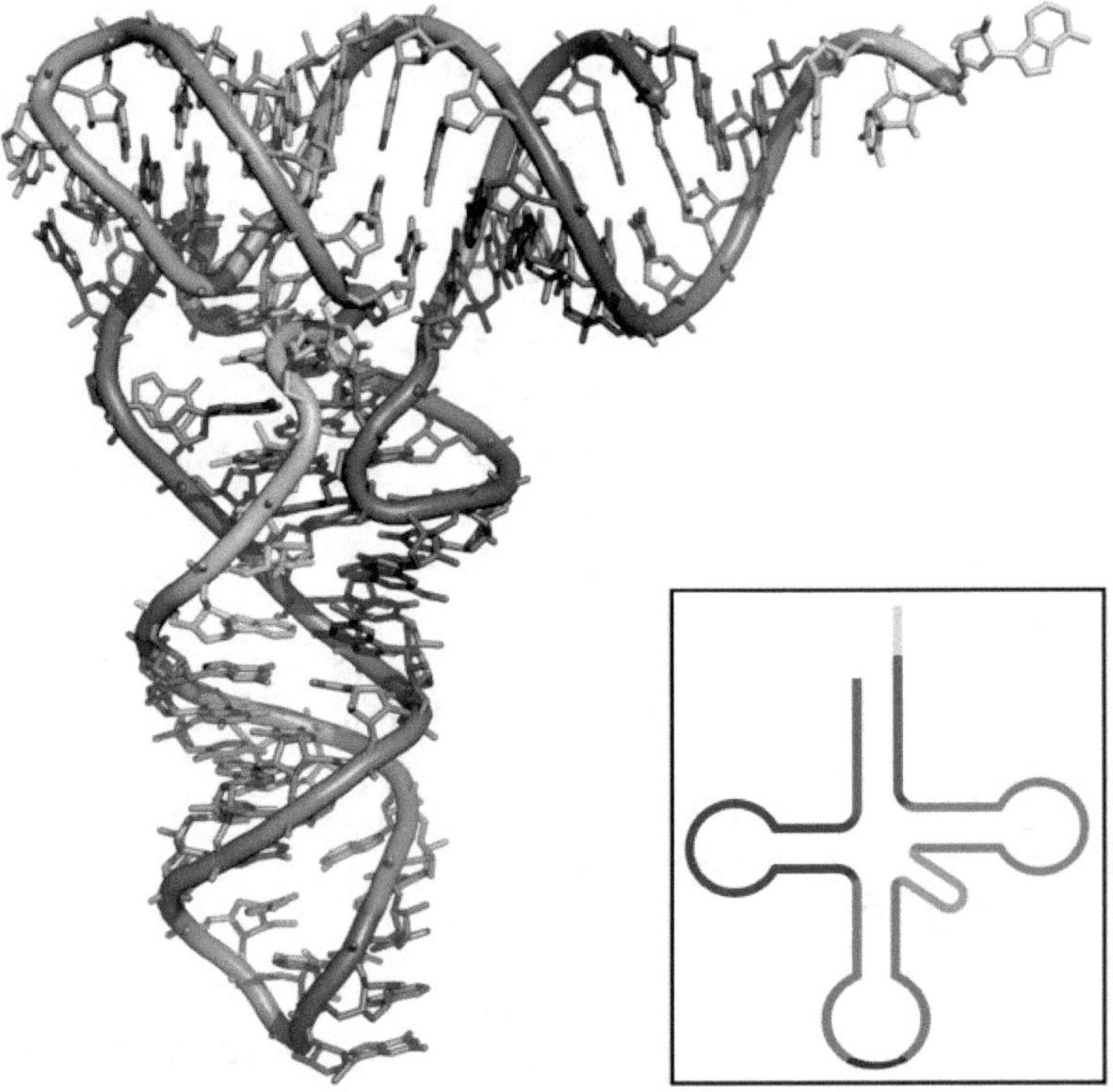

Tertiary structure of tRNA. CCA tail *in yellow,* Acceptor stem *in purple,* Variable loop *in orange,* D arm *in red,* Anticodon arm *in blue* with Anticodon *in black,* T arm *in green.*

example in mitochondria.[14]

To provide a one-to-one correspondence between tRNA molecules and codons that specify amino acids, 61 types of tRNA molecules would be required per cell. However, many cells contain fewer than 61 types of tRNAs because the wobble base is capable of binding to several, though not necessarily all, of the codons that specify a particular amino acid. A minimum of 31 tRNA are required to translate, unambiguously, all 61 sense codons of the standard genetic code.[3][15]

29.3 Aminoacylation

Aminoacylation is the process of adding an aminoacyl group to a compound. It produces a tRNA molecule with its CCA 3' end covalently linked to an amino acid.

Each tRNA is aminoacylated (or *charged*) with a specific amino acid by an aminoacyl tRNA synthetase. There is normally a single aminoacyl tRNA synthetase for each amino acid, despite the fact that there can be more than one tRNA, and more than one anticodon, for an amino acid. Recognition of the appropriate tRNA by the synthetases is not mediated solely by the anticodon, and the acceptor stem often plays a prominent role.[16]

Reaction:

1. amino acid + ATP → aminoacyl-AMP + PPi

2. aminoacyl-AMP + tRNA → aminoacyl-tRNA + AMP

Certain organisms can have one or more aminoacyl tRNA synthetases missing. This leads to charging of the tRNA by a chemically related amino acid. An enzyme or enzymes modify the charged amino acid to the final one. For example, *Helicobacter pylori* has glutaminyl tRNA synthetase missing. Thus, glutamate tRNA synthetase charges tRNA-glutamine(tRNA-Gln) with glutamate. An amidotransferase then converts the acid side chain of the glutamate to the amide, forming the correctly charged gln-tRNA-Gln.

29.4 Binding to ribosome

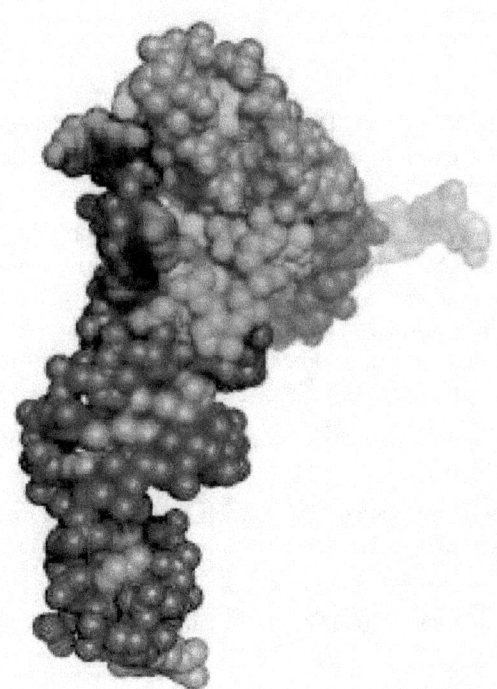

The range of conformations adopted by tRNA as it transits the A/T through P/E sites on the ribosome. The Protein Data Bank (PDB) codes for the structural models used as end points of the animation are given. Both tRNAs are modeled as phenylalanine-specific tRNA from Escherichia coli, *with the A/T tRNA as a homology model of the deposited coordinates. Color coding as shown for tRNA tertiary structure. Adapted from.*[17]

The ribosome has three binding sites for tRNA molecules that span the space between the two ribosomal subunits: the A (aminoacyl),[18] P (peptidyl), and E (exit) sites. In addition, the ribosome has two other sites for tRNA binding that are used during mRNA decoding or during the initiation of protein synthesis. These are the T site (named elongation factor

Tu) and I site (initiation).[19][20] By convention, the tRNA binding sites are denoted with the site on the small ribosomal subunit listed first and the site on the large ribosomal subunit listed second. For example, the A site is often written A/A, the P site, P/P, and the E site, E/E.[19] The binding proteins like L27, L2, L14, L15, L16 at the A- and P- sites have been determined by affinity labeling by A.P. Czernilofsky et al. (Proc. Natl. Acad. Sci, USA, pp 230–234, 1974).

Once translation initiation is complete, the first aminoacyl tRNA is located in the P/P site, ready for the elongation cycle described below. During translation elongation, tRNA first binds to the ribosome as part of a complex with elongation factor Tu (EF-Tu) or its eukaryotic (eEF-1) or archaeal counterpart. This initial tRNA binding site is called the A/T site. In the A/T site, the A-site half resides in the small ribosomal subunit where the mRNA decoding site is located. The mRNA decoding site is where the mRNA codon is read out during translation. The T-site half resides mainly on the large ribosomal subunit where EF-Tu or eEF-1 interacts with the ribosome. Once mRNA decoding is complete, the aminoacyl-tRNA is bound in the A/A site and is ready for the next peptide bond to be formed to its attached amino acid. The peptidyl-tRNA, which transfers the growing polypeptide to the aminoacyl-tRNA bound in the A/A site, is bound in the P/P site. Once the peptide bond is formed, the tRNA in the P/P site is deacylated, or has a free 3' end, and the tRNA in the A/A site carries the growing polypeptide chain. To allow for the next elongation cycle, the tRNAs then move through hybrid A/P and P/E binding sites, before completing the cycle and residing in the P/P and E/E sites. Once the A/A and P/P tRNAs have moved to the P/P and E/E sites, the mRNA has also moved over by one codon and the A/T site is vacant, ready for the next round of mRNA decoding. The tRNA bound in the E/E site then leaves the ribosome.

The P/I site is actually the first to bind to aminoacyl tRNA, which is delivered by an initiation factor called IF2 in bacteria.[20] However, the existence of the P/I site in eukaryotic or archaeal ribosomes has not yet been confirmed. The P-site protein L27 has been determined by affinity labeling by E. Collatz and A.P. Czernilofsky (FEBS Lett., Vol. 63, pp 283–286, 1976).

29.5   tRNA genes

Organisms vary in the number of tRNA genes in their genome. The nematode worm *C. elegans*, a commonly used model organism in genetics studies, has 29,647 [21] genes in its nuclear genome, of which 620 code for tRNA.[22][23] The budding yeast *Saccharomyces cerevisiae* has 275 tRNA genes in its genome.

In the human genome, which, according to January 2013 estimates, has about 20,848 protein coding genes [24] in total, there are 497 nuclear genes encoding cytoplasmic tRNA molecules, and 324 tRNA-derived pseudogenes—tRNA genes thought to be no longer functional[25] (although pseudo tRNAs have been shown to be involved in antibiotic resistance in bacteria).[26] Regions in nuclear chromosomes, very similar in sequence to mitochondrial tRNA genes, have also been identified (tRNA-lookalikes).[27] These tRNA-lookalikes are also been considered as part of the nuclear mitochondrial DNA (genes transferred from the mitochondria to the nucleus).[27][28]

As with all eukaryotes, there are 22 mitochondrial tRNA genes[29] in humans. Mutations in some of these genes have been associated with severe diseases like the MELAS syndrome.

Cytoplasmic tRNA genes can be grouped into 49 families according to their anticodon features. These genes are found on all chromosomes, except 22 and Y chromosome. High clustering on 6p is observed (140 tRNA genes), as well on 1 chromosome.[25]

The HGNC, in collaboration with the Genomic tRNA Database (GtRNAdb) and experts in the field, has approved unique names for human genes that encode tRNAs.

29.5.1   Evolution

Genomic tRNA content is a differentiating feature of genomes among biological domains of life: Archaea present the simplest situation in terms of genomic tRNA content with a uniform number of gene copies, Bacteria have an intermediate situation and Eukarya present the most complex situation.[30] Eukarya present not only more tRNA gene content than the other two kingdoms but also a high variation in gene copy number among different isoacceptors, and this complexity seem to be due to duplications of tRNA genes and changes in anticodon specificity .

Evolution of the tRNA gene copy number across different species has been linked to the appearance of specific tRNA modification enzymes (uridine methyltransferases in Bacteria, and adenosine deaminases in Eukarya), which increase the decoding capacity of a given tRNA.[30] As an example, tRNAAla encodes four different tRNA isoacceptors (AGC, UGC, GGC and CGC). In Eukarya, AGC isoacceptors are extremely enriched in gene copy number in comparison to the rest of isoacceptors, and this has been correlated with its A-to-I modification of its wobble base. This same trend has been shown for most amino acids of eukaryal species. Indeed, the effect of these two tRNA modifications is also seen in codon usage bias. Highly expressed genes seem to be enriched in codons that are exclusively using codons that will be decoded by these modified tRNAs, which suggests a possible role of these codons—and consequently of these tRNA modifications—in translation efficiency. [30]

29.6   tRNA biogenesis

In eukaryotic cells, tRNAs are transcribed by RNA polymerase III as pre-tRNAs in the nucleus.[31] RNA polymerase III recognizes two highly conserved downstream promoter sequences: the 5' intragenic control region (5'-ICR, D-control region, or A box), and the 3'-ICR (T-control region or B box) inside tRNA genes.[2][32][33] The first promoter begins at +8 of mature tRNAs and the second promoter is located 30-60 nucleotides downstream of the first promoter. The transcription terminates after a stretch of four or more thymidines.[2][33]

Pre-tRNAs undergo extensive modifications inside the nucleus. Some pre-tRNAs contain introns that are spliced, or cut, to form the functional tRNA molecule;[34] in bacteria these self-splice, whereas in eukaryotes and archaea they are removed by tRNA-splicing endonucleases.[35] The 5' sequence is removed by RNase P,[36] whereas the 3' end is removed by the tRNase Z enzyme.[37] A notable exception is in the archaeon *Nanoarchaeum equitans,* which does not possess an RNase P enzyme and has a promoter placed such that transcription starts at the 5' end of the mature tRNA.[38] The non-templated 3' CCA tail is added by a nucleotidyl transferase.[39] Before tRNAs are exported into the cytoplasm by Los1/Xpo-t,[40][41] tRNAs are aminoacylated.[42] The order of the processing events is not conserved. For example, in yeast, the splicing is not carried out in the nucleus but at the cytoplasmic side of mitochondrial membranes.[43]

29.7   History

The existence of tRNA was first hypothesized by Francis Crick, based on the assumption that there must exist an adapter molecule capable of mediating the translation of the RNA alphabet into the protein alphabet. Significant research on structure was conducted in the early 1960s by Alex Rich and Don Caspar, two researchers in Boston, the Jacques Fresco group in Princeton University and a United Kingdom group at King's College London.[44] In 1965, Robert W. Holley of Cornell University reported the primary structure and suggested three secondary structures.[45] tRNA was first crystallized in Madison, Wisconsin, by Robert M. Bock.[46] The cloverleaf structure was ascertained by several other studies in the following years[47] and was finally confirmed using X-ray crystallography studies in 1974. Two independent groups, Kim Sung-Hou working under Alexander Rich and a British group headed by Aaron Klug, published the same crystallography findings within a year.[48][49]

29.8   See also

- Cloverleaf model of tRNA

- Kim Sung-Hou

- Kissing stem-loop

- mRNA

- non-coding RNA and introns

- Slippery sequence

- tmRNA

- Transfer RNA-like structures

- translation

- tRNADB

- Wobble hypothesis

29.9   References

[1] Plescia, O J; Palczuk, N C; Cora-Figueroa, E; Mukherjee, A; Braun, W (October 1965). "Production of antibodies to soluble RNA (sRNA)". *Proc. Natl. Acad. Sci. USA* **54** (4): 1281–1285. doi:10.1073/pnas.54.4.1281. PMC 219862. PMID 5219832.

[2] Sharp, Stephen J; Schaack, Jerome; Cooley, Lynn; Burke, Deborah J; Soll, Dieter (1985). "Structure and Transcription of Eukaryotic tRNA Genes". *CRC Critical Reviews in Biochemistry* **19** (2): 107–144. doi:10.3109/10409238509082541. PMID 3905254. Retrieved 23 November 2014.

[3] Crick F (1968). "The origin of the genetic code". *J Mol Biol* **38** (3): 367–379. doi:10.1016/0022-2836(68)90392-6. PMID 4887876.

[4] Stryer L, Berg JM, Tymoczko JL (2002). *Biochemistry* (5th ed.). San Francisco: W.H. Freeman. ISBN 0-7167-4955-6.

[5] Itoh, Yuzuru; Sekine, Shun-ichi Sekine; Suetsugu, Shiro; Yokoyama, Shigeyuki (6 May 2013). "Tertiary structure of bacterial selenocysteine tRNA". *Nucleic Acids Research* **41** (13): 6729–6738. doi:10.1093/nar/gkt321. Retrieved 23 November 2014.

[6] Goodenbour, J. M.; Pan, T. (29 October 2006). "Diversity of tRNA genes in eukaryotes" (PDF). *Nucleic Acids Research* **34** (21): 6137–6146. doi:10.1093/nar/gkl725. Retrieved 23 November 2014.

[7] Jahn, Martina; Rogers, M. John; Söll, Dieter (18 July 1991). "Anticodon and acceptor stem nucleotides in tRNAGln are major recognition elements for E. coli glutaminyl-tRNA synthetase". *Nature* **352** (6332): 258–260. doi:10.1038/352258a0. Retrieved 23 November 2014.

[8] Ibba, Michael; Söll, Dieter (June 2000). "Aminoacyl-tRNA Synthesis". *Annual Review of Biochemistry* **69** (1): 617–650. doi:10.1146/annurev.biochem.69.1.617. PMID 10966471. Retrieved 23 November 2014.

[9] Sprinzl, M., and Cramer, F. (1979) *Prog. Nucleic Acids Res. Mol. Biol.* 22, 1–16

[10] Green, R., and Noller, H. F. (1997) *Annu. Rev. Biochem.* 66, 679–716

[11] Aebi M, Kirchner G, Chen JY; et al. (September 1990). "Isolation of a temperature-sensitive mutant with an altered tRNA nucleotidyltransferase and cloning of the gene encoding tRNA nucleotidyltransferase in the yeast Saccharomyces cerevisiae". *J. Biol. Chem* **265** (27): 16216–16220. PMID 2204621.

[12] McCloskey, James A.; Nishimura, Susumu (November 1977). "Modified nucleosides in transfer RNA". *Accounts of Chemical Research* **10** (11): 403–410. doi:10.1021/ar50119a004. Retrieved 23 November 2014.

[13] Felsenfeld G, Cantoni G; Cantoni (1964). "Use of thermal denaturation studies to investigate the base sequence of yeast serine sRNA". *Proc Natl Acad Sci USA* **51** (5): 818–26. Bibcode:1964PNAS...51..818F. doi:10.1073/pnas.51.5.818. PMC 300168. PMID 14172997.

[14] Suzuki, T; Suzuki, T (June 2014). "A complete landscape of post-transcriptional modifications in mammalian mitochondrial tRNAs.". *Nucleic acids research* **42** (11): 7346–57. PMID 24831542.

[15] Lodish H, Berk A, Matsudaira P, Kaiser CA, Krieger M, Scott MP, Zipursky SL, Darnell J. (2004). Molecular Biology of the Cell. WH Freeman: New York, NY. 5th ed.

[16] Schimmel P, Giege R, Moras D, Yokoyama S; Giege; Moras; Yokoyama (1993). "An operational RNA code for amino acids and possible relationship to genetic code". *Proc. Natl. Acad. Sci. USA* **90** (19): 8763–876. Bibcode:1993PNAS...90.8763S. doi:10.1073/pnas.90.19.8763.

[17] Dunkle JA, Wang L, Feldman MB, Pulk A, Chen VB, Kapral GJ, Noeske J, Richardson JS, Blanchard SC, Cate JH; Wang; Feldman; Pulk; Chen; Kapral; Noeske; Richardson; Blanchard; Cate (2011). "Structures of the bacterial ribosome in classical and hybrid states of tRNA binding". *Science* **332** (6032): 981–984. Bibcode:2011Sci...332..981D. doi:10.1126/science.1202692. PMC 3176341. PMID 21596992.

[18] Konevega, AL; Soboleva, NG; Makhno, VI; Semenkov, YP; Wintermeyer, W; Rodnina, MV; Katunin, VI (Jan 2004). "Purine bases at position 37 of tRNA stabilize codon-anticodon interaction in the ribosomal A site by stacking and Mg2+-dependent interactions". *RNA* **10** (1): 90–101. doi:10.1261/rna.5142404. PMC 1370521. PMID 14681588.

[19] Agirrezabala X, Frank J; Frank (2009). "Elongation in translation as a dynamic interaction among the ribosome, tRNA, and elongation factors EF-G and EF-Tu". *Q Rev Biophys* **42** (3): 159–200. doi:10.1017/S0033583509990060. PMC 2832932. PMID 20025795.

[20] Allen GS, Zavialov A, Gursky R, Ehrenberg M, Frank J; Zavialov; Gursky; Ehrenberg; Frank (2005). "The cryo-EM structure of a translation initiation complex from Escherichia coli". *Cell* **121** (5): 703–712. doi:10.1016/j.cell.2005.03.023. PMID 15935757.

[21] WormBase web site, http://www.wormbase.org, release WS187, date 25-Jan-2008.

[22] Spieth, J; Lawson, D (Jan 2006). "Overview of gene structure". *WormBook*: 1–10. doi:10.1895/wormbook.1.65.1. PMID 18023127.

[23] Hartwell LH, Hood L, Goldberg ML, Reynolds AE, Silver LM, Veres RC. (2004). *Genetics: From Genes to Genomes* 2nd ed. McGraw-Hill: New York, NY. p 264.

[24] Ensembl release 70 - Jan 2013 http://www.ensembl.org/Homo_sapiens/Info/StatsTable?db=core

[25] Lander E.; et al. (2001). "Initial sequencing and analysis of the human genome". *Nature* **409** (6822): 860–921. doi:10.1038/350 PMID 11237011.

[26] Rogers Theresa E.; et al. (2012). "A Pseudo-tRNA Modulates Antibiotic Resistance in Bacillus cereus". *PLoS ONE* **7** (7): e41248. doi:10.1371/journal.pone.0041248. PMID 22815980.

[27] Telonis Aristeidis G.; et al. (2014). "Nuclear and Mitochondrial tRNA-lookalikes in the Human Genome". *Frontiers in Genetics* **5**: 00344. doi:10.3389/fgene.2014.00344. PMID 25339973.

[28] Ramos A.; et al. (2011). "Nuclear Insertions of Mitochondrial Origin: Database Updating and Usefulness in Cancer Studies". *Mitochondrion* **11** (6): 946–53. doi:10.1016/j.mito.2011.08.009. PMID 21907832.

[29] *Ibid.* p 529.

[30] Novoa, Eva Maria; Pavon-Eternod, Mariana; Pan, Tao; Ribas de Pouplana, Lluís (March 2012). "A Role for tRNA Modifications in Genome Structure and Codon Usage". *Cell* **149** (1): 202–213. doi:10.1016/j.cell.2012.01.050. Retrieved 23 November 2014.

[31] White RJ (1997). "Regulation of RNA polymerases I and III by the retinoblastoma protein: a mechanism for growth control?". *Trends in Biochemical Sciences* **22** (3): 77–80. doi:10.1016/S0968-0004(96)10067-0. PMID 9066256.

[32] Sharp, Stephen; Dingermann, Theodor; Söll, Dieter (1982). "The minimum intragenic sequences required for promotion of eukaryotic tRNA gene transcription" (PDF). *Nucleic Acids Research* **10** (18): 5393–5406. doi:10.1093/nar/10.18.5393. Retrieved 23 November 2014.

[33] Dieci G, Fiorino G, Castelnuovo M, Teichmann M, Pagano A; Fiorino; Castelnuovo; Teichmann; Pagano (December 2007). "The expanding RNA polymerase III transcriptome". *Trends Genet.* **23** (12): 614–22. doi:10.1016/j.tig.2007.09.001. PMID 17977614.

[34] Tocchini-Valentini, Giuseppe D.; Fruscoloni, Paolo; Tocchini-Valentini, Glauco P. (12 November 2009). "Processing of multiple-intron-containing pretRNA". *Proceedings of the National Academy of Sciences* **106** (48): 20246–20251. doi:10.1073/pnas.0911658106.PMC2787110. PMID 19910528.

[35] Abelson J, Trotta CR, Li H; Trotta; Li (1998). "tRNA Splicing". *J Biol Chem* **273** (21): 12685–12688. doi:10.1074/jbc.273.21. PMID 9582290. 12685.

[36] Frank DN, Pace NR; Pace (1998). "Ribonuclease P: unity and diversity in a tRNA processing ribozyme". *Annu. Rev. Biochem.* **67** (1): 153–80. doi:10.1146/annurev.biochem.67.1.153. PMID 9759486.

[37] Ceballos M, Vioque A; Vioque (2007). "tRNase Z". *Protein Pept. Lett.* **14** (2): 137–45. doi:10.2174/092986607779816050. PMID 17305600.

[38] Randau L, Schröder I, Söll D; Schröder; Söll (May 2008). "Life without RNase P". *Nature***453**(7191): 120–3. Bibcode:2008N doi:10.1038/nature06833. PMID 18451863.

[39] Weiner AM (October 2004). "tRNA maturation: RNA polymerization without a nucleic acid template". *Curr. Biol.* **14** (20): R883–5. doi:10.1016/j.cub.2004.09.069. PMID 15498478.

[40] Kutay, U. .; Lipowsky, G. .; Izaurralde, E. .; Bischoff, F. .; Schwarzmaier, P. .; Hartmann, E. .; Görlich, D. . (1998). "Identification of a tRNA-Specific Nuclear Export Receptor". *Molecular Cell* **1** (3): 359–369. doi:10.1016/S1097-2765(00)80036-2. PMID 9660920.

[41] Arts, G. J.; Fornerod, M. .; Mattaj, L. W. (1998). "Identification of a nuclear export receptor for tRNA". *Current Biology* **8** (6): 305–314. doi:10.1016/S0960-9822(98)70130-7. PMID 9512417.

[42] Arts, G. -J.; Kuersten, S.; Romby, P.; Ehresmann, B.; Mattaj, I. W. (1998). "The role of exportin-t in selective nuclear export of mature tRNAs". *The EMBO Journal* **17** (24): 7430–7441. doi:10.1093/emboj/17.24.7430. PMC 1171087. PMID 9857198.

[43] Yoshihisa, T.; Yunoki-Esaki, K.; Ohshima, C.; Tanaka, N.; Endo, T. (2003). "Possibility of cytoplasmic pre-tRNA splicing: the yeast tRNA splicing endonuclease mainly localizes on the mitochondria". *Molecular Biology of the Cell* **14** (8): 3266–3279. doi:10.1091/mbc.E02-11-0757. PMC 181566. PMID 12925762.

[44] Brian F.C. Clark (October 2006). "The crystal structure of tRNA" (PDF). *J. Biosci.* **31** (4): 453–7. doi:10.1007/BF02705184. PMID 17206065.

[45] HOLLEY RW; APGAR J; EVERETT GA; et al. (March 1965). "STRUCTURE OF A RIBONUCLEIC ACID". *Science* **147** (3664): 1462–5. Bibcode:1965Sci...147.1462H. doi:10.1126/science.147.3664.1462. PMID 14263761. Retrieved 2010-09-03.

[46] http://www.nytimes.com/1991/07/04/obituaries/robert-m-bock-67-biologist-and-a-dean.html

[47] "The Nobel Prize in Physiology or Medicine 1968". Nobel Foundation. Retrieved 2007-07-28.

[48] Ladner JE; Jack A; Robertus JD; et al. (November 1975). "Structure of yeast phenylalanine transfer RNA at 2.5 A resolution". *Proc. Natl. Acad. Sci. U.S.A.* **72** (11): 4414–8. Bibcode:1975PNAS...72.4414L. doi:10.1073/pnas.72.11.4414. PMC 388732. PMID 1105583.

[49] Kim SH; Quigley GJ; Suddath FL; et al. (1973). "Three-dimensional structure of yeast phenylalanine transfer RNA: folding of the polynucleotide chain". *Science* **179** (4070): 285–8. Bibcode:1973Sci...179..285K. doi:10.1126/science.179.4070.285. PMID 4566654.

29.10   External links

- tRNAdb (updated and completely restructured version of Spritzls tRNA compilation)

- original Sprinzl tRNA compilation

- tRNA link to heart disease and stroke

- GtRNAdb: Collection of tRNAs identified from complete genomes

- HGNC: Gene nomenclature of human tRNAs

- Molecule of the Month © RCSB Protein Data Bank:

  - Transfer RNA

  - Aminoacyl-tRNA Synthetases

  - Elongation Factors

- Rfam entry for tRNA

Chapter 30

Proteinogenic amino acid

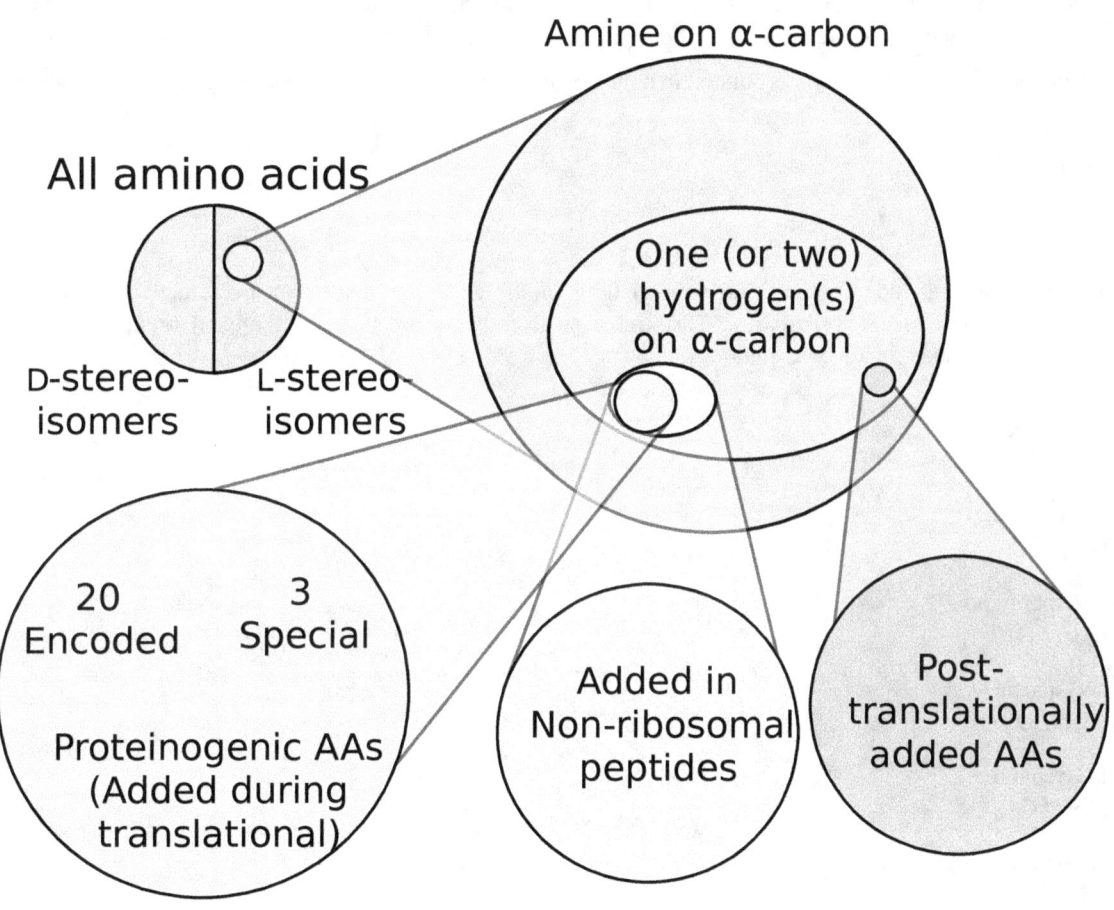

Proteinogenic amino acids are a small fraction of all amino acids

Proteinogenic amino acids are amino acids that are precursors to proteins, and are incorporated into proteins co-translationally — that is, during translation.[1] There are 23 proteinogenic amino acids in prokaryotes (including N-Formylmethionine, mainly used to initiate protein synthesis and often removed afterward), but only 21 are encoded by the nuclear genes of eukaryotes. Of the 23, pyrrolysine (O/Pyl) is incorporated into proteins by distinct post-translational biosynthetic mechanisms; all the other 21 are directly encoded by the genetic code, including selenocysteine (U/Sec), that uses a special case of insertion during the translational incorporation, but that is not considered a post-translational modification . Humans can synthesize 11 of these 20 from each other or from other molecules of intermediary metabolism. The other nine must be consumed (usually as their protein derivatives), and so they are called essential amino acids. The essential amino acids are histidine, isoleucine, leucine, lysine, methionine, phenylalanine, threonine, tryptophan, and valine (i.e. H I L K M F T W V).

The word "proteinogenic" means "protein creating". Proteinogenic amino acids can be condensed into a polypeptide (the subunit of a protein) through a process called translation (the second stage of protein biosynthesis, part of the overall process of gene expression).

In contrast, nonproteinogenic amino acids are either not incorporated in proteins (like GABA, L-DOPA, or triiodothyronine), or are not produced directly and in isolation by standard cellular machinery (like hydroxyproline and selenomethionine). The latter often results from post-translational modification of proteins.

The proteinogenic amino acids have been found to be related to the set of amino acids that can be recognized by ribozyme autoaminoacylation systems.[2] Thus, nonproteinogenic amino acids would have been excluded by the contingent evolutionary success of nucleotide-based life forms. Other reasons have been offered to explain why certain specific nonproteinogenic amino acids are not generally incorporated into proteins; for example, ornithine and homoserine cyclize against the peptide backbone and fragment the protein with relatively short half-lives, while others are toxic because they can be mistakenly incorporated into proteins, such as the arginine analog canavanine.

Nonproteinogenic amino acids are incorporated in nonribosomal peptides, which are not produced by the ribosome during translation.

30.1   Structures

The following illustrates the structures and abbreviations of the 21 amino acids that are directly encoded for protein synthesis by the genetic code of eukaryotes. The structures given below are standard chemical structures, not the typical zwitterion forms that exist in aqueous solutions.

- L-Alanine
  (Ala / A)

- L-Arginine
  (Arg / R)

- L-Asparagine
  (Asn / N)

- L-Aspartic acid
  (Asp / D)

- L-Cysteine
  (Cys / C)

- L-Glutamic acid
  (Glu / E)

- L-Glutamine
  (Gln / Q)

- Glycine
  (Gly / G)

- L-Histidine
  (His / H)

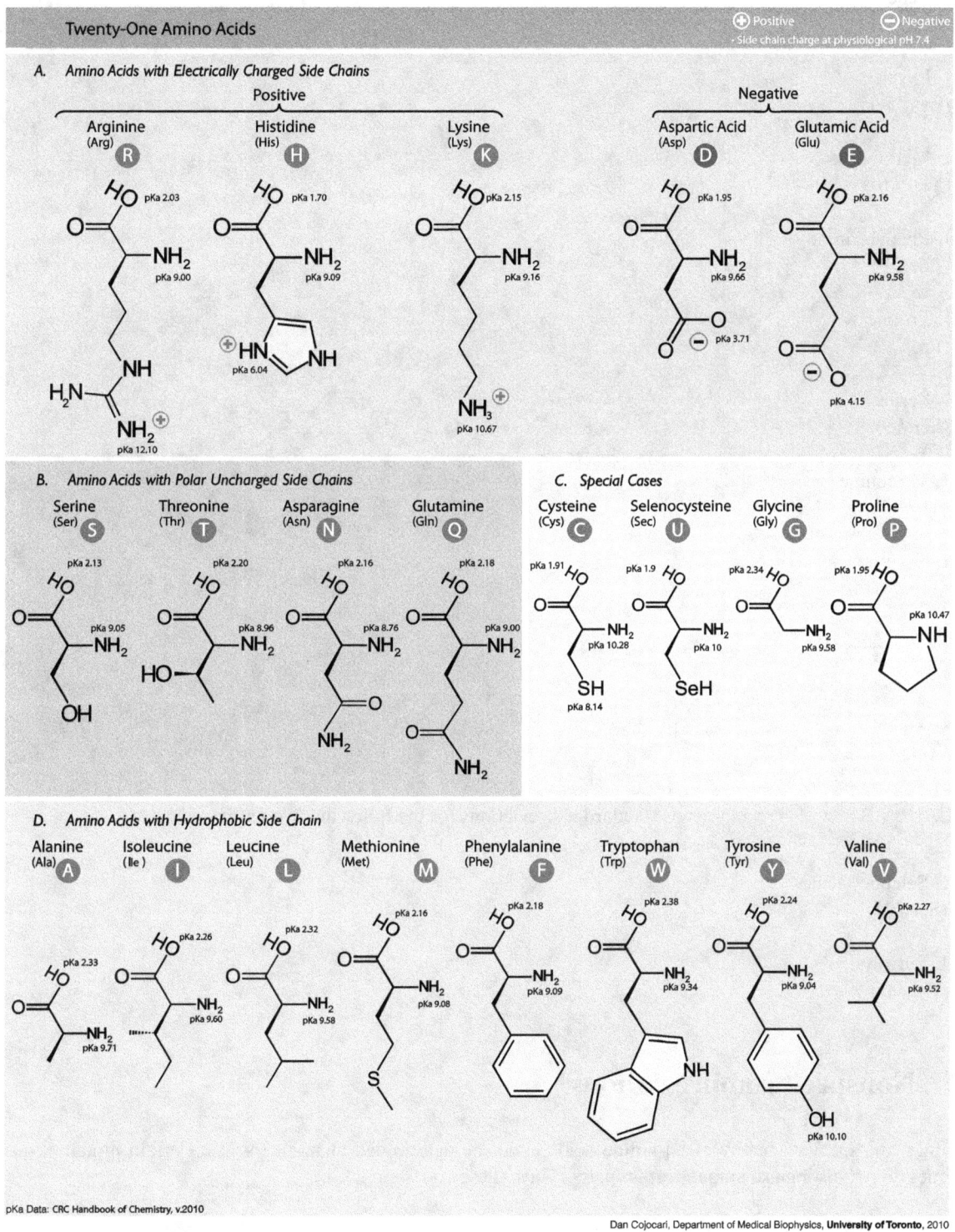

Grouped table of 21 amino acids' structures, nomenclature, and their side groups' pKa values

- L-Isoleucine
  (Ile / I)

- L-Leucine
  (Leu / L)

- L-Lysine
  (Lys / K)

- L-Methionine
  (Met / M)

- L-Phenylalanine
  (Phe / F)

- L-Proline
  (Pro / P)

- L-Serine
  (Ser / S)

- L-Threonine
  (Thr / T)

- L-Tryptophan
  (Trp / W)

- L-Tyrosine
  (Tyr / Y)

- L-Valine
  (Val / V)

IUPAC/IUBMB now also recommends standard abbreviations for the following two amino acids:

- L-Selenocysteine
  (Sec / U)

- L-Pyrrolysine
  (Pyl / O)

30.2   Nonspecific abbreviations

Sometimes, the specific identity of an amino acid cannot be determined unambiguously. Certain protein sequencing techniques do not distinguish among certain pairs. Thus, these codes are used:

- Asx (B) is "asparagine or aspartic acid"

- Glx (Z) is "glutamic acid or glutamine"

- Xle (J) is "leucine or isoleucine"

In addition, the symbol X is used to indicate an amino acid that is completely unidentified.

30.3 Chemical properties

Following is a table listing the one-letter symbols, the three-letter symbols, and the chemical properties of the side chains of the standard amino acids. The masses listed are based on weighted averages of the elemental isotopes at their natural abundances. Forming a peptide bond results in elimination of a molecule of water, so the mass of an amino acid unit within a protein chain is reduced by 18.01524 Da.

General chemical properties

30.3.1 Side chain properties

Note: The pKa values of amino acids are typically slightly different when the amino acid is inside a protein. Protein pKa calculations are sometimes used to calculate the change in the pKa value of an amino acid in this situation.

30.3.2 Gene expression and biochemistry

* UAG is normally the amber stop codon, but encodes pyrrolysine if a PYLIS element is present.
** UGA is normally the opal (or umber) stop codon, but encodes selenocysteine if a SECIS element is present.
† The stop codon is not an amino acid, but is included for completeness.
†† UAG and UGA do not always act as stop codons (see above).
‡ An essential amino acid cannot be synthesized in humans and must, therefore, be supplied in the diet. Conditionally essential amino acids are not normally required in the diet, but must be supplied exogenously to specific populations that do not synthesize it in adequate amounts.

30.3.3 Mass spectrometry

In mass spectrometry of peptides and proteins, knowledge of the masses of the residues is useful. The mass of the peptide or protein is the sum of the residue masses plus the mass of water.[3]

§ Monoisotopic mass

30.3.4 Stoichiometry and metabolic cost in cell

The table below lists the abundance of amino acids in *E.coli* cells and the metabolic cost (ATP) for synthesis the amino acids. Negative numbers indicate the metabolic processes are energy favorable and do not cost net ATP of the cell.[4] The abundance of amino acids includes amino acids in free form and in polymerization form (proteins).

30.3.5 Remarks

30.3.6 Catabolism

30.4 Life based on alternative proteinogenic sets

The proteinogenic set used by known life on Earth appears to be arbitrarily selected by evolution, according to current knowledge, from many hundreds of possible alpha-type amino acids. Xenobiology studies hypothetical life forms that could be constructed using alternative sets using expanded genetic codes. Miller-type experiments on artificial abiogenesis show that alpha-type amino acids predominate in water-based 'primordial soups', but beta-type amino acids dominate when less water is present. Both alpha- and beta-based sets could form the basis for alternative protein constructions and life forms.

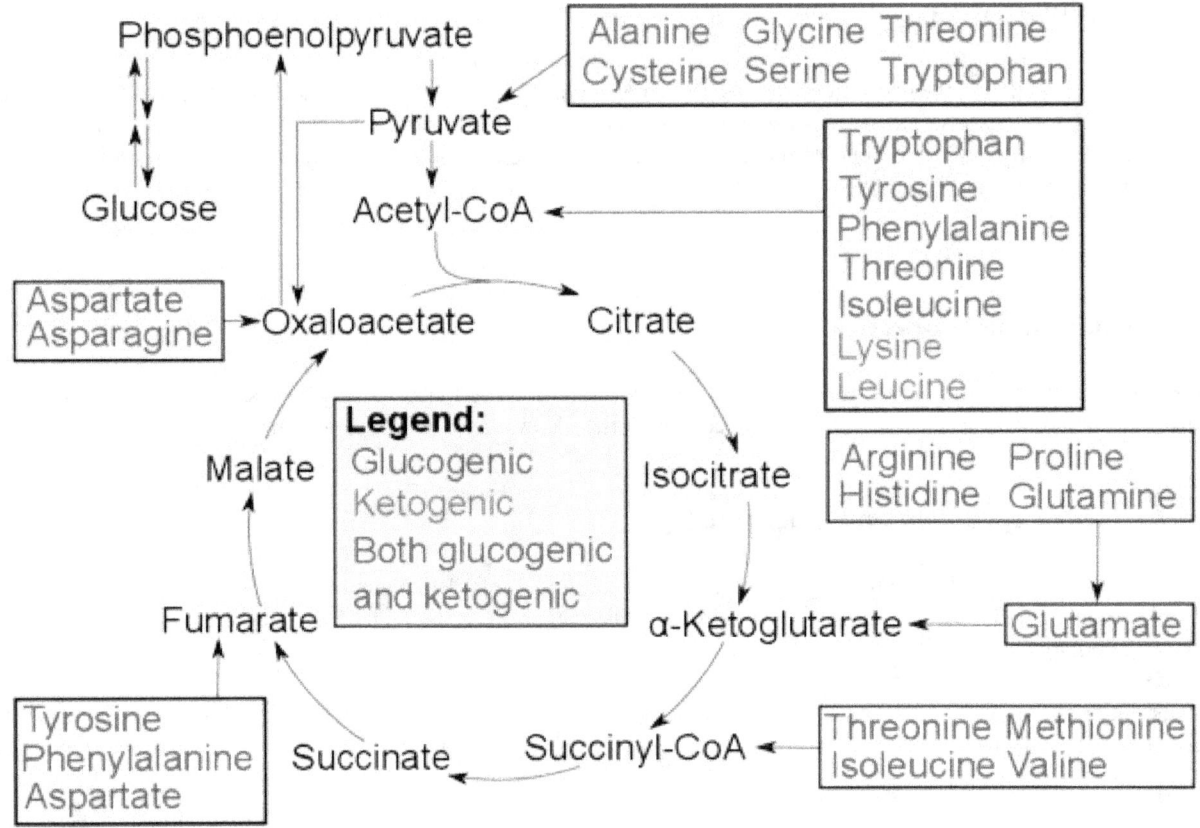

Amino acids can be classified according to the properties of their main products as either of:[5]

- *Glucogenic, with the products having the ability to form glucose by gluconeogenesis*

- *Ketogenic, with the products not having the ability to form glucose: These products may still be used for ketogenesis or lipid synthesis.*

- *Amino acids catabolized into both glucogenic and ketogenic products.*

30.5   References

[1] Ambrogelly A, Palioura S, Söll D (Jan 2007). "Natural expansion of the genetic code". *Nat Chem Biol* **3**(1): 29–35. doi:10.1038 PMID 17173027.

[2] Erives A (2011). "A Model of Proto-Anti-Codon RNA Enzymes Requiring L-Amino Acid Homochirality". *J Molecular Evolution* **73**: 10–22. doi:10.1007/s00239-011-9453-4. PMC 3223571. PMID 21779963.

[3] "The amino acid masses". ExPASy. Retrieved 2009-01-06.

[4] *Physical Biology of the Cell* (Garland Science) p. 178

[5] Chapter 20 (Amino Acid Degradation and Synthesis) in: Denise R., PhD. Ferrier. *Lippincott's Illustrated Reviews: Biochemistry (Lippincott's Illustrated Reviews)*. Hagerstwon, MD: Lippincott Williams & Wilkins. ISBN 0-7817-2265-9.

- Nelson, David L.; Cox, Michael M. (2000). *Lehninger Principles of Biochemistry* (3rd ed.). Worth Publishers. ISBN 1-57259-153-6.

- Kyte, J.; Doolittle, R. F. (1982). "A simple method for displaying the hydropathic character of a protein". *J. Mol. Biol.* **157** (1): 105–132. doi:10.1016/0022-2836(82)90515-0. PMID 7108955.

- Meierhenrich, Uwe J. (2008). *Amino acids and the asymmetry of life* (1st ed.). Springer. ISBN 978-3-540-76885-2.

- Biochemistry, Harpers (2015). *Harpers Illustrated Biochemistry* (30st ed.). Lange. ISBN 978-0-07-182534-4.

30.6 See also

- Glucogenic amino acid
- Ketogenic amino acid

Chapter 31

DNA-binding protein

DNA-binding proteins [3][4][5] are proteins that are composed of DNA-binding domains and thus have a specific or general affinity for either single or double stranded DNA. Sequence-specific DNA-binding proteins generally interact with the major groove of B-DNA, because it exposes more functional groups that identify a base pair. However there are some known minor groove DNA-binding ligands such as netropsin,[6] distamycin, Hoechst 33258, pentamidine, DAPI and others.[7]

31.1 Examples

DNA-binding proteins include transcription factors which modulate the process of transcription, various polymerases, nucleases which cleave DNA molecules, and histones which are involved in chromosome packaging and transcription in the cell nucleus. DNA-binding proteins can incorporate such domains as the zinc finger, the helix-turn-helix, and the leucine zipper (among many others) that facilitate binding to nucleic acid. There are also more unusual examples such as transcription activator like effectors.

31.2 Non-specific DNA-protein interactions

Structural proteins that bind DNA are well-understood examples of non-specific DNA-protein interactions. Within chromosomes, DNA is held in complexes with structural proteins. These proteins organize the DNA into a compact structure called chromatin. In eukaryotes this structure involves DNA binding to a complex of small basic proteins called histones, while in prokaryotes multiple types of proteins are involved.[8][9] The histones form a disk-shaped complex called a nucleosome, which contains two complete turns of double-stranded DNA wrapped around its surface. These non-specific interactions are formed through basic residues in the histones making ionic bonds to the acidic sugar-phosphate backbone of the DNA, and are therefore largely independent of the base sequence.[10] Chemical modifications of these basic amino acid residues include methylation, phosphorylation and acetylation.[11] These chemical changes alter the strength of the interaction between the DNA and the histones, making the DNA more or less accessible to transcription factors and changing the rate of transcription.[12] Other non-specific DNA-binding proteins in chromatin include the high-mobility group proteins, which bind to bent or distorted DNA.[13] These proteins are important in bending arrays of nucleosomes and arranging them into the larger structures that make up chromosomes.[14]

31.3 DNA-binding proteins that specifically bind single-stranded DNA

(See Single-stranded binding protein) A distinct group of DNA-binding proteins are the DNA-binding proteins that specifically bind single-stranded DNA. In humans, replication protein A is the best-understood member of this family and

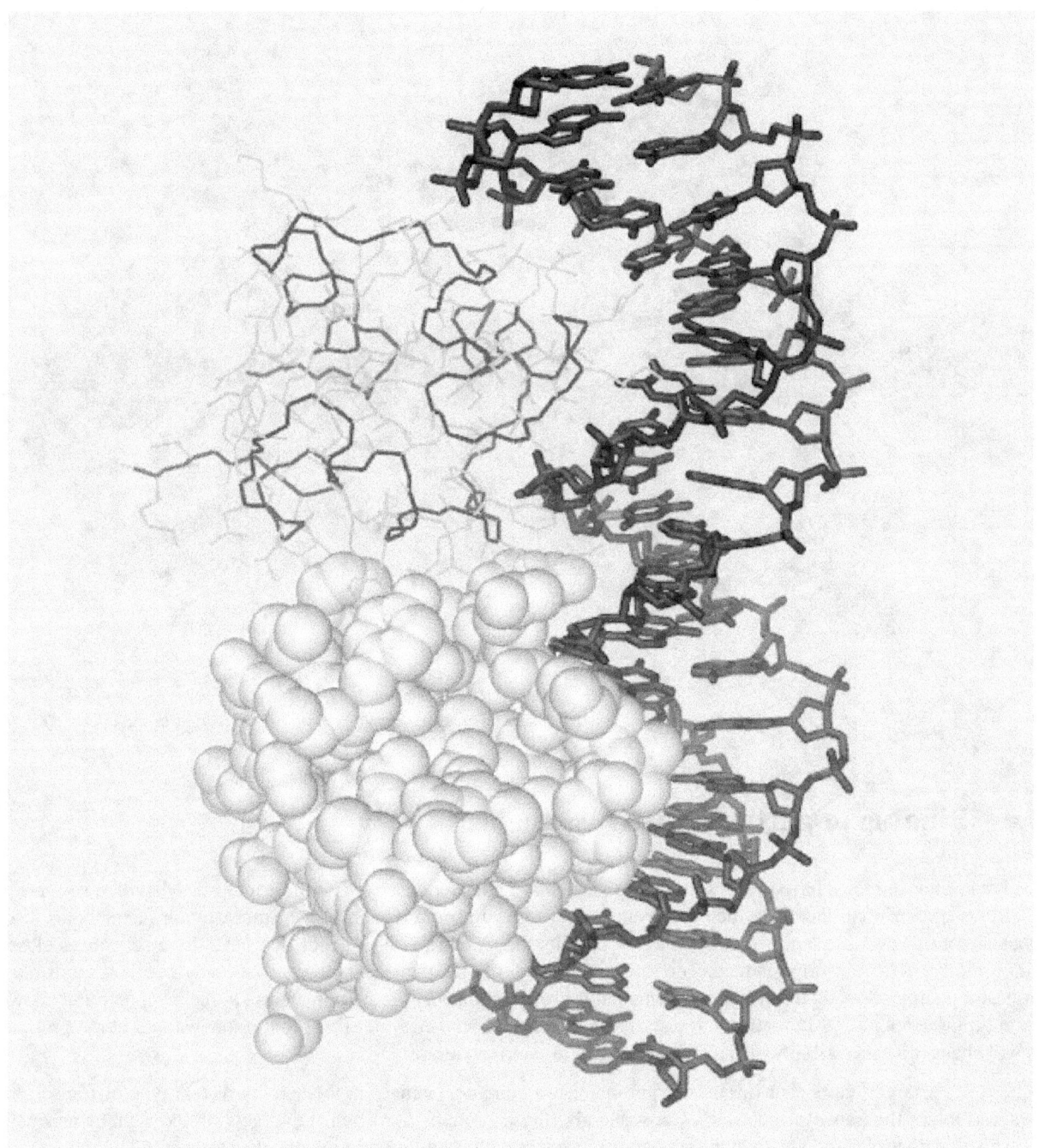

Cro *protein complex with DNA*

is used in processes where the double helix is separated, including DNA replication, recombination and DNA repair.[15] These binding proteins seem to stabilize single-stranded DNA and protect it from forming stem-loops or being degraded by nucleases.

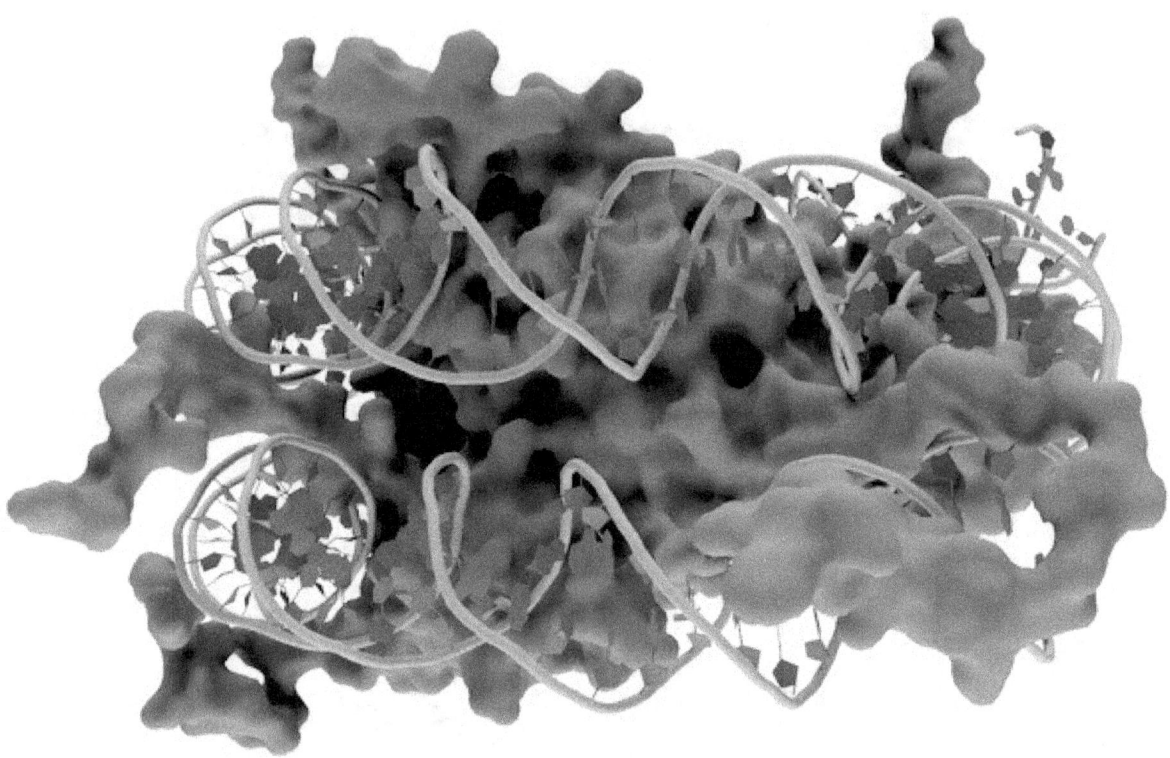

Interaction of DNA (shown in orange) with histones (shown in blue). These proteins' basic amino acids bind to the acidic phosphate groups on DNA.

31.4   Binding to particular DNA sequences

In contrast, other proteins have evolved to bind to particular DNA sequences. The most intensively studied of these are the various transcription factors, which are proteins that regulate transcription. Each transcription factor binds to one particular set of DNA sequences and activates or inhibits the transcription of genes that have these sequences close to their promoters. The transcription factors do this in two ways. Firstly, they can bind the RNA polymerase responsible for transcription, either directly or through other mediator proteins; this locates the polymerase at the promoter and allows it to begin transcription.[16] Alternatively, transcription factors can bind enzymes that modify the histones at the promoter; this will change the accessibility of the DNA template to the polymerase.[17]

As these DNA targets can occur throughout an organism's genome, changes in the activity of one type of transcription factor can affect thousands of genes.[18] Consequently, these proteins are often the targets of the signal transduction processes that control responses to environmental changes or cellular differentiation and development. The specificity of these transcription factors' interactions with DNA come from the proteins making multiple contacts to the edges of the DNA bases, allowing them to "read" the DNA sequence. Most of these base-interactions are made in the major groove, where the bases are most accessible.[19] Mathematical descriptions of protein-DNA binding taking into account sequence-specificity, competitive and cooperative binding of proteins of different types are usually performed with the help of the lattice models.[20] Computational methods to identify the DNA binding sequence specificity have been proposed to make a good use of the abundant sequence data in the post-genomic era.[21]

31.5   See also

- bZIP domain

- DNA-binding domain

- Helix-loop-helix

- Helix-turn-helix

- HMG-box

- Leucine zipper

- Lexitropsin

- Nucleic acid simulations

- Single-strand binding protein

- Zinc finger

31.6 References

[1] Created from PDB 1LMB

[2] Created from PDB 1RVA

[3] Travers, A. A. (1993). *DNA-protein interactions*. London: Springer. ISBN 978-0-412-25990-6.

[4] Pabo CO, Sauer RT (1984). "Protein-DNA recognition". *Annu. Rev. Biochem.* **53**(1): 293–321. doi:10.1146/annurev.bi.53.0 PMID 6236744.

[5] Dickerson R.E. (1983). "The DNA helix and how it is read". *Sci Am* **249** (6): 94–111. doi:10.1038/scientificamerican1283-94.

[6] Zimmer C, Wähnert U (1986). "Nonintercalating DNA-binding ligands: specificity of the interaction and their use as tools in biophysical, biochemical and biological investigations of the genetic material". *Prog. Biophys. Mol. Biol.* **47** (1): 31–112. doi:10.1016/0079-6107(86)90005-2. PMID 2422697.

[7] Dervan PB (April 1986). "Design of sequence-specific DNA-binding molecules". *Science* **232**(4749): 464–71. doi:10.1126/sc PMID 2421408.

[8] Sandman K, Pereira S, Reeve J (1998). "Diversity of prokaryotic chromosomal proteins and the origin of the nucleosome". *Cell Mol Life Sci* **54** (12): 1350–64. doi:10.1007/s000180050259. PMID 9893710.

[9] Dame RT (2005). "The role of nucleoid-associated proteins in the organization and compaction of bacterial chromatin". *Mol. Microbiol.* **56** (4): 858–70. doi:10.1111/j.1365-2958.2005.04598.x. PMID 15853876.

[10] Luger K, Mäder A, Richmond R, Sargent D, Richmond T (1997). "Crystal structure of the nucleosome core particle at 2.8 A resolution". *Nature* **389** (6648): 251–60. doi:10.1038/38444. PMID 9305837.

[11] Jenuwein T, Allis C (2001). "Translating the histone code". *Science* **293** (5532): 1074–80. doi:10.1126/science.1063127. PMID 11498575.

[12] Ito T (2003). "Nucleosome assembly and remodelling". *Curr Top Microbiol Immunol* **274**: 1–22. doi:10.1007/978-3-642-55747-7_1. PMID 12596902.

[13] Thomas J (2001). "HMG1 and 2: architectural DNA-binding proteins". *Biochem Soc Trans* **29**(Pt 4): 395–401. doi:10.1042/B PMID 11497996.

[14] Grosschedl R, Giese K, Pagel J (1994). "HMG domain proteins: architectural elements in the assembly of nucleoprotein structures". *Trends Genet* **10** (3): 94–100. doi:10.1016/0168-9525(94)90232-1. PMID 8178371.

[15] Iftode C, Daniely Y, Borowiec J (1999). "Replication protein A (RPA): the eukaryotic SSB". *Crit Rev Biochem Mol Biol* **34** (3): 141–80. doi:10.1080/10409239991209255. PMID 10473346.

[16] Myers L, Kornberg R (2000). "Mediator of transcriptional regulation". *Annu Rev Biochem* **69**(1): 729–49. doi:10.1146/annur PMID 10966474.

[17] Spiegelman B, Heinrich R (2004). "Biological control throughs regulated transcriptional coactivators". *Cell* **119** (2): 157–67. doi:10.1016/j.cell.2004.09.037. PMID 15479634.

[18] Li Z, Van Calcar S, Qu C, Cavenee W, Zhang M, Ren B (2003). "A global transcriptional regulatory role for c-Myc in Burkitt's lymphoma cells". *Proc Natl Acad Sci USA* **100** (14): 8164–9. doi:10.1073/pnas.1332764100. PMC 166200. PMID 12808131.

[19] Pabo C, Sauer R (1984). "Protein-DNA recognition". *Annu Rev Biochem* **53**(1): 293–321. doi:10.1146/annurev.bi.53.070184 PMID 6236744.

[20] Teif V.B., Rippe K. (2010). "Statistical-mechanical lattice models for protein-DNA binding in chromatin.". *Journal of Physics: Condensed Matter* **22** (41): 414105. arXiv:1004.5514. doi:10.1088/0953-8984/22/41/414105. PMID 21386588.

[21] Wong KC, Chan TM, Peng C., Li Y., and Zhang Z. "DNA Motif Elucidation using belief propagation" *Nucleic Acids Research* Advanced Online June 2013; doi:10.1093/nar/gkt574 PMID 23814189

31.7 External links

- Abalone tool for modeling DNA-ligand interactions.

- DBD database of predicted transcription factors Uses a curated set of DNA-binding domains to predict transcription factors in all completely sequenced genomes

- DNA-Binding Proteins at the US National Library of Medicine Medical Subject Headings (MeSH)

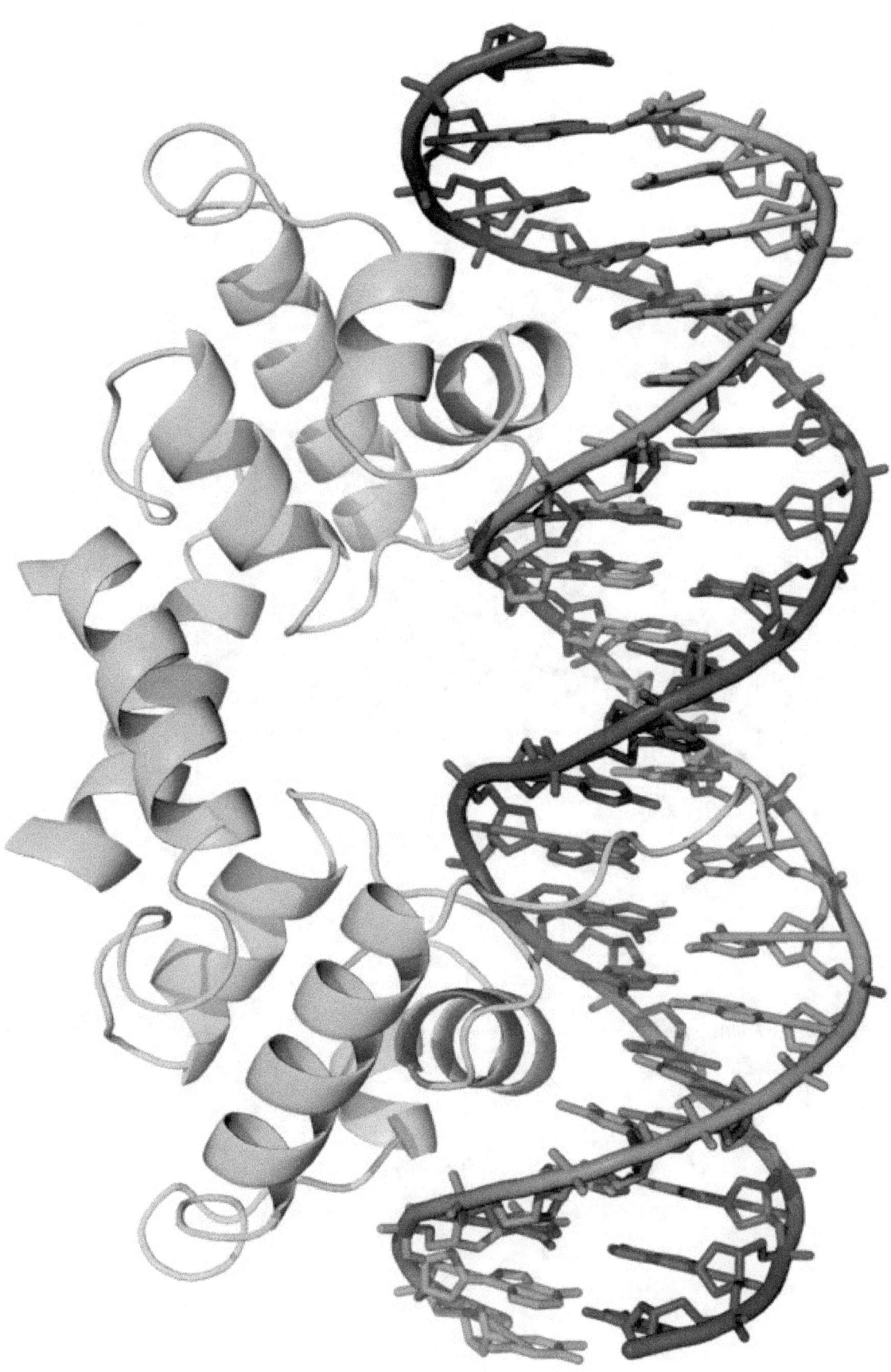

The lambda repressor helix-turn-helix transcription factor bound to its DNA target

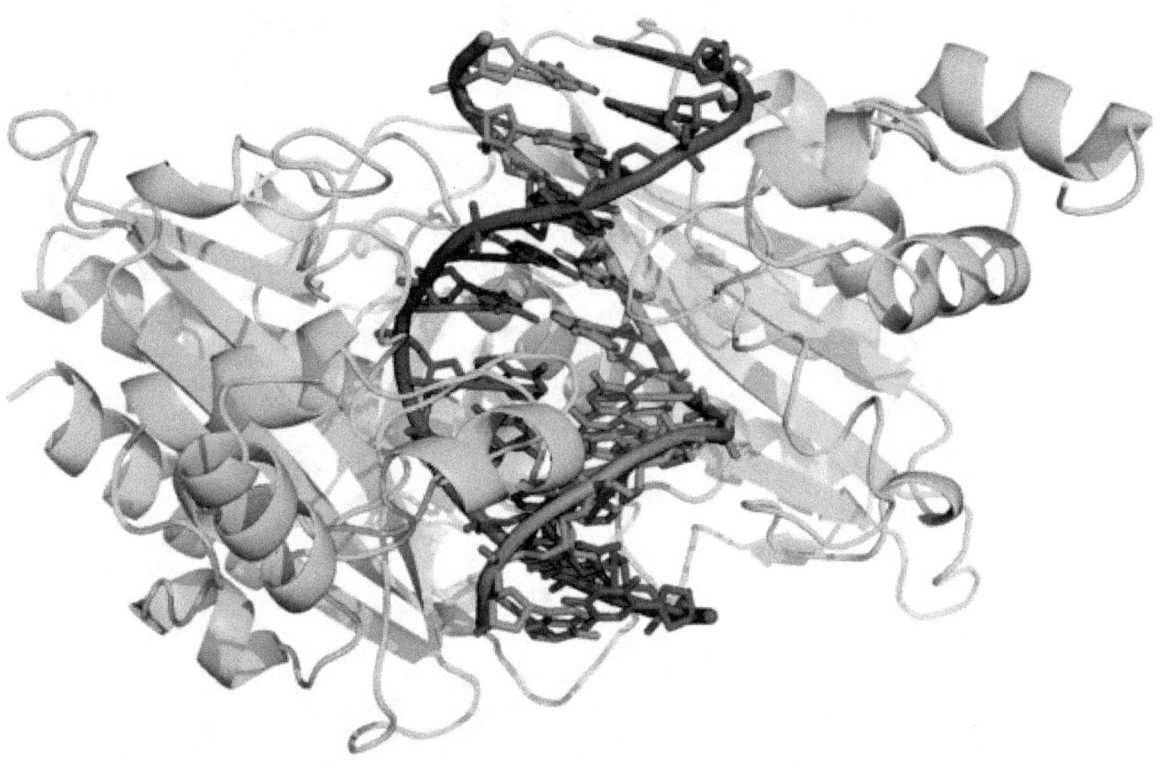

The restriction enzyme EcoRV (green) in a complex with its substrate DNA[2]

Chapter 32

Nuclease

A **nuclease** is an enzyme capable of cleaving the phosphodiester bonds between the nucleotide subunits of nucleic acids. Older publications may use terms such as "polynucleotidase" or "nucleodepolymerase".[1]

Nucleases are usually further divided into endonucleases and exonucleases, although some of the enzymes may fall in both categories. Well known nucleases are deoxyribonuclease and ribonuclease.

32.1 Introduction

In the late 1960s, scientists Stuart Linn and Werner Arber isolated examples of the two types of enzymes responsible for phage growth restriction in Escherichia coli (E. coli) bacteria.[2][3] One of these enzymes added a methyl group to the DNA, generating methylated DNA, while the other cleaved unmethylated DNA at a wide variety of locations along the length of the molecule. The first type of enzyme was called a "methylase" and the other a "restriction nuclease". These enzymatic tools were important to scientists who were gathering the tools needed to "cut and paste" DNA molecules. What was then needed was a tool that would cut DNA at specific sites, rather than at random sites along the length of the molecule, so that scientists could cut DNA molecules in a predictable and reproducible way.

32.2 Numerical Classification System

Most nucleases are classified by the Enzyme Commission number of the "Nomenclature Committee of the International Union of Biochemistry and Molecular Biology" as hydrolases (EC-number 3). The nucleases belong just like phosphodiesterase, lipase and phosphatase to the esterases (EC-number 3.1), a subgroup of the hydrolases. The esterases to which nucleases belong are classified with the EC-numbers 3.1.11 - EC-number 3.1.31.

32.3 Site-specific nuclease

32.3.1 Structure specific nuclease

For details see flap endonuclease.

32.3.2 Sequence specific nuclease

This important development came when H.O. Smith, K.W. Wilcox, and T.J. Kelley, working at Johns Hopkins University in 1968, isolated and characterized the first restriction nuclease whose functioning depended on a specific DNA nucleotide sequence. Working with Haemophilus influenzae bacteria, this group isolated an enzyme, called *Hind*II, that always cut DNA molecules at a particular point within a specific sequence of six base pairs.

They found that the *Hind*II enzyme always cuts directly in the center of this sequence. Wherever this particular sequence of six base pairs occurs unmodified in a DNA molecule, *Hind*II will cleave both DNA strands between the 3rd and 4th base pairs of the sequence. Moreover, *Hind*II will only cleave a DNA molecule at this particular site. For this reason, this specific base sequence is known as the "recognition sequence" for *Hind*II.

*Hind*II is only one example of the class of enzymes known as restriction nucleases. In fact, more than 900 restriction enzymes, some sequence specific and some not, have been isolated from over 230 strains of bacteria since the initial discovery of *Hind*II. These restriction enzymes generally have names that reflect their origin—The first letter of the name comes from the genus and the second two letters come from the species of the prokaryotic cell from which they were isolated. For example *Eco*RI comes from Escherichia coli RY13 bacteria, while HindII comes from Haemophilus influenzae strain Rd. Numbers following the nuclease names indicate the order in which the enzymes were isolated from single strains of bacteria: *Eco*RI, *Eco*RII. Nucleases are further described by addition of the prefix "endo" or "exo" to the name: The term "endonuclease" applies to nucleases that break nucleic acid chains somewhere in the interior, rather than at the ends, of the molecule. A nuclease that functions by removing nucleotides from the ends of the DNA molecule is called an **exonuclease.**

32.4   Endonucleases and DNA fragments

A restriction endonuclease functions by "scanning" the length of a DNA molecule. Once it encounters its particular specific recognition sequence, it will bind to the DNA molecule and makes one cut in each of the two sugar-phosphate backbones. The positions of these two cuts, both in relation to each other, and to the recognition sequence itself, are determined by the identity of the restriction endonuclease. Different endonucleases yield different sets of cuts, but one endonuclease will always cut a particular base sequence the same way, no matter what DNA molecule it is acting on. Once the cuts have been made, the DNA molecule will break into fragments.

32.5   Endonucleases and sticky ends

Not all restriction endonucleases cut symmetrically and leave blunt ends like *Hind*II described above. Many endonucleases cleave the DNA backbones in positions that are not directly opposite each other, creating overhangs. For example, the nuclease EcoRI has the following recognition sequence:

When the enzyme encounters this sequence, it cleaves each backbone between the G and the closest A base residues. Once the cuts have been made, the resulting fragments are held together only by the relatively weak hydrogen bonds that hold the complementary bases to each other. The weakness of these bonds allows the DNA fragments to separate from each other. Each resulting fragment has a protruding 5' end composed of unpaired bases. Other enzymes create cuts in the DNA backbone which result in protruding 3' ends. Protruding ends—both 3' and 5'—are sometimes called "sticky ends" because they tend to bond with complementary sequences of bases. In other words, if an unpaired length of bases (5' A A T T 3') encounters another unpaired length with the sequence (3' T T A A 5') they will bond to each other—they are "sticky" for each other. Ligase enzyme is then used to join the phosphate backbones of the two molecules. The cellular origin, or even the species origin, of the sticky ends does not affect their stickiness. Any pair of complementary sequences will tend to bond, even if one of the sequences comes from a length of human DNA, and the other comes from a length of bacterial DNA. In fact, it is this quality of stickiness that allows production of recombinant DNA molecules, molecules which are composed of DNA from different sources, and which has given birth to the genetic engineering technology.

32.6   Meganucleases

Main article: meganuclease

The frequency at which a particular nuclease will cut a given DNA molecule depends on the complexity of the DNA and the length of the nuclease's recognition sequence; due to the statistical likelihood of finding the bases in a particular order by chance, a longer recognition sequence will result in less frequent digestion. For example, a given four-base sequence (corresponding to the recognition site for a hypothetical nuclease) would be predicted to occur every 256 base pairs on average (where $4^4=256$), but any given six-base sequence would be expected to occur once every 4,096 base pairs on average ($4^6=4096$).

One unique family of nucleases is the meganucleases, which are characterized by having larger, and therefore less common, recognition sequences consisting of 12 to 40 base pairs. These nucleases are particularly useful for genetic engineering and Genome engineering applications in complex organisms such as plants and mammals, where typically larger genomes (numbering in the billions of base pairs) would result in frequent and deleterious site-specific digestion using traditional nucleases.

32.7 See also

- Nuclease protection assay

- Micrococcal nuclease

- S1 nuclease

- P1 nuclease

- HindIII

- PIN domain

32.8 References

[1] Avery, O.T., MacLeod, C.M., McCarty, M. (1944). Studies on the chemical nature of the substance inducing transformation of pneumococcal types: Induction of transformation by a desoxyribonucleic acid fraction isolated from Pneumococcus type III. J. Exp. Med. 79: 137-158.

[2] Linn S., Arber, W. (1968). Host specificity of DNA produced by Escherichia coli, X. In vitro restriction of phage fd replicative form. Proc. Natl. Acad. Sci. USA. 59:1300-1306

[3] Arber, W., Linn S. (1969) DNA modification and restriction. Annu. Rev. Biochem. 38:467-500

32.9 External links

- Examples of Restriction Enzymes Chart

- Restriction Enzyme Action of EcoRI

- Enzyme glossary

- Nucleases (Main source of the page...)

Chapter 33

Enzyme

"Biocatalyst" redirects here. For the use of natural catalysts in organic chemistry, see Biocatalysis.

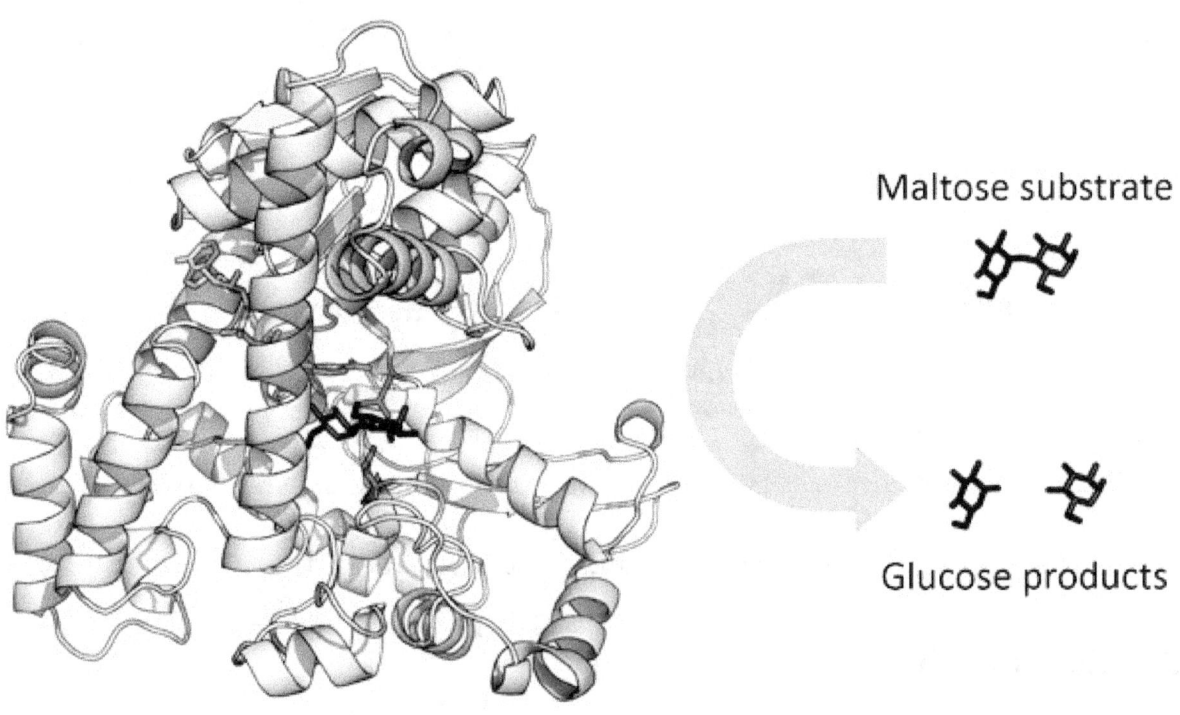

Maltose substrate

Glucose products

The enzyme glucosidase converts sugar maltose to two glucose sugars. Active site residues in red, maltose substrate in black, and NAD cofactor in yellow. (PDB: 1OBB)

Enzymes /ˈɛnzaɪmz/ are macromolecular biological catalysts. Enzymes accelerate, or catalyze, chemical reactions. The molecules at the beginning of the process are called substrates and the enzyme converts these into different molecules, called products. Almost all metabolic processes in the cell need enzymes in order to occur at rates fast enough to sustain life.[1]:8.1 The set of enzymes made in a cell determines which metabolic pathways occur in that cell. The study of enzymes is called *enzymology*.

Enzymes are known to catalyze more than 5,000 biochemical reaction types.[2] Most enzymes are proteins, although a few are catalytic RNA molecules. Enzymes' specificity comes from their unique three-dimensional structures.

Like all catalysts, enzymes increase the rate of a reaction by lowering its activation energy. Some enzymes can make their conversion of substrate to product occur many millions of times faster. An extreme example is orotidine 5'-phosphate decarboxylase, which allows a reaction that would otherwise take millions of years to occur in milliseconds.[3][4] Chemically, enzymes are like any catalyst and are not consumed in chemical reactions, nor do they alter the equilibrium of a reaction. Enzymes differ from most other catalysts by being much more specific. Enzyme activity can be affected by other molecules: inhibitors are molecules that decrease enzyme activity, and activators are molecules that increase activity. Many drugs and poisons are enzyme inhibitors. An enzyme's activity decreases markedly outside its optimal temperature and pH.

Some enzymes are used commercially, for example, in the synthesis of antibiotics. Some household products use enzymes to speed up chemical reactions: enzymes in biological washing powders break down protein, starch or fat stains on clothes, and enzymes in meat tenderizer break down proteins into smaller molecules, making the meat easier to chew.

33.1 Etymology and history

By the late 17th and early 18th centuries, the digestion of meat by stomach secretions[5] and the conversion of starch to sugars by plant extracts and saliva were known but the mechanisms by which these occurred had not been identified.[6]

French chemist Anselme Payen was the first to discover an enzyme, diastase, in 1833.[7] A few decades later, when studying the fermentation of sugar to alcohol by yeast, Louis Pasteur concluded that this fermentation was caused by a vital force contained within the yeast cells called "ferments", which were thought to function only within living organisms. He wrote that "alcoholic fermentation is an act correlated with the life and organization of the yeast cells, not with the death or putrefaction of the cells."[8]

In 1877, German physiologist Wilhelm Kühne (1837–1900) first used the term *enzyme*, which comes from Greek ἔνζυμον, "leavened", to describe this process.[9] The word *enzyme* was used later to refer to nonliving substances such as pepsin, and the word *ferment* was used to refer to chemical activity produced by living organisms.[10]

Eduard Buchner submitted his first paper on the study of yeast extracts in 1897. In a series of experiments at the University of Berlin, he found that sugar was fermented by yeast extracts even when there were no living yeast cells in the mixture.[11] He named the enzyme that brought about the fermentation of sucrose "zymase".[12] In 1907, he received the Nobel Prize in Chemistry for "his discovery of cell-free fermentation". Following Buchner's example, enzymes are usually named according to the reaction they carry out: the suffix *-ase* is combined with the name of the substrate (e.g., lactase is the enzyme that cleaves lactose) or to the type of reaction (e.g., DNA polymerase forms DNA polymers).[13]

The biochemical identity of enzymes was still unknown in the early 1900s. Many scientists observed that enzymatic activity was associated with proteins, but others (such as Nobel laureate Richard Willstätter) argued that proteins were merely carriers for the true enzymes and that proteins *per se* were incapable of catalysis.[14] In 1926, James B. Sumner showed that the enzyme urease was a pure protein and crystallized it; he did likewise for the enzyme catalase in 1937. The conclusion that pure proteins can be enzymes was definitively demonstrated by John Howard Northrop and Wendell Meredith Stanley, who worked on the digestive enzymes pepsin (1930), trypsin and chymotrypsin. These three scientists were awarded the 1946 Nobel Prize in Chemistry.[15]

The discovery that enzymes could be crystallized eventually allowed their structures to be solved by x-ray crystallography. This was first done for lysozyme, an enzyme found in tears, saliva and egg whites that digests the coating of some bacteria; the structure was solved by a group led by David Chilton Phillips and published in 1965.[16] This high-resolution structure of lysozyme marked the beginning of the field of structural biology and the effort to understand how enzymes work at an atomic level of detail.[17]

33.2 Structure

See also: Protein structure

Enzymes are generally globular proteins, acting alone or in larger complexes. Like all proteins, enzymes are linear chains of amino acids that fold to produce a three-dimensional structure. The sequence of the amino acids specifies the structure which in turn determines the catalytic activity of the enzyme.[18] Although structure determines function, a novel enzyme's activity cannot yet be predicted from its structure alone.[19] Enzyme structures unfold (denature) when heated or exposed to chemical denaturants and this disruption to the structure typically causes a loss of activity.[20] Enzyme denaturation is normally linked to temperatures above a species' normal level; as a result, enzymes from bacteria living in volcanic environments such as hot springs are prized by industrial users for their ability to function at high temperatures, allowing enzyme-catalysed reactions to be operated at a very high rate.

Enzymes are usually much larger than their substrates. Sizes range from just 62 amino acid residues, for the monomer of 4-oxalocrotonate tautomerase,[21] to over 2,500 residues in the animal fatty acid synthase.[22] Only a small portion of their structure (around 2–4 amino acids) is directly involved in catalysis: the catalytic site.[23] This catalytic site is located next to one or more binding sites where residues orient the substrates. The catalytic site and binding site together comprise the enzyme's active site. The remaining majority of the enzyme structure serves to maintain the precise orientation and dynamics of the active site.[24]

In some enzymes, no amino acids are directly involved in catalysis; instead, the enzyme contains sites to bind and orient catalytic cofactors.[24] Enzyme structures may also contain allosteric sites where the binding of a small molecule causes a conformational change that increases or decreases activity.[25]

A small number of RNA-based biological catalysts called ribozymes exist, which again can act alone or in complex with proteins. The most common of these is the ribosome which is a complex of protein and catalytic RNA components.[1]:2.2

33.3   Mechanism

33.3.1   Substrate binding

Enzymes must bind their substrates before they can catalyse any chemical reaction. Enzymes are usually very specific as to what substrates they bind and then the chemical reaction catalysed. Specificity is achieved by binding pockets with complementary shape, charge and hydrophilic/hydrophobic characteristics to the substrates. Enzymes can therefore distinguish between very similar substrate molecules to be chemoselective, regioselective and stereospecific.[26]

Some of the enzymes showing the highest specificity and accuracy are involved in the copying and expression of the genome. Some of these enzymes have "proof-reading" mechanisms. Here, an enzyme such as DNA polymerase catalyzes a reaction in a first step and then checks that the product is correct in a second step.[27] This two-step process results in average error rates of less than 1 error in 100 million reactions in high-fidelity mammalian polymerases.[1]:5.3.1 Similar proofreading mechanisms are also found in RNA polymerase,[28] aminoacyl tRNA synthetases[29] and ribosomes.[30]

Conversely, some enzymes display enzyme promiscuity, having broad specificity and acting on a range of different physiologically relevant substrates. Many enzymes possess small side activities which arose fortuitously (i.e. neutrally), which may be the starting point for the evolutionary selection of a new function.[31][32]

"Lock and key" model

To explain the observed specificity of enzymes, in 1894 Emil Fischer proposed that both the enzyme and the substrate possess specific complementary geometric shapes that fit exactly into one another.[33] This is often referred to as "the lock and key" model.[1]:8.3.2 This early model explains enzyme specificity, but fails to explain the stabilization of the transition state that enzymes achieve.[34]

Induced fit model

In 1958, Daniel Koshland suggested a modification to the lock and key model: since enzymes are rather flexible structures, the active site is continuously reshaped by interactions with the substrate as the substrate interacts with the enzyme.[35] As a result, the substrate does not simply bind to a rigid active site; the amino acid side-chains that make up the active site are molded into the precise positions that enable the enzyme to perform its catalytic function. In some cases, such as glycosidases, the substrate molecule also changes shape slightly as it enters the active site.[36] The active site continues to change until the substrate is completely bound, at which point the final shape and charge distribution is determined.[37] Induced fit may enhance the fidelity of molecular recognition in the presence of competition and noise via the conformational proofreading mechanism.[38]

33.3.2   Catalysis

See also: Enzyme catalysis

Enzymes can accelerate reactions in several ways, all of which lower the activation energy ($\Delta G^{\ddagger}$, Gibbs free energy)[39]

1. By stabilizing the transition state:

    - Creating an environment with a charge distribution complementary to that of the transition state to lower its energy.[40]

2. By providing an alternative reaction pathway:

    - Temporarily reacting with the substrate, forming a covalent intermediate to provide a lower energy transition state.[41]

3. By destabilising the substrate ground state:

    - Distorting bound substrate(s) into their transition state form to reduce the energy required to reach the transition state.[42]

    - By orienting the substrates into a productive arrangement to reduce the reaction entropy change.[43] The contribution of this mechanism to catalysis is relatively small.[44]

Enzymes may use several of these mechanisms simultaneously. For example, proteases such as trypsin perform covalent catalysis using a catalytic triad, stabilise charge build-up on the transition states using an oxyanion hole, complete hydrolysis using an oriented water substrate.

33.3.3   Dynamics

See also: Protein dynamics

Enzymes are not rigid, static structures; instead they have complex internal dynamic motions – that is, movements of parts of the enzyme's structure such as individual amino acid residues, groups of residues forming a protein loop or unit of secondary structure, or even an entire protein domain. These motions give rise to a conformational ensemble of slightly different structures that interconvert with one another at equilibrium. Different states within this ensemble may be associated with different aspects of an enzyme's function. For example, different conformations of the enzyme dihydrofolate reductase are associated with the substrate binding, catalysis, cofactor release, and product release steps of the catalytic cycle.[45]

33.3.4   Allosteric modulation

Main article: Allosteric regulation

Allosteric sites are pockets on the enzyme, distinct from the active site, that bind to molecules in the cellular environment. These molecules then cause a change in the conformation or dynamics of the enzyme that is transduced to the active site and thus affects the reaction rate of the enzyme.[46] In this way, allosteric interactions can either inhibit or activate enzymes. Allosteric interactions with metabolites upstream or downstream in an enzyme's metabolic pathway cause feedback regulation, altering the activity of the enzyme according to the flux through the rest of the pathway.[47]

33.4   Cofactors

Main article: Cofactor (biochemistry)

Some enzymes do not need additional components to show full activity. Others require non-protein molecules called cofactors to be bound for activity.[48] Cofactors can be either inorganic (e.g., metal ions and iron-sulfur clusters) or organic compounds (e.g., flavin and heme). Organic cofactors can be either coenzymes, which are released from the enzyme's active site during the reaction, or prosthetic groups, which are tightly bound to an enzyme. Organic prosthetic groups can be covalently bound (e.g., biotin in enzymes such as pyruvate carboxylase).[49]

An example of an enzyme that contains a cofactor is carbonic anhydrase, which is shown in the ribbon diagram above with a zinc cofactor bound as part of its active site.[50] These tightly bound ions or molecules are usually found in the active site and are involved in catalysis.[1]:8.1.1 For example, flavin and heme cofactors are often involved in redox reactions.[1]:17

Enzymes that require a cofactor but do not have one bound are called *apoenzymes* or *apoproteins*. An enzyme together with the cofactor(s) required for activity is called a *holoenzyme* (or haloenzyme). The term *holoenzyme* can also be applied to enzymes that contain multiple protein subunits, such as the DNA polymerases; here the holoenzyme is the complete complex containing all the subunits needed for activity.[1]:8.1.1

33.4.1   Coenzymes

Coenzymes are small organic molecules that can be loosely or tightly bound to an enzyme. Coenzymes transport chemical groups from one enzyme to another.[51] Examples include NADH, NADPH and adenosine triphosphate (ATP). Some coenzymes, such as riboflavin, thiamine and folic acid, are vitamins, or compounds that cannot be synthesized by the body and must be acquired from the diet. The chemical groups carried include the hydride ion (H^-) carried by NAD or $NADP^+$, the phosphate group carried by adenosine triphosphate, the acetyl group carried by coenzyme A, formyl, methenyl or methyl groups carried by folic acid and the methyl group carried by S-adenosylmethionine.[51]

Since coenzymes are chemically changed as a consequence of enzyme action, it is useful to consider coenzymes to be a special class of substrates, or second substrates, which are common to many different enzymes. For example, about 1000 enzymes are known to use the coenzyme NADH.[52]

Coenzymes are usually continuously regenerated and their concentrations maintained at a steady level inside the cell. For example, NADPH is regenerated through the pentose phosphate pathway and *S*-adenosylmethionine by methionine adenosyltransferase. This continuous regeneration means that small amounts of coenzymes can be used very intensively. For example, the human body turns over its own weight in ATP each day.[53]

33.5   Thermodynamics

Main articles: Activation energy, Thermodynamic equilibrium and Chemical equilibrium

As with all catalysts, enzymes do not alter the position of the chemical equilibrium of the reaction. In the presence of an enzyme, the reaction runs in the same direction as it would without the enzyme, just more quickly.[1]:8.2.3 For example, carbonic anhydrase catalyzes its reaction in either direction depending on the concentration of its reactants:[54]

$$CO_2 + H_2O \xrightarrow{\text{Carbonic anhydrase}} H_2CO_3 \text{ (in tissues; high } CO_2 \text{ concentration)}$$

$$H_2CO_3 \xrightarrow{\text{Carbonic anhydrase}} CO_2 + H_2O \text{ (in lungs; low } CO_2 \text{ concentration)}$$

The rate of a reaction is dependent on the activation energy needed to form the transition state which then decays into products. Enzymes increase reaction rates by lowering the energy of the transition state. First, binding forms a low energy enzyme-substrate complex (ES). Secondly the enzyme stabilises the transition state such that it requires less energy to achieve compared to the uncatalyzed reaction (ES‡). Finally the enzyme-product complex (EP) dissociates to release the products.[1]:8.3

Enzymes can couple two or more reactions, so that a thermodynamically favorable reaction can be used to "drive" a thermodynamically unfavourable one so that the combined energy of the products is lower than the substrates. For example, the hydrolysis of ATP is often used to drive other chemical reactions.[55]

33.6 Kinetics

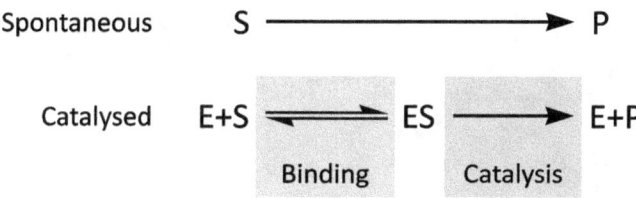

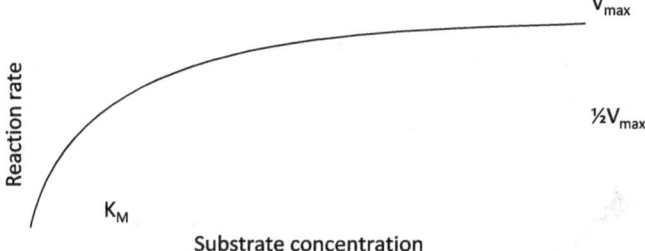

A chemical reaction mechanism with or without enzyme catalysis. The enzyme (E) binds substrate (S) to produce product (P).

Saturation curve for an enzyme reaction showing the relation between the substrate concentration and reaction rate. Main article: Enzyme kinetics

Enzyme kinetics is the investigation of how enzymes bind substrates and turn them into products. The rate data used in kinetic analyses are commonly obtained from enzyme assays. In 1913 Leonor Michaelis and Maud Leonora Menten proposed a quantitative theory of enzyme kinetics, which is referred to as Michaelis–Menten kinetics.[56] The major contribution of Michaelis and Menten was to think of enzyme reactions in two stages. In the first, the substrate binds reversibly to the enzyme, forming the enzyme-substrate complex. This is sometimes called the Michaelis-Menten complex in their honor. The enzyme then catalyzes the chemical step in the reaction and releases the product. This work was further developed by G. E. Briggs and J. B. S. Haldane, who derived kinetic equations that are still widely used today.[57]

Enzyme rates depend on solution conditions and substrate concentration. To find the maximum speed of an enzymatic reaction, the substrate concentration is increased until a constant rate of product formation is seen. This is shown in the saturation curve on the right. Saturation happens because, as substrate concentration increases, more and more of the free enzyme is converted into the substrate-bound ES complex. At the maximum reaction rate (V_{max}) of the enzyme, all the enzyme active sites are bound to substrate, and the amount of ES complex is the same as the total amount of enzyme.[1]:8.4

V_{max} is only one of several important kinetic parameters. The amount of substrate needed to achieve a given rate of reaction is also important. This is given by the Michaelis-Menten constant (K_m), which is the substrate concentration required for an enzyme to reach one-half its maximum reaction rate; generally, each enzyme has a characteristic K_m for a given substrate. Another useful constant is k_{cat}, also called the *turnover number*, which is the number of substrate molecules handled by one active site per second.[1]:8.4

The efficiency of an enzyme can be expressed in terms of k_{cat}/K_m. This is also called the specificity constant and incorporates the rate constants for all steps in the reaction up to and including the first irreversible step. Because the specificity constant reflects both affinity and catalytic ability, it is useful for comparing different enzymes against each other, or the same enzyme with different substrates. The theoretical maximum for the specificity constant is called the diffusion limit and is about 10^8 to 10^9 (M^{-1} s^{-1}). At this point every collision of the enzyme with its substrate will result in catalysis, and the rate of product formation is not limited by the reaction rate but by the diffusion rate. Enzymes with this property are called *catalytically perfect* or *kinetically perfect*. Example of such enzymes are triose-phosphate isomerase, carbonic anhydrase, acetylcholinesterase, catalase, fumarase, β-lactamase, and superoxide dismutase.[1]:8.4.2 The turnover of such enzymes can reach several million reactions per second.[1]:9.2

Michaelis–Menten kinetics relies on the law of mass action, which is derived from the assumptions of free diffusion and thermodynamically driven random collision. Many biochemical or cellular processes deviate significantly from these conditions, because of macromolecular crowding and constrained molecular movement.[58] More recent, complex extensions of the model attempt to correct for these effects.[59]

33.7 Inhibition

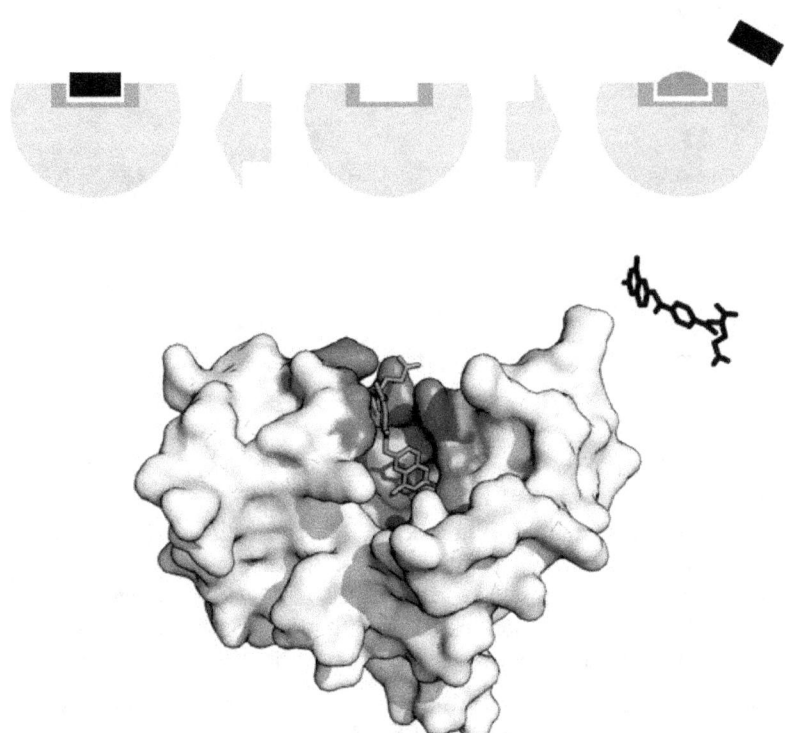

An enzyme binding site that would normally bind substrate can alternatively bind a competitive inhibitor, preventing substrate access. Dihydrofolate reductase is inhibited by methotrexate which prevents binding of its substrate, folic acid. Binding site in blue, inhibitor in green, and substrate in black. (PDB: 4QI9)

The coenzyme folic acid (left) and the anti-cancer drug methotrexate (right) are very similar in structure (differences show in green). As a result, methotrexate is a competitive inhibitor of many enzymes that use folates.
Main article: Enzyme inhibitor

Enzyme reaction rates can be decreased by various types of enzyme inhibitors.[60]:73–74

33.7.1   Types of inhibition

Competitive   A competitive inhibitor and substrate cannot bind to the enzyme at the same time.[61] Often competitive inhibitors strongly resemble the real substrate of the enzyme. For example, the drug methotrexate is a competitive inhibitor of the enzyme dihydrofolate reductase, which catalyzes the reduction of dihydrofolate to tetrahydrofolate. The similarity between the structures of dihydrofolate and this drug are shown in the accompanying figure. This type of inhibition can be overcome with high substrate concentration. In some cases, the inhibitor can bind to a site other than the binding-site of the usual substrate and exert an allosteric effect to change the shape of the usual binding-site.

Non-competitive   A non-competitive inhibitor binds to a site other than where the substrate binds. The substrate still binds with its usual affinity and hence K_m remains the same. However the inhibitor reduces the catalytic efficiency of the enzyme so that V_{max} is reduced. In contrast to competitive inhibition, non-competitive inhibition cannot be overcome with high substrate concentration.[60]:76–78

Uncompetitive   An uncompetitive inhibitor cannot bind to the free enzyme, only to the enzyme-substrate complex; hence, these types of inhibitors are most effective at high substrate concentration. In the presence of the inhibitor, the enzyme-substrate complex is inactive.[60]:78 This type of inhibition is rare.[62]

Mixed   A mixed inhibitor binds to an allosteric site and the binding of the substrate and the inhibitor affect each other. The enzyme's function is reduced but not eliminated when bound to the inhibitor. This type of inhibitor does not follow the Michaelis-Menten equation.[60]:76–78

Irreversible   An irreversible inhibitor permanently inactivates the enzyme, usually by forming a covalent bond to the protein. Penicillin[63] and aspirin[64] are common drugs that act in this manner.

33.7.2   Functions of inhibitors

In many organisms, inhibitors may act as part of a feedback mechanism. If an enzyme produces too much of one substance in the organism, that substance may act as an inhibitor for the enzyme at the beginning of the pathway that produces it, causing production of the substance to slow down or stop when there is sufficient amount. This is a form of negative feedback. Major metabolic pathways such as the citric acid cycle make use of this mechanism.[1]:17.2.2

Since inhibitors modulate the function of enzymes they are often used as drugs. Many such drugs are reversible competitive inhibitors that resemble the enzyme's native substrate, similar to methotrexate above; other well-known examples include statins used to treat high cholesterol,[65] and protease inhibitors used to treat retroviral infections such as HIV.[66] A common example of an irreversible inhibitor that is used as a drug is aspirin, which inhibits the COX-1 and COX-2 enzymes that produce the inflammation messenger prostaglandin.[64] Other enzyme inhibitors are poisons. For example, the poison cyanide is an irreversible enzyme inhibitor that combines with the copper and iron in the active site of the enzyme cytochrome c oxidase and blocks cellular respiration.[67]

33.8   Biological function

Enzymes serve a wide variety of functions inside living organisms. They are indispensable for signal transduction and cell regulation, often via kinases and phosphatases.[68] They also generate movement, with myosin hydrolyzing ATP to generate muscle contraction, and also transport cargo around the cell as part of the cytoskeleton.[69] Other ATPases in the cell membrane are ion pumps involved in active transport. Enzymes are also involved in more exotic functions, such as luciferase generating light in fireflies.[70] Viruses can also contain enzymes for infecting cells, such as the HIV integrase and reverse transcriptase, or for viral release from cells, like the influenza virus neuraminidase.[71]

An important function of enzymes is in the digestive systems of animals. Enzymes such as amylases and proteases break down large molecules (starch or proteins, respectively) into smaller ones, so they can be absorbed by the intestines. Starch molecules, for example, are too large to be absorbed from the intestine, but enzymes hydrolyze the starch chains into smaller molecules such as maltose and eventually glucose, which can then be absorbed. Different enzymes digest different food substances. In ruminants, which have herbivorous diets, microorganisms in the gut produce another enzyme, cellulase, to break down the cellulose cell walls of plant fiber.[72]

33.8.1 Metabolism

Several enzymes can work together in a specific order, creating metabolic pathways.[1]:30.1 In a metabolic pathway, one enzyme takes the product of another enzyme as a substrate. After the catalytic reaction, the product is then passed on to another enzyme. Sometimes more than one enzyme can catalyze the same reaction in parallel; this can allow more complex regulation: with, for example, a low constant activity provided by one enzyme but an inducible high activity from a second enzyme.[73]

Enzymes determine what steps occur in these pathways. Without enzymes, metabolism would neither progress through the same steps and could not be regulated to serve the needs of the cell. Most central metabolic pathways are regulated at a few key steps, typically through enzymes whose activity involves the hydrolysis of ATP. Because this reaction releases so much energy, other reactions that are thermodynamically unfavorable can be coupled to ATP hydrolysis, driving the overall series of linked metabolic reactions.[1]:30.1

33.8.2 Control of activity

There are five main ways that enzyme activity is controlled in the cell.[1]:30.1.1

Regulation Enzymes can be either activated or inhibited by other molecules. For example, the end product(s) of a metabolic pathway are often inhibitors for one of the first enzymes of the pathway (usually the first irreversible step, called committed step), thus regulating the amount of end product made by the pathways. Such a regulatory mechanism is called a negative feedback mechanism, because the amount of the end product produced is regulated by its own concentration.[74]:141–48 Negative feedback mechanism can effectively adjust the rate of synthesis of intermediate metabolites according to the demands of the cells. This helps with effective allocations of materials and energy economy, and it prevents the excess manufacture of end products. Like other homeostatic devices, the control of enzymatic action helps to maintain a stable internal environment in living organisms.[74]:141

Post-translational modification Examples of post-translational modification include phosphorylation, myristoylation and glycosylation.[74]:149–69 For example, in the response to insulin, the phosphorylation of multiple enzymes, including glycogen synthase, helps control the synthesis or degradation of glycogen and allows the cell to respond to changes in blood sugar.[75] Another example of post-translational modification is the cleavage of the polypeptide chain. Chymotrypsin, a digestive protease, is produced in inactive form as chymotrypsinogen in the pancreas and transported in this form to the stomach where it is activated. This stops the enzyme from digesting the pancreas or other tissues before it enters the gut. This type of inactive precursor to an enzyme is known as a zymogen.[74]:149–53

Quantity Enzyme production (transcription and translation of enzyme genes) can be enhanced or diminished by a cell in response to changes in the cell's environment. This form of gene regulation is called enzyme induction. For example, bacteria may become resistant to antibiotics such as penicillin because enzymes called beta-lactamases are induced that hydrolyse the crucial beta-lactam ring within the penicillin molecule.[76] Another example comes from enzymes in the liver called cytochrome P450 oxidases, which are important in drug metabolism. Induction or inhibition of these enzymes can cause drug interactions.[77] Enzyme levels can also be regulated by changing the rate of enzyme degradation.[1]:30.1.1

Subcellular distribution Enzymes can be compartmentalized, with different metabolic pathways occurring in different cellular compartments. For example, fatty acids are synthesized by one set of enzymes in the cytosol, endoplasmic reticulum and golgi and used by a different set of enzymes as a source of energy in the mitochondrion, through β-oxidation.[78] In addition, trafficking of the enzyme to different compartments may change the degree of protonation (cytoplasm neutral and lysosome acidic) or oxidative state [e.g., oxidized (periplasm) or reduced (cytoplasm)] which in turn affects enzyme activity.[79]

Organ specialization In multicellular eukaryotes, cells in different organs and tissues have different patterns of gene expression and therefore have different sets of enzymes (known as isozymes) available for metabolic reactions. This provides a mechanism for regulating the overall metabolism of the organism. For example, hexokinase, the first enzyme in the glycolysis pathway, has a specialized form called glucokinase expressed in the liver and pancreas that has a lower affinity for glucose yet is more sensitive to glucose concentration.[80] This enzyme is involved in sensing blood sugar and regulating insulin production.[81]

33.8.3 Involvement in disease

See also: Genetic disorder

Since the tight control of enzyme activity is essential for homeostasis, any malfunction (mutation, overproduction, underproduction or deletion) of a single critical enzyme can lead to a genetic disease. The malfunction of just one type of enzyme out of the thousands of types present in the human body can be fatal. An example of a fatal genetic disease due to enzyme insufficiency is Tay-Sachs disease, in which patients lack the enzyme hexosaminidase.[82][83]

One example of enzyme deficiency is the most common type of phenylketonuria. Many different single amino acid mutations in the enzyme phenylalanine hydroxylase, which catalyzes the first step in the degradation of phenylalanine, result in build-up of phenylalanine and related products. Some mutations are in the active site, directly disrupting binding and catalysis, but many are far from the active site and reduce activity by destabilising the protein structure, or affecting correct oligomerisation.[84][85] This can lead to intellectual disability if the disease is untreated.[86] Another example is pseudocholinesterase deficiency, in which the body's ability to break down choline ester drugs is impaired.[87] Oral administration of enzymes can be used to treat some functional enzyme deficiencies, such as pancreatic insufficiency[88] and lactose intolerance.[89]

Another way enzyme malfunctions can cause disease comes from germline mutations in genes coding for DNA repair enzymes. Defects in these enzymes cause cancer because cells are less able to repair mutations in their genomes. This causes a slow accumulation of mutations and results in the development of cancers. An example of such a hereditary cancer syndrome is xeroderma pigmentosum, which causes the development of skin cancers in response to even minimal exposure to ultraviolet light.[90][91]

33.9 Naming conventions

An enzyme's name is often derived from its substrate or the chemical reaction it catalyzes, with the word ending in -ase.[1]:8.1.3 Examples are lactase, alcohol dehydrogenase and DNA polymerase. Different enzymes that catalyze the same chemical reaction are called isozymes.[1]:10.3

The International Union of Biochemistry and Molecular Biology have developed a nomenclature for enzymes, the EC numbers; each enzyme is described by a sequence of four numbers preceded by "EC". The first number broadly classifies the enzyme based on its mechanism.[92]

The top-level classification is:

- EC 1, Oxidoreductases: catalyze oxidation/reduction reactions

- EC 2, Transferases: transfer a functional group (*e.g.* a methyl or phosphate group)

- EC 3, Hydrolases: catalyze the hydrolysis of various bonds

- EC 4, Lyases: cleave various bonds by means other than hydrolysis and oxidation

- EC 5, Isomerases: catalyze isomerization changes within a single molecule

- EC 6, Ligases: join two molecules with covalent bonds.

These sections are subdivided by other features such as the substrate, products, and chemical mechanism. An enzyme is fully specified by four numerical designations. For example, hexokinase (EC 2.7.1.1) is a transferase (EC 2) that adds a phosphate group (EC 2.7) to a hexose sugar, a molecule containing an alcohol group (EC 2.7.1).[93]

33.10   Industrial applications

Enzymes are used in the chemical industry and other industrial applications when extremely specific catalysts are required. Enzymes in general are limited in the number of reactions they have evolved to catalyze and also by their lack of stability in organic solvents and at high temperatures. As a consequence, protein engineering is an active area of research and involves attempts to create new enzymes with novel properties, either through rational design or *in vitro* evolution.[94][95] These efforts have begun to be successful, and a few enzymes have now been designed "from scratch" to catalyze reactions that do not occur in nature.[96]

33.11   See also

- List of enzymes

- Enzyme databases

  - BRENDA

  - ExPASy

  - IntEnz

  - KEGG

  - MetaCyc

33.12   References

[1] Stryer L, Berg JM, Tymoczko JL (2002). *Biochemistry* (5th ed.). San Francisco: W.H. Freeman. ISBN 0-7167-4955-6.

[2] Schomburg I, Chang A, Placzek S, Söhngen C, Rother M, Lang M, Munaretto C, Ulas S, Stelzer M, Grote A, Scheer M, Schomburg D (January 2013). "BRENDA in 2013: integrated reactions, kinetic data, enzyme function data, improved disease classification: new options and contents in BRENDA". *Nucleic Acids Research* **41** (Database issue): D764–72. doi:10.1093/nar/gks1049. PMID 23203881.

[3] Radzicka A, Wolfenden R (January 1995). "A proficient enzyme". *Science* **267** (5194): 90–931. doi:10.1126/science.7809611. PMID 7809611.

[4] Callahan BP, Miller BG (December 2007). "OMP decarboxylase—An enigma persists". *Bioorganic Chemistry* **35** (6): 465–9. doi:10.1016/j.bioorg.2007.07.004. PMID 17889251.

[5] de Réaumur RA (1752). "Observations sur la digestion des oiseaux". *Histoire de l'academie royale des sciences* **1752**: 266, 461.

[6] Williams HS (1904). *A History of Science: in Five Volumes. Volume IV: Modern Development of the Chemical and Biological Sciences*. Harper and Brothers.

[7] Payen A, Persoz JF (1833). "Mémoire sur la diastase, les principaux produits de ses réactions et leurs applications aux arts industriels" [Memoir on diastase, the principal products of its reactions, and their applications to the industrial arts]. *Annales de chimie et de physique*. 2nd (in French) **53**: 73–92.

[8] Manchester KL (December 1995). "Louis Pasteur (1822–1895)–chance and the prepared mind". *Trends in Biotechnology* **13** (12): 511–5. doi:10.1016/S0167-7799(00)89014-9. PMID 8595136.

[9] Kühne coined the word "enzyme" in: Kühne W (1877). "Über das Verhalten verschiedener organisirter und sog. unge-formter Fermente" [On the behavior of various organized and so-called unformed ferments]. *Verhandlungen des naturhistorisch-medicinischen Vereins zu Heidelberg.* new series (in German) **1** (3): 190–193. The relevant passage occurs on page 190: *"Um Missverständnissen vorzubeugen und lästige Umschreibungen zu vermeiden schlägt Vortragender vor, die ungeformten oder nicht organisirten Fermente, deren Wirkung ohne Anwesenheit von Organismen und ausserhalb derselben erfolgen kann, als* Enzyme *zu bezeichnen."* (Translation: In order to obviate misunderstandings and avoid cumbersome periphrases, [the author, a university lecturer] suggests designating as "enzymes" the unformed or not organized ferments, whose action can occur without the presence of organisms and outside of the same.)

[10] Holmes FL (2003). "Enzymes". In Heilbron JL. *The Oxford Companion to the History of Modern Science.* Oxford: Oxford University Press. p. 270.

[11] "Eduard Buchner". *Nobel Laureate Biography.* Nobelprize.org. Retrieved 23 February 2015.

[12] "Eduard Buchner – Nobel Lecture: Cell-Free Fermentation". *Nobelprize.org.* 1907. Retrieved 23 February 2015.

[13] The naming of enzymes by adding the suffix "-ase" to the substrate on which the enzyme acts, has been traced to French scientist Émile Duclaux (1840–1904), who intended to honor the discoverers of diastase – the first enzyme to be isolated – by introducing this practice in his book Duclaux E (1899). *Traité de microbiologie: Diastases, toxines et venins* [*Microbiology Treatise: diastases , toxins and venoms*] (in French). Paris, France: Masson and Co. See Chapter 1, especially page 9.

[14] Willstätter R (1927). "Faraday lecture. Problems and methods in enzyme research". *Journal of the Chemical Society (Resumed):* 1359. doi:10.1039/JR9270001359. quoted in Blow D (April 2000). "So do we understand how enzymes work?" (pdf). *Structure (London, England : 1993)* **8** (4): R77–R81. doi:10.1016/S0969-2126(00)00125-8. PMID 10801479.

[15] "Nobel Prizes and Laureates: The Nobel Prize in Chemistry 1946". *Nobelprize.org.* Retrieved 23 February 2015.

[16] Blake CC, Koenig DF, Mair GA, North AC, Phillips DC, Sarma VR (May 1965). "Structure of hen egg-white lysozyme. A three-dimensional Fourier synthesis at 2 Ångström resolution". *Nature* **206** (4986): 757–61. doi:10.1038/206757a0. PMID 5891407.

[17] Johnson LN, Petsko GA (1999). "David Phillips and the origin of structural enzymology". *Trends Biochem. Sci.* **24** (7): 287–9. doi:10.1016/S0968-0004(99)01423-1. PMID 10390620.

[18] Anfinsen CB (July 1973). "Principles that govern the folding of protein chains". *Science* **181** (4096): 223–30. doi:10.1126/scienc PMID 4124164.

[19] Dunaway-Mariano D (November 2008). "Enzyme function discovery". *Structure (London, England : 1993)* **16** (11): 1599–600. doi:10.1016/j.str.2008.10.001. PMID 19000810.

[20] Petsko GA, Ringe D (2003). "Chapter 1: From sequence to structure". *Protein structure and function.* London: New Science. p. 27. ISBN 978-1405119221.

[21] Chen LH, Kenyon GL, Curtin F, Harayama S, Bembenek ME, Hajipour G, Whitman CP (September 1992). "4-Oxalocrotonate tautomerase, an enzyme composed of 62 amino acid residues per monomer". *The Journal of Biological Chemistry* **267** (25): 17716–21. PMID 1339435.

[22] Smith S (December 1994). "The animal fatty acid synthase: one gene, one polypeptide, seven enzymes". *FASEB Journal : Official Publication of the Federation of American Societies for Experimental Biology* **8** (15): 1248–59. PMID 8001737.

[23] "The Catalytic Site Atlas". The European Bioinformatics Institute. Retrieved 4 April 2007.

[24] Suzuki H (2015). "Chapter 7: Active Site Structure". *How Enzymes Work: From Structure to Function.* Boca Raton, FL: CRC Press. pp. 117–140. ISBN 978-981-4463-92-8.

[25] Krauss G (2003). "The Regulations of Enzyme Activity". *Biochemistry of Signal Transduction and Regulation* (3rd ed.). Weinheim: Wiley-VCH. pp. 89–114. ISBN 9783527605767.

[26] Jaeger KE, Eggert T (August 2004). "Enantioselective biocatalysis optimized by directed evolution". *Current Opinion in Biotechnology* **15** (4): 305–13. doi:10.1016/j.copbio.2004.06.007. PMID 15358000.

[27] Shevelev IV, Hübscher U (May 2002). "The 3' 5' exonucleases". *Nature Reviews Molecular Cell Biology* **3** (5): 364–76. doi:10.1038/nrm804. PMID 11988770.

[28] Zenkin N, Yuzenkova Y, Severinov K (July 2006). "Transcript-assisted transcriptional proofreading". *Science* **313** (5786): 518–20. doi:10.1126/science.1127422. PMID 16873663.

[29] Ibba M,Soll D. "Aminoacyl-tRNA synthesis". *Annual Review of Biochemistry* **69**: 617–50. doi:10.1146/annurev.biochem.69.1. PMID 10966471.

[30] Rodnina MV, Wintermeyer W. "Fidelity of aminoacyl-tRNA selection on the ribosome: kinetic and structural mechanisms". *Annual Review of Biochemistry* **70**: 415–35. doi:10.1146/annurev.biochem.70.1.415. PMID 11395413.

[31] Khersonsky O, Tawfik DS. "Enzyme promiscuity: a mechanistic and evolutionary perspective". *Annual Review of Biochemistry* **79**: 471–505. doi:10.1146/annurev-biochem-030409-143718. PMID 20235827.

[32] O'Brien PJ, Herschlag D (April 1999). "Catalytic promiscuity and the evolution of new enzymatic activities". *Chemistry & Biology* **6** (4): R91–R105. doi:10.1016/S1074-5521(99)80033-7. PMID 10099128.

[33] Fischer E (1894). "Einfluss der Configuration auf die Wirkung der Enzyme" [Influence of configuration on the action of enzymes]. *Berichte der Deutschen chemischen Gesellschaft zu Berlin* (in German) **27** (3): 2985–93. doi:10.1002/cber.18940270364. From page 2992: *"Um ein Bild zu gebrauchen, will ich sagen, dass Enzym und Glucosid wie Schloss und Schlüssel zu einander passen müssen, um eine chemische Wirkung auf einander ausüben zu können."* (To use an image, I will say that an enzyme and a glucoside [i.e., glucose derivative] must fit like a lock and key, in order to be able to exert a chemical effect on each other.)

[34] Cooper GM (2000). "Chapter 2.2: The Central Role of Enzymes as Biological Catalysts". *The Cell: a Molecular Approach* (2nd ed.). Washington (DC): ASM Press. ISBN 0-87893-106-6.

[35] Koshland DE (February 1958). "Application of a Theory of Enzyme Specificity to Protein Synthesis". *Proceedings of the National Academy of Sciences of the United States of America* **44** (2): 98–104. doi:10.1073/pnas.44.2.98. PMC 335371. PMID 16590179.

[36] Vasella A, Davies GJ, Böhm M (October 2002). "Glycosidase mechanisms". *Current Opinion in Chemical Biology* **6** (5): 619–29. doi:10.1016/S1367-5931(02)00380-0. PMID 12413546.

[37] Boyer R (2002). "Chapter 6: Enzymes I, Reactions, Kinetics, and Inhibition". *Concepts in Biochemistry* (2nd ed.). New York, Chichester, Weinheim, Brisbane, Singapore, Toronto.: John Wiley & Sons, Inc. pp. 137–8. ISBN 0-470-00379-0. OCLC 51720783.

[38] Savir Y, Tlusty T (2007). Scalas E, ed. "Conformational proofreading: the impact of conformational changes on the specificity of molecular recognition" (PDF). *PloS One* **2** (5): e468. doi:10.1371/journal.pone.0000468. PMC 1868595. PMID 17520027.

[39] Fersht A (1985). *Enzyme Structure and Mechanism*. San Francisco: W.H. Freeman. pp. 50–2. ISBN 0-7167-1615-1.

[40] Warshel A, Sharma PK, Kato M, Xiang Y, Liu H, Olsson MH (August 2006). "Electrostatic basis for enzyme catalysis". *Chemical Reviews* **106** (8): 3210–35. doi:10.1021/cr0503106. PMID 16895325.

[41] Cox MM, Nelson DL (2013). "Chapter 6.2: How enzymes work". *Lehninger Principles of Biochemistry* (6th ed.). New York, N.Y.: W.H. Freeman. p. 195. ISBN 978-1464109621.

[42] Benkovic SJ, Hammes-Schiffer S (August 2003). "A perspective on enzyme catalysis". *Science* **301** (5637): 1196–202. doi:10.1126/science.1085515. PMID 12947189.

[43] Jencks WP (1987). *Catalysis in Chemistry and Enzymology*. Mineola, N.Y: Dover. ISBN 0-486-65460-5.

[44] Villa J, Strajbl M, Glennon TM, Sham YY, Chu ZT, Warshel A (October 2000). "How important are entropic contributions to enzyme catalysis?". *Proceedings of the National Academy of Sciences of the United States of America* **97** (22): 11899–904. doi:10.1073/pnas.97.22.11899. PMC 17266. PMID 11050223.

[45] Ramanathan A, Savol A, Burger V, Chennubhotla CS, Agarwal PK (2014). "Protein conformational populations and functionally relevant substates". *Acc. Chem. Res.* **47** (1): 149–56. doi:10.1021/ar400084s. PMID 23988159.

[46] Tsai CJ, Del Sol A, Nussinov R (2009). "Protein allostery, signal transmission and dynamics: a classification scheme of allosteric mechanisms". *Mol Biosyst* **5** (3): 207–16. doi:10.1039/b819720b. PMC 2898650. PMID 19225609.

[47] Changeux JP, Edelstein SJ (June 2005). "Allosteric mechanisms of signal transduction". *Science* **308** (5727): 1424–8. doi:10.1 PMID 15933191.

[48] de Bolster MW (1997). "Glossary of Terms Used in Bioinorganic Chemistry: Cofactor". International Union of Pure and Applied Chemistry. Retrieved 30 October 2007.

[49] Chapman-Smith A, Cronan JE (1999). "The enzymatic biotinylation of proteins: a post-translational modification of exceptional specificity". *Trends Biochem. Sci.* **24** (9): 359–63. doi:10.1016/s0968-0004(99)01438-3. PMID 10470036.

[50] Fisher Z, Hernandez Prada JA, Tu C, Duda D, Yoshioka C, An H, Govindasamy L, Silverman DN, McKenna R (February 2005). "Structural and kinetic characterization of active-site histidine as a proton shuttle in catalysis by human carbonic anhydrase II". *Biochemistry* **44** (4): 1097–115. doi:10.1021/bi0480279. PMID 15667203.

[51] Wagner AL (1975). *Vitamins and Coenzymes.* Krieger Pub Co. ISBN 0-88275-258-8.

[52] "BRENDA The Comprehensive Enzyme Information System". Technische Universität Braunschweig. Retrieved 23 February 2015.

[53] Törnroth-Horsefield S, Neutze R (December 2008). "Opening and closing the metabolite gate". *Proceedings of the National Academy of Sciences of the United States* **105** (50): 19565–6. doi:10.1073/pnas.0810654106. PMC 2604989. PMID 19073922.

[54] McArdle WD, Katch F, Katch VL (2006). "Chapter 9: The Pulmonary System and Exercise". *Essentials of Exercise Physiology* (3rd ed.). Baltimore, Maryland: Lippincott Williams & Wilkins. pp. 312–3. ISBN 978-0781749916.

[55] Ferguson SJ, Nicholls D, Ferguson S (2002). *Bioenergetics 3* (3rd ed.). San Diego: Academic. ISBN 0-12-518121-3.

[56] Michaelis L, Menten M (1913). "Die Kinetik der Invertinwirkung" [The Kinetics of Invertase Action] (PDF). *Biochem. Z.* (in German) **49**: 333–369. doi:10.1021/bi201284u.; Michaelis L, Menten ML, Johnson KA, Goody RS (2011). "The original Michaelis constant: translation of the 1913 Michaelis-Menten paper". *Biochemistry* **50** (39): 8264–9. doi:10.1021/bi201284u. PMC 3381512. PMID 21888353.

[57] Briggs GE, Haldane JB (1925). "A Note on the Kinetics of Enzyme Action". *The Biochemical Journal* **19** (2): 339–339. PMC 1259181. PMID 16743508.

[58] Ellis RJ (October 2001). "Macromolecular crowding: obvious but underappreciated". *Trends in Biochemical Sciences* **26** (10): 597–604. doi:10.1016/S0968-0004(01)01938-7. PMID 11590012.

[59] Kopelman R (September 1988). "Fractal reaction kinetics". *Science* **241** (4873): 1620–26. doi:10.1126/science.241.4873.1620. PMID 17820893.

[60] Cornish-Bowden A (2004). *Fundamentals of Enzyme Kinetics* (3 ed.). London: Portland Press. ISBN 1-85578-158-1.

[61] Price NC (1979). "What is meant by 'competitive inhibition'?". *Trends in Biochemical Sciences* **4**(11): N272–N273. doi:10.101 0004(79)90205-6.

[62] Cornish-Bowden A (July 1986). "Why is uncompetitive inhibition so rare? A possible explanation, with implications for the design of drugs and pesticides". *FEBS Letters* **203** (1): 3–6. doi:10.1016/0014-5793(86)81424-7. PMID 3720956.

[63] Fisher JF, Meroueh SO, Mobashery S (February 2005). "Bacterial resistance to beta-lactam antibiotics: compelling opportunism, compelling opportunity". *Chemical Reviews* **105** (2): 395–424. doi:10.1021/cr030102i. PMID 15700950.

[64] Johnson DS, Weerapana E, Cravatt BF (June 2010). "Strategies for discovering and derisking covalent, irreversible enzyme inhibitors". *Future Medicinal Chemistry* **2** (6): 949–64. doi:10.4155/fmc.10.21. PMID 20640225.

[65] Endo A (1 November 1992). "The discovery and development of HMG-CoA reductase inhibitors" (PDF). *J. Lipid Res.* **33** (11): 1569–82. PMID 1464741.

[66] Wlodawer A, Vondrasek J (1998). "Inhibitors of HIV-1 protease: a major success of structure-assisted drug design". *Annual Review of Biophysics and Biomolecular Structure* **27**: 249–84. doi:10.1146/annurev.biophys.27.1.249. PMID 9646869.

[67] Yoshikawa S, Caughey WS (May 1990). "Infrared evidence of cyanide binding to iron and copper sites in bovine heart cytochrome c oxidase. Implications regarding oxygen reduction". *The Journal of Biological Chemistry* **265** (14): 7945–58. PMID 2159465.

[68] Hunter T (January 1995). "Protein kinases and phosphatases: the yin and yang of protein phosphorylation and signaling". *Cell* **80** (2): 225–36. doi:10.1016/0092-8674(95)90405-0. PMID 7834742.

[69] Berg JS, Powell BC, Cheney RE (April 2001). "A millennial myosin census". *Molecular Biology of the Cell* **12** (4): 780–94. doi:10.1091/mbc.12.4.780. PMC 32266. PMID 11294886.

[70] Meighen EA (March 1991). "Molecular biology of bacterial bioluminescence". *Microbiological Reviews* **55** (1): 123–42. PMC 372803. PMID 2030669.

[71] De Clercq E (2002). "Highlights in the development of new antiviral agents". *Mini Rev Med Chem* **2**(2): 163–75. doi:10.2174/ PMID 12370077.

[72] Mackie RI, White BA (October 1990). "Recent advances in rumen microbial ecology and metabolism: potential impact on nutrient output". *Journal of Dairy Science* **73** (10): 2971–95. doi:10.3168/jds.S0022-0302(90)78986-2. PMID 2178174.

[73] Rouzer CA, Marnett LJ (2009). "Cyclooxygenases: structural and functional insights". *J. Lipid Res.* 50 Suppl: S29–34. doi:10.1194/jlr.R800042-JLR200. PMC 2674713. PMID 18952571.

[74] Suzuki H (2015). "Chapter 8: Control of Enzyme Activity". *How Enzymes Work: From Structure to Function.* Boca Raton, FL: CRC Press. pp. 141–69. ISBN 978-981-4463-92-8.

[75] Doble BW, Woodgett JR (April 2003). "GSK-3: tricks of the trade for a multi-tasking kinase". *Journal of Cell Science* **116** (Pt 7): 1175–86. doi:10.1242/jcs.00384. PMC 3006448. PMID 12615961.

[76] Bennett PM, Chopra I (1993). "Molecular basis of beta-lactamase induction in bacteria" (PDF). *Antimicrob. Agents Chemother.* **37** (2): 153–8. doi:10.1128/aac.37.2.153. PMC 187630. PMID 8452343.

[77] Skett P, Gibson GG (2001). "Chapter 3: Induction and Inhibition of Drug Metabolism". *Introduction to Drug Metabolism* (3 ed.). Cheltenham, UK: Nelson Thornes Publishers. pp. 87–118. ISBN 978-0748760114.

[78] Faergeman NJ, Knudsen J (April 1997). "Role of long-chain fatty acyl-CoA esters in the regulation of metabolism and in cell signalling". *The Biochemical Journal* **323** (Pt 1): 1–12. PMC 1218279. PMID 9173866.

[79] Suzuki H (2015). "Chapter 4: Effect of pH, Temperature, and High Pressure on Enzymatic Activity". *How Enzymes Work: From Structure to Function.* Boca Raton, FL: CRC Press. pp. 53–74. ISBN 978-981-4463-92-8.

[80] Kamata K, Mitsuya M, Nishimura T, Eiki J, Nagata Y (March 2004). "Structural basis for allosteric regulation of the monomeric allosteric enzyme human glucokinase". *Structure* **12** (3): 429–38. doi:10.1016/j.str.2004.02.005. PMID 15016359.

[81] Froguel P, Zouali H, Vionnet N, Velho G, Vaxillaire M, Sun F, Lesage S, Stoffel M, Takeda J, Passa P (March 1993). "Familial hyperglycemia due to mutations in glucokinase. Definition of a subtype of diabetes mellitus". *The New England Journal of Medicine* **328** (10): 697–702. doi:10.1056/NEJM199303113281005. PMID 8433729.

[82] Okada S, O'Brien JS (August 1969). "Tay-Sachs disease: generalized absence of a beta-D-N-acetylhexosaminidase component". *Science* **165** (3894): 698–700. doi:10.1126/science.165.3894.698. PMID 5793973.

[83] "Learning About Tay-Sachs Disease". U.S. National Human Genome Research Institute. Retrieved 1 March 2015.

[84] Erlandsen H, Stevens RC (October 1999). "The structural basis of phenylketonuria". *Molecular Genetics and Metabolism* **68** (2): 103–25. doi:10.1006/mgme.1999.2922. PMID 10527663.

[85] Flatmark T, Stevens RC (August 1999). "Structural Insight into the Aromatic Amino Acid Hydroxylases and Their Disease-Related Mutant Forms". *Chemical Reviews* **99** (8): 2137–2160. doi:10.1021/cr980450y. PMID 11849022.

[86] "Phenylketonuria". *Genes and Disease [Internet].* Bethesda (MD): National Center for Biotechnology Information (US). 1998–2015. Retrieved 4 April 2007.

[87] "Pseudocholinesterase deficiency". U.S. National Library of Medicine. Retrieved 5 September 2013.

[88] Fieker A, Philpott J, Armand M (2011). "Enzyme replacement therapy for pancreatic insufficiency: present and future". *Clinical and Experimental Gastroenterology* **4**: 55–73. doi:10.2147/CEG.S17634. PMID 21753892.

[89] Misselwitz B, Pohl D, Frühauf H, Fried M, Vavricka SR, Fox M (June 2013). "Lactose malabsorption and intolerance: pathogenesis, diagnosis and treatment". *United European Gastroenterology Journal* **1** (3): 151–9. doi:10.1177/2050640613484463. PMID 24917953.

[90] Cleaver JE (May 1968). "Defective repair replication of DNA in xeroderma pigmentosum". *Nature* **218** (5142): 652–6. doi:10.1038/218652a0. PMID 5655953.

[91] James WD, Elston D, Berger TG (2011). *Andrews' Diseases of the Skin: Clinical Dermatology* (11th ed.). London: Saunders/ Elsevier. p. 567. ISBN 978-1437703146.

[92] Nomenclature Committee. "Classification and Nomenclature of Enzymes by the Reactions they Catalyse". *International Union of Biochemistry and Molecular Biology (NC-IUBMB)*. School of Biological and Chemical Sciences, Queen Mary, University of London.

[93] Nomenclature Committee. "EC 2.7.1.1". *International Union of Biochemistry and Molecular Biology (NC-IUBMB)*. School of Biological and Chemical Sciences, Queen Mary, University of London.

[94] Renugopalakrishnan V, Garduño-Juárez R, Narasimhan G, Verma CS, Wei X, Li P (November 2005). "Rational design of thermally stable proteins: relevance to bionanotechnology". *Journal of Nanoscience and Nanotechnology* **5** (11): 1759–1767. doi:10.1166/jnn.2005.441. PMID 16433409.

[95] Hult K, Berglund P (August 2003). "Engineered enzymes for improved organic synthesis". *Current Opinion in Biotechnology* **14** (4): 395–400. doi:10.1016/S0958-1669(03)00095-8. PMID 12943848.

[96] Jiang L, Althoff EA, Clemente FR, Doyle L, Röthlisberger D, Zanghellini A, Gallaher JL, Betker JL, Tanaka F, Barbas CF, Hilvert D, Houk KN, Stoddard BL, Baker D (March 2008). "De novo computational design of retro-aldol enzymes". *Science* **319** (5868): 1387–91. doi:10.1126/science.1152692. PMC 3431203. PMID 18323453.

[97] Sun Y, Cheng J (May 2002). "Hydrolysis of lignocellulosic materials for ethanol production: a review". *Bioresource Technology* **83** (1): 1–11. doi:10.1016/S0960-8524(01)00212-7. PMID 12058826.

[98] Kirk O, Borchert TV, Fuglsang CC (August 2002). "Industrial enzyme applications". *Current Opinion in Biotechnology* **13** (4): 345–351. doi:10.1016/S0958-1669(02)00328-2.

[99] Briggs DE (1998). *Malts and Malting* (1st ed.). London: Blackie Academic. ISBN 978-0412298004.

[100] Dulieu C, Moll M, Boudrant J, Poncelet D. "Improved performances and control of beer fermentation using encapsulated alpha-acetolactate decarboxylase and modeling". *Biotechnology Progress* **16** (6): 958–65. doi:10.1021/bp000128k. PMID 11101321.

[101] Tarté R (2008). *Ingredients in Meat Products Properties, Functionality and Applications*. New York: Springer. p. 177. ISBN 978-0-387-71327-4.

[102] "Chymosin – GMO Database". *GMO Compass*. European Union. 10 July 2010. Retrieved 1 March 2015.

[103] Molimard P, Spinnler HE (February 1996). "Review: Compounds Involved in the Flavor of Surface Mold-Ripened Cheeses: Origins and Properties". *Journal of Dairy Science* **79** (2): 169–184. doi:10.3168/jds.S0022-0302(96)76348-8.

[104] Guzmán-Maldonado H, Paredes-López O (September 1995). "Amylolytic enzymes and products derived from starch: a review". *Critical Reviews in Food Science and Nutrition* **35** (5): 373–403. doi:10.1080/10408399509527706. PMID 8573280.

[105] "Protease – GMO Database". *GMO Compass*. European Union. 10 July 2010. Retrieved 28 February 2015.

[106] Alkorta I, Garbisu C, Llama MJ, Serra JL (January 1998). "Industrial applications of pectic enzymes: a review". *Process Biochemistry* **33** (1): 21–28. doi:10.1016/S0032-9592(97)00046-0.

[107] Bajpai P (March 1999). "Application of enzymes in the pulp and paper industry". *Biotechnology Progress* **15** (2): 147–157. doi:10.1021/bp990013k. PMID 10194388.

[108] Begley CG, Paragina S, Sporn A (March 1990). "An analysis of contact lens enzyme cleaners". *Journal of the American Optometric Association* **61** (3): 190–4. PMID 2186082.

[109] Farris PL (2009). "Economic Growth and Organization of the U.S. Starch Industry". In BeMiller JN, Whistler RL. *Starch Chemistry and Technology* (3rd ed.). London: Academic. ISBN 9780080926551.

33.13   Further reading

Eduard Buchner

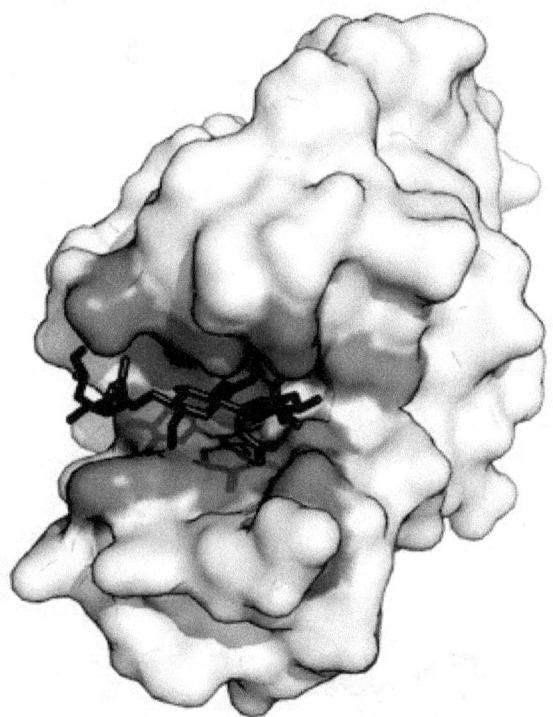

PROTEIN STRUCTURE
Scaffold to support and
position active site

ACTIVE SITE

BINDING SITES
Bind and orient
substrate(s)

CATALYTIC SITE
Reduce chemical
activation energy

Organisation of enzyme structure and lysozyme example. Binding sites in blue, catalytic site in red and peptidoglycan substrate in black. (PDB: 9LYZ)

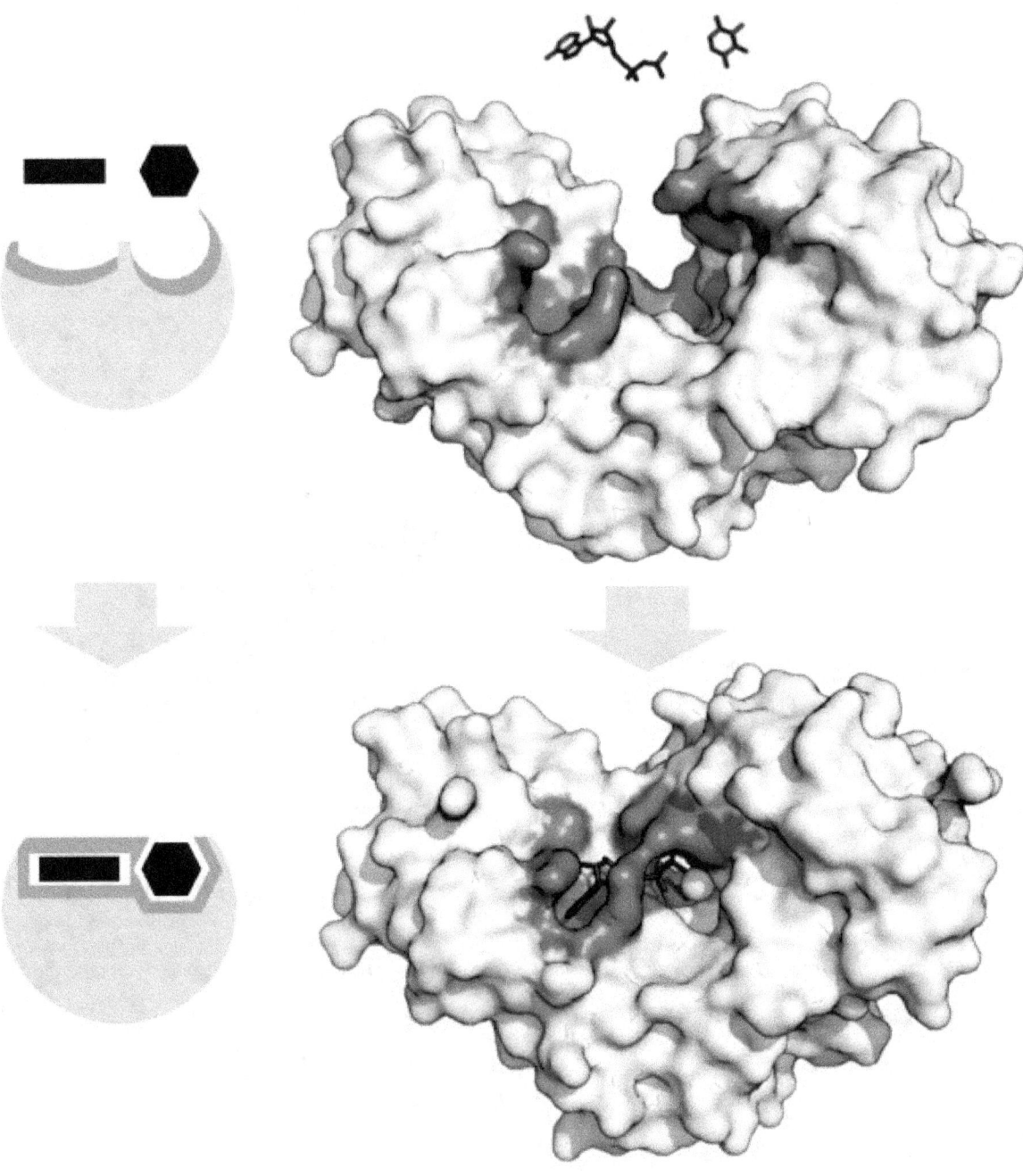

Enzyme changes shape by induced fit upon substrate binding to form enzyme-substrate complex. Hexokinase has a large induced fit motion that closes over the substrates adenosine triphosphate and xylose. Binding sites in blue, substrates in black and Mg^{2+} cofactor in yellow. (PDB: 2E2N, 2E2Q)

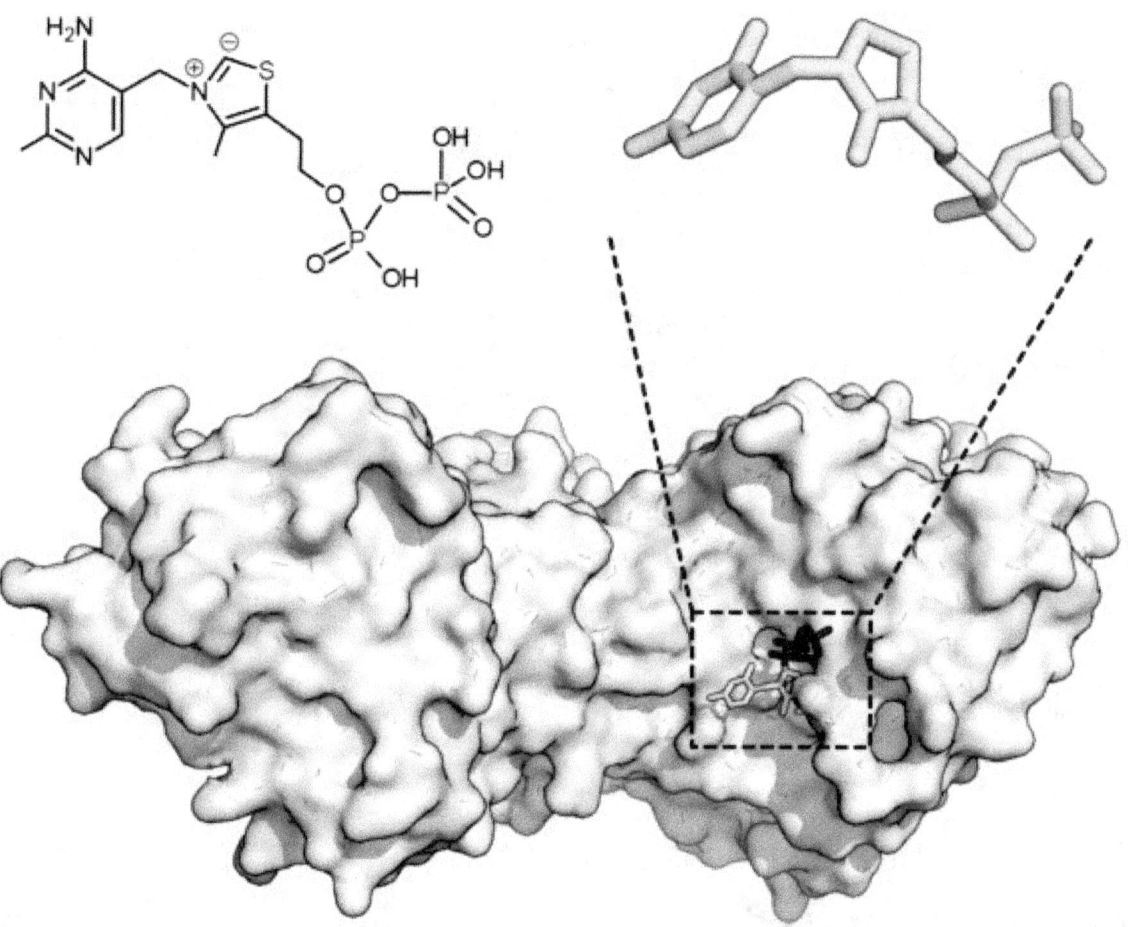

Chemical structure for thiamine pyrophosphate and protein structure of transketolase. Thiamine pyrophosphate cofactor in yellow and xylulose 5-phosphate substrate in black. (PDB: 4KXV)

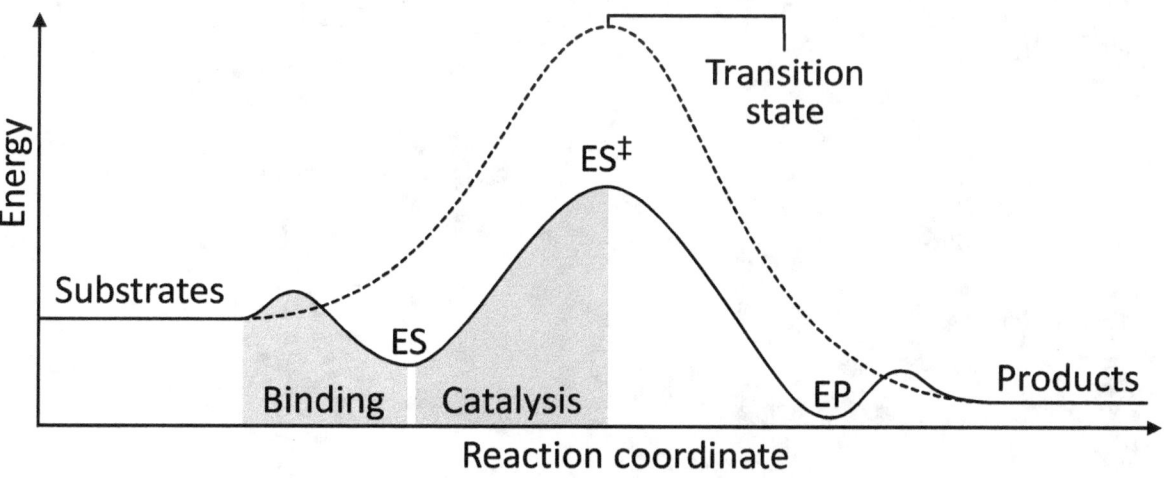

The energies of the stages of a chemical reaction. Uncatalysed (dashed line), substrates need a lot of activation energy to reach a transition state, which then decays into lower-energy products. When enzyme catalysed (solid line), the enzyme binds the substrates (ES), then stabilizes the transition state (ES‡) to reduce the activation energy required to produce products (EP) which are finally released.

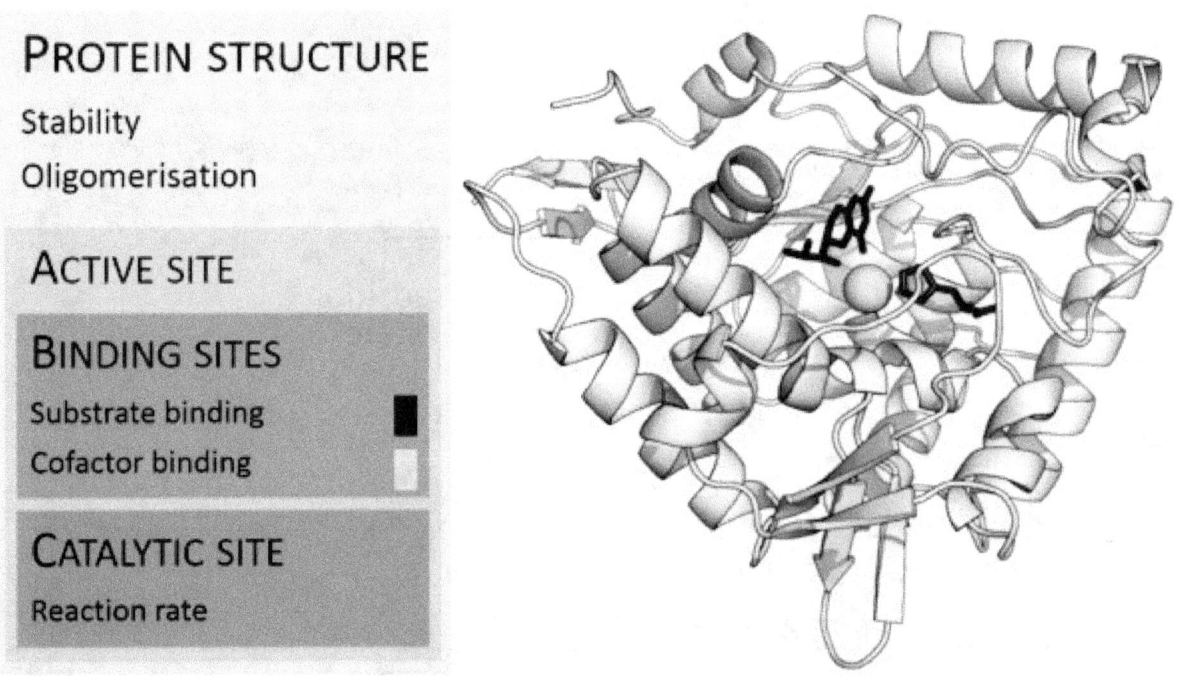

The metabolic pathway of glycolysis releases energy by converting glucose to pyruvate by via a series of intermediate metabolites. Each chemical modification (red box) is performed by a different enzyme.

PROTEIN STRUCTURE
Stability

Oligomerisation

ACTIVE SITE

BINDING SITES
Substrate binding

Cofactor binding

CATALYTIC SITE
Reaction rate

In phenylalanine hydroxylase over 300 different mutations throughout the structure cause phenylketonuria. Phenylalanine substrate and tetrahydrobiopterin coenzyme in black, and Fe^{2+} cofactor in yellow. (PDB: 1KW0)

Chapter 34

Topoisomerase

Topoisomerases are enzymes that regulate the overwinding or underwinding of DNA. The winding problem of DNA arises due to the intertwined nature of its double-helical structure. During DNA replication and transcription, DNA becomes overwound ahead of a replication fork. If left unabated, this tension would eventually stop the ability of the [enzymes] involved in these processes to continue down the DNA strand.

In order to prevent and correct these types of topological problems caused by the double helix, topoisomerases bind to either single-stranded or double-stranded DNA and cut the phosphate backbone of the DNA. This intermediate break allows the DNA to be untangled or unwound, and, at the end of these processes, the DNA backbone is resealed again. Since the overall chemical composition and connectivity of the DNA do not change, the tangled and untangled DNAs are chemical isomers, differing only in their global topology, thus their name. Topoisomerases are isomerase enzymes that act on the topology of DNA.[1]

Bacterial topoisomerase and human topoisomerase proceed via the same mechanism for replication and transcription.

34.1 Discovery

James C. Wang was the first to discover a topoisomerase when he identified *E. coli* topoisomerase I. Topo EC-codes are as follows: type I, EC 5.99.1.2; type II: EC 5.99.1.3. His discovery was made in the 1970s.

34.2 Function

The double-helical configuration that DNA strands naturally reside, makes them difficult to separate and yet they must be separated by helicase enzymes, if other enzymes are to transcribe the sequences that encode proteins, or if chromosomes are to be replicated. In so-called circular DNA, in which double-helical DNA is bent around and joined in a circle, the two strands are topologically linked, or knotted. Otherwise identical loops of DNA, having different numbers of twists, are topoisomers, and cannot be interconverted by any process that does not involve the breaking of DNA strands. Topoisomerases catalyze and guide the unknotting or unkinking of DNA[2] by creating transient breaks in the DNA using a conserved Tyrosine as the catalytic residue.[1]

The insertion of (viral) DNA into chromosomes and other forms of recombination can also require the action of topoisomerases.

34.3 Clinical significance

See also topoisomerase inhibitor

389

Many drugs operate through interference with the topoisomerases . The broad-spectrum fluoroquinolone antibiotics act by disrupting the function of bacterial type II topoisomerases. These small molecule inhibitors act as efficient anti-bacterial agents by hijacking the natural ability of topoisomerase to create breaks in chromosomal DNA.

Some chemotherapy drugs called topoisomerase inhibitors work by interfering with mammalian-type eukaryotic topoisomerases in cancer cells. This induces breaks in the DNA that ultimately lead to programmed cell death (apoptosis). This DNA-damaging effect, outside of its potential curative properties, may lead to secondary neoplasms in the patient.

Topoisomerase I is the antigen recognized by Anti Scl-70 antibodies in scleroderma.

34.4   Topological problems

There are three main types of topology: supercoiling, knotting, and catenation. Outside of the essential processes of replication or transcription, DNA must be kept as compact as possible, and these three states help this cause. However, when transcription or replication occurs, DNA must be free, and these states seriously hinder the processes. In addition, during replication, the newly replicated duplex of DNA and the original duplex of DNA become intertwined and must be completely separated in order to ensure genomic integrity as a cell divides. As a transcription bubble proceeds, DNA ahead of the transcription fork becomes overwound, or positively supercoiled, while DNA behind the transcription bubble becomes underwound, or negatively supercoiled. As replication occurs, DNA ahead of the replication bubble becomes positively supercoiled, while DNA behind the replication fork becomes entangled forming precatenanes. One of the most essential topological problems occurs at the very end of replication, when daughter chromosomes must be fully disentangled before mitosis occurs. Topoisomerase IIA plays an essential role in resolving these topological problems.

34.5   Classes

Topoisomerases can fix these topological problems and are separated into two types depending on the number of strands cut in one round of action:[3] Both these classes of enzyme utilize a conserved tyrosine. However these enzymes are structurally and mechanistically different. For a video of this process see: http://www.youtube.com/watch?v=EYGrElVyHnU& feature=related.

- A type I topoisomerase cuts one strand of a DNA double helix, relaxation occurs, and then the cut strand is re-annealed. Cutting one strand allows the part of the molecule on one side of the cut to rotate around the uncut strand, thereby reducing stress from too much or too little twist in the helix. Such stress is introduced when the DNA strand is "supercoiled" or uncoiled to or from higher orders of coiling. Type I topoisomerases are subdivided into two subclasses: type IA topoisomerases, which share many structural and mechanistic features with the type II topoisomerases, and type IB topoisomerases, which utilize a controlled rotary mechanism. Examples of type IA topoisomerases include topo I and topo III. In the past, type IB topoisomerases were referred to as eukaryotic topo I, but IB topoisomerases are present in all three domains of life. Like type II topoisomerases, type IA topoisomerases form a covalent intermediate with the 5' end of DNA, whereas the IB topoisomerases form a covalent intermediate with the 3' end of DNA. Recently, a type IC topoisomerase has been identified, called topo V. While it is structurally unique from type IA and IB topoisomerases, it shares a similar mechanism with type IB topoisomerase.

- A type II topoisomerase cuts both strands of one DNA double helix, passes another unbroken DNA helix through it, and then reanneals the cut strands. This class is also split into two subclasses: type IIA and type IIB topoisomerases, which possess similar structure and mechanisms. Examples of type IIA topoisomerases include eukaryotic topo II, E. coli gyrase, and E. coli topo IV. Examples of type IIB topoisomerase include topo VI. Type II topisomerases utilize ATP hydrolysis.

Both type I and type II topoisomerases change the linking number (L) of DNA. Type IA topoisomerases change the linking number by one, type IB and type IC topoisomerases change the linking number by any integer, whereas type IIA and type IIB topoisomerases change the linking number by two.

34.6 See also

- DNA topology

- Supercoil

- TOP1

- Type II topoisomerase

34.7 References

[1] Champoux JJ (2001). "DNA topoisomerases: structure, function, and mechanism". *Annu. Rev. Biochem.* **70**: 369–413. doi:10.1146/annurev.biochem.70.1.369. PMID 11395412.

[2] C.Michael Hogan. 2010. *Deoxyribonucleic acid*. Encyclopedia of Earth. National Council for Science and the Environment. eds. S.Draggan and C.Cleveland. Washington DC

[3] Wang JC (April 1991). "DNA topoisomerases: why so many?". *J. Biol. Chem.* **266** (11): 6659–62. PMID 1849888.

- Pommier, Yves (May 28, 2010). "DNA topoisomerases and their poisoning by anticancer and antibacterial drugs". *Chemistry & Biology*. Retrieved May 28, 2010.

34.8 Further reading

- James C. Wang (2009) *Untangling the Double Helix. DNA Entanglement and the Action of the DNA Topoisomerases*, Cold Spring Harbor Laboratory Press, Cold Spring Harbor, NY, 2009. 245 pp. ISBN 978-0-87969-879-9

34.9 External links

- DNA Topoisomerases at the US National Library of Medicine Medical Subject Headings (MeSH)

Chapter 35

Polymerase

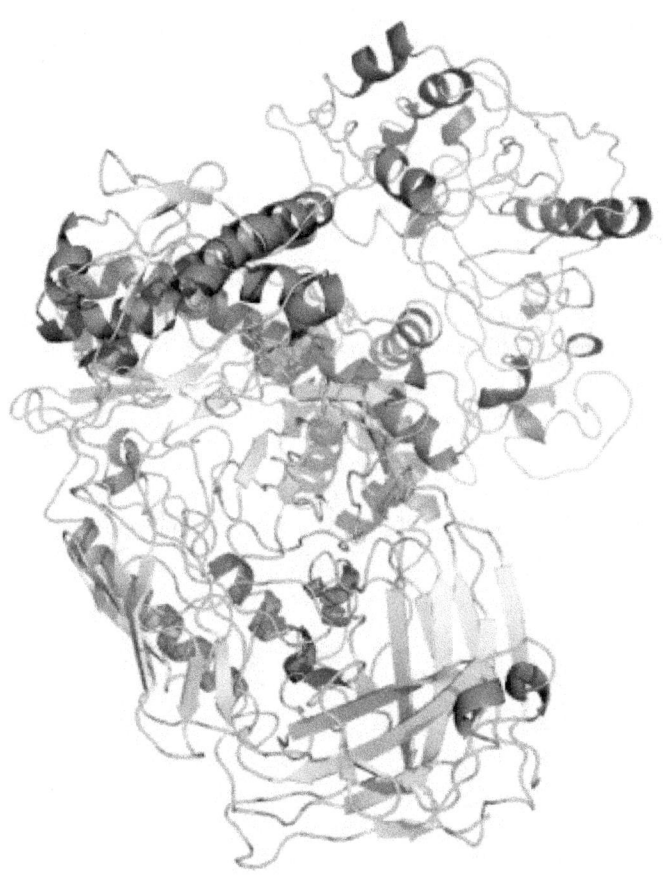

Structure of Taq DNA polymerase

A **polymerase** is an enzyme (EC 2.7.7.6/7/19/48/49) that synthesizes long chains or polymers of nucleic acids. DNA polymerase and RNA polymerase are used to assemble DNA and RNA molecules, respectively, by copying a DNA or RNA template strand using base-pairing interactions.

It is an accident of history that the enzymes responsible for the generation of other biopolymers are not also referred to as polymerases. For example, the enzymatic complex that assembles amino acids into proteins is termed the ribosome, rather than "protein polymerase".

A polymerase from the thermophilic bacterium, *Thermus aquaticus* (*Taq*) (PDB 1BGX, EC 2.7.7.7) is used in the polymerase chain reaction, an important technique of molecular biology.

Other well-known polymerases include:

- Terminal Deoxynucleotidyl Transferase (TDT), which lends diversity to antibody heavy chains

- Reverse Transcriptase, an enzyme used by RNA retroviruses like HIV, which is used to create a complementary strand to the preexisting strand of viral RNA before it can be integrated into the DNA of the host cell. It is also a major target for antiviral drugs.

35.1 See also

- DNA polymerase

  - DNA polymerase I
  - DNA polymerase II
  - DNA polymerase III holoenzyme
  - DNA polymerase IV (DinB) – SOS repair polymerase

- RNA polymerase

  - RNA polymerase I
  - RNA polymerase II
  - RNA polymerase III
  - T7 RNA polymerase

Chapter 36

Retrovirus

"Retrovirology" redirects here. For the journal, see Retrovirology (journal).
For a more accessible and less technical introduction to this topic, see Introduction to viruses.

Retroviridae is a family of enveloped viruses that replicate in a host cell through the process of reverse transcription. A **retrovirus** is a single-stranded positive-sense RNA virus with a DNA intermediate and, as an obligate parasite, targets a host cell. Once inside the host cell cytoplasm, the virus uses its own reverse transcriptase enzyme to produce DNA from its RNA genome — the reverse of the usual pattern, thus *retro* (backwards). This new DNA is then incorporated into the host cell genome by an integrase enzyme, at which point the retroviral DNA is referred to as a provirus. The host cell then treats the viral DNA as part of its own genome, translating and transcribing the viral genes along with the cell's own genes, producing the proteins required to assemble new copies of the virus. It is difficult to detect the virus until it has infected the host. At that point, the infection will persist indefinitely.

In most viruses, DNA is transcribed into RNA, and then RNA is translated into protein. However, retroviruses function differently – their RNA is reverse-transcribed into DNA, which is integrated into the host cell's genome (when it becomes a provirus), and then undergoes the usual transcription and translational processes to express the genes carried by the virus. So, the information contained in a retroviral gene is used to generate the corresponding protein via the sequence: RNA → DNA → RNA → polypeptide. This extends the fundamental process identified by Francis Crick (one gene-one peptide) in which the sequence is: DNA → RNA → peptide (proteins are made of one or more polypeptide chain; e.g. haemoglobin is a four-chain peptide).

Retroviruses are valuable research tools in molecular biology, and have been used successfully in gene delivery systems.[1]

36.1 Structure

Virions of retroviruses consist of enveloped particles about 100 nm in diameter. The virions also contain two identical single-stranded RNA molecules 7–10 kilobases in length. Although virions of different retroviruses do not have the same morphology or biology, all the virion components are very similar.[2]

The main virion components are:

- Envelope: composed of lipids (obtained from the host plasma membrane during budding process) as well as glycoprotein encoded by the env gene. The retroviral envelope serves three distinct functions: protection from the extracellular environment via the lipid bilayer, enabling the retrovirus to enter/exit host cells through endosomal membrane trafficking, and the ability to directly enter cells by fusing with their membranes.

- RNA: consists of a dimer RNA. It has a cap at the 5' end and a poly(A) tail at the 3' end. The RNA genome also has terminal noncoding regions, which are important in replication, and internal regions that encode virion proteins for gene expression. The 5' end includes four regions, which are R, U5, PBS, and L. The R region is a short repeated sequence at each end of the genome used during the reverse transcription to ensure correct end-to-end transfer in the growing chain. U5, on the other hand, is a short unique sequence between R and PBS. PBS (primer binding site) consists of 18 bases complementary to 3' end of tRNA primer. L region is an untranslated leader region that gives the signal for packaging of the genome RNA. The 3' end includes 3 regions, which are PPT (polypurine tract), U3, and R. The PPT is a primer for plus-strand DNA synthesis during reverse transcription. U3 is a sequence between PPT and R, which serves as a signal that the provirus can use in transcription. R is the terminal repeated sequence at 3' end.

- Proteins: consisting of gag proteins, protease (PR), pol proteins, and env proteins.

  - Group-specific antigen (gag) proteins are major components of the viral capsid, which are about 2000–4000 copies per virion.

  - Protease is expressed differently in different viruses. It functions in proteolytic cleavages during virion maturation to make mature gag and pol proteins.

  - Pol proteins are responsible for synthesis of viral DNA and integration into host DNA after infection.

  - Env proteins play a role in association and entry of virions into the host cell.[3] Possessing a functional copy of an env gene is what makes retroviruses distinct from retroelements.[4] The ability of the retrovirus to bind to its target host cell using specific cell-surface receptors is given by the surface component (SU) of the Env protein, while the ability of the retrovirus to enter the cell via membrane fusion is imparted by the membrane-anchored trans-membrane component (TM). Thus the Env protein is what enables the retrovirus to be infectious.

36.2 Multiplication

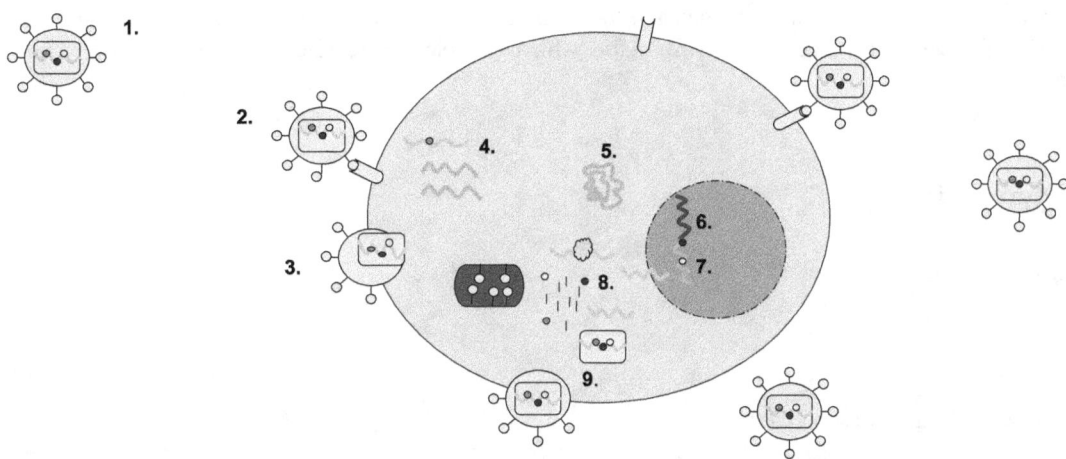

A retrovirus has a membrane containing glycoproteins, which are able to bind to a receptor protein on a host cell. There are two strands of RNA within the cell that have three enzymes; protease, reverse transcriptase, and integrase (1). The first step of replication is the binding of the glycoprotein to the receptor protein (2). Once these have been bound, the cell membrane degrades, becoming part of the host cell, and the RNA strands and enzymes go into the cell (3). Within the cell, reverse transcriptase creates a complementary strand of DNA from the retrovirus RNA and the RNA is degraded; this strand of DNA is known as cDNA (4). The cDNA is then replicated, and the two strands form a weak bond and go into the nucleus (5). Once in the nucleus, the DNA is integrated into the host cells DNA with the help of integrase (6). This cell can either stay dormant, or RNA may be synthesized from the DNA and used to create the proteins for a new retrovirus (7). Ribosome units are used to transcribe the mRNA of the virus into the amino acid sequences which can be made into proteins in the Rough Endoplasmic Reticulum. This step will also make viral enzymes and capsid proteins (8). Viral RNA will be made in the nucleus. These pieces are then gathered together and are pinched off of the cell membrane as a new retrovirus (9).

When retroviruses have integrated their own genome into the germ line, their genome is passed on to a following generation. These endogenous retroviruses (ERVs), contrasted with exogenous ones, now make up 5-8% of the human genome.[5] Most insertions have no known function and are often referred to as "junk DNA". However, many endogenous retroviruses play important roles in host biology, such as control of gene transcription, cell fusion during placental development in the course of the germination of an embryo, and resistance to exogenous retroviral infection. Endogenous retroviruses have also received special attention in the research of immunology-related pathologies, such as autoimmune diseases like multiple sclerosis, although endogenous retroviruses have not yet been proven to play any causal role in this class of disease.[6]

While transcription was classically thought to occur only from DNA to RNA, reverse transcriptase transcribes RNA into DNA. The term "retro" in retrovirus refers to this reversal (making DNA from RNA) of the central dogma of molecular biology. Reverse transcriptase activity outside of retroviruses has been found in almost all eukaryotes, enabling the generation and insertion of new copies of retrotransposons into the host genome. These inserts are transcribed by enzymes of the host into new RNA molecules that enter the cytosol. Next, some of these RNA molecules are translated into viral proteins. For example, the *gag* gene is translated into molecules of the capsid protein, the *pol* gene is translated into molecules of reverse transcriptase, and the *env* gene is translated into molecules of the envelope protein. It is important to note that a retrovirus must "bring" its own reverse transcriptase in its capsid, otherwise it is unable to utilize the enzymes of the infected cell to carry out the task, due to the unusual nature of producing DNA from RNA.

Industrial drugs that are designed as protease and reverse transcriptase inhibitors are made such that they target specific sites and sequences within their respective enzymes. However these drugs can quickly become ineffective due to the fact that the gene sequences that code for the protease and the reverse transcriptase quickly mutate. These changes in bases cause specific codons and sites with the enzymes to change and thereby avoid drug targeting by losing the sites that the drug actually targets.

Because reverse transcription lacks the usual proofreading of DNA replication, a retrovirus mutates very often. This enables the virus to grow resistant to antiviral pharmaceuticals quickly, and impedes the development of effective vaccines and inhibitors for the retrovirus.[7]

One difficulty faced with some retroviruses, such as the Moloney retrovirus, involves the requirement for cells to be actively dividing for transduction. As a result, cells such as neurons are very resistant to infection and transduction by retroviruses. This gives rise to a concern that insertional mutagenesis due to integration into the host genome might lead to cancer or leukemia. This is unlike *Lentivirus*, a genus of *Retroviridae*, which are able to integrate their RNA into the genome of non-dividing host cells.

36.3   Transmission

- Cell-to-cell[8]

- Fluids

- Airborne, like the Jaagsiekte sheep retrovirus.

36.4   Genes

Retrovirus genomes commonly contain these three open reading frames that encode for proteins that can be found in the mature virus:

- *group-specific antigen* (gag) codes for core and structural proteins of the virus;

- *polymerase* (pol) codes for reverse transcriptase, protease and integrase;

- *envelope* (env) codes for the retroviral coat proteins.

36.5 Provirus

This DNA can be incorporated into host genome as a provirus that can be passed on to progeny cells. The retrovirus DNA is inserted at random into the host genome. Because of this, it can be inserted into oncogenes. In this way some retroviruses can convert normal cells into cancer cells. Some provirus remains latent in the cell for a long period of time before it is activated by the change in cell environment.

36.6 Early evolution

Studies of retroviruses led to the first demonstrated synthesis of DNA from RNA templates, a fundamental mode for transferring genetic material that occurs in both eukaryotes and prokaryotes. It has been speculated that the RNA to DNA transcription processes used by retroviruses may have first caused DNA to be used as genetic material. In this model, the RNA world hypothesis, cellular organisms adopted the more chemically stable DNA when retroviruses evolved to create DNA from the RNA templates.

36.7 Gene therapy

Gammaretroviral and lentiviral vectors for gene therapy have been developed that mediate stable genetic modification of treated cells by chromosomal integration of the transferred vector genomes. This technology is of use, not only for research purposes, but also for clinical gene therapy aiming at the long-term correction of genetic defects, e.g., in stem and progenitor cells. Retroviral vector particles with tropism for various target cells have been designed. Gammaretroviral and lentiviral vectors have so far been used in more than 300 clinical trials, addressing treatment options for various diseases.[1][9] Retro viral mutations can be developed to make transgenic mouse models to study various cancers and their metastatic models.

36.8 Cancer

Retroviruses that cause tumor growth include *Rous sarcoma virus* and *Mouse mammary tumor virus*. Cancer can be triggered by proto-oncogenes that were mistakenly incorporated into proviral DNA or by the disruption of cellular proto-oncogenes. Rous sarcoma virus contains the src gene that triggers tumor formation. Later it was found that a similar gene in cells is involved in cell signaling, which was most likely excised with the proviral DNA. Nontransforming viruses can randomly insert their DNA into proto-oncogenes, disrupting the expression of proteins that regulate the cell cycle. The promoter of the provirus DNA can also cause over expression of regulatory genes.

36.9 Classification

36.9.1 Exogenous

These are infectious RNA-containing viruses which are transmitted from human to human.

The following genera are included here:

- Genus *Alpharetrovirus*; type species: *Avian leukosis virus*; others include *Rous sarcoma virus*

- Genus *Betaretrovirus*; type species: *Mouse mammary tumour virus*

- Genus *Gammaretrovirus*; type species: *Murine leukemia virus*; others include *Feline leukemia virus*

- Genus *Deltaretrovirus*; type species: *Bovine leukemia virus*; others include the cancer-causing *Human T-lymphotropic virus*

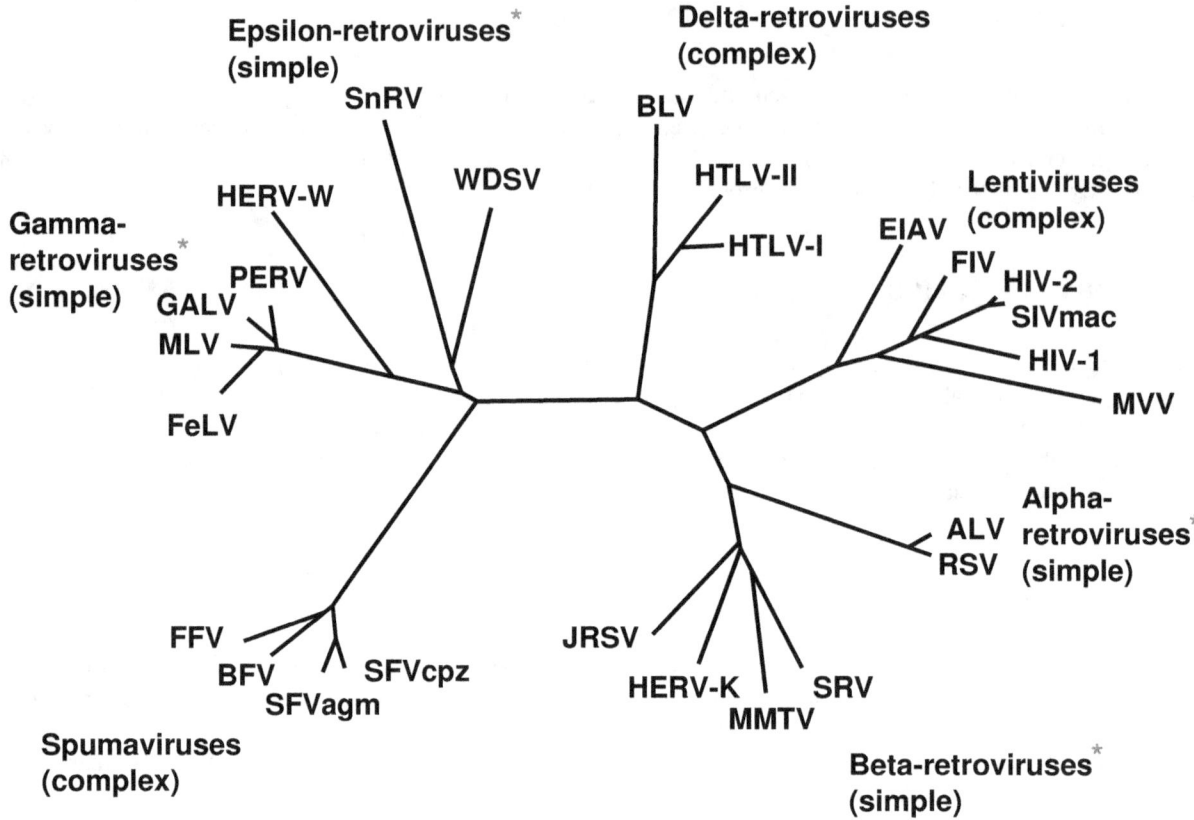

Phylogeny of Retroviruses

- Genus *Epsilonretrovirus*; type species: *Walleye dermal sarcoma virus*

- Genus *Lentivirus*; type species: *Human immunodeficiency virus 1*; others include *Simian, Feline* immunodeficiency viruses

- Genus *Spumavirus*; type species: *Simian foamy virus*

These were previously divided into three subfamilies (*Oncovirinae, Lentivirinae*, and *Spumavirinae*), but are now divided into two: *Orthoretrovirinae* and *Spumaretrovirinae*. The term oncovirus is now commonly used to describe a cancer-causing virus.

Retroviruses were in 2 groups of the Virus classification#Baltimore classification.

Group VI viruses

All members of Group VI use virally encoded reverse transcriptase, an RNA-dependent DNA polymerase, to produce DNA from the initial virion RNA genome. This DNA is often integrated into the host genome, as in the case of retroviruses and pseudoviruses, where it is replicated and transcribed by the host.

Group VI includes:

- Family *Metaviridae*

- Family *Pseudoviridae*

- Family *Retroviridae* — Retroviruses, e.g. *HIV*

Group VII viruses

Both families in Group VII have DNA genomes contained within the invading virus particles. The DNA genome is transcribed into both mRNA, for use as a transcript in protein synthesis, and pre-genomic RNA, for use as the template during genome replication. Virally encoded reverse transcriptase uses the pre-genomic RNA as a template for the creation of genomic DNA.

Group VII includes:

- Family *Hepadnaviridae* — e.g. *Hepatitis B virus*
- Family *Caulimoviridae* — e.g. *Cauliflower mosaic virus*

36.9.2 Endogenous

Main article: Endogenous retrovirus

Endogenous retroviruses are not formally included in this classification system, and are broadly classified into three classes, on the basis of relatedness to exogenous genera:

- Class I are most similar to the gammaretroviruses
- Class II are most similar to the betaretroviruses and alpharetroviruses
- Class III are most similar to the spumaviruses.

36.10 Treatment

Antiretroviral drugs are medications for the treatment of infection by retroviruses, primarily HIV. Different classes of antiretroviral drugs act on different stages of the HIV life cycle. Combination of several (typically three or four) antiretroviral drugs is known as highly active anti-retroviral therapy (HAART).[10]

36.11 Treatment of veterinary retroviruses

Feline leukemia virus and *Feline immunodeficiency virus* infections are treated with biologics, including the only immunomodulator currently licensed for sale in the United States, Lymphocyte T-Cell Immune Modulator (LTCI).[11]

36.12 References

[1] Kurth, Reinhard; Bannert, Norbert, eds. (2010). *Retroviruses: Molecular Biology, Genomics and Pathogenesis*. Horizon Scientific. ISBN 978-1-904455-55-4.

[2] Coffin, John M. (1992). "Structure and Classification of Retroviruses". In Levy, Jay A. *The Retroviridae* **1** (1st ed.). New York: Plenum. p. 20. ISBN 0-306-44074-1.

[3] Coffin 1992, pp. 26–34

[4] Kim FJ, Battini JL, Manel N, Sitbon M (January 2004). "Emergence of vertebrate retroviruses and envelope capture". *Virology* **318** (1): 183–91. doi:10.1016/j.virol.2003.09.026. PMID 14972546.

[5] Robert Belshaw; Pereira V; Katzourakis A; Talbot G; Paces J; Burt A; Tristem M. (April 2004). "Long-term reinfection of the human genome by endogenous retroviruses". *Proc Natl Acad Sci USA* **101** (14): 4894–9. doi:10.1073/pnas.0307800101. PMC 387345. PMID 15044706.

[6] Medstrand P, van de Lagemaat L, Dunn C, Landry J, Svenback D, Mager D (2005). "Impact of transposable elements on the evolution of mammalian gene regulation". *Cytogenet Genome Res* **110** (1-4): 342–52. doi:10.1159/000084966. PMID 16093686.

[7] Svarovskaia ES; Cheslock SR; Zhang WH; Hu WS; Pathak VK. (January 2003). "Retroviral mutation rates and reverse transcriptase fidelity.". *Front Biosci.* **8** (1-3): d117–34. doi:10.2741/957. PMID 12456349.

[8] Jolly C (March 2011). "Cell-to-cell transmission of retroviruses: Innate immunity and interferon-induced restriction factors.". *Virology* **411** (2): 251–9. doi:10.1016/j.virol.2010.12.031. PMC 3053447. PMID 21247613.

[9] Desport, M, ed. (2010). *Lentiviruses and Macrophages: Molecular and Cellular Interactions.* Caister Academic. ISBN 978-1-904455-60-8.

[10] Haddad M, Inch C, Glazier RH, et al. (2000). "Patient support and education for promoting adherence to highly active antiretroviral therapy for HIV/AIDS". *Cochrane Database of Systematic Reviews (Online)* (3): CD001442. doi:10.1002/14651858.CD001442.PMID10908497.

[11] Gingerich DA (2008). "Lymphocyte T-cell immunomodulator (LTCI): Review of the immunopharmacology of a new biologic" (PDF). *Intern J Appl Res Vet Med* **6** (2): 61–8.

36.13   External links

- *ViralZone* A Swiss Institute of Bioinformatics resource for all viral families, providing general molecular and epidemiological information (follow links for *"Retro-transcribing viruses"*)

- Retrovirus Animation (Flash Required)

- Retrovirology Scientific journal

- Retrovirus life cycle chapter From Kimball's *Biology* (online biology textbook pages)

- Coffin, John M; Hughes, Stephen H; Varmus, Harold E, eds. (1997). *Retroviruses.* Cold Spring Harbor Laboratory. ISBN 0-87969-571-4. NBK19376.

- Specter, Michael (3 December 2007). "Annals of Science: Darwin's Surprise". *The New Yorker.*

Chapter 37

Telomerase

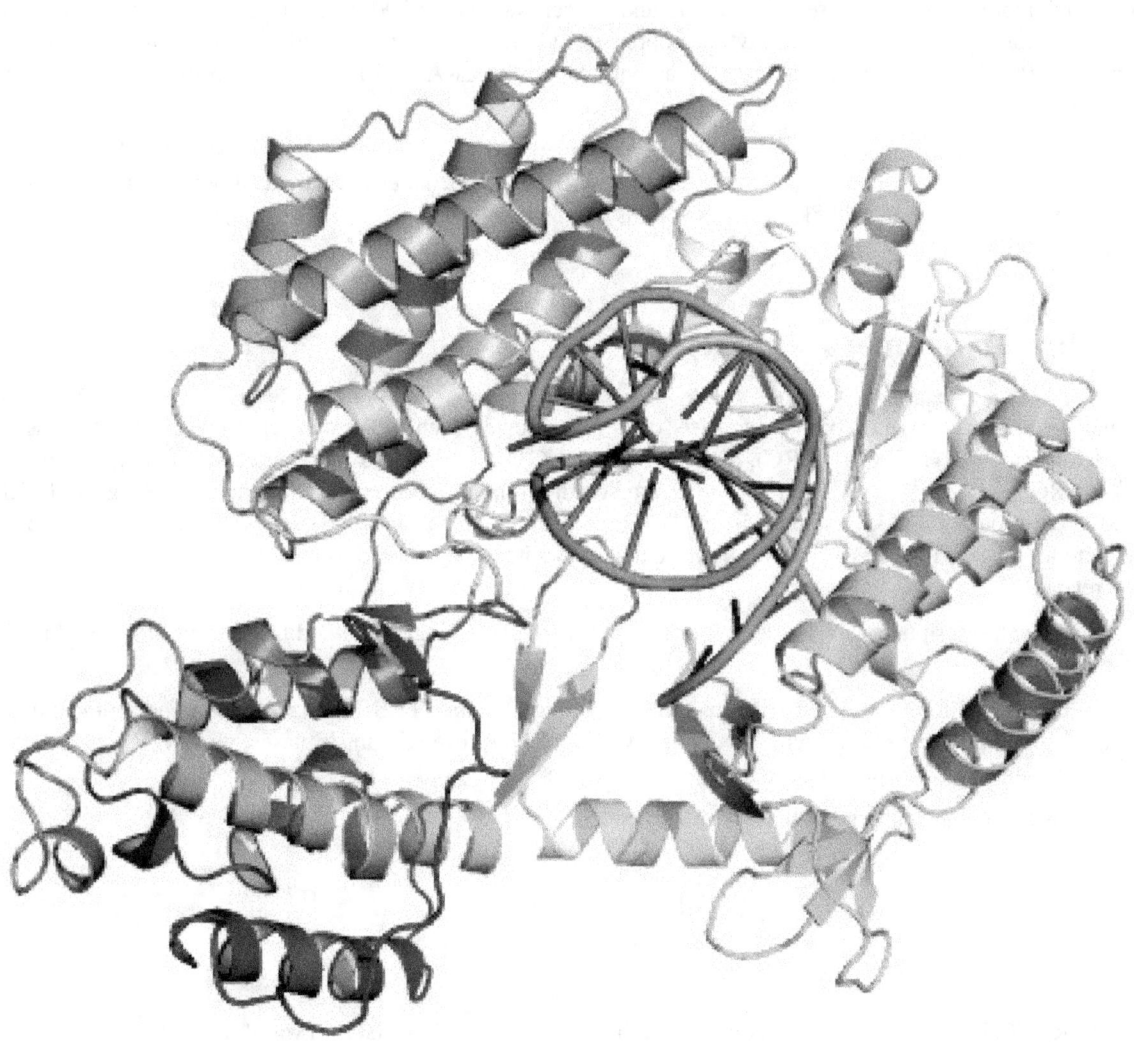

Tribolium castaneum *telomerase catalytic subunit, TERT, bound to putative RNA template and telomeric DNA (PDB 3KYL)*

Telomerase also called **telomere terminal transferase**[1] is a ribonucleoprotein that adds the polynucleotide "TTAGGG" to the 3' end of telomeres, which are found at the ends of eukaryotic chromosomes. A telomere is a region of repetitive sequences at each end of a chromatid, which protects the end of the chromosome from deterioration or from fusion with neighbouring chromosomes.

Telomerase is a reverse transcriptase enzyme that carries its own RNA molecule (with the pattern of "CCCAAUCCC" in vertebrates), which is used as a template when it elongates telomeres.

37.1   History

The existence of a compensatory mechanism for telomere shortening was first predicted by Soviet biologist Alexey Olovnikov in 1973,[2] who also suggested the telomere hypothesis of aging and the telomere's connections to cancer.

Telomerase was discovered by Carol W. Greider and Elizabeth Blackburn in 1984 in the ciliate *Tetrahymena*.[3] Together with Jack W. Szostak, Greider and Blackburn were awarded the 2009 Nobel Prize in Physiology or Medicine for their discovery.[4]

The role of telomeres and telomerase in cell aging and cancer was established by scientists at biotechnology company Geron with the cloning of the RNA and catalytic components of human telomerase[5] and the development of a polymerase chain reaction (PCR) based assay for telomerase activity called the TRAP assay, which surveys telomerase activity in multiple types of cancer.[6]

Its protein composition was identified in 2007 by Scott Cohen and his team at Children's Medical Research Institute (Australia).[7] The high-resolution protein structure of the *Tribolium castaneum* catalytic subunit of telomerase TERT was decoded in 2008 by Emmanuel Skordalakes and his team.[8]

37.2   Human telomerase structure

The human telomerase enzyme complex consists of two molecules each of human telomerase reverse transcriptase (TERT), telomerase RNA (TR or TERC), and dyskerin (DKC1).[7] The genes of telomerase subunits, which include TERT,[9] TERC,[10] DKC1[11] and TEP1,[12] are located on different chromosomes. The human TERT gene (hTERT) is translated into a protein of 1132 amino acids.[13] TERT polypeptide folds with (and carries) TERC, a non-coding RNA (451 nucleotides long). TERT has a 'mitten' structure that allows it to wrap around the chromosome to add single-stranded telomere repeats.

TERT is a reverse transcriptase, which is a class of enzyme that creates single-stranded DNA using single-stranded RNA as a template.

The protein consists of four conserved domains (RNA-Binding Domain (TRBD), fingers, palm and thumb), organized into a ring configuration that shares common features with retroviral reverse transcriptases, viral RNA polymerases and bacteriophage B-family DNA polymerases.

TERT proteins from many eukaryotes have been sequenced.[14]

37.3   Function

By using TERC, TERT can add a six-nucleotide repeating sequence, 5'-TTAGGG (in vertebrates, the sequence differs in other organisms) to the 3' strand of chromosomes. These TTAGGG repeats (with their various protein binding partners) are called telomeres. The template region of TERC is 3'-CAAUCCCAAUC-5'.[15]

Telomerase can bind the first few nucleotides of the template to the last telomere sequence on the chromosome, add a new telomere repeat (5'-GGTTAG-3') sequence, let go, realign the new 3'-end of telomere to the template, and repeat the process. Telomerase reverses telomere shortening.

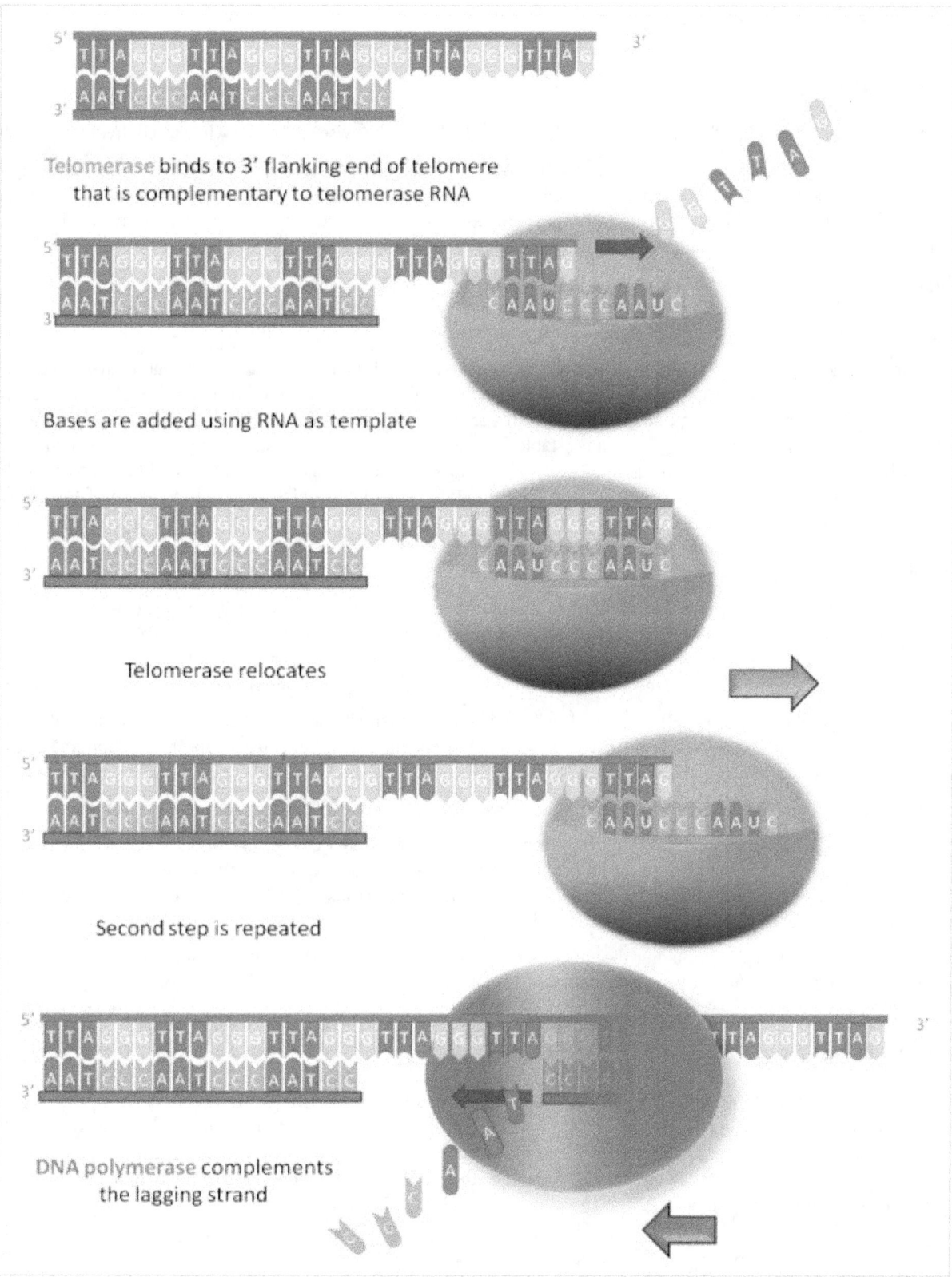

Telomerase binds to 3′ flanking end of telomere that is complementary to telomerase RNA

Bases are added using RNA as template

Telomerase relocates

Second step is repeated

DNA polymerase complements the lagging strand

An image illustrating how telomerase elongates telomere ends progressively.

37.4   Clinical implications

37.4.1   Aging

Telomerase replaces short bits of DNA known as telomeres, which are otherwise shortened when a cell divides via mitosis.

In normal circumstances, absent telomerase, if a cell divides recursively, at some point the progeny reach their Hayflick limit,[16] which is believed to be between 50–70 cell divisions. At the limit the cells become senescent and cell division stops.[17] Telomerase allows each offspring to replace the lost bit of DNA allowing the cell line to divide without ever reaching the limit. This same unbounded growth is a feature of cancerous growth.[18]

Embryonic stem cells express telomerase, which allows them to divide repeatedly and form the individual. In adults, telomerase is highly expressed only in cells that need to divide regularly especially in male germ cells but also in activated lymphocytes and certain adult stem cells, whereas other somatic cells do not express it.[19]

A comparative biology study of mammalian telomeres indicated that telomere length of some mammalian species correlates inversely, rather than directly, with lifespan, and concluded that the contribution of telomere length to lifespan is unresolved.[20] Telomere shortening does not occur with age in some postmitotic tissues, such as in the rat brain.[21] In humans, skeletal muscle telomere lengths remain stable from ages 23 –74.[22] In baboon skeletal muscle, which consists of fully differentiated post-mitotic cells, less than 3% of myonuclei contain damaged telomeres and this percentage does not increase with age.[23] Thus, telomere shortening does not appear to be a major factor in the aging of the differentiated cells of brain or skeletal muscle. In human liver, cholangiocytes and hepatocytes show no age-related telomere shortening.[24] Another study found little evidence that, in humans, telomere length is a significant biomarker of normal aging with respect to important cognitive and physical abilities.[25]

Some experiments have raised questions on whether telomerase can be used as an anti-aging therapy, namely, the fact that mice with elevated levels of telomerase have higher cancer incidence and hence do not live longer. Telomerase also favors tumorogenesis, which leads to questions about its potential as an anti-aging therapy.[26] On the other hand, one study showed that activating telomerase in cancer-resistant mice by overexpressing its catalytic subunit extended lifespan.[27]

Exposure of T lymphocytes from HIV-infected human donors to a small molecule telomerase activator (TAT2) retards telomere shortening, increases proliferative potential and enhances cytokine/chemokine production and antiviral activity.[28]

A study that focused on Ashkenazi Jews found that long-lived subjects inherited a hyperactive version of telomerase.[29]

Mice engineered to block the gene that produces telomerase, unless they are given a certain drug, aged at a much faster rate, and died at about six months, instead of reaching the average mouse lifespan, about three years. Administering the drug at 6 months turned on telomerase production and caused their organs to be "rejuvenated," restored fertility, and normalized their ability to detect or process odors.[30][31]

A 2012 study reported that introducing the TERT gene into healthy one-year-old mice using an engineered adeno-associated virus led to a 24% increase in lifespan, without any increase in cancer.[32]

Premature aging

Premature aging syndromes including Werner syndrome, Ataxia telangiectasia, Ataxia-telangiectasia like disorder, Bloom syndrome, Fanconi anemia and Nijmegen breakage syndrome are associated with short telomeres.[33] However, the genes that have mutated in these diseases all have roles in the repair of DNA damage and the increased DNA damage may, itself, be a factor in the premature aging (see DNA damage theory of aging). An additional role in maintaining telomere length is an active area of investigation.

37.4.2   Cancer

In vitro, when cells approach the Hayflick limit, the time to senescence can be extended by inactivating the tumor suppressor proteins - p53 and Retinoblastoma protein (pRb). Cells that have been so-altered eventually undergo an event termed a "crisis" when the majority of the cells in the culture die. Sometimes, a cell does not stop dividing once it reaches crisis. In a typical situation, the telomeres are shortened[34] and chromosomal integrity declines with every subsequent cell division. Exposed chromosome ends are interpreted as double-stranded breaks (DSB) in DNA; such damage is usually repaired by reattaching (religating) the broken ends together. When the cell does this due to telomere-shortening, the ends of different chromosomes can be attached to each other. This solves the problem of lacking telomeres, but during cell division anaphase, the fused chromosomes are randomly ripped apart, causing many mutations and chromosomal abnormalities. As this process continues, the cell's genome becomes unstable. Eventually, either fatal damage is done to the cell's chromosomes (killing it via apoptosis), or an additional mutation that activates telomerase occurs.

With telomerase activation some types of cells and their offspring become immortal (bypass the Hayflick limit), thus avoiding cell death as long as the conditions for their duplication are met. Many cancer cells are considered 'immortal' because telomerase activity allows them to divide indefinitely, which is why they can form tumors. A good example of immortal cancer cells is HeLa cells, which have been used in laboratories as a model cell line since 1951.

While this method of modeling human cancer in cell culture is effective and has been used for many years by scientists, it is also very imprecise. The exact changes that allow for the formation of the tumorigenic clones in the above-described experiment are not clear. Scientists addressed this question by the serial introduction of multiple mutations present in a variety of human cancers. This has led to the identification of mutation combinations that form tumorigenic cells in a variety of cell types. While the combination varies by cell type, the following alterations are required in all cases: TERT activation, loss of p53 pathway function, loss of pRb pathway function, activation of the Ras or myc proto-oncogenes, and aberration of the PP2A protein phosphatase. That is to say, the cell has an activated telomerase, eliminating the process of death by chromosome instability or loss, absence of apoptosis-induction pathways, and continued mitosis activation.

This model of cancer in cell culture accurately describes the role of telomerase in actual human tumors. Telomerase activation has been observed in ~90% of all human tumors,[35] suggesting that the immortality conferred by telomerase plays a key role in cancer development. Of the tumors without TERT activation,[36] most employ a separate pathway to maintain telomere length termed Alternative Lengthening of Telomeres (ALT).[37] The exact mechanism behind telomere maintenance in the ALT pathway is unclear, but likely involves multiple recombination events at the telomere.

Two telomerase vaccines have been developed: GRNVAC1 and GV1001. GRNVAC1 isolates dendritic cells and the RNA that codes for the telomerase protein and puts them back into the patient to make cytotoxic T cells that kill the telomerase-active cells. GV1001 comes from the active site of hTERT and is recognized by the immune system that reacts by killing the telomerase-active cells.[38]

Elizabeth Blackburn *et al.*, identified the upregulation of 70 genes known or suspected in cancer growth and spread through the body, and the activation of glycolysis, which enables cancer cells to rapidly use sugar to facilitate their programmed growth rate (roughly the growth rate of a fetus).[39]

Approaches to controlling telomerase and telomeres for cancer therapy include gene therapy, immunotherapy, small-molecule and signal pathway inhibitors.[38]

Drugs

The ability to maintain functional telomeres may be one mechanism that allows cancer cells to grow *in vitro* for decades.[40] Telomerase activity is necessary to preserve many cancer types and is inactive in somatic cells, creating the possibility that telomerase inhibition could selectively repress cancer cell growth with minimal side effects.[41] If a drug can inhibit telomerase in cancer cells, the telomeres of successive generations will progressively shorten, limiting tumor growth.[42]

Telomerase is a good biomarker for cancer detection because most human cancers cells express high levels of it. Telomerase activity can be identified by its catalytic protein domain (hTERT). This is the rate-limiting step in telomerase activity. It is associated with many cancer types. Various cancer cells and fibroblasts transformed with hTERT cDNA have high telomerase activity, while somatic cells do not. Cells testing positive for hTERT have positive nuclear signals. Epithelial stem cell tissue and its early daughter cells are the only noncancerous cells in which hTERT can be detected. Since hTERT expression is dependent only on the number of tumor cells within a sample, the amount of hTERT indicates the severity of a cancer.[43]

The expression of hTERT can also be used to distinguish benign tumors from malignant tumors. Malignant tumors have higher hTERT expression than benign tumors. Real-time reverse transcription polymerase chain reaction (RT-PCR) quantifying hTERT expression in various tumor samples verified this varying expression.[44]

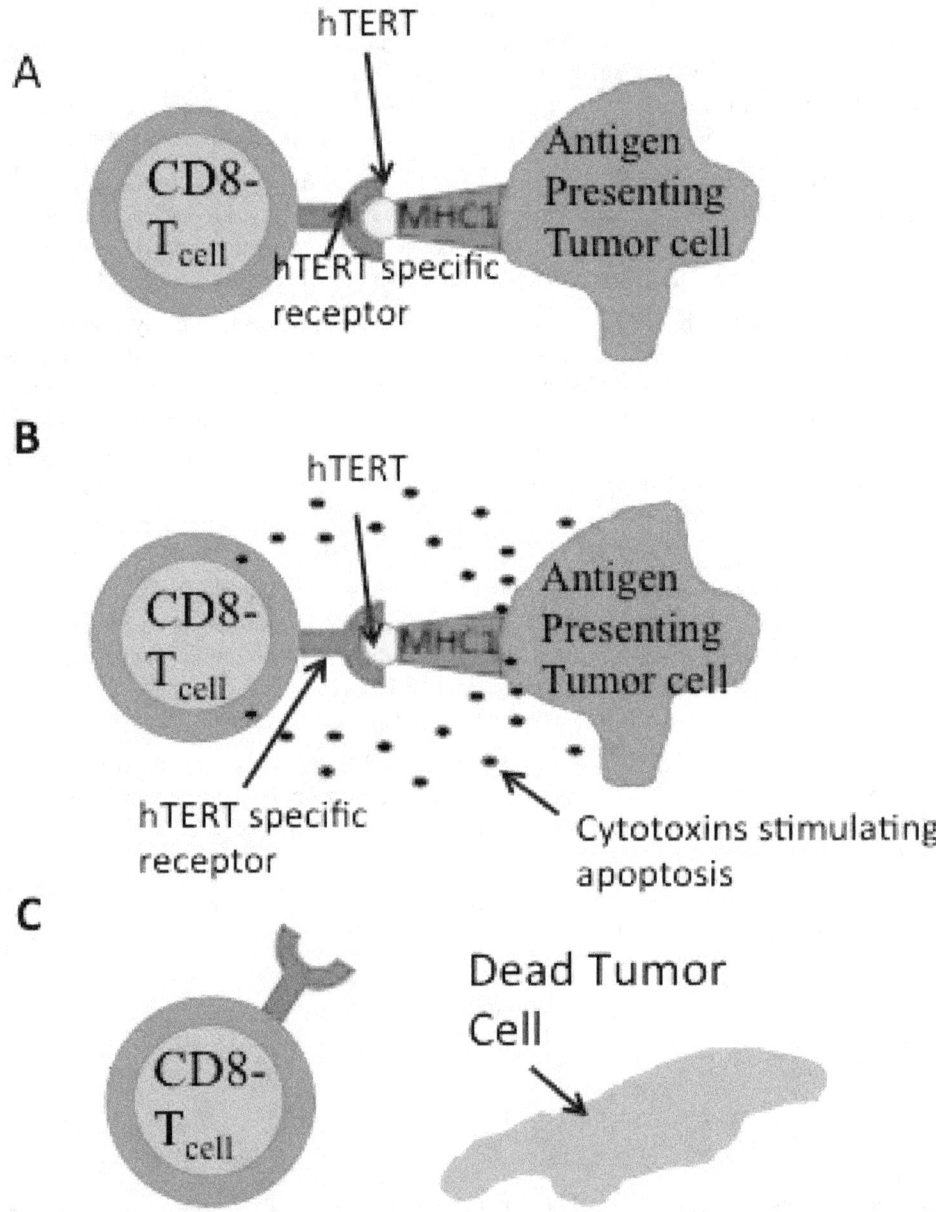

Figure 4:A) Tumor cells expressing hTERT will actively degrade some of the protein and process for presenting. The major histocompatibility complex 1(MHC1), can then present the hTERT epitote. CD8- T cells that have antibodies against hTERT will then bind to the presented epitote. B) As a result of the antigenic binding, the T cells will release cytotoxins, which can be absorbed by the affected cell. C) These cytotoxins induce multiple proteases and results in apoptosis (or cell death).

The lack of telomerase does not affect cell growth, until the telomeres are short enough to cause cells to "die or undergo growth arrest". However, inhibiting telomerase alone is not enough to destroy large tumors. It must be combined with surgery, radiation, chemotherapy or immunotherapy.[43]

Cells may reduce their telomere length by only 50-252 base pairs per cell division, which can lead to a long lag phase.[45][46]

Immunotherapy Immunotherapy successfully treats some kinds of cancer, such as melanoma. This treatment involves manipulating a human's immune system to destroy cancerous cells. Humans have two major antigen identifying molecules: cytotoxic T-lymphocytes (CTL) and CD4+ helper T-lymphocytes that can destroy cells. Antigen receptors on CTL can bind to a 9-10 amino acid chain that is presented by the major histocompatibility complex (MHC) as in Figure 4. HTERT is a potential target antigen. Immunotargeting should result in relatively few side effects since hTERT expression is associated only with telomerase and is not essential in almost all somatic cells.[47] GV1001 uses this pathway.[38] Experimental drug and vaccine therapies targeting active telomerase have been tested in mouse models, and clinical trials have begun.

In 2014 Geron Corporation received permission to resume a trial of its drug imetelstat for myelofibrosis after addressing FDA concerns over liver toxicity.[48][49] Geron licensee Merck had approval of an IND for one vaccine type. Imetelstat (GRN163L) binds directly to the telomerase's RNA template. One 2015 study reported that Imetelstat caused partial or complete remission in seven of 33 patients, while a second reported that it decreased blood platelet levels in all 18 study patients with essential thrombocythemia, a disorder in which the body overproduces blood platelets, increasing the risk of blood clots.[50]

Most of the harmful cancer-related effects of telomerase are dependent on an intact RNA template. Cancer stem cells that use an alternative method of telomere maintenance are still killed when telomerase's RNA template is blocked or damaged.

Targeted apoptosis Another independent approach is to use oligoadenylated anti-telomerase antisense oligonucleotides and ribozymes to target telomerase RNA, inducing dissociation and apoptosis (Figure 5). The fast induction of apoptosis through antisense binding may be a good alternative to the slower telomere shortening.[45]

37.4.3 Heart disease, diabetes and quality of life

Blackburn also discovered that mothers caring for very sick children have shorter telomeres when they report that their emotional stress is at a maximum and that telomerase was active at the site of blockages in coronary artery tissue, possibly accelerating heart attacks.

In 2009, it was shown that the amount of telomerase activity significantly increased following psychological stress. Across the sample of patients telomerase activity in increased peripheral blood mononuclear cellsby 18% one hour after the end of the stress.[51]

E. V. Gostjeva *et al.* found no differences between colon cancer stem cells and fetal colon stem cells.[52]

A study in 2010 found that there was "significantly greater" telomerase activity in participants than controls after a three-month meditation retreat.[53]

Telomerase deficiency has been linked to diabetes mellitus and impaired insulin secretion in mice, due to loss of pancreatic insulin-producing cells.[54]

37.4.4 Rare human diseases

Mutations in TERT have been implicated in predisposing patients to aplastic anemia, a disorder in which the bone marrow fails to produce blood cells, in 2005.[55]

Cri du chat syndrome (CdCS) is a complex disorder involving the loss of the distal portion of the short arm of chromosome 5. TERT is located in the deleted region, and loss of one copy of TERT has been suggested as a cause or contributing factor of this disease.[56]

Dyskeratosis congenita (DC) is a disease of the bone marrow that can be caused by some mutations in the telomerase subunits.[57] In the DC cases, about 35% cases are X-linked-recessive on the DKC1 locus[58] and 5% cases are autosomal dominant on the TERT[59] and TERC[60] loci.

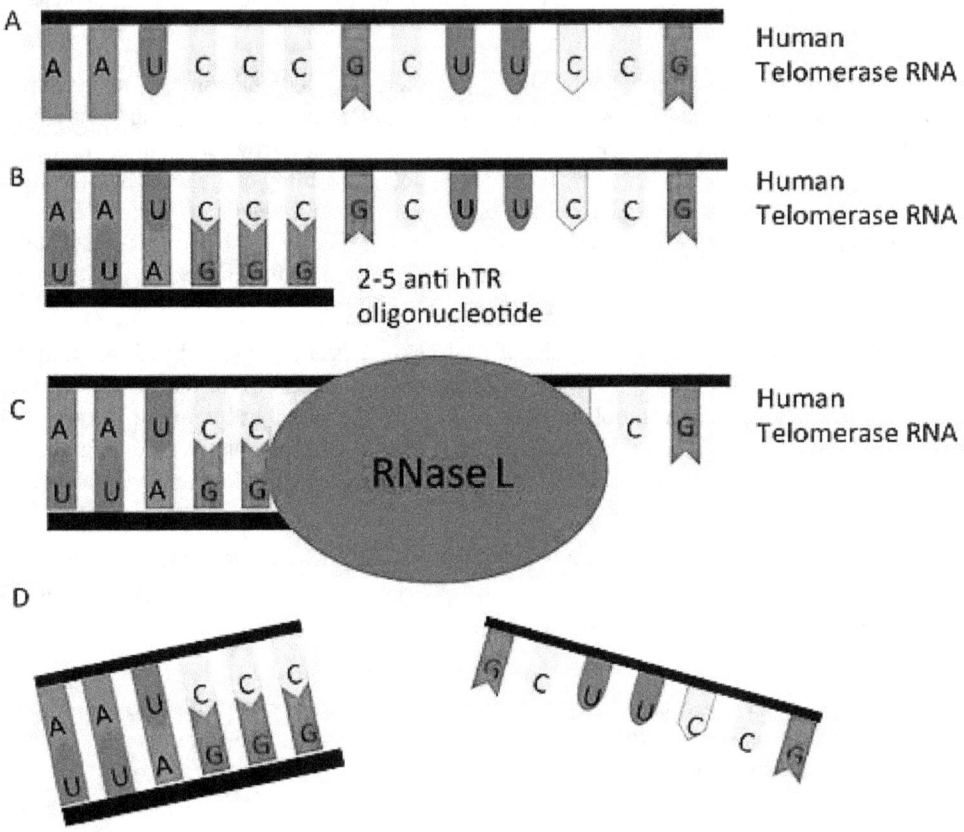

Figure 5: A) Human telomerase RNA (hTR) is present in the cell and can be targeted. B) 2-5 anti-hTR oligonucleotides is a specialized antisense oligo that can bind to the telomerase RNA. C) Once bound, the 2-5 anti-hTR oligonucleotide recruits RNase L to the sequence. Once recruited, the RNase L creates a single cleavage in the RNA (D) and causes dissociation of the RNA sequence.

Patients with DC have severe bone marrow failure manifesting as abnormal skin pigmentation, leucoplakia (a white thickening of the oral mucosa) and nail dystrophy, as well as a variety of other symptoms. Individuals with either TERC or DKC1 mutations have shorter telomeres and defective telomerase activity *in vitro* versus other individuals of the same age.[61]

In one family autosomal dominant DC was linked to a heterozygous TERT mutation.[62] These patients also exhibited an increased rate of telomere-shortening, and genetic anticipation (i.e., the DC phenotype worsened with each generation).

37.5   See also

- DNA repair

- TA-65

- Telomere

37.6   References

[1] What are telomeres and telomerase?

[2] Olovnikov AM (September 1973). "A theory of marginotomy. The incomplete copying of template margin in enzymic synthesis of polynucleotides and biological significance of the phenomenon". *J. Theor. Biol.* **41** (1): 181–90. doi:10.1016/0022-5193(73)90198-7. PMID 4754905.

[3] Greider CW, Blackburn EH (December 1985). "Identification of a specific telomere terminal transferase activity in Tetrahymena extracts". *Cell* **43** (2 Pt 1): 405–13. doi:10.1016/0092-8674(85)90170-9. PMID 3907856.

[4] "The Nobel Prize in Physiology or Medicine 2009". The Nobel Foundation. 2009-10-05. Retrieved 2010-10-23.

[5] Feng J, Funk WD, Wang SS, Weinrich SL, Avilion AA, Chiu CP, Adams RR, Chang E, Allsopp RC, Yu J (September 1995). "The RNA component of human telomerase". *Science* **269** (5228): 1236–41. doi:10.1126/science.7544491. PMID 7544491.

[6] Kim, N.; Piatyszek, M.; Prowse, K.; Harley, C.; West, M.; Ho, P.; Coviello, G.; Wright, W.; Weinrich, S. (1994). "Specific association of human telomerase activity with immortal cells and cancer". *Science* **266** (5193): 2011–5. doi:10.1126/science.7605428. PMID 7605428.

[7] Cohen S, Graham M, Lovrecz G, Bache N, Robinson P, Reddel R (2007). "Protein composition of catalytically active human telomerase from immortal cells". *Science* **315** (5820): 1850–3. doi:10.1126/science.1138596. PMID 17395830.

[8] Gillis AJ, Schuller AP, Skordalakes E (October 2008). "Structure of the Tribolium castaneum telomerase catalytic subunit TERT". *Nature* **455** (7213): 633–7. doi:10.1038/nature07283. PMID 18758444.

[9] "HGNC database of human gene names - HUGO Gene Nomenclature Committee". *genenames.org*.

[10] HGNC - TERC

[11] HGNC - DKC1

[12] HGNC - TEP1

[13] NCBI - telomerase reverse transcriptase isoform 1

[14] NCBI - telomerase reverse transcriptase

[15] Gavory G, Farrow M, Balasubramanian S (October 2002). "Minimum length requirement of the alignment domain of human telomerase RNA to sustain catalytic activity in vitro". *Nucleic Acids Res.* **30** (20): 4470–80. doi:10.1093/nar/gkf575. PMC 137139. PMID 12384594.

[16] Hayflick L, Moorhead PS (1961). "The serial cultivation of human diploid cell strains". *Exp Cell Res* **25** (3): 585–621. doi:10.1016/0014-4827(61)90192-6. PMID 13905658.

[17] Siegel, L (2013). Are Telomeres the Key to Aging and Cancer? The University of Utah. Retrieved 30 September 2013

[18] Hanahan D & Weinberg RA (March 2011). "Hallmarks of cancer: the next generation". *Cell* **144** (5): 646–74. doi:10.1016/j.cell.2 PMID 21376230.

[19] Cong YS (2002). "Human Telomerase and It's Regulation". *Microbiology and Molecular Biology Reviews* **66** (3): 407–425. doi:10.1128/MMBR.66.3.407-425.2002. PMC 120798. PMID 12208997.

[20] Gomes, NM; Ryder, OA; Houck, ML; Charter, SJ; Walker, W,; Forsyth, NR; Austad, SN; Venditti, C; Pagel, M; Shay, JW; Wright, WE (2011). "Comparative biology of mammalian telomeres: hypotheses on ancestral states and the roles of telomeres in longevity determination.". *Aging Cell* **105** (5): 761–768. doi:10.1111/j.1474-9726.2011.00718.x. PMID 21518243.

[21] Cherif H, Tarry JL, Ozanne SE, Hales CN (2003). "Ageing and telomeres: a study into organ- and gender-specific telomere shortening". *Nucleic Acids Res* **31** (5): 1576–1583. doi:10.1093/nar/gkg208. PMID 12595567.

[22] "Regenerative potential of human skeletal muscle during aging.". *Aging Cell* **1** (2): 132–9. Dec 2002. doi:10.1046/j.1474-9728.2002.00017.x. PMID 12882343.

[23] Jeyapalan JC, Ferreira M, Sedivy JM, Herbig U (2007). "Accumulation of senescent cells in mitotic tissue of aging primates". *Mech Ageing Dev* **128** (1): 36–44. doi:10.1016/j.mad.2006.11.008. PMID 17116315.

[24] Verma S, Tachtatzis P, Penrhyn-Lowe S, Scarpini C, Jurk D, Von Zglinicki T, Coleman N, Alexander GJ (2012). "Sustained telomere length in hepatocytes and cholangiocytes with increasing age in normal liver". *Hepatology* **56** (4): 1510–1520. doi:10.1002/hep.25787. PMID 22504828.

[25] "Telomere length and aging biomarkers in 70-year-olds: the Lothian Birth Cohort 1936.". *Neurobiol Aging* **33** (7): 1486.e3–8. Jul 2012. doi:10.1016/j.neurobiolaging.2010.11.013. PMID 21194798.

[26] de Magalhães JP, Toussaint O (2004). "Telomeres and telomerase: a modern fountain of youth?". *Rejuvenation Res* **7** (2): 126–33. doi:10.1089/1549168041553044. PMID 15312299.

[27] Tomás-Loba A, Flores I, Fernández-Marcos PJ, Cayuela ML, Maraver A, Tejera A, Borrás C, Matheu A, Klatt P, Flores JM, Viña J, Serrano M, Blasco MA (November 2008). "Telomerase reverse transcriptase delays aging in cancer-resistant mice". *Cell* **135** (4): 609–22. doi:10.1016/j.cell.2008.09.034. PMID 19013273.

[28] Fauce SR, Jamieson BD, Chin AC, Mitsuyasu RT, Parish ST, Ng HL, Kitchen CM, Yang OO, Harley CB, Effros RB (November 2008). "Telomerase-Based Pharmacologic Enhancement of Antiviral Function> of Human CD8+ T Lymphocytes". *J. Immunol.* **181** (10): 7400–6. doi:10.4049/jimmunol.181.10.7400. PMC 2682219. PMID 18981163.

[29] Atzmon G, Cho M, Cawthon RM, Budagov T, Katz M, Yang X, Siegel G, Bergman A, Huffman DM, Schechter CB, Wright WE, Shay JW, Barzilai N, Govindaraju DR, Suh Y (January 2010). "Genetic variation in human telomerase is associated with telomere length in Ashkenazi centenarians". *Proc. Natl. Acad. Sci. U.S.A.* 107 Suppl 1 (suppl_1): 1710–7. doi:10.1073/pnas.090619 1106.PMC2868292. PMID 19915151. Lay summary– *LiveScience*.

[30] Jaskelioff, M; Muller, FL; Paik, JH; Thomas, E; Jiang, S; Adams, AC; Sahin, E; Kost-Alimova, M; Protopopov, A; Cadiñanos, J; Horner, JW; Maratos-Flier, E; Depinho, RA (November 2010). "Telomerase reactivation reverses tissue degeneration in aged telomerase deficient mice". *Nature* **469** (7328): 102–6. doi:10.1038/nature09603. PMC 3057569. PMID 21113150. Lay summary – *news.discovery.com*.

[31] "Stop, rewind: the scientists slowing the ageing process". *BBC News.* 2011-01-26.

[32] Bernardes de Jesus, B; Vera, E; Schneeberger, K; Tejera, AM; Ayuso, E; Bosch, F; Blasco, MA (August 2012). "Telomerase gene therapy in adult and old mice delays aging and increases longevity without increasing cancer". *EMBO Molecular Medicine* **4** (8): 691–704. doi:10.1002/emmm.201200245. PMC 3494070. PMID 22585399.

[33] Blasco, MA (August 2005). "Telomeres and human disease: ageing, cancer and beyond". *Nature Reviews Genetics* **6** (8): 611–22. doi:10.1038/nrg1656. PMID 16136653.

[34] Skloot, Rebecca (2010). *The Immortal Life of Henrietta Lacks.* New York: Broadway Paperbacks. pp. 216, 217. ISBN 978-1-4000-5218-9.

[35] Shay, JW; Bacchetti, S (April 1997). "A survey of telomerase activity in human cancer". *Eur. J. Cancer* **33** (5): 787–91. doi:10.1016/S0959-8049(97)00062-2. PMID 9282118.

[36] Bryan, TM; Englezou, A; Gupta, J; Bacchetti, S; Reddel, RR (September 1995). "Telomere elongation in immortal human cells without detectable telomerase activity". *EMBO J.* **14** (17): 4240–8. PMC 394507. PMID 7556065.

[37] Henson, JD; Neumann, AA; Yeager, TR; Reddel, RR (January 2002). "Alternative lengthening of telomeres in mammalian cells". *Oncogene* **21** (4): 598–610. doi:10.1038/sj.onc.1205058. PMID 11850785.

[38] Tian, X; Chen, B; Liu, X (March 2013). "Telomere and Telomerase as Targets for Cancer Therapy". *Applied Biochemistry and Biotechnology* **160** (5): 906–21. doi:10.1007/s00018-007-6481-8. PMID 17310277.

[39] Blackburn, EH (February 2005). "Telomeres and telomerase: their mechanisms of action and the effects of altering their functions". *FEBS Lett.* **579** (4): 859–62. doi:10.1016/j.febslet.2004.11.036. PMID 15680963.

[40] Griffiths, Anthony J. F.; Wesslet, Susan R.; Carroll, Sean B.; Doebley, John (2008). *Introduction to Genetic Analysis.* W. H. Freeman. ISBN 978-0-7167-6887-6.

[41] Williams, SC (January 2013). "No end in sight for telomerase-targeted cancer drugs.". *Nat Med.* **19** (1): 6. doi:10.1038/nm0113-6. PMID 23295993.

[42] Blasco, MA (2001). "Telomeres in Cancer Therapy". *J Biomed Biotechnol.* **1** (1): 3–4. doi:10.1155/S1110724301000109. PMC 79678. PMID 12488618.

[43] Shay, Jerry W,; Ying, Zou; Hiyama, Eiso; Wright, Woodring E. (2001). "Telomerase and Cancer". *Human Molecular Genetics* **10** (7): 677–685. doi:10.1093/hmg/10.7.677. Retrieved June 2015.

[44] Gul, Ilhami; Dundar, Ozgur; Bodur, Serkan; Tunca, Yusuf; Tutuncu, Levent (2013). "The Status of Telomerase Enzyme Activity in Benign and Malignant Gynecologic Pathologies". *Balkan Medical Journal* **30**: 287–292. doi:10.5152/balkanmedj.2013.73. Retrieved June 2015.

[45] Saretzki, Gabriele (2003). "Telomerase inhibition as Cancer Therapy". *Cancer Letter* **194** (2): 209–219. PMID 12757979. Retrieved June 2015.

[46] Stoyanov, V (2009). "T-loop deletion factor showing speeding aging of Homo telomere diversity and evolution". *Rejuvenation Research* **12** (1): 52.

[47] Patel, Kunal P.; Robert H., Vonderheide (2004). "Telomerase as a tumor-associated antigen for cancer immunotherapy". *Cytotechnology* **45** (1-2): 91–99. PMID 11850795. Retrieved June 2015.

[48] "Geron's cancer drug shakes off one FDA hold but remains on pause". *FierceBiotech*. Retrieved 2015-06-28.

[49] "J&J bets up to $935M that Geron's drug can shake a checkered past". *FierceBiotech*. Retrieved 2015-06-28.

[50] Johnson, Steven Ross (September 2, 2015). "Experimental blood disorder therapy shows promise in new studies". *Modern Healthcare*. Retrieved September 2015.

[51] Epel, ES; Lin, J; Dhabhar, FS; Wolkowitz, OM; Puterman, E; Karan, L; Blackburn, EH (2010). "Dynamics of telomerase activity in response to acute psychological stress". *Brain Behav Immun* **24** (4): 531–9. doi:10.1016/j.bbi.2009.11.018. PMC 2856774. PMID 20018236.

[52] "Stem cell stages and the origins of colon cancer: a multidisciplinary perspective.". *Stem Cell Rev* **1** (3): 243–51. 2005. doi:10.1385/SCR:1:3:243. PMID 17142861.

[53] Jacobs TL, et al. "Intensive meditation training, immune cell telomerase activity, and psychological mediators.". *nih.gov*.

[54] Ristow, Michael (2010). "Telomerase deficiency impairs glucose metabolism and insulin secretion" (PDF). *Aging* **2** (10): 650–658. PMC 2993795. PMID 20876939.

[55] Yamaguchi, H; Calado, RT; Ly, H; Kajigaya, S; Baerlocher, GM; Chanock, SJ; Lansdorp, PM; Young, NS (April 2005). "Mutations in TERT, the gene for telomerase reverse transcriptase, in aplastic anemia". *N. Engl. J. Med.* **352** (14): 1413–24. doi:10.1056/NEJMoa042980. PMID 15814878.

[56] Zhang, A; Zheng, C; Hou, M; Lindvall, C; Li, KJ; Erlandsson, F; Björkholm, M; Gruber, A; Blennow, E; Xu, D (April 2003). "Deletion of the Telomerase Reverse Transcriptase Gene and Haploinsufficiency of Telomere Maintenance in Cri du Chat Syndrome". *Am. J. Hum. Genet.* **72** (4): 940–8. doi:10.1086/374565. PMC 1180356. PMID 12629597.

[57] Yamaguchi, H (June 2007). "Mutations of telomerase complex genes linked to bone marrow failures". *J Nippon Med Sch* **74** (3): 202–9. doi:10.1272/jnms.74.202. PMID 17625368.

[58] Heiss, NS; Knight, SW; Vulliamy, TJ; Klauck, SM; Wiemann, S; Mason, PJ; Poustka, A; Dokal, I (May 1998). "X-linked dyskeratosis congenita is caused by mutations in a highly conserved gene with putative nucleolar functions". *Nat. Genet.* **19** (1): 32–8. doi:10.1038/ng0598-32. PMID 9590285.

[59] Vulliamy, TJ; Walne, A; Baskaradas, A; Mason, PJ; Marrone, A; Dokal, I (2005). "Mutations in the reverse transcriptase component of telomerase (TERT) in patients with bone marrow failure". *Blood Cells Mol. Dis.* **34** (3): 257–63. doi:10.1016/j.bcmd.20 PMID15885610.

[60] Vulliamy, T; Marrone, A; Goldman, F; Dearlove, A; Bessler, M; Mason, PJ; Dokal, I (September 2001). "The RNA component of telomerase is mutated in autosomal dominant dyskeratosis congenita". *Nature* **413** (6854): 432–5. doi:10.1038/35096585. PMID 11574891.

[61] Marrone, A; Walne, A; Dokal, I (June 2005). "Dyskeratosis congenita: telomerase, telomeres and anticipation". *Current Opinion in Genetics & Development* **15** (3): 249–57. doi:10.1016/j.gde.2005.04.004. PMID 15917199.

[62] Armanios, M; Chen, JL; Chang, YP; Brodsky, RA; Hawkins, A; Griffin, CA; Eshleman, JR; Cohen, AR; Chakravarti, A; Hamosh, A; Greider, CW (November 2005). "Haploinsufficiency of telomerase reverse transcriptase leads to anticipation in autosomal dominant dyskeratosis congenita". *Proc. Natl. Acad. Sci. U.S.A.* **102** (44): 15960–4. doi:10.1073/pnas.0508124102. PMC 1276104. PMID 16247010.

37.7   Further reading

- *The Immortal Cell*, by Michael D. West, Doubleday (2003) ISBN 978-0-385-50928-2

37.8   External links

- The Telomerase Database - A Web-based tool for telomerase research.

- Three-dimensional model of telomerase at MUN

- Elizabeth Blackburn's seminars: Telomeres and Telomerase

- Telomerase at the US National Library of Medicine Medical Subject Headings (MeSH)

- Puri, Neelu; Girard, Jennifer (September 23, 2013). "Novel Therapeutics Targeting Telomerase and Telomeres" (PDF). *Journal of Cancer Science & Therapy*. doi:10.4172/1948-5956.1000e127. Retrieved June 2015.

Chapter 38

RNA world

For the general discussion about the origin of life, see Abiogenesis.

The **RNA world** refers to the self-replicating ribonucleic acid (RNA) molecules thought to have been precursors to all current life on Earth.[1][2][3] RNA stores genetic information like DNA, and catalyzes chemical reactions like an enzyme protein. It may, therefore, have played a major role in the evolution of cellular life.

The RNA world would have eventually been replaced by the DNA, RNA and protein world of today, likely through an intermediate stage of ribonucleoprotein enzymes such as the ribosome and ribozymes, since proteins large enough to self-fold and have useful activities would only have come about after RNA was available to catalyze peptide ligation or amino acid polymerization.[4] DNA is thought to have taken over the role of data storage due to its increased stability,[5] while proteins, through a greater variety of monomers (amino acids), replaced RNA's role in specialized biocatalysis.

The hypothesis that current life on Earth descends from an RNA world has wide support,[6][7] although several alternatives are also researched and discussed, including the hypothesis that RNA and peptides emerged together.[8][9] It has also been suggested that an RNA world came after a still earlier form of life based on different biochemistry — a pre-RNA world.[10][11]

The RNA world hypothesis is supported by many independent lines of evidence, such as the observations that RNA is central to the translation process and that small RNAs can catalyze all of the chemical group and information transfers required for life.[11][12] The structure of the ribosome has been called the "smoking gun," as it showed that the ribosome is a ribozyme, with a central core of RNA and no amino acid side chains within 18 angstroms of the active site where peptide bond formation is catalyzed.[10] Many of the most critical components of cells (those that evolve the slowest) are composed mostly or entirely of RNA. Also, many critical cofactors (ATP, Acetyl-CoA, NADH, etc.) are either nucleotides or substances clearly related to them. This would mean that the RNA and nucleotide cofactors in modern cells are an evolutionary remnant of an RNA-based enzymatic system that preceded the protein-based one seen in all extant life.

38.1 History

One of the challenges in studying abiogenesis is that the system of reproduction and metabolism utilized by all extant life involves three distinct types of interdependent macromolecules (DNA, RNA, and protein). This suggests that life could not have arisen in its current form, and mechanisms have then been sought whereby the current system might have arisen from a simpler precursor system. The concept of RNA as a primordial molecule[4] can be found in papers by Francis Crick[13] and Leslie Orgel,[14] as well as in Carl Woese's 1967 book *The Genetic Code*.[15] In 1962 the molecular biologist Alexander Rich, of the Massachusetts Institute of Technology, had posited much the same idea in an article he contributed to a volume issued in honor of Nobel-laureate physiologist Albert Szent-Györgyi.[16] Hans Kuhn in 1972 laid out a possible process by which the modern genetic system might have arisen from a nucleotide-based precursor, and this led Harold White in 1976 to observe that many of the cofactors essential for enzymatic function are either nucleotides or could have been derived from nucleotides. He proposed that these nucleotide cofactors represent "fossils of nucleic acid

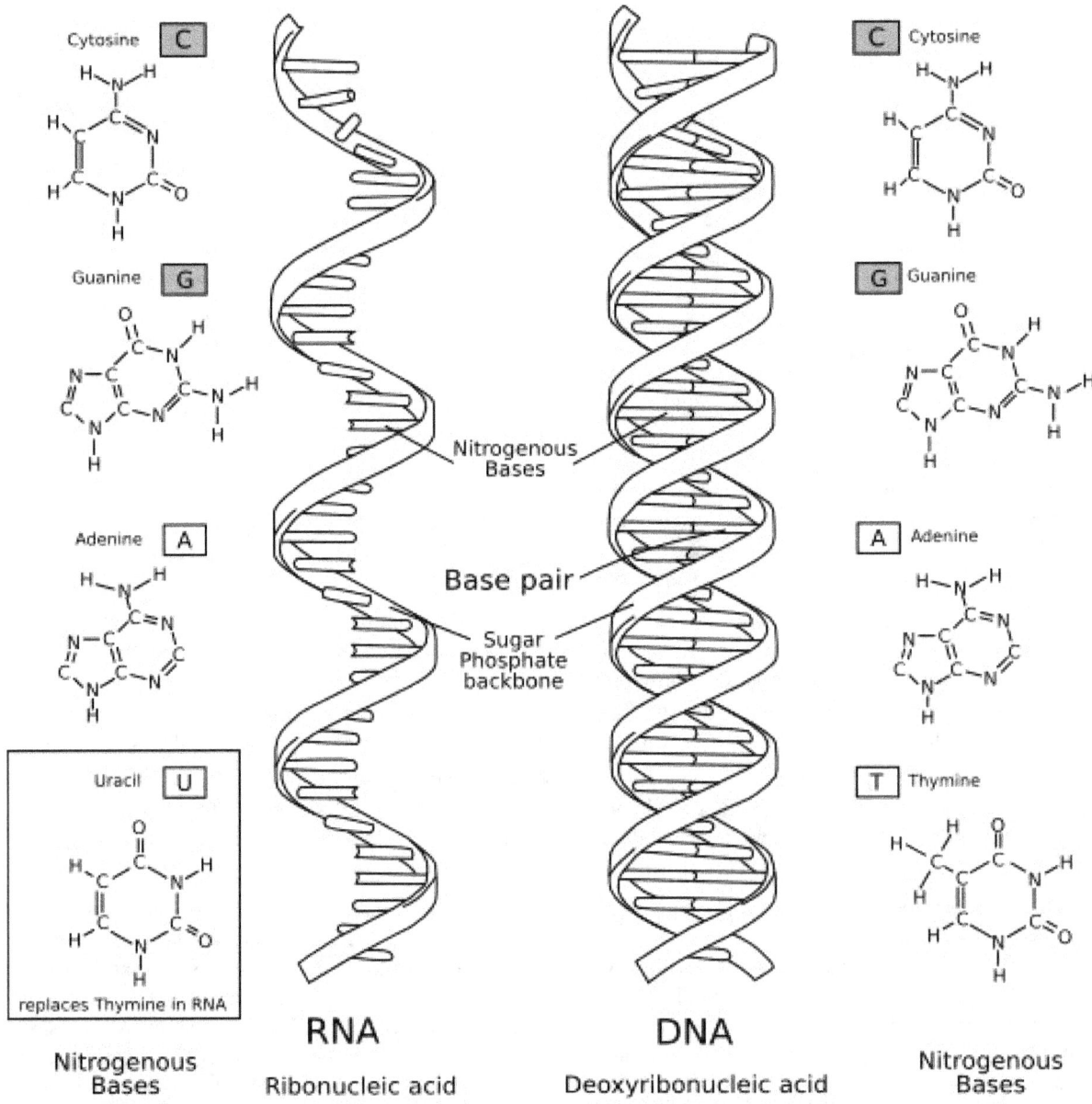

A comparison of RNA (left) with DNA (right), showing the helices and nucleobases each employs

enzymes".[17] The phrase "RNA World" was first used by Nobel laureate Walter Gilbert in 1986, in a commentary on how recent observations of the catalytic properties of various forms of RNA fit with this hypothesis.[18]

38.2 Properties of RNA

The properties of RNA make the idea of the RNA world hypothesis conceptually plausible, though its general acceptance as an explanation for the origin of life requires further evidence.[16] RNA is known to form efficient catalysts and its similarity to DNA makes its ability to store information clear. Opinions differ, however, as to whether RNA constituted the first autonomous self-replicating system or was a derivative of a still-earlier system.[4] One version of the hypothesis is that a different type of nucleic acid, termed *pre-RNA*, was the first one to emerge as a self-reproducing molecule, to be replaced by RNA only later. On the other hand, the recent finding that activated pyrimidine ribonucleotides can be synthesized under plausible prebiotic conditions[19] means that it is premature to dismiss the RNA-first scenarios.[4] Suggestions for 'simple' *pre-RNA* nucleic acids have included Peptide nucleic acid (PNA), Threose nucleic acid (TNA) or Glycol nucleic acid (GNA).[20][21] Despite their structural simplicity and possession of properties comparable with RNA, the chemically plausible generation of "simpler" nucleic acids under prebiotic conditions has yet to be demonstrated.[22]

38.2.1 RNA as an enzyme

Further information: Ribozyme

RNA enzymes, or ribozymes, are found in today's DNA-based life and could be examples of living fossils. Ribozymes play vital roles, such as those in the ribosome, which is vital for protein synthesis. Many other ribozyme functions exist; for example, the hammerhead ribozyme performs self-cleavage[23] and an RNA polymerase ribozyme can synthesize a short RNA strand from a primed RNA template.[24]

Among the enzymatic properties important for the beginning of life are:

- **Self-replication**. The ability to self-replicate, or synthesize other RNA molecules; relatively short RNA molecules that can synthesize others have been artificially produced in the lab. The shortest was 165-bases long, though it has been estimated that only part of the molecule was crucial for this function. One version, 189-bases long, had an error rate of just 1.1% per nucleotide when synthesizing an 11 nucleotide long RNA strand from primed template strands.[25] This 189 base pair ribozyme could polymerize a template of at most 14 nucleotides in length, which is too short for self replication, but a potential lead for further investigation. The longest primer extension performed by a ribozyme polymerase was 20 bases.[26]

- **Catalysis**. The ability to catalyze simple chemical reactions—which would enhance creation of molecules that are building blocks of RNA molecules (i.e., a strand of RNA which would make creating more strands of RNA easier). Relatively short RNA molecules with such abilities have been artificially formed in the lab.[27][28] A recent study showed that almost any nucleic acid can evolve into a catalytic sequence under appropriate selection. For instance, an arbitrarily chosen 50-nucleotide DNA fragment encoding for the *Bos taurus* (cattle) albumin mRNA was subjected to test-tube evolution to derive a catalytic DNA (DNAzyme) with RNA-cleavage activity. After only a few weeks, a DNAzyme with significant catalytic activity had evolved.[29] In general, DNA is much more chemically inert than RNA and hence much more resistant to obtaining catalytic properties. If in vitro evolution works for DNA it will happen much more easily with RNA.

- **Amino acid-RNA ligation**. The ability to conjugate an amino acid to the 3'-end of an RNA in order to use its chemical groups or provide a long-branched aliphatic side-chain.[30]

- **Peptide bond formation**. The ability to catalyse the formation of peptide bonds between amino acids to produce short peptides or longer proteins. This is done in modern cells by ribosomes, a complex of several RNA molecules known as rRNA together with many proteins. The rRNA molecules are thought responsible for its enzymatic activity, as no amino acid molecules lie within 18Å of the enzyme's active site,[16] and, when the majority of the amino acids in the ribosome were stringently removed, the resulting ribosome retained its full peptidyl transferase activity, fully able to catalyze the formation of peptide bonds between amino acids.[31] A much shorter RNA molecule has been synthesized in the laboratory with the ability to form peptide bonds, and it has been suggested that rRNA has evolved from a similar molecule.[32] It has also been suggested that amino acids may have initially been involved with RNA molecules as cofactors enhancing or diversifying their enzymatic capabilities, before evolving to more complex peptides. Similarly, tRNA is suggested to have evolved from RNA molecules that began to catalyze amino acid transfer.[33]

38.2.2 RNA in information storage

RNA is a very similar molecule to DNA, and only has two chemical differences. The overall structure of RNA and DNA are immensely similar—one strand of DNA and one of RNA can bind to form a double helical structure. This makes the storage of information in RNA possible in a very similar way to the storage of information in DNA. However RNA is less stable.

The major difference between RNA and DNA is the presence of a hydroxyl group at the 2'-position.

Comparison of DNA and RNA structure

Main articles: RNA and DNA

The major difference between RNA and DNA is the presence of a hydroxyl group at the 2'-position of the ribose sugar in RNA (*illustration, right*).[16] This group makes the molecule less stable because when not constrained in a double helix, the 2' hydroxyl can chemically attack the adjacent phosphodiester bond to cleave the phosphodiester backbone. The hydroxyl group also forces the ribose into the C3'-*endo* sugar conformation unlike the C2'-*endo* conformation of the deoxyribose sugar in DNA. This forces an RNA double helix to change from a B-DNA structure to one more closely resembling A-DNA.

RNA also uses a different set of bases than DNA—adenine, guanine, cytosine and uracil, instead of adenine, guanine, cytosine and thymine. Chemically, uracil is similar to thymine, differing only by a methyl group, and its production requires less energy.[34] In terms of base pairing, this has no effect. Adenine readily binds uracil or thymine. Uracil is, however, one product of damage to cytosine that makes RNA particularly susceptible to mutations that can replace a **GC** base pair with a **GU** (wobble) or **AU** base pair.

RNA is thought to have preceded DNA, because of their ordering in the biosynthetic pathways. The deoxyribonucleotides used to make DNA are made from ribonucleotides, the building blocks of RNA, by removing the 2'-hydroxyl group. As a consequence a cell must have the ability to make RNA before it can make DNA.

Limitations of information storage in RNA

The chemical properties of RNA make large RNA molecules inherently fragile, and they can easily be broken down into their constituent nucleotides through hydrolysis.[35][36] These limitations do not make use of RNA as an information storage system impossible, simply energy intensive (to repair or replace damaged RNA molecules) and prone to mutation. While this makes it unsuitable for current 'DNA optimised' life, it may have been acceptable for more primitive life.

38.2.3   RNA as a regulator

Main article: Riboswitch

Riboswitches have been found to act as regulators of gene expression, particularly in bacteria, but also in plants and archaea. Riboswitches alter their secondary structure in response to the binding of a metabolite. This change in structure can result in the formation or disruption of a terminator, truncating or permitting transcription respectively.[37] Alternatively, riboswitches may bind or occlude the Shine-Dalgarno sequence, affecting translation.[38] It has been suggested that these originated in an RNA-based world.[39] In addition, RNA thermometers regulate gene expression in response to temperature changes.[40]

38.3   Support and difficulties

The RNA world hypothesis is supported by RNA's ability to store, transmit, and duplicate genetic information, as DNA does. RNA can act as a ribozyme, a special type of enzyme. Because it can perform the tasks of both DNA and enzymes, RNA is believed to have once been capable of supporting independent life forms.[16] Some viruses use RNA as their genetic material, rather than DNA.[41] Further, while nucleotides were not found in Miller-Urey's origins of life experiments, their formation in prebiotically plausible conditions has now been reported, as noted above;[19] the purine base known as adenine is merely a pentamer of hydrogen cyanide. Experiments with basic ribozymes, like Bacteriophage Qβ RNA, have shown that simple self-replicating RNA structures can withstand even strong selective pressures (e.g., opposite-chirality chain terminators).[42]

Since there were no known chemical pathways for the abiogenic synthesis of nucleotides from pyrimidine nucleobases cytosine and uracil under prebiotic conditions, it is thought by some that nucleic acids did not contain these nucleobases seen in life's nucleic acids.[43] The nucleoside cytosine has a half-life in isolation of 19 days at 100 °C (212 °F) and 17,000 years in freezing water, which some argue is too short on the geologic time scale for accumulation.[44] Others have questioned whether ribose and other backbone sugars could be stable enough to find in the original genetic material,[45] and have raised the issue that all ribose molecules would have had to be the same enantiomer, as any nucleotide of the wrong chirality acts as a chain terminator.[46]

Pyrimidine ribonucleosides and their respective nucleotides have been prebiotically synthesised by a sequence of reactions that by-pass free sugars and assemble in a stepwise fashion by going against the dogma that nitrogenous and oxygenous chemistries should be avoided. In a series of publications, The *Sutherland Group* at the School of Chemistry, University of Manchester have demonstrated high yielding routes to cytidine and uridine ribonucleotides built from small 2 and 3 carbon fragments such as glycolaldehyde, glyceraldehyde or glyceraldehyde-3-phosphate, cyanamide and cyanoacetylene.

One of the steps in this sequence allows the isolation of enantiopure ribose aminooxazoline if the enantiomeric excess of glyceraldehyde is 60% or greater, of possible interest towards biological homochirality.[47] This can be viewed as a prebiotic purification step, where the said compound spontaneously crystallised out from a mixture of the other pentose aminooxazolines. Aminooxazolines can react with cyanoacetylene in a mild and highly efficient manner, controlled by inorganic phosphate, to give the cytidine ribonucleotides. Photoanomerization with UV light allows for inversion about the 1' anomeric centre to give the correct beta stereochemistry, one problem with this chemistry is the selective phosphorylation of alpha-cytidine at the 2' position.[48] However, in 2009 they showed that the same simple building blocks allow access, via phosphate controlled nucleobase elaboration, to 2',3'-cyclic pyrimidine nucleotides directly, which are known to be able to polymerise into RNA.[49] Organic chemist Donna Blackmond described this finding as "strong evidence" in favour of the RNA world.[50] However, John Sutherland said that while his team's work suggests that nucleic acids played an early and central role in the origin of life, it did not necessarily support the RNA world hypothesis in the strict sense, which he described as a "restrictive, hypothetical arrangement".[51]

The Sutherland group's 2009 paper also highlighted the possibility for the photo-sanitization of the pyrimidine-2',3'-cyclic phosphates.[49] A potential weakness of these routes is the generation of enantioenriched glyceraldehyde, or its 3-phosphate derivative (glyceraldehyde prefers to exist as its keto tautomer dihydroxyacetone).

On August 8, 2011, a report, based on NASA studies with meteorites found on Earth, was published suggesting building blocks of RNA (adenine, guanine and related organic molecules) may have been formed extraterrestrially in outer space.[52][53][54] On August 29, 2012, and in a world first, astronomers at Copenhagen University reported the detection of a specific sugar molecule, glycolaldehyde, in a distant star system. The molecule was found around the protostellar binary *IRAS 16293-2422*, which is located 400 light years from Earth.[55][56] Glycolaldehyde is needed to form ribonucleic acid, or RNA, which is similar in function to DNA. This finding suggests that complex organic molecules may form in stellar systems prior to the formation of planets, eventually arriving on young planets early in their formation.[57]

38.3.1 "Molecular biologist's dream"

"Molecular biologist's dream" is a phrase coined by Gerald Joyce and Leslie Orgel to refer to the problem of emergence of self-replicating RNA molecules, as any movement towards an RNA world on a properly modeled prebiotic early Earth would have been continuously suppressed by destructive reactions.[58] It was noted that many of the steps needed for the nucleotides formation do not proceed efficiently in prebiotic conditions.[59] Joyce and Orgel specifically referred the molecular biologist's dream to "a magic catalyst" that could "convert the activated nucleotides to a random ensemble of polynucleotide sequences, a subset of which had the ability to replicate".[58]

Joyce and Orgel further argued that nucleotides cannot link unless there is some activation of the phosphate group, whereas the only effective activating groups for this are "totally implausible in any prebiotic scenario", particularly adenosine triphosphate.[58] According to Joyce and Orgel, in case of the phosphate group activation, the basic polymer product would have 5',5'-pyrophosphate linkages, while the 3',5'-phosphodiester linkages, which are present in all known RNA, would be much less abundant.[58] The associated molecules would have been also prone to addition of incorrect nucleotides or to reactions with numerous other substances likely to have been present.[58] The RNA molecules would have been also continuously degraded by such destructive process as spontaneous hydrolysis, present on the early Earth.[58] Joyce and Orgel proposed to reject "the myth of a self-replicating RNA molecule that arose *de novo* from a soup of random polynucleotides"[58] and hypothesised a scenario where the prebiotic processes furnish pools of enantiopure beta-D-ribonucleosides.[60]

38.4   Prebiotic RNA synthesis

Nucleotides are the fundamental molecules that combine in series to form RNA. They consist of a nitrogenous base attached to a sugar-phosphate backbone. RNA is made of long stretches of specific nucleotides arranged so that their sequence of bases carries information. The RNA world hypothesis holds that in the primordial soup (or sandwich), there existed free-floating nucleotides. These nucleotides regularly formed bonds with one another, which often broke because the change in energy was so low. However, certain sequences of base pairs have catalytic properties that lower the energy of their chain being created, enabling them to stay together for longer periods of time. As each chain grew longer, it attracted more matching nucleotides faster, causing chains to now form faster than they were breaking down.

These chains have been proposed by some as the first, primitive forms of life.[61] In an RNA world, different sets of RNA strands would have had different replication outputs, which would have increased or decreased their frequency in the population, i.e. natural selection. As the fittest sets of RNA molecules expanded their numbers, novel catalytic properties added by mutation, which benefitted their persistence and expansion, could accumulate in the population. Such an autocatalytic set of ribozymes, capable of self replication in about an hour, has been identified. It was produced by molecular competition (*in vitro* evolution) of candidate enzyme mixtures.[62]

Competition between RNA may have favored the emergence of cooperation between different RNA chains, opening the way for the formation of the first protocell. Eventually, RNA chains developed with catalytic properties that help amino acids bind together (a process called peptide-bonding). These amino acids could then assist with RNA synthesis, giving those RNA chains that could serve as ribozymes the selective advantage. The ability to catalyze one step in protein synthesis, aminoacylation of RNA, has been demonstrated in a short (five-nucleotide) segment of RNA.[63]

One of the problems with the RNA world hypothesis is to discover the pathway by which RNA became upgraded to the DNA system. Ken Stedman of Portland State University in Oregon, may have found the solution. While filtering virus-sized particles from a hot acidic lake in Lassen Volcanic National Park, California, he discovered 400,000 pieces of viral DNA. Some of these, however, contained a protein coat of reverse transcriptase enzyme normally associated with RNA based retroviruses. This lack of respect for biochemical boundaries virologists like Luis Villareal of the University of California Irvine believe would have been a characteristic of a pre RNA virus world up to 4 billion years ago.[64][65] This finding bolsters the argument for the transfer of information from the RNA world to the emerging DNA world before the emergence of the Last Universal Common Ancestor. From the research, the diversity of this virus world is still with us.

In March 2015, NASA scientists reported that, for the first time, complex DNA and RNA organic compounds of life, including uracil, cytosine and thymine, have been formed in the laboratory under conditions found only in outer space, using starting chemicals, like pyrimidine, found in meteorites. Pyrimidine, like polycyclic aromatic hydrocarbons (PAHs), the most carbon-rich chemical found in the Universe, may have been formed in giant red stars or in interstellar dust and gas clouds, according to the scientists.[66]

38.5 Viroids

Additional evidence supporting the concept of an RNA world has resulted from research on viroids, the first representatives of a novel domain of "subviral pathogens."[67][68] Viroids are mostly plant pathogens, which consist of short stretches (a few hundred nucleobases) of highly complementary, circular, single-stranded, and non-coding RNA without a protein coat. Compared with other infectious plant pathogens, viroids are extremely small in size, ranging from 246 to 467 nucleobases. In comparison, the genome of the smallest known viruses capable of causing an infection are about 2,000 nucleobases long.[69]

In 1989, Diener proposed that, based on their characteristic properties, viroids are more plausible "living relics" of the RNA world than are introns or other RNAs then so considered.[70] If so, viroids have attained potential significance beyond plant pathology to evolutionary biology, by representing the most plausible macromolecules known capable of explaining crucial intermediate steps in the evolution of life from inanimate matter (see: abiogenesis).

Apparently, Diener's hypothesis lay dormant until 2014, when Flores et al. published a review paper, in which Diener's evidence supporting his hypothesis was summarized.[71] In the same year, a New York Times science writer published a popularized version of Diener's proposal, in which, however, he mistakenly credited Flores et al. with the hypothesis' original conception.[72]

Pertinent viroid properties listed in 1989 are:

1. their small size, imposed by error-prone replication; 2. their high guanine and cytosine content, which increases stability and replication fidelity; 3. their circular structure, which assures complete replication without genomic tags; 4. existence of structural periodicity, which permits modular assembly into enlarged genomes; 5. their lack of protein-coding ability, consistent with a ribosome-free habitat; and 6. replication mediated in some by ribozymes—the fingerprint of the RNA world.[71]

The existence, in extant cells, of RNAs with molecular properties predicted for RNAs of the RNA World constitutes an additional argument supporting the RNA World hypothesis.

38.6   Origin of sex

Eigen et al.[73] and Woese[74] proposed that the genomes of early protocells were composed of single-stranded RNA, and that individual genes corresponded to separate RNA segments, rather than being linked end-to-end as in present day DNA genomes. A protocell that was haploid (one copy of each RNA gene) would be vulnerable to damage, since a single lesion in any RNA segment would be potentially lethal to the protocell (e.g. by blocking replication or inhibiting the function of an essential gene).

Vulnerability to damage could be reduced by maintaining two or more copies of each RNA segment in each protocell, i.e. by maintaining diploidy or polyploidy. Genome redundancy would allow a damaged RNA segment to be replaced by an additional replication of its homolog. However for such a simple organism, the proportion of available resources tied up in the genetic material would be a large fraction of the total resource budget. Under limited resource conditions, the protocell reproductive rate would likely be inversely related to ploidy number. The protocell's fitness would be reduced by the costs of redundancy. Consequently, coping with damaged RNA genes while minimizing the costs of redundancy would likely have been a fundamental problem for early protocells.

A cost-benefit analysis was carried out in which the costs of maintaining redundancy were balanced against the costs of genome damage.[75] This analysis led to the conclusion that, under a wide range of circumstances, the selected strategy would be for each protocell to be haploid, but to periodically fuse with another haploid protocell to form a transient diploid. The retention of the haploid state maximizes the growth rate. The periodic fusions permit mutual reactivation of otherwise lethally damaged protocells. If at least one damage-free copy of each RNA gene is present in the transient diploid, viable progeny can be formed. For two, rather than one, viable daughter cells to be produced would require an extra replication of the intact RNA gene homologous to any RNA gene that had been damaged prior to the division of the fused protocell. The cycle of haploid reproduction, with occasional fusion to a transient diploid state, followed by splitting to the haploid state, can be considered to be the sexual cycle in its most primitive form.[75][76] In the absence of this sexual cycle, haploid protocells with a damage in an essential RNA gene would simply die.

This model for the early sexual cycle is hypothetical, but it is very similar to the known sexual behavior of the segmented RNA viruses, which are among the simplest organisms known. Influenza virus, whose genome consists of 8 physically separated single-stranded RNA segments,[77] is an example of this type of virus. In segmented RNA viruses, "mating" can occur when a host cell is infected by at least two virus particles. If these viruses each contain an RNA segment with a lethal damage, multiple infection can lead to reactivation providing that at least one undamaged copy of each virus gene is present in the infected cell. This phenomenon is known as "multiplicity reactivation". Multiplicity reactivation has been reported to occur in influenza virus infections after induction of RNA damage by UV-irradiation,[78] and ionizing radiation.[79]

38.7   Further developments

Patrick Forterre has been working on a novel hypothesis, called "three viruses, three domains":[80] that viruses were instrumental in the transition from RNA to DNA and the evolution of Bacteria, Archaea, and Eukaryota. He believes the last common ancestor (specifically, the "last universal cellular ancestor")[80] was RNA-based and evolved RNA viruses. Some of the viruses evolved into DNA viruses to protect their genes from attack. Through the process of viral infection into hosts the three domains of life evolved.[80][81] Another interesting proposal is the idea that RNA synthesis might have been driven by temperature gradients, in the process of thermosynthesis.[82] Single nucleotides have been shown to catalyze organic reactions.[83]

Steven Benner has argued that chemical conditions on the planet Mars, such as the presence of boron, molybdenum and oxygen, may have been better for initially producing RNA molecules than those on Earth. If so, life-suitable molecules, originating on Mars, may have later migrated to Earth via panspermia or similar process.[2][3]

38.8 Alternative hypotheses

The hypothesized existence of an RNA world does not exclude a "Pre-RNA world", where a metabolic system based on a different nucleic acid is proposed to pre-date RNA. A candidate nucleic acid is peptide nucleic acid (PNA), which uses simple peptide bonds to link nucleobases.[84] PNA is more stable than RNA, but its ability to be generated under prebiological conditions has yet to be demonstrated experimentally.

Threose nucleic acid (TNA) has also been proposed as a starting point, as has glycol nucleic acid (GNA), and like PNA, also lack experimental evidence for their respective abiogenesis.

An alternative — or complementary — theory of RNA origin is proposed in the PAH world hypothesis, whereby polycyclic aromatic hydrocarbons (PAHs) mediate the synthesis of RNA molecules.[85] PAHs are the most common and abundant of the known polyatomic molecules in the visible Universe, and are a likely constituent of the primordial sea.[86] PAHs, along with fullerenes (also implicated in the origin of life),[87] have been recently detected in nebulae.[88]

The iron-sulfur world theory proposes that simple metabolic processes developed before genetic materials did, and these energy-producing cycles catalyzed the production of genes.

Some of the difficulties of producing the precursors on earth are bypassed by another alternative or complementary theory for their origin, panspermia. It discusses the possibility that the earliest life on this planet was carried here from somewhere else in the galaxy, possibly on meteorites similar to the Murchison meteorite.[89] This does not invalidate the concept of an RNA world, but posits that this world or its precursors originated not on Earth but rather another, probably older, planet.

There are hypotheses that are in direct conflict to the RNA world hypothesis. The relative chemical complexity of the nucleotide and the unlikelihood of it spontaneously arising, along with the limited number of combinations possible among four base forms, as well as the need for RNA polymers of some length before seeing enzymatic activity, have led some to reject the RNA world hypothesis in favor of a metabolism-first hypothesis, where the chemistry underlying cellular function arose first, along with the ability to replicate and facilitate this metabolism.

38.8.1 RNA-peptide coevolution

Another proposal is that the dual-molecule system we see today, where a nucleotide-based molecule is needed to synthesize protein, and a protein-based molecule is needed to make nucleic acid polymers, represents the original form of life.[90] This theory is called RNA-peptide coevolution,[8] or the Peptide-RNA world, and offers a possible explanation for the rapid evolution of high-quality replication in RNA (since proteins are catalysts), with the disadvantage of having to postulate the formation of two complex molecules, an enzyme (from peptides) and a RNA (from nucleotides). In this Peptide-RNA World scenario, RNA would have contained the instructions for life, while peptides (simple protein enzymes) would have accelerated key chemical reactions to carry out those instructions.[91] The study leaves open the question of exactly how those primitive systems managed to replicate themselves — something neither the RNA World hypothesis nor the Peptide-RNA World theory can yet explain, unless polymerases (enzymes that rapidly assemble the RNA molecule) played a role.[91]

A research project completed in March 2015 by the Sutherland group found that a network of reactions beginning with hydrogen cyanide and hydrogen sulfide, in streams of water irradiated by UV light, could produce the chemical components of proteins and lipids, alongside those of RNA.[91][92] The researchers used the term "cyanosulfidic" to describe this network of reactions.[9]

38.9 Implications of the RNA world

The RNA world hypothesis, if true, has important implications for the definition of life. For most of the time that followed Watson and Crick's elucidation of DNA structure in 1953, life was largely defined in terms of DNA and proteins: DNA and proteins seemed the dominant macromolecules in the living cell, with RNA only aiding in creating proteins from the DNA blueprint.

The RNA world hypothesis places RNA at center-stage when life originated. This has been accompanied by many studies in the last ten years that demonstrate important aspects of RNA function not previously known—and supports the idea of a critical role for RNA in the mechanisms of life. The RNA world hypothesis is supported by the observations that ribosomes are ribozymes: the catalytic site is composed of RNA, and proteins hold no major structural role and are of peripheral functional importance. This was confirmed with the deciphering of the 3-dimensional structure of the ribosome in 2001. Specifically, peptide bond formation, the reaction that binds amino acids together into proteins, is now known to be catalyzed by an adenine residue in the rRNA.

Other interesting discoveries demonstrate a role for RNA beyond a simple message or transfer molecule.[93] These include the importance of small nuclear ribonucleoproteins (snRNPs) in the processing of pre-mRNA and RNA editing, RNA interference (RNAi), and reverse transcription from RNA in eukaryotes in the maintenance of telomeres in the telomerase reaction.[94]

38.10   See also

- GADV-protein world hypothesis
- *The Major Transitions in Evolution*

38.11   References

[1] Zimmer, Carl (September 25, 2014). "A Tiny Emissary From the Ancient Past". *New York Times*. Retrieved September 26, 2014.

[2] Zimmer, Carl (September 12, 2013). "A Far-Flung Possibility for the Origin of Life". *New York Times*. Retrieved September 12, 2013.

[3] Webb, Richard (August 29, 2013). "Primordial broth of life was a dry Martian cup-a-soup". *New Scientist*. Retrieved September 13, 2013.

[4] Cech, T.R. (2011). The RNA Worlds in Context. Source: Department of Chemistry and Biochemistry, University of Colorado, Boulder, Colorado 80309-0215. Cold Spring Harb Perspect Biol. 2011 Feb 16. pii: cshperspect.a006742v1. doi:10.1101/cshperspect.a006742.[Epub ahead of print]

[5] Garwood, Russell J. (2012). "Patterns In Palaeontology: The first 3 billion years of evolution". *Palaeontology Online* **2** (11): 1–14. Retrieved June 25, 2015.

[6] Wade, Nicholas (May 4, 2015). "Making Sense of the Chemistry That Led to Life on Earth". *New York Times*. Retrieved May 10, 2015.

[7]   • Copley SD, Smith E, Morowitz HJ (2007). "The origin of the RNA world: co-evolution of genes and metabolism.". *Bioorg Chem* **35** (6): 430–43. doi:10.1016/j.bioorg.2007.08.001. PMID 17897696. "The proposal that life on Earth arose from an RNA World is widely accepted."
  • Orgel LE (2003). "Some consequences of the RNA world hypothesis.". *Orig Life Evol Biosph* **33** (2): 211–8. PMID 12967268. "It now seems very likely that our familiar DNA/RNA/protein world was preceded by an RNA world"
  • Robertson MP, Joyce GF (2012). "The origins of the RNA world". *Cold Spring Harb Perspect Biol* **4** (5): a003608. doi:10.1101/cshperspect.a003608. PMC 3331698. PMID 20739415. "There is now strong evidence indicating that an RNA World did indeed exist before DNA- and protein-based life."
  • Neveu M, Kim HJ, Benner SA (2013). "The "strong" RNA world hypothesis: fifty years old.". *Astrobiology* **13** (4): 391–403. doi:10.1089/ast.2012.0868. PMID 23551238."[The RNA world's existence] has broad support within the community today."

[8] Pascal, Robert (2007), "A scenario starting from the first chemical building blocks", in Reisse, Jacques, *From Suns to Life: A Chronological Approach to the History of Life on Earth*, Springer Science & Business Media, pp. 163–166, ISBN 0-387-45083-1

[9] Patel, Bhavesh H.; Percivalle, Claudia; Ritson, Dougal J.; Duffy, Colm D.; Sutherland, John D. (April 2015). "Common origins of RNA, protein and lipid precursors in a cyanosulfidic protometabolism". *Nature Chemistry* (London: Nature Publishing Group) **7** (4): 301–307. Bibcode:2015NatCh...7..301P. doi:10.1038/nchem.2202. ISSN 1755-4330. PMID 25803468. Retrieved 2015-07-22.

[10] Robertson MP, Joyce GF (2012). "The origins of the RNA world". *Cold Spring Harb Perspect Biol*4(5): a003608. doi:10.1101/cs PMC 3331698. PMID 20739415.

[11] Cech TR (2012). "The RNA worlds in context.". *Cold Spring Harb Perspect Biol* 4 (7): a006742. doi:10.1101/cshperspect.a006742. PMC 3385955. PMID 21441585.

[12] Yarus M (2011). "Getting past the RNA world: the initial Darwinian ancestor". *Cold Spring Harb Perspect Biol* 3 (4): a003590. doi:10.1101/cshperspect.a003590. PMC 3062219. PMID 20719875.

[13] Crick FH (1968). "The origin of the genetic code". *J Mol Biol* **38** (3): 367–379. doi:10.1016/0022-2836(68)90392-6. PMID 4887876.

[14] Orgel LE (1968). "Evolution of the genetic apparatus". *J Mol Biol* **38** (3): 381–393. doi:10.1016/0022-2836(68)90393-8. PMID 5718557.

[15] Woese C.R. (1967). The genetic code: The molecular basis for genetic expression. p. 186. Harper & Row

[16] Atkins, John F.; Gesteland, Raymond F.; Cech, Thomas (2006). *The RNA world: the nature of modern RNA suggests a prebiotic RNA world*. Plainview, N.Y: Cold Spring Harbor Laboratory Press. ISBN 0-87969-739-3.

[17] White, HB III (1976). "Coenzymes as Fossils of an Earlier Metabolic State". *J Mol Evol*7(2): 101–104. doi:10.1007/BF017324 PMID 1263263.

[18] Gilbert, Walter(February 1986). "The RNA World". *Nature*319(6055): 618. Bibcode:1986Natur.319..618G.doi:10.1038/3

[19] Powner M.W., Gerland B, Sutherland J.D. (2009). "Synthesis of activated pyrimidine ribonucleotides in prebiotically plausible conditions". *Nature* **459** (7244): 239–242. Bibcode:2009Natur.459..239P. doi:10.1038/nature08013. PMID 19444213.

[20] Orgel, Leslie (November 2000). "A Simpler Nucleic Acid". *Science* **290** (5495): 1306–7. doi:10.1126/science.290.5495.1306. PMID 11185405.

[21] Nelson, K.E.; Levy, M.; Miller, S.L. (April 2000). "Peptide nucleic acids rather than RNA may have been the first genetic molecule". *Proc. Natl. Acad. Sci. USA* **97** (8): 3868–71. Bibcode:2000PNAS...97.3868N. doi:10.1073/pnas.97.8.3868. PMC 18108. PMID 10760258.

[22] Sutherland, J.D; Anastasi, C., Buchet F.F, Crower M.A, Parkes A.L, Powner M. W., Smith J.M. (April 2007). "RNA: Prebiotic Product, or Biotic Invention". *Chemistry & Biodiversity* **4** (4): 721–739. doi:10.1002/cbdv.200790060. PMID 17443885.

[23] Forster AC, Symons RH (1987). "Self-cleavage of plus and minus RNAs of a virusoid and a structural model for the active sites". *Cell* **49** (2): 211–220. doi:10.1016/0092-8674(87)90562-9. PMID 2436805.

[24] Johnston W, Unrau P, Lawrence M, Glasner M, Bartel D (2001). "RNA-catalyzed RNA polymerization: accurate and general RNA-templated primer extension" (PDF). *Science* **292** (5520): 1319–25. Bibcode:2001Sci...292.1319J. doi:10.1126/science.1 060786.PMID11358999.

[25] Johnston, W. K.; Unrau, P. J.; Lawrence, M. S.; Glasner, M. E.; Bartel, D. P. (2001). "RNA-Catalyzed RNA Polymerization: Accurate and General RNA-Templated Primer Extension". *Science* **292** (5520): 1319–25. doi:10.1126/science.1060786. PMID 11358999.

[26] Hani S. Zaher and Peter J. Unrau, Selection of an improved RNA polymerase ribozyme with superior extension and fidelity. RNA (2007), 13:1017-1026

[27] Huang, Yang, and Yarus, RNA enzymes with two small-molecule substrates. Chemistry & Biology, Vol 5, 669-678, November 1998

[28] Unrau, P. J.; Bartel, D. P. (1998). "RNA-catalysed nucleotide synthesis". *Nature*395(6699): 260–263. Bibcode:1998Natur.3 doi:10.1038/26193. PMID 9751052.

[29] Gysbers, R; Tram, K; Gu, J; Li, Y (2015). "Evolution of an Enzyme from a Noncatalytic Nucleic Acid Sequence". *Scientific Reports* **5**: 11405. doi:10.1038/srep11405. PMC 4473686. PMID 26091540.

[30] Erives A (2011). "A Model of Proto-Anti-Codon RNA Enzymes Requiring L-Amino Acid Homochirality". *J Molecular Evolution* **73** (1–2): 10–22. doi:10.1007/s00239-011-9453-4. PMC 3223571. PMID 21779963.

[31] Noller, H. F. , V. Hoffarth, and L. Zimniak (1992). Unusual resistance of peptidyl transferase to protein extraction procedures. Science 256:1416-1419.

[32] Zhang, Biliang; Cech, Thomas R. (1997). "Peptide bond formation by *in vitro* selected ribozymes". *Nature* **390** (6655): 96–100. Bibcode:1997Natur.390...96Z. doi:10.1038/36375. PMID 9363898.

[33] Szathmary, E. (1999). "The origin of the genetic code: amino acids as cofactors in an RNA world". *Trends in Genetics* **15** (6): 223–229. doi:10.1016/S0168-9525(99)01730-8. PMID 10354582.

[34] http://www.humpath.com/uracil

[35] Lindahl, T (April 1993). "Instability and decay of the primary structure of DNA". *Nature***362**(6422): 709–15. Bibcode:1993 doi:10.1038/362709a0. PMID 8469282.

[36] Pääbo, S (November 1993). "Ancient DNA". *Scientific American* **269** (5): 60–66. doi:10.1038/scientificamerican1193-86.

[37] Nudler E, Mironov AS (2004). "The riboswitch control of bacterial metabolism". *Trends Biochem Sci***29**(1): 11–7. doi:10.101 PMID 14729327.

[38] Tucker BJ, Breaker RR (2005). "Riboswitches as versatile gene control elements". *Current Opinion in Structural Biology* **15** (3): 342–8. doi:10.1016/j.sbi.2005.05.003. PMID 15919195.

[39] Switching the light on plant riboswitches. Samuel Bocobza and Asaph Aharoni Trends in Plant Science Volume 13, Issue 10, October 2008, Pages 526-533 doi:10.1016/j.tplants.2008.07.004 PMID 18778966

[40] Narberhaus F, Waldminghaus T, Chowdhury S (January 2006). "RNA thermometers". *FEMS Microbiol. Rev.* **30** (1): 3–16. doi:10.1111/j.1574-6976.2005.004.x. PMID 16438677. Retrieved 2011-04-23.

[41] Patton, John T. Editor (2008). Segmented Double-stranded RNA Viruses: Structure and Molecular Biology. Caister Academic Press. Editor's affiliation: Laboratory of Infectious Diseases, NIAID, NIH, Bethesda, MD 20892-8026. ISBN 978-1-904455-21-9

[42] Bell, Graham: The Basics of Selection. Springer, 1997.

[43] Orgel, L. (1994). "The origin of life on earth". *Scientific American* **271** (4): 81. doi:10.1038/scientificamerican1094-76. PMID 7524147.

[44] Levy, Matthew; Miller, Stanley L. (1998). "The stability of the RNA bases: Implications for the origin of life". *PNAS* **95** (14): 7933–7938. Bibcode:1998PNAS...95.7933L. doi:10.1073/pnas.95.14.7933. PMC 20907. PMID 9653118.

[45] Larralde, R.; Robertson, M. P.; Miller, S. L. (1995). "Rates of decomposition of ribose and other sugars: implications for chemical evolution". *PNAS* **92** (18): 8158–8160. Bibcode:1995PNAS...92.8158L. doi:10.1073/pnas.92.18.8158. PMC 41115. PMID 7667262.

[46] Joyce GF; et al. (1984). "Chiral selection in poly(C)-directed synthesis of oligo(G)". *Nature***310**(5978): 602–604. Bibcode:198 doi:10.1038/310602a0. PMID 6462250.

[47] Direct Assembly of Nucleoside Precursors from Two- and Three-Carbon Units Carole Anastasi, Michael A. Crowe, Matthew W. Powner, John D. Sutherland Angewandte Chemie International Edition Volume 45, Issue 37 , Pages 6176–79

[48] Potentially Prebiotic Synthesis of Pyrimidine β-D-Ribonucleotides by Photoanomerization/Hydrolysis of α-D-Cytidine-2'-Phosphate Matthew W. Powner, John D. Sutherland ChemBioChem Volume 9, Issue 15 , Pages 2386–87

[49] Powner MW, Gerland B, Sutherland JD (2009). "Synthesis of activated pyrimidine ribonucleotides in prebiotically plausible conditions". *Nature* **459** (7244): 239–242. Bibcode:2009Natur.459..239P. doi:10.1038/nature08013. PMID 19444213.

[50] Van Noorden R (2009). "RNA world easier to make". *Nature*. doi:10.1038/news.2009.471.

[51] Urquhart, James (13 May 2009), "Insight into RNA origins", *Chemistry World* (Royal Society of Chemistry)

[52] Callahan, M.P.; Smith, K.E.; Cleaves, H.J.; Ruzicka, J.; Stern, J.C.; Glavin, D.P.; House, C.H.; Dworkin, J.P. (11 August 2011). "Carbonaceous meteorites contain a wide range of extraterrestrial nucleobases". *Proceedings of the National Academy of Sciences* (PNAS) **108** (34): 13995–8. Bibcode:2011PNAS..10813995C. doi:10.1073/pnas.1106493108. PMC 3161613. PMID 21836052. Retrieved 2011-08-15.

[53] Steigerwald, John (8 August 2011). "NASA Researchers: DNA Building Blocks Can Be Made in Space". NASA. Retrieved 2011-08-10.

[54] ScienceDaily Staff (9 August 2011). "DNA Building Blocks Can Be Made in Space, NASA Evidence Suggests". ScienceDaily. Retrieved 2011-08-09.

[55] Than, Ker (August 29, 2012). "Sugar Found In Space". *National Geographic*. Retrieved August 31, 2012.

[56] Staff (August 29, 2012). "Sweet! Astronomers spot sugar molecule near star". AP News. Retrieved August 31, 2012.

[57] Jørgensen, J. K.; Favre, C.; Bisschop, S.; Bourke, T.; Dishoeck, E.; Schmalzl, M. (2012). "Detection of the simplest sugar, glyco-laldehyde, in a solar-type protostar with ALMA" (PDF). *The Astrophysical Journal Letters*. eprint **757**: L4. Bibcode:2012ApJ.. .757L...4J.doi:10.1088/2041-8205/757/1/L4.

[58] Gordon C. Mills, Dean Kenyon. "The RNA World: A Critique". Access Research Network. Retrieved 10 Sep 2011.

[59] Schopf, J. William (2002). *Life's origin: the beginnings of biological evolution*. University of California Press. p. 150. ISBN 0-520-23390-5.

[60] "Prebiotic RNA chemistry: realising the molecular biologist's dream". Engineering and Physical Sciences Research Council. Retrieved 10 Sep 2011.

[61] Villarreal LP, Witzany G (2013). The DNA Habitat and its RNA Inhabitants: At the Dawn of RNA Sociology 6: 1-12. doi:10.4137/GEI.S11490

[62] Lincoln, Tracey A.; Joyce, Gerald F. (January 8, 2009). "Self-Sustained Replication of an RNA Enzyme". *Science* (New York: American Association for the Advancement of Science) **323** (5918): 1229–32. Bibcode:2009Sci...323.1229L. doi:10.1126/scie nce.1167856.PMC2652413. PMID 19131595. Retrieved2009-01-13. Lay summary–*Medical News Today*(January12,2009).

[63] Rebecca M. Turk, Nataliya V. Chumachenko, and Michael Yarus (February 22, 2010). "Multiple translational products from a five-nucleotide ribozyme". *Proceedings of the National Academy of Sciences* **107** (10): 4585–89. Bibcode:2010PNAS..107.4585T. doi:10.1073/pnas.0912895107. ISSN 1091-6490. PMC 2826339. PMID 20176971. Lay summary – *ScienceDaily* (February 24, 2010).

[64] Holmes, Bob (2012) "First Glimpse at the birth of DNA" (New Scientist April 12, 2012)

[65] Diemer, Geoffrey; Stedman, Kenneth (19 April 2012). "A novel virus genome discovered in an extreme environment suggests recombination between unrelated groups of RNA and DNA viruses". *Biology Direct* (BioMed Central) **7** (1): 14. doi:10.1186/17 45-6150-7-13. Retrieved7February2015.

[66] Marlaire, Ruth (3 March 2015). "NASA Ames Reproduces the Building Blocks of Life in Laboratory". *NASA*. Retrieved 5 March 2015.

[67] Diener TO (August 1971). "Potato spindle tuber "virus". IV. A replicating, low molecular weight RNA". *Virology* **45** (2): 411–28. doi:10.1016/0042-6822(71)90342-4. PMID 5095900.

[68] "ARS Research Timeline – Tracking the Elusive Viroid". 2006-03-02. Retrieved 2007-07-18.

[69] Sänger first1=HL; Klotz, G; Riesner, D; Gross, HJ; Kleinschmidt, AK (1979). "viroids are single-stranded, covalently closed, circular RNA molecules, existing as highly base-paired rod-like structures". *Proc.Natl.Acad.Sci.USA* **73** (11): 3852–56.

[70] Diener, T.O.(1989). "Circular RNAs: Relics of precellular evolution?" Proc.Natl.Acad.Sci.USA **86** (23): 9370–9374. Bibcode: 1989PNAS...86.9370D. doi:10.1073/pnas.86.23.9370. PMC 298497. PMID 2480600. Retrieved November 1, 2014.

[71] Flores, R., Gago-Zachert, S., Serra, P., Sanjuan, R., Elena, S.F. "Viroids: Survivors from the RNA World?" Ann.Rev.Microbiol. **68**: 395–41, 2014. Retrieved November 1, 2014

[72] Zimmer, Carl (September 25, 2014). "A Tiny Emissary From the Ancient Past. url=http://www.nytimes.com/2014/09/25/ science/a-tiny-emissary-from-the-ancient-past.html?partner=rss&emc=rss". *New York Times*.

[73] Eigen M, Gardiner W, Schuster P, Winkler-Oswatitsch R (April 1981). "The origin of genetic information". *Sci. Am.* **244** (4): 88–92, 96, et passim. Bibcode:1981SciAm.244...88H. doi:10.1038/scientificamerican0481-88. PMID 6164094.

[74] Woese CR (1983). The primary lines of descent and the universal ancestor. Chapter in Bendall, D. S. (1983). *Evolution from molecules to men*. Cambridge, UK: Cambridge University Press. ISBN 0-521-28933-5. pp. 209-233.

[75] Bernstein H, Byerly HC, Hopf FA, Michod RE (October 1984). "Origin of sex". *J. Theor. Biol.* **110**(3): 323–51. doi:10.1016/5193(84)80178-2. PMID 6209512.

[76] Bernstein, Carol; Bernstein, Harris (1991). *Aging, sex, and DNA repair*. Boston: Academic Press. ISBN 0-12-092860-4. see pgs. 293-297

[77] Lamb RA, Choppin PW (1983). "The gene structure and replication of influenza virus". *Annu. Rev. Biochem.* **52**: 467–506. doi:10.1146/annurev.bi.52.070183.002343. PMID 6351727.

[78] BARRY RD (August 1961). "The multiplication of influenza virus. II. Multiplicity reactivation of ultraviolet irradiated virus". *Virology* **14** (4): 398–405. doi:10.1016/0042-6822(61)90330-0. PMID 13687359.

[79] Gilker JC, Pavilanis V, Ghys R (June 1967). "Multiplicity reactivation in gamma irradiated influenza viruses". *Nature* **214** (5094): 1235–57. Bibcode:1967Natur.214.1235G. doi:10.1038/2141235a0. PMID 6066111.

[80] Forterre, Patrick. "Three RNA cells for ribosomal lineages and three DNA viruses to replicate their genomes: A hypothesis for the origin of cellular domain"

[81] Zimmer C. (2006). "Did DNA come from viruses?". *Science* **312** (5775): 870–2. doi:10.1126/science.312.5775.870. PMID 16690855.

[82] Anthonie W.J. Muller (2005). "Thermosynthesis as energy source for the RNA World: a model for the bioenergetics of the origin of life". *Biosystems* **82** (1): 93–102. doi:10.1016/j.biosystems.2005.06.003. PMID 16024164.

[83] Kumar, Atul; Siddharth Sharma; Ram Awatar Maurya (2010). "Single Nucleotide-Catalyzed Biomimetic Reductive Amination". *Advanced Synthesis and Catalyst* **352** (13): 2227. doi:10.1002/adsc.201000178.

[84] Egholm M, Buchardt O, Christensen L, Behrens C, Freier SM, Driver DA, Berg RH, Kim SK, Nordén B, and Nielsen PE (1993). "PNA Hybridizes to Complementary Oligonucleotides Obeying the Watson-Crick Hydrogen Bonding Rules". *Nature* **365** (6446): 566–8. Bibcode:1993Natur.365..566E. doi:10.1038/365566a0. PMID 7692304.

[85] Platts, Simon Nicholas, "The PAH World – Discotic polynuclear aromatic compounds as a mesophase scaffolding at the origin of life"

[86] Allamandola, Louis et Al. "Cosmic Distribution of Chemical Complexity"

[87] Atkinson, Nancy (2010-10-27). "Buckyballs Could Be Plentiful in the Universe". Universe Today. Retrieved 2010-10-28.

[88] García-Hernández, D. A.; Manchado, A.; García-Lario, P.; Stanghellini, L.; Villaver, E.; Shaw, R. A.; Szczerba, R.; Perea-Calderón, J. V. (2010-10-28). "Formation Of Fullerenes In H-Containing Planetary Nebulae". *The Astrophysical Journal Letters* **724**: L39. arXiv:1009.4357. Bibcode:2010ApJ...724L..39G. doi:10.1088/2041-8205/724/1/L39.

[89] Bernstein MP, Sandford SA, Allamandola LJ, Gillette JS, Clemett SJ, Zare RN (February 1999). "UV irradiation of polycyclic aromatic hydrocarbons in ices: production of alcohols, quinones, and ethers". *Science* **283** (5405): 1135–8. Bibcode:1999Sci...283.1135B.doi:10.1126/science.283.5405.1135. PMID10024233.

[90] Kunin V. (October 2000). "A system of two polymerases--a model for the origin of life". *Origins of life and evolution of the biosphere* **30** (5): 459–466. Bibcode:2000OLEB...30..459K. doi:10.1023/A:1006672126867. PMID 11002892.

[91] "Challenging Assumptions About the Origin of Life". *Astrobiology Magazine*. 18 September 2013. Retrieved 2014-05-07.

[92] Service, Robert F. (16 March 2015). "Researchers may have solved origin-of-life conundrum". *Science* (News) (Washington, D.C.: American Association for the Advancement of Science). ISSN 1095-9203. Retrieved 2015-07-26.

[93] Witzany G (2014) RNA Sociology: Group Behavioral Motifs of RNA Consortia. Life 4:800-818. PMID 25426799

[94] Witzany, Guenther (2008). "The Viral Origins of Telomeres and Telomerases and their Important Role in Eukaryogenesis and Genome Maintenance". *Biosemiotics* **1** (2): 191. doi:10.1007/s12304-008-9018-0.

38.12 Further reading

- Cairns-Smith, A. G. (1993). *Genetic Takeover: And the Mineral Origins of Life*. Cambridge University Press. ISBN 0-521-23312-7.

- Orgel, L. E. (October 1994). "The origin of life on the Earth". *Scientific American* **271**(4): 76–83. doi:10.1038/sci 76. PMID 7524147.

- Orgel, L. E. (2004). "Prebiotic Chemistry and the Origin of the RNA World". *Critical Reviews in Biochemistry and Molecular Biology* **39** (2): 99–123. doi:10.1080/10409230490460765. ISSN 1549-7798. PMID 15217990.

- Woolfson, Adrian (September 2000). *Life Without Genes*. London: Flamingo. ISBN 978-0-00-654874-4.

- Vlassov, Alexander V.; Kazakov, Sergei A.; Johnston, Brian H.; Landweber, Laura F. (July 2005). "The RNA World on Ice: A New Scenario for the Emergence of RNA Information". *Journal of Molecular Evolution* **61** (2): 264–273. doi:10.1007/s00239-004-0362-7. PMID 16044244.

- Engelhart, Aaron E.; Hud, Nicholas V. (2011). "Primitive genetic polymers" (PDF). *Cold Spring Harb Perspect Biol* **2** (12): a002196. doi:10.1101/cshperspect.a002196. PMC 2982173. PMID 20462999.

- Bernhardt, Harold S. (2012). "The RNA world hypothesis: the worst theory of the early evolution of life (except for all the others)". *Biol Direct* **7** (1): 23. doi:10.1186/1745-6150-7-23. PMC 3495036. PMID 22793875.

- Sutherland, JD (April 2010). "Ribonucleotides". *Cold Spring Harb Perspect Biol.* **2**(4): a005439. doi:10.1101/csh PMC 2845210. PMID 20452951.

38.13 External links

- "The RNA world" (2001) by Sidney Altman, on the Nobel prize website

- "Exploring the new RNA world" (2004) by Thomas R. Cech, on the Nobel prize website

- "The Formation of the RNA World" by James P. Ferris

- "Exploring Life's Origins: a Virtual Exhibit"

- http://www.hhmi.org/bulletin/pdf/june2002/RNA.pdf HHMI bulletin

- Sutherland, J. D. (2010). "Ribonucleotides". *Cold Spring Harbor Perspectives in Biology* **2**(4): a005439. doi:10.1 PMC 2845210. PMID 20452951.

Chapter 39

DNA profiling

For DNA testing for inherited diseases, see Genetic testing.
Not to be confused with DNA barcoding or DNA phenotyping.

DNA profiling (also called **DNA fingerprinting**, **DNA testing**, or **DNA typing**) is a forensic technique used to identify individuals by characteristics of their DNA. A **DNA profile** is a small set of DNA variations that is very likely to be different in all unrelated individuals, thereby being as unique to individuals as are fingerprints (hence the alternate name for the technique). DNA profiling should not be confused with full genome sequencing.[1] First developed and used in 1985,[2] DNA profiling is used in, for example, parentage testing and criminal investigation, to identify a person or to place a person at a crime scene, techniques which are now employed globally in forensic science to facilitate police detective work and help clarify paternity and immigration disputes.[3]

Although 99.9% of human DNA sequences are the same in every person, enough of the DNA is different that it is possible to distinguish one individual from another, unless they are monozygotic ("identical") twins.[4] DNA profiling uses repetitive ("repeat") sequences that are highly variable,[4] called variable number tandem repeats (VNTRs), in particular short tandem repeats (STRs). VNTR loci are very similar between closely related humans, but are so variable that unrelated individuals are extremely unlikely to have the same VNTRs.

The DNA profiling technique nowadays used is based on technology developed in 1988.[5][6]

39.1   DNA profiling process

Developed by Professor of Genetics Sir Alec Jeffreys, the process begins with a sample of an individual's DNA (typically called a "reference sample"). The most desirable method of collecting a reference sample is the use of a buccal swab, as this reduces the possibility of contamination. When this is not available (e.g. because a court order is needed but not obtainable) other methods may need to be used to collect a sample of blood, saliva, semen, or other appropriate fluid or tissue from personal items (e.g. a toothbrush, razor) or from stored samples (e.g. banked sperm or biopsy tissue). Samples obtained from blood relatives (related by birth, not marriage) can provide an indication of an individual's profile, as could human remains that had been previously profiled.

A reference sample is then analyzed to create the individual's DNA profile using one of a number of techniques, discussed below. The DNA profile is then compared against another sample to determine whether there is a genetic match.

39.1.1   RFLP analysis

Main article: Restriction fragment length polymorphism

The first methods for finding out genetics used for DNA profiling involved **RFLP analysis**. DNA is collected from cells,

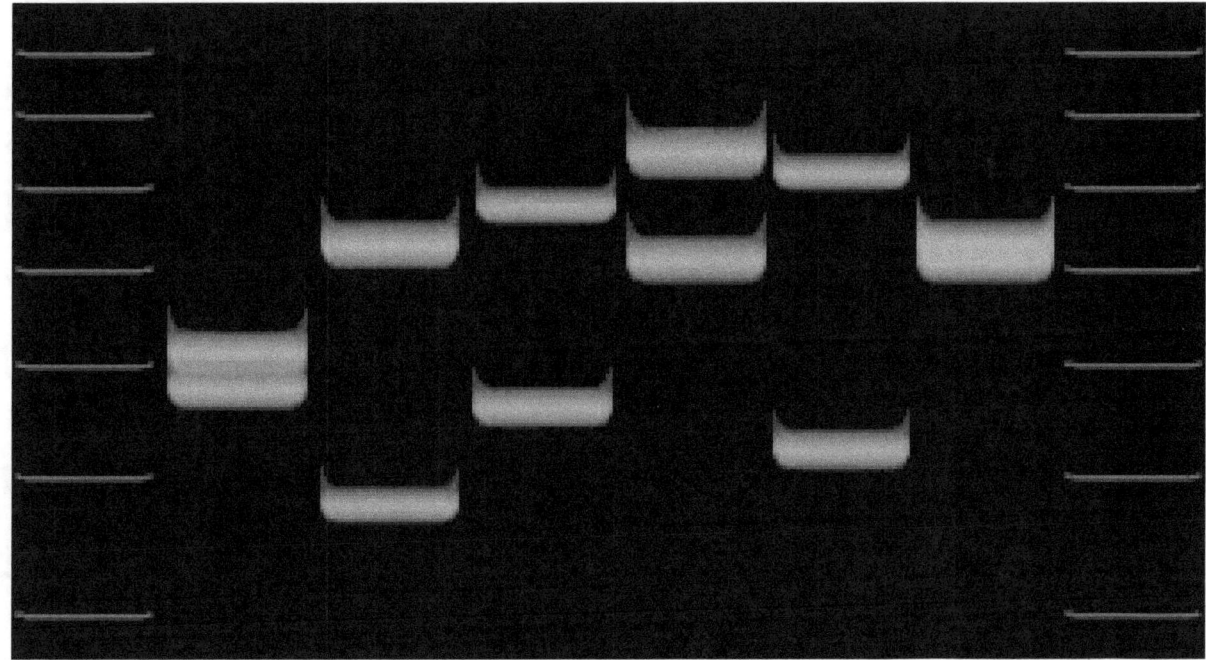

Variations of VNTR allele lengths in 6 individuals.

such as a blood sample, and cut into small pieces using a restriction enzyme (a restriction digest). This generates thousands of DNA fragments of differing sizes as a consequence of variations between DNA sequences of different individuals. The fragments are then separated on the basis of size using gel electrophoresis.

The separated fragments are then transferred to a nitrocellulose or nylon filter; this procedure is called a Southern blot. The DNA fragments within the blot are permanently fixed to the filter, and the DNA strands are denatured. Radiolabeled probe molecules are then added that are complementary to sequences in the genome that contain repeat sequences. These repeat sequences tend to vary in length among different individuals and are called variable number tandem repeat sequences or VNTRs. The probe molecules hybridize to DNA fragments containing the repeat sequences and excess probe molecules are washed away. The blot is then exposed to an X-ray film. Fragments of DNA that have bound to the probe molecules appear as dark bands on the film.

The Southern blot technique is laborious, and requires large amounts of undegraded sample DNA. Also, Karl Brown's original technique looked at many minisatellite loci at the same time, increasing the observed variability, but making it hard to discern individual alleles (and thereby precluding paternity testing). These early techniques have been supplanted by PCR-based assays.

39.1.2   PCR analysis

Main article: polymerase chain reaction

Developed by Kary Mullis in 1983, a process was reported by which specific portions of the sample DNA can be amplified almost indefinitely (Saiki et al. 1985, 1988). This has revolutionized the whole field of DNA study. The process, the polymerase chain reaction (PCR), mimics the biological process of DNA replication, but confines it to specific DNA sequences of interest. With the invention of the PCR technique, DNA profiling took huge strides forward in both discriminating power and the ability to recover information from very small (or degraded) starting samples.

PCR greatly amplifies the amounts of a specific region of DNA. In the PCR process, the DNA sample is denatured into the separate individual polynucleotide strands through heating. Two oligonucleotide DNA primers are used to hybridize

Alec Jeffreys, the pioneer of DNA profiling.

to two corresponding nearby sites on opposite DNA strands in such a fashion that the normal enzymatic extension of the active terminal of each primer (that is, the 3' end) leads toward the other primer. PCR uses replication enzymes that are tolerant of high temperatures, such as the thermostable Taq polymerase. In this fashion, two new copies of the sequence of interest are generated. Repeated denaturation, hybridization, and extension in this fashion produce an exponentially growing number of copies of the DNA of interest. Instruments that perform thermal cycling are now readily available from commercial sources. This process can produce a million-fold or greater amplification of the desired region in 2 hours or less.

Early assays such as the HLA-DQ alpha reverse dot blot strips grew to be very popular due to their ease of use, and the speed with which a result could be obtained. However, they were not as discriminating as RFLP analysis. It was also difficult to determine a DNA profile for mixed samples, such as a vaginal swab from a sexual assault victim.

However, the PCR method was readily adaptable for analyzing VNTR, in particular STR loci. In recent years, research in human DNA quantitation has focused on new "real-time" quantitative PCR (qPCR) techniques. Quantitative PCR methods enable automated, precise, and high-throughput measurements. Interlaboratory studies have demonstrated the importance of human DNA quantitation on achieving reliable interpretation of STR typing and obtaining consistent results across laboratories.

39.1.3   STR analysis

Main article: Short tandem repeats

The system of DNA profiling used today is based on PCR and uses simple sequences[7] or short tandem repeats (STR). This method uses highly polymorphic regions that have short repeated sequences of DNA (the most common is 4 bases repeated, but there are other lengths in use, including 3 and 5 bases). Because unrelated people almost certainly have different numbers of repeat units, STRs can be used to discriminate between unrelated individuals. These STR loci (locations on a chromosome) are targeted with sequence-specific primers and amplified using PCR. The DNA fragments that result are then separated and detected using electrophoresis. There are two common methods of separation and detection, capillary electrophoresis (CE) and gel electrophoresis.

Each STR is polymorphic, but the number of alleles is very small. Typically each STR allele will be shared by around 5 - 20% of individuals. The power of STR analysis comes from looking at multiple STR loci simultaneously. The pattern of alleles can identify an individual quite accurately. Thus STR analysis provides an excellent identification tool. The more STR regions that are tested in an individual the more discriminating the test becomes.

From country to country, different STR-based DNA-profiling systems are in use. In North America, systems that amplify the CODIS 13 core loci are almost universal, whereas in the United Kingdom the SGM+ 11 loci system (which is compatible with The National DNA Database) is in use. Whichever system is used, many of the STR regions used are the same. These DNA-profiling systems are based on multiplex reactions, whereby many STR regions will be tested at the same time.

The true power of STR analysis is in its statistical power of discrimination. Because the 13 loci that are currently used for discrimination in CODIS are independently assorted (having a certain number of repeats at one locus does not change the likelihood of having any number of repeats at any other locus), the product rule for probabilities can be applied. This means that, if someone has the DNA type of ABC, where the three loci were independent, we can say that the probability of having that DNA type is the probability of having type A times the probability of having type B times the probability of having type C. This has resulted in the ability to generate match probabilities of 1 in a quintillion (1×10^{18}) or more. However, DNA database searches showed much more frequent than expected false DNA profile matches.[8] Moreover, since there are about 12 million monozygotic twins on Earth, the theoretical probability is not accurate.

In practice, the risk of contaminated-matching is much greater than matching a distant relative, such as contamination of a sample from nearby objects, or from left-over cells transferred from a prior test. The risk is greater for matching the most common person in the samples: Everything collected from, or in contact with, a victim is a major source of contamination for any other samples brought into a lab. For that reason, multiple control-samples are typically tested in order to ensure that they stayed clean, when prepared during the same period as the actual test samples. Unexpected matches (or variations) in several control-samples indicates a high probability of contamination for the actual test samples. In a relationship test, the full DNA profiles should differ (except for twins), to prove that a person was not actually matched as being related to their own DNA in another sample.

39.1.4   AmpFLP

Main article: Amplified fragment length polymorphism

Another technique, AmpFLP, or amplified fragment length polymorphism was also put into practice during the early 1990s. This technique was also faster than RFLP analysis and used PCR to amplify DNA samples. It relied on variable number tandem repeat (VNTR) polymorphisms to distinguish various alleles, which were separated on a polyacrylamide gel using an allelic ladder (as opposed to a molecular weight ladder). Bands could be visualized by silver staining the gel. One popular locus for fingerprinting was the D1S80 locus. As with all PCR based methods, highly degraded DNA or very small amounts of DNA may cause allelic dropout (causing a mistake in thinking a heterozygote is a homozygote) or other stochastic effects. In addition, because the analysis is done on a gel, very high number repeats may bunch together at the top of the gel, making it difficult to resolve. AmpFLP analysis can be highly automated, and allows for easy creation of phylogenetic trees based on comparing individual samples of DNA. Due to its relatively low cost and ease of set-up and operation, AmpFLP remains popular in lower income countries.

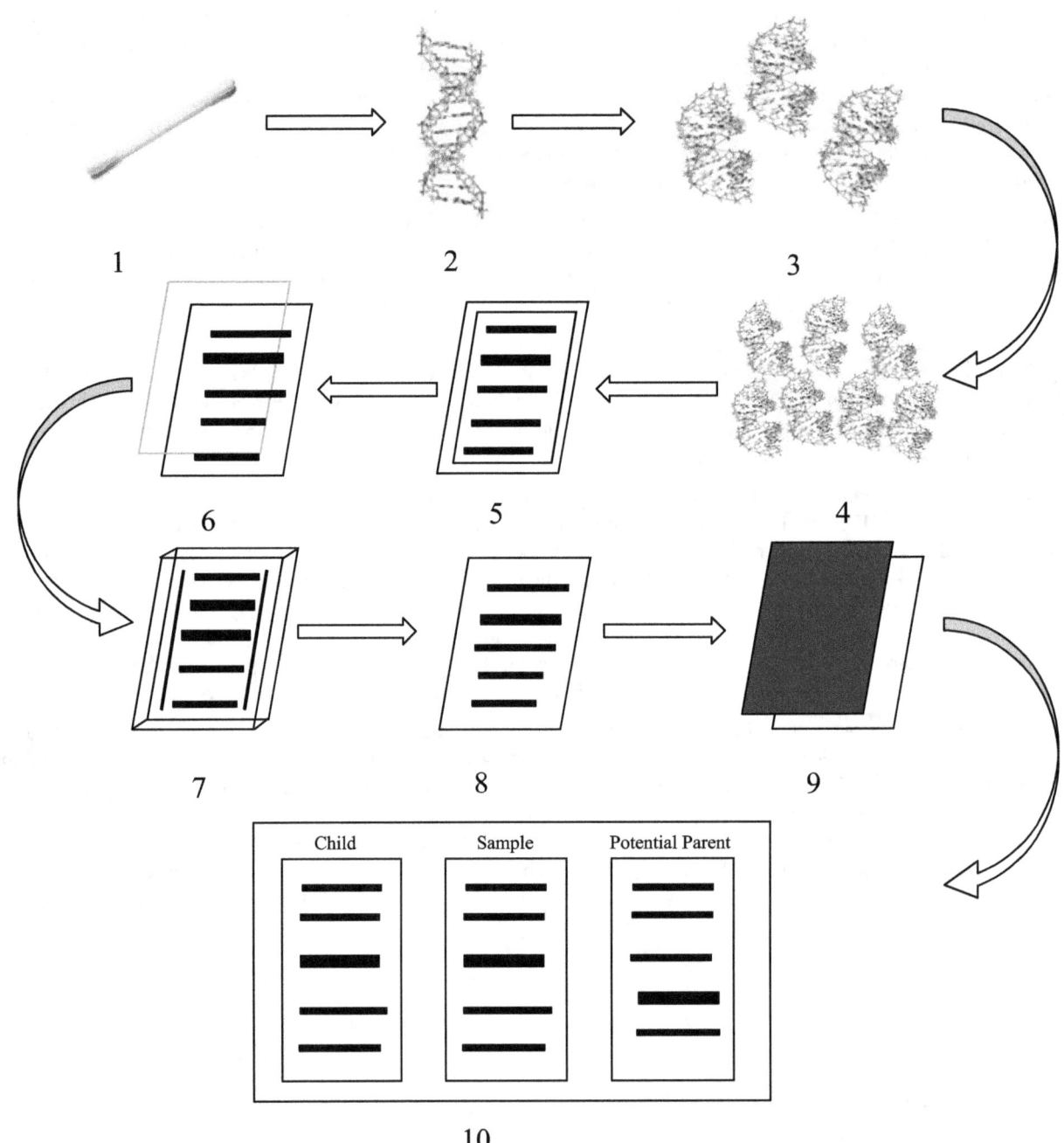

1: *A cell sample is taken- usually a cheek swab or blood test*
2: *DNA is extracted from sample*
3: *Cleavage of DNA by restriction enzyme- the DNA is broken into small fragments*
4: *Small fragments are amplified by the Polymerase Chain Reaction- results in many more fragments*
5: *DNA fragments are separated by electrophoresis*
6: *The fragments are transferred to an agar plate*
7: *On the Agar Plate specific DNA fragments are bound to a radioactive DNA probe*
8: *The Agar Plate is washed free of excess probe*
9: *An x-ray film is used to detect a radioactive pattern*
10: *The DNA is compared to other DNA samples*

39.1.5 DNA family relationship analysis

Using PCR technology, DNA analysis is widely applied to determine genetic family relationships such as paternity, maternity, siblingship and other kinships.

During conception, the father's sperm cell and the mother's egg cell, each containing half the amount of DNA found in other body cells, meet and fuse to form a fertilized egg, called a zygote. The zygote contains a complete set of DNA molecules, a unique combination of DNA from both parents. This zygote divides and multiplies into an embryo and later, a full human being.

At each stage of development, all the cells forming the body contain the same DNA—half from the father and half from the mother. This fact allows the relationship testing to use all types of all samples including loose cells from the cheeks collected using buccal swabs, blood or other types of samples.

There are predictable inheritance patterns at certain locations (called loci) in the human genome, which have been found to be useful in determining identity and biological relationships. These loci contain specific DNA markers that scientists use to identify individuals. In a routine DNA paternity test, the markers used are Short Tandem Repeats (STRs), short pieces of DNA that occur in highly differential repeat patterns among individuals.

Each person's DNA contains two copies of these markers—one copy inherited from the father and one from the mother. Within a population, the markers at each person's DNA location could differ in length and sometimes sequence, depending on the markers inherited from the parents.

The combination of marker sizes found in each person makes up his/her unique genetic profile. When determining the relationship between two individuals, their genetic profiles are compared to see if they share the same inheritance patterns at a statistically conclusive rate.

For example, the following sample report from this commercial DNA paternity testing laboratory Universal Genetics signifies how relatedness between parents and child is identified on those special markers:

The partial results indicate that the child and the alleged father's DNA match among these five markers. The complete test results show this correlation on 16 markers between the child and the tested man to enable a conclusion to be drawn as to whether or not the man is the biological father.

Each marker is assigned with a Paternity Index (PI), which is a statistical measure of how powerfully a match at a particular marker indicates paternity. The PI of each marker is multiplied with each other to generate the Combined Paternity Index (CPI), which indicates the overall probability of an individual being the biological father of the tested child relative to a randomly selected man from the entire population of the same race. The CPI is then converted into a Probability of Paternity showing the degree of relatedness between the alleged father and child.

The DNA test report in other family relationship tests, such as grandparentage and siblingship tests, is similar to a paternity test report. Instead of the Combined Paternity Index, a different value, such as a Siblingship Index, is reported.

The report shows the genetic profiles of each tested person. If there are markers shared among the tested individuals, the probability of biological relationship is calculated to determine how likely the tested individuals share the same markers due to a blood relationship.

39.1.6 Y-chromosome analysis

Recent innovations have included the creation of primers targeting polymorphic regions on the Y-chromosome (Y-STR), which allows resolution of a mixed DNA sample from a male and female or cases in which a differential extraction is not possible. Y-chromosomes are paternally inherited, so Y-STR analysis can help in the identification of paternally related males. Y-STR analysis was performed in the Sally Hemings controversy to determine if Thomas Jefferson had sired a son with one of his slaves. The analysis of the Y-chromosome yields weaker results than autosomal chromosome analysis. The Y male sex-determining chromosome, as it is inherited only by males from their fathers, is almost identical along the patrilineal line. This leads to a less precise analysis than if autosomal chromosomes were testing, because of the random matching that occurs between pairs of chromosomes as zygotes are being made.[9]

39.1.7 Mitochondrial analysis

Main article: Mitochondrial DNA

For highly degraded samples, it is sometimes impossible to get a complete profile of the 13 CODIS STRs. In these situations, mitochondrial DNA (mtDNA) is sometimes typed due to there being many copies of mtDNA in a cell, while there may only be 1-2 copies of the nuclear DNA. Forensic scientists amplify the HV1 and HV2 regions of the mtDNA, and then sequence each region and compare single-nucleotide differences to a reference. Because mtDNA is maternally inherited, directly linked maternal relatives can be used as match references, such as one's maternal grandmother's daughter's son. In general, a difference of two or more nucleotides is considered to be an exclusion. Heteroplasmy and poly-C differences may throw off straight sequence comparisons, so some expertise on the part of the analyst is required. mtDNA is useful in determining clear identities, such as those of missing people when a maternally linked relative can be found. mtDNA testing was used in determining that Anna Anderson was not the Russian princess she had claimed to be, Anastasia Romanov.

mtDNA can be obtained from such material as hair shafts and old bones/teeth. Control mechanism based on interaction point with data. This can be determined by tooled placement in sample.

39.2 DNA databases

Main article: National DNA database

An early application of a DNA database was the compilation of A Mitochondrial DNA Concordance,[10] prepared by Kevin W. P. Miller and John L. Dawson at the University of Cambridge from 1996 to 1998[11] from data collected as part of Miller's PhD thesis. There are now several DNA databases in existence around the world. Some are private, but most of the largest databases are government controlled. The United States maintains the largest DNA database, with the Combined DNA Index System (CODIS) holding over 5 million records as of 2007.[12] The United Kingdom maintains the National DNA Database (NDNAD), which is of similar size, despite the UK's smaller population. The size of this database, and its rate of growth, is giving concern to civil liberties groups in the UK, where police have wide-ranging powers to take samples and retain them even in the event of acquittal.[13]

The U.S. Patriot Act of the United States provides a means for the U.S. government to get DNA samples from other countries if they are either a division of or a head office of a company operating in the U.S. Under the act; the American offices of the company cannot divulge to their subsidiaries/offices in other countries the reasons that these DNA samples are sought or by whom.

When a match is made from a National DNA Databank to link a crime scene to an offender having provided a DNA Sample to a databank that link is often referred to as a *cold hit*. A cold hit is of value in referring the police agency to a specific suspect but is of less evidential value than a DNA match made from outside the DNA Databank.[14]

FBI agents cannot legally store DNA of a person not convicted of a crime. DNA collected from a suspect not later convicted must be disposed of and not entered into the database. In 1998, a man residing in the UK was arrested on accusation of burglary. His DNA was taken and tested, and he was later released. Nine months later, this man's DNA was accidentally and illegally entered in the DNA database. New DNA is automatically compared to the DNA found at cold cases and, in this case, this man was found to be a match to DNA found at a rape and assault case one year earlier. The government then prosecuted him for these crimes. During the trial the DNA match was requested to be removed from the evidence because it had been illegally entered into the database. The request was carried out.[15]

The DNA collected from victims of rape are often stored for years until matched with the perpetrator's, usually when committing another crime. In 2014, Congress extended a bill that helps states deal with "a backlog" of unexamined evidence.[16]

39.3   Considerations when evaluating DNA evidence

In the early days of the use of genetic fingerprinting as criminal evidence, juries were often swayed by spurious statistical arguments by defense lawyers along these lines: Given a match that had a 1 in 5 million probability of occurring by chance, the lawyer would argue that this meant that in a country of say 60 million people there were 12 people who would also match the profile. This was then translated to a 1 in 12 chance of the suspect's being the guilty one. This argument is not sound unless the suspect was drawn at random from the population of the country. In fact, a jury should consider how likely it is that an individual matching the genetic profile would also have been a suspect in the case for other reasons. Another spurious statistical argument is based on the false assumption that a 1 in 5 million probability of a match automatically translates into a 1 in 5 million probability of innocence and is known as the prosecutor's fallacy.

When using RFLP, the theoretical risk of a coincidental match is 1 in 100 billion (100,000,000,000), although the practical risk is actually 1 in 1000 because monozygotic twins are 0.2% of the human population. Moreover, the rate of laboratory error is almost certainly higher than this, and often actual laboratory procedures do not reflect the theory under which the coincidence probabilities were computed. For example, the coincidence probabilities may be calculated based on the probabilities that markers in two samples have bands in *precisely* the same location, but a laboratory worker may conclude that similar—but not precisely identical—band patterns result from identical genetic samples with some imperfection in the agarose gel. However, in this case, the laboratory worker increases the coincidence risk by expanding the criteria for declaring a match. Recent studies have quoted relatively high error rates, which may be cause for concern.[17] In the early days of genetic fingerprinting, the necessary population data to accurately compute a match probability was sometimes unavailable. Between 1992 and 1996, arbitrary low ceilings were controversially put on match probabilities used in RFLP analysis rather than the higher theoretically computed ones.[18] Today, RFLP has become widely disused due to the advent of more discriminating, sensitive and easier technologies.

Since 1998, the DNA profiling system supported by The National DNA Database in the UK is the SGM+ DNA profiling system that includes 10 STR regions and a sex-indicating test. STRs do not suffer from such subjectivity and provide similar power of discrimination (1 in 10^{13} for unrelated individuals if using a full SGM+ profile). Figures of this magnitude are not considered to be statistically supportable by scientists in the UK; for unrelated individuals with full matching DNA profiles a match probability of 1 in a billion is considered statistically supportable. However, with any DNA technique, the cautious juror should not convict on genetic fingerprint evidence alone if other factors raise doubt. Contamination with other evidence (secondary transfer) is a key source of incorrect DNA profiles and raising doubts as to whether a sample has been adulterated is a favorite defense technique. More rarely, chimerism is one such instance where the lack of a genetic match may unfairly exclude a suspect.

39.3.1   Evidence of genetic relationship

It is also possible to use DNA profiling as evidence of genetic relationship, although such evidence varies in strength from weak to positive. Testing that shows no relationship is absolutely certain.

While almost all individuals have a single and distinct set of genes, ultra-rare individuals, known as "chimeras", have at least two different sets of genes. There have been two cases of DNA profiling that falsely suggested that a mother was unrelated to her children.[19] This happens when two eggs are fertilized at the same time and fuse together to create one individual instead of twins.

39.4   Fake DNA evidence

In one case, a criminal even planted fake DNA evidence in his own body: John Schneeberger raped one of his sedated patients in 1992 and left semen on her underwear. Police drew what they believed to be Schneeberger's blood and compared its DNA against the crime scene semen DNA on three occasions, never showing a match. It turned out that he had surgically inserted a Penrose drain into his arm and filled it with foreign blood and anticoagulants.

The functional analysis of genes and their coding sequences (open reading frames [ORFs]) typically requires that each ORF be expressed, the encoded protein purified, antibodies produced, phenotypes examined, intracellular localization determined, and interactions with other proteins sought.[20] In a study conducted by the life science company Nucleix and published in the journal Forensic Science International, scientists found that an In vitro synthesized sample of DNA matching any desired genetic profile can be constructed using standard molecular biology techniques without obtaining any actual tissue from that person. Nucleix claims they can also prove the difference between non-altered DNA and any that was synthesized.[21]

In the case of the Phantom of Heilbronn, police detectives found DNA traces from the same woman on various crime scenes in Austria, Germany, and France—among them murders, burglaries and robberies. Only after the DNA of the "woman" matched the DNA sampled from the burned body of a *male* asylum seeker in France, detectives began to have serious doubts about the DNA evidence. In that case, DNA traces were already present on the cotton swabs used to collect the samples at the crime scene, and the swabs had all been produced at the same factory in Austria. The company's product specification said that the swabs were guaranteed to be sterile, but not DNA-free.

39.5   DNA evidence as evidence in criminal trials

39.5.1   Familial DNA searching

Familial DNA searching (sometimes referred to as "Familial DNA" or "Familial DNA Database Searching") is the practice of creating new investigative leads in cases where DNA evidence found at the scene of a crime (forensic profile) strongly resembles that of an existing DNA profile (offender profile) in a state DNA database but there is not an exact match.[22][23] After all other leads have been exhausted, investigators may use specially developed software to compare the forensic profile to all profiles taken from a state's DNA database to generate a list of those offenders already in the database who are most likely to be a very close relative of the individual whose DNA is in the forensic profile.[24] To eliminate the majority of this list when the forensic DNA is a man's, crime lab technicians conduct Y-STR analysis. Using standard investigative techniques, authorities are then able to build a family tree. The family tree is populated from information gathered from public records and criminal justice records. Investigators rule out family members' involvement in the crime by finding excluding factors such as sex, living out of state or being incarcerated when the crime was committed. They may also use other leads from the case, such as witness or victim statements, to identify a suspect. Once a suspect has been identified, investigators seek to legally obtain a DNA sample from the suspect. This suspect DNA profile is then compared to the sample found at the crime scene to definitively identify the suspect as the source of the crime scene DNA.

Familial DNA database searching was first used in an investigation leading to the conviction of Craig Harman of manslaughter in the United Kingdom on April 19, 2004.[25] Craig Harman was convicted using familial DNA because of the partial matches from Harman's brother. When the police questioned Harman's brother, the police noticed Harman lived very close to the original crime scene. Harman confessed when his DNA isolated from the DNA found on the brick, matched.[26] Currently, familial DNA database searching is not conducted on a national level in the United States. States determine their own policies and decision making processes for how and when to conduct familial searches. The first familial DNA search and subsequent conviction in the United States was conducted in Denver, Colorado, in 2008 using software developed under the leadership of Denver District Attorney Mitch Morrissey and Denver Police Department Crime Lab Director Gregg LaBerge.[27] California was the first state to implement a policy for familial searching under then Attorney General, now Governor, Jerry Brown.[28] In his role as consultant to the Familial Search Working Group of the California Department of Justice, former Alameda County Prosecutor Rock Harmon is widely considered to have been the catalyst in the adoption of familial search technology in California. The technique was used to catch the Los Angeles serial killer known as the "Grim Sleeper" in 2010.[29] It wasn't a witness or informant that tipped off law enforcement to the identity of the "Grim Sleeper" serial killer, who had eluded police for more than two decades, but DNA from the suspect's own son. The suspect's son was arrested and convicted in a felony weapons charge and swabbed for DNA last year. When his DNA was entered into the database of convicted felons, detectives were alerted to a partial match to evidence found at the "Grim Sleeper" crime scenes. David Franklin Jr., also known as the Grim Sleeper, was charged with ten counts of murder and one count of attempted murder.[30] More recently, familial DNA, led to the arrest of 21-year-old Elvis Garcia on charges of sexual assault and false imprisonment of a woman in Santa Cruz in 2008.[31] In March 2011 Virginia Governor Bob McDonnell announced that Virginia would begin using familial DNA searches.[32] Other states are expected to follow.

At a press conference in Virginia on March 7, 2011, regarding the East Coast Rapist, Prince William County prosecutor

Paul Ebert and Fairfax County Police Detective John Kelly said the case would have been solved years ago if Virginia had used familial DNA searching. Aaron Thomas, the suspected East Coast Rapist, was arrested in connection with the rape of 17 women from Virginia to Rhode Island, but familial DNA was not used in the case.[33]

Critics of familial DNA database searches argue that the technique is an invasion of an individual's 4th Amendment rights.[34] Privacy advocates are petitioning for DNA database restrictions, arguing that the only fair way to search for possible DNA matches to relatives of offenders or arrestees would be to have a population-wide DNA database.[15] Some scholars have pointed out that the privacy concerns surrounding familial searching are similar in some respects to other police search techniques,[35] and most have concluded that the practice is constitutional.[36] The Ninth Circuit Court of Appeals in *United States v. Pool* (vacated as moot) suggested that this practice is somewhat analogous to a witness looking at a photograph of one person and stating that it looked like the perpetrator, which leads law enforcement to show the witness photos of similar looking individuals, one of whom is identified as the perpetrator.[37] Regardless of whether familial DNA searching was the method used to identify the suspect, authorities always conduct a normal DNA test to match the suspect's DNA with that of the DNA left at the crime scene.

Critics also claim that racial profiling could occur on account of Familial DNA testing. In the United States, the conviction rates of racial minorities are much higher than that of the overall population. It is unclear whether this is due to discrimination from police officers and the courts, as opposed to a simple higher rate of offence among minorities. Arrest-based databases, which are found in the majority of the United States, lead to an even greater level of racial discrimination. An arrest, as opposed to conviction, relies much more heavily on police discretion.[15]

For instance, investigators with Denver District Attorney's Office successfully identified a suspect in a property theft case using a familial DNA search. In this example, the suspect's blood left at the scene of the crime strongly resembled that of a current Colorado Department of Corrections prisoner.[38] Using publicly available records, the investigators created a family tree. They then eliminated all the family members who were incarcerated at the time of the offense, as well as all of the females (the crime scene DNA profile was that of a male). Investigators obtained a court order to collect the suspect's DNA, but the suspect actually volunteered to come to a police station and give a DNA sample. After providing the sample, the suspect walked free without further interrogation or detainment. Later confronted with an exact match to the forensic profile, the suspect pled guilty to criminal trespass at the first court date and was sentenced to two years probation.

In Italy a familiar DNA search has been done to solve the case of the murder of Yara Gambirasio whose body was found in the bush three months after her disappearance. A DNA trace was found on the underwear of the murdered teenage near and a DNA sample was requested from a person who lived near the municipality of Brembate di Sopra and a common male ancestor was found in the DNA sample of a young man not involved in the murder. After a long investigation the father of the supposed killer was identified in Giuseppe Guerinoni a deceased man but his two sons born from his wife were not related with the DNA samples found on the body of Yara. After 3 and a half years the DNA found on the underwear of the deceased girl was matched with Massimo Giuseppe Bosetti who was arrested and accused of the murder of the 13-year-old girl. Now Bosetti is awaiting in jail his trial.

39.5.2 Partial matches

Partial DNA matches are not searches themselves, but are the result of moderate stringency CODIS searches that produce a potential match that shares at least one allele at every locus.[39] Partial matching does not involve the use of familial search software, such as those used in the UK and United States, or additional Y-STR analysis, and therefore often misses sibling relationships. Partial matching has been used to identify suspects in several cases in the UK and United States,[40] and has also been used as a tool to exonerate the falsely accused. Darryl Hunt was wrongly convicted in connection with the rape and murder of a young woman in 1984 in North Carolina.[41] Hunt was exonerated in 2004 when a DNA database search produced a remarkably close match between a convicted felon and the forensic profile from the case. The partial match led investigators to the felon's brother, Willard E. Brown, who confessed to the crime when confronted by police. A judge then signed an order to dismiss the case against Hunt.

39.5.3   Surreptitious DNA collecting

Police forces may collect DNA samples without the suspects' knowledge, and use it as evidence. Legality of this mode of proceeding has been questioned in Australia.

In the United States, it has been accepted, courts often claiming that there was no expectation of privacy, citing *California v. Greenwood* (1985), in which the Supreme Court held that the Fourth Amendment does not prohibit the warrantless search and seizure of garbage left for collection outside the curtilage of a home. Critics of this practice underline that this analogy ignores that "most people have no idea that they risk surrendering their genetic identity to the police by, for instance, failing to destroy a used coffee cup. Moreover, even if they do realize it, there is no way to avoid abandoning one's DNA in public."[42]

In the UK, the Human Tissue Act 2004 prohibited private individuals from covertly collecting biological samples (hair, fingernails, etc.) for DNA analysis, but excluded medical and criminal investigations from the offence.[43]

The U.S. Supreme Court ruled 5–4 on June 3, 2013, in the case of *Maryland v. King*, that DNA sampling of prisoners arrested for serious crimes is constitutional.[44][45][46]

39.5.4   England and Wales

Evidence from an expert who has compared DNA samples must be accompanied by evidence as to the sources of the samples and the procedures for obtaining the DNA profiles.[47] The judge must ensure that the jury must understand the significance of DNA matches and mismatches in the profiles. The judge must also ensure that the jury does not confuse the 'match probability' (the probability that a person that is chosen at random has a matching DNA profile to the sample from the scene) with the probability that a person with matching DNA committed the crime. In 1996 *R v. Doheny*[48] Phillips LJ gave this example of a summing up, which should be carefully tailored to the particular facts in each case:

> Members of the Jury, if you accept the scientific evidence called by the Crown, this indicates that there are probably only four or five white males in the United Kingdom from whom that semen stain could have come. The Defendant is one of them. If that is the position, the decision you have to reach, on all the evidence, is whether you are sure that it was the Defendant who left that stain or whether it is possible that it was one of that other small group of men who share the same DNA characteristics.

Juries should weigh up conflicting and corroborative evidence, using their own common sense and not by using mathematical formulae, such as Bayes' theorem, so as to avoid "confusion, misunderstanding and misjudgment".[49]

Presentation and evaluation of evidence of partial or incomplete DNA profiles

In *R v Bates*,[50] Moore-Bick LJ said:

> We can see no reason why partial profile DNA evidence should not be admissible provided that the jury are made aware of its inherent limitations and are given a sufficient explanation to enable them to evaluate it. There may be cases where the match probability in relation to all the samples tested is so great that the judge would consider its probative value to be minimal and decide to exclude the evidence in the exercise of his discretion, but this gives rise to no new question of principle and can be left for decision on a case by case basis. However, the fact that there exists in the case of all partial profile evidence the possibility that a "missing" allele might exculpate the accused altogether does not provide sufficient grounds for rejecting such evidence. In many there is a possibility (at least in theory) that evidence that would assist the accused and perhaps even exculpate him altogether exists, but that does not provide grounds for excluding relevant evidence that is available and otherwise admissible, though it does make it important to ensure that the jury are given sufficient information to enable them to evaluate that evidence properly[51]

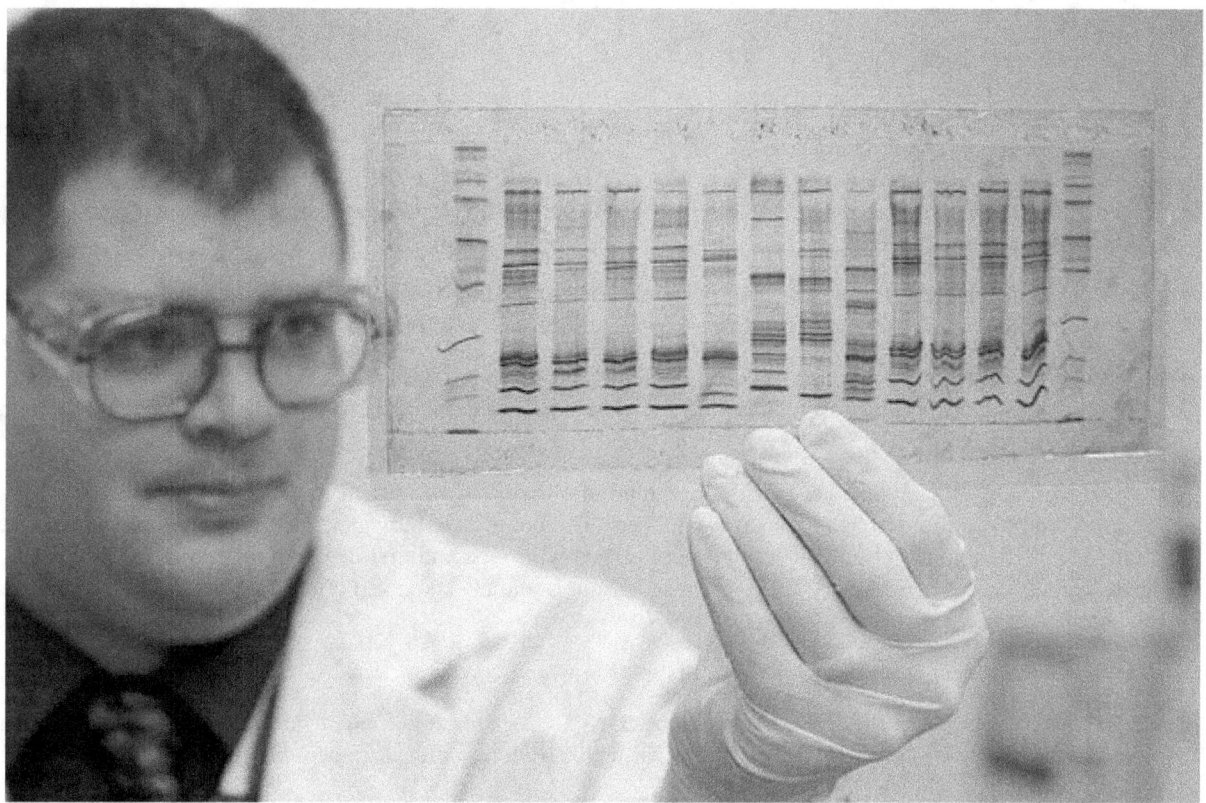

CBP chemist reads a DNA profile to determine the origin of a commodity.

39.5.5   DNA testing in the United States

There are state laws on DNA profiling in all 50 states of the United States.[52] Detailed information on database laws in each state can be found at the National Conference of State Legislatures website.[53]

39.5.6   Development of artificial DNA

In August 2009, scientists in Israel raised serious doubts concerning the use of DNA by law enforcement as the ultimate method of identification. In a paper published in the journal *Forensic Science International: Genetics*, the Israeli researchers demonstrated that it is possible to manufacture DNA in a laboratory, thus falsifying DNA evidence. The scientists fabricated saliva and blood samples, which originally contained DNA from a person other than the supposed donor of the blood and saliva.[54]

The researchers also showed that, using a DNA database, it is possible to take information from a profile and manufacture DNA to match it, and that this can be done without access to any actual DNA from the person whose DNA they are duplicating. The synthetic DNA oligos required for the procedure are common in molecular laboratories.[54]

The New York Times quoted the lead author, Daniel Frumkin, saying, "You can just engineer a crime scene...any biology undergraduate could perform this".[54] Frumkin perfected a test that can differentiate real DNA samples from fake ones. His test detects epigenetic modifications, in particular, DNA methylation. Seventy percent of the DNA in any human genome is methylated, meaning it contains methyl group modifications within a CpG dinucleotide context. Methylation at the promoter region is associated with gene silencing. The synthetic DNA lacks this epigenetic modification, which allows the test to distinguish manufactured DNA from genuine DNA.[54]

It is unknown how many police departments, if any, currently use the test. No police lab has publicly announced that it is using the new test to verify DNA results.[55]

39.6 Cases

- In 1986, Richard Buckland was exonerated, despite having admitted to the rape and murder of a teenager near Leicester, the city where DNA profiling was first developed. This was the first use of DNA fingerprinting in a criminal investigation.[56]

- In 1987, in the same case as Buckland, British baker Colin Pitchfork was the first criminal caught and convicted using DNA fingerprinting.[57]

- In 1987, genetic fingerprinting was used in criminal court for the first time in the trial of a man accused of unlawful intercourse with a mentally handicapped 14-year-old female who gave birth to a baby.[58]

- In 1987, Florida rapist Tommie Lee Andrews was the first person in the United States to be convicted as a result of DNA evidence, for raping a woman during a burglary; he was convicted on November 6, 1987, and sentenced to 22 years in prison.[59][60]

- In 1988, Timothy Wilson Spencer was the first man in Virginia to be sentenced to death through DNA testing, for several rape and murder charges. He was dubbed "The South Side Strangler" because he killed victims on the south side of Richmond, Virginia. He was later charged with rape and first-degree murder and was sentenced to death. He was executed on April 27, 1994. David Vasquez, initially convicted of one of Spencer's crimes, became the first man in America exonerated based on DNA evidence.

- In 1989, Chicago man Gary Dotson was the first person whose conviction was overturned using DNA evidence.

- In 1991, Allan Legere was the first Canadian to be convicted as a result of DNA evidence, for four murders he had committed while an escaped prisoner in 1989. During his trial, his defense argued that the relatively shallow gene pool of the region could lead to false positives.

- In 1992, DNA evidence was used to prove that Nazi doctor Josef Mengele was buried in Brazil under the name Wolfgang Gerhard.

- In 1992, DNA from a palo verde tree was used to convict Mark Alan Bogan of murder. DNA from seed pods of a tree at the crime scene was found to match that of seed pods found in Bogan's truck. This is the first instance of plant DNA admitted in a criminal case.[61][62][63]

- In 1993, Kirk Bloodsworth was the first person to have been convicted of murder and sentenced to death, whose conviction was overturned using DNA evidence.

- The 1993 rape and murder of Mia Zapata, lead singer for the Seattle punk band The Gits was unsolved nine years after the murder. A database search in 2001 failed, but the killer's DNA was collected when he was arrested in Florida for burglary and domestic abuse in 2002.

- The science was made famous in the United States in 1994 when prosecutors heavily relied on DNA evidence allegedly linking O. J. Simpson to a double murder. The case also brought to light the laboratory difficulties and handling procedure mishaps that can cause such evidence to be significantly doubted.

- In 1994, Royal Canadian Mounted Police (RCMP) detectives successfully tested hairs from a cat known as Snowball, and used the test to link a man to the murder of his wife, thus marking for the first time in forensic history the use of non-human animal DNA to identify a criminal (Plant DNA was used in 1992, see above).

- In 1994, the claim that Anna Anderson was Grand Duchess Anastasia Nikolaevna of Russia was tested after her death using samples of her tissue that had been stored at a Charlottesville, Virginia hospital following a medical procedure. The tissue was tested using DNA fingerprinting, and showed that she bore no relation to the Romanovs.[64]

- In 1994, Earl Washington, Jr., of Virginia had his death sentence commuted to life imprisonment a week before his scheduled execution date based on DNA evidence. He received a full pardon in 2000 based on more advanced testing.[65] His case is often cited by opponents of the death penalty.

- In 1995, the British Forensic Science Service carried out its first mass intelligence DNA screening in the investigation of the Naomi Smith murder case.

- In 1998, Richard J. Schmidt was convicted of attempted second-degree murder when it was shown that there was a link between the viral DNA of the human immunodeficiency virus (HIV) he had been accused of injecting in his girlfriend and viral DNA from one of his patients with AIDS. This was the first time viral DNA fingerprinting had been used as evidence in a criminal trial.

- In 1999, Raymond Easton, a disabled man from Swindon, England, was arrested and detained for seven hours in connection with a burglary. He was released due to an inaccurate DNA match. His DNA had been retained on file after an unrelated domestic incident some time previously.[66]

- In 2000 Frank Lee Smith was proved innocent by DNA profiling of the murder of an eight-year-old girl after spending 14 years on death row in Florida, USA. However he had died of cancer just before his innocence was proven.[67] In view of this the Florida state governor ordered that in future any death row inmate claiming innocence should have DNA testing.[65]

- In May 2000 Gordon Graham murdered Paul Gault at his home in Lisburn, Northern Ireland. Graham was convicted of the murder when his DNA was found on a sports bag left in the house as part of an elaborate ploy to suggest the murder occurred after a burglary had gone wrong. Graham was having an affair with the victim's wife at the time of the murder. It was the first time Low Copy Number DNA was used in Northern Ireland.[68]

- In 2001, Wayne Butler was convicted for the murder of Celia Douty. It was the first murder in Australia to be solved using DNA profiling.[69][70]

- In 2002, the body of James Hanratty, hanged in 1962 for the "A6 murder", was exhumed and DNA samples from the body and members of his family were analysed. The results convinced Court of Appeal judges that Hanratty's guilt, which had been strenuously disputed by campaigners, was proved "beyond doubt".[71] Paul Foot and some other campaigners continued to believe in Hanratty's innocence and argued that the DNA evidence could have been contaminated, noting that the small DNA samples from items of clothing, kept in a police laboratory for over 40 years "in conditions that do not satisfy modern evidential standards", had had to be subjected to very new amplification techniques in order to yield any genetic profile.[72] However, no DNA other than Hanratty's was found on the evidence tested, contrary to what would have been expected had the evidence indeed been contaminated.[73]

- In 2002, DNA testing was used to exonerate Douglas Echols, a man who was wrongfully convicted in a 1986 rape case. Echols was the 114th person to be exonerated through post-conviction DNA testing.

- In August 2002, Annalisa Vincenzi was shot dead in Tuscany. Bartender Peter Hamkin, 23, was arrested, in Merseyside, in March 2003 on an extradition warrant heard at Bow Street Magistrates' Court in London to establish whether he should be taken to Italy to face a murder charge. DNA "proved" he shot her, but he was cleared on other evidence.[74]

- In 2003, Welshman Jeffrey Gafoor was convicted of the 1988 murder of Lynette White, when crime scene evidence collected 12 years earlier was re-examined using STR techniques, resulting in a match with his nephew.[75] This may be the first known example of the DNA of an innocent yet related individual being used to identify the actual criminal, via "familial searching".

- In March 2003, Josiah Sutton was released from prison after serving four years of a twelve-year sentence for a sexual assault charge. Questionable DNA samples taken from Sutton were retested in the wake of the Houston Police Department's crime lab scandal of mishandling DNA evidence.

- In June 2003, because of new DNA evidence, Dennis Halstead, John Kogut and John Restivo won a re-trial on their murder conviction, their convictions were struck down and they were released.[76] The three men had already served eighteen years of their thirty-plus-year sentences.

- The trial of Robert Pickton (convicted in December 2003) is notable in that DNA evidence is being used primarily to identify the *victims*, and in many cases to prove their existence.

- In 2004, DNA testing shed new light into the mysterious 1912 disappearance of Bobby Dunbar, a four-year-old boy who vanished during a fishing trip. He was allegedly found alive eight months later in the custody of William Cantwell Walters, but another woman claimed that the boy was her son, Bruce Anderson, whom she had entrusted in Walters' custody. The courts disbelieved her claim and convicted Walters for the kidnapping. The boy was raised and known as Bobby Dunbar throughout the rest of his life. However, DNA tests on Dunbar's son and nephew revealed the two were not related, thus establishing that the boy found in 1912 was not Bobby Dunbar, whose real fate remains unknown.[77]

• In 2005, Gary Leiterman was convicted of the 1969 murder of Jane Mixer, a law student at the University of Michigan, after DNA found on Mixer's pantyhose was matched to Leiterman. DNA in a drop of blood on Mixer's hand was matched to John Ruelas, who was only four years old in 1969 and was never successfully connected to the case in any other way. Leiterman's defense unsuccessfully argued that the unexplained match of the blood spot to Ruelas pointed to cross-contamination and raised doubts about the reliability of the lab's identification of Leiterman.[78][79][80]

• In December 2005, Evan Simmons was proven innocent of a 1981 attack on an Atlanta woman after serving twenty-four years in prison. Mr. Clark is the 164th person in the United States and the fifth in Georgia to be freed using post-conviction DNA testing.

• In March 2009, Sean Hodgson—convicted of 1979 killing of Teresa De Simone, 22, in her car in Southampton—was released after tests proved DNA from the scene was not his. It was later matched to DNA retrieved from the exhumed body of David Lace. Lace had previously confessed to the crime but was not believed by the detectives. He served time in prison for other crimes committed at the same time as the murder and then committed suicide in 1988.[81]

• In November 2008, Anthony Curcio was arrested for masterminding one of the most elaborately planned armored car heists in history. DNA evidence linked Curcio to the crime.[82]

39.7   See also

• DNA barcoding
• DNA database
• National DNA database
• DNA paternity testing
• Capillary electrophoresis (CE)
• Forensic identification
• Full genome sequencing
• Gene mapping
• Genealogical DNA test
• *Harvey v. Horan*
• Identification (biology)
• Kinship analysis
• *Maryland v. King*
• Phantom of Heilbronn
• Project Innocence
• Restriction fragment length polymorphism (RFLP)
• Ribotyping
• Short tandem repeat (STR)
• International Society for Forensic Genetics

39.8 References

[1] *Kijk* magazine, 1 January 2009

[2] "The Guardian Interview With Sir Alec Jeffreys"http://www.theguardian.com/science/2009/may/24/dna-fingerprinting-alec-jef

[3] DNA pioneer's 'eureka' moment BBC. Retrieved 14 October 2011

[4] "Use of DNA in Identification". Accessexcellence.org. Retrieved 2010-04-03.

[5] Tautz, D. (1989). Hypervariability of simple sequences as a general source for polymorphic DNA markers. Nucleic Acids Research, 17, 6463-6471.

[6] Patent Jäckle H & Tautz D (1989) "Process For Analyzing Length Polymorphisms in DNA Regions" europäische Patent Nr. 0 438 512

[7] Tautz, D. (1989). Hypervariability of simple sequences as a general source for polymorphic DNA markers. Nucleic Acids Research, 17, 6463-6471.

[8] Felch, Jason; et al. (July 20, 2008). "FBI resists scrutiny of 'matches'". *Los Angeles Times*. pp. P8.

[9] "STR Analysis"

[10] Miller, Kevin. "Mitochondrial DNA Concordance".

[11] Miller, K.W.P.; Dawson, J.L.; Hagelberg, E. (1996). "A concordance of nucleotide substitutions in the first and second hyper-variable segments of the human mtDNA control region". *International Journal of Legal Medicine* (109): 107–113.

[12] "CODIS — National DNA Index System". Fbi.gov. Archived from the original on March 6, 2010. Retrieved 2010-04-03.

[13] "Restrictions on use and destruction of fingerprints and samples". Wikicrimeline.co.uk. 2009-09-01. Retrieved 2010-04-03.

[14] Rose & Goos. *DNA: A Practical Guide*. Toronto: Carswell Publications.

[15] "Double Helix Jeopardy"

[16] "Congress OKs bill to cut rape evidence backlog". Associated Press. Retrieved 18 September 2014.

[17] Nick Paton Walsh False result fear over DNA tests The Observer, Sunday 27 January 2002.

[18] The Evaluation of Forensic DNA Evidence 1996.

[19] "Two Women Don't Match Their Kids' DNA". Abcnews.go.com. 2006-08-15. Retrieved 2010-04-03.

[20] James L. Hartley, Gary F. Temple, and Michael A. Brasch (2000). "DNA Cloning Using In Vitro Site-Specific Recombination". *Cold Spring Harbor Laboratory Press.*

[21] A new test distinguishes between real and fake genetic evidence-Published by MIT Technology Review 2009-08-17, article by Emily Singer; Retrieved 2014-09-03

[22] Diamond, Diane. "Searching the Family DNA Tree to Solve Crime." The Huffington Post accessed April 17, 2011.

[23] Bieber Frederick; et al. (2006). "Finding Criminals Through DNA of Their Relatives". *Science* **312** (5778): 1315–16. doi:10.1126/science.1122655.

[24] "Science of the Future: Identifying Criminals Through Their Family Members".

[25] Bhattacharya, Shaoni "Killer Convicted Thanks to Relative's DNA." New Scientist accessed April 17, 2011.

[26] "Family Ties: the Use of DNA Databases to Catch Offenders' Kin".

[27] Pankratz, Howard. "Denver Uses 'Familial DNA Evidence' to Solve Car Break-Ins." The Denver Post accessed April 17, 2011.

[28] Steinhaur, Jennifer. "'Grim Sleeper' Arrest Fans Debate on DNA Use." The New York Times accessed April 17, 2011.

[29] Dolan, Maura. "A New Track in DNA Search." LA Times accessed April 17, 2011.

[30] New DNA Technique Led Police to 'Grim Sleeper' Serial Killer and Will 'Change Policing in America' - ABC News

[31] Dolan, Maura. "Familial DNA Search Used In Grim Sleeper Case Leads to Arrest of Santa Cruz Sex Offender." LA Times accessed April 17, 2011.

[32] Helderman, Rosalind. "McDonnell Approves Familial DNA for VA Crime Fighting." The Washington Post accessed April 17, 2011.

[33] Christoffersen, John and Barakat, Matthew. "Other victims of East Coast Rapist suspect sought." Associated Press. Accessed May 25, 2011.

[34] Murphy, Erin Elizabeth, (2009). "Relative Doubt: Familial Searches of DNA Databases" Michigan Law Review, Vol. 109, 291-348, 2010.

[35] Suter, Sonia. " All in The Family: Privacy and DNA Familial Searching" (2010). Harvard Journal of Law and Technology, Vol. 23,328, 2010.

[36] Kaye, David H., (2013). "The Genealogy Detectives: A Constitutional Analysis of Familial Searching" American Criminal Law Review, Vol. 51, No. 1, 109-163, 2013.

[37] "US v. Pool" Pool 621F .3d 1213.

[38] Pankratz, Howard. "Denver Uses 'Familial DNA Evidence' to Solve Car Break-Ins." The Denver Post accessed April 17, 2011.

[39] "Finding Criminals Through DNA Testing of Their Relatives" Technical Bulletin, Chromosomal Laboratories, Inc. accessed April 22, 2011.

[40] "Denver District Attorney DNA Resources" accessed April 20, 2011.

[41] "Darryl Hunt, The Innocence Project".

[42] Amy Harmon, "Lawyers Fight DNA Samples Gained on Sly", The New York Times, April 3, 2008.

[43] Human Tissue Act 2004, UK, available in PDF.

[44] "U.S. Supreme Court allows DNA sampling of prisoners". UPI. Retrieved 3 June 2013.

[45] http://www.supremecourt.gov/opinions/12pdf/12-207_d18e.pdf

[46] Samuels, J.E., E.H. Davies, and D.B. Pope. (2013). Collecting DNA at Arrest: Policies, Practices, and Implications, Final Technical Report. Washington, D.C.: Urban Institute, Justice Policy Center.

[47] R v. Loveridge, EWCA Crim 734 (2001).

[48] R v. Doheny [1996] EWCA Crim 728, [1997] 1 Cr App R 369 (31 July 1996), Court of Appeal

[49] R v. Adams [1997] EWCA Crim 2474 (16 October 1997), Court of Appeal

[50] R v Bates [2006] EWCA Crim 1395 (7 July 2006), Court of Appeal

[51] "WikiCrimeLine DNA profiling". Wikicrimeline.co.uk. Retrieved 2010-04-03.

[52] "Genelex: The DNA Paternity Testing Site". Healthanddna.com. 1996-01-06. Retrieved 2010-04-03.

[53] Donna Lyons — Posted by Glenda. "State Laws on DNA Data Banks". Ncsl.org. Retrieved 2010-04-03.

[54] Pollack, Andrew (August 18, 2009). "DNA Evidence Can Be Fabricated, Scientists Show". The New York Times. Retrieved April 1, 2010.

[55] "Elsevier". Fsigenetics.com. Retrieved 2010-04-03.

[56] "DNA pioneer's 'eureka' moment". BBC News. September 9, 2009. Retrieved April 1, 2010.

[57] Joseph Wambaugh, The Blooding (New York, New York: A Perigord Press Book, 1989), 369.

[58] Joseph Wambaugh, The Blooding (New York, New York: A Perigord Press Book, 1989), 316.

[59] "Gene Technology — Page 14". Txtwriter.com. 1987-11-06. Retrieved 2010-04-03.

[60] "frontline: the case for innocence: the dna revolution: state and federal dna database laws examined". Pbs.org. Retrieved 2010-04-03.

[61] "Court of Appeals of Arizona: Denial of Bogan's motion to reverse his conviction and sentence" (PDF). Denver DA: www.denverda.org. 2005-04-11. Retrieved 2011-04-21.

[62] "DNA Forensics: Angiosperm Witness for the Prosecution". Human Genome Project. Retrieved 2011-04-21.

[63] "Crime Scene Botanicals". Botanical Society of America. Retrieved 2011-04-21.

[64] Identification of the remains of the Romanov family by DNA analysis by Peter Gill, Central Research and Support Establishment, Forensic Science Service, Aldermaston, Reading, Berkshire, RG7 4PN, UK, Pavel L. Ivanov, Engelhardt Institute of Molecular Biology, Russian Academy of Sciences, 117984, Moscow, Russia, Colin Kimpton, Romelle Piercy, Nicola Benson, Gillian Tully, Ian Evett, Kevin Sullivan, Forensic Science Service, Priory House, Gooch Street North, Birmingham B5 6QQ, UK, Erika Hagelberg, University of Cambridge, Department of Biological Anthropology, Downing Street, Cambridge CB2 3DZ, UK -

[65] Murnaghan, Ian, (28 December 2012) Famous Trials and DNA Testing; Earl Washington Jr. Explore DNA, Retrieved 13 November 2014

[66] Jeffries, Stuart (2006-10-08). "Suspect Nation". London: The Guardian. Retrieved April 1, 2010.

[67] (June 2012) Frank Lee Smith The University of Michigan Law School, National Registry of Exonerations, Retrieved 13 November 2014

[68] Gordon, Stephen (2008-02-17). "Freedom in bag for killer Graham?". Belfasttelegraph.co.uk. Retrieved 2010-06-19.

[69] Dutter, Barbie (2001-06-19). "18 years on, man is jailed for murder of Briton in 'paradise'". London: The Telegraph. Retrieved 2008-06-17.

[70] McCutcheon, Peter (2004-09-08). "DNA evidence may not be infallible: experts". Australian Broadcasting Corporation. Retrieved 2008-06-17.

[71] Joshua Rozenberg,"DNA proves Hanratty guilt 'beyond doubt'", *Daily Telegraph*, London, 11 May 2002.

[72] John Steele, "Hanratty lawyers reject DNA 'guilt'", *Daily Telegraph*, London, 23 June 2001.

[73] "Hanratty: The damning DNA". *BBC News*. 10 May 2002. Retrieved 2011-08-22.

[74] "Mistaken identity claim over murder". *BBC News*. February 15, 2003. Retrieved April 1, 2010.

[75] Satish Sekar. "Lynette White Case: How Forensics Caught the Cellophane Man". Lifeloom.com. Retrieved 2010-04-03.

[76] (18 April 2014) Dennis Halstead The National Registry of Exonerations, University of Michigan Law School, Retrieved 12 January 2015

[77] "DNA clears man of 1914 kidnapping conviction", *USA Today*, (May 5, 2004), by Allen G. Breed, Associated Press.

[78] CBS News story on the Jane Mixer murder case; March 24, 2007.

[79] Another CBS News story on the Mixer case; July 17, 2007.

[80] An advocacy site challenging Leiterman's conviction in the Mixer murder.

[81] Booth, Jenny. "Police name David Lace as true killer of Teresa De Simone". *The Times*.

[82] Doughery, Phil. "D.B. Tuber". History Link.

39.9 Further reading

- Kaye, David H. (2010). *The Double Helix and the Law of Evidence*. Cambridge, Mass.: Harvard University Press. ISBN 9780674035881. OCLC 318876881.

39.10 External links

- McKie, Robin McKie (24 May 2009). "Eureka moment that led to the discovery of DNA fingerprinting". *The Observer* (London).

- Forensic Science, Statistics, and the Law—Blog that tracks scientific and legal developments pertinent to forensic DNA profiling

- Create a DNA Fingerprint—PBS.org

- In silico simulation of Molecular Biology Techniques—A place to learn typing techniques by simulating them

- National DNA Databases in the EU

- The Innocence Record, Winston & Strawn LLP/The Innocence Project

- Making Sense of DNA Backlogs, 2012: Myths vs. Reality United States Department of Justice

- France Tries Mass DNA Test in Hunt for School Rapist

Chapter 40

DNA digital data storage

DNA digital data storage refers to any scheme to store digital data in the base sequence of DNA. This technology uses artificial DNA made using commercially available oligonucleotide synthesis machines for storage and DNA sequencing machines for retrieval. This type of storage system is more compact than current magnetic tape or hard drive storage systems due to the data density of the DNA. It also has the capability for longevity, as long as the DNA is held in cold, dry and dark conditions, as is shown by the study of woolly mammoth DNA from up to 60,000 years ago, and for resistance to obsolescence, as DNA is a universal and fundamental data storage mechanism in biology. These features have led to researchers involved in their development to call this method of data storage "apocalypse-proof" because "after a hypothetical global disaster, future generations might eventually find the stores and be able to read them." [1] It is, however, a slow process, as the DNA needs to be sequenced in order to retrieve the data, and so the method is intended for uses with a low access rate such as long-term archival of large amounts of scientific data.[1][2]

40.1 History

The idea and the general considerations about the possibility of recording, storage and retrieval of information on DNA molecules were originally made by Mikhail Neiman and published in 1964–65 in the *Radiotekhnika* journal, USSR, and the technology may therefore be referred to as MNeimON (Mikhail Neiman OligoNucleotides).[3]

On August 16, 2012, the journal *Science* published research by George Church and colleagues at Harvard University, in which DNA was encoded with digital information that included an HTML draft of a 53,400 word book written by the lead researcher, eleven JPG images and one JavaScript program. Multiple copies for redundancy were added and 5.5 petabits can be stored in each cubic millimeter of DNA.[4] The researchers used a simple code where bits were mapped one-to-one with bases, which had the shortcoming that it led to long runs of the same base, the sequencing of which is error-prone. This research result showed that besides its other functions, DNA can also be another type of storage medium such as hard drives and magnetic tapes.[1]

An improved system was reported in the journal *Nature* in January 2013, in an article lead by researchers from the European Bioinformatics Institute (EBI) and submitted at around the same time as the paper of Church and colleagues. Over five million bits of data, appearing as a speck of dust to researchers, and consisting of text files and audio files, were successfully stored and then perfectly retrieved and reproduced. Encoded information consisted of all 154 of Shakespeare's sonnets, a twenty-six-second audio clip of the "I Have a Dream" speech by Martin Luther King, the well known paper on the structure of DNA by James Watson and Francis Crick, a photograph of EBI headquarters in Hinxton, United Kingdom, and a file describing the methods behind converting the data. All the DNA files reproduced the information between 99.99% and 100% accuracy.[2] The main innovations in this research were the use of an error-correcting encoding scheme to ensure the extremely low data-loss rate, as well as the idea of encoding the data in a series of overlapping short oligonucleotides identifiable through a sequence-based indexing scheme.[1] Also, the sequences of the individual strands of DNA overlapped in such a way that each region of data was repeated four times to avoid errors. Two of these four strands were constructed backwards, also with the goal of eliminating errors.[2] The costs per megabyte were estimated at $12,400 to encode data and $220 for retrieval. However, it was noted that the exponential decrease in DNA synthesis and sequencing costs, if it continues into the future, should make the technology cost-effective for long-term data storage within about ten years.[1]

The long-term stability of data encoded in DNA was reported in February 2015, in an article by researches from ETH Zurich. By adding redundancy via Reed–Solomon error correction coding and by encapsulating the DNA within silica glass spheres via Sol-gel chemistry, the researchers predict error-free information recovery after up to 1 million years at -18 °C and 2000 years if stored at 10 °C.[5][6] By adding the possibility of being able to handle errors, the research team could reduce the cost of DNA synthesis down to ~$500/MB by choosing a more error-prone DNA synthesis method. In a news article in the New Scientist the team stated that if they are able to further decrease the cost they would store an archive version of wikipedia in DNA.

40.2   See also

- DNA computing

- DNA nanotechnology

- Nanobiotechnology

- Natural computing

40.3   References

[1] Yong, E. (2013). "Synthetic double-helix faithfully stores Shakespeare's sonnets". *Nature*. doi:10.1038/nature.2013.12279.

[2] Goldman, N.; Bertone, P.; Chen, S.; Dessimoz, C.; Leproust, E. M.; Sipos, B.; Birney, E. (2013). "Towards practical, high-capacity, low-maintenance information storage in synthesized DNA". *Nature* **494** (7435): 77–80. doi:10.1038/nature11875. PMC 3672958. PMID 23354052.

[3] https://sites.google.com/site/msneiman1905/eng

[4] Church, G. M.; Gao, Y.; Kosuri, S. (2012). "Next-Generation Digital Information Storage in DNA". *Science* **337** (6102): 1628. doi:10.1126/science.1226355. PMID 22903519.

[5] Grass, R. N.; Heckel, R.; Puddu, M.; Paunescu, D.; Stark, W. J. (2015). "Robust Chemical Preservation of Digital Information on DNA in Silica with Error-Correcting Codes". *Angewandte Chemie International Edition***54**(8): 2552. doi:10.1002/anie.201

[6] Jacobs, Angelika (February 13, 2015). "Data-storage for eternity". Eidgenössische Technische Hochschule (ETH) Zürich. Archived from the original on March 15, 2015. Retrieved March 15, 2015.

40.4   Further reading

- Edwards, Lin (August 17, 2012). "DNA used to encode a book and other digital information". *Phys Org* (Phys Org). Retrieved 2013-01-28.

- Mardis, E. R. (2008). "Next-Generation DNA Sequencing Methods". *Annual Review of Genomics and Human Genetics* **9**: 387–402. doi:10.1146/annurev.genom.9.081307.164359. PMID 18576944.

- Cole, Adam (January 24, 2013). "Shall I Encode Thee In DNA? Sonnets Stored On Double Helix?" (Download article and audio is available). National Public Radio.

- Naik, Gautam (January 24, 2013). "Storing Digital Data in DNA". *The Wall Street Journal* (New York City: Dow Jones & Company). Retrieved 2012-01-25.

- Wall Street Journal article. "Storing Digital Data in DNA"

- Ewan Birney's Blog. "Using DNA as a digital archive media"

  - Also see "The 10,000 year archive"

- Ed Yong's National Geographic blog. "Shakespeare's Sonnets and MLK's Speech Stored in DNA Speck"

- DNA Sequencing Caught in Deluge of Data. The New York Times (NYTimes.com).

- Aron, Jacob (February 15, 2015). "Glassed-in DNA makes the ultimate time capsule". *New Scientist.* Retrieved February 19, 2015.

Chapter 41

DNA sequencing

DNA sequencing is the process of determining the precise order of nucleotides within a DNA molecule. It includes any method or technology that is used to determine the order of the four bases—adenine, guanine, cytosine, and thymine—in a strand of DNA. The advent of rapid DNA sequencing methods has greatly accelerated biological and medical research and discovery.

Knowledge of DNA sequences has become indispensable for basic biological research, and in numerous applied fields such as medical diagnosis, biotechnology, forensic biology, virology and biological systematics. The rapid speed of sequencing attained with modern DNA sequencing technology has been instrumental in the sequencing of complete DNA sequences, or genomes of numerous types and species of life, including the human genome and other complete DNA sequences of many animal, plant, and microbial species.

The first DNA sequences were obtained in the early 1970s by academic researchers using laborious methods based on two-dimensional chromatography. Following the development of fluorescence-based sequencing methods with a DNA sequencer,[1] DNA sequencing has become easier and orders of magnitude faster.[2]

41.1 Use of sequencing

DNA sequencing may be used to determine the sequence of individual genes, larger genetic regions (i.e. clusters of genes or operons), full chromosomes or entire genomes. Sequencing provides the order of individual nucleotides present in molecules of DNA or RNA isolated from animals, plants, bacteria, archaea, or virtually any other source of genetic information. This information is useful to various fields of biology and other sciences, medicine, forensics, and other areas of study.

41.1.1 Molecular biology

Sequencing is used in molecular biology to study genomes and the proteins they encode. Information obtained using sequencing allows researchers to identify changes in genes, associations with diseases and phenotypes, and identify potential drug targets.

41.1.2 Evolutionary biology

Since DNA is an informative macromolecule in terms of transmission from one generation to another, DNA sequencing is used in evolutionary biology to study how different organisms are related and how they evolved.

41.1.3 Metagenomics

Main article: Metagenomics

The field of metagenomics involves identification of organisms present in a body of water, sewage, dirt, debris filtered from the air, or swab samples from organisms. Knowing which organisms are present in a particular environment is critical to research in ecology, epidemiology, microbiology, and other fields. Sequencing enables researchers to determine which types of microbes may be present in a microbiome, for example.

41.1.4 Medicine

Medical technicians may sequence genes (or, theoretically, full genomes) from patients to determine if there is risk of genetic diseases. This is a form of genetic testing, though some genetic tests may not involve DNA sequencing.

41.1.5 Forensics

DNA sequencing may be used along with DNA profiling methods for forensic identification and paternity testing.

41.2 The four canonical bases

Main article: Nucleotide

The canonical structure of DNA has four bases: thymine (T), adenine (A), cytosine (C), and guanine (G). DNA sequencing is the determination of the physical order of these bases in a molecule of DNA. However, there are many other bases that may be present in a molecule. In some viruses (specifically, bacteriophage), cytosine may be replaced by hydroxy methyl or hydroxy methyl glucose cytosine.[3] In mammalian DNA, variant bases with methyl groups or phosphosulfate may be found.[4][5] Depending on the sequencing technique, a particular modification may or may not be detected, e.g., the 5mC (5 methyl cytosine) common in humans may or may not be detected.[6]

41.3 History

Deoxyribonucleic acid (DNA) was first discovered and isolated by Friedrich Miescher in 1869, but it remained understudied for many decades because proteins, rather than DNA, were thought to hold the genetic blueprint to life. This situation changed after 1944 as a result of some experiments by Oswald Avery, Colin MacLeod, and Maclyn McCarty demonstrated that purified DNA could change one strain of bacteria into another type. This was the first time that DNA was shown capable of transforming the properties of cells.

In 1953 James Watson and Francis Crick put forward their double-helix model of DNA which depicted DNA being made up of two strands of nucleotides coiled around each other, linked together by hydrogen bonds, in a spiral configuration. Each strand they argued was composed of four complementary nucleotides: adenine (A), cytosine (C), guanine (G) and thymine (T) and was oriented in opposite directions. Such a structure they proposed allowed each strand to reconstruct the other and was central to the passing on of hereditary information between generations.[7]

The foundation for sequencing DNA was first laid by the work of Fred Sanger who by 1955 had completed the sequence of all the amino acids in insulin, a small protein secreted by the pancreas. This provided the first conclusive evidence that proteins were chemical entities with a specific molecular pattern rather than a random mixture of material suspended in fluid. Sanger's success in sequencing insulin greatly electrified x-ray crystallographers, including Watson and Crick who by now were trying to understand how DNA directed the formation of proteins within a cell. Soon after attending a series of lectures given by Fred Sanger in October 1954, Crick began to develop a theory which argued that the arrangement of nucleotides in DNA determined the sequence of amino acids in proteins which in turn helped determine the function of a protein. He published this theory in 1958. [8]

41.3.1   RNA sequencing

RNA sequencing was one of the earliest forms of nucleotide sequencing. The major landmark of RNA sequencing is the sequence of the first complete gene and the complete genome of Bacteriophage MS2, identified and published by Walter Fiers and his coworkers at the University of Ghent (Ghent, Belgium), in 1972[9] and 1976.[10]

41.3.2   Early DNA sequencing methods

The first method for determining DNA sequences involved a location-specific primer extension strategy established by Ray Wu at Cornell University in 1970.[11] DNA polymerase catalysis and specific nucleotide labeling, both of which figure prominently in current sequencing schemes, were used to sequence the cohesive ends of lambda phage DNA[12][13][14] Between 1970 and 1973, Wu, R Padmanabhan and colleagues demonstrated that this method can be employed to determine any DNA sequence using synthetic location-specific primers.[15][16][17] Frederick Sanger then adopted this primer-extension strategy to develop more rapid DNA sequencing methods at the MRC Centre, Cambridge, UK and published a method for "DNA sequencing with chain-terminating inhibitors" in 1977.[18] Walter Gilbert and Allan Maxam at Harvard also developed sequencing methods, including one for "DNA sequencing by chemical degradation".[19][20] In 1973, Gilbert and Maxam reported the sequence of 24 basepairs using a method known as wandering-spot analysis.[21] Advancements in sequencing were aided by the concurrent development of recombinant DNA technology, allowing DNA samples to be isolated from sources other than viruses.

41.3.3   Sequencing of full genomes

The first full DNA genome to be sequenced was that of bacteriophage φX174 in 1977.[22] Medical Research Council scientists deciphered the complete DNA sequence of the Epstein-Barr virus in 1984, finding it contained 172,282 nucleotides. Completion of the sequence marked a significant turning point in DNA sequencing because it was achieved with no prior genetic profile knowledge of the virus.[23]

A non-radioactive method for transferring the DNA molecules of sequencing reaction mixtures onto an immobilizing matrix during electrophoresis was developed by Pohl and co-workers in the early 80's.[24][25] Followed by the commercialization of the DNA sequencer "Direct-Blotting-Electrophoresis-System GATC 1500" by GATC Biotech, which was intensively used in the framework of the EU genome-sequencing programme, the complete DNA sequence of the yeast *Saccharomyces cerevisiae* chromosome II.[26] Leroy E. Hood's laboratory at the California Institute of Technology announced the first semi-automated DNA sequencing machine in 1986.[27] This was followed by Applied Biosystems' marketing of the first fully automated sequencing machine, the ABI 370, in 1987 and by Dupont's Genesis 2000[28] which used a novel fluorescent labeling technique enabling all four dideoxynucleotides to be identified in a single lane. By 1990, the U.S. National Institutes of Health (NIH) had begun large-scale sequencing trials on *Mycoplasma capricolum*, *Escherichia coli*, *Caenorhabditis elegans*, and *Saccharomyces cerevisiae* at a cost of US$0.75 per base. Meanwhile, sequencing of human cDNA sequences called expressed sequence tags began in Craig Venter's lab, an attempt to capture the coding fraction of the human genome.[29] In 1995, Venter, Hamilton Smith, and colleagues at The Institute for Genomic Research (TIGR) published the first complete genome of a free-living organism, the bacterium *Haemophilus influenzae*. The circular chromosome contains 1,830,137 bases and its publication in the journal Science[30] marked the first published use of whole-genome shotgun sequencing, eliminating the need for initial mapping efforts.

By 2001, shotgun sequencing methods had been used to produce a draft sequence of the human genome.[31][32]

41.3.4   Next-generation sequencing methods

Several new methods for DNA sequencing were developed in the mid to late 1990s and were implemented in commercial DNA sequencers by the year 2000.

On October 26, 1990, Roger Tsien, Pepi Ross, Margaret Fahnestock and Allan J Johnston filed a patent describing stepwise ("base-by-base") sequencing with removable 3' blockers on DNA arrays (blots and single DNA molecules).[33] In 1996, Pål Nyrén and his student Mostafa Ronaghi at the Royal Institute of Technology in Stockholm published their method of pyrosequencing.[34]

On April 1, 1997, Pascal Mayer and Laurent Farinelli submitted patents to the World Intellectual Property Organization describing DNA colony sequencing.[35] The DNA sample preparation and random surface-PCR arraying methods described in this patent, coupled to Roger Tsien et al.'s "base-by-base" sequencing method, is now implemented in Illumina's Hi-Seq genome sequencers.

Lynx Therapeutics published and marketed "Massively parallel signature sequencing", or MPSS, in 2000. This method incorporated a parallelized, adapter/ligation-mediated, bead-based sequencing technology and served as the first commercially available "next-generation" sequencing method, though no DNA sequencers were sold to independent laboratories.[36]

In 2004, 454 Life Sciences marketed a parallelized version of pyrosequencing.[37] The first version of their machine reduced sequencing costs 6-fold compared to automated Sanger sequencing, and was the second of the new generation of sequencing technologies, after MPSS.[38]

The large quantities of data produced by DNA sequencing have also required development of new methods and programs for sequence analysis. Phil Green and Brent Ewing of the University of Washington described their phred quality score for sequencer data analysis in 1998.[39]

41.4 Basic methods

41.4.1 Maxam-Gilbert sequencing

Main article: Maxam-Gilbert sequencing

Allan Maxam and Walter Gilbert published a DNA sequencing method in 1977 based on chemical modification of DNA and subsequent cleavage at specific bases.[19] Also known as chemical sequencing, this method allowed purified samples of double-stranded DNA to be used without further cloning. This method's use of radioactive labeling and its technical complexity discouraged extensive use after refinements in the Sanger methods had been made.

Maxam-Gilbert sequencing requires radioactive labeling at one 5' end of the DNA and purification of the DNA fragment to be sequenced. Chemical treatment then generates breaks at a small proportion of one or two of the four nucleotide bases in each of four reactions (G, A+G, C, C+T). The concentration of the modifying chemicals is controlled to introduce on average one modification per DNA molecule. Thus a series of labeled fragments is generated, from the radiolabeled end to the first "cut" site in each molecule. The fragments in the four reactions are electrophoresed side by side in denaturing acrylamide gels for size separation. To visualize the fragments, the gel is exposed to X-ray film for autoradiography, yielding a series of dark bands each corresponding to a radiolabeled DNA fragment, from which the sequence may be inferred.[19]

41.4.2 Chain-termination methods

Main article: Sanger sequencing

The chain-termination method developed by Frederick Sanger and coworkers in 1977 soon became the method of choice, owing to its relative ease and reliability.[18][40] When invented, the chain-terminator method used fewer toxic chemicals and lower amounts of radioactivity than the Maxam and Gilbert method. Because of its comparative ease, the Sanger method was soon automated and was the method used in the first generation of DNA sequencers.

Sanger sequencing is the method which prevailed from the 80's until the mid-2000s. Over that period, great advances were made in the technique, such as fluorescent labelling, capillary electrophoresis, and general automation. These developments allowed much more efficient sequencing, leading to lower costs. The Sanger method, in mass production form, is the technology which produced the first human genome in 2001, ushering in the age of genomics. However, later in the decade, radically different approaches reached the market, bringing the cost per genome down from $100 million in 2001 to $10,000 in 2011.[41]

41.5 Advanced methods and *de novo* sequencing

Large-scale sequencing often aims at sequencing very long DNA pieces, such as whole chromosomes, although large-scale sequencing can also be used to generate very large numbers of short sequences, such as found in phage display. For longer targets such as chromosomes, common approaches consist of cutting (with restriction enzymes) or shearing (with mechanical forces) large DNA fragments into shorter DNA fragments. The fragmented DNA may then be cloned into a DNA vector and amplified in a bacterial host such as *Escherichia coli*. Short DNA fragments purified from individual bacterial colonies are individually sequenced and assembled electronically into one long, contiguous sequence. Studies have shown that adding a size selection step to collect DNA fragments of uniform size can improve sequencing efficiency and accuracy of the genome assembly. In these studies, automated sizing has proven to be more reproducible and precise than manual gel sizing.[42][43][44]

The term "*de novo* sequencing" specifically refers to methods used to determine the sequence of DNA with no previously known sequence. *De novo* translates from Latin as "from the beginning". Gaps in the assembled sequence may be filled by primer walking. The different strategies have different tradeoffs in speed and accuracy; shotgun methods are often used for sequencing large genomes, but its assembly is complex and difficult, particularly with sequence repeats often causing gaps in genome assembly.

Most sequencing approaches use an *in vitro* cloning step to amplify individual DNA molecules, because their molecular detection methods are not sensitive enough for single molecule sequencing. Emulsion PCR[45] isolates individual DNA molecules along with primer-coated beads in aqueous droplets within an oil phase. A polymerase chain reaction (PCR) then coats each bead with clonal copies of the DNA molecule followed by immobilization for later sequencing. Emulsion PCR is used in the methods developed by Marguilis et al. (commercialized by 454 Life Sciences), Shendure and Porreca et al. (also known as "Polony sequencing") and SOLiD sequencing, (developed by Agencourt, later Applied Biosystems, now Life Technologies).[46][47][48]

41.5.1 Shotgun sequencing

Main article: Shotgun sequencing

Shotgun sequencing is a sequencing method designed for analysis of DNA sequences longer than 1000 base pairs, up to and including entire chromosomes. This method requires the target DNA to be broken into random fragments. After sequencing individual fragments, the sequences can be reassembled on the basis of their overlapping regions.[49]

41.5.2 Bridge PCR

Another method for *in vitro* clonal amplification is bridge PCR, in which fragments are amplified upon primers attached to a solid surface[35][50][51] and form "DNA colonies" or "DNA clusters". This method is used in the Illumina Genome Analyzer sequencers. Single-molecule methods, such as that developed by Stephen Quake's laboratory (later commercialized by Helicos) are an exception: they use bright fluorophores and laser excitation to detect base addition events from individual DNA molecules fixed to a surface, eliminating the need for molecular amplification.[52]

41.6 Next-generation methods

Next-generation sequencing applies to genome sequencing, genome resequencing, transcriptome profiling (RNA-Seq), DNA-protein interactions (ChIP-sequencing), and epigenome characterization.[53] Resequencing is necessary, because the genome of a single individual of a species will not indicate all of the genome variations among other individuals of the same species.

The high demand for low-cost sequencing has driven the development of high-throughput sequencing (or next-generation sequencing) technologies that parallelize the sequencing process, producing thousands or millions of sequences concurrently. High-throughput sequencing technologies are intended to lower the cost of DNA sequencing beyond what is possible with standard dye-terminator methods.[38] In ultra-high-throughput sequencing as many as 500,000 sequencing-by-synthesis operations may be run in parallel.[56][57][58]

41.6.1   Massively parallel signature sequencing (MPSS)

The first of the next-generation sequencing technologies, massively parallel signature sequencing (or MPSS), was developed in the 1990s at Lynx Therapeutics, a company founded in 1992 by Sydney Brenner and Sam Eletr. MPSS was a bead-based method that used a complex approach of adapter ligation followed by adapter decoding, reading the sequence in increments of four nucleotides. This method made it susceptible to sequence-specific bias or loss of specific sequences. Because the technology was so complex, MPSS was only performed 'in-house' by Lynx Therapeutics and no DNA sequencing machines were sold to independent laboratories. Lynx Therapeutics merged with Solexa (later acquired by Illumina) in 2004, leading to the development of sequencing-by-synthesis, a simpler approach acquired from Manteia Predictive Medicine, which rendered MPSS obsolete. However, the essential properties of the MPSS output were typical of later "next-generation" data types, including hundreds of thousands of short DNA sequences. In the case of MPSS, these were typically used for sequencing cDNA for measurements of gene expression levels.[36]

41.6.2   Polony sequencing

Main article: Polony sequencing

The Polony sequencing method, developed in the laboratory of George M. Church at Harvard, was among the first next-generation sequencing systems and was used to sequence a full *E. coli* genome in 2005.[71] It combined an in vitro paired-tag library with emulsion PCR, an automated microscope, and ligation-based sequencing chemistry to sequence an *E. coli* genome at an accuracy of >99.9999% and a cost approximately 1/9 that of Sanger sequencing.[71] The technology was licensed to Agencourt Biosciences, subsequently spun out into Agencourt Personal Genomics, and eventually incorporated into the Applied Biosystems SOLiD platform. Applied Biosystems was later acquired by Life Technologies, now part of Thermo Fisher Scientific.

41.6.3   454 pyrosequencing

Main article: 454 Life Sciences § Technology

A parallelized version of pyrosequencing was developed by 454 Life Sciences, which has since been acquired by Roche Diagnostics. The method amplifies DNA inside water droplets in an oil solution (emulsion PCR), with each droplet containing a single DNA template attached to a single primer-coated bead that then forms a clonal colony. The sequencing machine contains many picoliter-volume wells each containing a single bead and sequencing enzymes. Pyrosequencing uses luciferase to generate light for detection of the individual nucleotides added to the nascent DNA, and the combined data are used to generate sequence read-outs.[46] This technology provides intermediate read length and price per base compared to Sanger sequencing on one end and Solexa and SOLiD on the other.[38]

41.6.4   Illumina (Solexa) sequencing

Main article: Illumina dye sequencing

Solexa, now part of Illumina, was founded by Shankar Balasubramanian and David Klenerman in 1998, and developed a sequencing method based on reversible dye-terminators technology, and engineered polymerases.[72] The terminated chemistry was developed internally at Solexa and the concept of the Solexa system was invented by Balasubramanian and Klenerman from Cambridge University's chemistry department. In 2004, Solexa acquired the company Manteia Predictive Medicine in order to gain a massivelly parallel sequencing technology invented in 1997 by Pascal Mayer and Laurent Farinelli.[35] It is based on "DNA Clusters" or "DNA colonies", which involves the clonal amplification of DNA on a surface. The cluster technology was co-acquired with Lynx Therapeutics of California. Solexa Ltd. later merged with Lynx to form Solexa Inc.

In this method, DNA molecules and primers are first attached on a slide or flow cell and amplified with polymerase so that local clonal DNA colonies, later coined "DNA clusters", are formed. To determine the sequence, four types of reversible terminator bases (RT-bases) are added and non-incorporated nucleotides are washed away. A camera takes images of the fluorescently labeled nucleotides. Then the dye, along with the terminal 3' blocker, is chemically removed from the DNA, allowing for the next cycle to begin. Unlike pyrosequencing, the DNA chains are extended one nucleotide at a time and image acquisition can be performed at a delayed moment, allowing for very large arrays of DNA colonies to be captured by sequential images taken from a single camera.

Decoupling the enzymatic reaction and the image capture allows for optimal throughput and theoretically unlimited sequencing capacity. With an optimal configuration, the ultimately reachable instrument throughput is thus dictated solely by the analog-to-digital conversion rate of the camera, multiplied by the number of cameras and divided by the number of pixels per DNA colony required for visualizing them optimally (approximately 10 pixels/colony). In 2012, with cameras operating at more than 10 MHz A/D conversion rates and available optics, fluidics and enzymatics, throughput can be multiples of 1 million nucleotides/second, corresponding roughly to 1 human genome equivalent at 1x coverage per hour per instrument, and 1 human genome re-sequenced (at approx. 30x) per day per instrument (equipped with a single camera).[73]

41.6.5   SOLiD sequencing

Main article: 2 base encoding

Applied Biosystems' (now a Life Technologies brand) SOLiD technology employs sequencing by ligation. Here, a pool of all possible oligonucleotides of a fixed length are labeled according to the sequenced position. Oligonucleotides are annealed and ligated; the preferential ligation by DNA ligase for matching sequences results in a signal informative of the nucleotide at that position. Before sequencing, the DNA is amplified by emulsion PCR. The resulting beads, each containing single copies of the same DNA molecule, are deposited on a glass slide.[74] The result is sequences of quantities and lengths comparable to Illumina sequencing.[38] This sequencing by ligation method has been reported to have some issue sequencing palindromic sequences.[70]

41.6.6   Ion Torrent semiconductor sequencing

Main article: Ion semiconductor sequencing

Ion Torrent Systems Inc. (now owned by Life Technologies) developed a system based on using standard sequencing chemistry, but with a novel, semiconductor based detection system. This method of sequencing is based on the detection of hydrogen ions that are released during the polymerisation of DNA, as opposed to the optical methods used in other sequencing systems. A microwell containing a template DNA strand to be sequenced is flooded with a single type of nucleotide. If the introduced nucleotide is complementary to the leading template nucleotide it is incorporated into the growing complementary strand. This causes the release of a hydrogen ion that triggers a hypersensitive ion sensor, which indicates that a reaction has occurred. If homopolymer repeats are present in the template sequence multiple nucleotides will be incorporated in a single cycle. This leads to a corresponding number of released hydrogens and a proportionally higher electronic signal.[75]

41.6.7 DNA nanoball sequencing

Main article: DNA nanoball sequencing

DNA nanoball sequencing is a type of high throughput sequencing technology used to determine the entire genomic sequence of an organism. The company Complete Genomics uses this technology to sequence samples submitted by independent researchers. The method uses rolling circle replication to amplify small fragments of genomic DNA into DNA nanoballs. Unchained sequencing by ligation is then used to determine the nucleotide sequence.[76] This method of DNA sequencing allows large numbers of DNA nanoballs to be sequenced per run and at low reagent costs compared to other next generation sequencing platforms.[77] However, only short sequences of DNA are determined from each DNA nanoball which makes mapping the short reads to a reference genome difficult.[76] This technology has been used for multiple genome sequencing projects and is scheduled to be used for more.[78]

41.6.8 Heliscope single molecule sequencing

Main article: Helicos single molecule fluorescent sequencing

Heliscope sequencing is a method of single-molecule sequencing developed by Helicos Biosciences. It uses DNA fragments with added poly-A tail adapters which are attached to the flow cell surface. The next steps involve extension-based sequencing with cyclic washes of the flow cell with fluorescently labeled nucleotides (one nucleotide type at a time, as with the Sanger method). The reads are performed by the Heliscope sequencer.[79][80] The reads are short, averaging 35 bp.[81] In 2009 a human genome was sequenced using the Heliscope, however in 2012 the company went bankrupt.[82]

41.6.9 Single molecule real time (SMRT) sequencing

Main article: Single molecule real time sequencing

SMRT sequencing is based on the sequencing by synthesis approach. The DNA is synthesized in zero-mode waveguides (ZMWs) – small well-like containers with the capturing tools located at the bottom of the well. The sequencing is performed with use of unmodified polymerase (attached to the ZMW bottom) and fluorescently labelled nucleotides flowing freely in the solution. The wells are constructed in a way that only the fluorescence occurring by the bottom of the well is detected. The fluorescent label is detached from the nucleotide upon its incorporation into the DNA strand, leaving an unmodified DNA strand. According to Pacific Biosciences (PacBio), the SMRT technology developer, this methodology allows detection of nucleotide modifications (such as cytosine methylation). This happens through the observation of polymerase kinetics. This approach allows reads of 20,000 nucleotides or more, with average read lengths of 5 kilobases.[65][83] In 2015, Pacific Biosciences announced the launch of a new sequencing instrument called the Sequel System, with 1 million ZMWs compared to 150,000 ZMWs in the PacBio RS II instrument.[84][85]

41.7 Methods in development

DNA sequencing methods currently under development include reading the sequence as a DNA strand transits through nanopores,[86][87] and microscopy-based techniques, such as atomic force microscopy or transmission electron microscopy that are used to identify the positions of individual nucleotides within long DNA fragments (>5,000 bp) by nucleotide labeling with heavier elements (e.g., halogens) for visual detection and recording.[88][89] Third generation technologies aim to increase throughput and decrease the time to result and cost by eliminating the need for excessive reagents and harnessing the processivity of DNA polymerase.[90]

41.7.1   Nanopore DNA sequencing

Main article: Nanopore sequencing

This method is based on the readout of electrical signals occurring at nucleotides passing by alpha-hemolysin pores covalently bound with cyclodextrin. The DNA passing through the nanopore changes its ion current. This change is dependent on the shape, size and length of the DNA sequence. Each type of the nucleotide blocks the ion flow through the pore for a different period of time. The method has a potential of development as it does not require modified nucleotides, however single nucleotide resolution is not yet available.[91]

Two main areas of nanopore sequencing in development are solid state nanopore sequencing, and protein based nanopore sequencing. Protein nanopore sequencing utilizes membrane protein complexes $\propto$-Hemolysin and MspA (Mycobacterium Smegmatis Porin A), which show great promise given their ability to distinguish between individual and groups of nucleotides.[92] Whereas, solid-state nanopore sequencing utilizes synthetic materials such as silicon nitride and aluminum oxide and it is preferred for its superior mechanical ability and thermal and chemical stability.[93] The fabrication method is essential for this type of sequencing given that the nanopore array can contain hundreds of pores with diameters smaller than eight nanometers.[92]

The concept originated from the idea that single stranded DNA or RNA molecules can be electrophoretically driven in a strict linear sequence through a biological pore that can be less than eight nanometers, and can be detected given that the molecules release an ionic current while moving through the pore. The pore contains a detection region capable of recognizing different bases, with each base generating various time specific signals corresponding to the sequence of bases as they cross the pore which are then evaluated.[93] When implementing this process it is important to note that precise control over the DNA transport through the pore is crucial for success. Various enzymes such as exonucleases and polymerases have been used to moderate this process by positioning them near the pore's entrance.[94]

Oxford Nanopore Technologies, a United Kingdom-based startup company, is currently developing products using nanopore sequencing. These products include the MinION, a handheld sequencer capable of generating more than 150 megabases of sequencing data in one run. The MinION is not yet available to the public and has been found to produce numerous errors, though further study may alleviate the issue.[95][96]

41.7.2   Tunnelling currents DNA sequencing

Another approach uses measurements of the electrical tunnelling currents across single-strand DNA as it moves through a channel. Depending on its electronic structure each base affects the tunnelling current differently, allowing differentiation between different bases.[97]

The use of tunnelling currents has the potential to sequence orders of magnitude faster than ionic current methods and the sequencing of several DNA oligomers and micro-RNA has already been achieved.[98]

41.7.3   Sequencing by hybridization

Sequencing by hybridization is a non-enzymatic method that uses a DNA microarray. A single pool of DNA whose sequence is to be determined is fluorescently labeled and hybridized to an array containing known sequences. Strong hybridization signals from a given spot on the array identifies its sequence in the DNA being sequenced.[99]

This method of sequencing utilizes binding characteristics of a library of short single stranded DNA molecules (oligonucleotides) also called DNA probes to reconstruct a target DNA sequence. Non-specific hybrids are removed by washing and the target DNA is eluted.[100] Hybrids are re-arranged such that the DNA sequence can be reconstructed. The benefit of this sequencing type is its ability to capture a large number of targets with a homogenous coverage.[101] Although a large number of chemicals and starting DNA is usually required. But, with the advent of solution-based hybridization much less equipment and chemicals are necessary.[100]

41.7.4 Sequencing with mass spectrometry

Mass spectrometry may be used to determine DNA sequences. Matrix-assisted laser desorption ionization time-of-flight mass spectrometry, or MALDI-TOF MS, has specifically been investigated as an alternative method to gel electrophoresis for visualizing DNA fragments. With this method, DNA fragments generated by chain-termination sequencing reactions are compared by mass rather than by size. The mass of each nucleotide is different from the others and this difference is detectable by mass spectrometry. Single-nucleotide mutations in a fragment can be more easily detected with MS than by gel electrophoresis alone. MALDI-TOF MS can more easily detect differences between RNA fragments, so researchers may indirectly sequence DNA with MS-based methods by converting it to RNA first.[102]

The higher resolution of DNA fragments permitted by MS-based methods is of special interest to researchers in forensic science, as they may wish to find single-nucleotide polymorphisms in human DNA samples to identify individuals. These samples may be highly degraded so forensic researchers often prefer mitochondrial DNA for its higher stability and applications for lineage studies. MS-based sequencing methods have been used to compare the sequences of human mitochondrial DNA from samples in a Federal Bureau of Investigation database[103] and from bones found in mass graves of World War I soldiers.[104]

Early chain-termination and TOF MS methods demonstrated read lengths of up to 100 base pairs.[105] Researchers have been unable to exceed this average read size; like chain-termination sequencing alone, MS-based DNA sequencing may not be suitable for large *de novo* sequencing projects. Even so, a recent study did use the short sequence reads and mass spectroscopy to compare single-nucleotide polymorphisms in pathogenic *Streptococcus* strains.[106]

41.7.5 Microfluidic Sanger sequencing

Main article: Sanger sequencing

In microfluidic Sanger sequencing the entire thermocycling amplification of DNA fragments as well as their separation by electrophoresis is done on a single glass wafer (approximately 10 cm in diameter) thus reducing the reagent usage as well as cost.[107] In some instances researchers have shown that they can increase the throughput of conventional sequencing through the use of microchips.[108] Research will still need to be done in order to make this use of technology effective.

41.7.6 Microscopy-based techniques

Main article: Transmission electron microscopy DNA sequencing

This approach directly visualizes the sequence of DNA molecules using electron microscopy. The first identification of DNA base pairs within intact DNA molecules by enzymatically incorporating modified bases, which contain atoms of increased atomic number, direct visualization and identification of individually labeled bases within a synthetic 3,272 base-pair DNA molecule and a 7,249 base-pair viral genome has been demonstrated.[109]

41.7.7 RNAP sequencing

This method is based on use of RNA polymerase (RNAP), which is attached to a polystyrene bead. One end of DNA to be sequenced is attached to another bead, with both beads being placed in optical traps. RNAP motion during transcription brings the beads in closer and their relative distance changes, which can then be recorded at a single nucleotide resolution. The sequence is deduced based on the four readouts with lowered concentrations of each of the four nucleotide types, similarly to the Sanger method.[110] A comparison is made between regions and sequence information is deduced by comparing the known sequence regions to the unknown sequence regions.[111]

41.7.8 *In vitro* virus high-throughput sequencing

A method has been developed to analyze full sets of protein interactions using a combination of 454 pyrosequencing and an *in vitro* virus mRNA display method. Specifically, this method covalently links proteins of interest to the mRNAs encoding them, then detects the mRNA pieces using reverse transcription PCRs. The mRNA may then be amplified and sequenced. The combined method was titled IVV-HiTSeq and can be performed under cell-free conditions, though its results may not be representative of *in vivo* conditions.[112]

41.8 Sample preparation

Successful DNA sequencing depends upon sample preparation. In any particular sample used for DNA sequencing, the percent by mass that is DNA may be well under 1%. Furthermore, most *in vivo* DNA molecules are millions of base pairs long, while existing sequencing technology typically can handle DNA that is no more than one kilobase (or, perhaps, tens of kilobases for some methods). Thus there is a significant amount of work that is needed to purify the DNA, chop it into small pieces, make the pieces sequence ready, and determine the sequence. Researchers use bioinformatics to reassemble the short pieces into a biologically relevant sequence.

41.9 Development initiatives

In October 2006, the X Prize Foundation established an initiative to promote the development of full genome sequencing technologies, called the Archon X Prize, intending to award $10 million to "the first Team that can build a device and use it to sequence 100 human genomes within 10 days or less, with an accuracy of no more than one error in every 100,000 bases sequenced, with sequences accurately covering at least 98% of the genome, and at a recurring cost of no more than $10,000 (US) per genome."[113]

Each year the National Human Genome Research Institute, or NHGRI, promotes grants for new research and developments in genomics. 2010 grants and 2011 candidates include continuing work in microfluidic, polony and base-heavy sequencing methodologies.[114]

41.10 Computational challenges

The sequencing technologies described here produce raw data that needs to be assembled into longer sequences such as complete genomes (sequence assembly). There are many computational challenges to achieve this, such as the evaluation of the raw sequence data which is done by programs and algorithms such as Phred and Phrap. Other challenges have to deal with repetitive sequences that often prevent complete genome assemblies because they occur in many places of the genome. As a consequence, many sequences may not be assigned to particular chromosomes. The production of raw sequence data is only the beginning of its detailed bioinformatical analysis.[115] Yet new methods for sequencing and correcting sequencing errors were developed.[116]

41.10.1 Read trimming

Sometimes, the raw reads produced by the sequencer are correct and precise only in a fraction of their length. Using the entire read may introduce artifacts in the downstream analyses like genome assembly, snp calling, or gene expression estimation. Two classes of trimming programs have been introduced, based on the window-based or the running-sum classes of algorithms.[117] This is a partial list of the trimming algorithms currently available, specifying the algorithm class they belong to:

41.11 Ethical issues

Main article: Bioethics

Human genetics have been included within the field of bioethics since the early 1970s[124] and the growth in the use of DNA sequencing (particularly high-throughput sequencing) has introduced a number of ethical issues. One key issue is the ownership of an individual's DNA and the data produced when that DNA is sequenced.[125] Regarding the DNA molecule itself, the leading legal case on this topic, *Moore v. Regents of the University of California* (1990) ruled that individuals have no property rights to discarded cells or any profits made using these cells (for instance, as a patented cell line). However, individuals have a right to informed consent regarding removal and use of cells. Regarding the data produced through DNA sequencing, *Moore* gives the individual no rights to the information derived from their DNA.[125]

As DNA sequencing becomes more widespread, the storage, security and sharing of genomic data has also become more important.[125][126] For instance, one concern is that insurers may use an individual's genomic data to modify their quote, depending on the perceived future health of the individual based on their DNA.[126][127] In May 2008, the Genetic Information Nondiscrimination Act (GINA) was signed in the United States, prohibiting discrimination on the basis of genetic information with respect to health insurance and employment.[128][129] In 2012, the US Presidential Commission for the Study of Bioethical Issues reported that existing privacy legislation for DNA sequencing data such as GINA and the Health Insurance Portability and Accountability Act were insufficient, noting that whole-genome sequencing data was particularly sensitive, as it could be used to identify not only the individual from which the data was created, but also their relatives.[130][131]

Ethical issues have also been raised by the increasing use of genetic variation screening, both in newborns, and in adults by companies such as 23andMe.[132][133] It has been asserted that screening for genetic variations can be harmful, increasing anxiety in individuals who have been found to have an increased risk of disease.[134] For example, in one case noted in *Time*, doctors screening an ill baby for genetic variants chose not to inform the parents of an unrelated variant linked to dementia due to the harm it would cause to the parents.[135] However, a 2011 study in *The New England Journal of Medicine* has shown that individuals undergoing disease risk profiling did not show increased levels of anxiety.[134]

41.12 See also

41.13 References

[1] Olsvik O, Wahlberg J, Petterson B, Uhlén M, Popovic T, Wachsmuth IK, Fields PI (January 1993). "Use of automated sequencing of polymerase chain reaction-generated amplicons to identify three types of cholera toxin subunit B in Vibrio cholerae O1 strains". *J. Clin. Microbiol.* **31** (1): 22–25. PMC 262614. PMID 7678018.

[2] Pettersson E, Lundeberg J, Ahmadian A (February 2009). "Generations of sequencing technologies". *Genomics* **93** (2): 105–11. doi:10.1016/j.ygeno.2008.10.003. PMID 18992322.

[3] Moréra, Solange; Larivière, Laurent; Kurzeck, Jürgen; Aschke-Sonnenborn, Ursula; Freemont, Paul S; Janin, Joël; Rüger, Wolfgang (August 2001). "High resolution crystal structures of T4 phage β-glucosyltransferase: induced fit and effect of substrate and metal binding". *Journal of Molecular Biology* **311** (3): 569–577. doi:10.1006/jmbi.2001.4905. PMID 11493010.

[4] Ehrlich, Melanie; Gama-Sosa, Miguel A.; Huang, Lan-Hsiang; Midgett, Rose Marie; Kuo, Kenneth C.; McCune, Roy A.; Gehrke, Charles (1982). "Amount and distribution of 5-methylcytosine in human DNA from different types of tissues or cells". *Nucleic Acids Research* **10** (8): 2709–2721. doi:10.1093/nar/10.8.2709. PMC 320645. PMID 7079182.

[5] Ehrlich, M; Wang, R. (19 June 1981). "5-Methylcytosine in eukaryotic DNA". *Science* **212**(4501): 1350–1357. Bibcode:1981 doi:10.1126/science.6262918.

[6] Song, Chun-Xiao; Clark, Tyson A; Lu, Xing-Yu; Kislyuk, Andrey; Dai, Qing; Turner, Stephen W; He, Chuan; Korlach, Jonas (20 November 2011). "Sensitive and specific single-molecule sequencing of 5-hydroxymethylcytosine". *Nature Methods* **9** (1): 75–77. doi:10.1038/nmeth.1779.

[7]   Watson JD, Crick FH (1953). "The structure of DNA". *Cold Spring Harb. Symp. Quant. Biol.* **18**: 123–31. doi:10.1101/SQB
       PMID 13168976.

[8]   Marks, L, The path to DNA sequencing: The life and work of Fred Sanger.

[9]   Min Jou W, Haegeman G, Ysebaert M, Fiers W (May 1972). "Nucleotide sequence of the gene coding for the bacteriophage
       MS2 coat protein". *Nature* **237** (5350): 82–8. Bibcode:1972Natur.237...82J. doi:10.1038/237082a0. PMID 4555447.

[10] Fiers W, Contreras R, Duerinck F, Haegeman G, Iserentant D, Merregaert J, Min Jou W, Molemans F, Raeymaekers A, Van den
       Berghe A, Volckaert G, Ysebaert M (April 1976). "Complete nucleotide sequence of bacteriophage MS2 RNA: primary and
       secondary structure of the replicase gene". *Nature* **260** (5551): 500–7. Bibcode:1976Natur.260..500F. doi:10.1038/260500a0.
       PMID 1264203.

[11] "Ray Wu Faculty Profile". Cornell University. Archived from the original on 2009-03-04.

[12] PADMANABHAN, R; Ray Wu; Ernest Jay (June 1974). "Chemical Synthesis of a Primer and Its Use in the Sequence Anal-
       ysis of the Lysozyme Gene of Bacteriophage T4". *Proceedings of the National Academy of Sciences* **71** (6): 2510–2514.
       Bibcode:1974PNAS...71.2510P. doi:10.1073/pnas.71.6.2510.

[13] Onaga LA (June 2014). "Ray Wu as Fifth Business: Demonstrating Collective Memory in the History of DNA Sequencing".
       Studies in the History and Philosophy of Science. Part C **46**: 1–14. doi:10.1016/j.shpsc.2013.12.006. PMID 24565976.

[14] Wu R (1972). "Nucleotide sequence analysis of DNA". *Nature New Biol.* **236** (68): 198–200. doi:10.1038/newbio236198a0.
       PMID 4553110.

[15] Padmanabhan R, Wu R (1972). "Nucleotide sequence analysis of DNA. IX. Use of oligonucleotides of defined sequence as
       primers in DNA sequence analysis". *Biochem. Biophys. Res. Commun.* **48** (5): 1295–302. doi:10.1016/0006-291X(72)90852-
       2. PMID 4560009.

[16] Wu R, Tu CD, Padmanabhan R (1973). "Nucleotide sequence analysis of DNA. XII. The chemical synthesis and sequence
       analysis of a dodecadeoxynucleotide which binds to the endolysin gene of bacteriophage lambda". *Biochem. Biophys. Res.
       Commun.* **55** (4): 1092–9. doi:10.1016/S0006-291X(73)80007-5. PMID 4358929.

[17] Jay E, Bambara R, Padmanabhan R, Wu R (March 1974). "DNA sequence analysis: a general, simple and rapid method for se-
       quencing large oligodeoxyribonucleotide fragments by mapping". *Nucleic Acids Research* **1** (3): 331–353. doi:10.1093/nar/1.3.3
       31.PMC344020. PMID10793670.

[18] Sanger F, Nicklen S, Coulson AR (December 1977). "DNA sequencing with chain-terminating inhibitors". *Proc. Natl. Acad.
       Sci. U.S.A.* **74** (12): 5463–7. Bibcode:1977PNAS...74.5463S. doi:10.1073/pnas.74.12.5463. PMC 431765. PMID 271968.

[19] Maxam AM, Gilbert W (February 1977). "A new method for sequencing DNA". *Proc. Natl. Acad. Sci. U.S.A.* **74** (2): 560–4.
       Bibcode:1977PNAS...74..560M. doi:10.1073/pnas.74.2.560. PMC 392330. PMID 265521.

[20] Gilbert, W. DNA sequencing and gene structure. Nobel lecture, 8 December 1980.

[21] Gilbert W, Maxam A (December 1973). "The Nucleotide Sequence of the lac Operator". *Proc. Natl. Acad. Sci. U.S.A.* **70**
       (12): 3581–4. Bibcode:1973PNAS...70.3581G. doi:10.1073/pnas.70.12.3581. PMC 427284. PMID 4587255.

[22] Sanger F, Air GM, Barrell BG, Brown NL, Coulson AR, Fiddes CA, Hutchison CA, Slocombe PM, Smith M (February
       1977). "Nucleotide sequence of bacteriophage phi X174 DNA". *Nature* **265** (5596): 687–95. Bibcode:1977Natur.265..687S.
       doi:10.1038/265687a0. PMID 870828.

[23] The path to DNA seqencing: The life and work of Fred Sanger.

[24] Beck S, Pohl FM (1984). "DNA sequencing with direct blotting electrophoresis". *EMBO J* **3** (12): 2905–2909. PMC 557787.
       PMID 6396083.

[25] United States Patent 4,631,122 (1986)

[26] Feldmann H, Aigle M, Aljinovic G, André B, Baclet MC, Barthe C, Baur A, Bécam AM, Biteau N, Boles E, Brandt T, Brendel M,
       Brückner M, Bussereau F, Christiansen C, Contreras R, Crouzet M, Ciepluch C, Démolis N, Delaveau T, Doignon F, Domdey
       H, Düsterhus S, Dubois E, Dujon B, El Bakkoury M, Entian KD, Feurmann M, Fiers W, Fobo GM, Fritz C, Gassenhuber H,
       Glandsdorff N, Goffeau A, Grivell LA, de Haan M, Hein C, Herbert CJ, Hollenberg CP, Holmstrøm K, Jacq C, Jacquet M,
       Jauniaux JC, Jonniaux JL, Kallesøe T, Kiesau P, Kirchrath L, Kötter P, Korol S, Liebl S, Logghe M, Lohan AJ, Louis EJ, Li ZY,

Maat MJ, Mallet L, Mannhaupt G, Messenguy F, Miosga T, Molemans F, Müller S, Nasr F, Obermaier B, Perea J, Piérard A, Piravandi E, Pohl FM, Pohl TM, Potier S, Proft M, Purnelle B, Ramezani Rad M, Rieger M, Rose M, Schaaff-Gerstenschläger I, Scherens B, Schwarzlose C, Skala J, Slonimski PP, Smits PH, Souciet JL, Steensma HY, Stucka R, Urrestarazu A, van der Aart QJ, van Dyck L, Vassarotti A, Vetter I, Vierendeels F, Vissers S, Wagner G, de Wergifosse P, Wolfe KH, Zagulski M, Zimmermann FK, Mewes HW, Kleine K (1994). "Complete DNA sequence of yeast chromosome II". *EMBO J.* **13** (24): 5795–809. PMC 395553. PMID 7813418.

[27] Smith LM, Sanders JZ, Kaiser RJ, Hughes P, Dodd C, Connell CR, Heiner C, Kent SB, Hood LE (12 June 1986). "Fluorescence Detection in Automated DNA Sequence Analysis". *Nature* **321** (6071): 674–79. Bibcode:1986Natur.321..674S. doi:10.1038/321674a0. PMID 3713851.

[28] Prober JM, Trainor GL, Dam RJ, Hobbs FW, Robertson CW, Zagursky RJ, Cocuzza AJ, Jensen MA, Baumeister K (16 Oct 1987). "A system for rapid DNA sequencing with fluorescent chain-terminating dideoxynucleotides". *Science* **238** (4825): 336–41. Bibcode:1987Sci...238..336P. doi:10.1126/science.2443975. PMID 2443975.

[29] Adams MD, Kelley JM, Gocayne JD, Dubnick M, Polymeropoulos MH, Xiao H, Merril CR, Wu A, Olde B, Moreno RF (June 1991). "Complementary DNA sequencing: expressed sequence tags and human genome project". *Science* **252** (5013): 1651–6. Bibcode:1991Sci...252.1651A. doi:10.1126/science.2047873. PMID 2047873.

[30] Fleischmann RD, Adams MD, White O, Clayton RA, Kirkness EF, Kerlavage AR, Bult CJ, Tomb JF, Dougherty BA, Merrick JM (July 1995). "Whole-genome random sequencing and assembly of *Haemophilus influenzae Rd*". *Science* **269** (5223): 496–512. Bibcode:1995Sci...269..496F. doi:10.1126/science.7542800. PMID 7542800.

[31] Lander ES, Linton LM, Birren B, Nusbaum C, Zody MC, Baldwin J, Devon K, Dewar K, Doyle M, FitzHugh W, Funke R, Gage D, Harris K, Heaford A, Howland J, Kann L, Lehoczky J, LeVine R, McEwan P, McKernan K, Meldrim J, Mesirov JP, Miranda C, Morris W, Naylor J, Raymond C, Rosetti M, Santos R, Sheridan A, Sougnez C, Stange-Thomann N, Stojanovic N, Subramanian A, Wyman D, Rogers J, Sulston J, Ainscough R, Beck S, Bentley D, Burton J, Clee C, Carter N, Coulson A, Deadman R, Deloukas P, Dunham A, Dunham I, Durbin R, French L, Grafham D, Gregory S, Hubbard T, Humphray S, Hunt A, Jones M, Lloyd C, McMurray A, Matthews L, Mercer S, Milne S, Mullikin JC, Mungall A, Plumb R, Ross M, Shownkeen R, Sims S, Waterston RH, Wilson RK, Hillier LW, McPherson JD, Marra MA, Mardis ER, Fulton LA, Chinwalla AT, Pepin KH, Gish WR, Chissoe SL, Wendl MC, Delehaunty KD, Miner TL, Delehaunty A, Kramer JB, Cook LL, Fulton RS, Johnson DL, Minx PJ, Clifton SW, Hawkins T, Branscomb E, Predki P, Richardson P, Wenning S, Slezak T, Doggett N, Cheng JF, Olsen A, Lucas S, Elkin C, Uberbacher E, Frazier M, Gibbs RA, Muzny DM, Scherer SE, Bouck JB, Sodergren EJ, Worley KC, Rives CM, Gorrell JH, Metzker ML, Naylor SL, Kucherlapati RS, Nelson DL, Weinstock GM, Sakaki Y, Fujiyama A, Hattori M, Yada T, Toyoda A, Itoh T, Kawagoe C, Watanabe H, Totoki Y, Taylor T, Weissenbach J, Heilig R, Saurin W, Artiguenave F, Brottier P, Bruls T, Pelletier E, Robert C, Wincker P, Smith DR, Doucette-Stamm L, Rubenfield M, Weinstock K, Lee HM, Dubois J, Rosenthal A, Platzer M, Nyakatura G, Taudien S, Rump A, Yang H, Yu J, Wang J, Huang G, Gu J, Hood L, Rowen L, Madan A, Qin S, Davis RW, Federspiel NA, Abola AP, Proctor MJ, Myers RM, Schmutz J, Dickson M, Grimwood J, Cox DR, Olson MV, Kaul R, Raymond C, Shimizu N, Kawasaki K, Minoshima S, Evans GA, Athanasiou M, Schultz R, Roe BA, Chen F, Pan H, Ramser J, Lehrach H, Reinhardt R, McCombie WR, de la Bastide M, Dedhia N, Blöcker H, Hornischer K, Nordsiek G, Agarwala R, Aravind L, Bailey JA, Bateman A, Batzoglou S, Birney E, Bork P, Brown DG, Burge CB, Cerutti L, Chen HC, Church D, Clamp M, Copley RR, Doerks T, Eddy SR, Eichler EE, Furey TS, Galagan J, Gilbert JG, Harmon C, Hayashizaki Y, Haussler D, Hermjakob H, Hokamp K, Jang W, Johnson LS, Jones TA, Kasif S, Kaspryzk A, Kennedy S, Kent WJ, Kitts P, Koonin EV, Korf I, Kulp D, Lancet D, Lowe TM, McLysaght A, Mikkelsen T, Moran JV, Mulder N, Pollara VJ, Ponting CP, Schuler G, Schultz J, Slater G, Smit AF, Stupka E, Szustakowski J, Thierry-Mieg D, Thierry-Mieg J, Wagner L, Wallis J, Wheeler R, Williams A, Wolf YI, Wolfe KH, Yang SP, Yeh RF, Collins F, Guyer MS, Peterson J, Felsenfeld A, Wetterstrand KA, Patrinos A, Morgan MJ, de Jong P, Catanese JJ, Osoegawa K, Shizuya H, Choi S, Chen YJ, Szustakowki J (February 2001). "Initial sequencing and analysis of the human genome". *Nature* **409** (6822): 860–921. Bibcode:2001Natur.409..860L. doi:10.1038/35057062. PMID 11237011.

[32] Venter JC, Adams MD, Myers EW, Li PW, Mural RJ, Sutton GG, Smith HO, Yandell M, Evans CA, Holt RA, Gocayne JD, Amanatides P, Ballew RM, Huson DH, Wortman JR, Zhang Q, Kodira CD, Zheng XH, Chen L, Skupski M, Subramanian G, Thomas PD, Zhang J, Gabor Miklos GL, Nelson C, Broder S, Clark AG, Nadeau J, McKusick VA, Zinder N, Levine AJ, Roberts RJ, Simon M, Slayman C, Hunkapiller M, Bolanos R, Delcher A, Dew I, Fasulo D, Flanigan M, Florea L, Halpern A, Hannenhalli S, Kravitz S, Levy S, Mobarry C, Reinert K, Remington K, Abu-Threideh J, Beasley E, Biddick K, Bonazzi V, Brandon R, Cargill M, Chandramouliswaran I, Charlab R, Chaturvedi K, Deng Z, Di Francesco V, Dunn P, Eilbeck K, Evangelista C, Gabrielian AE, Gan W, Ge W, Gong F, Gu Z, Guan P, Heiman TJ, Higgins ME, Ji RR, Ke Z, Ketchum KA, Lai Z, Lei Y, Li Z, Li J, Liang Y, Lin X, Lu F, Merkulov GV, Milshina N, Moore HM, Naik AK, Narayan VA, Neelam B, Nusskern D, Rusch DB, Salzberg S, Shao W, Shue B, Sun J, Wang Z, Wang A, Wang X, Wang J, Wei M, Wides R, Xiao C, Yan C, Yao A, Ye J, Zhan M, Zhang W, Zhang H, Zhao Q, Zheng L, Zhong F, Zhong W, Zhu S, Zhao S, Gilbert D, Baumhueter S, Spier G, Carter C, Cravchik A, Woodage T, Ali F, An H, Awe A, Baldwin D, Baden H, Barnstead M, Barrow I, Beeson K, Busam D,

Carver A, Center A, Cheng ML, Curry L, Danaher S, Davenport L, Desilets R, Dietz S, Dodson K, Doup L, Ferriera S, Garg N, Gluecksmann A, Hart B, Haynes J, Haynes C, Heiner C, Hladun S, Hostin D, Houck J, Howland T, Ibegwam C, Johnson J, Kalush F, Kline L, Koduru S, Love A, Mann F, May D, McCawley S, McIntosh T, McMullen I, Moy M, Moy L, Murphy B, Nelson K, Pfannkoch C, Pratts E, Puri V, Qureshi H, Reardon M, Rodriguez R, Rogers YH, Romblad D, Ruhfel B, Scott R, Sitter C, Smallwood M, Stewart E, Strong R, Suh E, Thomas R, Tint NN, Tse S, Vech C, Wang G, Wetter J, Williams S, Williams M, Windsor S, Winn-Deen E, Wolfe K, Zaveri J, Zaveri K, Abril JF, Guigó R, Campbell MJ, Sjolander KV, Karlak B, Kejariwal A, Mi H, Lazareva B, Hatton T, Narechania A, Diemer K, Muruganujan A, Guo N, Sato S, Bafna V, Istrail S, Lippert R, Schwartz R, Walenz B, Yooseph S, Allen D, Basu A, Baxendale J, Blick L, Caminha M, Carnes-Stine J, Caulk P, Chiang YH, Coyne M, Dahlke C, Mays A, Dombroski M, Donnelly M, Ely D, Esparham S, Fosler C, Gire H, Glanowski S, Glasser K, Glodek A, Gorokhov M, Graham K, Gropman B, Harris M, Heil J, Henderson S, Hoover J, Jennings D, Jordan C, Jordan J, Kasha J, Kagan L, Kraft C, Levitsky A, Lewis M, Liu X, Lopez J, Ma D, Majoros W, McDaniel J, Murphy S, Newman M, Nguyen T, Nguyen N, Nodell M, Pan S, Peck J, Peterson M, Rowe W, Sanders R, Scott J, Simpson M, Smith T, Sprague A, Stockwell T, Turner R, Venter E, Wang M, Wen M, Wu D, Wu M, Xia A, Zandieh A, Zhu X (February 2001). "The sequence of the human genome". *Science* **291** (5507): 1304–51. Bibcode:2001Sci...291.1304V. doi:10.1126/science.1058040. PMID 11181995.

[33] Tsien base-by-base sequencing patent

[34] Ronaghi M, Karamohamed S, Pettersson B, Uhlén M, Nyrén P (1996). "Real-time DNA sequencing using detection of py-rophosphate release". *Analytical Biochemistry* **242** (1): 84–9. doi:10.1006/abio.1996.0432. PMID 8923969.

[35] Kawashima, Eric H.; Laurent Farinelli; Pascal Mayer (2005-05-12). "Patent: Method of nucleic acid amplification". Retrieved 2012-12-22.

[36] Brenner S, Johnson M, Bridgham J, Golda G, Lloyd DH, Johnson D, Luo S, McCurdy S, Foy M, Ewan M, Roth R, George D, Eletr S, Albrecht G, Vermaas E, Williams SR, Moon K, Burcham T, Pallas M, DuBridge RB, Kirchner J, Fearon K, Mao J, Corcoran K (2000). "Gene expression analysis by massively parallel signature sequencing (MPSS) on microbead arrays". *Nature Biotechnology* (Nature Biotechnology) **18** (6): 630–634. doi:10.1038/76469. PMID 10835600.

[37] Stein RA (1 September 2008). "Next-Generation Sequencing Update". *Genetic Engineering & Biotechnology News* **28** (15).

[38] Schuster SC (January 2008). "Next-generation sequencing transforms today's biology". *Nat. Methods***5**(1): 16–8. doi:10.1038 PMID 18165802.

[39] Ewing B, Green P (March 1998). "Base-calling of automated sequencer traces using phred. II. Error probabilities". *Genome Res.* **8** (3): 186–94. doi:10.1101/gr.8.3.186 (inactive 2015-01-01). PMID 9521922.

[40] Sanger F, Coulson AR (May 1975). "A rapid method for determining sequences in DNA by primed synthesis with DNA polymerase". *J. Mol. Biol.* **94** (3): 441–8. doi:10.1016/0022-2836(75)90213-2. PMID 1100841.

[41] Wetterstrand, Kris. "DNA Sequencing Costs: Data from the NHGRI Genome Sequencing Program (GSP)". National Human Genome Research Institute. Retrieved 30 May 2013.

[42] Quail MA, Gu Y, Swerdlow H, Mayho M (2012). "Evaluation and optimisation of preparative semi-automated electrophoresis systems for Illumina library preparation". *Electrophoresis* **33** (23): 3521–8. doi:10.1002/elps.201200128. PMID 23147856.

[43] Duhaime MB, Deng L, Poulos BT, Sullivan MB (2012). "Towards quantitative metagenomics of wild viruses and other ultra-low concentration DNA samples: a rigorous assessment and optimization of the linker amplification method". *Environ. Microbiol.* **14** (9): 2526–37. doi:10.1111/j.1462-2920.2012.02791.x. PMC 3466414. PMID 22713159.

[44] Peterson BK, Weber JN, Kay EH, Fisher HS, Hoekstra HE (2012). "Double digest RADseq: an inexpensive method for de novo SNP discovery and genotyping in model and non-model species". *PLoS ONE* **7** (5): e37135. Bibcode:2012PLoSO...737135P. doi:10.1371/journal.pone.0037135. PMC 3365034. PMID 22675423.

[45] Williams R, Peisajovich SG, Miller OJ, Magdassi S, Tawfik DS, Griffiths AD (2006). "Amplification of complex gene libraries by emulsion PCR". *Nature methods* **3** (7): 545–550. doi:10.1038/nmeth896. PMID 16791213.

[46] Margulies M, Egholm M, Altman WE, Attiya S, Bader JS, Bemben LA, Berka J, Braverman MS, Chen YJ, Chen Z, Dewell SB, Du L, Fierro JM, Gomes XV, Godwin BC, He W, Helgesen S, Ho CH, Ho CH, Irzyk GP, Jando SC, Alenquer ML, Jarvie TP, Jirage KB, Kim JB, Knight JR, Lanza JR, Leamon JH, Lefkowitz SM, Lei M, Li J, Lohman KL, Lu H, Makhijani VB, McDade KE, McKenna MP, Myers EW, Nickerson E, Nobile JR, Plant R, Puc BP, Ronan MT, Roth GT, Sarkis GJ, Simons JF, Simpson JW, Srinivasan M, Tartaro KR, Tomasz A, Vogt KA, Volkmer GA, Wang SH, Wang Y, Weiner MP, Yu P, Begley RF, Rothberg JM (September 2005). "Genome Sequencing in Open Microfabricated High Density Picoliter Reactors". *Nature* **437** (7057): 376–80. Bibcode:2005Natur.437..376M. doi:10.1038/nature03959. PMC 1464427. PMID 16056220.

[47] Shendure J, Porreca GJ, Reppas NB, Lin X, McCutcheon JP, Rosenbaum AM, Wang MD, Zhang K, Mitra RD, Church GM (2005). "Accurate Multiplex Polony Sequencing of an Evolved Bacterial Genome". *Science* **309** (5741): 1728–32. Bibcode:2005Sci...309.1728S. doi:10.1126/science.1117389. PMID 16081699.

[48] Applied Biosystems' SOLiD technology

[49] Staden R (11 Jun 1979). "A strategy of DNA sequencing employing computer programs.". *Nucleic Acids Research* **6** (7): 2601–10. doi:10.1093/nar/6.7.2601. PMC 327874. PMID 461197.

[50] P. Mayer,L. Farinelli, G. Matton, C. Adessi, G. Turcatti, J. J. Mermod, E. Kawashima.DNA colony massively parallel sequencing ams98 presentation

[51] U.S. Patent 5,641,658

[52] Braslavsky I, Hebert B, Kartalov E, Quake SR (April 2003). "Sequence information can be obtained from single DNA molecules". *Proc. Natl. Acad. Sci. U.S.A.* **100** (7): 3960–4. Bibcode:2003PNAS..100.3960B. doi:10.1073/pnas.0230489100. PMC 153030. PMID 12651960.

[53] de Magalhães JP, Finch CE, Janssens G (2010). "Next-generation sequencing in aging research: emerging applications, problems, pitfalls and possible solutions". *Ageing Research Reviews* **9** (3): 315–323. doi:10.1016/j.arr.2009.10.006. PMC 2878865. PMID 19900591.

[54] Hall N (May 2007). "Advanced sequencing technologies and their wider impact in microbiology". *J. Exp. Biol.* **209** (Pt 9): 1518–1525. doi:10.1242/jeb.001370. PMID 17449817.

[55] Church GM (January 2006). "Genomes for all". *Sci. Am.* **294** (1): 46–54. doi:10.1038/scientificamerican0106-46. PMID 16468433.(subscription required)

[56] Kalb, Gilbert; Moxley, Robert (1992). *Massively Parallel, Optical, and Neural Computing in the United States*. IOS Press. ISBN 90-5199-097-9.

[57] ten Bosch JR, Grody WW (2008). "Keeping Up with the Next Generation". *The Journal of Molecular Diagnostics* **10** (6): 484–492. doi:10.2353/jmoldx.2008.080027. PMC 2570630. PMID 18832462.

[58] Tucker T, Marra M, Friedman JM (2009). "Massively Parallel Sequencing: The Next Big Thing in Genetic Medicine". *The American Journal of Human Genetics* **85** (2): 142–154. doi:10.1016/j.ajhg.2009.06.022. PMC 2725244. PMID 19679224.

[59] Quail MA, Smith M, Coupland P, Otto TD, Harris SR, Connor TR, Bertoni A, Swerdlow HP, Gu Y (1 January 2012). "A tale of three next generation sequencing platforms: comparison of Ion Torrent, Pacific Biosciences and illumina MiSeq sequencers". *BMC Genomics* **13** (1): 341. doi:10.1186/1471-2164-13-341. PMC 3431227. PMID 22827831.

[60] Liu L, Li Y, Li S, Hu N, He Y, Pong R, Lin D, Lu L, Law M (1 January 2012). "Comparison of Next-Generation Sequencing Systems". *Journal of Biomedicine and Biotechnology* (Hindawi Publishing Corporation) **2012**: 1–11. doi:10.1155/2012/251364. PMID 22829749.

[61] New Products: PacBio's RS II; Cufflinks | In Sequence | Sequencing | GenomeWeb

[62] "After a Year of Testing, Two Early PacBio Customers Expect More Routine Use of RS Sequencer in 2012". GenomeWeb. 10 January 2012.(registration required)

[63] Pacific Biosciences Introduces New Chemistry With Longer Read Lengths

[64] Chin CS, Alexander DH, Marks P, Klammer AA, Drake J, Heiner C, Clum A, Copeland A, Huddleston J, Eichler EE, Turner SW, Korlach J (2013). "Nonhybrid, finished microbial genome assemblies from long-read SMRT sequencing data". *Nat. Methods* **10** (6): 563–9. doi:10.1038/nmeth.2474. PMID 23644548.

[65] De novo bacterial genome assembly: a solved problem? | In between lines of code

[66] Rasko DA, Webster DR, Sahl JW, Bashir A, Boisen N, Scheutz F, Paxinos EE, Sebra R, Chin CS, Iliopoulos D, Klammer A, Peluso P, Lee L, Kislyuk AO, Bullard J, Kasarskis A, Wang S, Eid J, Rank D, Redman JC, Steyert SR, Frimodt-Møller J, Struve C, Petersen AM, Krogfelt KA, Nataro JP, Schadt EE, Waldor MK (25 August 2011). "Origins of the Strain Causing an Outbreak of Hemolytic–Uremic Syndrome in Germany". *N Engl J Med* **365** (8): 709–717. doi:10.1056/NEJMoa1106920. PMID 21793740.

[67] Tran B, Brown AM, Bedard PL, Winquist E, Goss GD, Hotte SJ, Welch SA, Hirte HW, Zhang T, Stein LD, Ferretti V, Watt S, Jiao W, Ng K, Ghai S, Shaw P, Petrocelli T, Hudson TJ, Neel BG, Onetto N, Siu LL, McPherson JD, Kamel-Reid S, Dancey JE (1 January 2012). "Feasibility of real time next generation sequencing of cancer genes linked to drug response: Results from a clinical trial". *Int. J. Cancer* **132** (7): 1547–1555. doi:10.1002/ijc.27817. PMID 22948899.(subscription required)

[68] Murray IA, Clark TA, Morgan RD, Boitano M, Anton BP, Luong K, Fomenkov A, Turner SW, Korlach J, Roberts RJ (2 October 2012). "The methylomes of six bacteria". *Nucleic Acids Research* **40** (22): 11450–62. doi:10.1093/nar/gks891. PMC 3526280. PMID 23034806.

[69] van Vliet AH (1 January 2010). "Next generation sequencing of microbial transcriptomes: challenges and opportunities". *FEMS Microbiology Letters* **302** (1): 1–7. doi:10.1111/j.1574-6968.2009.01767.x. PMID 19735299.

[70] Huang YF, Chen SC, Chiang YS, Chen TH, Chiu KP (2012). "Palindromic sequence impedes sequencing-by-ligation mechanism". *BMC Systems Biology*. 6 Suppl 2: S10. doi:10.1186/1752-0509-6-S2-S10. PMID 23281822.

[71] Shendure J, Porreca GJ, Reppas NB, Lin X, McCutcheon JP, Rosenbaum AM, Wang MD, Zhang K, Mitra RD, Church GM (9 Sep 2005). "Accurate multiplex polony sequencing of an evolved bacterial genome.". *Science* **309** (5741): 1728–32. Bibcode:2005Sci...309.1728S. doi:10.1126/science.1117389. PMID 16081699.

[72] Bentley DR, Balasubramanian S, Swerdlow HP, Smith GP, Milton J, Brown CG, Hall KP, Evers DJ, Barnes CL, Bignell HR, Boutell JM, Bryant J, Carter RJ, Keira Cheetham R, Cox AJ, Ellis DJ, Flatbush MR, Gormley NA, Humphray SJ, Irving LJ, Karbelashvili MS, Kirk SM, Li H, Liu X, Maisinger KS, Murray LJ, Obradovic B, Ost T, Parkinson ML, Pratt MR, Rasolonjatovo IM, Reed MT, Rigatti R, Rodighiero C, Ross MT, Sabot A, Sankar SV, Scally A, Schroth GP, Smith ME, Smith VP, Spiridou A, Torrance PE, Tzonev SS, Vermaas EH, Walter K, Wu X, Zhang L, Alam MD, Anastasi C, Aniebo IC, Bailey DM, Bancarz IR, Banerjee S, Barbour SG, Baybayan PA, Benoit VA, Benson KF, Bevis C, Black PJ, Boodhun A, Brennan JS, Bridgham JA, Brown RC, Brown AA, Buermann DH, Bundu AA, Burrows JC, Carter NP, Castillo N, Chiara E Catenazzi M, Chang S, Neil Cooley R, Crake NR, Dada OO, Diakoumakos KD, Dominguez-Fernandez B, Earnshaw DJ, Egbujor UC, Elmore DW, Etchin SS, Ewan MR, Fedurco M, Fraser LJ, Fuentes Fajardo KV, Scott Furey W, George D, Gietzen KJ, Goddard CP, Golda GS, Granieri PA, Green DE, Gustafson DL, Hansen NF, Harnish K, Haudenschild CD, Heyer NI, Hims MM, Ho JT, Horgan AM, Hoschler K, Hurwitz S, Ivanov DV, Johnson MQ, James T, Huw Jones TA, Kang GD, Kerelska TH, Kersey AD, Khrebtukova I, Kindwall AP, Kingsbury Z, Kokko-Gonzales PI, Kumar A, Laurent MA, Lawley CT, Lee SE, Lee X, Liao AK, Loch JA, Lok M, Luo S, Mammen RM, Martin JW, McCauley PG, McNitt P, Mehta P, Moon KW, Mullens JW, Newington T, Ning Z, Ling Ng B, Novo SM, O'Neill MJ, Osborne MA, Osnowski A, Ostadan O, Paraschos LL, Pickering L, Pike AC, Pike AC, Chris Pinkard D, Pliskin DP, Podhasky J, Quijano VJ, Raczy C, Rae VH, Rawlings SR, Chiva Rodriguez A, Roe PM, Rogers J, Rogert Bacigalupo MC, Romanov N, Romieu A, Roth RK, Rourke NJ, Ruediger ST, Rusman E, Sanches-Kuiper RM, Schenker MR, Seoane JM, Shaw RJ, Shiver MK, Short SW, Sizto NL, Sluis JP, Smith MA, Ernest Sohna Sohna J, Spence EJ, Stevens K, Sutton N, Szajkowski L, Tregidgo CL, Turcatti G, Vandevondele S, Verhovsky Y, Virk SM, Wakelin S, Walcott GC, Wang J, Worsley GJ, Yan J, Yau L, Zuerlein M, Rogers J, Mullikin JC, Hurles ME, McCooke NJ, West JS, Oaks FL, Lundberg PL, Klenerman D, Durbin R, Smith AJ (2008). "Accurate whole human genome sequencing using reversible terminator chemistry". *Nature* **456** (7218): 53–59. Bibcode:2008Natur.456...53B. doi:10.1038/nature07517. PMC 2581791. PMID 18987734. Vancouver style error (help)

[73] Mardis ER(2008). "Next-generation DNA sequencing methods". *Annu Rev Genomics Hum Genet***9**: 387–402. doi:10.1146/59. PMID 18576944.

[74] Valouev A, Ichikawa J, Tonthat T, Stuart J, Ranade S, Peckham H, Zeng K, Malek JA, Costa G, McKernan K, Sidow A, Fire A, Johnson SM (July 2008). "A high-resolution, nucleosome position map of C. elegans reveals a lack of universal sequence-dictated positioning". *Genome Res.* **18** (7): 1051–63. doi:10.1101/gr.076463.108. PMC 2493394. PMID 18477713.

[75] Rusk N (2011). "Torrents of sequence". *Nat Meth* **8** (1): 44–44. doi:10.1038/nmeth.f.330.

[76] Drmanac R, Sparks AB, Callow MJ, Halpern AL, Burns NL, Kermani BG, Carnevali P, Nazarenko I, Nilsen GB, Yeung G, Dahl F, Fernandez A, Staker B, Pant KP, Baccash J, Borcherding AP, Brownley A, Cedeno R, Chen L, Chernikoff D, Cheung A, Chirita R, Curson B, Ebert JC, Hacker CR, Hartlage R, Hauser B, Huang S, Jiang Y, Karpinchyk V, Koenig M, Kong C, Landers T, Le C, Liu J, McBride CE, Morenzoni M, Morey RE, Mutch K, Perazich H, Perry K, Peters BA, Peterson J, Pethiyagoda CL, Pothuraju K, Richter C, Rosenbaum AM, Roy S, Shafto J, Sharanhovich U, Shannon KW, Sheppy CG, Sun M, Thakuria JV, Tran A, Vu D, Zaranek AW, Wu X, Drmanac S, Oliphant AR, Banyai WC, Martin B, Ballinger DG, Church GM, Reid CA (2010). "Human Genome Sequencing Using Unchained Base Reads in Self-Assembling DNA Nanoarrays". *Science* **327** (5961): 78–81. Bibcode:2010Sci...327...78D. doi:10.1126/science.1181498. PMID 19892942.

[77] Porreca GJ (2010). "Genome Sequencing on Nanoballs". *Nature Biotechnology* **28** (1): 43–44. doi:10.1038/nbt0110-43. PMID 20062041.

[78] Complete Genomics Press release, 2010

[79] HeliScope Gene Sequencing / Genetic Analyzer System : Helicos BioSciences

[80] Thompson JF, Steinmann KE (October 2010). "Single molecule sequencing with a HeliScope genetic analysis system.". *Current Protocols in Molecular Biology*. Chapter 7: Unit7.10. doi:10.1002/0471142727.mb0710s92. PMC 2954431. PMID 20890904.

[81] "tSMS SeqLL Technical Explanation". SeqLL. Retrieved 9 Aug 2015.

[82] Sara El-Metwally, Osama M. Ouda, Mohamed Helmy. *New Horizons in Next-Generation Sequencing. Next Generation Sequencing Technologies and Challenges in Sequence Assembly, SpringerBriefs in Systems Biology Volume 7, 2014, pp 51-59.*

[83] PacBio Sales Start to Pick Up as Company Delivers on Product Enhancements I In Sequence I Sequencing I GenomeWeb

[84] http://www.bio-itworld.com/2015/9/30/pacbio-announces-sequel-sequencing-system.aspx

[85] https://www.genomeweb.com/business-news/pacbio-launches-higher-throughput-lower-cost-single-molecule-sequencing

[86] "The Harvard Nanopore Group". Mcb.harvard.edu. Retrieved 2009-11-15.

[87] "Nanopore Sequencing Could Slash DNA Analysis Costs".

[88] US patent 20060029957, ZS Genetics, "Systems and methods of analyzing nucleic acid polymers and related components", issued 2005-07-14

[89] Xu M, Fujita D, Hanagata N (December 2009). "Perspectives and challenges of emerging single-molecule DNA sequencing technologies". *Small* **5** (23): 2638–49. doi:10.1002/smll.200900976. PMID 19904762.

[90] Schadt EE, Turner S, Kasarskis A (2010). "A window into third-generation sequencing". *Human Molecular Genetics* **19** (R2): R227–40. doi:10.1093/hmg/ddq416. PMID 20858600.

[91] Stoddart D, Heron AJ, Mikhailova E, Maglia G, Bayley H (12 May 2009). "Single-nucleotide discrimination in immobilized DNA oligonucleotides with a biological nanopore". *Proceedings of the National Academy of Sciences of the United States of America* **106** (19): 7702–7. Bibcode:2009PNAS..106.7702S. doi:10.1073/pnas.0901054106. PMC 2683137. PMID 19380741.

[92] dela Torre R, Larkin J, Singer A, Meller A (2012). "Fabrication and characterization of solid-state nanopore arrays for high-throughput DNA sequencing". *Nanotechnology* **23** (38): 385308. Bibcode:2012Nanot..23L5308D. doi:10.1088/0957-4484/23/38/385308. PMC 3557807. PMID 22948520.

[93] Pathak, B., Lofas, H., Prasongkit, J., Grigoriev, A., Ahuja, R., & Scheicher, R. H. (9 January 2012). Double-functionalized nanopore-embedded gold electrodes for rapid DNA sequencing. Applied Physics Letters, 100, 2.)

[94] Korlach J, Marks PJ, Cicero RL, Gray JJ, Murphy DL, Roitman DB, Pham TT, Otto GA, Foquet M, Turner SW (2008). "Selective aluminum passivation for targeted immobilization of single DNA polymerase molecules in zero-mode waveguide nanostructures". *Proceedings of the National Academy of Sciences* **105** (4): 1176–1181. Bibcode:2008PNAS..105.1176K. doi:10.1073/pnas.0710982105. PMC 2234111. PMID 18216253.

[95] Mikheyev, Alexander S.; Tin, Mandy M. Y. (November 2014). "A first look at the Oxford Nanopore MinION sequencer". *Molecular Ecology Resources* **14** (6): 1097–1102. doi:10.1111/1755-0998.12324.

[96] Jain, Miten; Fiddes, Ian T; Miga, Karen H; Olsen, Hugh E; Paten, Benedict; Akeson, Mark (16 February 2015). "Improved data analysis for the MinION nanopore sequencer". *Nature Methods* **12** (4): 351–356. doi:10.1038/nmeth.3290.

[97] Di Ventra M (2013). "Fast DNA sequencing by electrical means inches closer". *Nanotechnology* **24**(34): 342501. Bibcode:2013 doi:10.1088/0957-4484/24/34/342501. PMID 23899780.

[98] Ohshiro T, Matsubara K, Tsutsui M, Furuhashi M, Taniguchi M, Kawai T (2012). "Single-molecule electrical random resequencing of DNA and RNA". *Sci Rep* **2**: 501. Bibcode:2012NatSR...2E.501O. doi:10.1038/srep00501. PMC 3392642. PMID 22787559.

[99] Hanna GJ, Johnson VA, Kuritzkes DR, Richman DD, Martinez-Picado J, Sutton L, Hazelwood JD, D'Aquila RT (1 July 2000). "Comparison of Sequencing by Hybridization and Cycle Sequencing for Genotyping of Human Immunodeficiency Virus Type 1 Reverse Transcriptase". *J. Clin. Microbiol.* **38** (7): 2715–21. PMC 87006. PMID 10878069.

[100] Morey M, Fernández-Marmiesse A, Castiñeiras D, Fraga JM, Couce ML, Cocho JA (2013). "A glimpse into past, present, and future DNA sequencing". *Molecular Genetics and Metabolism* **110** (1–2): 3–24. doi:10.1016/j.ymgme.2013.04.024. PMID 23742747.

[101] Qin Y, Schneider TM, Brenner MP (2012). Gibas C, ed. "Sequencing by Hybridization of Long Targets". *PLoS ONE* **7** (5): e35819. Bibcode:2012PLoSO...735819Q. doi:10.1371/journal.pone.0035819. PMC 3344849. PMID 22574124.

[102] Edwards JR, Ruparel H, Ju J (2005). "Mass-spectrometry DNA sequencing". *Mutation Research***573**(1–2): 3–12. doi:10.1016/ PMID 15829234.

[103] Hall TA, Budowle B, Jiang Y, Blyn L, Eshoo M, Sannes-Lowery KA, Sampath R, Drader JJ, Hannis JC, Harrell P, Samant V, White N, Ecker DJ, Hofstadler SA (2005). "Base composition analysis of human mitochondrial DNA using electrospray ionization mass spectrometry: A novel tool for the identification and differentiation of humans". *Analytical Biochemistry* **344** (1): 53–69. doi:10.1016/j.ab.2005.05.028. PMID 16054106.

[104] Howard R, Encheva V, Thomson J, Bache K, Chan YT, Cowen S, Debenham P, Dixon A, Krause JU, Krishan E, Moore D, Moore V, Ojo M, Rodrigues S, Stokes P, Walker J, Zimmermann W, Barallon R (15 Jun 2011). "Comparative analysis of human mitochondrial DNA from World War I bone samples by DNA sequencing and ESI-TOF mass spectrometry". *Forensic Science International: Genetics* **7** (1): 1–9. doi:10.1016/j.fsigen.2011.05.009. PMID 21683667.

[105] Monforte JA, Becker CH (1 March 1997). "High-throughput DNA analysis by time-of-flight mass spectrometry". *Nature Medicine* **3** (3): 360–362. doi:10.1038/nm0397-360. PMID 9055869.

[106] Beres SB, Carroll RK, Shea PR, Sitkiewicz I, Martinez-Gutierrez JC, Low DE, McGeer A, Willey BM, Green K, Tyrrell GJ, Goldman TD, Feldgarden M, Birren BW, Fofanov Y, Boos J, Wheaton WD, Honisch C, Musser JM (8 February 2010). "Molecular complexity of successive bacterial epidemics deconvoluted by comparative pathogenomics". *Proceedings of the National Academy of Sciences* **107** (9): 4371–4376. Bibcode:2010PNAS..107.4371B. doi:10.1073/pnas.0911295107. PMC 2840111. PMID 20142485.

[107] Kan CW, Fredlake CP, Doherty EA, Barron AE (1 November 2004). "DNA sequencing and genotyping in miniaturized electrophoresis systems". *Electrophoresis* **25** (21–22): 3564–3588. doi:10.1002/elps.200406161. PMID 15565709.

[108] Chen YJ, Roller EE, Huang X (2010). "DNA sequencing by denaturation: experimental proof of concept with an integrated fluidic device". *Lab on Chip* **10** (9): 1153–1159. doi:10.1039/b921417h. PMC 2881221. PMID 20390134.

[109] Bell DC, Thomas WK, Murtagh KM, Dionne CA, Graham AC, Anderson JE, Glover WR (9 Oct 2012). "DNA Base Identification by Electron Microscopy". *Microscopy and microanalysis : the official journal of Microscopy Society of America, Microbeam Analysis Society, Microscopical Society of Canada* **18** (5): 1–5. Bibcode:2012MiMic..18.1049B. doi:10.1017/S1431927612012 615.PMID23046798.

[110] Pareek CS, Smoczynski R, Tretyn A (November 2011). "Sequencing technologies and genome sequencing". *Journal of applied genetics* **52** (4): 413–35. doi:10.1007/s13353-011-0057-x. PMC 3189340. PMID 21698376.

[111] Pareek CS, Smoczynski R, Tretyn A (2011). "Sequencing technologies and genome sequencing". *Journal of Applied Genetics* **52** (4): 413–435. doi:10.1007/s13353-011-0057-x. PMC 3189340. PMID 21698376.

[112] Fujimori S, Hirai N, Ohashi H, Masuoka K, Nishikimi A, Fukui Y, Washio T, Oshikubo T, Yamashita T, Miyamoto-Sato E (2012). "Next-generation sequencing coupled with a cell-free display technology for high-throughput production of reliable interactome data". *Scientific reports* **2**: 691. Bibcode:2012NatSR...2E.691F. doi:10.1038/srep00691. PMC 3466446. PMID 23056904.

[113] "PRIZE Overview: Archon X PRIZE for Genomics"

[114] Genome.gov - Grant Information

[115] Severin J, Lizio M, Harshbarger J, Kawaji H, Daub CO, Hayashizaki Y, Bertin N, Forrest AR (2014). "Interactive visualization and analysis of large-scale sequencing datasets using ZENBU". *Nat. Biotechnol.* **32** (3): 217–9. doi:10.1038/nbt.2840. PMID 24727769.

[116] Shmilovici A,Ben-Gal I (2007). "Using a VOM model for reconstructing potential coding regions in EST sequences" (PDF). *Computational Statistics* **22** (1): 49–69. doi:10.1007/s00180-007-0021-8.

[117] Del Fabbro C, Scalabrin S, Morgante M, Giorgi FM (2013). "An Extensive Evaluation of Read Trimming Effects on Illumina NGS Data Analysis". *PLoS ONE* **8** (12): e85024. Bibcode:2013PLoSO...885024D. doi:10.1371/journal.pone.0085024. PMC 3871669. PMID 24376861.

[118] Martin, Marcel (2 May 2011). "Cutadapt removes adapter sequences from high-throughput sequencing reads". *EMBnet.journal* **17** (1): 10. doi:10.14806/ej.17.1.200.

[119] Smeds, Linnéa; Künstner, Axel; Donlin, Maureen J. (19 October 2011). "ConDeTri - A Content Dependent Read Trimmer for Illumina Data". *PLoS ONE* **6** (10): e26314. Bibcode:2011PLoSO...626314S. doi:10.1371/journal.pone.0026314.

[120] Spandow, O; Hellström, S; Schmidt, SH; De Paoli, Emanuale; Policriti, Alberto (2012). "ERNE-BS5: Aligning BS-treated Sequences by Multiple Hits on a 5-letters Alphabet". *Proceedings of the ACM Conference on Bioinformatics, Computational Biology and Biomedicine* **12**: 12–19. doi:10.1145/2382936.2382938.

[121] Schmieder, R.; Edwards, R. (28 January 2011). "Quality control and preprocessing of metagenomic datasets". *Bioinformatics* **27** (6): 863–864. doi:10.1093/bioinformatics/btr026.

[122] Bolger, A. M.; Lohse, M.; Usadel, B. (1 April 2014). "Trimmomatic: a flexible trimmer for Illumina sequence data". *Bioinformatics* **30** (15): 2114–2120. doi:10.1093/bioinformatics/btu170.

[123] Cox, Murray P; Peterson, Daniel A; Biggs, Patrick J (2010). "SolexaQA: At-a-glance quality assessment of Illumina second-generation sequencing data". *BMC Bioinformatics* **11** (1): 485. doi:10.1186/1471-2105-11-485.

[124] Murray, TH (January 1991). "Ethical issues in human genome research.". *FASEB journal : official publication of the Federation of American Societies for Experimental Biology* **5** (1): 55–60. PMID 1825074.

[125] Robertson, John A. (August 2003). "The $1000 Genome: Ethical and Legal Issues in Whole Genome Sequencing of Individuals". *The American Journal of Bioethics* **3** (3): 35–42. doi:10.1162/152651603322874762.

[126] Henderson, Mark. "Human genome sequencing: the real ethical dilemmas". *The Guardian*. Retrieved 20 May 2015.

[127] Harmon, Amy (24 February 2008). "Insurance Fears Lead Many to Shun DNA Tests". *The New York Times*. Retrieved 20 May 2015.

[128] Statement of Administration policy, Executive Office of the President, Office of Management and Budget, April 27, 2007

[129] National Human Genome Research Institute (May 21, 2008). "President Bush Signs the Genetic Information Nondiscrimination Act of 2008". Retrieved Feb 17, 2014.

[130] Baker, Monya. "US ethics panel reports on DNA sequencing and privacy". *Nature New Blog*. Retrieved 20 May 2015.

[131] "Privacy and Progress in Whole Genome Sequencing" (PDF). Presidential Commission for the Study of Bioethical Issues. Retrieved 20 May 2015.

[132] Goldenberg, Aaron J.; Sharp, Richard R. (1 February 2012). "The Ethical Hazards and Programmatic Challenges of Genomic Newborn Screening". *JAMA* **307** (5): 461. doi:10.1001/jama.2012.68.

[133] Hughes, Virginia. "It's Time To Stop Obsessing About the Dangers of Genetic Information". *Slate Magazine*. Retrieved 22 May 2015.

[134] Bloss, Cinnamon S.; Schork, Nicholas J.; Topol, Eric J. (10 February 2011). "Effect of Direct-to-Consumer Genomewide Profiling to Assess Disease Risk". *New England Journal of Medicine* **364** (6): 524–534. doi:10.1056/NEJMoa1011893.

[135] Rochman, Bonnie (25 October 2012). "What Your Doctor Isn't Telling You About Your DNA". *Time.com*. Retrieved 22 May 2015.

41.14 External links

- A wikibook on next generation sequencing

- A free didactic directory for DNA sequencing analysis.

- A The path to DNA sequencing: The life and work of Fred Sanger

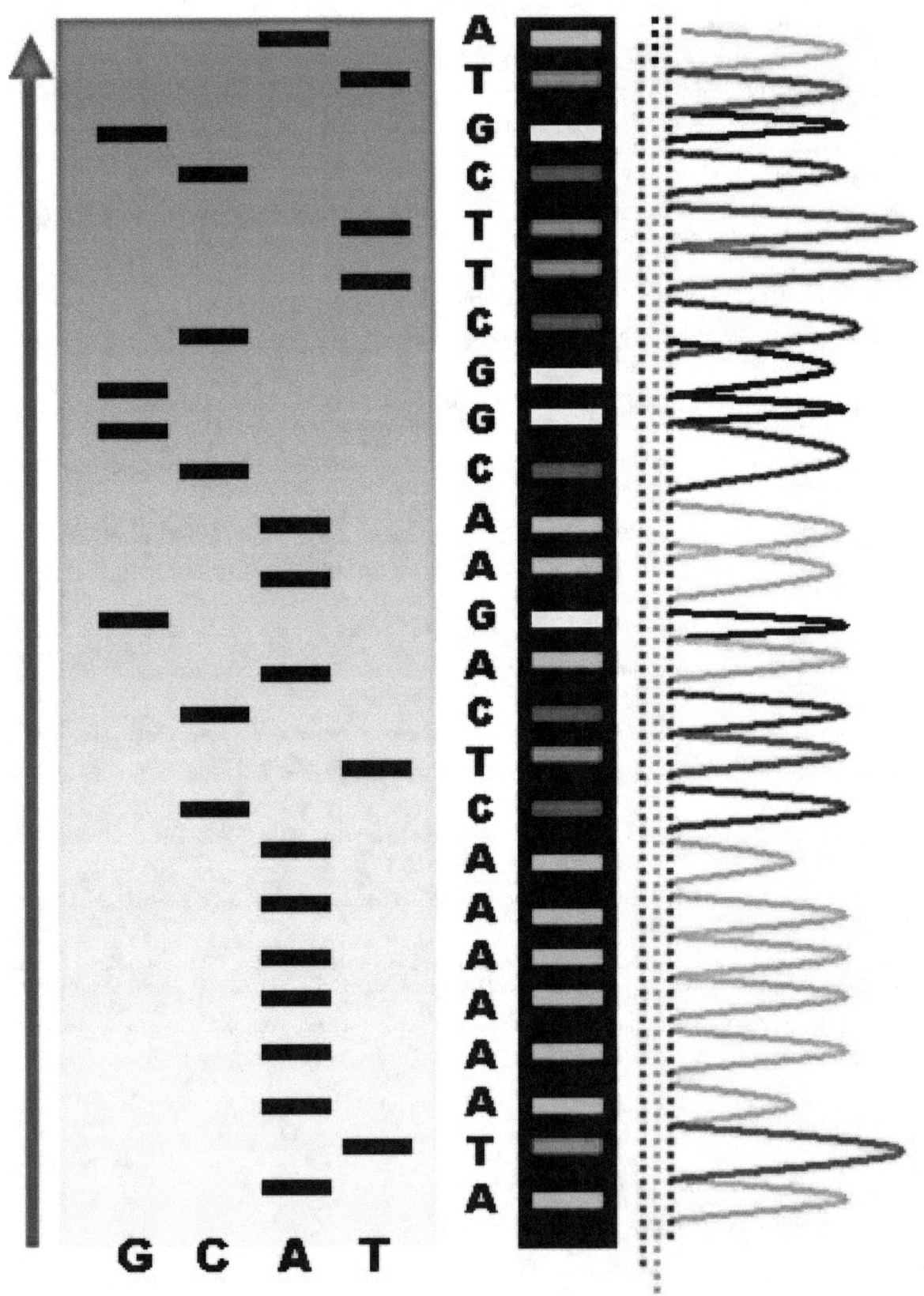

An example of the results of automated chain-termination DNA sequencing.

Genomic DNA is fragmented into random pieces and cloned as a bacterial library. DNA from individual bacterial clones is sequenced and the sequence is assembled by using overlapping DNA regions.(click to expand)

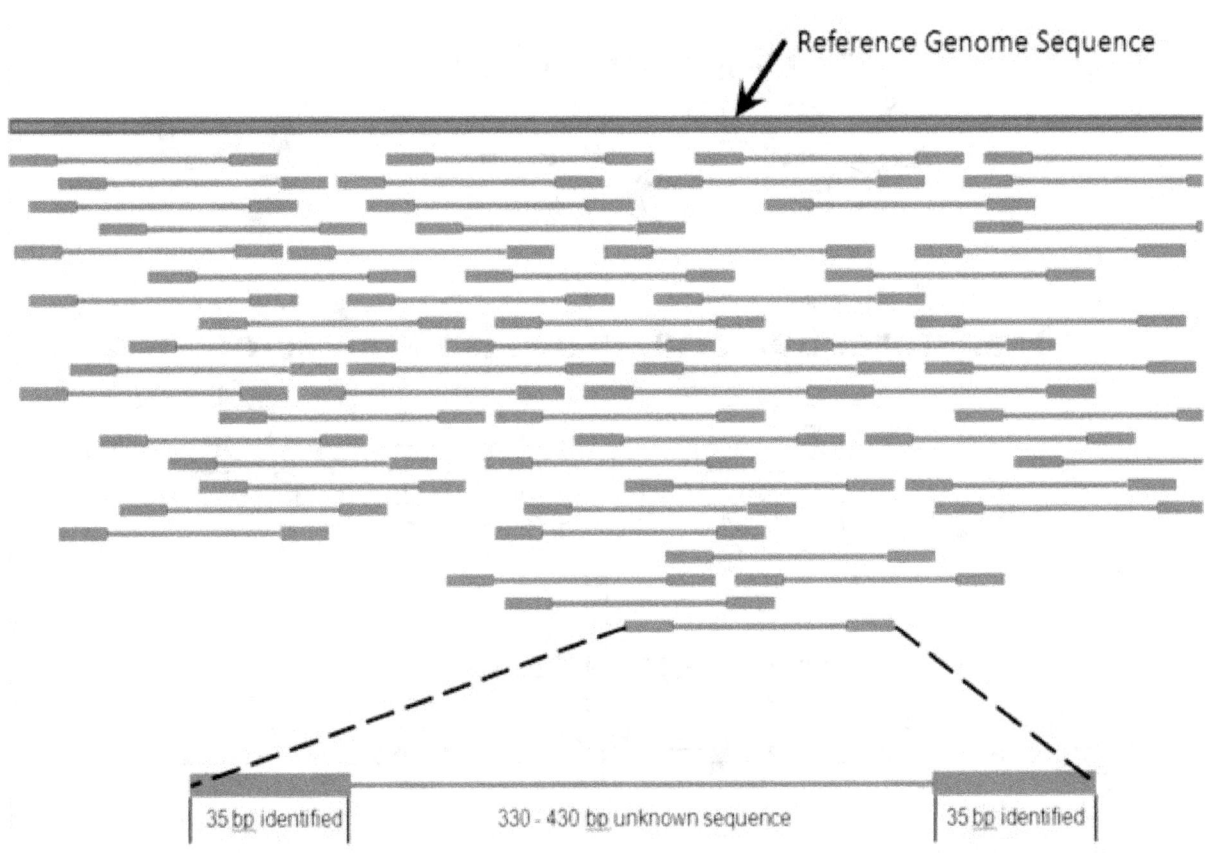

Reference Genome Sequence

| 35 bp identified | 330 - 430 bp unknown sequence | 35 bp identified |

Multiple, fragmented sequence reads must be assembled together on the basis of their overlapping areas.

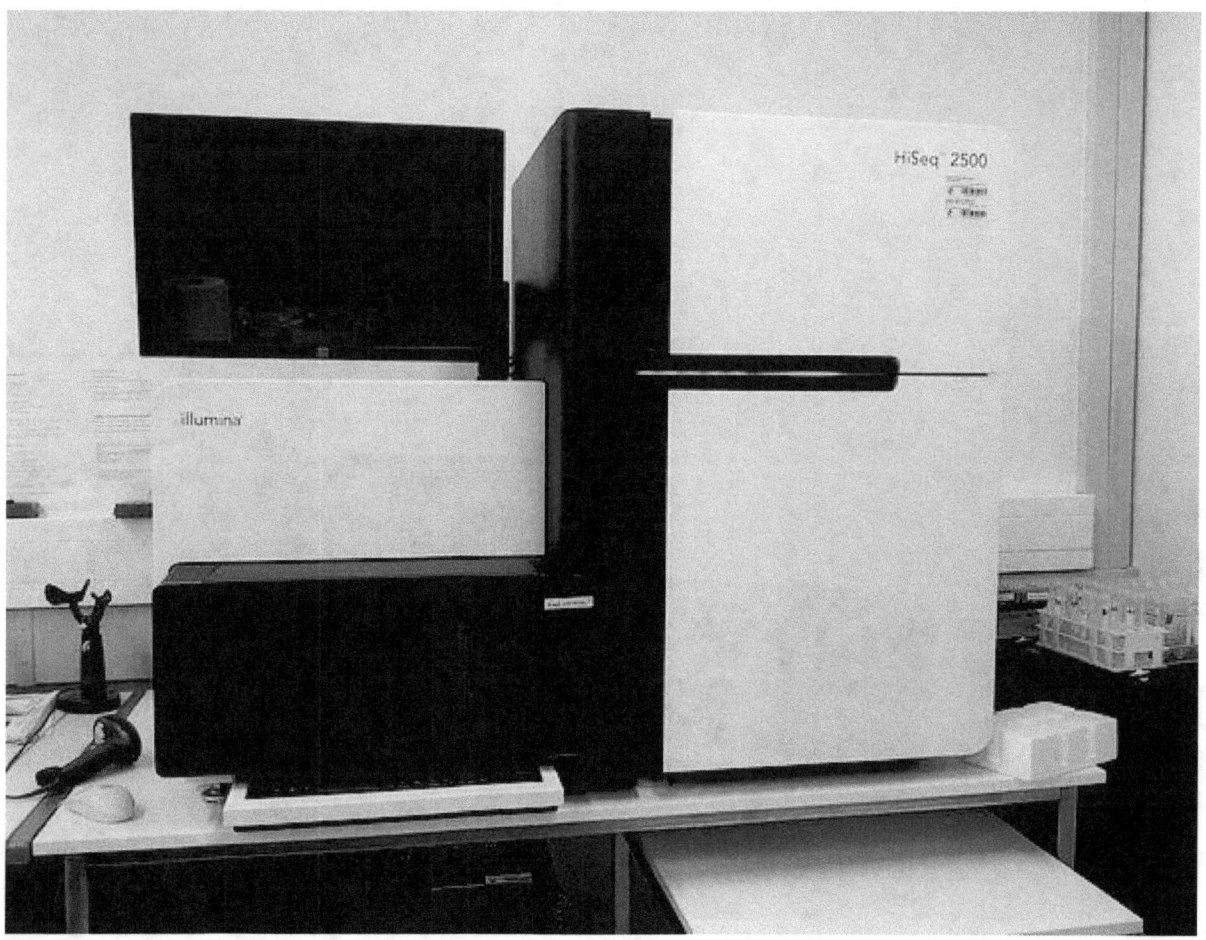

An Illumina HiSeq 2500 sequencer

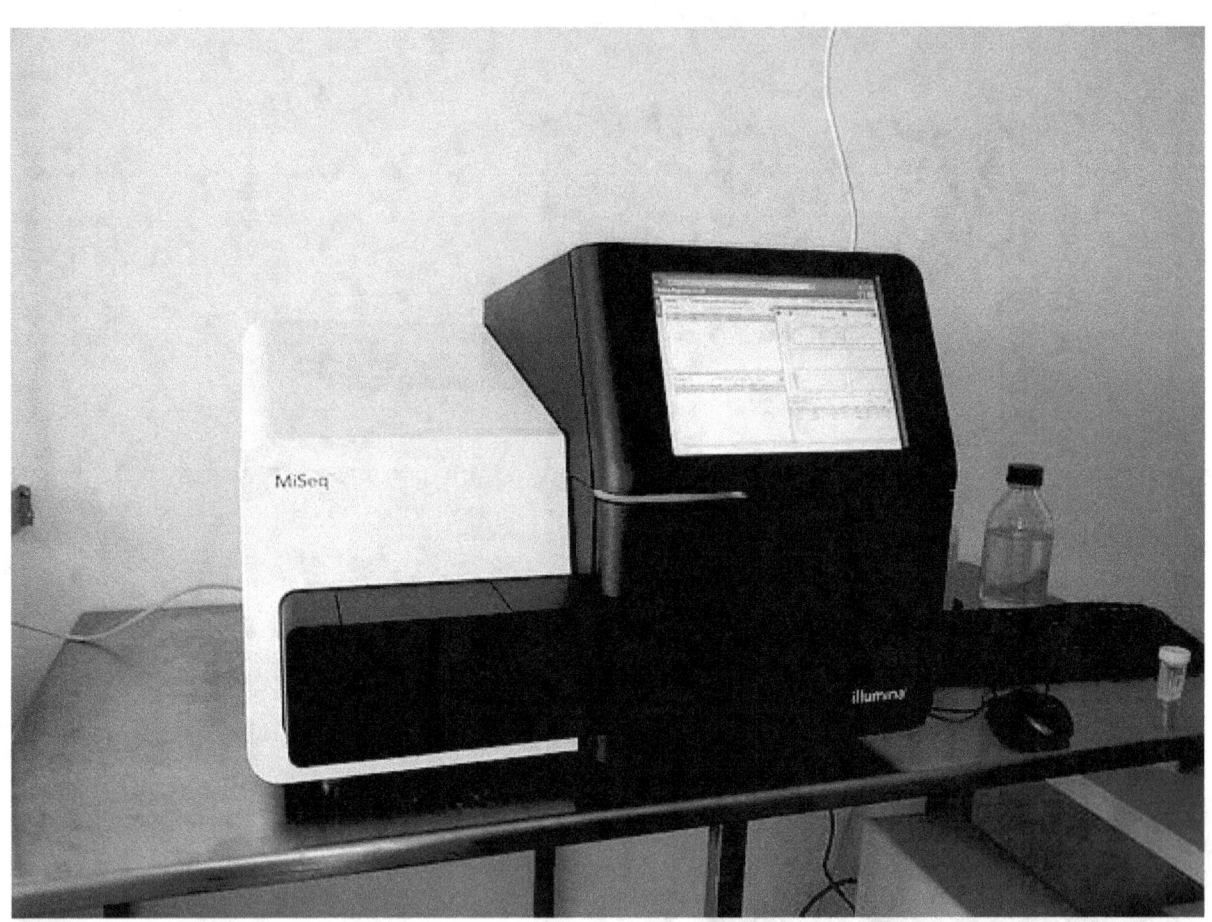

An Illumina MiSeq sequencer

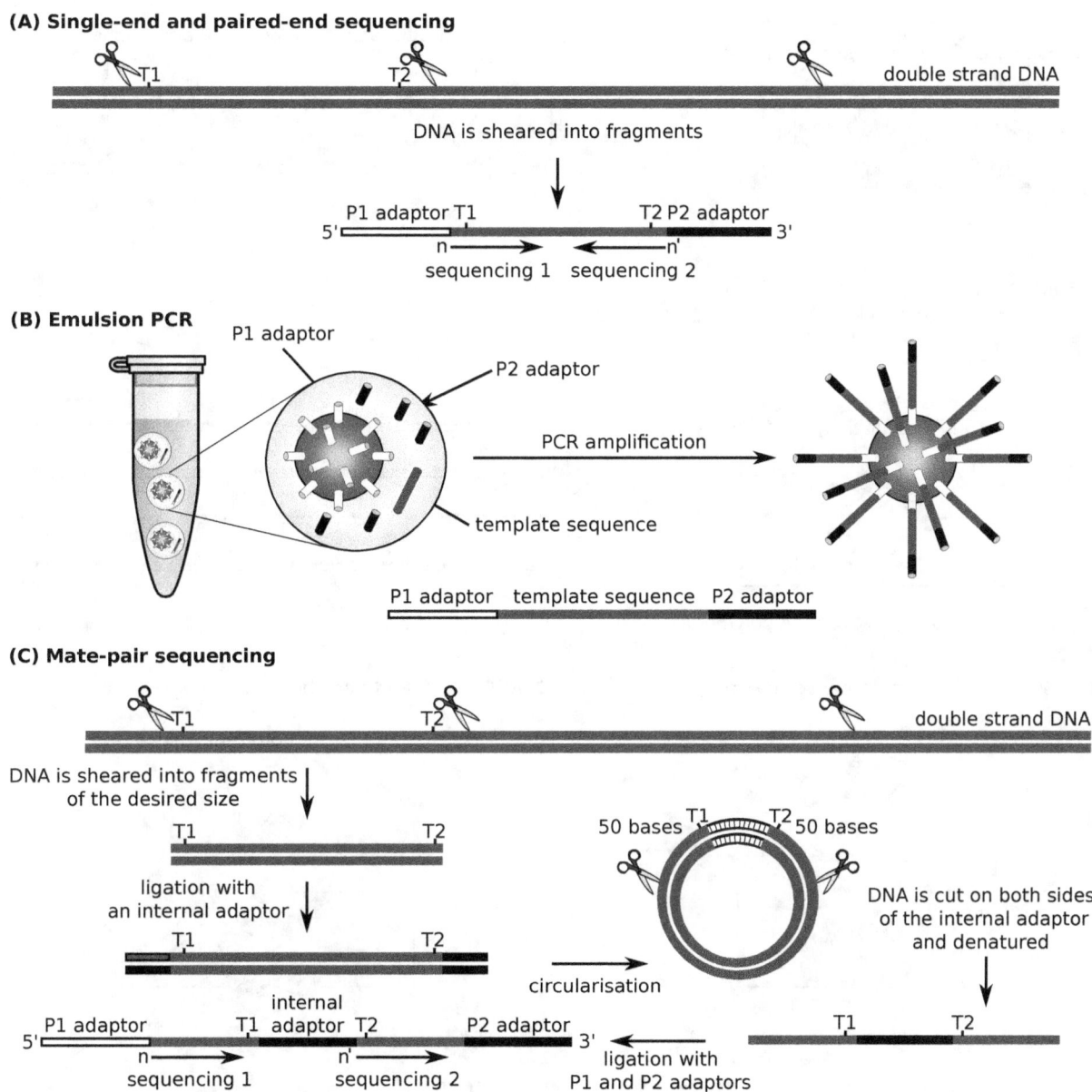

Library preparation for the SOLiD platform

template: TAGGCT

(A) Ion Torrent PGM

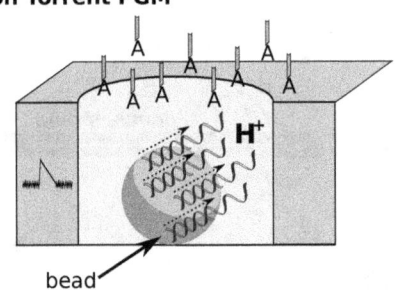

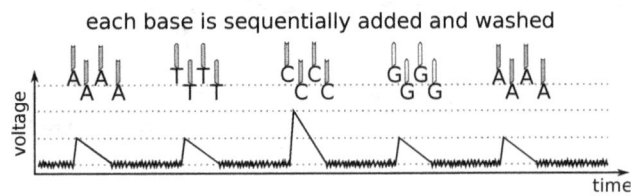

each base is sequentially added and washed

(B) PacBio RS

fluorochromes

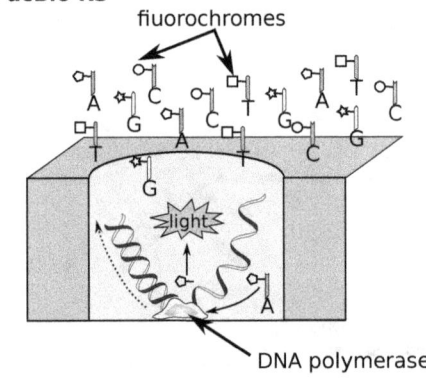

DNA polymerase

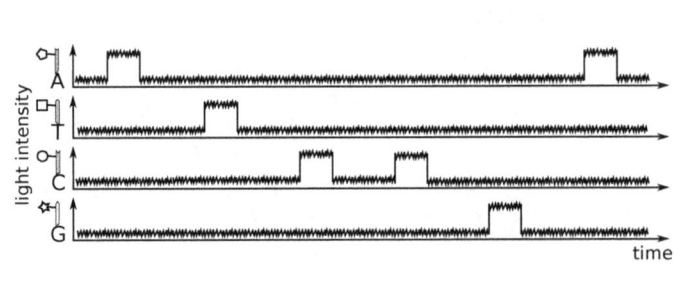

(C) GridION - exonuclease sequencing   **(D) GridION - strand sequencing**

exonuclease   Motor enzyme

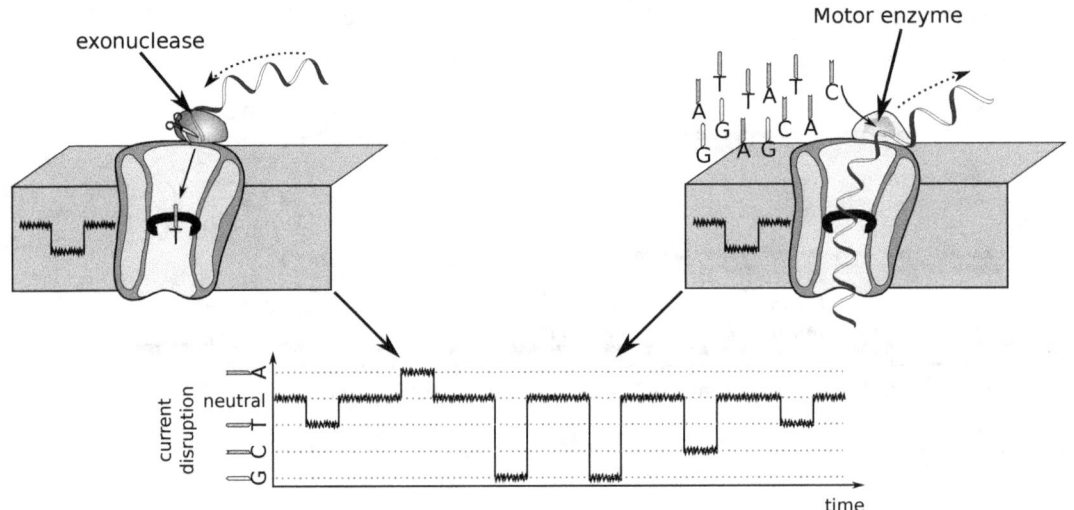

Sequencing of the TAGGCT template with IonTorrent, PacBioRS and GridION

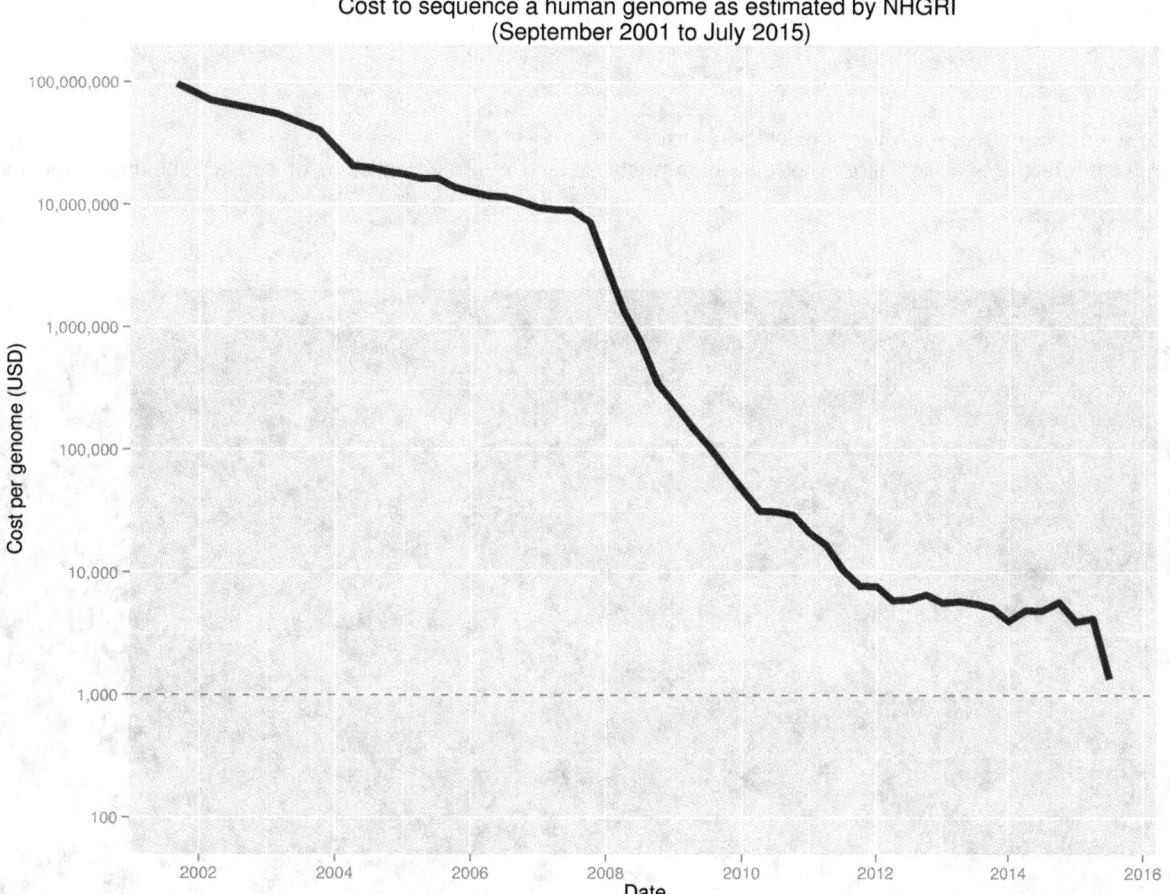

Total cost of sequencing a human genome over time as calculated by the NHGRI.

Chapter 42

Macromolecule

For the scientific journal, see Macromolecules (journal).

A **macromolecule** is a very large molecule commonly created by polymerization of smaller subunits (monomers).

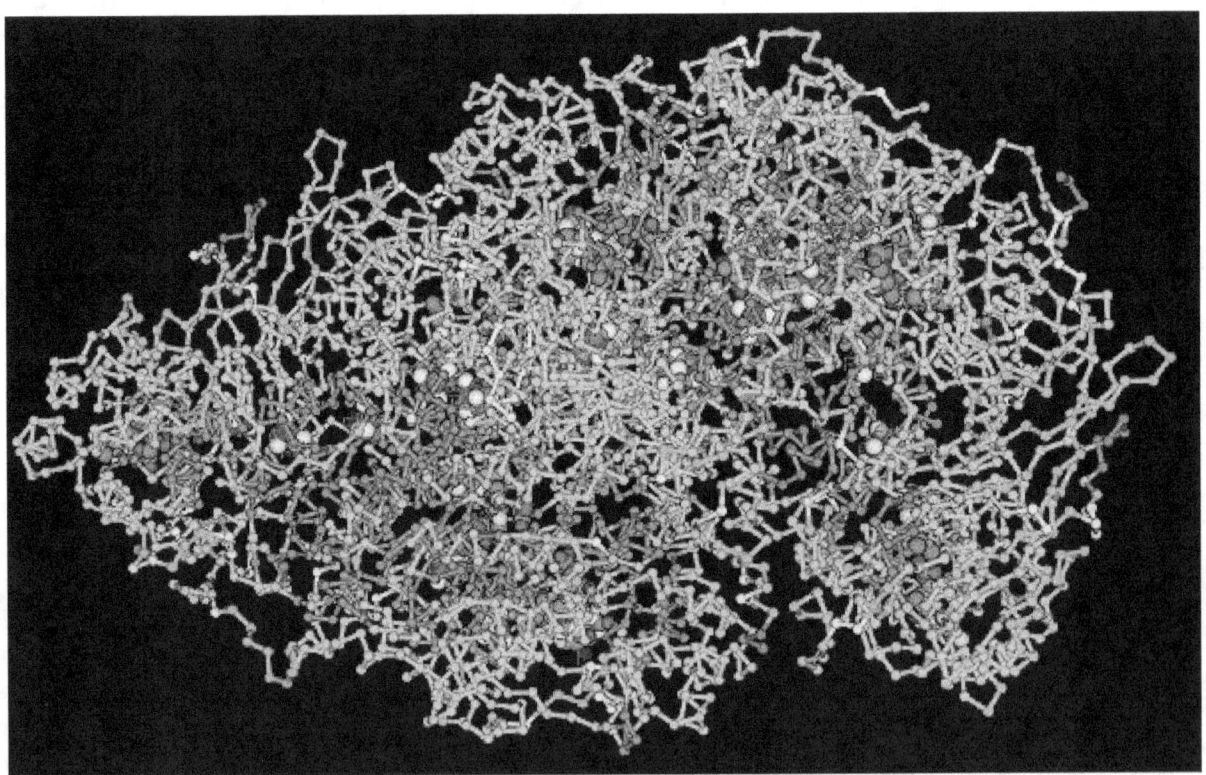

Chemical structure of a polypeptide macromolecule

They are typically composed of thousands or more atoms. The most common macromolecules in biochemistry are biopolymers (nucleic acids, proteins, carbohydrates and polyphenols) and large non-polymeric molecules (such as lipids and macrocycles).[1] Synthetic macromolecules include common plastics and synthetic fibres as well as experimental materials such as carbon nanotubes.[2][3]

42.1 Definition

IUPAC definition

Macromolecule
Polymer molecule

A molecule of high relative molecular mass, the structure of which essentially comprises the multiple repetition of units derived, actually or conceptually, from molecules of low relative molecular mass.

Notes

1. In many cases, especially for synthetic polymers, a molecule can be regarded as having a high relative molecular mass if the addition or removal of one or a few of the units has a negligible effect on the molecular properties. This statement fails in the case of certain macromolecules for which the properties may be critically dependent on fine details of the molecular structure.
2. If a part or the whole of the molecule fits into this definition, it may be described as either *macromolecular* or *polymeric*, or by *polymer* used adjectivally.[4]

The term *macromolecule* (*macro- + molecule*) was coined by Nobel laureate Hermann Staudinger in the 1920s, although his first relevant publication on this field only mentions *high molecular compounds* (in excess of 1,000 atoms).[5] At that time the phrase *polymer*, as introduced by Berzelius in 1833, had a different meaning from that of today: it simply was another form of isomerism for example with benzene and acetylene and had little to do with size.[6]

Usage of the term to describe large molecules varies among the disciplines. For example, while biology refers to macromolecules as the four large molecules comprising living things, in chemistry, the term may refer to aggregates of two or more molecules held together by intermolecular forces rather than covalent bonds but which do not readily dissociate.[7]

According to the standard IUPAC definition, the term *macromolecule* as used in polymer science refers only to a single molecule. For example, a single polymeric molecule is appropriately described as a "macromolecule" or "polymer molecule" rather than a "polymer", which suggests a substance composed of macromolecules.[8]

Because of their size, macromolecules are not conveniently described in terms of stoichiometry alone. The structure of simple macromolecules, such as homopolymers, may be described in terms of the individual monomer subunit and total molecular mass. Complicated biomacromolecules, on the other hand, require multi-faceted structural description such as the hierarchy of structures used to describe proteins. In British English, the word "macromolecule" tends to be called "**high polymer**". [9]

42.2 Properties

Macromolecules often have unusual physical properties that do not occur for smaller molecules.

For example, DNA in a solution can be broken simply by sucking the solution through an ordinary straw because the physical forces on the molecule can overcome the strength of its covalent bonds. The 1964 edition of Linus Pauling's *College Chemistry* asserted that DNA in nature is never longer than about 5,000 base pairs.[10] This error arose because biochemists were inadvertently breaking their samples into fragments. In fact, the DNA of chromosomes can be hundreds of millions of base pairs long, packaged into chromatin.

Another common macromolecular property that does not characterize smaller molecules is their relative insolubility in water and similar solvents, instead forming colloids. Many require salts or particular ions to dissolve in water. Similarly, many proteins will denature if the solute concentration of their solution is too high or too low.

High concentrations of macromolecules in a solution can alter the rates and equilibrium constants of the reactions of other macromolecules, through an effect known as macromolecular crowding.[11] This comes from macromolecules excluding other molecules from a large part of the volume of the solution, thereby increasing the effective concentrations of these molecules.

42.3   Linear biopolymers

All living organisms are dependent on three essential biopolymers for their biological functions: DNA, RNA and Proteins.[12] Each of these molecules is required for life since each plays a distinct, indispensable role in the cell.[13] The simple summary is that DNA makes RNA, and then RNA makes proteins.

DNA, RNA and proteins all consist of a repeating structure of related building blocks (nucleotides in the case of DNA and RNA, amino acids in the case of proteins). In general, they are all unbranched polymers, and so can be represented in the form of a string. Indeed, they can be viewed as a string of beads, with each bead representing a single nucleotide or amino acid monomer linked together through covalent chemical bonds into a very long chain.

In most cases, the monomers within the chain have a strong propensity to interact with other amino acids or nucleotides. In DNA and RNA, this can take the form of Watson-Crick base pairs (G-C and A-T or A-U), although many more complicated interactions can and do occur.

42.3.1   Structural features

Because of the double-stranded nature of DNA, essentially all of the nucleotides take the form of Watson-Crick base pairs between nucleotides on the two complementary strands of the double-helix.

In contrast, both RNA and proteins are normally single-stranded. Therefore, they are not constrained by the regular geometry of the DNA double helix, and so fold into complex three-dimensional shapes dependent on their sequence. These different shapes are responsible for many of the common properties of RNA and proteins, including the formation of specific binding pockets, and the ability to catalyse biochemical reactions.

DNA is optimised for encoding information

DNA is an information storage macromolecule that encodes the complete set of instructions (the genome) that are required to assemble, maintain, and reproduce every living organism.[14]

DNA and RNA are both capable of encoding genetic information, because there are biochemical mechanisms which read the information coded within a DNA or RNA sequence and use it to generate a specified protein. On the other hand, the sequence information of a protein molecule is not used by cells to functionally encode genetic information.[1]:5

DNA has three primary attributes that allow it to be far better than RNA at encoding genetic information. First, it is normally double-stranded, so that there are a minimum of two copies of the information encoding each gene in every cell. Second, DNA has a much greater stability against breakdown than does RNA, an attribute primarily associated with the absence of the 2'-hydroxyl group within every nucleotide of DNA. Third, highly sophisticated DNA surveillance and repair systems are present which monitor damage to the DNA and repair the sequence when necessary. Analogous systems have not evolved for repairing damaged RNA molecules. Consequently, chromosomes can contain many billions of atoms, arranged in a specific chemical structure.

Proteins are optimised for catalysis

Proteins are functional macromolecules responsible for catalysing the biochemical reactions that sustain life.[1]:3 Proteins carry out all functions of an organism, for example photosynthesis, neural function, vision, and movement.[15]

The single-stranded nature of protein molecules, together with their composition of 20 or more different amino acid building blocks, allows them to fold in to a vast number of different three-dimensional shapes, while providing binding pockets through which they can specifically interact with all manner of molecules. In addition, the chemical diversity of the different amino acids, together with different chemical environments afforded by local 3D structure, enables many proteins to act as Enzymes, catalyzing a wide range of specific biochemical transformations within cells. In addition, proteins have evolved the ability to bind a wide range of cofactors and Coenzymes, smaller molecules that can endow the protein with specific activities beyond those associated with the polypeptide chain alone.

RNA is multifunctional

RNA is multifunctional, its primary function is to encode proteins, according to the instructions within a cell's DNA.[1]:5 They control and regulate many aspects of protein synthesis in eukaryotes.

RNA encodes genetic information that can be translated into the amino acid sequence of proteins, as evidenced by the messenger RNA molecules present within every cell, and the RNA genomes of a large number of viruses. The single-stranded nature of RNA, together with tendency for rapid breakdown and a lack of repair systems means that RNA is not so well suited for the long-term storage of genetic information as is DNA.

In addition, RNA is a single-stranded polymer that can, like proteins, fold into a very large number of three-dimensional structures. Some of these structures provide binding sites for other molecules and chemically-active centers that can catalyze specific chemical reactions on those bound molecules. The limited number of different building blocks of RNA (4 nucleotides vs >20 amino acids in proteins), together with their lack of chemical diversity, results in catalytic RNA (ribozymes) being generally less-effective catalysts than proteins for most biological reactions.

42.4   Branched biopolymers

Raspberry ellagitannin, a tannin composed of core of glucose units surrounded by gallic acid esters and ellagic acid units

Carbohydrate macromolecules (polysaccharides) are formed from polymers of monosaccharides.[1]:11 Because monosaccharides have multiple functional groups, polysaccharides can form linear polymers (e.g. cellulose) or complex branched structures (e.g. glycogen). Polysaccharides perform numerous roles in living organisms, acting as energy stores (e.g. Starch) and as structural components (e.g.chitin in arthropods and fungi). Many carbohydrates contain modified monosaccharide units that have had functional groups replaced or removed.

Polyphenols consist of a branched structure of multiple phenolic subunits. They can perform structural roles (e.g. lignin) as well as roles as secondary metabolites involved in signalling, pigmentation and defense.

42.5 Synthetic macromolecules

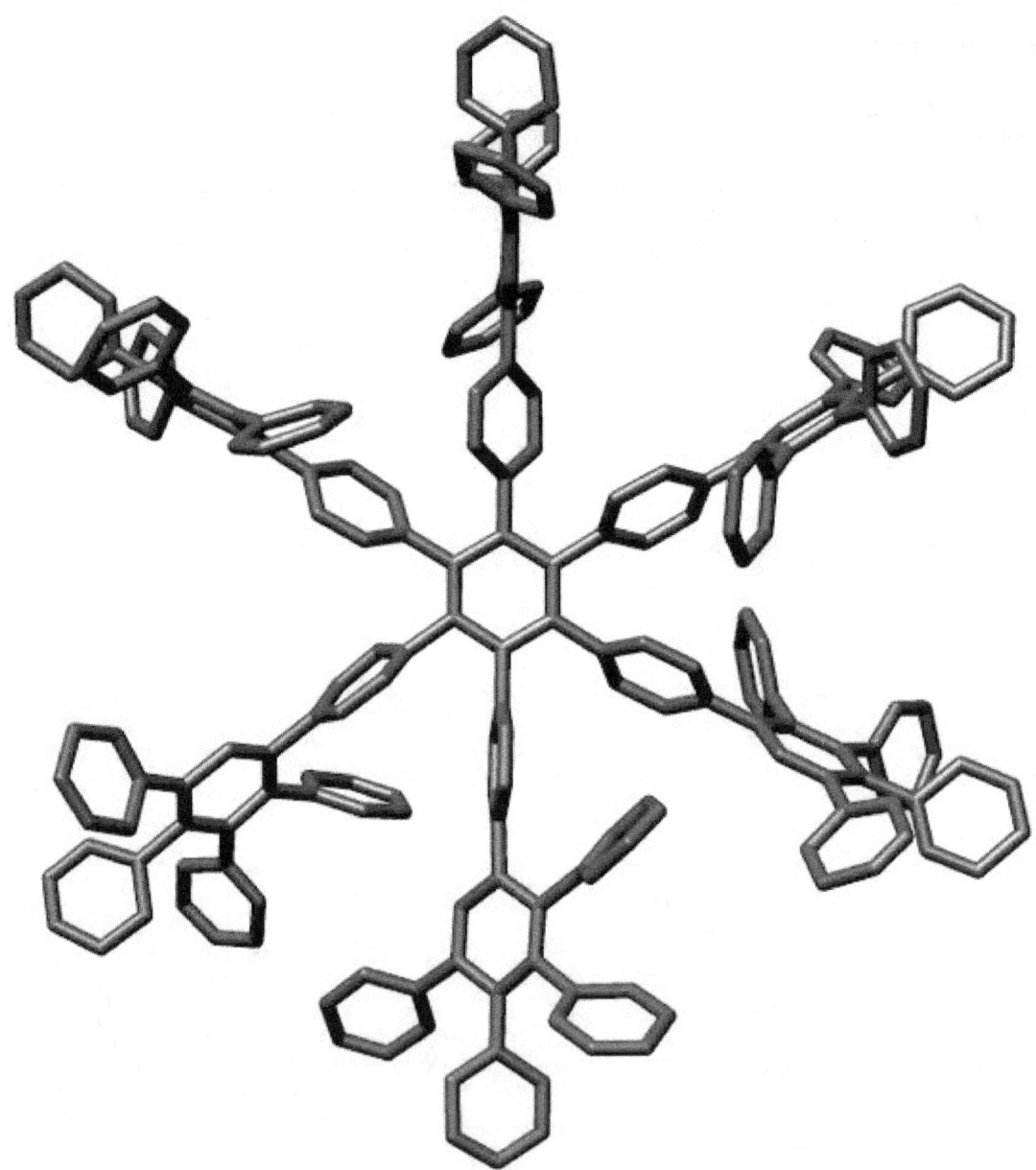

Structure of a polyphenylene dendrimer macromolecule reported by Müllen, et al.[16]

Some examples of macromolecules are synthetic polymers (plastics, synthetic fibers, and synthetic rubber), graphene, and carbon nanotubes.

42.6 References

[1] Stryer L, Berg JM, Tymoczko JL (2002). *Biochemistry* (5th ed.). San Francisco: W.H. Freeman. ISBN 0-7167-4955-6.

[2] Life cycle of a plastic product. Americanchemistry.com. Retrieved on 2011-07-01.

[3] Gullapalli, S.; Wong, M.S. (2011). "Nanotechnology: A Guide to Nano-Objects" (PDF). *Chemical Engineering Progress* **107** (5): 28–32.

[4] "Glossary of basic terms in polymer science (IUPAC Recommendations 1996)" (PDF). *Pure and Applied Chemistry* **68** (12): 2287–2311. 1996. doi:10.1351/pac199668122287.

[5] Staudinger, H.; Fritschi, J. (1922). "Über Isopren und Kautschuk. 5. Mitteilung. Über die Hydrierung des Kautschuks und über seine Konstitution". *Helvetica Chimica Acta* **5** (5): 785. doi:10.1002/hlca.19220050517.

[6] Jensen, William B. (2008). "The Origin of the Polymer Concept". *Journal of Chemical Education* **85**(5): 624. Bibcode:2008J doi:10.1021/ed085p624.

[7] van Holde, K.E. (1998) *Principles of Physical Biochemistry* Prentice Hall: New Jersey, ISBN 0-13-720459-0

[8] Jenkins, A. D.; Kratochvíl, P.; Stepto, R. F. T.; Suter, U. W. (1996). "Glossary of Basic Terms in Polymer Science" (PDF). *Pure and Applied Chemistry* **68** (12): 2287. doi:10.1351/pac199668122287.

[9] High Polymer Research Group

[10] Pauling, Linus (1964). *College Chemistry*. W.H. Feeman and Company.

[11] Minton AP (2006). "How can biochemical reactions within cells differ from those in test tubes?". *J. Cell. Sci.* **119** (Pt 14): 2863–9. doi:10.1242/jcs.03063. PMID 16825427.

[12] Berg, Jeremy Mark; Tymoczko, John L.; Stryer, Lubert (2010). *Biochemistry, 7th ed. (Biochemistry (Berg))*. W.H. Freeman & Company. ISBN 1-4292-2936-5. Fifth edition available online through the NCBI Bookshelf: link

[13] Walter, Peter; Alberts, Bruce; Johnson, Alexander S.; Lewis, Julian; Raff, Martin C.; Roberts, Keith (2008). *Molecular Biology of the Cell (5th edition, Extended version)*. New York: Garland Science. ISBN 0-8153-4111-3.. Fourth edition is available online through the NCBI Bookshelf: link

[14] Golnick, Larry; Wheelis, Mark. *The Cartoon Guide to Genetics*. Collins Reference. ISBN 978-0-06-273099-2.

[15] Takemura, Masaharu (2009). *The Manga Guide to Molecular Biology*. No Starch Press. ISBN 978-1-59327-202-9.

[16] Roland E. Bauer, Volker Enkelmann, Uwe M. Wiesler, Alexander J. Berresheim, Klaus Müllen (2002). "Single-Crystal Structures of Polyphenylene Dendrimers". *Chemistry – A European Journal* **8** (17): 3858. doi:10.1002/1521-3765(20020902)8:17<3 858::AID-CHEM3858>3.0.CO;2-5.

- Tanford, Charles (1961). *Physical Chemistry of Macromolecules*. New York, NY: John Wiley & Sons.

42.7 External links

- Synopsis of Chapter 5, Campbell & Reece, 2002

- Lecture notes on the structure and function of macromolecules

- Several (free) introductory macromolecule related internet-based courses

- Giant Molecules! by Ulysses Magee, *ISSA Review* Winter 2002–2003, ISSN 1540-9864. Cached HTML version of a missing PDF file. Retrieved March 10, 2010. The article is based on the book, *Inventing Polymer Science: Staudinger, Carothers, and the Emergence of Macromolecular Chemistry* by Yasu Furukawa.

42.8   Text and image sources, contributors, and licenses

42.8.1   Text

- **DNA** *Source:* https://en.wikipedia.org/wiki/DNA?oldid=691528201 *Contributors:* AxelBoldt, Magnus Manske, Peter Winnberg, Marj Tiefert, Lee Daniel Crocker, Eloquence, Mav, Bryan Derksen, Slrubenstein, RK, Andre Engels, Fnielsen, Ted Longstaffe, LA2, Danny, Aldie, Unukorno, Toby Bartels, PierreAbbat, Ortolan88, Ben-Zin~enwiki, Anthere, Ellmist, Graft, Heron, Hephaestos, Someone else, Stevertigo, Spiff~enwiki, Lir, Erik Zachte, Lexor, Wwwwolf, Ixfd64, Fruge~enwiki, TakuyaMurata, (, Pde, Pcb21, Goatasaur, Egil, 168..., Looxix~enwiki, Ellywa, Mortene, Ahoerstemeier, Fcrick, Mac, Docu, Snoyes, CatherineMunro, JWSchmidt, Kingturtle, Glenn, Cyan, Poor Yorick, Kwekubo, Rotem Dan, Llull, Samuel~enwiki, Mxn, Raven in Orbit, Quickbeam, Hashar, Mulad, Crusadeonilliteracy, Adam Bishop, Timwi, Dcoetzee, Nohat, Wikiborg, Reddi, Lfh, Jwrosenzweig, Gutza, Rednblu, Wik, Zoicon5, Steinsky, Patrick0Moran, Tpbradbury, Tom Allen, Samsara, Thue, Bevo, Shizhao, Dbabbitt, Raul654, Gakrivas, Bcorr, Pakaran, Jerzy, Lumos3, Donarreiskoffer, Robbot, Astronautics~enwiki, Chris 73, Schutz, Vyasa, Peak, Stewartadcock, Sverdrup, Academic Challenger, Timrollpickering, Bkell, Factual, Moink, Hadal, Jstech, Anthony, Neckro, Pifactorial, Diberri, Cyrius, Dmn, Dina, Ancheta Wis, Giftlite, JamesMLane, Graeme Bartlett, DocWatson42, Nunh-huh, Kapow, Netoholic, Fastfission, MSGJ, Obli, Everyking, No Guru, Dratman, Curps, Michael Devore, Bensaccount, Guanaco, Jorge Stolfi, Pascal666, Horatio, Luigi30, Solipsist, Ojl, Avala, SWAdair, Bobblewik, Alan Au, Delta G, Wmahan, Stevietheman, Adenosine, Utcursch, Pgan002, Andycjp, Dullhunk, CryptoDerk, Gazibara, Kums, Antandrus, Onco p53, MisfitToys, G3pro, PDH, Jossi, Rdsmith4, Mikko Paananen, OwenBlacker, Kevin B12, PFHLai, Magadan, Icairns, Troels Arvin, Figure, Asbestos, Neutrality, Golnazfotohabadi, JohnArmagh, Jh51681, Sonett72, Adashiel, Thorwald, Mike Rosoft, Alkivar, D6, Freakofnurture, Rdb, A-giau, William Pietri, ElTyrant, Rich Farmbrough, Guanabot, Ffirehorse, Cacycle, Qutezuce, Vsmith, EliasAlucard, Mgtoohey, Mjpieters, Zazou, Bender235, ESkog, Kbh3rd, Swid, Loren36, Danny B-), Brian0918, Charm, Ben Webber, Kwamikagami, Mwanner, Phoenix Hacker, Aude, Shanes, Susvolans, RoyBoy, EurekaLott, Andreww, Causa sui, Bobo192, Kghose, Infocidal, R. S. Shaw, Brim, Dungodung, Arcadian, Redquark, Timl, Tomgally, La goutte de pluie, Jojit fb, Malcolm rowe, Vanished user 19794758563875, Kierano, Hagerman, Bijee~enwiki, HasharBot~enwiki, Sam Burne James, Emoticon, Jumbuck, Danski14, Mithent, Gary, Anthony Appleyard, Chino, Borisblue, Atlant, SemperBlotto, Ricky81682, Loris, Benjah-bmm27, Wouterstomp, AzaToth, Yamla, Water Bottle, Echuck215, Seans Potato Business, Kocio, InShaneee, Hu, Malo, VladimirKorablin, Snowolf, PaePae, Wtmitchell, Melaen, Schapel, BaronLarf, ClockworkSoul, Unconventional, KingTT, Knowledge Seeker, Cburnett, Evil Monkey, Cal 1234, Max Naylor, CloudNine, TenOfAllTrades, Sciurinæ, Inge-Lyubov, Lerdsuwa, LFaraone, Gene Nygaard, Alai, Mattbrundage, Netkinetic, Johntex, Kitch, Adrian.benko, RyanGerbil10, Falcorian, Tariqabjotu, Ashujo, DeAceShooter, Stemonitis, Natarajanganesan, Gmaxwell, Zanaq, Kelly Martin, OwenX, Woohookitty, Mindmatrix, Katyare, TigerShark, Yansa, Rocastelo, Carcharoth, Nitecrawler, Lincher, Colorajo, JeremyA, Duncan.france, MONGO, Nirajrm, Astrowob, Eleassar777, Bbatsell, GregorB, SCEhardt, TheAlphaWolf, Junes, Palica, Turnstep, Marudubshinki, Mandarax, Frostyservant, RichardWeiss, Yasha~enwiki, Graham87, Deltabeignet, Magister Mathematicae, 168.., 169, GoldRingChip, Buxtehude, BD2412, Patrick2480, FreplySpang, Lion Wilson, Yurik, RxS, Effeietsanders, Whoutz, Crzrussian, Drbogdan, Rjwilmsi, Biriwilg, Koavf, Virtualphtn, Collins.mc, Vary, JHMM13, Bruce1ee, Jmcc150, Salix alba, JonMoulton, Oblivious, DonSiano, Ligulem, Gjuggler, SeanMack, Bubba73, Brighterorange, Bhadani, Ucucha, AySz88, Sango123, Yamamoto Ichiro, Kevmitch, Wobble, Ravidreams, FlaBot, Tufflaw, RobertG, Latka, Nihiltres, Crazycomputers, Chanting Fox, Chill Pill Bill, RexNL, Takometer, Karrmann, Jrtayloriv, DevastatorIIC, Wikipedia Administration, ZScout370, Alphachimp, McDogm, Daycd, BMF81, GordonWatts, Smithbrenon, Mstroeck, King of Hearts, Chobot, Nauseam, Antilived, Bornhj, Bjwebb, Mhking, Bgwhite, Digitalme, Whosasking, The Rambling Man, Wavelength, Cathalgarvey, Sceptre, WAvegetarian, Jumbo Snails, Spaully, DNA EDIT WAR, Editing DNA, Anarchy on DNA, DNA is shyt, I hate DNA, Zell Miller's DNA, Trent Lott's DNA, RJC, Pigman, Bernie Sanders' DNA, Orrin Hatch's DNA, Bill Nelson's DNA, Tom Harkin's DNA, Chuck Grassley's DNA, Richard Durbin's DNA, Max Baucus' DNA, Splette, SpuriousQ, Stephenb, Wimt, GeeJo, Tavilis, Anomalocaris, Canadaduane, Alcides, Sentausa, Fabhcún, EWS23, Alzhaid~enwiki, Dysmorodrepanis~enwiki, Wiki alf, Chick Bowen, Jaxl, Tmueck, Psora, Taco325i, Yoninah, Djm1279, Ragesoss, Retired username, Sangwine, Robdurbar, Bert Macklin, PhilipO, Misza13, Grafikm fr, Fs, Tony1, Dbfirs, Lockesdonkey, Roy Brumback, DeadEyeArrow, Bota47, Biolinker, Cosmotron, Supspirit, Scope creep, Caerwine, Chris84~enwiki, Somoza, Wknight94, Leptictidium, WAS 4.250, FF2010, Alarob, Calaschysm, 2over0, Theodolite, BenBildstein, Theda, Spondoolicks, Hurricanehink, Jolt76, GraemeL, Acer, JoanneB, Heathhunnicutt, Zahiri, ArielGold, Curpsbot-unicodify, Michigan user, RunOrDie, Stuhacking, Kungfuadam, Banus, Airconswitch, Samuel Blanning, BiH, DVD R W, Jer ome, Luk, RichG, Blastwizard, Itub, Twilight Realm, A bit iffy, SmackBot, Eperotao, Spongebobsqpants, Tarret, Chodges, Prodego, KnowledgeOfSelf, TestPilot, Royalguard11, Martin.Budden, Melchoir, Nitramrekcap, Shoy, Joconnol, Jacek Kendysz, Pkirlin, Delldot, Blondtraillite, Zephyris, Yamaguchi⯑⯑, Cool3, Aksi great, Peter Isotalo, Gilliam, Julian Diamond, Chaojoker, Jimwong, ERcheck, Carbon-16, Yaser al-Nabriss, Kazkaskazkasako, Izehar, Unint, Keegan, Persian Poet Gal, RDBrown, MidgleyDJ, Uthbrian, Adamstevenson, Dlohcierekim's sock, DHN-bot~enwiki, Terraguy, Zven, Darth Panda, Firetrap9254, Gracenotes, Lightspeedchick, Mikker, Hgrosser, Zsinj, Can't sleep, clown will eat me, Jorvik, Danielkueh, Snowmanradio, OOODDD, Roadnottaken, TheKMan, Anita1988, GVnayR, Grover cleveland, Aldaron, NoIdeaNick, Spectrogram, Iapetus, Nakon, Savidan, Ne0Freedom, Jiddisch~enwiki, Rezecib, RandomP, Smokefoot, SteveHopson, Drphilharmonic, Portugue6927, Fagstein, Kendrick7, Where, Jls043, Pilotguy, Madeleine Price Ball, SashatoBot, Nishkid64, Visium, Rory096, Zahid Abdassabur, Kuru, General Ization, Tazmaniacs, Kahlfin~enwiki, Epingchris, Statsone, Shadowlynk, AstroChemist, JoshuaZ, Edwy, Coredesat, Mgiganteus1, MichaelHa, IronGargoyle, Fernando S. Aldado~enwiki, The Man in Question, Loadmaster, Digger3000, Alexandremas~enwiki, Hvn0413, Mantissa128, Echo park00, Munita Prasad, Mr Stephen, Dicklyon, Serephine, 4u1e, SandyGeorgia, Sir.Loin, Midnightblueowl, Ryulong, Elb2000, Carlo.milanesi, Squirepants101, RoyLaurie, Autonova, Glen Hunt's DNA, PaulGS, SimonD, Vanished user, Paul venter, Ariel Pontes, Younusporteous, Paul Foxworthy, ScottBS, Esurnir, Adambiswanger1, Tawkerbot2, Alegoo92, Ouishoebean, Davidbspalding, Redneckjimmy, Fvasconcellos, Switchercat, JForget, RSido, Friendly Neighbour, Gatortpk, CmdrObot, Mattbr, Wafulz, Zarex, Lavateraguy, SupaStarGirl, Leevanjackson, CWY2190, GHe, STAN SWANSON, Mintman16, Outriggr (2006-2009), Cerberus lord, Logical2u, Nthornberry, Moorice~enwiki, RoddyYoung, GeoMor, Sopoforic, A D 13, WillowW, Bryan, Steel, YanWong~enwiki, Michaelas10, Gogo Dodo, Travelbird, Red Director, Trd300gt, Hughdbrown, Sloth monkey, Studerby, Shekharsuman, B, Tawkerbot4, Carstensen, Doug Weller, DumbBOT, Narayanese, ErrantX, Omicronpersei8, Daniel Olsen, Gimmetrow, Casliber, Sabbre, Thijs!bot, Bloger, Opabinia regalis, Ante Aikio, Russ47025, Headbomb, West Brom 4ever, NorwegianBlue, Tellyaddict, Peter K., GAThrawn22, EdJohnston, Eddycrai, Nick Number, S77914767, Sean William, Natalie Erin, Escarbot, David D., KrakatoaKatie, Cyclonenim, Ialsoagree, Mattjblythe, AntiVandalBot, Mr Bungle, Ty146Ty146, Majorly, Luna Santin, Jeka911, Why My Fleece?, Opelio, Jayron32, 17Drew, Pro crast in a tor, TimVickers, Isilanes, Darklilac, Davegrupp, Storkk, Sjollema, Lklundin, Figma, Canadian-Bacon, JAnDbot,

Deflective, Leuko, MER-C, Janejellyroll, DigitalGhost, Db099221, Alpinu, Andonic, Hut 8.5, Tstrobaugh, Clivedelmonte, Yahel Guhan, Efbfweborg, Sangak, Magioladitis, Pigmietheclub, WolfmanSF, Pedro, Bongwarrior, Hb2019, Carlwev, NighthawkJ, Bhar100101, Hullaballoo Wolfowitz, Jlh29, CattleGirl, Docjames, Scincesociety, GODhack~enwiki, Tobogganoggin, ThoHug, Avicennasis, WhatamIdoing, BatteryIncluded, Torchiest, David Eppstein, Emw, Gurko, Doctor Faust, Mr Meow Meow, Mika293, TheRanger, Stolsvik, Squidonius, Patstuart, E. Wayne, NatureA16, PsyMar, Yobol, MartinBot, NochnoiDozor, Sjjupadhyay~enwiki, ShaunL, Jonrunles, Pierceno, Dudewheresmywallet, Motley Crue Rocks, Rettetast, Fishingpal99, Zouavman Le Zouave, R'n'B, Test100000, CommonsDelinker, Nono64, Armored Ear, Impamiizgraa, Hairchrm, Fconaway, Jarhed, WelshMatt, Grandegrandegrande, 3dscience, Cinnamon colbert, DrKay, CFCF, Trusilver, Nate1028, Spathaky, UBeR, Nbauman, Boghog, Shmee47, Maurice Carbonaro, Vandelizer, WarthogDemon, Tdadamemd, Andrew wilson, BikA06, Ellis O'Neill, TheChrisD, Bmtbomb, DJRafe, Tonyrenploki, Nemo bis, Dr d12, MrErku, Lhenslee, Kemyou, Pyrospirit, (jarbarf), WHeimbigner, Aquaplus, NewEnglandYankee, Antony-22, SmilesALot, ArazZeynili, Master dingley, Touch Of Light, MKoltnow, Aatomic1, Potatoswatter, Scoterican, Cornacchia123, FOTEMEH, Aj123456, WJBscribe, YOUR DNA, Mleefs7, Dr.Kerr, Cainer91, Treisijs, Vaernnond, Etanol~enwiki, Idioma-bot, Hammersoft, VolkovBot, Preston47, DrMicro, Seldon1, Midoriko, Coolawesome~enwiki, Mstislavl, Daniel987600, Hersfold, Orthologist, Lia Todua, AlnoktaBOT, Fences and windows, The WikiWhippet, Loginbuddy, Philip Trueman, Director, TXiKiBoT, Muro de Aguas, Billmcgn189, CupOBeans, ElinorD, Qxz, Ocolon, Harianto, Sintaku, Michael H 34, Toninu, Chanora, Seb az86556, Ilia Kr., UnitedStatesian, Pristontale111, Chuck02, Luuva, Johanvs, Hockey21dude, DJAX, Nighthawk380, Rajwiki123, Gilisa, Hannes Röst, Gwsrinme, Mentalmaniac07, MarvPaule, Flavaflav1005, Yomama9753, Madhero88, Amboo85, Ashnard, JamesMt1984, Usergreatpower, Synthebot, CephasE, Northfox, AlleborgoBot, Sly G, Timewatcher, Pediaknowledge, Isis07, EmxBot, Gustav von Humpelschmumpel, Kbrose, Safwan40, SieBot, Cubskrazy29, ShiftFn, Tiddly Tom, Graham Beards, BotMultichill, Sakkura, Gerakibot, Dawn Bard, Cwkmail, RJaguar3, Triwbe, Brockett, Execvator, Keilana, Aaaxlp, Crowstar, Prestonmag, Hzh, Artoasis, Lightmouse, Nanettea, Escape Artist Swyer, Hairwheel, BenoniBot~enwiki, Cradlelover123, Diego Grez-Cañete, Vividonset, Juneythomas, Andrij Kursetsky, Taulant23, Bogwhistle, Randomblue, Paulinho28, Florentino floro, Lascorz, Forluvoft, SallyForth123, Stuart7m, Jimriz, ClueBot, GorillaWarfare, Artichoker, Wikievil666, Yikrazuul, Enthusiast01, TobyWilson1992, Niceguyedc, GoEThe, Llongland, DragonBot, Excirial, Alexbot, Pumpkingrower05, Gulmammad, ThreeDaysGraceFan101, Estirabot, Coinmanj, NuclearWarfare, Jotterbot, Medos2, Highfly3442, Jackrm, OekelWm, Heyheyhack, Br shadow, Ardyn, Muro Bot, Gaara san, Aitias, Jonverve, Versus22, Josq, Johnuniq, Wnt, Wkboonec, Psymier, Rishi.bedi, DumZiBoT, Jetsetpainter, Mandyj61596, Simultaneous, SilvonenBot, Priscilla 95925, Lemchesvej, Addbot, Giftiger wunsch, DOI bot, Medessec, TutterMouse, OBloodyHell, Low-frequency internal, Keepweek~enwiki, NjardarBot, Bernstein0275, DFS454, LinkFABot, Kisbesbot, 84user, Numbo3-bot, Tide rolls, Luckas Blade, Zorrobot, WikiDreamer Bot, Ettrig, Legobot, Drpickem, Luckas-bot, Yobot, Marcus.aerlous, Aquilla34, Nallimbot, KamikazeBot, LightFlare, AnomieBOT, Tryptofish, 1exec1, Jim1138, Galoubet, JWSurf, Mahmudmasri, Materialscientist, Citation bot, Bci2, ArthurBot, Quebec99, Xqbot, Sciencechick, Chaboura, Timir2, Kholdstare99, Khajidha, JWBE, P99am, GrouchoBot, Ute in DC, Wizardist, RibotBOT, Wiki emma johnson, 7434be, Briland, GhalyBot, AustralianRupert, Hersfold tool account, Jack B108, Legobot III, FrescoBot, Paine Ellsworth, Dogposter, Rohitsuratekar, Jc3s5h, KerryO77, HJ Mitchell, Steve Quinn, Citation bot 1, SebastianHawes, Scarce, Javert, Chenopodiaceous, Redrose64, DrilBot, Pinethicket, I dream of horses, HRoestBot, Jonesey95, Tom.Reding, Supreme Deliciousness, RedBot, Curehd, Jujutacular, DixonDBot, Jann, Dinamik-bot, Clarkcj12, Sgt. R.K. Blue, HayleyJohnson21, Jynto, Tbhotch, Jesse V., Minimac, DARTH SIDIOUS 2, Mean as custard, RjwilmsiBot, Cloakedyoshi, Salvio giuliano, Mandolinface, Toto Azéro, Mashin6, EmausBot, Orphan Wiki, Acather96, WikitanvirBot, Never give in, Jodon1971, Joeywallace9, GoingBatty, Black Yoshi, Winner 42, Dcirovic, TeleComNasSprVen, Hhhippo, JSquish, Otterinfo, Empty Buffer, Dffgd, John Mackenzie Burke, SporkBot, AManWithNoPlan, Ocaasi, Thine Antique Pen, Hccc, JZuehlke, Brandmeister, Pkank, L Kensington, Perseus, Son of Zeus, Rr2wiki, Donner60, Dnacond, Nanodance, Scientific29, Puffin, Nofatlandshark, LikeLakers2, Davidartois, Woodsrock, Mikhail Ryazanov, Will Beback Auto, ClueBot NG, Jnorton7558, Ds2207, Gareth Griffith-Jones, MelbourneStar, Prathfig, Kaushlendratripathi, Zynwyx, O.Koslowski, Argionember, Username 772, Theopolisme, Vogel2014, Helpful Pixie Bot, Iyentra Rasonica, Aaronstonestrom, Calabe1992, Gob Lofa, Bibcode Bot, Lowercase sigmabot, BG19bot, BOK602, Piguy101, Midnight Green, Astpurcell, JasonK33, Jazzlw, Min.neel, Snow Blizzard, Uluru345, Achowat, Djihinne1, AndroidOS, Biosthmors, TuringMachine17, Stigmatella aurantiaca, ChrisGualtieri, Saxophilist, Khazar2, IjonTichyIjonTichy, Dexbot, Webclient101, Mogism, Oliverrichardson, Laxative Brownies, THFC1996, Fox2k11, Zziccardi, Hopefuldonor, Mshamza112, Phil2793, Fjozk, RandomLittleHelper, Cor Ferrum, Joeinwiki, ProDawg5, 9FireStar, Diegomanzana, Keiyashi, Dbzhero5000, IncredibleWondersYes1, Youngdro2, AmericanLemming, ManofQueens, Evolution and evolvability, Bever, BruceBlaus, Anrnusna, Meteor sandwich yum, Andrewmhhs, Abitslow, Csutric, Chaya5260, Zettek95, Mahusha, Giancarlobasile, Monkbot, Virion123, Owais Khursheed, Acagastya, SusithCM, Entitymasterblaster, HMSLavender, Cyntiamaspian, KH-1, Bhootrina, Soldier of the Empire, DiscantX, MicroPaLeo, Jensberzelius, Forscienceonly, Azealia911, Craftwerker, 3 of Diamonds, COOL ANKUR CHOUDHURY, ProprioMe OW, Unmaterial scientist, MANEVIL 187, Rrwanga17, Maddog9962002, Christianmorasco, Underlyingboss3, CAPTAIN RAJU, S281305, Nn9888, Ryanross123, MoonmanMCD, Stabila711, Matiar Chris Brown, Mirenoula, Sweatypalmsggg, Joshhassweatylankles, Mikemorris123456789, Nafi172, ProWiki21, PubSci, Lordmitchimus, CutiePie420, Milkdudstruth, Alexander Rohner, Milkdudstruthback, Milkdudstruthagian, MariusOrion and Anonymous: 1043

- **Base pair** *Source:* https://en.wikipedia.org/wiki/Base_pair?oldid=688739219 *Contributors:* Magnus Manske, The Anome, SimonP, Ewen, Axel Driken, Edward, Michael Hardy, Lexor, MichaelJanich, Cyan, Alf, Tpbradbury, J D, Donarreiskoffer, Hadal, Nagelfar, Giftlite, DocWatson42, Obli, Bensaccount, Ferren~enwiki, Delta G, Nova77, Antandrus, G3pro, PDH, Ary29, Sam Hocevar, MRSC, Thorwald, Discospinster, Rich Farmbrough, Guanabot, Cacycle, Deelkar, Bender235, Truthflux, Cmdrjameson, Jerryseinfeld, Alansohn, Gary, Seans Potato Business, Wtmitchell, Gene Nygaard, Kgrr, Graham87, BD2412, Effeietsanders, Drbogdan, Rjwilmsi, JHMM13, Badhotra, Kazrak, Wragge, FlaBot, Vilcxjo, King of Hearts, TAKE, YurikBot, Borgx, Ejl, Divide, Allens, GrinBot~enwiki, SmackBot, Martin.Budden, Jrockley, BiT, Keegan, SchfiftyThree, Polyhedron, Hgrosser, Sephiroth BCR, Roadnottaken, Drphilharmonic, Wizardman, Sjester, Vina-iwbot~enwiki, SvenskaJohannes, JorisvS, Ben Moore, Dcflyer, Pqrstuv, Kaarel, Martious, Dumeinar~enwiki, Matt26, Ppgardne, Was a bee, Tawkerbot4, Narayanese, Freak in the bunnysuit, Mauroesguerroto, Thijs!bot, Epbr123, Opabinia regalis, CopperKettle, Lauranrg, Dfrg.msc, AntiVandalBot, Seaphoto, Isilanes, JAnDbot, Andonic, Johnman239, MSBOT, VoABot II, Nyq, BatteryIncluded, Adrian J. Hunter, Emw, Squidonius, MartinBot, CommonsDelinker, Tgeairn, Hierophantasmagoria, Antony-22, Tsutsuonda, Pdcook, RJASE1, Tekkaman~enwiki, VolkovBot, Paranoid600, A4bot, Rei-bot, Valencerian, Jaimepeschiera, Littlealien182, BrandonKloppenburgPh.D, PaulTanenbaum, Hannes Röst, AlleborgoBot, Petergans, SieBot, BotMultichill, Flyer22 Reborn, Mrgnomes, Rocknrollsuicide, Nopetro, Yerpo, Oxymoron83, Vanished user kijsdion3i4jf, StaticGull, Anchor Link Bot, Asikhi, Forluvoft, WikipedianMarlith, ClueBot, Saatwik Katiha, CounterVandalismBot, NuclearWarfare, Wiki libs, BOTarate, Kakofonous, Aitias, DumZiBoT, RMFan1, Jordanp, Trabelsiismail, DOI bot, CanadianLinuxUser, AndersBot, West.andrew.g, Wgpautoplay, Flakinho, Tide rolls, Legobot, Luckas-bot, Yobot, Materialscientist, Carlsotr, Franciscosp2, TheAMmollusc, Abce2, Citation bot 1, Flecha2, Tom.Reding, Techyactor15, MirankerAD, BertinoSA, YuanH, Vinnyzz, WikitanvirBot, Mordgier, Ecorahul, JSquish, John Mackenzie Burke, Donner60, ClueBot NG, This lousy T-shirt, Catlemur, HMSSolent, BG19bot, NotWith, ChrisGualtieri, Webclient101, TortoiseWrath, Rhanzhou, Xxkitsune, Sosthenes12, Liam Neal, Monkbot, Noatpoiasioguio, CV9933 and Anonymous: 189

- **Sense (molecular biology)** *Source:* https://en.wikipedia.org/wiki/Sense_(molecular_biology)?oldid=687622947 *Contributors:* Vik-Thor, Alan Liefting, Jeremiah, Graeme Bartlett, Pgan002, PDH, JeffreyN, Rich Farmbrough, Guettarda, Iamunknown, Benbest, Rjwilmsi, Feydey, Jon-Moulton, Wavelength, Dysmorodrepanis~enwiki, Lesotho, Leptictidium, WAS 4.250, SmackBot, Telliott, Gilliam, Radagast83, Drphilharmonic, DMacks, Clicketyclack, FrozenMan, Ben Moore, Beetstra, Serephine, InfoCan, Opabinia regalis, Rosarinagazo, Adrian J. Hunter, Sabedon, CommonsDelinker, Nono64, Abecedare, Xris0, Lantonov, Mikael Häggström, Dciitkgp, VolkovBot, Jeff G., TXiKiBoT, Luuva, Michaeldsuarez, Temporaluser, SylviaStanley, Graham Beards, Phe-bot, Juan-Nafrán, Happysailor, ClueBot, TableManners, Pysar, Schreiber-Bike, Dana boomer, XLinkBot, Addbot, Fyrael, Leszek Jańczuk, SpBot, Yobot, Ptbotgourou, JackieBot, Almabot, MerlLinkBot, FrescoBot, Flecha2, Jusses2, Jesse V., Seekerojustice, RjwilmsiBot, Dcirovic, Sjoosse, FeatherPluma, Thepigdog, Helpful Pixie Bot, Iyentra Rasonica, BG19bot, AKalsbeek, DemonRipper, MusikAnimal, BattyBot, PICAWN, Philipl1221, ComfyKem, Benjamin.hepp, Quenhitran, Monkbot, Dsbhong, Kkimeric and Anonymous: 55

- **Messenger RNA** *Source:* https://en.wikipedia.org/wiki/Messenger_RNA?oldid=689415723 *Contributors:* Magnus Manske, Sodium, Bryan Derksen, Edward, Lexor, Axeloide, Ahoerstemeier, Andres, Mxn, Stismail, Jph, Phil Boswell, Robbot, Sverdrup, Puckly, Hadal, ManuelGR, Brona, Bensaccount, Manuel Anastácio, G3pro, PDH, PFHLai, Icairns, Adashiel, Discospinster, Cacycle, Vsmith, Brian0918, Plociam, Wisdom89, Tmh, La goutte de pluie, TheProject, HasharBot~enwiki, Alansohn, Zsero, Wtmitchell, Ceyockey, Jef-Infojef, OwenX, Bitshifter10, Kyleca, Rjwilmsi, OneWeirdDude, Sdornan, Klortho, AySz88, Tangram-Abacus, Sheldrake, FlaBot, Tobiasmayr, Gurch, Sairen42, Chobot, Roboto de Ajvol, YurikBot, Hairy Dude, Wimt, Sentausa, Aeusoes1, Jaxl, Nick C, Bota47, Orioane, Lt-wiki-bot, Peter, Slavakion, Banus, Axfangli, Kf4bdy, SmackBot, Martin.Budden, Edgar181, Apers0n, Zephyris, Gilliam, Izehar, Bluebot, NCurse, Jprg1966, SchfiftyThree, DHN-bot~enwiki, Wisden17, CheesyPuffs144, Can't sleep, clown will eat me, Nick Levine, Radagast83, TedE, Richard001, Drphilharmonic, Spiritia, Nishkid64, IronGargoyle, Ben Moore, Michael Greiner, Hu12, Kaarel, Vermiculus, Agathman, Robotsintrouble, FlyingToaster, ONUnicorn, Ppgardne, Was a bee, Narayanese, Thijs!bot, Marek69, Prontoo, AntiVandalBot, MECU, .anacondabot, TransControl, Magioladitis, VoABot II, Pvosta, Pupster21, Tgeairn, Mike.lifeguard, Rod57, Katalaveno, Oceanflynn, Gurchzilla, Tanaats, KylieTastic, Belasco, Dorftrottel, Tomas.Persson, ELLusKa 86, Lalvers, Squids and Chips, 28bytes, VolkovBot, Msweany, Adhesionguy, Tameeria, Antoni Barau, Kriak, Stephen.floor, Broadbot, McM.bot, Charles Kinbote, Dirkbb, Logan, Fredweis, SieBot, Gerakibot, Midknightr, Yintan, Keilana, Flyer22 Reborn, Enti342, Faradayplank, Louismaddox, Anchor Link Bot, Forluvoft, ClueBot, LonelyBeacon, Mrscappell, Excirial, Schreiber-Bike, Connah0047, BlownSugar, Kpok3019, Addbot, DOI bot, DougsTech, Bastion Monk, Favonian, LinkFA-Bot, Tide rolls, ماني, Math Champion, Luckas-bot, Yobot, Andreasmperu, II MusLiM HyBRiD II, AnomieBOT, Citation bot, ArthurBot, Xqbot, TheAMmollusc, GrouchoBot, SassoBot, Immatt96, Shadowjams, SchnitzelMannGreek, FrescoBot, Elocute, Cannolis, Citation bot 1, Bristi13, DrilBot, Pinethicket, Sideamongst, Flecha2, Zhernovoi, Dinamik-bot, LilyKitty, NerdyScienceDude, EmausBot, Dcirovic, JSquish, Traxs7, Lateg, John Mackenzie Burke, Hccc, Ibenami, Avatar9n, SamSnowden, Llightex, 28bot, Sonicyouth86, Will Beback Auto, ClueBot NG, Carandraug, Amin s 26, BG19bot, CitationCleanerBot, Arianjalali, Smettems, Achowat, Baemms, BattyBot, Biosthmors, Lophostrix, ChrisGualtieri, Kelvinsong, Dexbot, Mogism, Callum123343, XXJigsawXx, JakobSteenberg, Xxkitsune, Ajakis1, Msrp02, Tommy a. bacon, Seppi333, Jitan007, Slj758, Dodgeb2, Elephantsofearth, Johnfranciscollins, Monkbot, Vieque, Teun.Zijp, Enigmatore, LeGuybrush, Graehare, Rodrigo.duarte88, Johnrayphd, NugoMn and Anonymous: 283

- **DNA supercoil** *Source:* https://en.wikipedia.org/wiki/DNA_supercoil?oldid=682023930 *Contributors:* Rubik-wuerfel, Thorwald, Cwolfsheep, Redquark, BryanD, Oleg Alexandrov, Rjwilmsi, Jakob Suckale, Hede2000, Chuck Carroll, Hellbus, Gaius Cornelius, Vatassery, AndrewW-Taylor, SmackBot, TestPilot, Zephyris, Kazkaskazkasako, UberMD, Rigadoun, Jim.belk, Jac16888, GRBerry, Comcc, Adam B, Nbauman, Lantonov, Antony-22, Pdcook, Daimore, LokiClock, A4bot, Rei-bot, Anonymous Dissident, Sintaku, Luuva, AlleborgoBot, Cvf-ps, Forluvoft, WurmWoode, Excirial, Heathmoor, Ericci8996, Chrismichener1, Addbot, DOI bot, Noosh-X, LaaknorBot, CarTick, Yobot, Sz-iwbot, Materialscientist, Citation bot, Xqbot, RibotBOT, Anooranamarie, Citation bot 1, Trappist the monk, EmausBot, Ikerus, EleferenBot, Wikipelli, Dcirovic, Ebrambot, Theopolisme, Helpful Pixie Bot, Plantdrew, Estevezj and Anonymous: 38

- **Transcription (genetics)** *Source:* https://en.wikipedia.org/wiki/Transcription_(genetics)?oldid=689935410 *Contributors:* Mav, The Anome, Michael Hardy, Zashaw, Lexor, Kku, Menchi, Ronz, JWSchmidt, Александър, Mxn, Ec5618, Fuzheado, Steinsky, Phil Boswell, Kzhr, Timemutt, VanishedUser kfljdfjsg33k, Giftlite, Michael Devore, Bensaccount, Duncharris, Antandrus, G3pro, Oneiros, PFHLai, Grunt, Archer3, DanielCD, Discospinster, Mgtoohey, Pabloes, Perfecto, AKGhetto, Tmh, Arcadian, Sriram sh, Srlasky, Haham hanuka, Alansohn, Terrycojones, Sl, Wouterstomp, Riana, Seans Potato Business, Batmanand, Helixblue, Tycho, Amorymeltzer, GabrielF, Ceyockey, RyanGerbil10, Brookie, Woohookitty, Mindmatrix, EnSamulili, Fbv65edel, Mms, Jclemens, Dpv, Sjakkalle, Rjwilmsi, FlaBot, Margosbot~enwiki, Vossman, Fenoxielo, Roboto de Ajvol, Wavelength, Postglock, WAvegetarian, Hede2000, Rosieredfield, Shanel, Rmky87, Bucketsofg, WAS 4.250, JoanneB, Curpsbot-unicodify, Teply, Mengxu, Hughitt1, SmackBot, TestPilot, Martin.Budden, Hydrogen Iodide, Geno-Supremo, Apers0n, Zephyris, Yamaguchi⁇, Gilliam, BrotherGeorge, Kazkaskazkasako, Evandrix, Aaadddaaammm, MalafayaBot, Uthbrian, Oxhop, Miguel Andrade, DHN-bot~enwiki, Ribrob, Vidric, Khoikhoi, Totophe64~enwiki, Richard001, Drphilharmonic, Kshieh, Ifan160, Kukini, CoeurDeLion, The undertow, Nishkid64, ArglebargleIV, AThing, Kuru, Epingchris, Bloodpack, Ben Moore, MarkSutton, Slakr, Noah Salzman, Iridescent, Gentlemaan, Peter M Dodge, Kaarel, Cph3992, JForget, Liam Skoda, CmdrObot, OverlordKain, Bonás, Agathman, Sameerbau, Spottydog3, Neelix, Opus118, Yided, Tomjc, Nuplex, Nick.wiebe, RelentlessRecusant, Was a bee, Corpx, Smelissali, Narayanese, Vanished User jdksfajlasd, Phi*n!x, Thijs!bot, Opabinia regalis, Mojo Hand, John254, NERIUM, Kickassso, AntiVandalBot, MoogleDan, BokicaK, Eltanin, NightwolfAA2k5, TimVickers, Smartse, MDG38, Neur0X, Legolost, Kswenson, Dcooper, Jullag, Greensburger, .anacondabot, VoABot II, LeaHazel, Antorjal, Squidonius, NunoAgostinho, MartinBot, Cvd5012, Rettetast, Anaxial, R'n'B, Qrex123, Huzzlet the bot, HoergerJ, SU Linguist, Lantonov, Rod57, KDSKDS, Jmajeremy, Mikael Häggström, Martyn Axon, Antony-22, Vanished user 39948282, Useight, Zamftb, Llorenzi, G. Völcker, Hammersoft, Saurabh523, Nburden, AlnoktaBOT, Scresawn, Philip Trueman, Segabud, Z.E.R.O., Ilia Kr., Mishlai, Furfurfur, Gillyweed, Enviroboy, Jcdietz03, RaseaC, Pjoef, AlleborgoBot, Kehrbykid, PGWG, ASDZXCQWE, SieBot, Nubiatech, Sophos II, Sakkura, Lucasbfrbot, Flyer22 Reborn, Teethies563, Yerpo, Sunrise, Iknowyourider, DaDrought3, Altzinn, Neta90, Brettbarbaro, Forluvoft, ClueBot, Binksternet, Paulabek, DragonBot, Excirial, PixelBot, Skyuppercutt, Cbailey7, Sydney3803, Frigginacky, SchreiberBike, LeighClesterMolar, Thingg, BVBede, Qwfp, Johnuniq, Local hero, XLinkBot, LostLucidity, Feinoha, Vojtěch Dostál, WikiHead, Yvorez1274, Drosilia, Hilwhale, Alohascott, Geir.overland, Addbot, DOI bot, CanadianLinuxUser, R-skin, Lehtv, ChenzwBot, Tassedethe, Tide rolls, Gail, Tedtoal, Luckas-bot, Yobot, Ptbotgourou, Berkay0652, TaBOT-zerem, Amirobot, KamikazeBot, The Flying Spaghetti Monster, Legendre17,

AnomieBOT, Kerfuffler, Jim1138, Kingpin13, Hpswimmer, Materialscientist, Citation bot, Maxis ftw, Quebec99, Marshallsumter, Xqbot, GrouchoBot, Brandon5485, Vikky2904, E0steven, Thehelpfulbot, Rgocs, FrescoBot, Brianwatson94, S73v3n, Citation bot 1, I dream of horses, Foureyes915, Flecha2, A8UDI, RedBot, Mexican9493, Orenburg1, Bursting74, Περίεργος, Ferrari430man, Amkilpatrick, Jcorry10, Gamingmaster125, Tbhotch, Jesse V., Galneon23, RjwilmsiBot, Pjshort42, J36miles, EmausBot, WikitanvirBot, RA0808, Transitiveinstance, Wikipelli, K6ka, 20Lukianto, Grunny, John Mackenzie Burke, EWikist, Wayne Slam, Tropicalpurplekitty, Donner60, BioPupil, Theislikerice, ChuispastonBot, Jeffpkamp, AnnaJune, ClueBot NG, This lousy T-shirt, Rida97, Liney22, Alex-engraver, Mesoderm, Jogmiers, Brynedal, Ajdavis5, Yasmeh, Jacobso4, Helpful Pixie Bot, Miguelferig, BG19bot, Bths83Cu87Aiu06, CatPath, Dhp4, MusikAnimal, Cncmaster, Mdebortoli, Barton1234, Cmcclean22, Lolzpvp, Roleren, Klilidiplomus, SIMONCOHEN, BattyBot, TuringMachine17, Jimw338, Sermadison, Iamozy, Kelvinsong, Dexbot, Saba irshad, Mohamed 151995, Avijitarya64, Guma44, Dreesem, Xxkitsune, Limelightmk, FallingGravity, Lemnaminor, Melonkelon, Aadharm, Wally 84, A Certain Lack of Grandeur, Quenhitran, APsCollegeEditor, Slj758, ChelseaE, Mjt3727, Wikicology, Iwilsonp, Rai hamid raza kharal, Cyrej, Zhirzh, Mmm053, KasparBot, Husbro and Anonymous: 513

- **DNA replication** *Source:* https://en.wikipedia.org/wiki/DNA_replication?oldid=690485970 *Contributors:* AxelBoldt, The Anome, Vanderesch, Graft, Heron, Lexor, Skysmith, Alfio, Ahoerstemeier, Habj, Samuel~enwiki, Fibonacci, Robbot, Baldhur, ZimZalaBim, Whiteniko, Hadal, Lupo, Oddharmonic, Giftlite, Bensaccount, FriedMilk, Adenosine, Pgan002, GD~enwiki, Jossi, Taka, Ary29, Neutrality, Julianonions, Imjustmatthew, Mike Rosoft, Rich Farmbrough, Guanabot, Vsmith, Xezbeth, Bender235, Kaisershatner, El C, RoyBoy, Adambro, Bobo192, Stesmo, Whosyourjudas, Smalljim, Cmdrjameson, Arcadian, Timl, Dennis Valeev, La goutte de pluie, Nhandler, Nsaa, Alansohn, Arthena, Wouterstomp, Stillnotelf, Snowolf, Ravenhull, Velella, ClockworkSoul, Tycho, Netkinetic, RyanGerbil10, Jackhynes, Before My Ken, Fbv65edel, MONGO, LadyofHats, Plrk, Dysepsion, Snafflekid, Rjwilmsi, SMC, Thisismikesother, Alveolate, The wub, Mortice, FlaBot, Cless Alvein, RobertG, Nihiltres, Alphachimp, BradBeattie, Chobot, Whosasking, YurikBot, Borgx, WAvegetarian, Chris Capoccia, GLaDOS, SpuriousQ, Thiseye, Malcolma, Daniel Mietchen, Jpbowen, InvaderJim42, Raven4x4x, RUL3R, Wangi, Jhinman, Closedmouth, Maristoddard, Pb30, Xaxafrad, Svetlana Miljkovic~enwiki, Mfedder, JoanneB, CIreland, Rystic, SmackBot, Andthendougsaid, Ariliand, InverseHypercube, KnowledgeOfSelf, Royalguard11, Bomac, Stepa, Eskimbot, CapitalSasha, Bozartas, Edgar181, Blondtraillite, Zephyris, Ashermadan, Freddy S., Gilliam, DividedByNegativeZero, Skizzik, ERcheck, Andy M. Wang, Bluebot, Persian Poet Gal, JDCMAN, Tree Biting Conspiracy, Hichris, MalafayaBot, Celarnor, Zrulli, Bowlhover, Richard001, Drphilharmonic, Madeleine Price Ball, SashatoBot, Harryboyles, Scientizzle, Gobonobo, JulianBlow, Thegathering, Edwy, Steipe, Mgiganteus1, MichaelHa, PseudoSudo, A. Parrot, Stwalkerster, Noah Salzman, Wrlampe, Romanticcynic, TastyPoutine, Artman40, Sasata, Hu12, Dan Gluck, JayZ, Twas Now, Igoldste, Yukaxu, Pranay Biswas, Tawkerbot2, Dave Runger, Yashgaroth, Falconus, Jpeguero, Superandomness, JForget, CmdrObot, Agathman, Harej bot, AKaK, Dgw, Tex, Tomjc, Ppgardne, Meno25, RelentlessRecusant, Was a bee, Macduy, Narayanese, Biomedeng, Thijs!bot, Epbr123, Pstanton, Ucanlookitup, Nonagonal Spider, Marek69, Crazysunshineboy, Sturm55, Pfranson, CTZMSC3, Dantheman531, Mentifisto, David D., AntiVandalBot, Majorly, BokicaK, TimVickers, Shajilvt, TexMurphy, Tlabshier, Farosdaughter, Bio-queen, Instinct, Jonemerson, Hello32020, Jullag, Tstrobaugh, MSBOT, Freedomlinux, VoABot II, Sedmic, Arthmelow, SparrowsWing, Midgrid, Olavrg, Catgut, Cyktsui, Emw, DerHexer, Squidonius, MartinBot, STBot, Rettetast, Anaxial, CommonsDelinker, Player 03, Victor Blacus, Exarion, J.delanoy, Harsha112, Maurice Carbonaro, Captain Infinity, Ignacio Icke, MrErku, Mikael Häggström, WebHamster, Jcwf, Omes, NewEnglandYankee, Firelement85, Mufka, Hanacy, Iduggin, MishaPan, Useight, VolkovBot, DrMicro, AlnoktaBOT, Pkakaris, Soliloquial, Aesopos, Philip Trueman, Tameeria, A4bot, Monkey Bounce, JhsBot, Leafyplant, Im emo tastic, Archdevil75, LoneSeeker, Dr. Anton Funk PhD, Sylent, Guelphie, Twooars, Guelphie85, PGWG, Sandymit, Rafis v, Nubiatech, Work permit, BotMultichill, ToePeu.bot, Sakkura, Eganio, Radon210, Yerpo, Oxymoron83, Erick880, Faradayplank, Nancy, BSoD, Capitalismojo, Mygerardromance, Ucphilsc, Gloss, Nornour, ClueBot, Mstuddert, Medfreak, The Thing That Should Not Be, Voxpuppet, Unbuttered Parsnip, Kusb, Blanchardb, Neverquick, Cirt, Arunsingh16, Canis Lupus, Alexbot, Cookatoo.ergo.ZooM, Tassos Kan., Gregoots, Duxenaz, 101louise101, Kakofonous, Aitias, ForestDim, Burner0718, Johnuniq, DumZiBoT, Residenthalo, Jovianeye, Vojtěch Dostál, SilvonenBot, Jimhsu77479, Jadtnr1, Calito00000001, Addbot, DOI bot, Chasturn, CanadianLinuxUser, Ka Faraq Gatri, Download, LaaknorBot, Coveman999, Bernstein0275, Debresser, Favonian, LinkFA-Bot, Rtz-bot, Numbo3-bot, Tide rolls, Gail, Megaman en m, Legobot, Luckas-bot, Yobot, Andreasmperu, Nallimbot, Azcolvin429, AnomieBOT, LLDMart, Nutriveg, EryZ, Bluerasberry, FangedFaerie, Citation bot, ArthurBot, Xqbot, JimVC3, RibotBOT, WaysToEscape, Brianwatson94, Media1312, AstaBOTh15, Pinethicket, I dream of horses, 10metreh, Robinhaw, Σ, Emply shell China dry, Dac04, Saintonge235, Trappist the monk, Jonkerz, TBloemink, DragonofFire, Reaper Eternal, 564dude, Skakkle, Kidwet, Jessi bob, DARTH SIDIOUS 2, Whisky drinker, Onel5969, RjwilmsiBot, DASHBot, John of Reading, Immunize, Lotez, Bob100077, Golfandme, Wikipelli, K6ka, Dusanman1, PJinBoston, JSquish, Access Denied, OnePt618, Nipun.chamodh, Cyberdog958, Flightx52, Scientific29, Puffin, Ego White Tray, Orange Suede Sofa, DemonicPartyHat, Chatmonregina, Petrb, ClueBot NG, Dwc89, Rsc227, Mesoderm, Widr, Bosborne21212, Theopolisme, Wbm1058, BG19bot, Joydeep, Syeda Hassan Rabia, NotWith, Smettems, Lsanman, Shaun, Thermodynamic, BattyBot, Zhaofeng Li, Stigmatella aurantiaca, Rinkle gorge, Deathlasersonline, Ricochet Jones, Dexbot, Saba irshad, Mohamed 151995, YurTru, Kfh123, Everything Is Numbers, Churn and change, Kevin12xd, GabeIglesia, Epicgenius, FallingGravity, U1012738, Melonkelon, Tentinator, Epic123456789, MedicalStudentOxford, Triolysat, Jlmalcos, JaconaFrere, Bonob, Stormmeteo, Monkbot, Vieque, Freddieyao167, AmericanSocialist, Paul Badillo, Dai Pritchard, RationalBlasphemist, Manal Adil, KasparBot, Alannahmattice, Asdfghj12h2b and Anonymous: 752

- **Molecular models of DNA** *Source:* https://en.wikipedia.org/wiki/Molecular_models_of_DNA?oldid=680259920 *Contributors:* Michael Hardy, Topbanana, Bender235, MassGalactusUniversum, Rjwilmsi, Koavf, Thiseye, SmackBot, Oscarthecat, Chris the speller, RDBrown, Clive Delmonte, Headbomb, David Eppstein, Nanodetails, R'n'B, CommonsDelinker, Nono64, Antony-22, Plindenbaum, ShiftFn, Erythromycin, The Thing That Should Not Be, Boing! said Zebedee, Dthomsen8, Addbot, Boomur, LinkFA-Bot, Tassedethe, Lightbot, Yobot, Fraggle81, KRLS, Citation bot, Bci2, LilHelpa, P99am, Miym, Nmedard, Citation bot 1, Trappist the monk, Jonkerz, RjwilmsiBot, Dcirovic, ClueBot NG, Bibcode Bot, Dontneedtoknow....2013, Monkbot and Anonymous: 6

- **Nucleic acid structure** *Source:* https://en.wikipedia.org/wiki/Nucleic_acid_structure?oldid=690730028 *Contributors:* Giraffedata, Rjwilmsi, Rsrikanth05, Kazkaskazkasako, LadyofShalott, Ayzmo, EagleFan, CommonsDelinker, Boghog, Antony-22, Staylor71, Jbening, Yikrazuul, Therewasaguy, Addbot, Yobot, Amirobot, AnomieBOT, Citation bot, RibotBOT, Shadowjams, Jcc, John of Reading, Dcirovic, Donner60, ClueBot NG, BattyBot, Illia Connell, Lugia2453, Myaworsk, Monicamcoulter, Ravinder2012, Joeinwiki, Evolution and evolvability, Cantukarol, Monkbot, SashaBolshoy, Zeke1940, Katgreenly, Fractalsplash and Anonymous: 24

- **Molecular Structure of Nucleic Acids: A Structure for Deoxyribose Nucleic Acid** *Source:* https://en.wikipedia.org/wiki/Molecular_Structure_of_Nucleic_Acids%3A_A_Structure_for_Deoxyribose_Nucleic_Acid?oldid=684352907 *Contributors:* Patrick, Gaurav, JWSchmidt, Selket, Jnc, Wetman, Jonth, AJim, Thorwald, Rich Farmbrough, Swid, Lycurgus, Easyer, Danshil, John Vandenberg, Shenme, La goutte de pluie, Sam Korn, Mithent, Arthena, Wouterstomp, Ferrierd, ClockworkSoul, Grenavitar, Camw, Bkkbrad, Carcharoth, MFH, Jacob Finn, Casey Abell, Rjwilmsi, Tim!, Koavf, Jmcc150, Erkcan, RobertG, Intgr, Bubbachuck, Wavelength, Mushin, RussBot, NawlinWiki, Kvn8907, Howcheng, SMA11784, Zzuuzz, Petri Krohn, Snalwibma, SmackBot, SchfiftyThree, Addshore, Schnarr, Mgiganteus1, N6~enwiki, ChazYork, Kaarel, FairuseBot, Switchercat, WongFeiHung, Cydebot, Neilio 23, Vanished User jdksfajlasd, Thunder, Headbomb, Linksfuss, WolfmanSF, VoABot II, WhatamIdoing, DerHexer, DGG, Collinf, Dave Rado, CommonsDelinker, Petersec, Lizzie Harrison, Someguy1221, Eldaran~enwiki, Afireinside13t, Northfox, Cj1340, Tiddly Tom, Matthew Yeager, RJaguar3, LeadSongDog, GlassCobra, Bentogoa, Oxymoron83, Fratrep, ClueBot, Jacek FH, Abrech, Martin998877, LieAfterLie, Dthomsen8, Good Olfactory, Snapperman2, Addbot, Willking1979, Tcncv, Favonian, Nurasko, Luckas-bot, Yobot, TaBOT-zerem, Legobot II, Tempodivalse, AnomieBOT, Materialscientist, Citation bot, Maxis ftw, Bci2, LilHelpa, P99am, PimRijkee, Rotideypoc41352, Citation bot 1, RedBot, SW3 5DL, Trappist the monk, Jeffrd10, Wikipelli, Dcirovic, Traxs7, Autismdoctor, Chris857, Palaeozoic99, ClueBot NG, SeekingAnswers, Helpful Pixie Bot, Bibcode Bot, Mark Arsten, Glacialfox, DaHuzyBru, Stayncutelol, Dr. K. Uangry, Trillig, Epicgenius, Zhantongz, Monkbot, Nick468, Trittman, Trolled seriouse man, EvilLair and Anonymous: 126

- **Genetic code** *Source:* https://en.wikipedia.org/wiki/Genetic_code?oldid=691497002 *Contributors:* Magnus Manske, Bryan Derksen, The Anome, Taw, Slrubenstein, Josh Grosse, SimonP, AdamRetchless, Graft, Ryguasu, Bdesham, Michael Hardy, Lexor, Stw, Ahoerstemeier, CatherineMunro, Glenn, Evercat, Rob Hooft, Tobias Conradi, Timwi, Dysprosia, Zoicon5, Peregrine981, Krithin, Taxman, Mackensen, Robbot, Peak, Yelyos, Romanm, Sverdrup, Bkell, Hadal, Anthony, Diberri, Buster2058, Ancheta Wis, Giftlite, Average Earthman, Gracefool, Alvestrand, Utcursch, Pgan002, Garrett~enwiki, Ruy Lopez, Zeimusu, G3pro, PFHLai, Sam Hocevar, Sayeth, Gscshoyru, B.d.mills, Lazarus666, Imjustmatthew, Qef, Zro, ClockworkTroll, JTN, Discospinster, Rich Farmbrough, Guanabot, Pjacobi, Vsmith, Kbh3rd, Chewie, CanisRufus, Charm, Joanjoc~enwiki, Bobo192, John Vandenberg, Wisdom89, .:Ajvol:., Kevin Myers, Arcadian, Rajah, Alansohn, Anthony Appleyard, Keenan Pepper, Wouterstomp, Zephirum, ClockworkSoul, *Kat*, Amorymeltzer, Ceyockey, Eleassar777, Allen3, Rjwilmsi, DonSiano, TheIncredibleEdibleOompaLoompa, Drpaule, Anchemis, Wobble, FlaBot, Pruneau, Vietbio~enwiki, Takometer, TheDJ, Chobot, Hall Monitor, YurikBot, Wavelength, Anuran, Chris Capoccia, Stephenb, Tavilis, Dysmorodrepanis~enwiki, Welsh, Mccready, Ragesoss, Rmky87, Bota47, Kiddo54, Wknight94, Alx bio, Sandstein, Knotnic, Жованв6, Petri Krohn, HereToHelp, Allens, Katieh5584, Dabs, Sardanaphalus, SmackBot, Haza-w, Andrekaur, Reedy, Martin.Budden, KocjoBot~enwiki, Midway, Jrockley, Gaff, Kurykh, Persian Poet Gal, RDBrown, MK8, Thumperward, Hichris, Baa, Brinerustle, Scray, Smoken Flames, Valich, Snowmanradio, Geoffrey Gibson, Homestarmy, Richard001, Drphilharmonic, DMacks, Acdx, Ohconfucius, GoldenTorc, Madeleine Price Ball, Scientizzle, JH-man, Sir Nicholas de Mimsy-Porpington, Gareth Palidwor, Bjankuloski06en~enwiki, Ben Moore, Loadmaster, Slakr, Beetstra, SQGibbon, Kliqjaw, Gondooley, RMHED, Vanished user, Kaarel, Frank Lofaro Jr., Yashgaroth, CmdrObot, Agathman, Leevanjackson, Neelix, Jmcgloth, MC10, Was a bee, Carstensen, Comcc, Narayanese, Kozuch, Savitr, Thijs!bot, Epbr123, Mojo Hand, MrXow, Dgies, CarbonX, Dawnseeker2000, Mentifisto, AntiVandalBot, Luna Santin, QuiteUnusual, TimVickers, Smartse, Qwerty Binary, JAnDbot, Issaacc, IronArjen, Dcooper, Greensburger, East718, Acroterion, DRHagen, TransControl, Jaysweet, Bongwarrior, VoABot II, SineWave, Bulbeck, Catgut, Antorjal, Adrian J. Hunter, Emw, GetAgrippa, DerHexer, Doctor Faust, Squidonius, MartinBot, Arjun01, Chaotic rach, Anaxial, Cdcdoc, Fconaway, Nbauman, Sp3000, Boghog, SlowJog, Maproom, Bobo the Talking Clown, James A. Stewart, Imalipusram~enwiki, Dr d12, Mikael Häggström, Tarotcards, Eldang, NewEnglandYankee, GGenov, Secleinteer, AndreasJSbot, Xiahou, Speciate, Lights, DrMicro, Tbill92, Philip Trueman, Oshwah, TraumB~enwiki, Anna Lincoln, Elphion, Jackfork, BotKung, Dprust, Maxim, Alexbateman, L Yampolsky, Turgan, BrownBot, Xecoli, Ceranthor, Dmcq, Kehrbykid, Logan, Kosigrim, Radagast3, Givegains, Lesterama, ShiftFn, Graham Beards, Flyer22 Reborn, Yerpo, Jruderman, Sunrise, Anchor Link Bot, Copypasteusa, Naturespace, Explicit, Forluvoft, Agabirhei, ClueBot, Konamaiki, Cwilhoit, Polyamorph, Mspraveen, Copyeditor42, PixelBot, Eeekster, Thingg, Aitias, Galapah, DBodor, DumZiBoT, XLinkBot, Gnowor, Shadow600, WikHead, Arapacana, SkyLined, Addbot, DOI bot, Jojhutton, Diptanshu.D, Bernstein0275, DanaStreet, LinkFA-Bot, Doug youvan~enwiki, Tide rolls, QuadrivialMind, Arxiloxos, Luckas-bot, Yobot, Fraggle81, TaBOT-zerem, AnomieBOT, Jim1138, Galoubet, Boleyn2, Wenlongtian, Kingpin13, Shogatetus, Flewis, Materialscientist, Citation bot, ArthurBot, SchfiftyThree (Public), 8ennett, Aa77zz, Dave15o1, Richard.decal, Ghost3794, Jezhotwells, Maikfr, Miyagawa, Dan Wylie-Sears 2, Killdec, VI, Kwiki, Citation bot 1, DrilBot, Pinethicket, Tom.Reding, Jeyamalini, Dina.Vacca, Aspstren, Wikididact, Double sharp, Minhduc.cao, Jonkerz, Oktanyum, Franfemu, Gamingmaster125, Tbhotch, Jesse V., DARTH SIDIOUS 2, RjwilmsiBot, TjBot, EmausBot, WikitanvirBot, Dutilh, Jjechris, Tommy2010, Wikipelli, Vindicata, Sjoosse, Ida Shaw, 30Ikra, Skopetz, John Mackenzie Burke, Señor Aluminio, Scotware, Flightx52, B4swine, Beta cafe, CountMacula, ChuispastonBot, Bobthefish2, Codexbiogenesis, Vogelbe, 28bot, Teaktl17, Will Beback Auto, ClueBot NG, KookBot2, Natedog1300, Lanthanum-138, Frietjes, CasualVisitor, Bibcode Bot, BG19bot, Biolog75, Avocato-Bot, Yournamenothere, Bradlake10, NotWith, HPBiochemie, Bopiotrowski, Samein50, BattyBot, ChrisGualtieri, Bsbbs, Mogism, Pro pusey, Mozrael2357, CopyrightX, John54100, 拝承殿, Gaurav.du, Frank Layden, Anrnusna, Ryftstarr, Jamal Ahmed GM, DOBBY77, Monkbot, BethNaught, VeniVidiVicipedia, CV9933, CCevol2015 and Anonymous: 412

- **Conformational isomerism** *Source:* https://en.wikipedia.org/wiki/Conformational_isomerism?oldid=682993213 *Contributors:* Michael Hardy, Smjg, Christopherlin, H Padleckas, Cacycle, Martpol, Mashford, Aranel, El C, Renice, Irrbloss, Snarfevs, Gene Nygaard, V8rik, Syndicate, FlaBot, Kolbasz, WijzeWillem, Roboto de Ajvol, Shaddack, Cholmes75, Itub, SmackBot, Victor M. Vicente Selvas, Reedy, M stone, Richard001, Smokefoot, DMacks, Darktemplar, JorisvS, Peterlewis, RSido, Odie5533, Christian75, Thijs!bot, Dougher, JAnDbot, Japo, R'n'B, Boghog, Rod57, AlnoktaBOT, Nono le petit robot~enwiki, Luuva, Duncan.Hull, SieBot, Asdfasdf12345, Breakyunit, Flyer22 Reborn, OK-Bot, Ryocharlesyang, PipepBot, DragonBot, DumZiBoT, Mahmutuludag, Jujubot~enwiki, Addbot, Wickey-nl, Julia W, AnomieBOT, Mcpazzo, LilHelpa, P99am, RibotBOT, Dcrjsr, Sms1371, FrescoBot, Cleanthis, RjwilmsiBot, EmausBot, Dcirovic, Math-ghamhainn, Mentibot, ChuispastonBot, Shivsagardharam, Mohamed CJ, Smbanik, BattyBot, ChrisGualtieri, Dexbot, TOMMITOMMI12, Mr.Holmium, Monkbot, Ternary.pulsar, Azzozzz, Bananas5511, JamesP and Anonymous: 25

- **A-DNA** *Source:* https://en.wikipedia.org/wiki/A-DNA?oldid=681021220 *Contributors:* Thorwald, Mike Rosoft, Rich Farmbrough, Triona, Stemonitis, Mushin, Phantomsteve, Flsxx, Zephyris, Jjalexand, Harold f, Mauroesguerroto, Thijs!bot, Magioladitis, D-rew, VolkovBot, Safemariner, TXiKiBoT, Luuva, ClueBot, Thingg, Little Mountain 5, Addbot, DOI bot, Xocolatanegra, Luckas-bot, Citation bot 1, I dream of horses, EmausBot, Dcirovic, ZéroBot, Salaei, ClueBot NG, Liney22, Jeffluo1011, BattyBot, Monkbot, Zeke1940 and Anonymous: 15

- **Z-DNA** *Source:* https://en.wikipedia.org/wiki/Z-DNA?oldid=688981395 *Contributors:* Bryan Derksen, Lexor, Llull, Thorwald, Rich Farmbrough, MITalum, Grutness, Gene Nygaard, Drbogdan, Rjwilmsi, Kerowyn, Raviriyer, SmackBot, TestPilot, Zephyris, Chris the speller, Bluebot, OrphanBot, Npho, Smokefoot, Harold f, CmdrObot, Kupirijo, Carstensen, Mauroesguerroto, Bot-maru, JAnDbot, TAnthony, Appraiser, Amhantar, D-rew, Ben Ram, VolkovBot, Luuva, Hannes Röst, Forluvoft, Atif.t2, Luke4545, Dhugot, Addbot, DOI bot, Luckas-bot, Yobot, Jim1138, Citation bot, Xqbot, Citation bot 1, I dream of horses, RedBot, Adithya Sagar, Amkilpatrick, RjwilmsiBot, EmausBot,

Dcirovic, ZéroBot, ClueBot NG, Bibcode Bot, Zdrman, Dexbot, Spider55555, Faskal, HelgeUK, Anrnusna, Monkbot, Ghiutun, CV9933 and Anonymous: 31

- **Telomere** *Source:* https://en.wikipedia.org/wiki/Telomere?oldid=691283496 *Contributors:* AxelBoldt, Magnus Manske, Bryan Derksen, Josh Grosse, PierreAbbat, Ark~enwiki, Netesq, Gabbe, Ahoerstemeier, Timwi, IceKarma, Samsara, Jeffq, Phil Boswell, Robbot, Paranoid, Ke4roh, Yakisoba, Sbisolo, Litefantastic, SoLando, Giftlite, KMT, Dp, Bkonrad, Brona, Electric goat, Pascal666, Foobar, Peter Ellis, Wmahan, Tipiac, Mr d logan, PDH, Thincat, Sayeth, Creidieki, Chris Howard, AAAAA, 4pq1injbok, Rich Farmbrough, Bender235, Hrodrik, Lycurgus, R. S. Shaw, Giraffedata, La goutte de pluie, Alansohn, Ferrierd, Hohum, Melaen, RJII, Amorymeltzer, Spellcheck, Mindmatrix, StradivariusTV, Benbest, Isnow, Eras-mus, BorisTM, Rjwilmsi, Nandesuka, Fred Bradstadt, FlaBot, Themalau, Margosbot~enwiki, Intgr, Iridos~enwiki, Chobot, Karch, Mhking, Bgwhite, Banaticus, YurikBot, Wavelength, Matanya (renamed), Epolk, Splette, Dysmorodrepanis~enwiki, Dtrebbien, Redtails, Jrissman, Leptictidium, Pjprecze, Arthur Rubin, Paul White, Rchillyard, SmackBot, Slashme, Hydrogen Iodide, Masparasol, Od Mishehu, Giraldusfaber, RedSpruce, Bmord, BiT, Kazkaskazkasako, Chris the speller, Bluebot, Thom2002, RDBrown, Mdwh, Deli nk, Ggenoar, Darth Panda, JonHarder, Dr. Oki, Monotonehell, Dcteas17, WIIfred the Mack~enwiki, Drphilharmonic, Scootdive, Nishkid64, Tim bates, Mgiganteus1, Vpfritz, Rock4arolla, Murdochious, Axcelis555, Nathancor, Spinnick597, RekishiEJ, Mozozzy, CmdrObot, Leevanjackson, Pathh, JamesAGreen, Cydebot, Ntsimp, Ppgardne, Kanags, A876, Was a bee, Arrowned, Smelissali, Headbomb, Semiclear, Seaphoto, TimVickers, Smartse, Rsocol, Apparent Logic, Db099221, Dream Focus, Magioladitis, VoABot II, Harelx, Sugar128, J.B., A3nm, Emw, Zoidstar, Gjd001, Flandrensis, TechnoFaye, Nono64, Fconaway, Nbauman, Hisagi, Rsecor, Rod57, Shendar, Belovedfreak, Sprunk, STBotD, Ncrabb, Bonadea, Pdcook, Speciate, Fainites, Benstrider, VolkovBot, Larryisgood, Mabidex, TXiKiBoT, Myles325a, A4bot, Malljaja, Bjman, Someguy1221, Pwestep, Earthdirt, Theleopard, CarinaT, RYAN IS VERY WISE AND GREAT, Cmcnicoll, Hellodearcancer, CIBike6, Demize, Zenlax, Brenont, Jab416171, Prakash Nadkarni, Noisy Hand, Callipides~enwiki, Roesser, Formerly the IP-Address 24.22.227.53, Flyer22 Reborn, Jdaloner, Sharpray, Johndheathcote, Jack the Stripper, OKBot, Iknowyourider, Phillip K. H., Wpac5, ClueBot, NickCT, Mild Bill Hiccup, Piledhigheranddeeper, Korey1989, Ashashyou, Rui-kun, Frigginacky, SchreiberBike, Thehelpfulone, Galapah, Qwfp, DumZiBoT, Eve basenko, Sierrasciadmin, KarakasaObake, Ramatata, Dfoxvog, Vikash2k, Addbot, Some jerk on the Internet, DOI bot, Friginator, Metagraph, Murnane, Ronhjones, Renmanov, Bernstein0275, Omnipedian, Numbo3-bot, Venerdì17, Tide rolls, Icons789, Blah28948, Luckas-bot, Yobot, Fraggle81, TaBOT-zerem, Glscience, Dlwv, AnomieBOT, Piano non troppo, Citation bot, Trilliumroots, ArthurBot, The Firewall, Xqbot, Flavius.Aettius, Lloydsd, Jsharpminor, Andrewthomasj, Adentong, Aduron umd.edu, Altruistic Egotist, X5dna, SEASONnmr, SchnitzelMannGreek, Dogposter, Sillicia, Citation bot 1, Micromesistius, Jonesey95, RedBot, Σ, Veneventura, Aaabbbzzz, Double sharp, Jfu12450, Jeffrd10, Adi4094, Minimac, Aaharvey, Watcher0911, RjwilmsiBot, Becritical, Androstachys, Thergrarg, AManWithNoPlan, Inamorata666, Peteb4, Dawn08, ChuispastonBot, ClueBot NG, Horoporo, Pengortm, Kaushlendratripathi, Samibarna, 336, Ciechomskih, Shivsagardharam, BG19bot, Vokesk, Stevetihi, Lechu502, Stamford347, Dormantfreedom, Smatlin, NotWith, BattyBot, Cyberbot II, ChrisGualtieri, Blueheavan, Mogism, Everything Is Numbers, Tommy Pinball, Joeinwiki, Faizan, Daniel 82, BradYard, Dustin V. S., Divergence5, Anrnusna, Stamptrader, Bcwikiprg, AlmaMer, Pdavis9396, RpBoell, Monkbot, Vieque, Whitneym89, Sadegh76, Pivoter, Counteragingwise, Hbialic, Ybeigima, Dan Eisenberg and Anonymous: 300

- **Branched DNA assay** *Source:* https://en.wikipedia.org/wiki/Branched_DNA_assay?oldid=596423932 *Contributors:* Darkwind, Alan Liefting, PDH, Arcadian, Rjwilmsi, Chris Capoccia, Leptictidium, SB Johnny, Peter K., TimVickers, JBKramer, Sabedon, ClueBot, Snapperman2, Addbot, DOI bot, TomRau, FrescoBot, Citation bot 1, Jesse V., EmausBot, Dcirovic, Mohamed-Ahmed-FG, Mon3oturf and Anonymous: 6

- **DNA nanotechnology** *Source:* https://en.wikipedia.org/wiki/DNA_nanotechnology?oldid=686916149 *Contributors:* Shizhao, Giftlite, Stijn Ghesquiere, Thorwald, Iridia, Zlite, ZayZayEM, Anthony Appleyard, BD2412, Melesse, Rjwilmsi, Tony1, Kkmurray, Leptictidium, Nikkimaria, MaNeMeBasat, SmackBot, ShawnDouglas, Chris the speller, Epbr123, Gioto, Sangak, VoABot II, David Eppstein, Emw, Nanodetails, Schmloof, DrKay, Antony-22, Alnokta, Brvman, Izno, Malik Shabazz, Graham Beards, Cyfal, EoGuy, Niceguyedc, Piledhigheranddeeper, DragonBot, Aurelius173, Another Believer, Amaling, Dank, EEng, Addbot, Anthonydelaware, DOI bot, Baffle gab1978, Ettrig, AnomieBOT, 0x38I9J*, Kingpin13, Materialscientist, Citation bot, LilHelpa, P99am, Mouagip, Tolosthemagician, GrouchoBot, Pwkr, Mattimussi, Dcrjsr, Carlog3, Citation bot 1, Redrose64, Jonesey95, Specdude, Orenburg1, Trappist the monk, Jonkerz, RjwilmsiBot, Noommos, EmausBot, Beatnik8983, Slightsmile, Dcirovic, Hsleiman, ZéroBot, Bformhelix, Chalik1, Cberlind, Teapeat, ClueBot NG, Kirill Borisenko, Ankuzyk, Morgankevinj huggle, DeeperQA, Helpful Pixie Bot, Bibcode Bot, Razorbliss, Ginger Maine Coon, Aisteco, Justincheng12345-bot, Jicutler, Ling.Nut3, ChrisGualtieri, Lucquessoy, Khazar2, Dexbot, Br'er Rabbit, Leafonesky, JamesMadison Pres, Duckduckstop, Monkbot, Ittakesavillage2, Lilm345, Ashaul3 and Anonymous: 52

- **DNA methylation** *Source:* https://en.wikipedia.org/wiki/DNA_methylation?oldid=688702773 *Contributors:* Taw, Axel Driken, Lexor, AlexR, Charles Matthews, Peak, Lproven, St3vo, Gadfium, Eef (A), Pgan002, Piotrus, Julianonions, Thorwald, Mike Rosoft, Nina Gerlach, Vsmith, Polypompholyx, Bobo192, Arcadian, Scapermoya, Physicistjedi, Sbpreuss, Seans Potato Business, ClockworkSoul, Cangrejoinmortal, Ceyockey, Woohookitty, Dandv, Benbest, Jonnabuz, Mandarax, JIP, Sjö, Rjwilmsi, Ligulem, Mohawkjohn, Daycd, YurikBot, Wavelength, Chrispounds, NawlinWiki, Smowton, Arthur Rubin, SmackBot, InvictaHOG, Deniz Feneri, Chris the speller, Trikmc, Uthbrian, Miguel Andrade, JonHarder, Addshore, BobJones, Memming, TedE, Smokefoot, Drphilharmonic, Raetzsch, Yashgaroth, CmdrObot, Kodiak71, A876, Kweeket, Rifleman 82, Was a bee, Narayanese, Crana, Mikewax, Thijs!bot, Jsnover, Laportechicago, Elicsiegel, TimVickers, Nipisiquit, Tlabshier, Dokidok, BrotherE, Magioladitis, Adrian J. Hunter, David Eppstein, WLU, Boghog, Rod57, Plindenbaum, Specter01010, Jgreally, TraumB~enwiki, Malljaja, Luuva, Messengercrow, Wikipolonius, SieBot, WereSpielChequers, EditorInTheRye, Bigskiyesme, Touchstone42, The Thing That Should Not Be, Nursebhayes, DragonBot, Little Mountain 5, Evan.morien, WikHead, MystBot, Kembangraps, Addbot, MikePittsburgh, Aceofhearts1968, Guillaume Filion, DOI bot, Wickey-nl, Fluffernutter, Yobot, Amirobot, AnomieBOT, Jacobglass, Materialscientist, Citation bot, Shellymah, Cow4prez, Xqbot, Omnipaedista, RibotBOT, FrescoBot, Raylim34, Mikipedia2, Wiki1911, Citation bot 1, Fcrivera22, Skyerise, Tim1357, Trappist the monk, Omics, Akahst, Milotoor, Karin sandby, Daniel Cliff, EmausBot, John of Reading, Wikipelli, Dcirovic, Clementine2009, SBabovic, Mjbmrbot, ClueBot NG, Plantdrew, BG19bot, Stevetihi, Austinprince, Christoph-bock, Duxwing, BattyBot, Stigmatella aurantiaca, Khazar2, Dexbot, Dmitry Dzhagarov, Hopefuldonor, Hongbo919, Munkitchoy, Achriner, Tentinator, Enniscath, Hmliuiuc, Tessi87, Von Juvalt, Teddybme, Monkbot, Thetunicgod, Ca2james, Courtney Tait, CV9933, Szaina, Mariuswalter, Nkbbm, Sonam Sapra and Anonymous: 142

- **DNA damage (naturally occurring)** *Source:* https://en.wikipedia.org/wiki/DNA_damage_(naturally_occurring)?oldid=685588371 *Contributors:* Bearcat, LukeSurl, Rjwilmsi, Bgwhite, Danielkueh, Alan G. Archer, Cydebot, Bernstein0275, Yobot, AnomieBOT, Rflynn2, Chaya5260, Mama meta modal and Anonymous: 2

- **Mutation** *Source:* https://en.wikipedia.org/wiki/Mutation?oldid=690613054 *Contributors:* Magnus Manske, Kpjas, Marj Tiefert, Derek Ross, Mav, Bryan Derksen, Taw, Malcolm Farmer, Ed Poor, Rgamble, PierreAbbat, Mjb, Gog, Axel Driken, Michael Hardy, Palnatoke, Lexor, Gabbe, Ixfd64, Cyde, Skysmith, Ahoerstemeier, Stevenj, TUF-KAT, Plop, JWSchmidt, Nikai, Andres, Mxn, Nikola Smolenski, Quizkajer, Timwi, Steinsky, Markhurd, Furrykef, Omegatron, Samsara, Thue, Bevo, Pakaran, Robbot, Sander123, Romanm, Chopchopwhitey, Sverdrup, Academic Challenger, Hadal, Wikibot, Lupo, JerryFriedman, Pengo, Giftlite, Kim Bruning, Mintleaf~enwiki, Nunh-huh, Brian Kendig, Bensaccount, Cantus, Duncharris, Mboverload, Bobblewik, Wmahan, Adenosine, Andycjp, Quadell, Antandrus, PDH, PhDP, Oneiros, Joyous!, Ukexpat, Rahul sig, Stephenpratt, Mike Rosoft, D6, ClockworkTroll, Ocon, DanielCD, Jiy, Discospinster, Oliver Lineham, LindsayH, Ivan Bajlo, Bender235, ESkog, Mashford, Violetriga, Brian0918, Livajo, Gilgamesh he, Jpgordon, Guettarda, Bobo192, Smalljim, Arcadian, ParticleMan, Tgr, Nhandler, 4v4l0n42, Orangemarlin, Abstraktn, Alansohn, Etxrge, Plumbago, Wouterstomp, ClockworkSoul, Shogun~enwiki, Cal 1234, Amorymeltzer, Jon Cates, RainbowOfLight, LFaraone, Ndteegarden, MIT Trekkie, Embryomystic, Ceyockey, Andygainey, Camw, TomTheHand, Urod, WadeSimMiser, MONGO, Zzyzx11, Turnstep, GSlicer, RichardWeiss, BorisTM, Ketiltrout, Rjwilmsi, Koavf, Rogerd, Wikibofh, Stardust8212, Crazynas, Cww, Ian Dunster, Sango123, Yamamoto Ichiro, Dinosaurdarrell, FlaBot, Dantecubed, RexNL, Gurch, Zsingaya, Nabarry, Terrace4, Daycd, Chobot, DVdm, Gwernol, Whosasking, YurikBot, Wavelength, Sceptre, Phantomsteve, Hede2000, Pigman, Kodemage, Ansell, Shell Kinney, Eleassar, Wimt, Shanel, NawlinWiki, EWS23, Dysmorodrepanis~enwiki, Wiki alf, Dureo, Irishguy, Ragesoss, Dhollm, Misza13, Semperf, DeadEyeArrow, Bota47, Phenz, Juicy fisheye, Wknight94, WAS 4.250, Zargulon, Johndburger, Deville, Ninly, Bayerischermann, Closedmouth, JoanneB, Alasdair, Guillom, RenamedUser jaskldjslak904, ArielGold, Johnpseudo, Allens, Sancassania, Kungfuadam, Roke, Jungenbergs, DVD R W, Brentt, SmackBot, Reedy, TestPilot, Hydrogen Iodide, Bomac, ScaldingHotSoup, Jrockley, Delldot, Hanmi74, Zephyris, JosephCCampana, Yamaguchi図図, Gilliam, Rmosler2100, Sveika, Kazkaskazkasako, Chris the speller, Keegan, RDBrown, JohnRobertMartin, Jprg1966, Miquonranger03, MalafayaBot, Wedian, Zven, Gracenotes, John Reaves, Can't sleep, clown will eat me, Nixeagle, Voyajer, Rrburke, Tommyjb, Xyzzyplugh, GVnayR, Radagast83, Makemi, Jiddisch~enwiki, Dreadstar, Smokefoot, DMacks, LeoNomis, Alan G. Archer, Madeleine Price Ball, Dncarley, LestatdeLioncourt, Robert Stevens, Mgiganteus1, Seb951, Dingopup, Waggers, Ryulong, Hogyn Lleol, Novangelis, MTSbot~enwiki, Ryanjunk, Hu12, White Ash, Roland Deschain, Poechalkdust, JoeBot, Jason7825, Toastybread, Courcelles, Frank Lofaro Jr., Tawkerbot2, Dlohcierekim, Geezerbill, JForget, CmdrObot, Tanthalas39, SimonSayz, Fedir, Lavateraguy, Agathman, ShelfSkewed, Metzenberg, Standonbible, Seven of Nine, Equendil, Icek~enwiki, Kupirijo, Clappingsimon, Ppgardne, Thewall2, Carifio24, Gogo Dodo, Was a bee, Khatru2, Anthonyhcole, Flowerpotman, Corpx, Tawkerbot4, Clovis Sangrail, Chrislk02, Narayanese, Sharonlees, InfoCan, Michael Johnson, Oleksii0, Thijs!bot, Epbr123, Peachiezworld, Mojo Hand, Marek69, James086, Nezzadar, Tellyaddict, Davidhorman, SineCurve, Sithu.Win, Capeo, Escarbot, Mentifisto, David D., AntiVandalBot, Guy Macon, Seaphoto, Gusgould, Quintote, TimVickers, Scepia, Danger, Credema, Myanw, Joycierules, Res2216firestar, Ioeth, Erxnmedia, Solenoozerec, Gcm, MER-C, Plantsurfer, Sheitan, Instinct, Arch dude, Alex tudor, Agreene175, Dcooper, Frankie816, Savant13, .anacondabot, Somnathroy, Magioladitis, Bongwarrior, VoABot II, DWIII, Tupeliano~enwiki, Culverin, WhatamIdoing, C.-ting Wu, Faustnh, Cgingold, CodeCat, Ciar, Daarznieks, Allstarecho, Emw, DerHexer, Saganaki-, Peter coxhead, 0612, MartinBot, STBot, Arjun01, Lid6, Tuganax~enwiki, Middlenamefrank, Mike6271, R'n'B, Tgeairn, Anastacie, J.delanoy, Pharaoh of the Wizards, Filll, Sikatriz, Synapomorphy, Uncle Dick, Jmjanzen, Netnuevo, Gr8niladri, Katalaveno, Dr d12, Mikael Häggström, AntiSpamBot, Chiswick Chap, Hut 6.5, Plindenbaum, Kdawson66, Gcrossan, Bioephemera, Aled 12345, VolkovBot, Jeff G., Pparazorback, Philip Trueman, DoorsAjar, TXiKiBoT, Oshwah, Zurrr, Tameeria, PizzaBox, Ajoust, Rei-bot, Sk741~enwiki, Naohiro19 revertvandal, Vanished user ikijeirw34iuaeolaseriffic, Lradrama, Sintaku, Abdullais4u, Cremepuff222, Watchdogb, Meepmoop, Daveheathr, Meters, ITxT, Enviroboy, Ceranthor, HiDrNick, AlleborgoBot, Karajade, SMC89, SieBot, VK35, Chaosof99, Dawn Bard, Eagleal, Ncfuzzy101, LeadSongDog, Thebestlaidplans, Happysailor, Flyer22 Reborn, Radon210, Oda Mari, Hzh, Biskot~enwiki, Oxymoron83, Genenome, Faradayplank, Peternewell, Techman224, Hobartimus, OKBot, Chain funds, Alikingofkings, Anchor Link Bot, Alchemes, Rabo3, Denisarona, Forluvoft, SallyForth123, Rustygin, Ratemonth, ClueBot, Snigbrook, The Thing That Should Not Be, Matdrodes, Lewa.27, YassineMrabet~enwiki, Drmies, Boing! said Zebedee, CounterVandalismBot, Harland1, Excirial, Jusdafax, Leonard^Bloom, Lartoven, Rhododendrites, ParisianBlade, NuclearWarfare, Arjayay, Micha, Fungusnunchuck, ChrisHodgesUK, Thingg, Aitias, Christomaniac76, Johnuniq, Egmontaz, DumZiBoT, Graevemoore, Ktfreese, MeSoDark, Crazy Boris with a red beard, Rror, Mouse555~enwiki, Hylobius, TFOWR, Firebat08, WikiDao, Aunt Entropy, KnightofZion, Klgroom, Elvin fimel, Addbot, DOI bot, Jojhutton, Wickey-nl, Atethnekos, Jason.Rafe.Miller, Shirtwaist, CanadianLinuxUser, Imarockstarfoo, Jakeisbaked, Thedude212, Cst17, Bernstein0275, Quercus solaris, Lordlosss2, Ehrenkater, Tide rolls, Botanist3, Krano, Luckas Blade, Ngsmart, Gail, Zorrobot, AdamXtreme18, Ettrig, LuK3, Planettelex23, Ben Ben, Legobot, Luckas-bot, Tohd8BohaithuGh1, Julia W, Mauler90, Anypodetos, Ayrton Prost, Azcolvin429, AnomieBOT, Haetmachine, Phlyght, Jim1138, Piano non troppo, AdjustShift, Flewis, Materialscientist, The High Fin Sperm Whale, Citation bot, Brightgalrs, Swastikde, ArthurBot, WaffleMaster44, LovesMacs, Gsmgm, Sionus, Capricorn42, Kahkooi, Millahnna, Oxwil, Pb0823, Gatorgirl7563, Shirik, SassoBot, Methcub, Joejoeftw, SchnitzelMannGreek, Dante mars, FrescoBot, DoctorDNA, Riventree, Lothar von Richthofen, Ifiber, HJ Mitchell, Letatcestmoi94, Drew R. Smith, Matthew Ackerman, Citation bot 1, Trichodinamitra, Pinethicket, I dream of horses, Naif1989, Rushbugled13, Meaghan, Curehd, Jauhienij, Trappist the monk, MiRroar, Ooikahkooi0507, Specs112, Jcorry10, Bhawani Gautam, GoodScienceForYou, Bradshej, Hajatvrc, Wintonian, EmausBot, Liseranius, Orphan Wiki, WikitanvirBot, 478jjjz, Brendanology, Super48paul, RA0808, Teerickson, Christinebenson58, Winner 42, Jakirfirozkamal, Wikipelli, ChrisTheBrown, WhatTheBlack, Tuxedo junction, Josve05a, Martboy722, John Mackenzie Burke, Wayne Slam, Aidarzver, Abergabe, Standinguptoit, Donner60, Orange Suede Sofa, Iltoffier, TYelliot, DASH-BotAV, Rocketrod1960, PTaschner, Woodsrock, Will Beback Auto, ClueBot NG, Gareth Griffith-Jones, Jack Greenmaven, MelbourneStar, Ur moma123456789, Frietjes, Mesoderm, Go Phightins!, Widr, Scaple10, Helpful Pixie Bot, Excerpted31, Titodutta, Plantdrew, Jaba96, Stacyroy, RobLandau, Scratlikesacorns, MusikAnimal, AwamerT, Mark Arsten, P'tit Pierre, MrBill3, Glacialfox, Klilidiplomus, Achowat, Biosthmors, Icodemachine, MyNameIsLukeMonet, ChrisGualtieri, Professorrus, Fiona126, BrightStarSky, Mogism, Lugia2453, Sfgiants1995, 図図図図図, The Anonymouse, CsDix, Shahram.shirazi, LHSaavedra, Xinyukiwi, Realguy12, DavidLeighEllis, Hi my names bobt42, CensoredScribe, Nightlight77, GirlWithWings, Zenibus, Spgoggles, Jianhui67, Elizabeth sunny, OccultZone, Fundon1, StefanieStrohl, Jmmy2013, Csutric, Drericklagos, Chaya5260, Monkbot, JRElings, AKS.9955, NewEnglandDr, Dsprc, AdamoulasA, Squad1234, Amortias, Tetucarlos, HMSLavender, Hello1234566, Mike Swagger, !0CaterPillar0!, Cavefish777, Giraffesyrup, YeOldeGentleman, LarryBoy79, ScienceGuru123, Somecoolman7888, K scheik, GeneralizationsAreBad, Krmeyer17, CREASHUNIST, Proyecto genómica, Diptanshu mandal, FallenAngelXXXX, Duchamp 7 and Anonymous: 825

- **DNA damage theory of aging** *Source:* https://en.wikipedia.org/wiki/DNA_damage_theory_of_aging?oldid=674580767 *Contributors:* Edward, Topbanana, Bender235, PrometheusOne, Benbest, Rjwilmsi, Nihiltres, Bgwhite, Epipelagic, Edgar181, Addshore, GVnayR, SMasters, Ellis2ca, Cydebot, Smartse, Magioladitis, AtticusX, Fabrictramp, Maurice Carbonaro, Mkper9, Fyyer, Vanished 45kd09la13, Snapperman2, Addbot, Ka Faraq Gatri, Bernstein0275, Favonian, AV3000, FrescoBot, Citation bot 1, Bgpaulus, Trappist the monk, RjwilmsiBot, Ohsoseri-

Hangjian, CKlunck, Dmanning, Nsaa, JohnyDog, Alansohn, Sheehan, Atlant, Bart133, Wtmitchell, Yogi de, Tycho, Sciurinæ, Gene Nygaard, Blaxthos, Dan100, Isfisk, WadeSimMiser, Isnow, LadyofHats, Achim Raschka, CharlesC, Kralizec!, Gimboid13, Banpei~enwiki, Mandarax, Graham87, Kbdank71, Snafflekid, Sjö, Rjwilmsi, Wikibofh, Lordkinbote, JonMoulton, Scorpiuss, Yamamoto Ichiro, FuelWagon, FlaBot, Cless Alvein, Nihiltres, Gurch, Jrtayloriv, Vossman, Nsteinberg, Chobot, DVdm, Stoive, Bgwhite, WriterHound, Gwernol, The Rambling Man, YurikBot, Wavelength, Mushin, RobotE, Jimp, Flameviper, Severa, Hydrargyrum, Stephenb, Chaos, Rsrikanth05, Tavilis, Anomalocaris, NawlinWiki, Haranoh, Xdenizen, William R. Buckley, Zzuuzz, Lt-wiki-bot, Terry Longbaugh, Closedmouth, Petri Krohn, MaNeMeBasat, GraemeL, Chrishmt0423, Elfalem, Guillom, Banus, Dkasak, DVD R W, Kf4bdy, PVSpud, Trevorloflin, Crystallina, SmackBot, MattieTK, Tom Lougheed, KnowledgeOfSelf, TestPilot, Unyoyega, Michaelfavor, Alksub, Zephyris, Gilliam, Betacommand, Rmosler2100, Kazkaskazkasako, Amatulic, Jjalexand, Quinsareth, NCurse, Master of Puppets, Hichris, MalafayaBot, Papa November, DHN-bot~enwiki, Darth Panda, Rama's Arrow, Plasmid, Can't sleep, clown will eat me, OrphanBot, Rrburke, Khoikhoi, Nakon, MichaelBillington, Akriasas, Drphilharmonic, Jklin, DMacks, KeithB, Jls043, Thistheman, Kukini, Themadness, The undertow, SashatoBot, Rory096, Euchiasmus, ClarkFreifeld, Gobonobo, Ywith, Ben Moore, Munita Prasad, Mr Stephen, Ryulong, Novangelis, Cerealkiller13, Hu12, Iridescent, AAM, Kaarel, Newone, Cph3992, StephenBuxton, Cela~enwiki, Twas Now, Nubzor, Igoldste, Az1568, Tawkerbot2, JForget, Ale jrb, Sir Vicious, Agathman, Scirocco6, Dgw, Arnoldlcl, Chrisahn, MrFish, Brianmurphy22, Nilfanion, Elewton, Cydebot, Ppgardne, Kanags, WillowW, RelentlessRecusant, Gogo Dodo, Pascal.Tesson, Julian Mendez, Tawkerbot4, Carstensen, DumbBOT, Narayanese, Omicronpersei8, Thijs!bot, Epbr123, Nonagonal Spider, Marek69, John254, Misterstark, A3RO, James086, Orthank, RFerreira, Dfrg.msc, Benzenetoaster, Will Bradshaw, Sean William, Karin D. E. Everett, Escarbot, AntiVandalBot, Majorly, Quintote, ArchDaemon, TimVickers, Jdclevenger, Znkp, Res2216firestar, JAnDbot, Kaobear, MER-C, Plantsurfer, Igodard, Hut 8.5, Greensburger, PhilKnight, Acroterion, Magioladitis, Connormah, Mattb112885, VoABot II, Meredyth, Twsx, Alexllew, Xnurfz, DerHexer, Wikianon, Otvaltak, Climax Void, MartinBot, Treyd500, Rettetast, Strasserk08, R'n'B, CommonsDelinker, Pekaje, PrestonH, J.delanoy, Nbauman, Bluesquareapple, Hodja Nasreddin, Gzkn, Katalaveno, McSly, ROFLCOPTERone1!, Nwbeeson, SmilesALot, Tanaats, Sukkoth Qulmos, Thatwchduznotfly, Radioactivebloke, Megaprof, Scwerllguy, Xaxx, Pdcook, JavierMC, Halmstad, ThePointblank, CardinalDan, Speciate, Waffles209, VolkovBot, Have fun with this, Hersfold, Philip Trueman, TXiKiBoT, Oshwah, Anonymous Dissident, Gen. Quon, Qxz, Warrush, Aznshark4, Xaraikex, Mauror, Mishlai, Universewik, Falcon8765, Enviroboy, Burntsauce, Jones1337, Billybutterworthreborn, HiDrNick, AlleborgoBot, Blood sliver, Satish suryan, NHRHS2010, EmxBot, D. Recorder, SieBot, Mikemoral, Graham Beards, Hacker312, Bentogoa, Tiptoety, Ipodamos, Oxymoron83, Byrialbot, Smilesfozwood, Boppet, Techman224, Jack the Stripper, Sunrise, Mike2vil, Jmschrad, Drgarden, WikipedianMarlith, Atif.t2, ClueBot, Jbening, NickCT, Jackollie, Foxj, The Thing That Should Not Be, Rodhullandemu, KawaiiApocalypse, Arakunem, Sanjeev.singh3, CounterVandalismBot, Rotational, Peteruetz, Tim-larry, Excirial, Andy pyro, Biodoc546, Bendover12345, Tyler, NuclearWarfare, Razorflame, Syko66613, Walterdavidsmith, EdwardLawrence, Aitias, SoxBot III, BarretB, Jed 20012, PseudoOne, Little Mountain 5, Alexius08, Noctibus, ZooFari, Addbot, Some jerk on the Internet, DOI bot, Mr. Wheely Guy, Vishnava, CanadianLinuxUser, Download, Glane23, Debresser, Quercus solaris, Numbo3-bot, Tide rolls, Lightbot, Zorrobot, Alfie66, Math Champion, Luckas-bot, Tamiasciurus, Maxí, South Bay, Eric-Wester, AnomieBOT, Andrewrp, Jim1138, Piano non troppo, Keithbob, Ipatrol, AdjustShift, Kingpin13, Duhmutton, Citation bot, Neurolysis, Xqbot, Petercoccusfuriosus, Dcrjsr, IcedNut, Erik9, Dougofborg, Thehelpfulbot, SAVE US.L2P, Ion496, GT5162, Fingerz, FrescoBot, Ryryrules100, Dogposter, Sirtywell, Lambmeat, Galorr, DivineAlpha, HamburgerRadio, Citation bot 1, Cypher3c, Intelligentsium, Pinethicket, I dream of horses, Jonesey95, A8UDI, LukeGoodsell, SpaceFlight89, RandomStringOfCharacters, Caladont, Dashed, PiRSquared17, Vrenator, Capt. James T. Kirk, Aiken drum, Firsthuman, Breckp, Jd Tendril, Euland, Reach Out to the Truth, Brambleclawx, YuanH, Hyyfgk101, RjwilmsiBot, Jaredwinkler, TjBot, Gould363, NerdyScienceDude, Felix Tritscher, Skamecrazy123, LcawteHuggle, DASHBot, EmausBot, Never give in, RA0808, Rajesh822, Jayreimer, Evilninja180, Wikipelli, John Cline, Mantha.satish, StereoTypo, Lalawasabi, John Mackenzie Burke, Sky380, WeigelaPen, Beta cafe, Scientific29, Cstillabower, ChuispastonBot, Staticd, LikeLakers2, FeatherPluma, Petrb, ClueBot NG, CocuBot, This lousy T-shirt, Aram-van, Frietjes, Rezabot, MorganRJ, Widr, HMSSolent, Strike Eagle, BZTMPS, Lowercase sigmabot, BG19bot, Vokesk, MusikAnimal, Mark Arsten, EmadIV, Czernilofsky, Warriorsrule2478, MrBill3, Pablhern, Tdk327, Westie28, Jscudier, Roleren, BattyBot, Illia Connell, Dexbot, Captaincollect1970, Mogism, Frosty, Electronsaregreen, Epicgenius, FallingGravity, Frickson, Ananthkamath1995, Andrey.a.mitin, Origamite, Fremantle99, Monkbot, Vieque, Smhorseshoe123, Trackteur, Ziff xl 200, Cljolly, TheWorldsGreatestScientist, Nathen418, Pitchcapper, Hcnfur, Zortwort, CV9933, Crackstack22, Matthew4671, KasparBot, Asutosh.wikipedia, Phillips.h.hamilton, XJustina, CCevol2015 and Anonymous: 878

- **Transfer RNA** *Source:* https://en.wikipedia.org/wiki/Transfer_RNA?oldid=686459751 *Contributors:* Lexor, Chris~enwiki, Timwi, Taoster, Josh Cherry, Csibert, Mintleaf~enwiki, Bensaccount, Ferren~enwiki, Maximaximax, PFHLai, B.d.mills, Sorgenfrei, Mike Rosoft, Discospinster, Rich Farmbrough, Cacycle, Wk muriithi, ArnoldReinhold, Violetriga, Phoenix Hacker, Deicas, Wee Jimmy, Bobo192, Arcadian, Speedy-Gonsales, Ramujana, HasharBot~enwiki, Etxrge, Craigy144, Riana, Esrob, Ceyockey, WadeSimMiser, MarcoTolo, Rjwilmsi, Thisismikesother, FlaBot, Margosbot~enwiki, Vossman, Kinglz, Bgwhite, YurikBot, Wavelength, RobotE, Beethoven's DNA, Tavilis, Dbfirs, Whitejay251, Bkwrm18, Banus, Thorney¿?, SmackBot, InverseHypercube, Martin.Budden, Zephyris, Gilliam, Hichris, Jerrch, TedE, Michael-Billington, Drphilharmonic, Spiritia, Pjlmac, Akendall, Physis, Ben Moore, Dhp1080, Dan Gluck, Kaarel, Lord E, Agathman, Estanton, WeggeBot, Tex, Ppgardne, Was a bee, Narayanese, JodyB, Crum375, Talgalili, Thijs!bot, Epbr123, Opabinia regalis, CopperKettle, Headbomb, AntiVandalBot, Tmjlowe, Storkk, JAnDbot, Husond, MSBOT, VoABot II, Askari Mark, Mbc362, BernardP, Antorjal, -VL-, Su-no-G, Foregone conclusion, Wylve, J.delanoy, Tlim7882, Tarotcards, Nwbeeson, Plindenbaum, Steel1943, Idioma-bot, VolkovBot, Thedjatclubrock, TXiKiBoT, Parker007, Martin451, Alexbateman, Logan, Active contributor, Macdonald-ross, BotMultichill, Gerakibot, Lynnk20201, Jennifer Rfm, Sunrise, Denisarona, Niplet226, ClueBot, Yikrazuul, DragonBot, Lihmwiki, Kvongunten, Walterdavidsmith, Versus22, SoxBot III, Allsaints23, Marchije, Asrghasrhiojadrhr, Addbot, DOI bot, Binary TSO, OmgItsTheSmartGuy, Diptanshu.D, FiriBot, Favonian, Samat-Bot, JLSussman, Luckas Blade, Jarble, Yobot, Mirabellen, The Tanker, AnomieBOT, Kristen Eriksen, Citation bot, Xqbot, Capricorn42, Mike Chelen, Racknack7824, Captain-n00dle, Daleang, Dogposter, Blueronin44, BenzolBot, Biochemwiki, Citation bot 1, Pinethicket, SW3 5DL, 777sms, Diannaa, Minimac, RoseAE, NerdyScienceDude, John of Reading, WikitanvirBot, Racerx11, Tommy2010, Dcirovic, 30Ikra, Chelsea898, Ethaniel, John Mackenzie Burke, TeapotAgnostic, RE73, Scientific29, Morgis, Teaktl17, ClueBot NG, Masssly, MerlIwBot, Calabe1992, Bibcode Bot, BG19bot, TRNAtest, Vokesk, Egocentrism, CritiK1LLz, Czernilofsky, BattyBot, Ajaxfiore, Tyersome, Dexbot, Makecat-bot, JammyDodgers72, Sswan010, Ananthkamath1995, MisiC, JakeOliger, Jlmalcos, Elephantsofearth, Jcfrommn, Monkbot, Vpliatsika, Wikicology, Sweepy and Anonymous: 169

- **Proteinogenic amino acid** *Source:* https://en.wikipedia.org/wiki/Proteinogenic_amino_acid?oldid=689008617 *Contributors:* Selket, Robbot, Chris Roy, Nagelfar, Pgan002, Kaldari, Rich Farmbrough, Cacycle, Phoenix Hacker, Arcadian, Keenan Pepper, BrentN, Mindmatrix, Qchristensen, Rjwilmsi, Gurch, Takometer, Benlisquare, Kkmurray, Leptictidium, ChipperGuy, Itub, SmackBot, Edgar181, Eloil, Kazkaskazkasako, Chris the speller, Hichris, Uthbrian, Radagast83, Smokefoot, Drphilharmonic, DMacks, Grothmag, Fat64~enwiki, Stifynsemons, Kupirijo, Christian75, Thijs!bot, Erechtheus, BenJWoodcroft, TimVickers, Xact, Squidonius, Mikael Häggström, Phlounder, Cpt ricard, TXiKiBoT, Kumorifox, Gregogil, Zoasterboy, Agur bar Jacé, Myceteae, Mlaffs, MystBot, Addbot, Willking1979, Tanhabot, Flakinho, Luckas-bot, Amirobot, Microball, Carolina wren, Wiki007wiki, Citation bot, Xqbot, Capricorn42, Aa77zz, RibotBOT, L-Tyrosine, Jackroven, I dream of horses, Scoutfreak, RjwilmsiBot, Dancojocari, EmausBot, Dcirovic, MRandazzUCSD, GenyAncalagon, Whoop whoop pull up, Teaktl17, Lanthanum-138, Frietjes, HPBiochemie, ChrisGualtieri, Ivan trus, SophieAthena, Hieu nguyentrung12, CsDix, Doniitoo, Monkbot, WamSam, MGaloni and Anonymous: 60

- **DNA-binding protein** *Source:* https://en.wikipedia.org/wiki/DNA-binding_protein?oldid=665683337 *Contributors:* Timc, Josh Cherry, Diberri, MacGyverMagic, Icairns, Wisdom89, Arcadian, Jag123, Seans Potato Business, Marudubshinki, Rjwilmsi, FlaBot, Margosbot~enwiki, Wavelength, Epolk, Splette, SmackBot, Sergeymk, Edgar181, Betacommand, Rlevse, FrozenMan, Heimstern, Elb2000, DabMachine, Arfy900, ShelfSkewed, Skittleys, Hwttdz, Alaibot, Headbomb, .anacondabot, T@nn, CommonsDelinker, Alexbateman, JerrySteal, ClueBot, Hojo 10 19, Thehelpfulone, Deepraine, Addbot, CBCR, Ettrig, Yobot, Citation bot, Proconsoph, P99am, D'ohBot, Citation bot 1, WikitanvirBot, NGPriest, Chemberlen, ClueBot NG, BG19bot, Pascoe J Harvey, Monkbot and Anonymous: 24

- **Nuclease** *Source:* https://en.wikipedia.org/wiki/Nuclease?oldid=689312716 *Contributors:* DanKeshet, Ojigiri~enwiki, Bensaccount, Gadfium, HaNcI, Vsmith, Shanes, Arcadian, Aegis Maelstrom, Strangethingintheland, FlaBot, YurikBot, Splette, Gaius Cornelius, Slarson, Gilliam, Bluebot, Kukini, Agathman, John Riemann Soong, WinBot, Smartse, Magioladitis, Oven, Freddyd945, STBot, VolkovBot, Kaboytes, A4bot, Reibot, Alexbateman, England Expects, AlleborgoBot, Anabolic1, Redeemer079, Vojtěch Dostál, Addbot, LaaknorBot, Luckas-bot, Bunnyhop11, İnfoCan, Clark89, Xqbot, Mkosterv, Kathleenoj, Mitchctaylor, EmausBot, Dcirovic, ZéroBot, ClueBot NG, ChrisGualtieri, Makecat-bot, The Anonymouse, AKS.9955, KasparBot, Dcbennett2 and Anonymous: 31

- **Enzyme** *Source:* https://en.wikipedia.org/wiki/Enzyme?oldid=689656720 *Contributors:* AxelBoldt, Magnus Manske, Marj Tiefert, Derek Ross, Bryan Derksen, Taw, Malcolm Farmer, Andre Engels, Danny, Gianfranco, Tempel, Zoe, Graham, Netesq, Ewen, Olivier, Dwmyers, Lir, D, JohnOwens, Michael Hardy, GABaker, Lexor, DIG~enwiki, Ixfd64, Qaz, 168..., Ahoerstemeier, Ronz, Elano, 5ko, Den fjättrade ankan~enwiki, JWSchmidt, Glenn, Tristanb, TonyClarke, Rob Hooft, Hashar, Jengod, Gingekerr, Wikiborg, Stone, Ike9898, Wolfgang Kufner, Tpbradbury, John NRW, Taxman, Ed g2s, Mperkins, Shizhao, Exaton, Fvw, Pietrosperoni, Secretlondon, Jusjih, Lumos3, Gentgeen, Robbot, Chris 73, Yelyos, Romanm, Academic Challenger, Hadal, JesseW, Jpbrenna, Holeung, Lupo, Diberri, MilkMiruku, GreatWhiteNortherner, Giftlite, Christopher Parham, Fennec, Jyril, Nunh-huh, Ævar Arnfjörð Bjarmason, Doctorcherokee, Everyking, Chihowa, Michael Devore, Bensaccount, Maver1ck, Duncharris, FrYGuY, Mboverload, Eequor, Jackol, Pne, Delta G, Wmahan, ChicXulub, Gzuckier, GeneMosher, Onco p53, OverlordQ, MisfitToys, G3pro, PDH, Vina, MacGyverMagic, Bumm13, PFHLai, Neutrality, Phpham31, Joyous!, Clemwang, Fabrício Kury, Zondor, Trevor MacInnis, DanielCD, Johan Elisson, Discospinster, Rich Farmbrough, Guanabot, Cacycle, Narsil, Mjpieters, Mani1, Bender235, ESkog, Kbh3rd, Kjoonlee, Danny B-), CanisRufus, Pt, Bookofjude, DavidRader, CDN99, Causa sui, Bobo192, Smalljim, Xevious, Arcadian, Guiltyspark, Giraffedata, Nk, TheProject, Samulili, Nhandler, Haham hanuka, Krellis, Jumbuck, Siim, Alansohn, Anthony Appleyard, PopUpPirate, Petemorris, Arthena, Atlant, Riana, Seans Potato Business, Walkerma, Ynhockey, Hu, Stillnotelf, Bart133, Theyeti, PaePae, Mcy jerry, Knowledge Seeker, Evil Monkey, Jobe6, Dirac1933, IMeowbot, Arakin, Ajpratt56, Ianblair23, Redvers, Drummstikk, Capecodeph, Instantnood, RyanGerbil10, Jwanderson, Woohookitty, Dalmoz~enwiki, TigerShark, Tabletop, Dolfrog, SCEhardt, M412k, Wayward, Marco-Tolo, JohnJohn, Palica, Turnstep, Magister Mathematicae, BD2412, Chun-hian, Mendaliv, Josh Parris, Rjwilmsi, Tizio, Саша Стефановиħ, Quale, Dpark, Vary, Harry491, HappyCamper, Brighterorange, The wub, Bhadani, Sango123, DirkvdM, FlaBot, Ian Pitchford, SchuminWeb, RobertG, Shultzc, Jak123, Mister Matt, Nihiltres, Crazycomputers, Nivix, RexNL, Nige111, Takometer, Intgr, Stevenfruitsmaak, Daycd, BradBeattie, Smartsmith, Scimitar, Chobot, Jaraalbe, Jdhowens90, Bornhj, DVdm, Bgwhite, Gwernol, The Rambling Man, YurikBot, Wavelength, Mdemian, Phantomsteve, John Quincy Adding Machine, Conscious, Bcebul, Captaindan, DanMS, Stephenb, Bill52270, Gaius Cornelius, K.C. Tang, NawlinWiki, Wiki alf, Msikma, Grafen, Icelight, Nutiketaiel, Aaron Brenneman, Xdenizen, Matticus78, Ruhrfisch, Hv, BOT-Superzerocool, DeadEyeArrow, Bota47, Superiority, Brisvegas, Typer 525, Nlu, Nick123, Wknight94, Leptictidium, FF2010, Donbert, Zero1328, Albert109, Zzuuzz, Syd Midnight, Lt-wiki-bot, TimBarrel, Closedmouth, KGasso, Sean Whitton, BorgQueen, GraemeL, JoanneB, Amren, Anclation~enwiki, F. Cosoleto, Tvd12110, Mejor Los Indios, DVD R W, ChemGardener, Sardanaphalus, Attilios, Smack-Bot, Michael%Sappir, Saravask, Jamesters, KnowledgeOfSelf, Chazz88, unyoyega, David Shear, Giraldusfaber, Blue520, WookieInHeat, EncycloPetey, Delldot, Jab843, Veesicle, Frymaster, Edgar181, Alsandro, Zephyris, Gaff, Aksi great, Gilliam, Hmains, Modulus86, Yaser al-Nabriss, SergeantBolt, MrDrBob, SlimJim, RDBrown, NCurse, Fuzzform, Hichris, MalafayaBot, Moshe Constantine Hassan Al-Silverburg, DHN-bot~enwiki, Darth Panda, Oatmeal batman, Rlevse, Langbein Rise, RoyArnon, Tsca.bot, Can't sleep, clown will eat me, Vanished User 0001, NoahElhardt, Yaksha, JonHarder, EvelinaB, TKD, Addshore, SundarBot, Khukri, Decltype, Nakon, Savidan, Jiddisch~enwiki, Richard001, Weregerbil, Iridescence, Drphilharmonic, Hammer1980, DMacks, Sturm, Mion, Ck lostsword, Pilotguy, Kukini, Clicketyclack, Baoilleach, Dono, LtPowers, Harryboyles, Mouse Nightshirt, Kuru, John, J. Finkelstein, AmiDaniel, FrozenMan, Kipala, SilkTork, DrNixon, Minna Sora no Shita, Tlesher, IronGargoyle, Physis, Clone1, Mathel, Ben Moore, Knights who say ni, Beetstra, Serephine, Uuhuų, SandyGeorgia, Dalstadt, Mets501, Dcflyer, Artman40, Jose77, Stuckonempty, Iridescent, Paul venter, JoeBot, Shoeofdeath, 10014derek, Blackprince, CapitalR, Blehfu, Tawkerbot2, Dlohcierekim, Daniel5127, TWUChemLS, Ioannes Pragensis, Fvasconcellos, JForget, CmdrObot, Fedir, Van helsing, Robotsintrouble, John Riemann Soong, SupaStarGirl, Drinibot, MiShogun, THF, ShelfSkewed, WeggeBot, Casper2k3, WillowW, Steel, Mato, Michaelas10, Gogo Dodo, Lugnuts, Jedonnelley, Tawkerbot4, Carstensen, DumbBOT, Chrislk02, FastLizard4, Smellysam524, Kozuch, Omicronpersei8, Gourgeousladyy, Tunheim, Gimmetrow, הסרפד, Chris CII, Thijs!bot, Epbr123, Opabinia regalis, Tomharris~enwiki, N5iln, Young Pioneer, Sopranosmob781, Luigifan, Pjvpjv, Marek69, (3ucky(3all, Harlequin2134, West Brom 4ever, Elem3nt, James086, Miller17CU94, Dfrg.msc, CharlotteWebb, Wikidenizen, Tomos ANTIGUA Tomos~enwiki, Northumbrian, Escarbot, Eleuther, David D., KrakatoaKatie, AntiVandalBot, Luna Santin, Aljasm, Prolog, Kheiron~enwiki, Jj137, TimVickers, Gdo01, TJClark, JAnDbot, Deflective, MER-C, Nthep, Correctist, Hello32020, Hut 8.5, NEUROtiker, FaerieInGrey, Magioladitis, Bongwarrior, VoABot II, Wikidudeman, James-BWatson, Swpb, Think outside the box, Aerographer1981, Emhale, Twisted86, Pax deorum, Recurring dreams, WhatamIdoing, Fallschirmjäger, Catherine-maguire, Mikhail Garouznov, Economo, Dirac66, Adrian J. Hunter, Emw, User A1, ArmadilloFromHell, Glen, DerHexer, JaGa, Wdflake, Lelkesa, Lenticel, Squidonius, Patstuart, Fishyfishy, DancingPenguin, Sinharaja2002, MartinBot, Arjun01, ChemNerd, Rettetast, Curiousgeorgie, Anaxial, CommonsDelinker, Dr. St-Amant, Flannelgraf, Leyo, CBSB, AmH2~enwiki, Fondls, J.delanoy, Pharaoh of the Wizards, Jroenick2797, Timberman~enwiki, Cyclosa, DrKay, Trusilver, Awg usc, Dreamm, Hans Dunkelberg, Boghog, Shane wallmann, Cybercyclone3993, Jorwisgar, Hector kevin, Jstew87, Articles for analfuckbaggage-!, Sefog, Bot-Schafter, Kataleveno, Ghjfdhgdju, Victuallers, Jonnylyman, Rocket71048576, Chriswiki, AntiSpamBot, Flugs~enwiki, Plasticup, LittleHow, NewEnglandYankee, Evan57~enwiki, JohnnyRush10, Hellfire68, Cmichael, Suhail174, Benwd, RB972, Ichbintux, Pdcook, AndreasJSbot, Useight, Matthiascarrigan, Cardinal-Dan, Idioma-bot, Wikieditor06, PeaceNT, VolkovBot, Tkenna, Jeff G., Wolfnix, Philip Trueman, DoorsAjar, TXiKiBoT, Zidonuke, Bob888, Pkarp11, Sk741~enwiki, Qxz, Xxpsycho37xx, Indy 900, Tantal-ja, DennyColt, DocteurCosmos, Gregogil, Psyche825, Jitu jain, Scottiedog15, Katimawan2005, Iamstaurtmayle, Stuartmayle, Synthebot, Altermike, Falcon8765, CephasE, Sfmammamia, Kosigrim, Petergans, Hazel77, EmxBot, 01roshay, Cwhitt817, SieBot, Sonicology, Ttony21, Jauerback, Gerakibot, Viskonsas, Caltas, Matthew Yeager, Cwkmail, Yintan, Purbo T, Maphyche, Arbor to SJ, Monkeymox, Wmpearl, Oxymoron83, Kochipoik, Lightmouse, Poindexter Propellerhead, Mbols, Hobartimus, OKBot, Bobadillo, StaticGull, Cppgx, GiveItSomeThought, Maralia, Hezzy140, Ebuxbaum, Pinkadelica, M2Ys4U, Denisarona, Buzybeez, Lascorz, Blabberfruit, Forluvoft, Webridge, Atif.t2, Elassint, ClueBot, PipepBot, Snigbrook, The Thing That Should Not Be, Vs-Bot, Yikrazuul, Damonchicken, WDavis1911, Mild Bill Hiccup, Regibox, CounterVandalismBot, Niceguyedc, OccamzRazor, Blanchardb, Michał Sobkowski, LizardJr8, Liempt, Rose090, Timhewitt101, Mcrkramer, Moguls, DragonBot, Alexbot, Jusdafax, Vanisheduser12345, Pmm530, Idontknowmyname, Tyler, Arjayay, Jotterbot, Huntthetroll, Saebjorn, Aitias, Yasitha2, Dheenkumar, TAKEN00, Redeemer079, Qwfp, Jamesscottbrown, Skunkboy74, XLinkBot, Jovianeye, Rror, Vojtěch Dostál, Little Mountain 5, Skarebo, NellieBly, Jeremyb650, Wiki-Dao, ZooFari, MystBot, HexaChord, CalumH93, Addbot, Some jerk on the Internet, DOI bot, Jojhutton, Marxspiro, Ronhjones, PandaRepublican, Fieldday-sunday, Adrian 1001, Low-frequency internal, Cst17, MrOllie, LaaknorBot, EconoPhysicist, Bassbonerocks, Omnipedian, Debresser, Doniago, LinkFA-Bot, 84user, Numbo3-bot, Tide rolls, Alfie66, Luckas-bot, Yobot, Fraggle81, II MusLiM HyBRiD II, Yoilish, Mmxx, Beeswaxcandle, Chios1127, Naipicnirp, AnomieBOT, Haetmachine, 1exec1, Neptune5000, Barliman Butterbur, AdjustShift, Scuzzer, Kingpin13, Bluerasberry, Materialscientist, Kasper90, Citation bot, Videot43, Felyza, Frankenpuppy, Tekks, Xqbot, Timir2, Intelati, Apothecia, Jeffrey Mall, DSisyphBot, Rlmrace, Texasrocks, Aa77zz, J04n, GrouchoBot, ChristopherKingChemist, ProtectionTaggingBot, Omnipaedista, Shirik, RibotBOT, The Wiki ghost, Moxy, GainLine, Shadowjams, AlimanRuna, Some standardized rigour, Timehigh, Captain-n00dle, GT5162, 10buttz, LucienBOT, HJ Mitchell, Simpler12, Kmdouglass, Citation bot 1, Intelligentsium, AstaBOTh15, Pinethicket, I dream of horses, Elockid, HRoestBot, Edderso, MimisHandler, Calmer Waters, A8UDI, Killacoll123, Σ, KES47, Papapu, Shanmugamp7, Jauhienij, Mj455972007, K1lla abbas, FoxBot, TobeBot, Trappist the monk, كاشف عقيل, Barneycleo, Stefano Garibaldi, Vrenator, Mewilliams95, Shosh3333, Amkilpatrick, Justine Hill, Canuckian89, Yorubaspartan, Suffusion of Yellow, Myplace345, Jynto, Tbhotch, Jrc2005, Reach Out to the Truth, EMacG, DARTH SIDIOUS 2, AXRL, Maaayowa, RjwilmsiBot, Q3w3e3, Beki Hellens, Mdznr, Androstachys, DASH-Bot, EmausBot, Cpeditorial, WikitanvirBot, Gfoley4, Snow storm in Eastern Asia, Talktomarie1, Gn96, Kerivoula, Matty616, Tommy2010, Ballboy070707, Wikipelli, 6AND5, Dspoel, Shuipzv3, SimBioUSA, John Mackenzie Burke, Michaeljwilson, Rcjohns1, H3llBot, EWikist, Jarodalien, Wayne Slam, U+003F, Sky380, Beastinbman, L Kensington, Loganloganlogand, Donner60, Beta cafe, ChuispastonBot, Snehalshekatkar, ClueBot NG, Gareth Griffith-Jones, Apdang, MelbourneStar, HistoricPandaBear, Gilderien, Eltronio, Eseosala, Bjhiggins, Widr, Antiqueight, Flomenbom, Robertohaas, Timflutre, Bradley1545, Helpful Pixie Bot, Lalo1121, HMSSolent, Lowercase sigmabot, BG19bot, Maxwell754, Happylama101, The Banner Turbo, Dakarai99, 13shepht, JacobTrue, Gautehuus, Mark Arsten, Dhruvrajgohil, FutureTrillionaire, Dr. Krishna Prasad Nooralabettu, Mrspkapila, KhoousesWiki, BattyBot, Johnmichaeloliver, Studlaff, Miszatomic, A2-25, Pratyya Ghosh, Mrt3366, Dexbot, Webclient101, Dannyboy600, Lugia2453, Andrew Murkin, Joeinwiki, Mark viking, Epicgenius, Forever elementary school student, Bdoc13, The Sackinator, Blythwood, Jakec, DavidLeighEllis, Evolution and evolvability, Ugog Nizdast, Seppi333, Ginsuloft, Amr94, IsaacOcean, PierreFG5, Suderpie, SimonWombat8, Anrnusna, HYH.124, Ziahlao, Chloemthomson, EngScott, JohnGormleyJG, Monkbot, Rakeshyashroy, PointsofNoReturn, Y-S.Ko, KasparBot and Anonymous: 1334

- **Topoisomerase** *Source:* https://en.wikipedia.org/wiki/Topoisomerase?oldid=688809499 *Contributors:* Youssefsan, Shyamal, 168..., Fuelbottle, Giftlite, Jfdwolff, G3pro, Thorwald, Discospinster, MBisanz, Arcadian, Passw0rd, Anthony Appleyard, Axl, ClockworkSoul, TenOfAll-Trades, Capecodeph, Japanese Searobin, Stemonitis, Woohookitty, Chochopk, FlaBot, Gurch, Intgr, Banus, SmackBot, Melchoir, Dauto, Chris the speller, Kunalmehta, Miguel Andrade, Greg carter, Kcordina, Wen D House, Drphilharmonic, Clicketyclack, FrozenMan, Jafield, DabMachine, CRGreathouse, Chrumps, A876, Thijs!bot, Gharmon, SGGH, Orthank, SenorKristobbal, Pixelface, Alan Meyer, East718, Einstein76, Antorjal, NunoAgostinho, Cdcdoc, Stevenjohnkiss, CommonsDelinker, Pdeitiker, Nbauman, Boghog, LordAnubisBOT, EquationDoc, Alexhlau, Pdcook, Ummel, VolkovBot, Linefeed, Enozkan, A4bot, Broadbot, Luuva, AlleborgoBot, Lorangriel, Phe-bot, Forluvoft, Alexbot, PixelBot, Ngebendi, DumZiBoT, Vojtěch Dostál, Addbot, DanaStreet, CarTick, Luckas-bot, Ptbotgourou, Berkay0652, Rubinbot, Materialscientist, Citation bot, ArthurBot, Xqbot, LittleWink, Wikididact, EmausBot, The Mysterious El Willstro, Truthortruth, Kuba fischer, Nwoonet, Pommiery, Helpful Pixie Bot, Iyentra Rasonica, PhishIsChochmah, Plantdrew, ADeck17-27, JonathonSimister, Sermadison, Freddiecrane1, FoCuSandLeArN, Earl Moss, Dmander05, Brainiacal, Monkbot, Barani raam and Anonymous: 81

- **Polymerase** *Source:* https://en.wikipedia.org/wiki/Polymerase?oldid=636253055 *Contributors:* Dwmyers, Lexor, JWSchmidt, Wilke, Erik Garrison, Arcadian, Jag123, Seans Potato Business, Lkinkade, GregorB, Eras-mus, Mohawkjohn, Pumeleon, YurikBot, Wavelength, Kantokano, Lijealso, KnightRider~enwiki, MattieTK, InverseHypercube, Ultramandk, Bluebot, UniPharm, Dreadstar, Ywith, JHunterJ, TastyPoutine, Citicat, RelentlessRecusant, DJBullfish, Thijs!bot, Mglg, JPG-GR, CommonsDelinker, Pekaje, Nbauman, STBotD, Speciate, TXiKiBoT, A4bot, AlleborgoBot, Miqademus, ClueBot, DeltaQuad, Blackdrago56, Legobot, TaBOT-zerem, JackieBot, El Mayimbe, Dcirovic, ZéroBot, John Mackenzie Burke, Pnolan1, Teaktl17, ClueBot NG and Anonymous: 32

- **Retrovirus** *Source:* https://en.wikipedia.org/wiki/Retrovirus?oldid=691332214 *Contributors:* Bryan Derksen, Ed Poor, Andre Engels, Josh Grosse, Youssefsan, William Avery, Azhyd, David spector, Someone else, Edward, Lexor, Shyamal, Nina, Gbleem, Muriel Gottrop~enwiki, Cyan, Andres, AhmadH, Ec5618, Fuzheado, Steinsky, Furrykef, Lumos3, Robbot, Wikibot, Xanzzibar, Giftlite, Nunh-huh, Marcika, Unconcerned, Gubbubu, PDH, DragonflySixtyseven, Trevor MacInnis, Rich Farmbrough, Drano, Vsmith, Spundun, Bender235, AyJay, RoyBoy, Bobo192, BrokenSegue, Arcadian, Joe Jarvis, La goutte de pluie, Dillee1, Allstarzero, Monado, Malo, Arag0rn, Ickle~enwiki, RyanGerbil10, JarlaxleArtemis, Oliphaunt, Vineet KewalRamani, EnSamulili, Schzmo, Eras-mus, SDC, Sci guy, Ketiltrout, Rjwilmsi, Smoe, Zambani, FlaBot, HenrikB, Margosbot~enwiki, Diza, Chobot, Ghismax, WriterHound, YurikBot, RobotE, Huw Powell, Eleassar, Brandon, Redtails, William Graham, Zwobot, DeadEyeArrow, Bota47, Ke6jjj, WAS 4.250, Light current, Closedmouth, Maristoddard, ASmartKid, TechBear, SmackBot, Goldfishbutt, Eskimbot, Chriswig, Quinsareth, RDBrown, Hichris, Can't sleep, clown will eat me, Cophus, MattOates, HeteroZellous, Aldaron, Krich, G716, SashatoBot, DA3N, Kuru, Mat8989, Seb951, Xdx~enwiki, Serephine, Dhp1080, Peyre, Beefyt, BranStark, TerryE, Yodin, Andreas Rejbrand, Richard75, Courcelles, Tawkerbot2, Im.a.lumberjack, WeggeBot, SeanMon, Thijs!bot, Epbr123, Kablammo, Mojo Hand, Marek69, Peter Znamenskiy, Doc Comic, AntiVandalBot, MrMarmite, Xuchilbara, Fayenatic london, Richiez, JAnDbot, Giler, Connormah, JNW, Jiejunkong, WhatamIdoing, Emw, DerHexer, A-Nottingham, TechnoFaye, Fconaway, RockMFR, Manticore, J.delanoy, Nbauman, Rod57, Mikael Häggström, Belovedfreak, Zumlin, Deor, VolkovBot, TXiKiBoT, Edward Bower, Earthdirt, MCTales, AlleborgoBot, Graham Beards, Yintan, DevOhm, Fratrep, Sunrise, Cyfal, Ram rottenly, RobinHood70, Touchstone42, Martarius, ClueBot, Gadiandi, Bubbletruble, Aaroncorey, MagyarFiatalember, Zack wadghiri, DragonBot, Abrech, Jerry Zhang, Ykhwong, 1ForTheMoney, Derekstm, Shadow600, Vojtěch Dostál, Addbot, DOI bot, Captain-tucker, Binary TSO, Wamerocity, MrOllie, Tide rolls, Sammy theeditor, Luckas-bot, Yobot, THEN WHO WAS PHONE?, Nallimbot, Götz, Eteklema-GMU, Materialscientist, Swithrow2546, Aff123a, Citation bot, Dkabban-GMU, Lapabc, Masterpra2002, Xqbot, Capricorn42, Km2452-GMU, ArcadianOnUnsecuredLoc, GrouchoBot, DVMresearcher, Omnipaedista, RibotBOT, Axelmctavish, FrescoBot, Citation bot 1, DrilBot, Pinethicket, The Last Baron, Adrians executive, Aytrus, Jesse V., RjwilmsiBot, Uanfala, Whywhenwhohow, EmlivOlcano, Xetijelut, TuHan-Bot, Rupertsciamenna, Savh, Mhahnel, Mjcdowling, Dafonzdcom, Obotlig, Goretexguy, ClueBot NG, Frietjes, Fjalnes, Gcc111, Jkaralis1925, BG19bot, Phageghost, Czernilofsky, Ccevo2011, Basketcase87, ChrisGualtieri, Lawenlerk, ComfyKem, Rjdodger, Adiecoly, Mrdavis21, Katn*ss, Jamesikim, Monkbot, Treacles, KasparBot and Anonymous: 246

- **Telomerase** *Source:* https://en.wikipedia.org/wiki/Telomerase?oldid=691221659 *Contributors:* AxelBoldt, Netesq, Michael Hardy, Lquilter, GTBacchus, Karada, MichaK, Slakhan, Kaal, Frazzydee, Phil Boswell, Giftlite, Gotanda, Alvestrand, AdamJacobMuller, Andycjp, DMG413, Xrchz, AAAAA, Discospinster, Rich Farmbrough, Smyth, Bender235, Cmdrjameson, Arcadian, La goutte de pluie, Nhandler, Ramujana, Zaxxon de, ClockworkSoul, Tycho, TenOfAllTrades, Memenen, Ron Ritzman, Marcelo1229, Woohookitty, LizardWizard, Mindmatrix, Benbest, Astrowob, Arrkhal, Eras-mus, Achim Raschka, BorisTM, Tbird20d, Rjwilmsi, Renamed user 8262690166681, Bhadani, Gsp, Dullfig, EnDumEn, Chobot, WriterHound, Whosasking, YurikBot, Bencze Gyenge, TangParadise, James Bedford, Dysmorodrepanis~enwiki, Welsh, Open2universe, Mumuwenwu, Katieh5584, SmackBot, Dakoman, Bmord, Ashermadan, Jjalexand, TheAntitype, Frap, OrphanBot, Rrburke, Madman2001, Monotonehell, Dcteas17, Drphilharmonic, FrozenMan, Serephine, Ambuj.Saxena, IvanLanin, Twas Now, ChrisCork, Protiek, Dia^, CmdrObot, Svetachakrabarti, Cydebot, Ppgardne, Hwttdz, Crum375, SomethingCatchy, Thijs!bot, JustAGal, Oleykin, MoogleDan, TimVickers, Tlabshier, Darklilac, Pixelface, Lfstevens, Ozperp, MER-C, VoABot II, Decembermouse, Balloonguy, Destynova, Enquire, Rickard Vogelberg, AstarothCY, R'n'B, Pekaje, Fconaway, Boghog, Xris0, Rod57, ElinneaG, Sprunk, STBotD, Davrids, VolkovBot, Mirrordor, Mabidex, TXiKiBoT, Berichard, JCRansom, AlleborgoBot, Adr11iano, SieBot, Mikemoral, Noisy Hand, Bgordski, Henk Poley, Cmale, Plastikspork, Niceguyedc, Dylan620, Alexbot, Tundrill, Cowwaw, Bproman, Dana boomer, Sierrasciadmin, KarakasaObake, Noctibus, Dfoxvog, Kace7, Good Olfactory, Snapperman2, Addbot, Piz d'Es-Cha, DOI bot, Fyrael, Freakonomicsfan, Bernstein0275, Deamon138, Bouncingball2, Tide rolls, Icons789, Saintjoep, ערן, زرشک, Ben Ben, Luckas-bot, Yobot, Tiddlypeep, Theserialcomma, Legobot II, Vinodksingh1111, AnomieBOT, Superpoopoo, Citation bot, Trilliumroots, Flavius.Aettius, Makeswell, Kylelovesyou, FrescoBot, Citation bot 1, Citation bot 4, Jonesey95, RedBot, Bohalrantipol, Jeangabin, Alexorella, Trappist the monk, NickVertical, Autumnox, Time9, Theo10011, Amkilpatrick, Vitaliy0001, Feduchin, Watcher0911, RjwilmsiBot, Adarakdjian, TjBot, EmausBot, Wiki.Tango.Foxtrot, Lotez, Gnulinux, Dcirovic, TYelliot, Patch101, FeatherPluma, Roanotto, ClueBot NG, Peergorm, BG19bot, I7laseral, Dormantfreedom, Pndanilov, Emskorda, BattyBot, Dexbot, Everything Is Numbers, Ivanjoksimovic, Alexwho314, BradYard, EtymAesthete, Anrnusna, AlmaMer, Pdavis9396, Maethes, Monkbot, MattLBeck, Ip1673, Nith117, Noora Al Balushi, Dr Scott Cohen, Mahfuzur rahman shourov and Anonymous: 201

- **RNA world** *Source:* https://en.wikipedia.org/wiki/RNA_world?oldid=691196286 *Contributors:* Tobias Hoevekamp, Magnus Manske, Joao, Bryan Derksen, Taw, Deb, Graft, Ewen, Someone else, Lexor, Bueller 007, Julesd, Timwi, Steinsky, Wetman, Donarreiskoffer, Robbot, Korath, Aetheling, Jooler, Timemutt, Giftlite, Fastfission, Dratman, Duncharris, Christopherlin, Gadfium, Pgan002, Julianonions, JohnArmagh, Deglr6328, Gazpacho, Jayjg, Vague Rant, Vsmith, Bender235, Reinyday, BrokenSegue, Shenme, Viriditas, R. S. Shaw, Mpvdm, Cohesion, I9Q79oL78KiL0QTFHgyc, KBi, Srlasky, Elchupachipmunk, SMesser, Gsandi, Hunter1084, Spellcheck, Ceyockey, Mel Etitis, Woohookitty, Havermayer, GeorgeOrr, Mandarax, Drbogdan, Rjwilmsi, Margosbot~enwiki, NawlinWiki, Seb35, Killdevil, DeadEyeArrow, Leptictidium, Light current, Petri Krohn, Ryanhupka, Xanin, SmackBot, Dale Keiger, Zephyris, Skizzik, Elagatis, IlliniWikipedian, Audriusa, Royboycrashfan, Makemi, John D. Croft, Tktktk, Ben Moore, Stevebritgimp, Novangelis, Dan Gluck, Ossipewsk, JoeBot, Twas Now, Amakuru, CmdrObot, Metzenberg, Ppgardne, Michael C Price, Narayanese, InfoCan, Greeneto, Wisebridge, Crum375, Oleksii0, Thijs!bot, Epbr123, Headbomb, Speedyboy, Z10x, BlytheG, WinBot, Stefanwikipedia~enwiki, TimVickers, Smartse, LinkinPark, Igodard, Kevin Langdon, Tstrobaugh, Magioladitis, Professor marginalia, Engineman, BatteryIncluded, Teebol, GetAgrippa, Talon Artaine, Edward321, DGG, Agricolae, CommonsDelinker, Knowledge Junkie, Ian.thomson, Notreallydavid, STBotD, Speciate, Joeoettinger, TXiKiBoT, Ychastnik APL, Haplochromis, Northfox, AlleborgoBot, Awils1, Graham Beards, Agesworth, Aillema, Mimihitam, Boppet, Jdaloner, Mtrinque, Sunrise, Kalidasa 777, Hilzofkikiro, Sfan00 IMG, Mild Bill Hiccup, Niceguyedc, Sid-Vicious, Peteruetz, Carba, Excirial, AC+79 3888, MKDolphin, Vojtěch Dostál, Avoided, SilvonenBot, Orrad, Addbot, DOI bot, RangeWriter, Griselquiroz, Mentisock, Bernstein0275, Glane23, 5 albert square, Lightbot, Jarble, Ettrig, Luckas-bot, Yobot, Legobot II, AnomieBOT, Sz-iwbot, Citation bot, LilHelpa, Obersachsebot, RibotBOT, Tales23, Shadowjams, Middle 8, HighFlyingFish, FrescoBot, Citation bot 1, Jonesey95, Mbellish, Lemmiwinks2, Trappist the monk, Sven Jähnichen, Lotez, Dcirovic, ZéroBot, AManWithNoPlan, Peterebun, Brandmeister, Spr3sso, Good Samarian, ClueBot NG, Helpful Pixie Bot, Bibcode Bot, BG19bot, Ashaels, HP-Biochemie, BattyBot, Peggy hopper, ChrisGualtieri, TippyGoomba, Rozi Sh, Strauss12, Theemathas, Comp.arch, SzostakJack, AioftheStorm, Anrnusna, Chaya5260, Monkbot, Lor, Quintar360, Paleolithicus, Dienerto, Nn9888, CCevol2015 and Anonymous: 164

- **DNA profiling** *Source:* https://en.wikipedia.org/wiki/DNA_profiling?oldid=691547237 *Contributors:* AxelBoldt, Mav, Ap, Hephaestos, Someone else, Twilsonb, Patrick, Michael Hardy, Zashaw, Lexor, Gabbe, Karada, 168..., Ahoerstemeier, DavidWBrooks, JWSchmidt, Julesd, Jengod, Feedmecereal, Emperorbma, Steinsky, DJ Clayworth, Freechild, Maximus Rex, Pakaran, Twang, Phil Boswell, Branddobbe, Robbot, Schutz, Rorro, Acegikmo1, Mervyn, SoLando, Pengo, DocWatson42, Inter, Fastfission, Obli, Peruvianllama, Everyking, Curps, Michael Devore, Duncharris, Frankthetank, AlistairMcMillan, Golbez, Christopherlin, Quadell, Beland, DragonflySixtyseven, Thincat, Arcturus, Snicklefritzdee, Klemen Kocjancic, Discospinster, Rich Farmbrough, Cacycle, Vsmith, Naive cynic, Spencer BOOTH, Serpens, LindsayH, ESkog, Charm, Aude, RoyBoy, Triona, Bobo192, Stesmo, Reinyday, Viriditas, Dungodung, Twobells, Alansohn, Guy Harris, Linmhall, AzaToth, Wtmitchell, TaintedMustard, Toytown Mafia, TenOfAllTrades, Dan100, Richwales, Shimeru, Woohookitty, Mindmatrix, Nuggetboy, Awostrack, Smor4860, MONGO, Philbarker, MarcoTolo, Mandarax, RichardWeiss, BorisTM, Josh Parris, Rjwilmsi, Tizio, Nightscream, Jake Wartenberg, The wub, Keimzelle, Ucucha, Titoxd, Ground Zero, Cherubino, Gurch, Alphachimp, Chobot, Gwernol, The Rambling Man, Wavelength, TexasAndroid, Personman, RattusMaximus, Sekiyu, Phantomsteve, RussBot, Petiatil, John Quincy Adding Machine, Icarus3, Koffieyahoo, Severa, Anonymous editor, Chaser, RadioFan2 (usurped), Scubafish, Gaius Cornelius, Ksyrie, Shaddack, Rsrikanth05, Wiki alf, Motor.on, Maverick Leonhart, Yoninah, Daniel Mietchen, Jpbowen, Syrthiss, DeadEyeArrow, Bota47, Maunus, Sandstein, Uncke Herb, Theda, E Wing, Abune, GraemeL, GIR, TBadger, Phil Holmes, Markbenecke, CIreland, Sardanaphalus, MacsBug, KnightRider~enwiki, SmackBot, Eperotao, Hyper84, Prodego, Martin.Budden, Pgk, Stepa, Piccadilly, Bugg42, Imzadi1979, HalfShadow, Apers0n, Gilliam, Chris the speller, Bidgee, Stubblyhead, Robth, BW95, Can't sleep, clown will eat me, Jefffire, DéRahier, Skidude9950, Abrahami, Dharmabum420, Iapetus, Sens08, TedE, Wjcollier07, DMacks, Soarhead77, Drunken Pirate, Ceoil, TenPoundHammer, Madeleine Price Ball, The undertow,

SashatoBot, Paternity Test, Kuru, Akendall, Sjock, Tazmaniacs, IronGargoyle, RomanSpa, Werdan7, Beetstra, Funkboy3, Dr.K., Caiaffa, ShakingSpirit, StuartMcIntyre, Asyndeton, Masem, Mgummess, Iridescent, Joseph Solis in Australia, JoeBot, Lcamtuf, Marysunshine, Courcelles, Tawkerbot2, Alexbrewer, JForget, CmdrObot, Tanthalas39, Ron Barker, Aerternum8, NickW557, Fordmadoxfraud, HalJor, Cydebot, Kupirijo, Abeg92, A876, MC10, YanWong~enwiki, SyntaxError55, Gogo Dodo, Red Director, Danorton, GRBerry, Chasingsol, Marketing250, Marandbon, Chrislk02, Bungle, Noahsachs, Click23, Mattisse, Anyo Niminus, Epbr123, Mercury~enwiki, Wikid77, Qwyrxian, Peter johnson4, Russ47025, Mojo Hand, Zé da Silva, Nick Number, Dkropf, Ct2288, LachlanA, AntiVandalBot, MelchiorZ, F-451, TimVickers, Smartse, Anikin3, North Shoreman, Qwerty Binary, CNicol, Figma, MER-C, Rsanaie, Hoomanator, Tstrobaugh, Yahel Guhan, Propaniac, TransControl, Shumdw, VoABot II, JNW, Appraiser, Think outside the box, Catgut, WhatamIdoing, Allstarecho, Patstuart, Flowanda, MartinBot, Rettetast, SuperMarioMan, Ravichandar84, CommonsDelinker, LedgendGamer, J.delanoy, Uncle Dick, Maurice Carbonaro, Justin, Ginsengbomb, DD2K, SlowJog, Manuel Yayo, Notreallydavid, Woodega, Mikael Häggström, Gill110951, Tubeyes, Nwbeeson, Mufka, FJPB, Olegwiki, Lonesomefighter, Jamesontai, Lucifero4, Jimokay, Zophra, Kvdveer, WLRoss, JBarno, CardinalDan, Hugo999, Nitroshockwave, VolkovBot, ABF, Lldancer91, James Callahan, Mgalle, Abberley2, TXiKiBoT, Tameeria, Rei-bot, Xavierschmit~enwiki, Qxz, Someguy1221, Cmcnicoll, Enviroboy, Blaaake, Doc James, Nagy, W4chris, NHRHS2010, SylviaStanley, GreaterWikiholic, SieBot, Tresiden, Bastiche, Tiddly Tom, Caltas, Paulbrock, Flyer22 Reborn, Vifsm, Yerpo, Hardware Hank, Steven Crossin, Lightmouse, Poindexter Propellerhead, Rhsimard, Macy, Junger01, CHallam, CharlesBrenner, Angelo De La Paz, ClueBot, Binksternet, Kangdawei, Kennvido, Narom, RashersTierney, Enthusiast01, Mild Bill Hiccup, Boing! said Zebedee, LizardJr8, Dubitante, Michaplot, Puchiko, DragonBot, Sun Creator, Islaammaged126, Keepmytubes, Arjayay, Iohannes Animosus, Domikins, ElectricLemon, Aitias, Versus22, The Baroness of Morden, Skipper500, Johnuniq, Oore, Tdslk, DumZiBoT, Jamesflint, Otr500, BarretB, PseudoOne, Jovianeye, Avoided, Ixymapoe, Thatguyflint, Kembangraps, Addbot, Spongeboblover124, Brumski, PaleWhaleGail, Doom7987, DougsTech, Ronhjones, TutterMouse, S-MorrisVP, Fieldday-sunday, Glane23, Debresser, 5 albert square, TheFreeloader, Tassedethe, Justpassin, PRL42, Unibond, Tide rolls, Captain Obvious and his crime-fighting dog, HerculeBot, आशीष भटनागर, Luckas-bot, Yobot, Jonathan Haas, Librsh, ArchonMagnus, SnowsCode, AnomieBOT, Momoricks, Jim1138, Galoubet, Kingpin13, Bluerasberry, Materialscientist, Wwc93, OllieFury, Eumolpo, Quebec99, LilHelpa, Xqbot, Anneman, Capricorn42, Bugo30, J04n, Wilsonchas, 7434be, Jetjaxon5, Moxy, A.amitkumar, BloodGrapefruit2, FrescoBot, Tobby72, Pepper, Skyk98, JMS Old Al, InspectorSands, Haeinous, McPavement, HamburgerRadio, Þjóðólfr, Pinethicket, Vicenarian, 123imran, Universalgenetics, Sarah00001, Pv86, Nish242, Serols, SpaceFlight89, Tamoxidiva, HarringtonSmith, Banej, Dnalabsindia, Jenwen28, Keri, TobeBot, Animalparty, Lotje, Dinamik-bot, Vrenator, Clarkcj12, 808OG33, PleaseStand, Mean as custard, VernoWhitney, EmausBot, WikitanvirBot, Ajraddatz, NinjaTazzyDevil, Vanished user zq46pw21, Solarra, Thecheesykid, JSquish, Josve05a, Dolovis, Rugbyboroughman, Lateg, Hotrodizfilthy, A930913, H3llBot, Wayne Slam, Teksus, Ventus55, Brandmeister, L Kensington, Carmichael, Overstreets, ClueBot NG, Pallap, Serasuna, Don.doe, EGR-Valve, Baconchopface, Widr, NellieBlyMobile, Theopolisme, Ejshear, Helpful Pixie Bot, HMSSolent, Spu2011, Da5id403, BG19bot, Northamerica1000, Francescamilanolingue, Cusop Dingle, Writ Keeper, Damonatorv31, Dezastru, KugelaP, Stigmatella aurantiaca, Ddcm8991, Cyberbot II, ChrisGualtieri, Mediran, Tankhead2, Ssbbplayer, TwoTwoHello, Joshtaco, Frosty, Darbyday, Lucjenkins, Taylordavison, Vanamonde93, I am One of Many, PhantomTech, Rachel Cornell, Angel360xx, SoftMachine, Manul, Mendisar Esarimar Desktrwaimar, Monkbot, NewEnglandDr, Username20014, Χρυσάνθη Λυκούση, Sechelonline, KH-1, Legislature856, CSIGUY82, Charlesdeman69, Daddy1022, Senecarocks, Craftwerker, Alexandritechrysoberyl, Sneptunebear16 and Anonymous: 752

- **DNA digital data storage** *Source:* https://en.wikipedia.org/wiki/DNA_digital_data_storage?oldid=679455630 *Contributors:* Edcolins, Alexf, Brianhe, Jérôme, Neo-Jay, Woodshed, Hebrides, Kudpung, Antony-22, Fences and windows, Oxfordwang, Duncan.Hull, Glst2, Cdessimoz, XLinkBot, Citation bot, The Interior, Steve Quinn, Ирина Реброва, Dewritech, BG19bot, ChrisGualtieri, Ssscienccce, Dexbot, Skahaggus, Ananthkamath1995, CensoredScribe, Openpaul, TheEpTic and Anonymous: 7

- **DNA sequencing** *Source:* https://en.wikipedia.org/wiki/DNA_sequencing?oldid=690950789 *Contributors:* AxelBoldt, Magnus Manske, Shyamal, Kku, Paul A, Glenn, Omegatron, Samsara, Ketil, Pgan002, Beland, PDH, NeffK, Gipelttil, Robin klein, Abdull, Discospinster, Rich Farmbrough, David Schaich, Bender235, RoyBoy, Fuxx, Pharos, Terrycojones, Loris, Seans Potato Business, Wtmitchell, Velella, ClockworkSoul, Yurivict, Ceyockey, Abanima, Woohookitty, 25or6to4, Benbest, Robert K S, Frankatca, MarcoTolo, Marudubshinki, Mandarax, RichardWeiss, Rjwilmsi, Eyu100, Bgwhite, YurikBot, Wavelength, Charles Gaudette, Kymacpherson, Rsrikanth05, Sentausa, NawlinWiki, Dysmorodrepanis~enwiki, Ondenc, Yoninah, JHCaufield, Kkmurray, Theda, Closedmouth, Dupz, Lundse, SmackBot, TestPilot, Gilliam, Ohnoitsjamie, ERcheck, Chris the speller, RDBrown, George Church, DHN-bot~enwiki, Wcbpolish, Furby100, Rrburke, Akriasas, MBCF, Schnarr, Abizar, Clicketyclack, Qwerty0, Madeleine Price Ball, Victor D, Park3r, Mgiganteus1, Ben Moore, Suirauqa, Smith609, Mathsci, Novangelis, Dl2000, ToastyMallows, Twas Now, Lenoxus, Tawkerbot2, Shrimp wong, Patho~enwiki, CmdrObot, Pathh, Markchiang, Pgr94, Krobison13, Peripitus, Robinatron, Fedra, Gogo Dodo, Narayanese, Crana, SomethingCatchy, Konrad Foerstner, HappyInGeneral, Headbomb, Trevyn, Marek69, Ejwong, Dkropf, Rees11, Pixelface, Dougher, Lklundin, Res2216firestar, Erxnmedia, JAnDbot, SNP, Magioladitis, VoABot II, Bhar100101, Cspenser, Avicennasis, DWeissman, BatteryIncluded, Adrian J. Hunter, Spellmaster, Su-no-G, Genometer, DerHexer, Vinylmeister, Pvosta, MartinBot, Nono64, Cinnamon colbert, Petter Bøckman, Tgeairn, CFCF, Kadoo, Adavidb, Boghog, Dai mingjie, Lantonov, Eimeim, Mikael Häggström, SteveChervitzTrutane, NewEnglandYankee, Nwbeeson, SJP, 83d40m, Jbingham, Thisisborin9, Mercurywoodrose, Malljaja, NPrice, Agricola44, Sintaku, Bjorn9800991, UnitedStatesian, Jvbishop, Cremepuff222, VanishedUserABC, Lmsutton1, SHL-at-Sv, ShiftFn, Noveltyghost, ConfuciusOrnis, God Emperor, RJaguar3, Yintan, Zaimon~enwiki, Tammyz06, Oxymoron83, Svick, Brice one, CultureDrone, Owlmonkey, Graminophile, Naturespace, ClueBot, NickCT, Rosyaraur, Niceguyedc, Xenon54, Peteruetz, RRphys, Howie Goodell, Excirial, Iohannes Animosus, Schtina~enwiki, Johnuniq, DumZiBoT, Duncan, MystBot, Dnacash, Destrii, Kembangraps, Red89011, Addbot, Jacopo Werther, DOI bot, Tsmith423, DutchDevil, MrOllie, NailPuppy, Akita86, Tide rolls, Luckas Blade, Ettrig, Abduallah mohammed, Luckas-bot, Yobot, TaBOT-zerem, Legobot II, Theropod, Zalumon, Anradt, Azcolvin429, Eric-Wester, AnomieBOT, İnfoCan, BaChev, Enzymes-GMU, Jim1138, Gawi, Kern3020, Materialscientist, Citation bot, Shcha, Xqbot, Hozelda, 5piggies, Sirozha, Keaka411, Akjosh, Hoe4sho, FrescoBot, DoctorDNA, Paine Ellsworth, Ace of Spades, Citation bot 1, Pinethicket, PrincessofLlyr, Jonesey95, MastiBot, Xeworlebi, Vyahhi, Trappist the monk, SchreyP, Xook1kai Choa6aur, HelenOnline, Jonkerz, Vrenator, LilyKitty, DadOfBeanAndBug, Manuelcorpas, ANSMID, Amkilpatrick, Drcsk, Minimac, Sideways713, Mikestorck, RjwilmsiBot, Helwr, J36miles, EmausBot, Cpeditorial, Lord of Ruin, Dewritech, RA0808, Minimac's Clone, Marscher23, Dcirovic, HiW-Bot, Mooiboys, Shuipzv3, NicatronTg, John Mackenzie Burke, Aeonx, Sengupta 98, AManWithNoPlan, Thine Antique Pen, Rcsprinter123, Bhumity, Sal9820, Bill william compton, Chuispaston-Bot, Javnad, Wakebrdkid, MacStep, Rocketrod1960, Sunmylondon10, Cdslipp, Sarahburge, Woodsrock, ClueBot NG, Ebudishorn, Mesoderm, Jmgrants, Suspencewl, Paulhmiller, Roberto2004t, Leon mei, Bibcode Bot, BG19bot, CatPath, Californiadreams, MusikAnimal, AvocatoBot, Hellohanxin, MGThomas01, Notallthere2day, Estevezj, BattyBot, PhilipWongHS, Trepan77, ChrisGualtieri, Saltwolf, Chialuchen, JesseAlan-

Gordon, JYBot, Mwsal, Warriormeek, Mikepetroff, Mogism, Leburnett, Tbuzz, Dmitry Dzhagarov, Mvschneider, Thusz, Szssdy, Rfassbind, Phupe1, CsDix, Mwahahahahqwert, Rybec, Cynthia MEDG, Ginsuloft, HelgeUK, LibraryStudent24, Arnauddesfeux, Anrnusna, Sankohm, Stubbelm, Mahusha, Monkbot, Leegrc, Poohlalong, Jpalme53, Svijayasriharsha, Jessie Thuswaldner, YankerWanker, Ruslana Tyt, Laravmarks, DMLapato and Anonymous: 332

• **Macromolecule** *Source:* https://en.wikipedia.org/wiki/Macromolecule?oldid=690065112 *Contributors:* Youssefsan, Heron, Dwmyers, D, Lexor, Dcljr, Alfio, 168..., Looxix~enwiki, Prometheus~enwiki, Evercat, Nufy8, Robbot, Peak, Stewartadcock, Meelar, Diberri, Marc Venot,Alan Au, Kandar, Utcursch, LiDaobing, Antandrus, Onco p53, G3pro, Karol Langner, Rdsmith4, Icairns, Discospinster, Bender235, ESkog,Rgdboer, RoyBoy, Gary123, Dirac1933, Miaow Miaow, V8rik, Rjwilmsi, SMC, Dar-Ape, FlaBot, Gurch, Chobot , Dj Capricorn, Gwernol,YurikBot, Borgx, Bhny, Stephenb, Wikimachine, NawlinWiki, Joel7687, Anetode, D. F. Schmidt, CWenger, DVD R W, Luk, TestPilot,Jacek Kendysz, Eskimbot, Jab843, BiT, M stone, Gilliam, Drn8, Tree Biting Conspiracy, Moshe Constantine Hassan Al-Silverburg, Steven-mitchell, Aicheung, DMacks, Mion, Petr Kopač, Kuru, John, AstroChemist, JHunterJ, SmokeyJoe, Aeons, Lazulilasher, Rifleman 82, JFree-man, Skittleys, Christian75, VPliousnine, Thijs!bot, Epbr123, Marek69, Dawnseeker2000, AntiVandalBot, Rcej, TimVickers, Blacksun1942,JAnDbot, MER-C, Belg4mit, Efbfweborg, VoABot II, Careless hx , Soulbot, Allstarecho, Middlenamefrank, CommonsDelinker, J.delanoy,Trusilver, Nigholith, NewEnglandYankee, Cometstyles, RB972 , Geekdiva, Squids and Chips, CardinalDan, Idioma-bot, VolkovBot, IreneRingworm, Chienlit, Philip Trueman, TXiKiBoT, JhsBot, Autodidactyl, Nacho007, RaseaC, NHRHS2010, Rhinosaur0220, Kbrose, Botev,SieBot, Yintan, Flyer22 Reborn, MaynardClark, OKBot, Ascidian, Ptr123, ClueBot, The Thing That Should Not Be, Cp111, Yoshi Cano-pus, MARKELLOS, Blanchardb, Lensicon~enwiki, Aua, Excirial, Vriend, Versus22, Jytdog, SilvonenBot, Alexius08, Addbot, Xp54321,Narayansg, AkhtaBot, Quercus solaris, Bouncingball2, Tide rolls, Lightbot, Legobot, Luckas-bot, Yobot, 2D, Fraggle81, KamikazeBot,AnomieBOT, Jim 1138, Piano non troppo, Sisyph, AdjustShift, Materialscientist, Citation bot, Srinivas, Xqbot, Schmutz MDPI, Zad68,Marcibieber , DSisyphBot, Champlax, Danacademic, Coffeenutter, DivineAlpha, Cannolis, Citation bot 1, Drasek Riven, Im4evrsmrt, Tom.Reding,RedBot, TobeBot, Ajalygirl, Vanished user aoiowaiuyr894isdik43, GoingBatty, RA0808, Wikipelli, Dcirovic, JSquish, 1234r00t, Amargnanda,Hccc, Δ, DeeMoneyy23, Whoop whoop pull up, ClueBot NG, Gareth Griffith-Jones, O.Koslowski, 336, Widr, Helpful Pixie Bot, Bibcode Bot,Trhgr22, NotWith, Vanischenu, AutisticCatnip, Makecat-bot, Lugia2453, Sriharsh1234, Acetotyce, Jakec, Evolution and evolvability, Amr94, DragonZMP, JoshXD, Snavesyhr, KasparBot and Anonymous: 244

42.8.2 Images

• **File:10_large_subunit.gif** *Source:* https://upload.wikimedia.org/wikipedia/commons/c/c6/10_large_subunit.gif *License:* Public domain *Contributors:* <a data-x-rel='nofollow' class='external text' href='http://www.pdb.org/pdb/static.do?p=education_discussion/molecule_of_the_month/ pdb10_1.html'>*Molecule of the Month*</a> at the RCSB Protein Data Bank *Original artist:* Animation by David S. Goodsell, RCSB Protein Data Bank

• **File:10_small_subunit.gif** *Source:* https://upload.wikimedia.org/wikipedia/commons/3/3d/10_small_subunit.gif *License:* Public domain *Contributors:* <a data-x-rel='nofollow' class='external text' href='http://www.pdb.org/pdb/static.do?p=education_discussion/molecule_of_the_month/ pdb10_1.html'>*Molecule of the Month*</a> at the RCSB Protein Data Bank *Original artist:* Animation by David S. Goodsell, RCSB Protein Data Bank

• **File:1axc_tricolor.png** *Source:* https://upload.wikimedia.org/wikipedia/commons/6/61/1axc_tricolor.png *License:* CC-BY-SA-3.0 *Contributors:* Self created from PDB entry 1AXC using the freely available visualization and analysis package VMD

Credits (as required by the license):
Original artist: Opabinia regalis

• **File:2ConfBoltzmannDist.png** *Source:* https://upload.wikimedia.org/wikipedia/commons/d/d2/2ConfBoltzmannDist.png *License:* Public domain *Contributors:* Own work *Original artist:* Mcpazzo

• **File:30nm_Chromatin_Structures.png** *Source:* https://upload.wikimedia.org/wikipedia/commons/6/6a/30nm_Chromatin_Structures.png *License:* CC-BY-SA-3.0 *Contributors:* Transferred from en.wikipedia *Original artist:* Original uploader was Zephyris at en.wikipedia

• **File:5-Methylcytosine.svg** *Source:* https://upload.wikimedia.org/wikipedia/commons/3/3a/5-Methylcytosine.svg *License:* Public domain *Contributors:* Own work *Original artist:* Yikrazuul (<a href='//commons.wikimedia.org/wiki/User_talk:Yikrazuul' title='User talk:Yikrazuul'>talk</a>)

• **File:7atoms_interaction_trans-tBu-cyclohexyl_Cl.png** *Source:* https://upload.wikimedia.org/wikipedia/commons/2/2b/7atoms_interaction_ trans-tBu-cyclohexyl_Cl.png *License:* CC BY-SA 3.0 *Contributors:* Own work *Original artist:* Mr.Holmium

• **File:A-B-Z-DNA_Side_View.png** *Source:* https://upload.wikimedia.org/wikipedia/commons/b/b9/A-B-Z-DNA_Side_View.png *License:* Public domain *Contributors:* Transferred from en.wikipedia *Original artist:* Original uploader was Thorwald at en.wikipedia

• **File:A-DNA,_B-DNA_and_Z-DNA.png** *Source:* https://upload.wikimedia.org/wikipedia/commons/b/b1/A-DNA%2C_B-DNA_and_Z-DNA. png *License:* GFDL *Contributors:* Originally from en.wikipedia; description page is/was here. *Original artist:* Original uploader was Richard Wheeler (Zephyris) at en.wikipedia

• **File:ABDNAxrgpj.jpg** *Source:* https://upload.wikimedia.org/wikipedia/commons/8/8a/ABDNAxrgpj.jpg *License:* CC-BY-SA-3.0 *Contributors:* "Physical Chemistry of Foods", vol.2, van Nostrand reinhold: New York, 1994 *Original artist:* I.C. Baianu et al.

• **File:ADN_animation.gif** *Source:* https://upload.wikimedia.org/wikipedia/commons/8/81/ADN_animation.gif *License:* Public domain *Contributors:* Own work *Original artist:* brian0918™

• **File:AT_DNA_base_pair.png** *Source:* https://upload.wikimedia.org/wikipedia/commons/a/a2/AT_DNA_base_pair.png *License:* CC-BY-SA-3.0 *Contributors:* Originally uploaded on en.wikipedia *Original artist:* Originally uploaded by Roadnottaken (Transferred by Edgar181)

• **File:Adna3.ogv** *Source:* https://upload.wikimedia.org/wikipedia/commons/1/14/Adna3.ogv *License:* CC BY-SA 4.0 *Contributors:* Own work *Original artist:* Mauroesguerroto

- **File:Enzyme_catalysis_energy_levels_2.svg***Source:* https://upload.wikimedia.org/wikipedia/commons/c/c0/Enzyme_catalysis_energy_le 2.svg *License:* CC BY-SA 4.0 *Contributors:* Own work *Original artist:* Thomas Shafee

- **File:Enzyme_mechanism_2.svg** *Source:* https://upload.wikimedia.org/wikipedia/commons/6/69/Enzyme_mechanism_2.svg *License:* CC BY-SA 4.0 *Contributors:* Own work *Original artist:* Thomas Shafee

- **File:Enzyme_structure.png** *Source:* https://upload.wikimedia.org/wikipedia/commons/2/22/Enzyme_structure.png *License:* CC BY-SA 4.0 *Contributors:* Own work *Original artist:* Thomas Shafee

- **File:Equillibrium_conformers.jpg** *Source:* https://upload.wikimedia.org/wikipedia/commons/9/97/Equillibrium_conformers.jpg *License:* CC BY-SA 3.0 *Contributors:* Own work *Original artist:* Mr.Holmium

- **File:Escher_Depth.jpg** *Source:* https://upload.wikimedia.org/wikipedia/en/7/7d/Escher_Depth.jpg *License:* Fair use *Contributors:* The Official M.C. Escher Website Picture gallery *Original artist:* ?

- **File:Essential_metabolic_genes_in_bacteria.png***Source:* https://upload.wikimedia.org/wikipedia/commons/c/cc/Essential_metabolic_g in_bacteria.png *License:* CC BY-SA 3.0 *Contributors:* Powerpoint **Previously published:** none *Original artist:* Peteruetz

- **File:Ethanol-3D-balls.png** *Source:* https://upload.wikimedia.org/wikipedia/commons/b/b0/Ethanol-3D-balls.png *License:* Public domain *Contributors:* ? *Original artist:* ?

- **File:EukPreRC.jpg** *Source:* https://upload.wikimedia.org/wikipedia/commons/4/45/EukPreRC.jpg *License:* CC BY-SA 3.0 *Contributors:* Own work *Original artist:* Lsanman

- **File:Eukaryote_DNA-en.svg** *Source:* https://upload.wikimedia.org/wikipedia/commons/e/e2/Eukaryote_DNA-en.svg *License:* CC BY-SA 3.0*Contributors:* This file was derived fromEukaryote DNA.svg: <a href='//commons.wikimedia.org/wiki/File:Eukaryote_DNA.svg' class=' image'><imgalt='Eukaryote DNA.svg' src='https://upload.wikimedia.org/wikipedia/commons/thumb/8/82/Eukaryote_DNA.svg/50px-Eukaryote_DNA.svg.png' width='50' height='31' srcset='https://upload.wikimedia.org/wikipedia/commons/thumb/8/82/Eukaryote_DNA.svg/ 75px-Eukaryote_DNA.svg.png 1.5x, https://upload.wikimedia.org/wikipedia/commons/thumb/8/82/Eukaryote_DNA.svg/100px-Eukaryote_ DNA.svg.png 2x' data-file-width='1189' data-file-height='734' /></a>*Original artist:* Eukaryote_DNA.svg: *Difference_DNA_RNA-EN.svg: *Difference_DNA_RNA-DE.svg: Sponk(talk)

- **File:EvolutionOfDuplicateGenes.png** *Source:* https://upload.wikimedia.org/wikipedia/commons/b/b3/EvolutionOfDuplicateGenes.png *License:* CC BY-SA 3.0 *Contributors:* Own work *Original artist:* Veryhuman

- **File:FACIL_genetic_code_logo.png** *Source:* https://upload.wikimedia.org/wikipedia/commons/e/e1/FACIL_genetic_code_logo.png *License:* CC BY 2.5 *Contributors:* doi: 10.1093/bioinformatics/btr316 *Original artist:* Bas E. Dutilh, Rasa Jurgelenaite, Radek Szklarczyk, Sacha A.F.T. van Hijum, Harry R. Harhangi, Markus Schmid, Bart de Wild, Kees-Jan Françoijs, Hendrik G. Stunnenberg, Marc Strous, Mike S.M. Jetten, Huub J.M. Op den Camp and Martijn A. Huynen

- **File:Figure_Role_of_initiators_for_initiation_of_DNA_replication.png** *Source:* https://upload.wikimedia.org/wikipedia/commons/d/d5/ Figure_Role_of_initiators_for_initiation_of_DNA_replication.png *License:* CC BY-SA 3.0 *Contributors:* Own work *Original artist:* Emply shell China dry

- **File:Fisher_iris_versicolor_sepalwidth.svg***Source:* https://upload.wikimedia.org/wikipedia/commons/4/40/Fisher_iris_versicolor_ sepalwidth.svg *License:* CC BY-SA 3.0 *Contributors:* en:Image:Fisher iris versicolor sepalwidth.png *Original artist:* en:User:Qwfp (original); Pbroks13(talk) (redraw)

- **File:Flemming1882Tafel1Fig14.jpg** *Source:* https://upload.wikimedia.org/wikipedia/commons/6/69/Flemming1882Tafel1Fig14.jpg *License:* Public domain *Contributors:* Zellsubstanz, Kern und Zelltheilung. Leipzig, Verlag von F.C.W. Vogel. Websource: [1] *Original artist:* Walther Flemming. Professor der Anatomie in Kiel.

- **File:Folder_Hexagonal_Icon.svg** *Source:* https://upload.wikimedia.org/wikipedia/en/4/48/Folder_Hexagonal_Icon.svg *License:* Cc-by-sa-3.0 *Contributors:* ? *Original artist:* ?

- **File:Foodlogo2.svg** *Source:* https://upload.wikimedia.org/wikipedia/commons/d/d6/Foodlogo2.svg *License:* CC-BY-SA-3.0 *Contributors:* Original *Original artist:* Seahen

- **File:From_second_to_fourth-generation_sequencing,_illustration_on_TAGGCT_template.svg** *Source:* https://upload.wikimedia.org/ wikipedia/commons/2/2e/From_second_to_fourth-generation_sequencing%2C_illustration_on_TAGGCT_template.svg *License:* CC BY-SA 3.0 *Contributors:* Emmanuel Barillot, Laurence Calzone, Philippe Hupé, Jean-Philippe Vert, Andrei Zinovyev, Computational Systems Biology of Cancer Chapman & Hall/CRC Mathematical & Computational Biology , 2012 *Original artist:* Philippe Hupé

- **File:Fullerene_Nanogears_-_GPN-2000-001535.jpg** *Source:* https://upload.wikimedia.org/wikipedia/commons/b/b6/Fullerene_Nanogears_ -_GPN-2000-001535.jpg *License:* Public domain *Contributors:* Great Images in NASA: Home - info - pic *Original artist:* NASA

- **File:Galactosidase_enzyme.png** *Source:* https://upload.wikimedia.org/wikipedia/commons/0/03/Galactosidase_enzyme.png *License:* CC BY-SA 4.0 *Contributors:* Own work *Original artist:* Thomas Shafee

- **File:Gene_numbers.svg** *Source:* https://upload.wikimedia.org/wikipedia/commons/8/89/Gene_numbers.svg *License:* CC BY-SA 4.0 *Contributors:* [#cite_ref-Watson_1-0 ↑] Watson, JD, Baker TA, Bell SP, Gann A, Levine M, Losick R. (2004). "Ch9-10", Molecular Biology of the Gene, 5th ed., Peason Benjamin Cummings; CSHL Press. *Original artist:* Thomas Shafee

- **File:Gene_structure_eukaryote_2_unannotated.svg***Source:* https://upload.wikimedia.org/wikipedia/commons/a/ac/Gene_structure_euk 2_unannotated.svg *License:* CC BY-SA 4.0 *Contributors:* Own work *Original artist:* Thomas Shafee

- **File:Gene_structure_prokaryote_2_unannotated.svg***Source:* https://upload.wikimedia.org/wikipedia/commons/d/d9/Gene_structure_pr 2_unannotated.svg *License:* CC BY-SA 4.0 *Contributors:* Own work *Original artist:* Thomas Shafee

- **File:Genes_and_base_pairs_on_chromosomes.svg** *Source:* https://upload.wikimedia.org/wikipedia/commons/6/6c/Genes_and_base_pairs_ on_chromosomes.svg *License:* CC BY-SA 3.0 *Contributors:* Own work *Original artist:* Friend of a friends

42.8.3 Content license

compliance